Primate Sexuality

Primate Sexuality

Comparative Studies of the Prosimians, Monkeys, Apes, and Human Beings

Alan F. Dixson

Medical Research Council,
Group on Development and Integration of Behaviour,
Sub-Department of Animal Behaviour,
University of Cambridge

Foreword by

John Bancroft

The Kinsey Institute for Research in
Sex, Gender and Reproduction,
Indiana University,
USA

Oxford   New York   Tokyo

OXFORD UNIVERSITY PRESS

1998

Oxford University Press, Great Clarendon Street, Oxford OX2 6DP

Oxford   New York

*Athens   Auckland   Bangkok   Bogota   Bombay Buenos Aires   Calcutta
Cape Town   Chennai   Dar es Salaam   Delhi   Florence   Hong Kong   Istanbul
Karachi   Kuala Lumpur   Madrid   Melbourne   Mexico City   Mumbai
Nairobi   Paris   São Paolo   Singapore   Taipei   Tokyo   Toronto   Warsaw*

and associated companies in
Berlin   Ibadan

Oxford is a trade mark of Oxford University Press

*Published in the United States
by Oxford University Press Inc., New York*

A catalogue record for this book is available from the British Library

Library of Congress Cataloging in Publication Data

*Dixson, A. F.
Primate sexuality: comparative studies of the prosimians,
monkeys, apes, and human beings / Alan F. Dixson.
Includes bibliographical references and index.
1. Sex (Biology)   2. Primates—Behaviour.   I. Title.*
QH481.D56   1998   599.8'1562—DC21   98-18400

*ISBN 0 19 850183 8 (Hbk)
ISBN 0 19850182 X (Pbk)*

*Typeset by Technical Typesetting Ireland, Belfast
Printed in Great Britain by
The Bath Press, Avon*

Foreword

The behaviour of non-human primates has held a fascination for many people because of the obvious comparison it invites with human behaviour. This is particularly so with the great apes, our closest relatives among primates. For many years much of the evidence on primate behaviour depended on studies of animals in captivity. Experiments with artifically arranged pair-tests of animals in the laboratory, what Frank Beach once called the 'prostitute paradigm', may have led to useful experimental conclusions, but the relevance of the sexual behaviour involved to normal sexuality in the species studied has always been questionable. What Alan Dixson's scholarly review of primate sexuality shows us is that over the past 20 to 30 years there has been a steady and very substantial increase in evidence of primate sexual behaviour in the natural habitat, and involving many species. There is now a wealth of comparative information.

As a student of human sexuality I am committed to an interdisciplinary approach; both biological and sociocultural determinants have to be taken into consideration. Fundamental to this is the comparative approach. Cross-cultural comparisons of human societies help us to get the sociocultural determinants into perspective. Cross-species comparisons also help us to get the biological determinants into perspective; even more important, we can look at non-human primates to see how different social structures, and the varying patterns of sexual behaviour that are part of them, can be understood in terms of adaptation to the environment. Here, there is the added advantage that the social structure is uncomplicated by the peculiar complexity of human cultures, whereas the parallels between certain non-human primate social structures and early human societies are readily apparent. The great apes are particularly informative in this respect. We can see striking contrasts, for example, between the gorilla, the orang-utan, and the chimpanzee, all closely related genetically to humans, in their social structures and in the patterns of sexual behaviour that accompany them. As the evidence builds up so it becomes more convincing to seek explanations of these different patterns in terms of environmental adaptation. We may use such evidence as the starting point for analysing human social structures.

Whatever comparison you may be interested in making, Alan Dixson's book is a gold mine. He reviews the literature with the precision and attention to detail that have become the hallmarks of his writing. He also boldly ventures into the human literature. This reminds me of Alfred Kinsey, who did the opposite. When seeking to interpret and explain the extraordinary variety of human sexual expression that he encountered, he turned to the animal literature to find that much of what his human subjects reported had been observed in other species. He used this to emphasize the 'naturalness' of human sexual behaviour, and to contrast this with the unnatural constraints on sexual expression that he saw society imposing. Fifty years on his use of animal evidence in this way looks naïve and simplistic—he overlooked cross-species differences and the various ways in which sexuality was constrained by social factors among other species. But the evidence available to him was much less than it is now. I have enough respect for Kinsey to believe that if Dixson's book had been available to him he would have reached different conclusions.

Alan Dixson's use of the human literature, while inevitably selective, is certainly not naïve. And in the process he implicitly issues a challenge to his readers to respond to these human–non-human primate comparisons. I believe it will provoke some fruitful debate.

John Bancroft
Director, The Kinsey Institute for Research in Sex, Gender and Reproduction, Indiana University
July 1998

Preface

Is there a fatherless book,
an orphan volume in this world?
A book that is not the descendant
of other books?

Carlos Fuentes

So many books and articles have been written about primates, that a few words of explanation are advisable, when adding another title to the list. In my defence, it is fair to say that no detailed comparative volume on the subject of primate sexuality has been published until now. This is despite the fact that any complete study of primate reproductive biology must include the subject of sexual behaviour. Information on the sexual lives of prosimians, monkeys and apes is scattered throughout a vast literature, so that more than 2000 sources have been cited in the present volume. In distilling and reviewing this information, I have kept two objectives in mind. Firstly, I have attempted to construct some picture of how sexual behaviour has been shaped by selective forces during primate evolution. This has involved making comparisons of mating systems and mating tactics, as well as an exploration of copulatory patterns, genitalic morphologies, and sperm competition across the various primate lineages. Secondly, I have examined the proximate, endocrine and neurological, mechanisms which control the development and integration of sexual behaviour in males and females. Human sexuality has been included within these goals and placed in proper comparative perspective. Man is but one of approximately 250 extant primate species; all of them are fascinating.

Congreave once remarked that he 'could never look upon a monkey without very mortifying reflections'. On the contrary, the more I have studied primate behaviour, the greater my sympathy for the human condition has become. This has been a gradual process, however, taking place over the past 28 years. Perhaps Wordsworth was nearer the mark when he said that he had learned to view nature 'not as in the hour of thoughtless youth; but hearing oftentimes the still, sad music of humanity.' Certainly, there is little room for complacency when viewing modern society's treatment of sexual matters. In Britain, for example, frequencies of divorce, under-age pregnancies, abortions, sexually-related crime, and sexually transmitted disease are all on the increase. Yet we invest relatively little, either in research on human sexuality, or in the provision of sexual education for our children.

I should like to thank several colleagues who have helped tremendously in the preparation of this book. Jane Armitage typed and retyped the many draft versions and Claire Nevison prepared over 100 of the graphs and histograms. My especial thanks go to Matthew Anderson for tireless work and assistance with all aspects of preparing the typescript, illustrations, and reference list. All three of these friends have a great sense of humour; they cheered me up at times when the task seemed too great. I am most grateful to staff at Oxford University Press, for encouragement and guidance in preparing this book, and to Michael Tiernan and Ian Davies for their painstaking contributions at the copy-editing stage. To John Bancroft, who has written the Foreword, my particular thanks, and best wishes for the future of the Kinsey Institute. I should also like to thank the organizations and individuals listed below, for permission to reproduce figures, tables, photographs and data, either in their original form or as redrawn and modified versions.

Academic Press: Figures 3.5, 4.2b, 4.4, 4.6, 4.14, 4.23, 5.38, 7.8, 7.12, 7.13, 7.14, 7.25, 7.30, 7.31, 7.32, 7.33, 8.20, 10.4, 10.6, 10.17, 11.5, 11.11, 11.15, 11.16, 12.4,

12.13, 12.16, 12.23, 13.1, 13.5, 13.11, 13.14, 13.16, 13.33, 13.48, and 14.20; **Aldine Press:** Figures 4.24, 4.25, and table 4.11; **Allen Press:** Figure 8.5; **American Association for the Advancement of Science:** Figures 4.22, 8.19, 10.11, 10.19, 10.23, 10.25, and 11.13; **American Journal of Physical Anthropology:** Figure 7.6; **American Psychiatric Association:** Figure 6.17; **American Psychological Association Inc.:** Figures 5.18 (lower left and lower right), 8.21, 11.25, 12.5, 13.2, and 14.21; **American Psychosomatic Society:** Figure 11.33; **American Urological Association, Inc.:** Figure 13.46; **Balliere Tindall:** Figure 8.12; **Dr Simon Bearder:** Figures 2.7, 2.8, 2.9, 2.26, and 2.30; **Belknap Press of Harvard University Press:** Table 4.12; **The Benjamin/Cummings Publishing Co. Inc.:** Figure 8.7; **Biologica Gabonica:** Figure 7.9; **Blackwell Publishers Ltd:** Figures 2.37, 6.6, 12.20, and 12.21; **Brill Publishers:** Figure 4.8 and 6.12; **Dr Gillian Brown:** Figures 6.11 and 10.15; **Butterworth Scientific:** Figure 3.1; **Cambridge University Press:** Figures 6.10 (lower left), 6.14, 7.1, 7.2, 8.2, 8.8, and 12.29; **Chapman and Hall:** Figure 2.36; **Chicago University Press:** Figures 3.10, 3.11, 5.4, 5.15, 5.29, 6.8, 6.10 (lower right), 7.5, 10.8 and 11.18; **Dr David Chivers:** Figures 2.32 and 2.33; **Churchill Livingstone:** Figures 5.28, 5.30, 9.19, 9.21, 10.3, and 13.6; **Dr R. da Rocha e Silva:** Figure 2.21; **Danish Medical Association:** Figures 5.32 and 11.12; **Professor Ralf Dittmann and Praeger Press:** Figure 10.18; **Gerald Duckworth and Co. Limited:** Figures 2.11, 3.12, 7.15, 11.3, 12.2, 12.11, and 14.8; **Edinburgh University Press:** Figures 7.7, 9.6, 9.15c, and 9.17; **Elsevier Inc.:** Figures 4.20, 5.33, 5.35, 6.1 (upper left and upper right), 6.13, 9.2, 9.14, 10.13, 10.14, 10.20, 10.21, 10.24, 10.26, 11.7, 11.8, 11.9, 12.7, 12.10, 12.14, 12.15, 12.18, 12.19, 12.22, 12.32, 12.33, 13.8, 13.9, 13.12, 13.13, 13.20, 13.21, 13.22, 13.23, 13.26, 13.27, 13.28, 13.29, 13.30, 13.34, 13.38, 13.39, 13.40, 13.43, 13.47, 13.49, 13.51, 14.10, 14.13, 14.14, 14.16 and 14.22; **Field Studies of New World Monkeys:** Figure 9.16; **George Banter Publishing:** Figure 11.4; **Gustav Fischer Verlag:** Figures 6.9 and 7.21 d, e and f; **Dr E. Haimoff:** Figure 7.7; **Harvard University Press:** Table 9.2; **Harwood Academic Publishers:** Figure 4.15; **Holt, Reinhart and Winston:** Figure 5.5; **Professor Julianne Imperato-McGinley:** Figure 10.10; **Indian National Science Academy:** Figure 11.14; **Japan Monkey Centre:** Figures 3.6, 3.7, 4.7, 4.16, 4.17, 4.18, 4.19, 7.5, 11.26, 13.3, and tables 6.2 and 6.3; **Johns-Hopkins University Press:** Figures

10.1, 10.2, 10.9 and tables 10.1, and 10.2; **Journal of Endocrinology:** Figures 10.12, 12.6, 13.45, and 13.50; **Dr W. Junk Publishers:** Figure 14.5; **Karger AG Ltd:** Figures 2.1, 4.3, 5.10, 5.20, 5.22, 5.31, 6.10 (upper left and upper right), 6.15, 7.4, 7.10, 7.17, 8.13, 8.14, 8.18, 9.10, 9.20, 11.17, 11.22, 11.23, 12.1, 12.30, 12.31, 12.34, 12.35, 13.4, 13.25, and 14.12; **The Lancet Limited:** Figure 13.44; **Macmillan Magazines Limited:** Figure 10.27; **Professor Bob Martin:** Figures 2.3, 2.4, 2.5, 2.6, and 2.29; **Dr Patrick McGinnis and the Gombe Stream Research Centre:** Figure 5.25; **Dr R. A. Mittermeier:** Figures 2.12, 2.13, 2.16, 2.17, 2.18, 2.19, 2.20, 2.22, 2.23, 2.24, and 3.4; **The late Dr Kurt Modahl:** Figure 2.28; **The New England Journal of Medicine:** Figure 11.31; **Dr C.P. Noguerra:** Figure 2.15; **Oxford University Press:** Figures 5.6 (middle), 5.24, 5.37, 7.19, 8.4, 11.1, 11.2, 11.30 and 12.17; **Palaeontologische Zeitschift:** Figure 9.11; **Plenum Publishing Co. Ltd:** Figures 5.9, 5.21, 5.34, 6.1 (lower left and lower right), 6.2, 6.7, 7.28, 7.29, 8.3, 9.13, 9.18, 11.28, 13.7, 13.17, 13.41, and tables 6.2 and 6.3; **Portland Press:** Figures 6.5, 7.3, 7.27, 8.11, 8.16, 8.17, 9.24, 11.6, 14.6, 14.11, and tables 8.2 and 14.1; **Raven Press:** Figures 9.22, 9.23, 13.10, and tables 9.5 and 9.6; **The Royal Society:** Figure 7.24, 12.24, and 12.25; **Dr Anthony Rylands:** For advice and assistance in obtaining photographs of New World primates; **Dr Robert Sapolsky:** Figure 14.2 and 14.3; **W.B. Saunders and Co.:** Figures 6.16, 11.32, 13.18, and 13.19; **Dr C. Schmidt:** Figure 5.26; **Dr Devendra Singh:** Figure 7.35; **Society for the Scientific Study of Sex:** Figures 5.1, 12.3, and 12.26; **Dr Volker Sommer:** Figures 2.25 and 3.8; **Springer Verlag GmbH:** Figures 4.5, 4.21, 7.23, and 14.15; **Stanford University Press:** Figures 4.2a and 6.4; **Professor Karen Strier:** Figure 5.8; **The New York Academy of Sciences:** Figure 14.4; **University of Tokyo Press:** Figures 11.10 and 13.15; **Veenman and Zonen B.V.:** Figures 5.23 and 5.39; **Wadworth Publishing Company:** Figure 4.3; **Waverley Press, Inc.:** Figures 5.11 and 13.42; **Weidenfeld and Nicolson:** Figures 2.2, 2.34, 2.35, 5.6 (Lower), 5.27, 7.11, 7.18, 7.20, 7.34, 11.27, and 11.29; **Wiley-Liss Inc.:** Figures 3.2, 4.9, 4.13, 4.26, 7.26, 8.6, 9.8, 9.9, 11.24, 12.27, 12.28, 13.32, 13.35, 13.36, 13.37, 14.1, 14.9, 14.18, 14.19, and table 8.7; **Dr A. Young:** Figure 2.14; **Zoological Society of London:** Figures 3.9, 5.2, 5.6 (upper), 5.7, 5.12, 5.14, 5.17, 5.18 (upper left and upper right), 6.2, 6.3, 7.16, 8.10, 9.1 (a–d), 9.3, 9.4, 9.5, 9.12, 9.15 (A and B), 11.19, 11.20, and 11.21.

Contents

1. **Darwin and friends**   1

2. **Primate classification and evolution**   5
   Evolutionary relationships and time-scales   18

3. **Mating systems**   22
   Primate social organization   22
   Primate mating systems   24
   Monogamy in primates   26
   Polandry in primates   29
   Polygyny in primates   29
   Multimale–multifemale mating systems   37
   Evidence that females mate with multiple partners   37
   Examples of multimale–multifemale societies   39
      Ringtailed lemur   39
      Rhesus monkey   41
      Chimpanzees and bonobos   42
   Mating seasonality and numbers of males in primate groups   44
   Dispersed mating systems   45
   The mating system of the orang-utan   49

4. **Mating tactics and reproductive success**   51
   Dominance, mating success, and reproductive success   52
      Dominance in males   52
   Coalitions, alliances, and male mating success   57
   Consortship, mate-guarding, and possessiveness   57
   Alternative mating tactics—a lifespan view   62
   Coercive matings   66
   Infanticide: the sexual selection hypothesis   67
   Harassment and interruption of copulation   71
   Genetic assessment of male reproductive success   74
   Female dominance, mating success, and reproductive success   80
   Female mate 'choice'   84
   Sexual preferences—favouritism and friendship   85
   Incestuous matings and incest avoidance   88
   The major histocompatibility complex and mate choice   90
   The problem of skewed birth sex ratios   90

5. **Sexual behaviour and sexual response**   93
   Female sexuality: concepts and definitions   93

The question of oestrus   93
Proceptivity   94
Sexual receptivity   101
Sexual attractiveness   102
Male sexuality: concepts and definitions   104
Precopulatory behaviour in male primates   107
Copulatory behaviour in male primates   110
The evolution of copulatory postures   110
The evolution of intromission and ejaculatory patterns   114
Classifying primate copulatory patterns   116
Communication during copulation   126
Facial communication during copulation   126
Vocal communication during copulation   127
Tactile stimulation and the problem of orgasm   129
Some conceptual issues in human sexuality   136
Gender identity and gender role   136
Auto-erotism, including masturbation   139
Abnormal sexual preferences: the human paraphilias   141

6. Sociosexual behaviour and homosexuality   146
Sociosexual patterns   146
Does sociosexual behaviour reflect dominance rank?   146
Penile erection and sociosexual communication   150
Erection in response to the sight or proximity of a female   150
Erection during sleep   151
Erection during play, grooming, or when seeking 'reassurance'   151
Erections during aggression and ritualized penile displays   151
Sociosexual presentation in New World primates   153
The development of sociosexual behaviour   153
Social stimuli and sociosexual development   154
Is sociosexual behaviour equivalent to homosexual behaviour?   159
Does sociosexual behaviour occur in man?   161
Human homosexual behaviour   164
Homosexual patterns and relationships   165
The biological basis of homosexuality   165
Genetic contributions to homosexual development   166
Homosexuality and birth order   167
Experiential contributions to homosexual development   168

7. Sexual selection and sexually dimorphic traits   170
Sexual dimorphism in body weight   171
Sexual dimorphism in canine tooth size   174
Sexual dimorphism in vocal anatomy and display   176
Sexual dimorphism in cutaneous glands and scent-marking displays   183
Sexual skin and other secondary sexual traits in adult males   192
Blue and red sexual skin   192
Capes of hair and facial adornments   195

The problem of fluctuating asymmetry 198
Secondary sexual characters in the adult female 201
    The problem of female sexual skin 201
The evolution of sexual skin morphology in various catarrhine lineages 208
Secondary sexual features of the human female 214

8. Sperm competition 217
Relative testes size and mating systems in primates 217
Further tests and applications of relative testis size in primates 218
    The prosimian primates 218
    The marmosets and tamarins (Callitrichidae) 220
    The mandrill and drill 221
Relative testis size and mating systems in non-primates 222
Testis size, sperm production, and ejaculate quality 222
    The compartments of the testis 222
    Rates of sperm production and storage 223
    Problems of ejaculate quality 223
Does sexual selection influence sperm morphology? 224
Has sexual selection influenced the evolution of sperm length? 227
The male accessory reproductive organs and sperm competition 228
    Structure and functions of the vasa deferentia 228
    The seminal vesicles and prostrate 231
    The structure and functions of primate copulatory plugs 235
Possible effects of repeated ejaculations 237
Does mating order influence male reproductive success? 239
Do social or sexual stimuli affect sperm counts? 243

9. Sexual selection and genitalic evolution 244
Sexual selection and the evolution of male genitalia 244
    The lock-and-key hypothesis 244
    The genitalic recognition hypothesis 247
    The pleiotropism hypothesis 247
    Mechanical conflict of interest hypothesis 247
Eberhard's hypothesis: sexual selection by female choice 247
Penile morphology and mating systems in primates 248
Descriptions of penile morphologies in dispersed and
    multimale–multifemale species 250
Descriptions of penile morphologies in monogamous and polygynous forms 251
    The evolution of the baculum 252
    The evolution of penile spines 260
    The evolution of distal penile complexity 263
The striated penile muscles and morphological changes in the
    glans penis during copulation 263
Sexual selection and the evolution of female genitalia 265
    Some observations on evolution of the clitoris 265
The vagina 267
The cervix 269

The uterus   271
The uterotubal junction and fallopian tubes   273

10. Sexual differential of the brain and behaviour   277
The organization hypothesis   277
Sexual differentiation at the genital level   278
The organization hypothesis and primate behavioural development   279
The organization hypothesis and psychosexual differentiation in man   284
Congenital adrenal hyperplasia   285
Androgen-insensitivity in males: the testicular feminization syndrome   288
5-Alpha-reductase deficiency and male pseudohermaphroditism   288
Prenatal exposure to exogenous progestagens and oestrogens   290
The postnatal testosterone surge and psychosexual differentiation   292
Hormones and homosexuality   295
The neuroanatomical basis of sexually dimorphic behaviour   301
Sex differences in vocal control systems in the brain   301
Spinal motor neurons innervating penile muscles   304
Sex differences in the primate brain, in comparative perspective   305
Sex differences in the corpus callosum   305
Sex differences in the preoptic area and anterior hypothalamus   306
Sex differences in other hypothalamic nuclei   311
Conclusions: sexual differentiation of the brain and behaviour   312

11. The ovarian cycle and sexual behaviour   315
Ovarian cycles: proceptivity, receptivity, and attractiveness   315
The ovarian cycle and sexual behaviour in prosimians   316
Conclusions: the ovarian cycle and sexual behaviour in prosimians   321
The ovarian cycle and sexual behaviour in New World monkeys   321
Conclusions: the ovarian cycle and sexual behaviour in New World monkeys   331
The menstrual cycle and sexual behaviour in catarrhine primates   332
The rhesus monkey   332
The menstrual cycle and sexual behaviour in other Old World monkeys   338
The menstrual cycle and sexual behaviour in the great apes   343
Sexual behaviour during the human menstrual cycle   348
Some remarks on the question of 'concealed ovulation'   352

12. The neuroendocrine regulation of sexual behaviour in the adult female 354
Peripheral effects of hormones upon female sexuality   354
Possible effects upon visual and tactile cues   354
Possible effects upon olfactory cues   355
Central effects of hormones upon female sexuality   357
Some comments on methods of behavioural observation and measurement   357
Central effects of oestrogens upon sexual behaviour   358
Central effects of progesterone upon sexual behaviour   361
Central effects of androgens upon sexual behaviour   362
The adrenal glands and sexual behaviour in female primates   363
The hypothalamic basis of sexual receptivity and proceptivity   365

A neural model of proceptivity in female primates 376
Neurotransmitters and sexual behaviour 380
   Monoaminergic neurotransmitters 380
   Neuroactive peptides 384
   Luteinizing-hormone-releasing hormone (LHRH) 384
   Oxytocin 385
   Opioid peptides 386

13. Hormones and sexual behaviour in the adult male 389
   Seasonal changes in hormones and sexual behaviour 389
   Effects of castration and testosterone replacement 392
   Antiandrogens and sexual behaviour 394
   Behavioural effects of metabolites of testosterone 397
   Sources of individual variability in sexual behaviour 401
      Individual differences in circulating androgen levels 401
      Adrenal androgens 404
      The role of previous sexual experience 404
      The role of the female partner 404
      Genetic variability between males 405
      Effects of age 406
   Peripheral versus central effects of androgens upon male sexuality 409
   Androgens and penile morphology 409
   The accessory sexual organs 413
   Secondary sexual adornments 413
   Central effects of androgens upon male sexuality 413
   The preoptic area and hypothalamus 418
   Influences that impinge upon hypothalamic mechanisms 423
      Amygdala and stria terminalis 423
      The septum 424
      The midbrain 425
   The neural control of pelvic thrusting 425
   The neural control of erection and ejaculation 430
      The innervation of the penis 430
      Neurotransmitters, erection, and detumescence 431
   Erectile dysfunction 434
   Emission and ejaculation 436
   Neurotransmitters and sexual behaviour 437
   Neuroactive peptides 437
      β-Endorphin 437
      Luteinizing-hormone-releasing hormone 439
      Oxytocin 440
      Prolactin 440
      Other pro-opiomelanocortin-derived peptides 440
   Monoaminergic neurotransmitters 441
      Dopamine 441
      Noradrenaline 442
      Serotonin (5-HT) 443

14. Socioendocrinology and sexual behaviour   444
    Social rank and neuroendocrine function in male primates   444
    Social rank and secondary sexual traits in male primates   446
    Female social rank, the ovarian cycle, and fertility   449
    Social environment and reproductive synchrony   453
    Effects of social stimuli upon pubertal development   460
    Short-term effects of copulatory stimuli upon gonadal function   464
        Male primates   464
        Female primates   466

References   467

Index   529

1 Darwin and friends

In the years that followed the publication of Darwin's 'Origin of Species' acceptance was gained, albeit reluctantly in some quarters, that man must be closely related to the primates and especially to the great apes. The case for an evolutionary link between the ancestors of the great apes and man was cogently argued by Huxley (1863) in his essays on 'Man's Place in Nature'. Darwin's (1871) later treatise on 'The Descent of Man and Selection in Relation to Sex' also contained comparative material on the non-human primates, although little reference was made to sexual behaviour. Two principal avenues of sexual selection were explored by Darwin; intrasexual competition for access to mates and intersexual selection for secondary sexual adornments and displays as enhancers of sexual attractiveness. Both these processes were thought to act primarily upon the male sex (Fig. 1.1). Thus, intrasexual competition favoured the evolution of greater aggressiveness, as well as increased body size and weaponry in males. Intersexual selection, on the other hand, accounted for the evolution of extravagant secondary sexual characteristics, such as the brilliant plumage of the peacock and the red and blue pigmentation of the adult male mandrill.

Although Darwin was preoccupied with the effects of sexual selection upon males there is, of course, no reason to exclude the possibility that females might compete among themselves for mating opportunities, as well as more indirectly for the resources required to nurture their offspring. Likewise, examples might exist of intersexual selection for physical adornments or behavioural patterns which enhance female sexual attractiveness (Fig. 1.1). Both these propositions are valid as far as the Order Primates is concerned and examples will be described and discussed throughout the present volume.

Darwin considered that the primary genitalia of both sexes had been moulded by evolution in response to natural selection, rather than as a result of sexual selection. He certainly recognized that some structures, such as specialized prehensile organs in males of various invertebrate groups, might have been affected by sexual selection. Thus 'if the

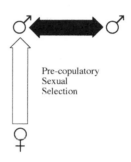

Traditional Darwinian View

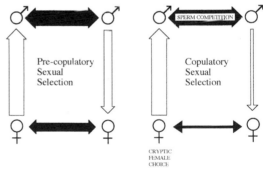

Modern Darwinian View

Intrasexual selection   Intersexual selection

Fig. 1.1 Traditional and modern views of the processes underlying sexual selection. The traditional Darwinian view emphasized the importance of intermale competition and female choice as precopulatory determinants of mating success. The modern Darwinian view acknowledges that intrasexual competition might sometimes involve females and that females can, in some circumstances, produce extravagant adornments to attract males (e.g. sexual skin in some female Old World monkeys). Most important has been the realization that sexual selection acts at the copulatory level and post-copulatory level, via sperm competition and cryptic female choice, to affect reproductive success (as indicated in lower part of diagram, on the right hand side).

chief service rendered to the male by his prehensile organs is to prevent the escape of the female before the arrival of other males, or when assaulted by them, these organs will have been perfected through sexual selection.' Yet he lamented that 'in most cases of this kind it is impossible to distinguish between the effects of natural and sexual selection.' Only in the last few decades has it become clear that sexual selection operates at the level of the gonads (via *sperm competition*: Parker 1970) and copulatory organs (via *cryptic female choice*: Eberhard 1985, 1996) as well as influencing the evolution of patterns of copulatory behaviour. This modern view of sexual selection, embracing copulatory and post-copulatory processes in addition to precopulatory competition and mate choice, is shown in schematic form on the right hand side of Fig. 1.1.

Given Darwin's genius as a biologist, it might seem surprising that he failed to register the importance of sexual selection at the genitalic level. His extensive taxonomic studies of barnacles, for example, should have alerted him to the possible effects of sexual selection upon masculine genital morphology and behaviour (Darwin 1851, 1854). Although barnacle species are usually hermaphroditic, cross-fertilization between individuals is the rule in this group. Being sessile, each barnacle is equipped with a greatly elongated penis; those with the longest penes gain a reproductive advantage, since they can reach and fertilize a larger number of their neighbours. Each individual may be inseminated by a number of conspecifics, and barnacles often have large testes, presumably because sperm competition favours the production of larger numbers of gametes under such conditions. It is possible that Darwin's reticence in dealing with the evolution of copulatory behaviour and genital anatomy was a result of the moral climate of his time. The 19th century was an era of prudery and prejudice regarding sexual behaviour. Biographers have attested to Darwin's retiring nature and his reluctance to participate in the heated public debates which greeted the publication of his ideas concerning natural selection (Bowlby 1990; Desmond and Moore 1991). One can only speculate as to what would have occurred if 'The Descent of Man' had included a detailed exposition on the evolution of the penis and testes or descriptions of the various copulatory postures and patterns employed by animals and human beings. It is not unreasonable to suggest that Darwin concentrated on the precopulatory aspects of sexual selection because they were less likely to cause offence. Even at the close of the 19th century the great sexologist Havelock Ellis records that his books had to be published in America since their content was viewed as obscene in England (Havelock Ellis 1902). It is quite possible therefore that Darwin was aware that

sexual selection might influence patterns of copulatory behaviour and genital morphology. Numerous examples of such effects of sexual selection will be dealt with in this book.

Setting aside the topic of sexual selection , it is essential also to consider the many proximate mechanisms that govern the expression of sexual behaviour. How do sex differences in behaviour develop during the lifetime of an individual and how do hormones and other physiological mechanisms influence sexual activities? In reviewing current knowledge of these fields it has been sobering to reflect how little was known about reproductive physiology or genetics in Darwin's time. Yet in the fields of reproductive physiology, behaviour, and genetics, contributions made by 19th century scientists have profoundly influenced the development of sexology during the 20th century. Mendel's experiments on inheritance in the garden pea provided the foundation for the science of genetics. His paper, published in the Proceedings of the Natural History Society of Brünn in 1866 went unnoticed by his contemporaries, but its importance was finally appreciated in 1900 when deVries, Correns, and Tschermak each independently 'rediscovered' it for science (Fisher 1965). In the field of reproductive physiology, the first proof of endocrine function is usually credited to Berthold (1849) who studied the effects of castration and testicular implantation in the cockerel. Berthold's single report was also a pioneering experiment in behavioural endocrinology. He noted that three cockerels which were castrated and implanted with a single testis remained normal 'so far as voice, sexual urge, belligerence, and growth of the comb and wattles were concerned.' The three untreated castrates became caponized, however. Since the transplanted testes lacked their original innervation, Berthold surmised that blood-born factors emanating from the transplants must be responsible, and that the nervous system might also be a target for such factors (Forbes 1949). In his survey of the history of behavioural endocrinology Beach (1981) notes that 'during the half century that followed Berthold's discovery, physiologists in France, Italy, Germany, and Russia undertook investigation of relationships between mating behaviour of various animals and functions of the sex glands.' Steinach (1894, 1910), for instance, conducted a variety of experiments using amphibians, birds, and rodents. His view was that gonadal secretions in some way stimulate, or inhibit, specialized centres in the brain to produce their behavioural effects. Bayliss and Starling (1902) are often credited with initiating the modern era of endocrinology; they demonstrated that secretin released by the duodenal mucosa stimulates the flow of pancreatic juice (Turner 1960; Turner and Bagnara 1976). Starling also coined the term 'hormone', deriving it from the Greek word *hormaein* meaning to 'impel or arouse to activity' (Barrington 1963). Yet it was the 19th century physiologists who had laid the foundations upon which endocrinology is based, and it is very much to their credit that behavioural observations formed an integral part of their studies of endocrine function.

If the sexual activities of cockerels or frogs were marginally acceptable as subjects for scientific contemplation in Darwin's time, human sexuality was quite another matter. As Hoenig (1977) says, 'the 19th century was a century of prudery, and to take up the study of sexual disorders and write about them required great courage.' Richard von Krafft-Ebing (1840–1902) was one such pioneer who established the study of sexual disorders as a distinct branch of psychiatry. His 'Psychopathia Sexualis: With Especial Reference to the Antipathic Sexual Instinct' was published in 1886 and ran to 12 editions. Despite criticisms of the book at the time and the fact that von Krafft-Ebing published other major treatises on psychiatry, 'Psychopathia Sexualis' remains his most enduring contribution. His detailed case histories and descriptions of paraphilic behaviour are still valuable; the reader will find them referred to in Chapter 5 of this book. Another, and less fortunate, student of human sexuality was Albert Moll (1862–1939) a private practitioner who lived in Berlin. Despite an active career during which he founded an International Society for Sex Research and organized the first ever International Congress on Sexology, Moll was to die penniless and unmourned by the scientific establishment. Havelock Ellis (1858–1939) who wrote widely on sexual issues, wryly commented that 'the pioneer in this field may well count himself happy if he meets with nothing worse than indifference'. Havelock Ellis was not an experimental scientist or collector of clinical histories; his contribution was made primarily by examining sexual problems in the light of information derived from studies in cultural anthropology. He referred occasionally to the sexual behaviour of animals, including monkeys and apes (Havelock Ellis 1902) but there was insufficient information upon which to base detailed comparisons.

Other pioneers in the field of sex research whose careers spanned the late 19th and early 20th centuries include Forel (1848–1931), Freud (1856–1939), Hirschfeld (1856–1935), and Bloch (1872–1922). Forel was a psychiatrist and an entomologist of renown whose interest in sexual matters was as an educator and campaigner for social reform. Freud,

the father of psychoanalytic theory, focused attention upon the phenomenon of infantile sexuality with the publication of 'Three Essays on the Theory of Sexuality' (1905). His contribution remains valuable, although few today would subscribe uncritically to his views regarding phases of infant erotosexual development (i.e. 'oral', 'anal' and, 'phallic' phases) or the existence of oedipal or castration complexes in human infants. Hirschfeld was more specifically a 'sex rescarcher' who coined the term *transvestite* with reference to men (usually of heterosexual disposition) who enjoy dressing in womens' clothes. Hirschfeld was a homosexual and campaigned for the rights of sexual minorities. His Centre for Sex Research, founded in 1919 and subsequently donated to the German nation, was closed and his books and collections were burned by the Nazis. His contemporary Iwan Bloch, who studied sexual abnormalities and originated the term 'sexology', wrote many of his early contributions under an assumed name (Eugene Dühren) in order to avoid persecution.

These early sexologists had little opportunity to compare their studies of human sexuality with information on the non-human primates. Indeed, very little was known about the sexual lives of monkeys and apes until the third and fourth decades of the 20th century, when important publications began to appear, both in the USA and in Great Britain. The realization that Old World monkeys and apes have a menstrual cycle which is physiologically very close to that of the human female, led to a number of studies to explore relationships between the menstrual cycle and sexual behaviour (in baboons: Zuckerman 1932; rhesus monkeys: Ball and Hartman 1935; and chimpanzees: Yerkes and Elder 1936; Young and Orbison 1944). In the USA, the Committee for Research in Problems of Sex, under the chairmanship of Robert Yerkes, did a great deal to support research in many aspects of sexual physiology and behaviour. William Young, who helped to lay the foundations of modern behavioural endocrinology with his experiments on the organization and activation of sexual behaviour by gonadal hormones, received support from this Committee. The first large-scale surveys of human sexual behaviour, carried out by Alfred Kinsey and his colleagues at Indiana University (Kinsey *et al.* 1948, 1953) and the field researches of Clarence Ray Carpenter on primate behaviour (Carpenter 1934, 1935, 1940, 1942) also benefited from the support of Yerkes and the Committee for Research in Problems of Sex.

Special mention must be made concerning Carpenter's contributions to primatology, because his comparative approach to studies of primate social organization and sexual behaviour was unique and well ahead of its time. Carpenter's work led to the appreciation that primate mating systems are variable, and include monogamy (e.g., in the gibbon) as well as multimale-multifemale groups (e.g., in howler and rhesus monkeys). He obtained the first useful information on the periodicity of sexual activity during ovarian cycles in female monkeys living under natural conditions. It became apparent that, although sexual receptivity is not limited to the peri-ovulatory period in species such as howlers and rhesus monkeys, there are, none the less, marked increases in sexual activity at this time. Further, these increases are due in part to changes in the females' invitational behaviour to males (what Beach (1976) would later call *proceptivity* in female mammals) and not simply to the males' greater willingness to copulate with females at mid-cycle.

Although a few scientists attempted to carry out fieldwork on primate behaviour (e.g., Nissen (1931) on chimpanzees and Bingham (1932) on the gorilla) none achieved either the depth or comparative breadth displayed by Carpenter. Not until the 1950s and 1960s was there a resurgence of interest in primate fieldwork which has continued until the present day. By 1967, Altmann estimated that field research on primate ecology and behaviour was doubling every five years. At that time the majority of work involved the Old World monkeys and apes. However, research interests have broadened over the years and there has been a gratifying increase in fieldwork both on the New World primates and on prosimians, including some of the nocturnal forms such as bushbabies and pottos. Although the majority of reports are not specifically concerned with sexual behaviour, the primate field literature as a whole constitutes an invaluable source of information on reproductive biology, mating systems, mating tactics, and patterns of sexual behaviour. A recent census of current field projects, compiled by Casperd (1996) for the Primate Society of Great Britain lists no fewer than 144 field studies, including 45 in Africa, 45 in Asia, 16 in Madagascar, and 38 in Central and South America. One of the tasks undertaken in preparing this book was to read as much field biology as possible, in an attempt to integrate information concerning sexual behaviour observed under natural conditions with observations and experiments on the reproductive physiology and behaviour of captive primates.

Mankind is fortunate to have the opportunity to study over 200 species of prosimians, monkeys, and apes; a priceless resource upon which to base comparative and evolutionary arguments concerning the nature and origins of primate sexuality. Less fortunate are our non-human primate relatives, many of which face extinction as the world's rainforests are destroyed. As the millennium approaches, fascinating creatures such as the aye-aye, woolly spider monkey, and mountain gorilla may be numbered in their hundreds, whilst the human population of our planet approaches six billion. The delusion is still widely prevalent that human population problems can be resolved simply by the provision of better contraceptive techniques. Yet, without acceptance of contraception as part of a healthy and happy sexual life, the miseries of overpopulation and disease increase with each passing year. If sexual intolerance and bigotry were problems during Darwin's era, it must be admitted that these attitudes still exist and impede man's quest for sexual health in many parts of the world. Recognition of mankind's sexual nature and our kinship with other primates as sexual beings is thus a matter of some importance. In this book, the human species is treated as just one member of the Order Primates; a remarkable primate it is true, but in many ways no more remarkable than the aye-aye, woolly spider monkey or mountain gorilla.

2   Primate classification and evolution

The extant primates may be placed into six major groups or superfamilies, as listed below:-

1   The lemurs of Madagascar . . . .
                    Superfamily Lemuroidea
2   The galagos and lorises of Africa and Asia . . . . .
                    Superfamily Lorisoidea
3   The Tarsiers of S.E. Asia . . . .
                    Superfamily Tarsioidea

4   The New World Monkeys . . . .
                    Superfamily Ceboidea
5   The Old World Monkeys . . . .
                    Superfamily Cercopithecoidea
6   The apes and human beings . . . .
                    Superfamily Hominoidea.

These six superfamilies comprise approximately

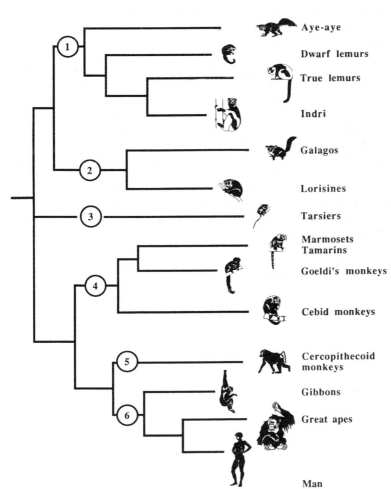

Aye-aye

Dwarf lemurs

True lemurs

Indri

Galagos

Lorisines

Tarsiers

Marmosets
Tamarins

Goeldi's monkeys

Cebid monkeys

Cercopithecoid
monkeys

Gibbons

Great apes

Man

Fig. 2.1 Phylogeny of the extant primates, showing relationships between the six superfamilies (1. Lemuroidea, 2. Lorisoidea, 3. Tarsioidea, 4. Ceboidea, 5. Cercopithecoidea, and 6. Hominoidea). The Tarsiers occupy an intermediate position between superfamilies 1+2 and other primates. Traditionally the tarsiers have been included with superfamilies of 1+2 in the suborder Prosimii, whereas superfamilies 4, 5, and 6 (the monkeys, apes, and man) have been placed in the suborder Anthropoidea (Simpson 1945). An alternative arrangement, discussed in the text, is to place the lemuroids and lorisoids together in the Strepsirhini, whilst uniting the tarsiers with the monkeys, apes and man in the Haplorhini. (Redrawn and modified from Martin 1995.)

Table 2.1 The extant primates: alphabetical list of the genera belonging to each of the six superfamilies, together with their common names

Genus	Common name	Number of species
1. Lemuroidea: Madagascar lemurs		
Allocebus	Hairy-eared dwarf lemur	1
Avahi	Avahi	1
Cheirogaleus	Dwarf lemurs	2
Daubentonia	Aye-aye	1
Hapalemur	Gentle lemurs	3
Indri	Indri	1
Lemur[a]	True lemurs	5
Lepilemur[b]	Sportive lemurs	7?
Microcebus	Mouse lemurs	4
Phaner	Fork-crowned lemur	1
Propithecus	Sifakas	3
Varecia[c]	Variegated lemur	1?
2. Lorisoidea: Lorises and Galagos		
Arctocebus[d]	Angwantibo	1 (2?)
Euoticus	Needle-claw bushbabies	2
Galago[e]	Bushbabies or galagos	5?
Galagoides[e]	Dwarf bushbabies	7
Loris	Slender loris	1
Nycticebus	Slow lorises	2?
Otolemur[f]	Greater galagos	2?
Perodicticus[g]	Potto	1?
3. Tarsioidea: Tarsiers		
Tarsius	Tarsiers	5?
4. Ceboidea: New World monkeys		
Alouatta	Howler monkeys	7
Aotus[h]	Owl monkeys	9?
Ateles[i]	Spider monkeys	4–6?
Brachyteles[j]	Woolly spider monkey	1
Cacajao	Uakaris	2
Callicebus[k]	Titi monkeys	3–13?
Callimico	Goeldi's monkey	1
Callithrix[l]	Marmosets	9?
Cebuella	Pygmy marmoset	1
Cebus	Capuchin monkeys	4
Chiropotes	Bearded sakis	2
Lagothrix	Woolly monkeys	2
Leontopithecus	Golden lion tamarins	4
Pithecia	Sakis	2–5?
Saguinus	Tamarins	12?
Saimiri	Squirrel monkeys	4–5?
5. Cercopithecoidea: Old World Monkeys		
Allenopithecus	Allen's swamp monkey	1
Cercocebus	Terrestrial mangabeys	3
Cercopithecus	Guenons	27?
Colobus[m]	Colobus/guerezas	6?
Erythrocebus	Patas monkey	1
Lophocebus[n]	Arboreal mangabeys	2
Macaca[o]	Macaques	19?
Mandrillus	Mandrill/drill	2
Miopithecus[p]	Talapoin	1 (2?)
Nasalis	Proboscis monkey	1
Papio[q]	Baboons	5?
Presbytis[r]	Langurs and leaf monkeys	19?
Pygathrix	Douc langur	1
Rhinopithecus[s]	Snub-nosed langurs	2–4?

Table 2.1—*Continued*

Genus	Common name	Number of species
Simias	Pagai Island Langur	1
Theropithecus	Gelada baboon	1
6. Hominoidea: Apes and Human Beings		
Gorilla[t]	Gorilla	1–2?
Homo	Human beings	1
Hylobates[u]	Gibbons/siamang	9
Pan[v]	Chimpanzee/bonobo	2–3?
Pongo[w]	Orang-utan	1–2?

Based upon Hershkovitz 1977; Napier and Napier 1985; Gautier-Hion *et al.* 1988; Mittermeier *et al.* 1988; Martin 1990; Alterman *et al.* 1995 and Rowe 1996.
? The exact number of species in the genus is debated.
[a] Some taxonomists place *L. catta* in the genus *Lemur*, and remaining species in a separate genus, *Eulemur*.
[b] The original species, *Lepilemur mustelinus* has been split into 7 spp. on the basis of genetic evidence.
[c] Two very distinct subspecies of *Varecia variegata* are known.
[d] Some authorities recognize two species of *Arctocebus*; *A. calabarensis* and *A. aureus*.
[e] Further species of *Galago* and *Galagoides* probably remain to be described.
[f] *Otelemur crassicaudatus* and *O. garnettii* are referred to as Galago spp. in this book.
[g] A second potto species *Pseudopotto martini* is known only from skeletal material.
[h] Two species groups (red-necked and grey-necked owl monkeys) are distinguishable; the exact number of species is debated.
[i] *Ateles chamek* and *A. marginatus* are only accorded subspecific rank by some taxonomists.
[j] Northern and southern forms of *Brachyteles* are ranked as separate species by some authorities.
[k] The status of many *Callicebus* species is still unresolved.
[l] Some *Callithrix* 'species' are ranked only as subspecies by some authorities. Hershkovitz recognizes five subspecies of *C. jacchus*, for example, whereas some taxonomists split these into five species.
[m] More species are recognized by some workers; the red colobus for example has been classified as a single species with 14 subspecies, or as four distinct species.
[n] The arboreal mangabeys are referred to as members of the genus *Cercocebus* (e.g., *C. albigena*) in this book.
[o] The exact number of Sulawesian macaque species varies in different texts.
[p] The Northern and Southern forms of the talapoin may be separate species, but this remains unresolved.
[q] Some authorities place all forms in the single species *P. hamadryas*. Five baboon species are recognized in this book.
[r] *Presbytis* is a large and complex genus; some of these monkeys are placed in the genera *Semnopithecus* and *Trachypithecus* by various taxonomists.
[s] *Rhinopithecus avunculus* and *R. roxellana* are recognized. Some authorities separate the three subspecies of R. roxellana into distinct species.
[t] The more western of the three *Gorilla* subspecies (*G. g. gorilla*) may warrant specific rank.
[u] The siamang is traditionally placed in its own genus, *symphalangus*, and is referred to by its original name in this book.
[v] The most western of the three *Pan* subspecies (*P. troglodytes verus*) may warrant specific rank.
[w] Genetic differences between Bornean and Sumatran orang-utans lead some taxonomists to place them in two separate species.

58 genera containing 229–252 species of primates (Table 2.1). I say 'approximately' because primate

taxonomy is a restless subject and the exact numbers of species and genera are still debated. Just occasionally new species are discovered; testimony to the fact that the world's shrinking rainforests still contain some surprises for mankind. Examples include the golden bamboo lemur (*Hapalemur aureus*: Meier *et al.* 1987) and Tattersall's sifaka (*Propithecus tattersalli*: Simons 1988), both from Madagascar, the suntailed monkey (*Cercopithecus solatus*: Harrison 1988) from Gabon, Diana's tarsier (*Tarsius dianae*: Niemitz *et al.* 1991) from Central Sulawesi, new marmosets from the Amazonian rainforest (e.g., *Callithrix mauesi*: Mittermeier *et al.* 1992), and an array of new and still unclassified African galagos (Bearder *et al.* 1995).

Phylogenetic relationships between the six primate superfamilies are represented in simplified form in Fig. 2.1, which incorporates the two major arrangements used in various texts. Firstly, the Lemuroidea, Lorisoidea, and Tarsioidea have traditionally been included within a single suborder, the Prosimii, whilst the remaining three superfamilies are grouped together as the simians or anthropoids (suborder Anthropoidea: Simpson 1945). The prosimians are anatomically more similar to stem forms which gave rise to the modern primates and for this reason they are sometimes (misleadingly) referred to as 'lower primates'. The terms *prosimian* and *anthropoid* will be used throughout this book. However, a second valid phylogenetic arrangement depicted in Fig. 2.1 takes account of the peculiar taxonomic position of the superfamily Tarsioidea. The tarsiers exhibit a spectrum of anatomical features, some of which they share with the extant prosimians and others with the anthropoids. For instance, in common with monkeys and apes, the tarsiers lack the moist rhinarium which characterizes the nasal structure of prosimians (Fig. 2.2). The tarsiers are accordingly sometimes included with the anthropoids as members of the suborder Haplorhini, whilst the malagasy lemurs and the lorisoids are placed in the suborder Strepsirhini.

Brief descriptions of the major features of each of the six major groups of primates are provided below, together with illustrations of some of the many species which will be referred to in the chapters which follow. These sections are provided for the benefit of those readers who are not specialists in primatology.

The modern-day lemurs represent the survivors of a remarkable adaptive radiation of prosimians which

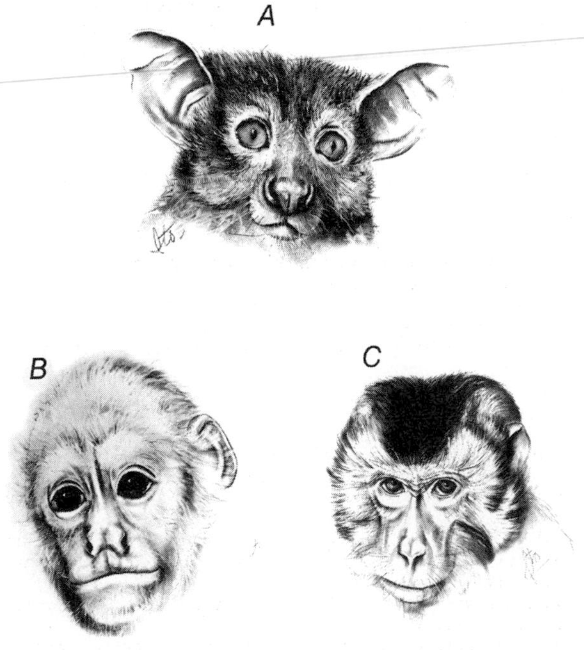

A

B   *C*

Fig. 2.2 Portraits of typical prosimian and anthropoid primates. A. Greater bushbaby (*Galago crassicaudatus*), a prosimian, showing the moist rhinarium typical of the lorisoids and lemuroids. B. Capuchin monkey (*Cebus*) with the widely-spaced, laterally directed nostrils found in New World primates (*Platyrrhines*). C. Pigtailed macaque (*Macaca nemestrina*) with nostrils situated close together and pointing downwards as in other Old World monkeys (*catarrhines*). (From Dixson 1981.)

Fig. 2.3 Fat-tailed dwarf lemur (*Cheirogaleus medius*). (Photograph by Professor Bob Martin.)

Fig. 2.4 Sportive lemur (*Lepilemur leucopus*) photographed at night and showing the effects of the reflecting layer or *tapetum lucidum* situated between the choroid and retina in the eye. (Photograph by Professor Bob Martin.)

Fig. 2.5 Female ringtailed lemur (*Lemur catta*) with young. (Photograph by Professor Bob Martin.)

important role (Bearder 1987; Bearder *et al.* 1995; Charles-Dominique 1977). Like other prosimians, mating is restricted to a limited period during the

took place in Madagascar. Secure from competition from mammals of the African mainland, the Malagasy lemurs produced a variety of arboreal and terrestrial forms, many of which have become extinct since Man first occupied the region 2000 years ago (Tattersall 1982). The survivors include the diminutive mouse lemurs, the dwarf lemurs, sportive lemurs, and the bizarre aye-aye, all of which are nocturnal. Diurnal species include the ringtail lemur, brown lemur, and the sifaka, which live in multi-male-multifemale groups, and the indri which is a pair-living species. Some examples of the Malagasy lemurs are shown in Figs 2.3–2.6.

The lorisoidea includes the galagos or 'bushbabies' as well as the slower-moving potto and angwantibo of the African rainforest and their relatives, the slow loris and slender loris of South East Asia. All are non-gregarious and nocturnal. The sexes usually occupy separate home ranges and female ranges overlap with those of males quite extensively. Social and sexual communication is complex in these animals and the secretions of specialized cutaneous glands, urinary cues and vocalizations all play an

Fig. 2.6 The sifaka (*Propithecus verreauxi*). (Photograph by Professor Bob Martin.)

Fig. 2.7 Lesser galagos (*Galago moholi*) feeding on the gum of an acacia tree. (Photographed in S. Africa by Dr Simon Bearder.)

peri-ovulatory phase of the female's oestrous cycle. This rigid control of sexual receptivity by ovarian hormones represents an important physiological distinction between the prosimians and the anthropoids. Prosimians also exhibit other specializations, such as possession of a 'toilet claw' on the second toe of each foot and a reflecting layer or *tapetum lucidum* behind the retina, (see Fig. 2.4), which enables them to make maximum use of the poor illumination available during nocturnal activities. Some typical lorisoids are shown in Figs 2.7–2.10. It will be noted that they display fundamental adapta-

tions found in all primates, including man. Thus, the eyes point forwards, enabling stereoscopic vision and most of the digits on the hands and feet bear flattened nails rather than claws. Such features are viewed as an inheritance from the nocturnal ancestors of primates and advantageous both for a life spent in the trees and also for hunting small prey. Cartmill (1974) has argued that just as selection has favoured the evolution of forward-pointing eyes in nocturnal predators such as owls and felids, so it may have encouraged the development of stereoscopic vision in ancestral primates. Modern prosimi-

Fig. 2.8 Greater galago (*Galago crassicaudatus*). (Photographed in S. Africa by Dr Simon Bearder.)

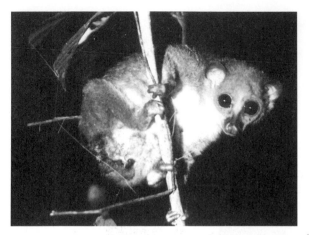

Fig. 2.9 Angwantibo (*Arctocebus calabarensis*). (Photographed in Cameroon by Dr Simon Bearder.)

ans such as the bushbabies or angwantibos feed upon insects and other invertebrates as well as upon a vegetarian diet. They lack colour vision, however, and the retina contains only rods, rather than cones and rods as in the diurnal monkeys and apes.

The final prosimian Superfamily, the Tarsioidea, contains a single surviving family; the tarsiers (Fig. 2.11). These small nocturnal creatures are found only on Borneo, Sumatra, Sulawesi, and the Philippine islands. They are non-gregarious in their social organization, although one species (*Tarsius spectrum*) has been reported to live in monogamous pairs. As stated previously, the tarsiers share some anatomical features with prosimians and others with anthropoids. However, as we shall see, the ancestors of tarsiers diverged very early from the line which gave rise to the anthropoids. Modern tarsiers exhibit many prosimian features, such as possession of large

Fig. 2.11 Horsfield's tarsier (*Tarsius bancanus*). (After Fogden 1976.)

forward-pointing eyes (albeit lacking a tapetum lucidum), large mobile ears and toilet claws (on two toes). In this book, the tarsiers are included in the suborder Prosimii.

Fig. 2.10 Slow loris (*Nycticebus coucang*).

Fig. 2.12 Bald uakari (*Cacajao calvus*). The skin on the head reddens and the hair is lost as these South American monkeys become sexually mature. (Photograph by Dr R. A. Mittermeier.)

Fig. 2.13 The rare black-faced saki (*Chiropotes satanas*). The head hair of these South American monkeys is strikingly arranged and both sexes also possess a well-developed beard. (Photograph by Dr R. A. Mittermeier.)

The New World monkeys (Superfamily Ceboidea) may be divided into two families; the Cebidae and the Callitrichidae. The family Cebidae includes many of the more monkey-like forms such as howlers, spider monkeys, woolly monkeys, capuchins, uakaris, and squirrel monkeys (Figs 2.12–2.17). Many cebids live in multimale–multifemale social groups but one-male groups have been recorded in some howler species and in several cases monogamous family groups are the norm (e.g., owl monkeys, titi monkeys, and white-faced sakis). The family Callitrichidae contains the smaller-bodied marmosets and tamarins (see Figs 2.18–2.23), ranging from 80 g in body weight (pygmy marmoset) to over 700 g (golden lion tamarins). These monkeys live in monogamous family groups, although a polyandrous mating system also occurs in some species. Females give birth to twins, and adult males and older siblings assist the breeding female in carrying and caring for the offspring. The New World genus *Callimico* (Goeldi's monkey, Fig. 2.24) is exceptional in sharing anatomical features with both the Callitrichidae and the Cebidae. Despite its disputed taxonomic position, however, modern molecular evidence indicates that Goeldi's monkey is most closely related to the marmosets and pygmy marmosets. Goeldi's monkey lives in monogamous family groups. A single offspring is produced, rather than twins, and the adult male assists the female in transporting the infant. In some wild groups of *Callimico* a polygynous mating system

Fig. 2.14 Woolly monkey (*Lagothrix lagotricha*). Four subspecies of this South American monkey have been described; this example is Poeppig's woolly monkey (*L. l. poeppigii*). (Photograph by A. Young, supplied by courtesy of Dr R. A. Mittermeier.)

may occur, two breeding females being present within a single group.

All the New World monkeys are arboreal and several genera of the Cebidae (e.g., Spider monkeys, woolly, and howler monkeys) possess prehensile tails (see Fig. 2.15). Apart from the owl monkey, which is nocturnal, all the New World monkeys are diurnal. Cutaneous scent-marking glands occur in the sternal or epigastric area as well as the circumgenital region. Many New World monkeys also scent-mark using urine. Olfactory communication is important to these monkeys, but in addition they employ facial expressions and postures for visual communication. Piloerection displays are especially prevalent among the marmosets and tamarins.

The Old World monkeys (Superfamily Cercopithecoidea) represent an extensive adaptive radiation of terrestrial and arboreal forms, distributed across Africa, Asia, and South East Asia. All the Old World monkeys belong to the single family

Fig. 2.16 Black howler (*Alouatta pigra*) from Belize in Central America. (Photograph by Dr R. A. Mittermeier.)

Fig. 2.15 The remarkable woolly spider monkey, or muriqui, (*Brachyteles arachnoides*) is the largest South American monkey. Isolated populations are found in surviving pockets of the Brazilian Atlantic coastal forest. (Photograph by C. P. Noguerra, supplied by courtesy of Professor Karen Strier.)

arboreal rainforest species. Some of the largest monkeys belong to this group, such as the baboons, geladas, and mandrills of Africa and the macaques of Asia and South East Asia (Figs 2.26–2.28). In addition to the better known macaque species such as the rhesus monkey, stumptail, cynomolgus monkey, and Japanese macaque, which have been exten-

Cercopithecidae which is divisible into two subfamilies; the Colobinae and the Cercopithecinae. Colobines include the langurs, such as the Hanuman langur of India which is a semiterrestrial species (Fig. 2.25), as well as an array of arboreal Asiatic rainforest leaf-monkeys, proboscis monkeys, and the African colobus species. These monkeys have complex stomachs to aid digestion of their predominantly folivorous diet. Many form one-male units; a single breeding male associates with a small group of females as in the black and white colobus of Africa or the proboscis monkey of Borneo. Others are found in large multimale-multifemale groups as in the red colobus monkeys of Africa and some populations of the Hanuman langur. In one case, *Presbytis potenziani*, adults live in pairs and the mating system is believed to be monogamous. Cercopithecines are represented by a vast array of terrestrial or semi-terrestrial monkeys, as well as by

Fig. 2.17 Brown capuchin (*Cebus apella*). (Photograph by Dr R. A. Mittermeier.)

Fig. 2.18 The common marmoset (*Callithrix jacchus*), which occurs in Northern Brazil. (Photograph by Dr R. A. Mittermeier.)

Fig. 2.20 Red-crested bare-face tamarin (*Saguinus oedipus geoffroyi*). Adult carrying twin offspring. In the closely-related cotton top tamarin (*S. o. oedipus*), the head hair is much longer. (Photograph by Dr R. A. Mittermeier.)

sively used for biomedical research, one species, the Barbary macaque (*Macaca sylvanus*), occurs in North Africa in the Atlas mountains and has been secondarily introduced to the rock of Gibraltar (Fig. 2.29). The more exclusively arboreal cercopithecines in-

clude the brightly coloured guenons (*Cercopithecus spp*), the talapoins (*Miopithecus*), and Allen's swamp monkey (*Allenopithecus*). Together with one closely related terrestrial genus, the patas monkey (*Erythrocebus*) , these forms are grouped into a single 'tribe', the Cercopithecini (Figs 2.30 and 2.31). All

Fig. 2.19 A pair of buffy-tufted-ear marmosets (*Callithrix flaviceps*). (Photograph by Dr R. A. Mittermeier.)

Fig. 2.21 Pied tamarin (*Saguinus bicolor*). (Photograph by R. da Rocha e Silva, supplied by courtesy of Dr R. A. Mittermeier.)

Fig. 2.22 Emperor tamarin (*Saguinus imperator*). (Photograph by Dr R. A. Mittermeier.)

the remaining cercopithecines (i.e., baboons, macaques, mangabeys, geladas, etc.) are placed in a second tribe, the Papionini.

Some of the cercopithecines live in large multi-male-multifemale social groups. This is the case in macaques, chacma baboons, mandrills, talapoins, and vervet monkeys. Others, such as the patas monkey and most guenon species, form one male groups, although influxes of additional males may occur during the annual mating season in some of these species.

Old World monkeys are sometimes referred to as catarrhines because the nostrils point downwards and are narrow by contrast with the widely-separated and laterally directed nostrils of the New World platyrrhines (Fig. 2.2). Old World monkeys also possess hard pads of skin covering the ischial bones of the pelvic girdle; these sitting pads are lacking in New World monkeys. Aside from these anatomical features, and certain differences in dentition between Old World and New World monkeys, there are numerous other differences which involve the social lives and reproductive biology of these two anthropoid groups. In all the Old World monkeys studied to date, females exhibit menstrual cycles, as is also the case in the apes and man. Very few New World primates menstruate, and in those cases where menstrual bleeding occurs it is minor and difficult to detect externally. Whilst the external genitalia of female New World monkeys do not undergo marked morphological changes during the ovarian cycle, many female Old World monkeys exhibit oedema and reddening of the perineal *sexual skin* during the follicular and peri-ovulatory phases of the menstrual cycle. Sexual skin swellings are sexually attractive to males and they provide an example of the greater importance of visual signals, rather than olfactory signals, in the social and sexual lives of the Old World anthropoids. In contrast to the New World primates very few Old World monkeys possess cutaneous scent-marking glands and likewise urine-marking is exceedingly unusual. Whereas callitrichids and some other New World

Fig. 2.23 Golden-headed lion tamarin (*Leontopithecus rosalia chysomelas*). Found only in a small area of forest in Eastern Brazil and sometimes considered as a distinct species, rather than as a subspecies of the golden lion tamarin (*L. r. rosalia*). ((Photograph by Dr R. A. Mittermeier, supplied by courtesy of Dr Anthony Rylands.)

Fig. 2.24 Goeldi's monkey (*Callimico goeldii*). (Photograph by Dr R. A. Mittermeier, supplied by courtesy of Dr Anthony Rylands.)

Fig. 2.25 Hanuman langurs (*Presbytis entellus*). Members of an all male band, photographed at Jodhpur in N. India by Dr Volker Sommer.

monkeys have a functional vomeronasal organ and accessory olfactory system, the vomeronasal organ is usually lacking in Old World anthropoids except during foetal development (some exceptions to this generalization occur, however).

The sixth primate superfamily, the Hominoidea, contains three families; the lesser apes (Hylobatidae), great apes (Pongidae), and man (Hominidae). The lesser apes include the gibbons and the larger siamang, all of which are found in the rainforests of S.E. Asia. The lesser apes have sitting pads (ischial callosities) like those of Old World monkeys. However, even at first glance they are morphologically quite distinct from monkeys (Fig. 2.32). The chest is broad and the vertebral column is shorter than in monkeys. A tail is lacking and the arms are long in relation to the length of the legs. These apes are beautifully adapted for rapid arm-over-arm suspensory locomotion through the trees; they are the masters of *brachiation*. Although the great apes are also capable of moving in this way they lack the speed and grace of the gibbons. All the lesser apes live in small family groups consisting of an adult male, an adult female, and up to 3 or 4 offspring.

Fig. 2.26 A female chacma baboon (*Papio ursinus*); with her infant. Photographed in S. Africa by Dr Simon Bearder.

Fig. 2.27 Mandrill (*Mandrillus sphinx*). Adult female grooming the alpha male in a semi-free-ranging mandrill group in Gabon.

Fig. 2.28 Celebes (Sulawesi) macaques (*Macaca nigra*). Four females and a juvenile offspring; two of the females are inviting grooming by adopting a distinctive presentation posture (Photograph taken at Oregon Primate Research Centre by the late Dr Kurt Modahl.)

These primarily monogamous families are also territorial and adults of both sexes perform complex vocal duets as intergroup spacing displays.

The great apes include the orang-utan (*Pongo pymaeus*) of Borneo and Northern Sumatra and three African species—the chimpanzee (*Pan troglodytes*), the pygmy chimpanzee or bonobo (*Pan paniscus*), and the gorilla (*Gorilla gorilla*). Orang-utans are massive, primarily arboreal, and frugivorous apes. Adult males may weigh 70–80 kg, approximately twice the weight of the adult female. Such males are non-gregarious and territorial in their behaviour; they possess striking fatty flanges on each side of the face and a pendulous laryngeal sac. Adult males emit 'long-calls' which can be heard for over a kilometre and serve to communicate their positions to conspecifics. Females are also non-gregarious (Fig. 2.33) and when they are in breeding condition, they preferentially approach and consort with the larger territorial males. Smaller males, lacking the prominent secondary sexual adornments of the territorial males, attempt to mate opportunistically and coercively with females.

The African apes lead a partly terrestrial and partly arboreal existence and all make use of a specialized *knuckle-walking* form of quadrupedal locomotion. The fingers are flexed and the weight is born on the knuckles and second carpals which are covered with thick pads of skin. Although chimpanzees and gorillas can brachiate, they do so more slowly and cautiously than the lesser apes whereas they can move rapidly at ground level by knuckle-walking. The gorilla is the most massive of all the primates (Fig. 2.34). Adult males weight 160 kg or more and adult females approximately half this weight. Gorillas do climb, but much of their time is spent on the ground and, like all the great apes, they construct nests from branches and leaves which are used for sleeping purposes. They are primarily vegetarian, although the western lowland subspecies also includes substantial amounts of fruit in its diet. Social groups of gorillas usually consist of a small number of females and offspring accompanied by an adult male, referred to as a *silverback* because the hair on the dorsal surface of the body is white in contrast to the brownish or black colour of the

Fig. 2.29 Barbary macaques (*Macaca sylvanus*). Females and offspring. (Photograph taken on the rock of Gibraltar by Professor Bob Martin.)

Fig. 2.30 Vervet monkeys (*Cero-pithecus aethiops*). Several members of a free-ranging group, photographed in S. Africa by Dr Simon Bearder.

remaining pelage. In the mountain gorilla, two or three silverbacks occasionally co-exist within a single group, but the mating system is primarily polygynous —a single dominant male copulates with the females and is presumed to sire their offspring.

The chimpanzee is much smaller than the gorilla; adult males weigh approximately 40 kg and are only about 10% heavier than females (Fig. 2.35). Both the larger chimpanzee (*Pan troglodytes*) and the pygmy chimpanzee or bonobo (*P. paniscus*) live in large *communities* consisting of about 60 males, females, and offspring. These are not permanent multimale–multifemale groups, instead the commu-

nity fragments into smaller subgroups which forage separately and meet and mingle from time to time. This has been referred to as a *fusion–fission* type of social organization, quite unlike that of other apes, but not dissimilar to the social organization of the New World spider monkeys. Chimpanzees are frugivores and subgroups are able to move rapidly at ground level through the forest in search of suitable fruiting trees. Loud *pant-hoots* facilitate communication between members of different subgroups. These apes are extraordinary in terms of their cognitive development. Tool use occurs in the wild, and although the level of technology may be simple, such

Fig. 2.31 The talapoin (*Miopithecus talapoin*) is the smallest Old World monkey. Here an adult female is shown grooming a male.

as the use of grass stems to fish for termites or rocks to crack nuts, it is clear that chimpanzees possess unusual intellectual abilities. Their mating system is multimale–multifemale in structure; females exhibit sexual skin swelling during the follicular phase of the menstrual cycle and may mate with most of the adult males in their community when the swelling is largest and ovulation is most likely to occur. In the pigmy chimpanzee, sexual activities are especially frequent; patterns of sociosexual activity and copulatory postures are also more variable than those exhibited by chimpanzees.

Evolutionary relationships and time-scales

Figure 2.36 shows the evolutionary relationships of the living and extinct primates, together with time-scales for divergences of the various lineages. According to this schema, which follows Martin (1990, 1993) the treeshrews are excluded from the Order Primates. This contrasts with some earlier classification schemes (e.g., Clark 1959; Napier and Napier 1967). The earliest primates are thought to have lived during the Cretaceous period approximately 90–100 million years ago. In Fig. 2.36 an offshoot of the line which gave rise to modern primates is shown on the left hand side of the evolutionary tree. This side branch contained archaic primates including the plesiadapiform group, all of which had become extinct by the end of the Palaeocene (55 million years ago) or during the Eocene. During the late Cretaceous, the line which gave rise to modern primates had already diverged into two stocks, the prosimians and the tarsiers/anthropoids. It will be noted, however, that ancestors of the modern tarsiers had already diverged from the anthropoid stock during the Palaeocene between 55–66 million years ago. Hence, although tarsiers share various anatomical features with members of the Anthropoidea, their prosimian affinities are pronounced and it is valid to regard them as members of that suborder. The split between the ancestral stocks of New World and Old World monkeys is thought to have taken place during the late Eocene. Forerunners of the New World primates reached South America either via North America, by using land bridges which were available at times during the Eocene, or from Africa which was less separated from the South American

Fig. 2.32 Siamang (*Hylobates (Symphalangus) syndactylus*). Suspensory postures and terminal branch feeding. The male siamang has a prominent preputial tuft of black hair, which is clearly visible in the example on the right. (Photographs taken in Malaysia by Dr David Chivers.)

Fig. 2.34 Adult male western lowland gorilla (*Gorilla g. gorilla*). This is an old silverback, and white hairs have spread from his back onto the flanks, buttocks and thighs. (After Dixson, 1981).

Fig. 2.33 Orang-utan (*Pongo pygmaeus*). An adult female, climbing quadrumanously and carrying a small infant on her hip. (Photographed in Sumatra by Dr David Chivers.)

land mass at that time. Zoogeographers and palaeontologists still debate this issue (Ciochon and Chiarelli 1980; Martin 1990). The last common ancestors of the Old World monkey and ape lineages existed at the end of the Eocene, approximately 40 million years ago. At about that time, the line which gave rise to the lesser apes, great apes, and man diverged from the Old World anthropoid stock. Subsequently this gave rise to the lesser apes, the orang-utan lineage, and, later, to the ancestors of the African apes and man. Man is therefore viewed by most authorities as sharing a more recent common ancestry with the African chimpanzees and gorillas, rather than with the Asiatic genus *Pongo* (Fig. 2.37).

Evolutionary trees and time-scales serve to remind us of mankind's comparatively recent descent from ape-like hominid ancestors. Foley (1995) proposes that hominid evolution can best be appreciated 'not as a ladder, nor even as a bush of phylogenetic branches, but as a series of adaptive radiations taking place over the last 5 million years' (Fig. 2.37). The first of these radiations consisted of small bipedal, ape-like australopithecines, including *Aus-*

tralopithecus ramidus, *A. afarensis* and *A. africanus* whose fossils are known from eastern and southern Africa. Beginning approximately 2 million years ago, two further adaptive radiations emerge from these early bipeds—the robust australopithecines and the larger-brained hominids. Robust australopithecines included *Australopithecus robustus* and *A. boisei* which occurred in southern and eastern Africa, respectively. They were heavily built apes with large jaws and molar teeth suitable for coping with a fibrous vegetarian diet, but with comparatively small brains. By contrast, early representatives of the genus *Homo* were characterized by having larger brains, a smaller dentition, and more human bodily proportions. Although some representatives are known only from Africa (*Homo habilis*; *H. rudolfensis*) one of these larger-brained hominids, *Homo erectus*, dispersed as far as Java and China. Finally, there is the radiation consisting of archaic forms of *homo sapiens* culminating in the appearance of modern *Homo sapiens* at around 100 000 years ago.

The evolutionary schemata embodied in Figs 2.36 and 2.37 are important in the present context because they provide a background to comparative studies of sexual behaviour and reproductive physiology in extant prosimians and anthropoids. Sexual behaviour leaves no fossil record, but a sound knowledge of the evolutionary history of the primates can help us to suggest possible pathways for the evolution of behaviour and its physiological con-

Fig. 2.35 Chimpanzees (*Pan troglodytes*). Left: adult male; Right: adult female and infant. (After Dixson 1981.)

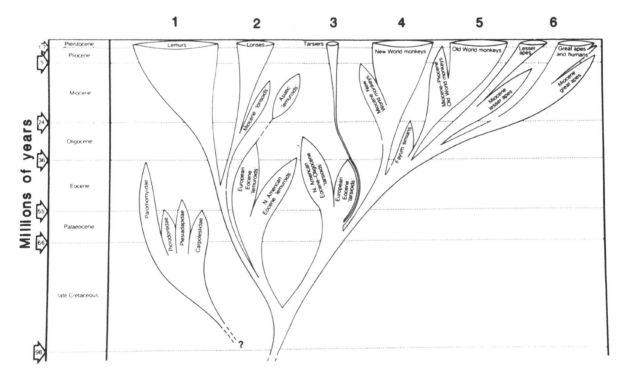

Fig. 2.36 Provisional outline phylogenetic tree for the Order Primates. This schema dates the ancestral stock which gave rise to the extant primates at 90–100 million years ago. Further discussion of the tree is provided in the text. (After Martin 1990.)

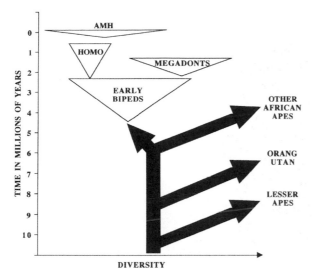

Fig. 2.37 Evolutionary relationships of the apes and hominids. Man is thought to have shared a common ancestry with the forerunners of the African apes; the divergence between hominids and the ancestral African apes is placed at between 5–6 million years ago. Hominid evolution may then be viewed a series of adaptive radiations. The first radiation was of early apes (australopithecines). The second involved the specialization of some australopithecines for a diet of coarse vegetable foods, resulting in megadontic adaptations. The third radiation saw the emergence of the genus *Homo*, becoming more encephalized and spreading beyond Africa. Finally, there was the appearance and dispersal of modern human beings. (Modified from Foley 1995.)

trol. This will become apparent, for example, as we consider the mating systems and mating tactics employed by primates (Chapters 3 and 4), their patterns of copulatory and sociosexual behaviour (Chapters 5–9) or the hormonal control sexual behaviour in prosimians and anthropoids (Chapters 10–14).

3 Mating systems

Primate social organization

The vast majority of primate species lives in complex social groups whose members stay together all the year round. This generalization applies with particular force to the anthropoids, for among the monkeys and apes only the orang-utan has a non-gregarious social organization. Among the prosimian primates, by contrast, the nocturnal forms include many examples of non-gregarious social organizations. This is not to say that these animals are solitary, for they occupy complex social networks and communicate with one another by means of olfactory and vocal displays (Bearder 1987). For the monkeys, apes, and man, however, group living is the norm and social groups range in size from small families (e.g. in *Aotus*, *Callicebus*, *Hylobates*, and *Symphalangus*) to multimale–multifemale units numbering scores or even hundreds of individuals (e.g. in *Macaca*, *Papio*, and *Mandrillus*). In an early attempt to explain this phenomenon, Zuckerman (1932) suggested that the propensity of anthropoid primates to engage in sexual activity throughout the year might act as a proximate mechanism to cement social relationships between the sexes and to maintain social groups. In its simplistic form, this provocative hypothesis is now known to be incorrect. Many Old World and New World monkeys are seasonal breeders, yet they continue to associate in groups at all times and not just during the annual mating season (Lancaster and Lee 1965; Lindburg 1986). Furthermore, in primates which form small family groups, copulation is infrequent once a pair is well-established (e.g. in *Symphalangus*: Chivers 1974; *Callicebus*: Mason 1968; *Aotus*: Dixson 1983a). Indeed, Kleiman (1977) has pointed out that monogamous primates are notable for their relatively low frequencies of mating behaviour, so that it seems unlikely that copulation alone could account for the maintenance of social groups. Modern field research has also demonstrated that major differences often exist between the social system and the mating system of individual primate species. As an example, groups of mountain gorillas (*Gorilla g. beringei*) might contain

up to three or (exceptionally) four sexually mature, silverback males in addition to females and offspring (Schaller 1963). Yet the great majority of copulations are performed by the single, highest-ranking, silverback so that the mating system is one-male and not multimale (Harcourt *et al.* 1980). An interesting contrast is provided by social groups of forest-living *Cercopithecus* monkeys, such as *C. mitis* and *C. ascanius*. These monkeys usually occur in groups containing a single large adult male, a number of adult females and offspring of various ages. At a superficial level such groups appear to be invariably one-male units and polygynous in the sense that the resident male has exclusive access to females for sexual purposes. Yet detailed field studies have revealed that additional influxes of sexually active males can occur during the annual mating season (Cords 1988; Rowell 1988). Hence a 'one-male group' can convert to a multimale mating system in certain circumstances. Examples could be multiplied to strengthen the argument that the social organization of a primate group is not necessarily synonymous with its mating system. Whilst Zuckerman was incorrect in explaining group cohesion in terms of year-round mating activity, it is correct to say that sexual behaviour can result in profound changes in social organization and especially so in primates which are seasonal breeders. In the West African talapoin (*Miopithecus talapoin*), for example, the monkeys live all year round in large multimale–multifemale groups numbering as many as 60–80 individuals. For most of the year, however, the adult males occupy satellite positions on the fringes of the troop (Rowell 1973). It is only during the annual mating season that adult males move into the central core of the troop to associate and copulate with sexually attractive females (Rowell and Dixson 1975). These shifts in social organization, triggered as they are by hormonal changes and associated increases in sexual activity, are not accomplished without some conflict. Indeed, talapoins are extremely aggressive during their annual mating season, particularly the adult males. Increased intermale aggression during mating seasons has also been recorded in ringtailed lemurs (stink fights within and between groups; Jolly 1966), rhesus macaques (increased wounding and mortality; Wilson and Boelkins 1970) and in many other species (Dixson 1980).

Given that mating activity is patterned upon a pre-existing social organization in primate groups, the first question to ask when considering primate mating systems is 'why do most primates live in groups and what determines group size and composition?' A valid answer to this question must explain how group membership favours survival and reproductive success for its individual members. An amalgam of benefits might exist which outweighs the costs of group-living in terms of feeding competition or sexual conflict. Behavioural ecologists have stressed the advantages of group membership in combating predation pressures (Alexander 1974; Van Schaik 1983) or obtaining food (Wrangham 1979, 1980; Raemakers and Chivers 1980). Some workers recognize that both these selective forces have played a role in the evolution of primate groups (e.g. Jolly 1972; Clutton-Brock 1974) but the relative importance of predation and dietary factors has been much debated (Dunbar 1988). The search for a unitary causation of primate social groups may prove to be sterile since a combination of factors is almost certainly involved. In lions, for instance, foraging success alone does not account for group size. Female lions also form groups to provide a communal crèche for their cubs and to repel neighbouring prides during territorial disputes (Packer *et al.* 1990). Among the primates, a strong case has been made by Wrangham (1979, 1980) that females form the focus of many groups and that females band together primarily to exploit food resources in the most efficient manner. Males, by contrast, compete amongst themselves and are distributed with respect to the female groups; 'the entwined distribution of females and males yields the social system' (Fig. 3.1).

Wrangham's hypothesis rests heavily upon the theory of sexual selection, as propounded originally by Darwin (1871) and refined by Hamilton (1964), Williams (1966), and Trivers (1972). As noted in Chapter 1, Darwin envisaged two main selective processes; intrasexual selection, which he thought of principally in terms of competition between males for access to females, and intersexual selection, which he viewed as operating via female preferences for the most attractive males. In addition to this traditional view of sexual selection, the possibility of *competition between females* and of *male mate choice* as selective forces must also be considered, as these processes can also influence sexuality in primates. The fundamental reason why sexual selection might be biased towards intermale competition and female

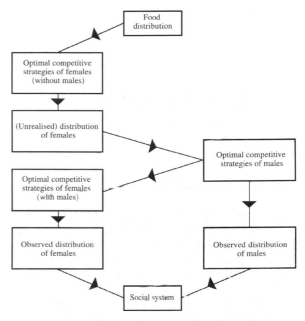

Fig. 3.1 Framework for the analysis of social (and mating) systems in anthropoid primates. Females compete for nutritional resources and this leads to a given distribution of females. Males compete for access to females and the entwined distributions of the two sexes yield the social system. (Redrawn from Wrangham 1979.)

choosiness was explained by Trivers (1972) in terms of different levels of 'parental investment' by the two sexes. Females typically invest far more in their offspring than males do. Female mammals nurture the developing offspring *in utero* and then provide milk and maternal care once birth has occurred. With the exception of certain monogamous species, males contribute little more than their spermatozoa to the reproductive equation. Under these circumstances females become a limiting resource for males as the latter compete to maximize their reproductive success (Williams 1966; Trivers 1972). This line of reasoning led Wrangham (1979) to suggest that female apes or monkeys might distribute themselves in the most advantageous manner to utilize food resources, and that kin-selection should be important in this process. Thus many groups are *female-bonded* in the sense that 'females maintain affiliative bonds with other females in their group and normally spend their lives in the group where they are born'. Indeed, the evidence then available indicated that in most social groups of primates, females are philopatric and that male-biased dispersal from natal groups is the norm. However, in four species (red colobus, hamadryas baboon, mountain gorilla, and chimpanzee) it was clear that females transfer between groups more readily than males do. With

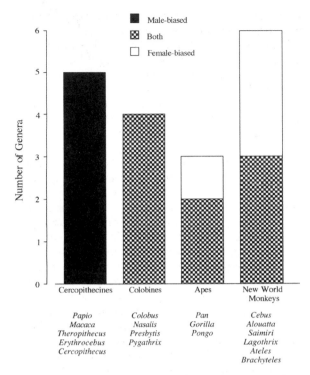

Fig. 3.2 Dispersal patterns in anthropoid primates. (Redrawn from Strier 1994.)

the passage of time, still more examples have come to light of anthropoid species in which both sexes disperse or in which female-biased dispersal is the usual pattern (Fig. 3.2). Strier (1994) has rightly cautioned that it is incorrect to regard female-bonded groups as the rule among anthropoids; they are far more common among Old World cercopithecines than among New World primates, for instance (Fig. 3.2). It is worth mentioning, however, that among many Old World prosimians, which often have non-gregarious social organizations, females tend to remain in their natal areas, whilst males emigrate and undergo a 'vagabond' stage of development as sexual maturity approaches (Charles-Dominique 1977; Bearder 1987). Thus, it appears that female pottos, lorises or galagos within a particular area of forest, are likely to be more closely related genetically than is the case for adult males.

The notion that females form alliances in order to exploit nutritional resources, and that such alliances are more common among related individuals and involve kin selection, is an attractive one. It allows a conceptual framework to be constructed for the evolution of female-bonded primate societies. Females lie at the core of such primate societies, whether they involve large groups of baboons or

macaques or collections of individual female home-ranges, as in galagos or lorisines. They represent a limiting resource for males, and males distribute themselves in the most advantageous way to ensure their own survival and to obtain mating opportunities. Where primate groups are not female-bonded, however (as in the gorilla or chimpanzee) is it not still the case that females form an important core to such groups, even though they are not always close relatives? Parish (1996), for example, notes that in the bonobo 'females are remarkably skilful in establishing strong affiliative bonds with each other despite being unrelated... The immediate advantage to females in such alliances is increased control over food, the main resource on which their reproductive success depends.' Females may thus represent a limiting resource in non-female-bonded groups and males still compete and distribute themselves in order to maximize their potential for reproduction. With these thoughts in mind we shall now consider the types of mating systems which occur in the Order Primates.

Primate mating systems

Animal mating systems are diverse and complex; a number of authors have attempted to classify these systems and to explain their evolutionary bases (birds: Emlen and Oring 1977; insects: Thornhill and Alcock 1983; mammals: Clutton-Brock 1989; vertebrates in general: Alcock 1984; Rees and Harvey 1991; Davies 1991). Some mating systems which have been well-described and studied in a variety of animals are not known to occur in the primates. As an example, we may consider the *lek* mating system, in which males aggregate in order to defend small individual territories; females visit these male display grounds in order to choose mates. Leks occur in various bird species (e.g. sage grouse: Bradbury and Gibson 1983), ungulates (e.g. Uganda kob: Buechner and Schloeth 1965), in the hammer-headed bat (*Hypsignathus monstrosus*: Bradbury 1977), in certain frogs (Wells 1977), and insects (Thornhill and Alcock 1983). Males aggregate on lekking grounds solely for sexual purposes and their small display territories provide no food or other resources for prospective mates. Some males are exceedingly successful under these circumstances. In hammer-headed bats, for example Bradbury (1977) found that 6% of males at one riverside lek accounted for 79% of probable copulations. Leks have therefore been likened to a polygynous mating system in which females exert considerable mate choice.

Although polygyny occurs in primates, as in many other mammals, nothing comparable to a lek-type mating system has been described. Given the fact that so many primates live in permanent groups with complex social infrastructures, it seems unlikely that leks would occur. In many cases, male monkeys remain associated with females in groups all the year round, even in those species which are seasonal breeders. Polygyny, when it occurs, involves a single male defending access to a small group of females, the so-called *female-defence* polygynous mating system.

Rather than review all possible animal mating systems and then attempt to place the primates within some universally acceptable scheme, I have chosen to define only those mating systems which occur in prosimians, monkeys, and apes. Human mating systems may also be set in context within this framework. Two important considerations are, firstly, whether a female usually mates with one male, or with more than one male, during the fertile phase of her ovarian cycle and, secondly, whether her sexual relationships are long-term and relatively exclusive or short-term and non-exclusive. This line of reasoning results in the recognition of five mating systems (see Table 3.1): 1. monogamy, 2. polygyny, 3. polyandry, 4. multimale–multifemale, and 5. dispersed or non-gregarious. The types of mating interaction which typify these five systems are represented diagramatically in Fig. 3.3. In monogamous, polygynous and, polyandrous primate groups, females have longer-term sexual relationships, either with a single male or, in the case of polyandry, with two or more partners. By contrast, in multimale–multifemale and dispersed mating sys-

tems females mate with a number of partners in a more labile, non-exclusive manner.

Rees and Harvey (1991) have cautioned that 'any species-specific mating system classification is likely to be an unfair representation of the true gametic pattern'. Thus, it might be incorrect to strait-jacket each primate species into one of the five mating systems depicted in Fig. 3.3. Firstly, we must acknowledge that more than one mating system can occur within a single species (e.g. monogamy and polyandry in certain marmosets or tamarins (family *Callitrichidae*), or polygyny, monogamy and polyandry in *Homo sapiens*). To address this problem I shall adopt the approach that a given species has a *primary mating system* and one or more *secondary systems*. Further, it must be kept in mind that any particular mating system is not impermeable to external pressures which affect sexual interactions. Extra-pair copulations occur outwith monogamous pairings, for instance, as has been amply documented in many avian species (Birkhead and Møller 1992). Extra-pair copulations have been described for the monogamous titi monkey (*Callicebus moloch*: Mason 1966), the white handed gibbon (*Hylobates lar*: Palombit 1994; Reichard and Sommer 1997) and the siamang (*Symphalangus syndactylus*: Palomit 1994) but we have little notion of their frequencies among monogamous primate species in general, or their consequences in terms of reproductive success. The occurrence of such additional matings does not alter the fact that the primary system is one of pair-formation and monogamy. Likewise in the case of female defence polygyny, resident males may be ousted during 'takeovers' by non-group males (e.g. in the Hanuman langur: Sugiyama 1967 or

Table 3.1 A classification scheme for primate mating systems

Number of males mating per female cycle	Type of Sexual relationship	Mating system	Examples
One	Long-term, exclusive	Monogamy	*Indri, Aotus, Callicebus, Symphalangus Hylobates*
One	Long-term, exclusive[†] but other females also mate with the resident male	Polygyny	*Theropithecus, Papio hamadryas, Nasalis, Gorilla*
Two or more	Long-term, exclusive	Polyandry	*Saguinus fuscicollis*[*]: *Callithrix humeralifer*[*]
Two or more	Short-term, not exclusive, gregarious	Multimale–multifemale	*Propithecus, Macaca,* most *Papio spp, Cercocebus* S*aimiri, Lagothrix, Pan*
Two or more	Short-term, not exclusive, non-gregarious	Dispersed	*Microcebus, Daubentonia,* most *Galago, spp., Perodicticus.*

[†] The single male mates with other females in the 'harem' unit.
[*] Monogamy also occurs in these callitrichid species, and is probably the primary mating system in many cases. From Dixson (1997a).

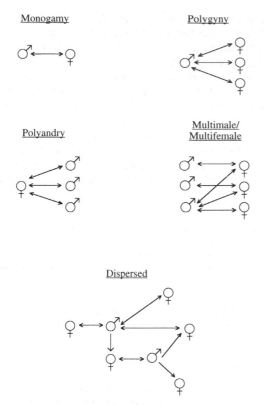

Monogamy

Polygyny

Polyandry

Multimale/
Multifemale

Dispersed

Fig. 3.3 Types of sexual interaction which occur in the five primary mating systems found among primates. In monogamous and polygynous mating systems, females mate mainly (but not necessarily exclusively), with a single male. Sexual relationships are usually of a long-term nature. In multimale–multifemale and dispersed mating systems, females more commonly mate with a number of males. Sexual relationships are more labile but they are not necessarily indiscriminate or 'promiscuous'. Polyandry is rare among primates; when it occurs a single female maintains long term relationships with two (or more) males rather than with a single partner.

gelada: Dunbar 1984). Lone males, living outside multimale–multifemale groups, may enter groups and mate with resident females (Japanese macaque: Tokuda 1961). Young males in such multimale–multifemale groups may emigrate and attempt to gain acceptance in neighbouring groups; females exhibit some preference for mating with these new males (Pusey and Packer 1987). Within any given mating system sexual selection favours those individuals which maximize their lifetime reproductive success. In this regard a bewildering array of *mating tactics* are observed in various primates. Males and females often pursue different tactical paths and, given the differences in parental investment between the two sexes (Trivers 1972), we should expect differences and conflicts to occur. As an example,

the multipartner matings which occur in multimale–multifemale primate groups are far from being totally indiscriminate or promiscuous. Temporary consortships can occur between particular partners during the fertile phase of the female's cycle (e.g. baboons: Seyfarth 1978; chimpanzees: Tutin 1979), and females sometimes exhibit sexual preferences for certain males (pigtailed macaque: Eaton 1973) and vice-versa (rhesus macaque: Herbert 1968). Long-term alliances between particular males and females have been described in yellow baboons (Smuts 1985) and these might influence mate choice by females when they enter breeding condition. Social rank can also have powerful effects upon mating frequency and reproductive success. In some species low-ranking males copulate less frequently and sire fewer offspring than more dominant conspecifics (e.g. in *Macaca fascicularis*: De Ruiter *et al.* 1992; *Mandrillus sphinx*: Dixson *et al.* 1993). Despite the differences in mating tactics exhibited by males and females I do not consider it useful to attempt a classification of 'male mating systems' and 'female mating systems'. For all its faults the scheme outlined in Table 3.1 has much to recommend it when considering effects of sexual selection upon the evolution of copulatory behaviour, genital morphology, sperm competition, and secondary sexual characteristics. In the following sections, the major features of each of the five mating systems are described in more detail.

Monogamy in primates

The tendency for primates to live in small family groups, in which a long-term, monogamous relationship exists between an adult male and adult female, has occurred repeatedly during the evolution of the Order. Monogamy occurs in prosimians such as *Indri indri* (Pollock 1979) and *Tarsius spectrum* (Mackinnon and Mackinnon 1980), in New World monkeys of the family *Callitrichidae* (Stevenson and Rylands 1988) as well as in some cebids (e.g. *Aotus*: Wright 1981), in very few Old World monkeys (e.g. *Presbytis potenziani*: Tilson and Tenaza 1976), and in the lesser apes (*Hylobates lar*: Carpenter 1940; *Symphalangus syndactylus*: Chivers 1974). A listing of monogamous primate species is provided in Table 3.2, together with information on group sizes and adult body weights. In some cases it is not known whether particular species are monogamous, due to lack of sufficient field data on their sexual behaviour. The ruffed lemur (*Varecia variegata*) has been classified as pair-living by some authors

(Tattersall 1982; White 1991) but it has been suggested that small groups encountered in the wild are parts of a larger fusion−fission type of social organization (Pereira *et al.* 1988). Several nocturnal prosimian species have been observed feeding and associating in pairs or small groups (e.g. *Avahi laniger* and *Phaner furciver*: Pollock 1979; *Galago zanzibaracus*: Harcourt and Nash 1984). However, in the absence of information on copulatory behaviour and on the biological relatedness of group members a monogamous classification is provisional at best. Better information exists for the spectral tarsier (*Tarsius spectrum*) in which adults forage as male−female pairs, exhibit complex vocal duets, and sleep together during the daytime (Mackinnon and Mackinnon 1980).

All the monogamous primates are arboreal and inhabit quite small home ranges or territories. White-handed gibbons (*Hylobates lar*) live in family groups of three or four individuals with a territory size of approximately 30−50 hectares. Both sexes produce loud vocalizations, such choruses being most frequent and intense during the early morning hours. Groups often approach the margins of their territo-

Table 3.2 Group sizes and body weights of monogamous primates

Species	Group size	Adult body weights (g) Male	Female	Source
Prosimians				
Indri indri	2+ offspring	10 500	10 500	Pollock 1979
Lemur mongoz	2+ offspring	1800	1800	Tattersall 1982
Tarsius spectrum	2	200	200	Mackinnon and Mackinnon 1980
New World monkeys				
Cebuella pygmaea	2−11 (6)	122	110	Soini 1988
Callithrix jacchus	3−13 (8.6)	310	290	Hubrecht 1984
Callithrix humeralifer	8−15 (11.5)	300	300	Rylands 1981, 1982
Callithrix argentata	5−14			Stallings and Mittermeier 1983
Saguinus oedipus geoffroyi	3−9 (6.8)	500	510	Dawson 1976, 1979
Saguinus fuscicollis	3−8 (5.8)	420	370	Snowdon and Soini 1988 Yoneda 1981
Saguinus mystax	3−9 (5.3)			Snowdon and Soini 1988
Saguinus midas	3−6 (5.9)	600	530	Mittermeier 1977
Saguinus labiatus	3−6 (4.2)			Yoneda 1981
Saguinus leucopus	2−8 (4.7)			
Leontopithecus rosalia	2−8 (3 or 4)	560	550	Kleiman *et al.* 1988
Callimico goeldii	(4 or 5)	400−555	481	Heltne *et al.* 1981
Aotus lemurinus[†]	(4)	923	1009	Wright 1981
Aotus azarae	(4)	1141	1091	Wright 1981
Callicebus moloch	2−5 (3.2)	1100	1050	Mason 1968; Kinzey 1981
Callicebus torquatus	(4)	1100	1100	Defler 1983
Pithecia pithecia	2−5 (2.67)	1600	1400	Buchanan *et al.* 1981
Pithecia monarchus	2−6 (2.4)	−	−	Buchanan *et al.* 1981
Old World monkeys				
Presbytis potenziani	2−6	6500	6400	Tilson and Tenaza 1976
Apes				
Hylobates lar	(3.3)	5700	5300	Ellefson 1968
Hylobates hoolock	(3.2)	6900	6500	Preuschoft *et al.* 1984
Hylobates klossii	(3.7)	5700	5900	Preuschoft *et al.* 1984
Hylobates agilis	(4.4)	6000	5700	Preuschoft *et al.* 1984
Symphalangus syndactylus	2−6 (4 or 5)	10 900	10 600	Chivers 1974

Group sizes are given as ranges and means (in parentheses) where known.
[†] There are nine species in the genus *Aotus*: all are thought to be monogamous.
Polyandry also occurs in some callitrichid species, as a secondary mating system.

ries in response to vocal displays and adults may
display by leaping rapidly through the trees and by
branch-shaking (Ellefson 1968). There is little evidence that males outrank females in gibbon groups,
or in groups of other monogamous primate species.
The two sexes fulfil broadly similar roles in many
aspects of the social life of the group. Maturing
offspring emigrate from the family unit, both by
active peripheralization and in response to aggression from their parents, especially the parent of the
same sex (Carpenter 1940; Ellefson 1974). Selection
for monogamy in gibbons is thought to have operated via ecological constraints. Small-bodied females
are territorial, feeding upon scattered clumps of
high-quality foods (primarily fruits of various kinds).
Males are distributed with respect to the females,
which represent a limiting resource in reproductive
terms (Gittins and Raemakers 1980). Hostility shown
by females towards each other, coupled with lack of
male dominance, perhaps prevents males from associating with more than one mate and limits the
mating system to monogamy rather than polygyny.

Intermale competition for mating opportunities is
less intense in monogamous mating systems than in
those primates which are polygynous or which live in
multimale–multifemale groups. Sexual selection for
greater body size, increased weaponry (e.g. canine
size) or for striking secondary sexual adornments is
minimal in monogamous primate species. Typically,
adult males and females are of similar body weight
(Table 3.2) and are usually very similar in general
appearance (see Fig. 3.4). Male canine size is likewise modest and the same as, or only slightly larger
than, in the adult female. A few exceptions occur to
this rule of monomorphism in monogamous primates and the most striking concerns the white-faced
saki (*Pithecia pithecia*) of South America. The adult
male white-faced saki has a striking face mask and
black muzzle, completely different from that of the
female (Fig. 3.4). The reasons for this dichromatism
are unknown. In swallows, which are monogamous,
the male has longer tail feathers than the female.
Møller (1988a) has shown experimentally that the
male's longer tail is advantageous in encouraging
female mate choice. Males with longer tails pair
earlier and are more likely to be able to pair again
once the first brood of chicks has been reared.
Whether the white facial mask of *P. pithecia* plays
any role in mate choice is not known, although it
has been observed that extra-pair copulations occur
in this species in the wild (Rosenberger and
Norconk 1996).

Monogamy involves considerable parental investment by the male partner as well as by the female.
Although there have been no DNA typing studies of

Fig. 3.4 Examples of primarily monogamous primates.
Upper. A family group of owl monkeys (*Aotus sp.*) consisting of an adult female (right), an adult male carrying
a young infant (left) and a juvenile. Lower. An adult male
white-faced saki (*Pithecia pithecia*) photographed in the
wild, in Surinam. In this species only the male has a
distinctive white face mask and black muzzle. (Photos by
Dr R. A. Mittermeier and by courtesy of Dr Anthony
Rylands.)

any monogamous primate species (with the exception of *Callithrix jacchus* in which it was not possible
to assign paternity on the basis of DNA fingerprints:
Dixson *et al.* 1992), it is usually inferred that males
in family groups are probably the sires of most (if
not all) the offspring produced. Selection for parental
care of offspring has therefore been high and males,
as well as females, carry the offspring in a number
of species (*Indri*: Pollock 1979; *Callithrix jacchus*:
Ingram 1977; other marmoset and tamarin species:
Goldizen 1987; *Aotus*: Dixson and Fleming 1981;
Callicebus: Fragaszy *et al.* 1982; *Symphalangus*:
Chivers 1974). In the marmosets and tamarins it is

not uncommon for older siblings to carry infants or for non-reproductive adults to act as helpers in this way (Goldizen 1987). It is usually the case that a single female breeds in many of these species whilst her adult daughters exhibit suppression of ovarian function (e.g. *C. jacchus*: Abbott *et al.* 1993). Kin selection may operate in such groups to favour animals which carry the offspring of the dominant female, which are their genetic relatives. In some monogamous species, by contrast, only the adult female transports her offspring (e.g. in gibbons or in the white-faced saki). In such cases it is possible (but by no means certain) that the male contributes to the offspring's well-being in some indirect way, such as by territorial defence.

Polyandry in primates

Polyandrous mating systems are rare in animals, occurring for instance in a certain bird species (spotted sandpiper: Oring and Knudson 1973; jacana: Jenni and Collier 1972). Among the primates, polyandry is found as a secondary mating system in certain callitrichid species which are primarily monogamous. With increasing knowledge of callitrichid field biology it has become apparent that the single reproductive female in a social group might mate with more than one partner (*Callithrix humeralifer*: Rylands 1982, 1986; *Saguinus fuscicollis*: Goldizen 1987; *Saguinus mystax*: Garber *et al.* 1991). It has been suggested that 'facultative polyandry' occurs in such species; i.e., that two or more males copulate with a female and co-operate in rearing her offspring (Terborgh and Goldizen 1985; Sussman and Garber 1987). Terborgh and Goldizen argue that polyandry is adaptive in *S. fusicollis* because, for a newly formed reproductive pair lacking helpers, the task of carrying and caring for twins might be impossible. Twins may represent as much as 50% of their mother's body weight by the time they are weaned. The acceptance of an extra adult male as a helper, and the formation of a polyandrous group is advantageous, although, from the male's perspective it has the disadvantage that paternity is not assured given that both males mate with the female during her fertile period. Goldizen (1987) advances arguments that ovulation might be *concealed* in callitrichid primates and hence that a number of males might mate with a single female without any one of them having a reproductive advantage. However, fine-grained studies of sexual behaviour in two callitrichid species have shown that peri-ovulatory peaks in copulatory activity do occur,

and that these peaks are correlated with changes in female proceptivity and/or attractiveness (*Callithrix jacchus*: Dixson and Lunn 1987; *Saguinus oedipus*: Ziegler *et al.* 1993). It is most interesting, however, that female common marmosets remain attractive and proceptive towards their male partners if conception occurs, so that copulations continue into early pregnancy (Dixson and Lunn 1987). This may be one mechanism by which females maintain close relationships with adult males which transport the resulting offspring. The paternity of offspring in free-ranging marmoset and tamarin groups is not known and the biological relatedness of adults in such groups has not been determined by genetic means. The details of sexual interactions in 'polyandrous' groups are poorly understood and it remains uncertain whether polyandry is typical of the Callitrichidae as a whole, or whether it is restricted to particular species or populations (Goldizen 1987; Dixson 1993a).

Polyandry occurs in a few human societies, but only rarely and under unusual conditions. In parts of Tibet, for instance, polyandry involves two or more brothers marrying the same woman and co-operating to farm the family lands (Crook and Crook 1988). Even in such societies, polyandry is not universal and monogamous marriages also occur.

Polygyny in primates

In this book, the term *polygyny* or *polygynous mating system* refers to an enduring relationship between a single male and a number of females for the purposes of mating and production of offspring (Table 3.1). This system of female defence polygyny is equivalent to a harem-type of mating arrangement in which a single male typically mates with all the females in his unit, whilst the females mate primarily with this male. It is not uncommon in the literature on animal behaviour, behavioural ecology, or reproductive biology to find the term polygyny applied to other mating systems where males mate with numbers of partners, irrespective of whether one male has a more exclusive relationship with such females (e.g. Daly and Wilson 1983; Alcock 1984; Davies 1991). Thus, multimale–multifemale primate groups, such as those of macaques, savannah baboons, or red colobus monkeys are referred to as polygynous by some authors. I believe that this approach is misleading, and causes confusion when effects of sexual selection in polygynous mating systems are considered. In a macaque or savannah baboon group, for example, females commonly mate with a number of males during a single ovarian cycle. The system could just as well be labelled polyandrous and to employ the term polygyny in this context places an incorrect bias on the males' contribution to multi-partner copulations. Polygyny may be better defined with respect to a single male's defence of a group of females. This is the sense, for instance, in which Short (1979) or Harcourt *et al.* (1981a) applied the term in their studies of sexual selection, relative testes size, and sperm competition in primates. Polygynous species such as the gorilla experience relatively little sperm competition and relative testes size is low. In macaques by contrast, sperm competition is an important issue, because females often mate with multiple partners, and males have large testes in relation to their body weight. Effects of sexual selection, via sperm competition, are quite different in polygynous one-male units from those which operate in a multimale–multifemale system. Hence, it is important at the outset not to confuse the two types of mating system by referring to males as 'polygynous' in both cases. This is not to say that polygynous, harem-type units are impervious to influxes of additional males or that a single male has lifetime 'tenure' within a female unit. However, it should be the case that the resident male in such a unit is the principal mating partner and that he has a reproductive advantage over non-unit males during the period when he defends a group of females.

Among the primates, a polygynous mating system is found in a number of catarrhine species, such as black and white colobus, proboscis monkeys, various guenons (*Cercopithecus* species), the patas monkey, gelada, hamadryas baboon, and the gorilla. In some cases, such as the Hanuman langur (*Presbytis entellus*) one-male units and multimale–multifemale social groups occur within the same species (Jay 1965; Sugiyama 1967). Polygyny is much rarer among the platyrrhine monkeys of the New World (some groups of howlers are one-male units for instance: for example, *Alouatta palliata*: Crockett and Eisenberg 1987) and I know of no case where one-male units occur in prosimians. Table 3.3 provides a listing of polygynous primate species together with data on adult body weights and on the numbers of females which occur in a typical unit. Comparison of Table 3.3 with Table 3.2 on monogamous primates reveals that sexual dimorphism in body weight is much more extreme in the polygynous species. Male proboscis

Table 3.3 Examples of primate species which have polygynous mating systems: group sizes, number of adult females per group, and adult body weights of both sexes

Species	Group size	No. of adult females per group	Body weights (kg) Adult male	Body weights (kg) Adult female	Source
*Alouatta seniculus**	7.8	2.6	8.1	6.4	Crockett and Eisenberg 1987
Colobus guereza	12.3	5.0	11.8	9.25	Struhsaker and Leland 1979
Presbytis johnii	8.9	2.4 / 1–5	14.8	12.0	Poirier 1979
P. senex	8.4	4.6 / 3–7	8.5	7.8	Rudran 1973
P. cristatus	14–19	10–12	8.6	8.1	Wolf and Fleagle 1977
*P. entellus**	15.6	8.8 / 6–13	18.4	11.4	Sugiyama 1967
Nasalis larvatus	12.6	5.0	20.3	9.9	Yeager 1990a
*Cercopithecus mitis***	18.25	7.25	7.6	4.4	Struhsaker and Leland 1979
*C. ascanius***	15–50	10.1 / 5–15.5	4.2	2.9	Struhsaker and Leland 1988
*Erythrocebus patas***	20.3	7.1	10.0	5.6	Hall 1965
Theropithecus gelada	10.0	3.7 / 1–10	20.5	13.6	Dunbar 1984
Papio hamadryas	7.3	1–9	21.5	9.4	Sigg *et al.* 1982
*Gorilla g. beringei****	16.9	6.2 / 3–10	160.0	93.0	Schaller 1963

Data are means and/or ranges. Body weights are from Harvey *et al.* 1987. * Multimale–multifemale groups also occur in these species.
** Influxes of extra-group males sometimes occur during annual mating seasons in these species, as is described in the text.
*** Although as many as 3 or 4 silverbacks can co-exist in mountain gorilla groups, the mating system is polygynous as (virtually) all copulations are performed by a single, dominant, resident male.

monkeys or gorillas, for instance, might weigh two or three times as much as an average adult female. The effects of sexual selection on male body size, weaponry (canine size), and secondary sexual characteristics in the various mating systems, will be dealt with in detail in Chapter 7. It is sufficient at this point to note that intermale competition (and perhaps also intersexual selection) has favoured polygynous males which are larger, more formidable, and often more brightly adorned than males of other mating systems. Secondary sexual adornments include the capes of hair of male hamadryas baboons and geladas, the brilliant red sexual skin on the chest of the gelada, the extraordinarily prominent nose of the male proboscis monkey, and the silver saddle of hair on the back of the male gorilla.

In the majority of polygynous primates, one-male units occupy spatially separate home ranges. This is the case, for example, in the patas monkey, black and white colobus, or gorilla. However, in three species, the hamadryas baboon, the gelada, and the proboscis monkey, one-male units are small components of much larger and more complex social groups. One-male units typically forage separately for at least part of the day but, as evening approaches, they re-assemble and sleep together as a single troop. In the hamadryas baboon of Ethiopia and Arabia troops commonly number 100 or so individuals, with a maximum size of 750. As Fig. 3.5 shows, the troop is in reality composed of a number of *bands* each of which contains several *clans*; each clan contains between 1–4 one-male units (Kummer 1968, 1990; Sigg *et al.* 1982; Abegglen 1984). Troops gather at sleeping cliffs each evening since these provide security from nocturnal predators. Bands do not mix on the sleeping cliffs and in the morning the members of each band depart in the same direction from the sleeping site. Thereafter the band splits into its component clans and these forage separately. Kummer (1990) comments that 'clans can split into one-male units for an hour or so but do not normally lose contact. There are no specific cohesive mechanisms. Females often walk in an approximate line between their one-male unit leader and a second one-male unit leader'. The females in these units are not closely related and as such hamadryas society is not a female-bonded system. Males 'herd' the females in their units and may 'neck bite' a female if she strays. These are long-lasting associations; units contain from 1–10 adult and sub-adult females which sometimes remain together for over 5.5 years (in 7% of cases studied by Sigg *et al.* 1982). Maturing females typically transfer from their natal unit and adult females may also

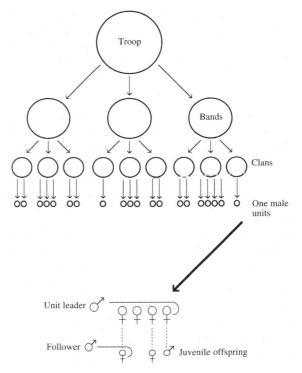

Fig. 3.5 Multi-level troop structure in the hamadryas baboon (*Papio hamadryas*). The troop is composed of bands, each of which contains a number of clans. Clans contain one-male units. A typical one-male unit is represented, comprising a leader male, a follower male, and four adult females with their offspring. Hooks indicate male herding or courting behaviour. (Based upon Kummer 1984, 1990.)

transfer. Males leave the unit in which they were born, but although they may transfer to another clan or band for up to several months they usually return and remain in their natal clans.

Two tactics are adopted by males in order to found their one-male units. A bachelor might join an established unit as a *follower*, initially subordinate to the leader male but with the possibility of taking over as the sexually active male once the leader ages. Alternatively, the bachelor may attempt to start a harem by capturing and herding a juvenile female—in which case he must wait until her ovarian cycles commence at puberty. The propensity of male hamadryas baboons to herd and control females is partially under genetic control. Thus, naturally occurring hybrids between *Papio hamadryas* and *P. anubis* show an intermediate type of behaviour, in which the brief consortships typical of *P. anubis* are extended and might last for several months (Nagel 1973). Whilst a male hamadryas will attempt to herd any additional females which become available, established relationships are rarely challenged, so that males do not tend to fight one another or attempt harem takeovers (Bachmann and Kummer 1980).

At a superficial level the social organization of the Ethiopian gelada closely resembles that of the hamadryas baboon. Thus, geladas live in large troops, which assemble on sleeping cliffs but then disperse during the day as one-male units (Crook 1966). A number of these units might continue to forage together during the day; they form *bands* of between 2–27 units (Kawai *et al.* 1983). However, the gelada one-male unit is female-bonded. Females rarely transfer from these units and, in contrast to female hamadryas baboons, they form close affiliations and grooming relationships. Each unit consists of an adult male and 4 or 5 adult females (as many as 12 females in the largest units) with their young offspring. Males born in these units emigrate as subadults and join an all-male group for 2–4 years (Dunbar and Dunbar 1975). A bachelor male may join an established unit as a follower and attempt to acquire one or more of its lower-ranking peripheral females. These will provide the starting point for a new unit, and the male will depart, taking the females with him. Alternatively, and in contrast to hamadryas baboons, bachelor males may attempt direct aggressive takeovers of established units. Females in these units play a key role in determining whether a new male will be accepted, intruders must not only fight the resident male but must solicit grooming and proximity with the females in the unit (Dunbar 1984, 1993). Dunbar (1993) comments that 'takeover contests involve the most serious fighting ever observed among the gelada, and the protagonists often incur injuries from razor sharp canines'. Like the male hamadryas baboon, the adult male gelada is heavily built and with an impressive array of secondary sexual adornments. Intermale competition is intense and, in the case of the gelada especially, the possibility of female choice, favouring particular males during takeovers, is apparent.

The third species in which one-male units form part of a larger troop is the proboscis monkey. In contrast to geladas and hamadryas baboons, the proboscis monkeys of Borneo are arboreal, being found mainly in the forests and mangroves which border tidal creeks or along river banks. Troops often sleep by rivers and all the group members will share 1–3 adjacent trees (Yeager 1989). Although it was formerly thought that proboscis monkeys live in multimale–multifemale groups (e.g. Kawabe and Mano 1972), more detailed fieldwork by Bennett and Sebastian (1988) and Yeager (1990a, 1995) has shown that groups are composed of distinct one-male units. Extra-group males form all-male groups containing 10 or more monkeys. One-male groups are stable entities varying in size from 3–23 individuals (mean group size: 12.6). In Kalimantan, Yeager (1990a) found that units consisted, on average, of 1 adult male, 5 adult females, 2.6 adolescents and juveniles, and 4 infants. Bennett and Sebastian (1988) recorded slightly smaller one-male units in Sarawak ($N = 6$, range 5–16, mean = 9 individuals with 1 adult male, 3.6 adult females, 2.4 adolescents, and 2.0 infants per unit). In both studies, approximately 18% of the population consisted of additional males, living in all-male groups (the ratio of units to all-male groups averaged 6:1). As juveniles, males become peripheralized and emigrate from their natal units. However, transfers of adult females between units have also occasionally been observed (Bennett and Sebastian 1988). Since one-male units and all-male groups frequently sleep less than 100 m apart, this might have lead to the earlier impression that the proboscis monkey lives in stable multimale–multifemale groups. However, during the day the various units range inland and forage separately. It is not known whether higher-level social assemblages such as bands or clans occur in the proboscis monkey. Nor is it known how long males are able to defend female units or precisely how new units are formed. Aggressive *takeovers* by males in all-male groups seem likely, but have not been reported. Within the units agonistic interactions are apparently infrequent. Adult females groom one another, but the unit male is rarely involved in this behaviour. Neither one-male nor all-male groups are territorial. However, intergroup displays have been observed during which adult males vocalize (deep 'snorking' sounds are produced), leap through the trees, and branch-shake.

The mating systems of the proboscis monkey, gelada, and hamadryas baboon share the characteristic that troops are composed of one-male units. There is insufficient information to discuss how multi-unit troops of proboscis monkeys might have evolved. However, in the case of the gelada and the hamadryas baboon various authors have suggested that ancestral forms lived in multimale–multifemale groups such as occur in present-day savannah baboons (Struhsaker and Leland 1979; Dixson 1983b; Kummer 1990). Kummer (1990) argues that, in the case of the hamadryas baboon, ancestral multimale–multifemale troops began to split into smaller foraging units in response to ecological conditions. Female relatives tended to band together in such foraging units and a mating system based on female defence polygyny developed. The tendency for males in one-male units to 'herd' females might have been selected for because units continued to congregate at scarce resources such as waterholes and sleeping cliffs. Large and powerful males were beneficial in such circumstances, to protect females and offspring from predators. Kummer comments that 'this stage is comparable with gelada one-male units'. However, in the case of ancestral hamadryas baboons the further development of herding and neck-biting techniques involved males assembling units of unrelated females. In contrast to the gelada, therefore, the units of hamadryas baboons are not female-bonded but they are probably secondarily derived from a female-bonded society.

Since man also lives in complex societies in which polygyny has played a significant role, it is appropriate to consider some aspects of human mating systems at this point. Sexual dimorphism in body weight and the occurrence of various secondary sexual characteristics in human males (e.g. muscular build, beard, enlarged larynx, and deeper voice) are indicative of a polygynous ancestry (Short 1980). Although marriages are intended to be monogamous in many modern societies, cross-cultural studies indicate that most human societies allow polygyny. Thus Ford and Beach (1952) found that polygyny occurred in 84% of 185 human societies for which they could obtain sufficient information. Later surveys indicate that polygyny occurs in 83% of human societies, whilst 16% are monogamous and <1% practise polyandry. Yet, even in polygynous societies many marriages are monogamous and polygyny is allowed when a man is sufficiently wealthy or powerful to support more than one wife. In the Trobriand islands, for example, Malinowski (1932) found polygyny to be prevalent among the village chiefs. In Papua New Guinea a number of tribes are polygynous to the extent that respected and powerful men in the community (so-called 'big-men') often have several wives, whereas most men have a single wife. These examples serve as reminders that it is incorrect to assume that a particular primate species will have a single mating system. It might be surmised that our prehuman ancestors were not as deeply polygynous as gorillas, hamadryas baboons, or geladas. Thus sexual dimorphism in body weight, canine size, and other features is not so pronounced in man as in many polygynous primates.

For many polygynous primate species, one-male units are spatially separate and do not form part of larger social groups or sleeping assemblages. The gorilla affords a particularly clear example of this type of polygynous system. Gorilla groups vary in size from 2–27 individuals and western lowland gorilla groups tend to be smaller than those of the mountain gorilla (*Gorilla g. gorilla* in Rio Muni: Range 2–12, mean group size = 6.8, $N = 13$, Jones and Sabater PI 1971. *G. g. beringei* in the Virungas Park: Range 5–27, mean group size = 16.9, N = 10, Schaller 1963). Schaller found that mountain gorilla groups contained an average of 1.7 silverbacks, 1.5 younger, black-backed males, 6.2 adult females, 2.9 juveniles, and 4.6 infants. Lone silverback males also occurred in the population and sometimes these joined established groups and were tolerated for varying periods. Subsequent research on the mountain gorilla has shown that not only do maturing males often emigrate from their natal group but that females tend to transfer to another unit once they become reproductively mature (Harcourt *et al.* 1976). Given that only one silverback is sexually active in a group (Harcourt *et al.* 1980) and that his tenure can last for many years, female transfers as well as male emigration may be important in preventing inbreeding. Formation of a new group occurs when a lone male acquires one or more females from an established unit. Such female transfers are accompanied by a good deal of aggression and silverbacks may hoot, chest beat, and bluff charge at one another (Caro 1976; Harcourt *et al.* 1976). Although no genetic studies have been undertaken to establish paternity in gorilla groups, it seems likely that the dominant silverback sires most of the offspring. Female gorillas have long interbirth intervals (4–5 years) and are sexually active mainly during the peri-ovulatory period of the menstrual cycle. With only 5–6 females in a unit it appears that a silverback is able to monopolize sexual access to them and unit takeovers or multimale incursions (such as occur in geladas or some *Cercopithecus* species) have not been recorded in gorillas. Sperm competition is probably minimal in the gorilla, therefore, since females typically mate with just one male during the fertile period of the cycle. The gorilla has the smallest testes, both absolutely and in relation to its body weight, of any of the hominoids.

In several examples of polygynous mating systems discussed so far it is clear that the so-called 'one-male units' are not invariably uni-male. A follower may attach himself to a one-male unit in the hamadryas baboon or gelada, for example, and takeovers occur in the gelada, so that the unit male is replaced by an outsider from an all-male group. In the gorilla, lone males may join established groups or seek to precipitate a female transfer. Since the very term 'female defence polygyny' implies a high degree of intermale competition, it should not surprise us if male tenure is limited or that non-unit males adopt counter tactics to obtain mating opportunities (Rowell 1988). The concept of polygyny is still perfectly valid, however, provided that information is sought on male tenure, on various counter tactics, and on lifetime reproductive success. Several further examples will be discussed to illustrate this proposition. Firstly, we shall consider the polygynous mating systems of forest guenons (*Cercopithecus* spp) and of the more terrestrial patas monkey (*Erythrocebus patas*). In these cases, multimale influxes sometimes alter the picture of a rigidly maintained one-male unit mating system. Then we shall consider the existence of one-male groups and multimale–multifemale groups within a single species, the Hanuman langur (*Presbytis entellus*). In this case, takeovers of one-male units sometimes involve the new male in attacks upon young infants. The significance of these *infanticides* has been much debated.

The forest guenons of the genus *Cercopithecus* comprise a colourful adaptive radiation in the rainforests of central and western Africa. Most of the 20 or so species are arboreal and include the redtail monkey (*C. ascanius*), the moustached monkey (*C. cephus*), and the blue monkey (*C. mitis*). With the exception of the vervet (*C. aethiops*) these guenons have traditionally been considered as living in one-male units. Cords (1987), who has reviewed much of the information pertaining to this subject, points out that most earlier studies were of limited duration and showed that *Cercopithecus* groups contained a single adult male, a number of adult females and offspring (e.g. Struhsaker 1969; Quris 1976). These group-associated males were antagonistic towards one another and towards lone males (Struhsaker and Gartlan 1970; Struhsaker and Leland 1979). It was known that adult males typically gave loud vocalizations, amplified by means of a larygeal sac, and these vocalizations appeared to be relevant to intergroup spacing and territoriality (Gautier 1971; Gautier and Gautier 1977). Longer term studies of redtail and blue monkeys then showed that at least some males remain in their groups for lengthy peri-

ods. Cords (1987) cites instances where male redtails in the Kibale forest (Uganda) and Kakamega forest (Kenya) are known to have remained in their same groups for 13–39 months. In these forests blue monkeys have been followed for 17–62 months as members of the same groups. However, longer term field studies also revealed that multimale influxes sometimes occur into redtail and blue monkey groups and that a number of males are sexually active in these circumstances (Table 3.4). Many of the forest-living guenons are seasonal breeders (Butynski 1988) and, in the Kakamega forest at least, multimale influxes of redtail and blue monkeys coincide with the mating season. The crucial question of how these events influence male reproductive success remains unanswered. A reliable assessment of the problem would require genetic analyses (e.g. DNA typing) to determine the parentage of resulting offspring and accurate data on copulatory frequencies of the various males. Multimale influxes in blue monkeys are accompanied by aggression and competition for mating opportunities (Tsingalia and Rowell 1984). In that study, resident males sometimes also mated with females in neighbouring groups. Cords (1987) expresses the view that in the Kakamega forest, resident male blue monkeys 'probably sired few, if any, of the offspring conceived during multimale influxes when they were absent or not seen to mate'. Kakamega and Kibale resident male redtails 'may have sired some of the offspring but other males were observed mating concurrently'. Yet, since it is exceedingly difficult to observe copulation in free-living *Cercopithecus* monkeys, little weight can be given to reports that certain males were not observed to mate. Further, in the absence of genetic data, the reproductive success of resident versus intruder males cannot be measured. Of particular interest is the problem of *lifetime reproductive success* given that certain males remain with the females in a given group for years

Table 3.4 Influxes of males into free-ranging groups of redtail and blue monkeys during the annual mating season

Species and year studied	Months when mating occurred	Number of males in group	Number of males seen simultaneously (range)	Male tenure (range in days)	Number of males seen mating	Number of breeding females	Number of resulting births
Redtail monkey							
L. group (1980)	Feb–Sept.	8	1–4	1–153	7	9–10	7
Blue monkey							
T. group (1981)	June–Nov.	16	2–10	5–105	9	18	3
T. group (1983)	June–Aug.	19	2–11	1–44	12	18	unknown

Data are from Cords (1987).

Fig. 3.6 A resident male patas monkey (*Erythrocebus patas*) copulates with a female, immediately after an intruder male (on the right) has mated with her. (After Ohsawa *et al.* 1993).

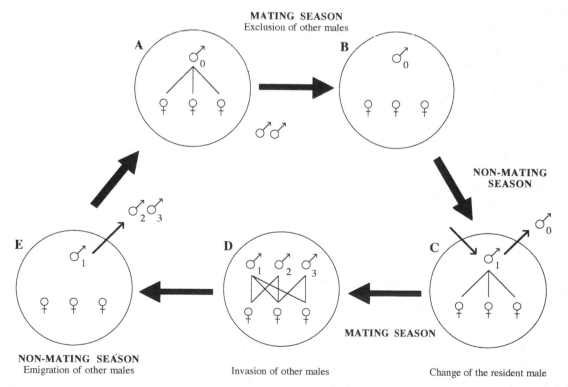

Fig. 3.7 Shifts in social organization and mating interactions in a free-ranging group of patas monkeys (*Erythrocebus patas*) during successive non-mating and mating seasons. O=the former resident male; 1=the current resident male; 2, 3=intruder males. Lines between individuals indicate mating relations. (Redrawn from Ohsawa *et al.* 1993.)

and that this tactic might confer some long-term reproductive advantage.

Closely related to the guenons is the patas monkey (*Erythrocebus patas*), a large but lightly built terrestrial monkey which occurs in groups of 5–31 individuals (mean group size = 15) containing a single adult male, several adult females, and offspring (Hall 1965). Hall's pioneering fieldwork on the patas monkey in Uganda also revealed the existence of 'bachelor bands' and lone males in the population. Home ranges of patas groups were found to be extensive (51.8 km² in the case of the largest group studied). However, little was learned concerning sexual behaviour or the dynamics of the mating system, and it was assumed that patas one-male units were exclusively polygynous. Subsequent fieldwork revealed that multimale influxes also occur into some patas groups and that females mate with a number of partners (Harding and Olson 1986; Chism and Rowell 1986; Ohsawa 1991; Ohsawa *et al*. 1993). The study of Ohsawa *et al*. in Cameroon is especially interesting because these authors succeeded in trapping groups and in conducting DNA typing studies to determine parentage of offspring and hence to measure male reproductive success. In two out of four cases where it was possible to assign parentage to offspring born as a result of a uni-male situation, the unit male was not the sire. Ohsawa *et al*. thus concluded that 'sneak copulations', involving extra-group males, were occurring, although these were never witnessed. When multimale influxes occurred, the resident continued to mate and he accounted for 31% of observed copulations, whilst the remaining 69% of sexual activity was distributed between five intruder males. The resident male often responded to an opportunistic copulation by promptly remating with the female concerned (Fig. 3.6). Such behaviour may be significant in terms of sperm competition between rival males. Four out of five progeny born as a result of multimale influxes, for which DNA typing was successful, were found to be the offspring of the group's resident male. Ohsawa *et al*. considered that many of the additional matings with intruders or extra-group copulations were the result of female initiations. Indeed the female patas has a distinctive repertoire of proceptive displays (including approaches to the male, puffing out the cheeks and gurgling vocalizations) by which she invites copulation. Multimale influxes typically occur in periods after a unit male has been displaced and when a new male is present in the group. Under these conditions females might be more likely to solicit copulations from intruders. Ohsawa *et al*. have produced a model of the patas mating system in which the one-male unit alternates with a multimale system, as depicted in Fig. 3.7. Most interesting is their finding that the resident male of the unit sired six out of nine (67%) of offspring born under both conditions. Although these genetic data are limited, they support the view that female defence polygyny confers a reproductive advantage upon the resident male and that this is the primary mating system of the patas.

The final example of a polygynous primate I wish to discuss is the Hanuman langur (*Presbytis entellus*). This monkey is widely distributed throughout the Indian subcontinent, as far south as the island of Sri Lanka. It is an adaptable species occurring in open habitats as well as in forests. Group sizes are also variable—one-male units, multimale–multifemale groups and all-male groups have been described (Jay 1965; Sugiyama 1967). Sugiyama (1967) conducted the bulk of his studies on *P. entellus* at Dharwar in the state of Mysore in western India, where the animals inhabit open-scrub forest. He observed 44 troops of langurs there; 38 contained both sexes whilst 6 were all-male groups. Most of the bisexual troops were one-male units (almost 74%) but multimale–multifemale troops containing from two to seven adult males also occurred. In some cases Sugiyama found that some multimale troops contained a single large adult male and a number of smaller individuals. These might therefore be equivalent to what Eisenberg *et al*. (1972) have called *age-graded male units*. In age-graded units, only one male is sexually active and younger male relatives are subordinate to this individual. It is unknown how far this structure might apply to Hanuman langurs; some groups are truly multimale–multifemale and a number of males are sexually active. Multimale–multifemale units are in the majority in parts of Northern India and Nepal as well as in the far south and in Sri Lanka. One-male units tend to form the majority of the population in more central and north western areas of the continent (Fig. 3.8). Perhaps this is related to the dryer and harsher nature of the climate at sites such as Jodhpur and Mt. Abu. Sugiyama (1967) described the average troop size for Hanuman langurs as 15.1 individuals, including 8.0 adult females. All-male groups were more variable or 'loose gatherings' varying in size from 2–60 monkeys. The bisexual troops occupied overlapping home ranges and males in neighbouring groups sometimes displayed aggressively towards one another. In Sri Lanka, Ripley (1967) observed such intergroup displays which involved loud 'whooping' vocalizations by resident males and leaping and branch-shaking behaviour.

Sugiyama recorded a number of male incursions and takeovers of one-male units. He was the first to describe how the new leader male sometimes attacked and bit young infants in the group after such

SOCIAL ORGANIZATION OF HANUMAN-LANGURS

Fig. 3.8 Variations in the social organization of Hanuman langur (*Presbytis entellus*) troops in India and Sri Lanka. The percentages of one-male troops (dark shading) and multimale–multifemale troops in each area are shown by pie charts. (Figure by kind permission of Dr Volker Sommer.)

takeover episodes. He surmised that infanticidal behaviour favoured the new leader male because females give birth only every 2 or 3 years and 'loss of the infant has the effect of advancing the oestrus of the female'. Sugiyama also suggested that 'this astonishing attack may show to the females of the troop the leader's power for organizing them without injury to themselves'. In later years other cases of infanticide (either observed directly or suspected from circumstantial evidence), were reported in Hanuman langur groups (Mohnot 1971; Hrdy 1974, 1977). These reports prompted the search for examples of similar behaviour in other primate species. A critique of the evidence pertaining to infant-killing and its possible adaptive significance will be provided in the next chapter which deals with mating tactics.

Multimale–multifemale mating systems

In a multimale–multifemale mating system females

usually mate with a number of partners and long-term sexual relationships with a single male do not occur. Temporary associations or consortships between a male and a female are seen in some species, especially during the fertile phase of the ovarian cycle, but these are rarely of long duration (e.g. *Papio ursinus*, 1–8 days: Seyfarth 1978; *Macaca fascicularis*, from less than 1 h to 2 or 3 weeks: Van Noordwijk 1985; *Pan troglodytes*, from 3 h to 28 days: Tutin 1979). Consortship between a male and a female does not guarantee exclusivity of sexual activities between the two animals. Thus in *M. fascicularis*, Van Noordwijk (1985) found that 'both the female and the male copulated occasionally with other partners, returning to the original consort partner immediately afterwards'. In *Papio ursinus*, 35% of adult females studied by Hausfater (1975) consorted with a series of males during the peri-ovulatory period of the menstrual cycle and changed partners during a single day. Mating is rarely indiscriminate in multimale–multifemale primate groups, however. A variety of factors, including kinship ties, social rank, age, sexual attractiveness and individual sexual preferences might influence mate choice in both sexes. It is, therefore, incorrect to label such mating systems as *promiscuous*. None the less, females in multimale–multifemale primate groups are usually more sexually active, in terms of the numbers of their mating partners, than is the case for the monogamous or polygynous mating systems considered so far. To reinforce this point I shall review evidence relating to occurrences of multipartner copulations by females in multimale–multifemale social groups.

Evidence that females mate with multiple partners

It might be argued that multimale–multifemale groups are functionally polygynous, and that a high-ranking male might monopolize access to all females, or that such an individual might exclude all other males during the peri-ovulatory phase of each female's cycle. Alternatively, it might be claimed that some groups containing a number of adult males are *age-graded* in the sense that they contain a single mature and sexually active male as well as his younger male relatives (Eisenberg *et al.* 1972). Whilst not denying these possibilities, since they are believed to occur in some groups of *Presbytis entellus* (Sugiyama 1967) or *Alouatta* spp. (Crockett and Eisenberg 1987; Pope 1990) there is now ample evidence that females copulate with multiple partners during a single ovarian cycle in many primates

Table 3.5 Evidence for multiple-partner matings by females in multimale–multifemale primate groups

Species	Type of study	Numbers of males mated per female	Source
Prosimians			
Lemur catta	Field study	3–5	Koyama 1988
Propithecus verreauxi	Field study	1–2	Richard 1976
New World monkeys			
Alouatta palliata	Field study	>1	Jones 1985
Ateles paniscus	Field study	>1	Roosmalen and Klein 1988
Brachyteles arachnoides	Field study	4	Milton 1985a
Cebus apella	Field study	1–7	Janson 1984
Old World monkeys			
Cercopithecus aethiops	Field study	>1	Andelman 1987
Miopithecus talapoin	Field study	>1	Rowell and Dixson 1975
Papio ursinus	Field study	1–3	Hall 1962
Papio cynocephalus	Field study	>1	Hausfater 1975
Papio anubis	Field study	>1	Scott 1984
		1–3	Smuts 1985
Cercocebus albigena	Field study	>1	Wallis 1983
Mandrillus sphinx	Semi-free ranging	1–3	Dixson *et al.* 1993
Macaca mulatta	Field study	1–11 (mean 3.2)	Conoway and Koford 1965
	Field study	mean 3.8	Loy 1971
	Field study	>1	Lindburg 1983
	Field study	1–9 (mean 3 or 4)	Manson 1992
Macaca fuscata	Semi-free ranging	>1	Wolfe 1984*b*
	Captive group	3–19 (mean 10.7)*	Hanby *et al.* 1971
Macaca fascicularis	Field study	>1	Van Noordwijk 1985
Macaca radiata	Captive group	>1	Glick 1980
Macaca sylvanus	Field study	3–11 (mean 6.0)	Taub 1980
	Field study	5–11 (mean 7.12)	Menard *et al.* 1992
Macaca nemestrina	Captive group	1–5 (mean 2.5)	Tokuda *et al.* 1968
Apes			
Pan troglodytes	Field study	8+	Goodall 1986
	Field study	>1	Hasegawa and Hiraiwa-Hasegawa 1990
Pan paniscus	Field study	>1	Furuichi 1987
	Field study	>1	Kano 1992

Cases where multipartner matings occur but the number of males involved is unknown are indicated by >1.
Data in this table refer to matings during a single ovarian cycle, including the presumptive period of ovulation. Copulations occurring during pregnancy are not considered.
* These data refer to numbers of male partners during a single mating season and not necessarily during a single ovarian cycle (from Dixson 1997a).

which live in multimale–multifemale societies. Table 3.5 cites records of multipartner matings by females of 21 primate species, including observations of such behaviour in the wild for 17 of the species listed. Field studies of primate sexual behaviour contain graphic accounts of copulatory behaviour involving a single female with a succession of partners, as the following examples show:

Papio ursinus (Saayman 1970) 'The roving appetitive be-

haviour of inflating females* from one male to another was conspicuous. It was not uncommon for an inflating female to present to, and be mounted by, as many as three males within the space of two or three minutes.' [*Here he refers to enlargement of the female's sexual skin swelling during the follicular phase of the menstrual cycle].

Macaca mulatta (Lindburg 1983) 'Over a period of about two hours the female shifted at least ten times among the four males of her group, apparently gaining acceptance with all but the alpha male.'

Brachyteles arachnoides (Milton 1985a) 'Sexual receptivity in the female was indicated by the arrival in her foraging area of a large number (7–9) of males. These males followed the female constantly, copulating with her whenever possible, over a period of some 36–48 hours. The receptive female was seen to mate as many as eleven times with at least four males in a single 12-hour observation period.'

Pan troglodytes (Goodall 1986) 'Once a female's swelling reaches full size, particularly if she is undergoing a fertile cycle, she typically becomes the nucleus of a party comprising many or all of the community males. In this situation... her sexual partners will probably include every male in the party with the exception (almost always) of any adult son or (sometimes) adult maternal sibling.'

In the woolly monkey (*Lagothrix lagotricha*) females have been observed mating with up to four males during a single day (Nishimura 1988). Menard *et al.* (1992), who studied Barbary Macaques in the Atlas Mountains of North Africa, found that all females copulated with more than one male and most mated 'with at least half of the available sexually mature males.' In some cases direct observations of the multipartner matings are lacking and yet the circumstantial evidence for their occurrence is very strong indeed. In the talapoin monkey (*Miopithecus talapoin*) of western central Africa, males and females do not engage in consortships during the annual mating season. Instead, 'a male would approach a female from 20 or 30 yards and, after mating, each would leave in a different direction' (Rowell and Dixson 1975). Copulations occurred frequently at the height of the mating season and could be seen (or heard) in several subgroups simultaneously. Indeed, 'it seemed that females probably mated with several males in rapid succession.'

For some species with multimale–multifemale mating systems, individual females engage in very large numbers of copulations before conception. Under these circumstances sexual selection has favoured males which can produce large numbers of motile sperm per ejaculate (see Chapter 8 on Sperm Competition) as well as males which have the most efficient copulatory patterns and penile morphologies (see Chapters 5 and 9). In a few cases quantitative data on ejaculatory frequencies per female have been gathered under field conditions. In the ringtailed lemur, for instance, Koyama (1988) records that one female mated with five males and received 27 ejaculations during the 4 h for which she remained sexually receptive. Among the anthropoid primates, in which there is no rigid restriction of receptivity to a peri-ovulatory oestrus, males may compete and copulate at even higher frequencies

over an extended period. Manson (1992) calculates that female rhesus monkeys on Cayo Santiago may copulate 50–90 times during a 14-day period which includes the putative day of ovulation. Hasegawa and Hiraiwa-Hasegawa (1990) estimate that middle-aged female chimpanzees in the Mahale mountains of Tanzania usually exhibit three cycles of sexual skin tumescence and copulate 'approximately 135 times before each conception.'

Examples of multimale–multifemale societies

Although scores of primate species have multimale–multifemale mating systems (see Table 3.6) their societies are diverse and present numerous variations upon the central theme. To illustrate this I shall briefly describe the social organization and mating system of four representative species; the ringtailed lemur (*Lemur catta*), rhesus monkey (*Macaca mulatta*), chimpanzee (*Pan troglodytes*), and bonobo (*Pan paniscus*).

Ringtailed lemur

Ringtailed lemurs occur in southern and southwestern Madagascar where they inhabit both deciduous forest and brush/scrub forests. This adaptable species spends approximately 70% of the time in the trees, but groups often descend to ground level to travel about the home range, which varies in size from 5.7–8.8 hectares (Jolly 1966; Sussman 1976). The diet consists of fruit, flowers, leaves, sap, and bark. Groups range in size from 12–24 individuals (mean group size = 18.8) and commonly contain slightly more females than males (Sussman 1976). Females form the core of ringtailed lemur groups; males by contrast often transfer between groups. Intermale aggression and dominance relationships are marked and subordinate males tend to occupy peripheral positions compared with other group members (Budnitz and Dainis 1975; Sussman 1976). Ringtails are seasonal breeders and the mating season lasts for just 2 weeks, during April. Intermale aggression and scent-marking activity increases markedly during this annual mating period. Males give aggressive 'spat calls', cuff one another with their hands, or attempt to slash opponents with their canines during 'jump fights' and chases. Palmar marking, tail marking, and tail-waving displays all increase in frequency during the mating season (Fig. 3.9); thus males are said to engage in 'stink fights' (Jolly 1966). Dominance relationships between males tend to be more labile during the annual mating season. Koyama (1988) found no simple relationship between mating success and dominance rank in free ranging male ringtails. However, most of the ejaculatory mounts he observed (13 out of a total of 38 ejaculations) involved the second ranking out of 10 males. This male also emitted loud howls at dusk ('Kehon-Kehon' calls) whilst associating with females at the sleeping site. Koyama suggests that as

Table 3.6 Primate species with multimale–multifemale mating systems: group sizes, numbers of adult males and females, and body weights of adults

Species	Mean group size	Number of adults		Body weights (kg)		Source
		Males	Females	Males	Females	
Prosimians						
Lemur catta (S)	20 (1)	8	5	2.9	2.5	Koyama 1988
	18 (3)	6	6.3			Sussman 1976
L. fulvus (S)	9.5 (17)	3.4	4.3	2.5	1.9	Sussman 1976
L. macaco						
Propithecus verreauxi (S)	5.75 (4)	2.25	2.25	3.7	3.5	Richard 1976
New World monkeys						
Saimiri sciureus (S)	24 (3)	2.3	6	0.75	0.58	Baldwin and Baldwin 1981
Cebus capucinus	24 (2)	4	6.5	3.8	2.7	Freese and Oppenheimer 1981
Cebus nigrivittatus	19 (1)	2	8	2.9	2.3	Freese and Oppenheimer 1981
Cebus albifrons	7 (1)	2	2	2.6	2.6	Freese and Oppenheimer 1981
Cebus apella	6–7 (1)	1–2	2	2.86	2.1	Freese and Oppenheimer 1981
Alouatta palliata	17.3 (23)*	2.7	7.4	7.4	5.7	Neville *et al.* 1988; Carpenter 1934
	18.9 (8)	3.9	8.0			Neville *et al.* 1988
Brachyteles arachnoides	19.2 (6)+	6.2	6.8	15.0	9.5	Nishimura *et al.* 1988
Ateles belzebuth	23.5 (2)	4	11.5	6.2	5.8	Van Roosmalen and Klein 1988
Lagothrix lagotricha	22.5 (2)	6	7	6.8	5.8	Nishimura 1988
Old World monkeys						
Presbytis entellus	27.5 (4)	4.25	9.25	18.4	11.4	Jay 1965
Colobus badius tephrosceles	37.6 (8)	3.6	9.2	10.5	5.8	Struhsaker 1975
Cercopithecus aethiops (S)	17 (2)	2	4	4.75	3.56	Struhsaker 1967a
Miopithecus talapoin (S)	86 (2)	13	20.5	1.4	1.1	Rowell and Dixson 1975
Cercocebus albigena	21(2)	4.5	8.0	9.0	6.4	Chalmers 1968a
Papio anubis	25.8 (5)	3	7.4	21.0	12.0	DeVore and Hall 1965
P. cynocephalus	39 (1)	8.4	12.6	20.0	15.0	Altmann and Altmann 1970
P. ursinus	42.7 (3)	4	13.3	20.4	16.8	Hall and DeVore 1965
Macaca mulatta (S)	49.8 (7)	5.6	19.2	11.2	8.2	Southwick *et al.* 1965
M. fuscata (S)	81 (1)	7	24	11.7	9.1	Kurland 1977
M. fascicularis	29.1 (1)	5.7	9.9	5.9	4.1	Wheatley 1980
M. sylvanus	18.3 (6)	2.7	5.2	11.2	10.0	Deag and Crook 1971
M. silenus	21 (4)	2	7.8	6.8	5.0	Green and Minkowski 1977
M. nemestrina	50 (1)	2	18	10.4	7.8	Caldecott 1981
M. sinica	24.8 (18)	2.7	6.2	6.5	3.4	Dittus 1975
M. radiata	30 (12)	8.0	9.7	6.6	3.7	Sugiyama 1971
M. thibetana	39.2 (6)	5.2	10	–	–	Zhao 1993
Apes						
Pan paniscus	63**	15	16	45.0	33.2	Kano 1982
P. troglodytes	41.4**	8.8	11.6	41.6	31.1	Goodall 1986

Bodyweight data are from Harvey *et al.* 1987 and Martin and Dixson (unpublished). (S)=species with discrete mating seasons.
* Includes four one-male groups. One-male groups also occur in the four other *Alouatta* spp. so that this genus is not purely multimale–multifemale (see text).
+ Milton (1985a) suggests that a fusion–fission type of organization occurs in *Brachyteles*, with a total group size of 45 in her study.
** Unit groups or communities occur in both *Pan spp* and a fusion–fission society consists of changeable sub-groups. This arrangement may also occur in *Ateles* and *Lagothrix*: see text.

Fig. 3.9 Adult male ringtailed lemur (*Lemur catta*). The tail is marked with secretions from specialized glands situated on the shoulder and wrist. The male then distributes the odours by flicking the tail. (Author's drawings: after Evans and Goy 1968.)

the brief period of sexual receptivity of the female (approximately 4 h in duration) may extend into the night, these vocalizations may serve to attract such females to a sexually active male. Jolly (1966) by contrast thought that these howling vocalizations might serve for inter-troop communication. Koyama found that receptive female ringtails mated with a number of partners and suggested that female choice, and not simply male dominance, determined the pattern of sexual encounters. Indeed male ringtails are not necessarily dominant over females; so that females, for instance, may outrank males in feeding situations (Jolly 1984). Although the timing of oestrus within a ringtail group is quite tightly co-ordinated among the various females (Pereira 1991) there is variability between groups. Males transfer actively between groups during the mating season and copulations occur between immigrants and resident females. Koyama (1988) proposed that resident males experience a reproductive advantage in terms of their familiarity with females and their established agonistic relationships with other males.

DNA-typing studies of free-ranging ringtailed lemurs will be required to enable some assessment of the effects of male rank or mobility between groups upon reproductive success.

Rhesus monkey

The rhesus monkey is the most widely distributed of all the macaque species, being found in Northern India (including Kashmir), Nepal, Pakistan and Assam. This is also an adaptable species, occurring in forests, in more open habitats, and even in urban areas, such as the precincts of Indian temples or close to native villages (Southwick *et al.* 1965). In the forests of Uttar Pradesh, Southwick *et al.* (1965) recorded that seven rhesus monkey troops ranged in size from around 40–68 individuals (mean group size was 49.8). Troops inhabiting temple precincts were also large (range 17–78, mean = 41.9 for five troops studied) and more than twice the size of those living near villages or towns where animals received less protection. In all these cases rhesus monkeys formed multimale–multifemale groups and the same holds true for rhesus monkeys transplanted to seminaturalistic environments such as the island of Cayo Santiago in Puerto Rico (Carpenter 1942; Koford 1965). Seasonality of reproduction is the norm for this species; in N. India (Uttar Pradesh), for instance, mating occurs between November and January, with a corresponding birth season between March and May. Also consistent across studies of rhesus monkeys in a wide range of habitats is the basic social structure of their groups. At the core of the troop are females and offspring. Matrilineal relationships have important effects upon the development of offspring. The daughters of high-ranking females are more likely to achieve high rank themselves, and females remain in their natal groups. Males by contrast tend to become peripheralized and to emigrate from their natal groups as sexual maturity approaches (at 3.5–4.5 years of age). In the rhesus monkey, as in other primates which form female-bonded social groups, it is males which disperse and enter other groups in order to breed (Pusey and Packer 1987). Male transfers mainly occur during the annual mating season and may be due in part to the sexual attractiveness of non-related females in neighbouring groups (Boelkins and Wilson 1972; Drickamer and Vessey 1973). A number of studies (but not all), report positive correlations between male social rank and mating success in rhesus monkeys (Cowlishaw and Dunbar 1991). Males are intensely aggressive during the annual mating season and the frequency of wounds and deaths increases at this time (Wilson and Boelkins 1970). Since male rank tends to increase with the length of time spent in a social group, it may be advantageous for male rhesus monkeys to emigrate early from their natal groups (for instance, at three and a half years when puberty begins) to gain acceptance in a new group. It is noteworthy, however, that the sons of high-ranking females tend to migrate later and sometimes stay in their natal groups and breed there for several years (Chapais 1983). Mortality rates tend to be higher in younger males (aged 4–6 years) and in adult male rhesus monkeys, perhaps because of predation during the migratory and solitary stages of development and increased frequencies of fighting among adult males, as compared with adult females (Koford 1965). This circumstance might contribute to the disparity in the sex ratio in rhesus monkey groups, since adult females usually outnumber adult males, as is the case in many primates which live in multimale–multifemale societies (see Table 3.6).

As the annual mating season approaches, the sexual skin which covers the perineal area and scrotum of the male rhesus monkey begins to redden under the influence of testosterone, secreted by the testes (Vandenbergh 1965; Gordon *et al.* 1976). These changes coincide with marked enlargement of the testes and increased spermatogenesis (Conoway and Sade 1965) as well as the onset of masturbatory behaviour (Koford 1965) and copulation (Gordon *et al.* 1976). Among pubescent males, dominance rank correlates with testicular development, so that the testes are largest in high-ranking individuals (Bercovitch 1993) and in sons of high-ranking females (Dixson and Nevison 1997).

Copulatory behaviour in the rhesus monkey involves a series of mounts and intromissions, culminating in ejaculation. Copulations often occur as part of close associations (consortships) between females and males during periods which include the peri-ovulatory phase of the female's cycle. Females may consort with a number of males at such times; Kaufman (1965) observed females consorting with 1–4 partners and similar data have been obtained by other observers (e.g. three or four males on average: Carpenter 1942; Loy 1971; Manson 1992). Thus, as in the ringtailed lemur, female rhesus monkeys mate with a number of males during the peri-ovulatory period, but copulations are spread over a much larger time interval, lasting for days or weeks in some cases, and do not occur during a limited peri-ovulatory 'oestrus' as in prosimian primates. The dynamics of consortships are complex and involve effects of female rank (highest ranking females consort with more partners) and sexual preferences for particular partners or partners of similar age (Loy 1971). Female choice as well as male rank influences consortship and copulatory frequency. Manson

(1992) observed female rhesus monkeys actively soliciting copulations from lower-ranking males as well as from high-ranking individuals in groups on Cayo Santiago. However, male dominance rank was positively correlated with copulatory rates with fertile females, and in one group females maintained proximity more actively to males which, in turn, exhibited the highest copulatory rates. In captive rhesus monkey groups, seeking proximity to the male represents female invitational (proceptive) behaviour and peaks during the peri-ovulatory phase of the cycle as levels of circulating oestradiol increase (Wallen *et al.* 1984). It may be, therefore, that whilst female rhesus monkeys are receptive and can mate at other times during the menstrual cycle, their proceptivity increases before ovulation and is directed towards specific, preferred males. Carpenter (1942) also recorded that females tended to associate with males most frequently during the mid-phase of putative 'oestrous' periods, when ovulation was most likely to occur. As regards the choice of a mate, Manson (1992) concluded that 'female choice for relatively stable (presently unknown) male traits, independent of male dominance rank, is apparently a component of a mixed female mating strategy that incorporates both choice and promiscuity'.

Chimpanzees and bonobos

The recent distribution range of the three chimpanzee subspecies is vast, extending from Sierra Leone in West Africa (*Pan troglodytes verus*) through Central Africa (*P.t. troglodytes*) to Tanzania in the east (*P.t. schweinfurthii*). The majority of our information about chimpanzee social and sexual behaviour derives from long-term studies in Tanzania conducted at the Gombe Stream Reserve (Goodall 1986) and by Japanese fieldworkers in the Mahale Mountains (Nishida 1990). Chimpanzees occur in temporary parties or subgroups, which form part of a much larger *unit-group* (Nishida 1968) or *community* (Goodall 1973). This variant of the multimale–multifemale social organization has been called a *fusion–fission society* (Kummer 1971). The name is apt, since chimpanzee subgroups frequently meet, mingle and divide once more as they travel through the forest in search of fruiting trees. A community may contain between 20–100+ individuals (Nishida and Hiraiwa-Hasegawa 1987); females tend to outnumber males (see Table 3.6). Subgroups usually contain fewer than seven individuals; 82% of subgroups observed at Gombe and 55% of subgroups at Mahale were of this type (Goodall 1968; Nishida 1968). Females often travel with their offspring and range over a smaller area than do adult males. Male offspring continue to associate with their mothers until adolescence, and they leave at approximately 12 years of age to join all-male subgroups (Nishida 1979). Chimpanzees, in contrast to ringtailed lemurs or rhesus monkeys, provide an example of a non-female-bonded society. Thus, whilst males remain in their natal communities, maturing females usually transfer to a neighbouring community before having their first offspring (Goodall 1986). These female transfers commonly occur when a female develops a sexual skin swelling and is sexually attractive to males. Nishida (1979) records that although rank relationships are not pronounced among female chimpanzees, newly transferring females tend to be subordinate to the resident animals. By contrast, there is a distinct hierarchy among the adult males in a community and males invariably outrank females in their social group.

Chimpanzees, unlike ringtailed lemurs or rhesus monkeys, are not seasonal breeders. Females may develop their sexual skin swellings, ovulate and conceive at any time of year. In practice such occurrences are comparatively rare in a chimpanzee community because females spend most of their adult lives either pregnant or lactating; interbirth intervals last for 4–5 years on average (Harcourt *et al.* 1980). When a female develops her prominent pink anogenital swelling, males alter their ranging patterns to seek proximity with her. Three types of mating tactic have been recorded in free-ranging chimpanzees (Tutin 1979). Firstly, males may copulate one after another with a female during the height of her swelling (approximately 11–12 days). As described previously, all the non-related adult males in a community may mate with such a female, so the occurrence of sperm competition is marked under these circumstances. Copulations are typically brief, involving a single intromission with pelvic thrusting to attain ejaculation. Either sex may initiate copulation but most mounts occur in response to penile displays and other precopulatory invitations by the adult male (Goodall 1968). Tutin (1979) recorded two further types of mating tactic employed by chimpanzees, in addition to the multimale copulations described above. Firstly, the alpha male in a community may display an extreme form of mate-guarding, or *possessiveness* towards a female, preventing all other males from approaching or copulating with her during the height of sexual skin swelling. Similar behaviour has been described for chimpanzees of the Mahale Mountains by Hasegawa and Hiraiwa-Hasegawa (1990). Secondly, consortships occur, during which a male attempts to lead a female away from the group and to remain associated with her for periods ranging from a few hours, to several weeks. Females are by no means passive in these activities and consortships are only successful if they co-operate with their male partners. By examining data on birth dates and putative timing of conceptions among the Gombe chimpanzees, Tutin (1979) proposed that consortship was the most effective male tactic in terms of reproductive success. Conversely, Hasegawa and Hiraiwa-Hasegawa (1990) found that, out of 18 conceptions for which data were available, opportunistic matings and resultant sperm competition occurred in at least 10 cases whereas consortship could be confirmed for just one case. In reality, only one study has thus far combined behavioural data with DNA typing in free-ranging chimpanzees to determine paternity of offspring and to measure male reproductive success. Gagneux *et al.* (1997) have reported that in a group of chimpanzees (*P.t. verus*) of the Tai forest, Ivory Coast, seven out of thirteen infants tested were sired by males from other communities. Dominant males in the home community and those engaging in consortships did not experience enhanced reproductive success. These surprising findings require confirmation and until further studies are conducted arguments about the relative success of the various male tactics in chimpanzees must remain speculative.

Like the chimpanzee, the bonobo or pygmy chimpanzee (*Pan paniscus*) lives in unit groups or communities consisting of changeable subgroups. It has a fusion–fission type of social organization resembling that of its larger relative. Bonobos have evolved in isolation from the common stock which also gave rise to the chimpanzee; the bonobo is restricted to an enclave bounded on its northern side by the Rivers Zaire and Lualaba. An adult bonobo weighs, on average, 84.5% of the weight of an adult chimpanzee and has a more rounded skull and relatively longer lower limbs (Kano 1992). Sexual dimorphism in cranial measurements is less pronounced in bonobos than in chimpanzees (Cramer 1977), and there are also differences between the two species in the anatomy of the genitalia of both sexes (Dahl 1985; Izor *et al.* 1981). Significant differences also exist between chimpanzees and bonobos as regards social organization and the mating system. Bonobo subgroups are usually larger than those of chimpanzees. The majority (74.1%) of bonobo subgroups contain 11 or more individuals (Kano 1982). Moreover, adults of both sexes occupy the same subgroups more commonly than is the case amongst chimpanzees. The 'E group' of bonobos studied by Kano at Wamba in Zaire, consisted of 65 individuals ranging over an area of 40–50 km^2 of rainforest. In E group, 96% of subgroups contained adults of both sexes and these often included lactating and non-pregnant females and those with sexual skin swellings. Kano (1982,

1992) has reported that bonobo females retain their sexual skin swellings for far longer than do female chimpanzees. Breeding is non-seasonal, as in the chimpanzee, and there is an effect of the menstrual cycle upon copulatory frequency in the bonobo. Most copulations occur when the swelling is firmest, at mid cycle, although it remains large and prominent at other times and copulations can occur at any stage (Furuichi 1987). Kano ascribes the larger sub-group sizes of bonobos and more common occurrence of mixed-sex subgroupings to the greater frequency of swollen females and greater copulatory activity of these apes, when compared with the chimpanzee. Certainly, sexual behaviour and socio-sexual behaviour plays a most important role in bonobo society, as will be described in a later section (see Chapter 6 pages 146–8). Female bonobos mate with a number of males, in an equivalent manner to the opportunistic matings described for chimpanzees at Gombe and at Mahale Mountains. Ventro-ventral copulatory postures occur among bonobos, particularly among younger animals, whereas chimpanzees invariably mate in a dorso-ventral position. Although male dominance rank is positively correlated with mating success in the bonobo (Kano 1992), possessive mate-guarding by alpha male bonobos has not been described. Nor have consortships been described for bonobos, so that the mating tactics of this species differ significantly from those of its larger relative, and sperm competition may be even more intense due to the frequency of opportunistic, multi-male copulations.

Mating seasonality and numbers of males in primate groups

At the beginning of this chapter it was pointed out that females lie at the core of many primate social groups. Females band together in order to exploit resources and males distribute themselves with respect to females (Wrangham 1979, 1980). The number of males in a primate group is thus believed to reflect the defensibility of the females which constitute a limiting resource as far as mating opportunities are concerned. Predation risk might also be a factor in determining how many males are present. Van Schaik, who favours the latter view, has recently pointed out that small groups sometimes contain several adult males and that this might be related to high predation pressure. Van Schaik and Horstermann (1994) compared group compositions in arboreal colobus monkeys, howlers, and langurs.

Colobus monkeys and howlers are subject to predation by monkey-eating eagles, whereas langurs are not at risk from these birds of prey. Groups of howlers and colobus monkeys contain multiple males whereas arboreal langurs typically form one-male groups. This correlates with the higher predation risk in howler and colobus groups; males are thought to be more vigilant and extra males are of value in protecting females and offspring.

It seems unlikely that a single factor can explain the occurrence of multiple males in such a wide range of primate species. Predation may be relevant in certain cases, but not in all. It is even thought that the overall size of primate groups is related to neocortical development; the larger the ratio of neocortex to hind-brain weight in primates, the larger are their social groups (Dunbar 1992). If we accept the biologically sound argument that females band together to exploit food resources and that males associate with females for mating purposes (Wrangham 1980) then multimale–multifemale groups might occur because a single male would be unable to exclude others from mating with females in the group. In this context, it is important to consider not just the sex ratio of adult females versus adult males in the group, but also the *operational sex ratio*, i.e., the number of females which are in breeding condition versus the number of males which are sexually active at any given time. If a primate species has a limited mating season, and all the females ovulate within a restricted period, as in the ringtailed lemurs described above, then a single male might be unable to mate with all the receptive females or to exclude other males from gaining copulations. Multimale groups might then be favoured in seasonally breeding primate species, whereas one-male groups might be more likely to occur in primates which mate and give birth at all times of year. This hypothesis was examined by Ridley (1986), who found that birth seasons (and hence by inference mating seasons) tend to be shorter in multimale–multifemale primate groups than in one-male (polygynous) groups. The sex ratio of adults in a group is likewise correlated with the length of the breeding season; a higher proportion of males occurs in groups of those species with short birth seasons.

Ridley's hypothesis has various shortcomings and he was well aware of some of these. Thus, there is no doubt that some multimale–multifemale species are not seasonal breeders. Chimpanzees and bonobos give birth at all times of year, for instance, as do baboon (*Papio*) species. Ridley divided primates into those with 'short' breeding seasons (75% of births occur within a 2-month period) and those with 'long' breeding seasons in order to facilitate statistical comparisons. His methods have been criticized on the basis that the important variable is the number of females per group which are simultaneously fertile and likely to conceive; this measurement should be used rather than numbers of males and duration of the birth season (Rees and Harvey 1991; Altmann 1990). In practice, however, it is exceedingly difficult to obtain information on the numbers of females per group which ovulate within the same time period for a sufficiently large sample of primate species and genera. Ridley's attempts to circumvent these problems seem entirely reasonable. Since the publication of his original report, data have continued to accumulate from field studies, and these still indicate a tendency for one-male groups (in which the ratio of adult females: males is very high) to have extended breeding seasons or to be non-seasonal breeders. By contrast far more species and genera which have a multimale–multifemale mating system also have restricted birth seasons. Exceptions to this general principle, such as chimpanzees, bonobos, chacma and savannah baboons, and Celebes macaques, are not understood. It is not sufficient in such cases to claim that adequate data are lacking or that there is something 'exceptional' about their social systems. Chimpanzees and bonobos have a fusion–fission organization accompanied by female emigration for instance, but this is not the case in *Papio ursinus* or *P. cynocephalus* which are also non-seasonal breeders. If these unusual cases run counter to Ridley's hypothesis, however, there are others which tend to support it. The one-male units of various forest guenons (*Cercopithecus* spp.) are also markedly seasonal in their reproduction (Butynski 1988). During their mating season, influxes of additional sexually active males may occur, so that a multimale–multifemale system is produced (e.g. *C. mitis*, *C. ascanius*: Cords 1987). The same phenomenon occurs in some one-male units of patas monkeys, as discussed previously (e.g. Chism and Rowell 1986). In these cases, where short mating seasons occur in polygynous primate species, there is a strong tendency for multimale influxes to transform the mating system in at least some groups.

Ridley's hypothesis is not without important exceptions, therefore, and cannot, in isolation, explain the occurrence of multiple males in primate groups. In reality there is probably an unidentified mixture of factors which determines the numbers of males present. Mitani *et al.* (1996) conducted a detailed comparative analysis of this problem and were able to confirm that a positive correlation exists between female group sizes and numbers of adult males. They were unable to confirm any relationship between mating season duration and the numbers of males in social groups, however.

Dispersed mating systems

The monogamous, polygynous, and multimale–multifemale mating systems considered far are not confined to any particular taxonomic grouping among the primates. Monogamy occurs in prosimians (*Indri*), in certain New World and Old World monkeys, and among the lesser apes. Multimale–multifemale groups are to be found in some diurnal prosimians, in monkeys of the New and Old World, and in chimpanzees. However, the 'dispersed' or non-gregarious type of mating system is found almost exclusively among nocturnal prosimians, such as galagos, lorises, and the aye-aye (Table 3.7). I say 'almost exclusively' because, among the anthropoids, the orang-utan is unique in having a non-gregarious social system and a specialized form of dispersed mating system. Primates which have dispersed mating systems do not live in permanent social groups, and this type of mating system is prevalent among nocturnal prosimians because nocturnality predisposes these animals towards a non-gregarious mode of life. Ancestral primates, like early mammals, were nocturnal (Martin 1990). Studies of extant nocturnal prosimians are therefore valuable in providing some clues about the social organization and sexual behaviour of the common ancestors of prosimians and anthropoids. Field research has shown that the simplistic view of nocturnal prosimians as 'solitary' primates leading impoverished social lives is incorrect. Non-gregarious prosimians, such as pottos, galagos, and mouse lemurs, occupy complex social networks which are maintained by means of olfactory and vocal communication, as well as by direct encounters during nocturnal activities (Charles-Dominique 1977; Schilling 1979; Bearder 1987). Bearder (1987) has defined five types of social organization among nocturnal primates, all of which involve varying degrees of home-range overlap and range-size differentials between the sexes (Fig. 3.10). The Type-1 system shown in Fig. 3.10 is typical of galagos (bushbabies) with the possible exception of *Galago zanzibaricus*. In the greater and lesser bushbabies or Allen's and Demidoff's bushbabies, the home ranges of adult males exhibit minimal overlap and are larger than those of adult females. Adult females in a particular area are often inter-related and a matriarchal system occurs in which female home ranges are tightly grouped. The example shown in Fig. 3.11 is from Bearder and Martin's (1979) field study of

Table 3.7 Home-range sizes and body weights of some primates which have dispersed mating systems

Species	Range size (hectares)		Body weight (kg)		Source
	Male	Female	Male	Female	
Microcebus murinus	3.2	1.8	0.08	0.08	Pagés-Feuillade 1988
Daubentonia madagascariensis	200[a]	40[a]	2.8	2.8	Andriamasimanana 1994
Galago alleni	30, 50	8–16	0.23	0.27	Charles-Dominique 1977
G. moholi	9.5–22.9	4.4–11.7	0.21	0.193	Bearder and Martin 1979
G. garnettii	17.9	11.6	0.822	0.721	Harcourt 1984
G. crassicaudatus	10	7	1.51	1.258	Bearder and Doyle 1974
Galagoides demidoff	0.5–2.7	0.6–1.4	0.063	0.062	Charles-Dominique 1977
Euoticus elegantulus	–	–	0.29	0.28	–
Perodicticus potto	9–40	6–9	1.02	1.08	Charles-Dominique 1977
Arctocebus calabarensis	–	0.32	0.31	–	
Nycticebus coucang	–	–	0.679*	0.626*	–
Loris tardigradus	–	–	0.29	0.26	
Pongo pygmaeus	v. large, limits unknown	>1.5 km² 5–6 km²	70.0	37.0	Rodman and Mitani 1987 Galdikas 1978

Body weights are from Harvey *et al.* 1987 and Bearder 1987.
[a] Personal communication from Sterling to Andriamasimanana 1994.
* Weights vary in this species; some specimens weigh >1 kg.

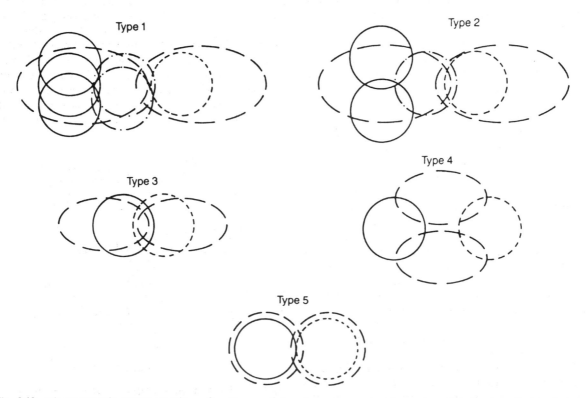

Fig. 3.10 Diagrammatic representation of patterns of home-range overlap in nocturnal primates. Circles=female ranges; Ovals=male ranges (except in Type 5 where both sexes are represented by circles). Type 1. for example *Galago senegalensis*; *G. Alleni*; Type 2. for example *Perodicticus potto*; Type 3. for example *Tarsius bancanus*; Type 4. for example *Microcebus murinus*; Type 5. for example *Aotus spp.* (After Bearder 1987.)

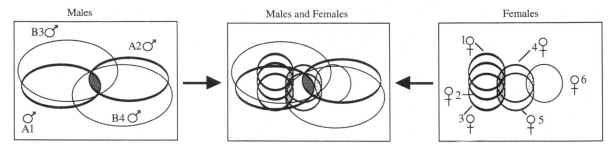

Fig. 3.11 Social organization and home-range overlap in *Galago senegalensis moholi* (Now usually considered as a distinct species—*G. moholi*). The ranges of six females are shown, including two matriarchies (1+2+3; 4+5) in which related females frequently sleep together. Two A-males (1 and 2) exclude each other from their territories, but tolerate the presence of smaller, wide-ranging, B-males (3+4). (Redrawn from Bearder 1987.)

the lesser bushbaby (*G. senegalensis*). Here there are two matriarchies, consisting of female numbers 1 + 2 + 3 and 4 + 5. Related females have extensively overlapping ranges whereas the range of the 6th and unrelated female is spatially more separate from the others. Female ranges also overlap extensively with the ranges of two dominant 'A-males' (Nos. 1 and 2). These males are older and heavier (mean weight 226 g) than younger non-territorial 'B-males' (mean weight 211 g) in the population. The B-males (Nos. 3 and 4 in Fig. 3.11) are tolerated by the more dominant, territorial individuals. A-males, by contrast, are antagonistic to one another and they have greater access to females, being found in sleeping associations with them during the daytime, as well as sharing a greater degree of home-range overlap at night.

Female lesser bushbabies usually give birth to twins and initially these are cared for in a nest or in a tree hollow. After a few days the mother leaves her infants in the branches of a tree whilst she forages at night and she returns to suckle them. The offspring differ markedly in their social development depending upon their sex. Females begin to show territorial behaviour (scent-marking and vocalizations) from about 200 days of age onwards, but they usually remain within their mother's home range unless a neighbouring area becomes vacant. Males, by contrast, emigrate from their natal ranges, moving 1–2 km before settling into large peripheral home ranges as B-males and awaiting an opportunity to establish themselves and gain access to the ranges of reproductive females.

The Type 2 social system depicted in Fig. 3.10, occurs in the potto (*Perodicticus potto*) and perhaps also in the other lorisines (*Arctocebus*, *Loris*, and *Nycticebus*). Bearder (1987) distinguishes this system from the Type-1 social organization of bushbabies, because matriarchies are not thought to occur, although genetic evidence on relatedness of females

has not yet been obtained. Nor do pottos employ territorial vocalizations or alarm calls as bushbabies do; indirect communication by means of scent-marking plays the most important role in lorisine social and sexual behaviour. Male pottos have larger home ranges (9–40 hectares) than females (6–9 hectares) and male ranges overlap very little, in contrast with the ranges of females. An example is shown in Fig. 3.12, from the field studies of Charles-Dominique (1977) of home ranges of two adult male pottos (A and B) and three adult females. Male A shares his range with females 1 and 2 whilst male B shares his range with Female 3. Yet on closer examination of Fig. 3.12 it becomes clear that female 2's range includes an extension into male B's range whilst male A also shares a small area of his range with female 3. The point to be

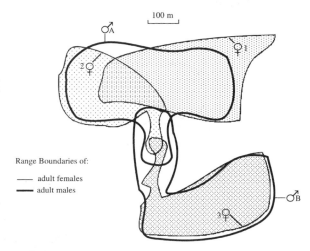

Fig. 3.12 Home range overlap and social organization in the potto (*Perodicticus potto*). The home ranges of adult males (A and B) exhibit minimal overlap. The smaller ranges of three adult females (1, 2, and 3) overlap extensively with individual male ranges. (Redrawn from Charles-Dominique 1977.)

made, which I shall return to in more detail below, is that even in these slower-moving nocturnal primates the potential exists for females to interact with a number of males in adjacent home ranges and to communicate indirectly with them (e.g. via olfactory cues). Maturing male pottos emigrate from their natal area and pass through what Charles-Dominique (1977) describes as a vagabond stage of development. Vagabonds are a little lighter and have smaller testes, on average, than established territorial males. Territorial males visit and court females in neighbouring ranges; the frequency of these visits increases in the period preceding the female's oestrus.

Very few direct observations of copulatory behaviour in nocturnal prosimians have been made under field conditions. We do not know if just one or a number of male pottos visit a female as she approaches oestrus. The structure of potto territories makes multimale visits a distinct possibility, however. In several nocturnal species, observers have witnessed a number of males congregating around a single oestrous female (in. *G. crassicaudatus*: Clark 1985; *G. senegalensis*: Bearder 1987, and *N. Coucang*: Barrett 1985). In Coquerel's dwarf lemur adult males expand their home ranges fourfold during the annual mating season, and their ranges may overlap with those of more than ten females (Kappeler 1997a). In the aye-aye, several males have been found sleeping close to the nest of an oestrous female during the daylight hours (Sterling 1993) and there is now evidence that females mate with more than one male (Sterling and Richard 1995). The true extent to which females of these, or of other, prosimian species mate with multiple partners during the peri-ovulatory period remains to be established. Because the home range of a territorial male potto, galago or mouse lemur can overlap several female ranges, there has been a tendency in the literature to equate the mating systems of these prosimians with the polygynous 'one-male units' of some diurnal anthropoids. It is presumed that a territorial male has priority or exclusivity of access in sexual relationships with neighbouring females but there is little evidence for this (Martin 1973; Bearder 1987; Izard 1990).

Charles-Dominique (1977), who studied five sympatric nocturnal prosimian species in Gabon (*G. demidoff*; *G. alleni*; *Euoticus elegantulus*, *P. potto*, and *Arctocebus calabarensis*), concluded that 'there is active competition between males giving rise to sexual selection for the most vigorous among them. The smallest and weakest males are hence excluded to various degrees from contact with the female home ranges'. Intermale competition is probably intense in many of these species and there is evidence that males fight with and wound one another. Charles-Dominique examined adults and immature individuals of *G. demidoff*, *G. alleni*, *E. elegantulus*, and *A. calabarensis* for signs of healed wounds. Damaged ears, tails, digits, eyes and body scars were present in 13–23% of adults examined and were two to four times more common in males than in females. Immature animals, by contrast, showed no signs of involvement in intraspecific fights. It is certainly true that physically larger and more dominant males, whose ranges occupy central positions vis à vis the ranges of females, have been identified in several species (e.g. *G. senegalensis*: Bearder and Martin 1979; *G. demidoff*: Charles Dominique 1977; *Microcebus murinus*: Martin 1973). In experiments on captive *M. murinus*, Perret (1992) has shown that dominant males are indeed aggressive towards subordinates, perform most copulations, and exhibit higher plasma testosterone levels. Perret has also shown that chemical cues in the urine of dominant males cause significant decreases in circulating testosterone in subordinates. Hence, the scent-marking behaviour used by so many nocturnal prosimian species may play a role in intermale competition and in the physiological suppression of lower ranking males.

Whilst all these observations indicate that intermale competition is pronounced in these nocturnal species, they do not show that their mating systems are polygynous. Given their non-gregarious social organization and the difficulties of mate-guarding spatially separated females throughout the hours of darkness, it seems most unlikely that male bushbabies, pottos, and other species could be strictly polygynous. Also, we shall see in later sections of this book that many of these prosimians have very large testes relative to their body weights (indicative of sperm competition, see Chapter 8) as well as complex penile morphologies and copulatory patterns such as result from sexual selection in a multimale mating context (see Chapters 5 and 9). It is for these reasons that I refer to the mating systems of many nocturnal prosimians as being 'dispersed' and equivalent, in terms of sexual interactions, to the multimale–multifemale mating systems of certain diurnal anthropoids. This is not meant to imply that all nocturnal primates have dispersed mating systems. Pair-formation and monogamy has been described in some cases (e.g. *Tarsius spectrum*: and *Lemur mongoz*) and possible monogamy has been claimed for at least one galago species (*G. zanzibaricus*: Harcourt and Nash 1984; Bearder 1987). In no case has DNA-typing been applied to a nocturnal prosimian population to determine the biological relatedness of adults or parentage of offspring. Clearly, this would be a most valuable exercise.

The mating system of the orang-utan

The orang-utan is remarkable among the diurnal primates in being the only species which has a non-gregarious social system. (Mackinnon 1974; Galdikas 1981; Rodman and Mitani 1987). Adult males are usually encountered as solitary individuals, as are adult females, although the latter may be accompanied by a dependent infant or a juvenile offspring. Adult males are massive and, at approximately 70–80 kg, they weigh more then twice as much as adult females (Table 3.7). Such males are territorial and the true extent of their ranges is unknown, since they travel far afield through the middle storey of the rainforest, as well as descending to ground level from time to time. The area of the home range may shift seasonally and over the course of time. Females occupy smaller ranges (1.5 km^2: Rodman and Mitani 1987; 5–6 km^2: Galdikas 1978) and rarely descend to ground level. Female ranges overlap one another and several females live within the range of an adult male. This situation is not dissimilar to that described previously for certain nocturnal prosimians.

Younger male orang-utans also resemble the vagabonds described for pottos and some other nocturnal *Lorisidae*. Sub-adult male orang-utans range widely but are not territorial. They mate opportunistically with females and their forceful mating tactics have been likened to rapes (Mackinnon 1974; Mitani 1985; Galdikas 1985). Sub-adult males are much smaller than territorial individuals and they lack the large cheek flanges, laryngeal sac, and long hair of fully developed adults. Nonetheless, these non-flanged males are sexually mature, since their testes are fully developed, and they are capable of impregnating females. In captivity male orang-utans reach this stage by as early as 6–7 years of age (Dixson *et al.* 1982). Development of the secondary sexual characters is affected by social stimuli, however, and the presence of a large, flanged male orang-utan in a captive colony can retard secondary sexual development in younger males for some years (Kingsley 1982). The same might be true in the wild, but it is not known how long the so-called 'sub-adult' males in a wild population take to achieve complete physical development. The territorial or flanged males advertise their presence by means of the *long call*. This consists of a series of loud moaning vocalizations, interspersed with 'bubbling' sounds pro-duced by inhalation. The male's laryngeal sac is inflated and his lips protruded during the call which lasts 1–2 min and carries for up to 1 km in the rainforest. Such calls are apparently spontaneous, but are also given in response to the sound of a falling tree, after an inter-male encounter, or in association with copulatory behaviour (Schurmann 1982; Galdikas 1983). Long calls are believed to function for intermale communication and spacing (Rodman and Mitani 1987) although they may also be significant in intersexual communication. In the wild, female orang-utans may approach and consort with fully-flanged adult males more frequently when conception is likely. There is some evidence that younger, nulliparous females are most persistent in their attempts to consort with flanged males (Schurmann 1982; Galdikas 1995). Thus, the secondary sexual adornments of adult males might have some significance in enhancing male sexual attractiveness. However, their potential role in intermale display is better documented; if flanged males meet they often behave aggressively, swaying trees and branches, piloerecting and glaring at one another. Fights may occur between such males, but they usually avoid contact (Horr 1972; Galdikas 1985). Flanged males form consortships with females which last for several days, during which time the male follows the female and copulations occur (Galdikas 1981, 1985). Sub-adult males, by contrast, are much less successful at forming consortships and females usually resist their copulatory attempts. Such resisted matings are probably less likely to result in conceptions than those which occur between fully developed adult males and females during consortships. It appears that fully flanged territorial male orang-utans are more successful reproductively, whilst non-flanged males adopt opportunistic/coercive tactics and are ousted from proximity to females by the larger males (Galdikas 1985).

How then, should one interpret the mating system of the orang-utan? Orangs are slow moving, quadrumanous climbers which travel around their large home ranges in search of fruiting trees. Their non-gregarious way of life is probably an adaptation to their feeding ecology; permanent social groups of orang-utans would probably be unable to sustain themselves as arboreal frugivores. In their review on orang-utan social and sexual behaviour, Rodman and Mitani (1987) conclude that the mating system 'tends to be promiscuous, with each female mating with more than one male and each male mating with more than one female'. This might make the orang-utan's mating system equivalent to the dispersed mating systems of many nocturnal prosimians. However, I think this is unlikely for the following rea-

sons. Unlike pottos, galagos, and many other nocturnal prosimians, male orang-utans do not have large
testes in relation to body weight (see Chapter 8,
pages 217–19). It is unlikely that sperm competition
has played a major role during the evolution of the
orang-utan, and such competition would be expected in a dispersed, 'promiscuous' mating system.
Perhaps the key to recognizing differences between
the orang-utan and other non-gregarious primates
lies in a combination of factors; its huge home
ranges, slow locomotion, long interbirth intervals
and formation of consortships for mating purposes.
Female orang-utans give birth every 3–4 years in
captivity (Lippert 1977) and interbirth intervals might
be longer in the wild. Flanged males consort with
females during the presumptive period of ovulation
and, given that a male's range contains those of just
a few females, it is likely that territorial males are
better able to guard and copulate with their partners
during the crucial fertile phase of the female's cycle
than is the case among nocturnal prosimians. In
some cases females preferentially approach and associate with flanged males at this time (Schurmann
1982) and laboratory studies have also shown that
female proceptivity (sexual initiating behaviour) increases during the peri-ovulatory phase of the menstrual cycle (Nadler 1982). In contrast with pottos or
bushbabies, male orang-utans have the advantage of
a diurnal lifestyle, enabling them to communicate by
visual displays (e.g. tree-swaying and 'snag crashing')
as well as by long-distance vocalizations. I am not
suggesting that multimale matings never occur in
the orang-utan, merely that they are much less common than in multimale–multifemale mating systems
(e.g. the chimpanzee) or dispersed systems (e.g.
bushbabies) where sperm competition has resulted
in the evolution of large relative testes size. Male
orang-utans are twice the size of females and possess striking secondary sexual adornments. These
are the hallmarks of primates in which there is
intermale competition and female defence polygyny,
as in geladas and hamadryas baboons. Thus, although the orang-utan is non-gregarious, its mating
system probably lies somewhere between polygnyous
and dispersed. Decisive evidence could be obtained
by conducting DNA-typing studies of free-ranging
orang-utans, perhaps by making use of hair samples
collected from their night nests. Such studies are
currently in progress (De Ruiter and Bruford: personal communication).

4   Mating tactics and reproductive success

Many aspects of an animal's environment might be expected to act as selective forces and to shape the evolution of behaviour. Normally when we consider environmental factors it is the physical make-up of the animal's world which springs to mind; temperature, rainfall, light, and availability of food throughout the year as well as the dangers imposed by predators and parasites. A species might be well-adapted to survive and reproduce within a particular habitat or 'niche' within the ecosystem. However, there is a very real sense in which the organism itself might contribute to the conditions under which natural selection and sexual selection operate (Lewontin 1983; Odling-Smee *et al.* 1996). Some animals alter their physical surroundings and change the environment in ways which influence evolution. Earthworms, for instance, alter the characteristics of the topsoil in which they live (Darwin 1881) and so influence conditions experienced by future generations of their own species. Industrialized human societies have brought about enormous changes in the physical environment so that, in the fullness of time, these also might have evolutionary consequences. The burrows produced by numerous animals, the 'lodges' constructed by beavers, the complex nests of social insects, and the webs of spiders, all constitute specialized environments and potential agents for evolutionary change. For male spiders of many species the female's web presents an important challenge. The smaller male must pluck and vibrate the female's web in particular ways as a courtship display; should he fail to communicate effectively he is likely to be eaten. Presumably both natural and sexual selection have refined these male displays; the most successful males survive, mate, and sire the largest numbers of offspring. How could such courtship displays have been perfected during evolution without the environment provided by the female's web?

And so, from the analogy of sex and the spider's web, let us turn to webs of social and sexual relationships within groups of monkeys or apes. Since so many primates are, par excellence, social animals it is not unreasonable to suggest that the social group

constitutes an important vehicle for evolutionary change. Selection should, for example, favour mechanisms of social communication which assist an individual (and its kin) to survive, compete with others of its own sex, attract mates and raise the resulting offspring. The sexual tactics pursued by males and females might be expected to differ in important ways. In those social groups which are female–bonded it is usual for female offspring to remain in their natal group; daughters of dominant females tend to achieve higher rank and this, in turn, can have a positive influence upon their reproductive success (e.g., Harcourt 1987; Gomendio, 1990). The core of a macaque or baboon group, for example, consists of a number of matrilines. Males, by contrast, usually emigrate as sexual maturity approaches (e.g. *Papio*, Pusey and Packer 1987). Such males must transfer to other groups and compete for mating opportunities. In the case of polygynous primates, emigrating males must either acquire females to found a new one-male unit (as in the gorilla) or take over an existing unit by ousting the resident male (as in geladas or Hanuman langurs). A wide variety of mating tactics exists in primate societies and it is the purpose of this chapter to review the major types. No attempt will be made to classify mating tactics, for example, in terms of how far a given tactic is genetically or environmentally determined or whether it is either reversible or adhered to across the lifespan (Caro and Bateson 1986; Moore 1990). Nor shall I make any distinction between mating tactics and mating strategies; in practice both terms are used interchangeably by various authors. In correct usage, it is the case that when alternative mating tactics result in a similar level of reproductive success for the individuals concerned, then the combination of these tactics represents an evolutionary stable strategy (ESS) in the population (Maynard Smith 1982). In reality, we have only limited knowledge of how the various mating tactics employed by prosimians, monkeys and apes translate into quantitative measurements of lifetime reproductive success. Evidence is accumulating on these questions, especially since the advent of DNA typing techniques has made it possible to measure biological relatedness of individuals and parentage of offspring in primate groups (Martin *et al.* 1992; De Ruiter and Inoue 1993). Progress in relating behaviour to genetic measurements of reproductive success will also be reviewed in this chapter.

Dominance, mating success, and reproductive success
Dominance in males

The concept of dominance has a chequered history as applied to studies of primate social organization. Dominance has been used to indicate priority of access to food, water, and sexual and grooming partners, as well as superiority in aggressive and sociosexual encounters (Zuckerman 1932; Maslow 1936–1937; Wickler 1967). Quantitative tests of these propositions provide little support for a unitary definition of dominance; rank orders constructed by measuring agonistic interactions frequently do not agree with those arrived at by measuring grooming, sociosexual activity, or feeding behaviour (Bernstein 1970; Dixson *et al.* 1975; Smith, M. *et al.* 1977). Gartlan (1964) suggested that the use of the term should be dropped, therefore. Yet there are correlations between an animal's position in its social group and patterns of aggression, sexual behaviour, grooming, and other activities. In talapoins, for example, a clear-cut aggressive hierarchy is usually measurable in captive groups. Animals adjacent in rank groom one another much more frequently than would be expected by chance. Further, if sociosexual presentations and mounts occur after an aggressive episode, it is usually the recipient of aggression which presents and the dominant individual which mounts (Dixson *et al.* 1975). Low-ranking talapoins glance at high-ranking individuals more frequently than vice-versa; measurements of this 'looking at' behaviour correlate closely with dominance in aggressive encounters (Keverne *et al.* 1978). In baboons, measurements of approach/avoidance behaviour correlate well with measurements of agonistic rank (Rowell 1967a) as is true for avoidance/displacement behaviour in male mandrills (Wickings and Dixson 1992a).

What concerns us here is whether a male's position in the dominance framework of the group, as reflected by agonistic relationships (or correlated behavioural measurements), affects mating success. Beyond that, we wish to know if mating success translates into reproductive success as reflected by numbers of offspring sired. Although there are co-

gent reasons for believing that intermale competition and mating success are linked (Darwin 1871), the degree to which male dominance rank influences mating success in primates has been much debated (Bernstein 1981; Fedigan 1983; Bercovitch 1986, 1992; Ellis 1995). A comparative analysis of relationships between male rank and mating success in 75 groups of 14 anthropoid species led Cowlishaw and Dunbar (1991) to conclude that there is 'a reliable positive relationship between male dominance rank and mating success amongst animals of the same age class'. In Table 4.1 I have assembled examples from studies of free-ranging and captive groups of monkeys and apes where male rank and mating success have been found to correlate in a positive way. The list includes a preponderance of species which have a multimale–multifemale mating system, since the presence of a number of males in a group is required in order to measure their dominance relationships. In those species which are polygynous it is usually the case that the single male defending a group of females has a mating advantage during the period of his 'tenure' (e.g., in gorillas: Schaller 1963; hamadryas baboons: Kummer 1968; or geladas: Dunbar 1984). However, it is necessary to consider effects of male 'take-overs', such as occur in geladas and Hanuman langurs, and the effect of influxes of additional males into one-male units, such as occur in the patas monkey and certain *Cercopithecus* species (Cords 1987, 1988). Likewise, among nocturnal prosimians it is often assumed that male bushbabies, pottos, or mouse lemurs which occupy favoured territories, overlapping those of reproductive females, should experience a mating and reproductive advantage (Martin 1973; Charles-Dominique 1977; Bearder 1987). The evidence is weak, however, since we have almost no quantitative data on copulatory interactions in these non-gregarious primates.

Table 4.2 and Figs 4.1 and 4.2 provide examples of relationships between male dominance rank and copulatory frequencies in monkeys and apes. Table 4.2 refers to the talapoin, and here the effect of dominance is extreme, for only the highest-ranking males in three captive groups were observed to mount females. Talapoins are also unusual, because adult females consistently rank above males. Female choice, as well as male rank, influences copulatory frequencies since females are more proceptive towards the most dominant male (Dixson *et al.* 1975; Dixson and Herbert 1977). In the wild, a number of male talapoins are sexually active during the annual mating season and effects of male dominance are unknown (Rowell and Dixson 1975). The importance of female proceptivity in partner choice and copulatory frequency has been established in field studies of various primate species, however (e.g., in *Cebus apella*: invitational behaviour towards the dominant male: Janson 1984; *Brachyteles arachnoides*: female vocalization, increased ranging behaviour and scent marks attract a 'mating aggregation' of males: Milton 1985a).

A second example of effects of male dominance upon mating success is shown in Fig. 4.1 and concerns the mandrill (*Mandrillus sphinx*). In this case, a pronounced linear hierarchy occurred among six male mandrills (three adults and three sub-adults) in a semi-free-ranging heterosexual group. The aggressive hierarchy correlated with the avoidance/displacement hierarchy among these males. Outside the group, three further adult males lived a solitary/semi-solitary existence. A hierarchy was also discernible among these males, and the most dominant solitary male also outranked the majority of the group-associated males, with the notable exception of the alpha individual. Copulatory success in these male mandrills was clearly related to their dominance ranks. The alpha, group-associated, male succeeded in mating with all the adult females and

Table 4.1 Dominance and mating success in adult male anthropoid primates: some examples where rank and mating frequencies are positively correlated

Species	Type of study	Sources
New World monkeys		
Cebuella pygmaea*	Field study	Soini 1988
Alouatta palliata[+]	Field study	Clarke 1983; Jones 1985
A. caraya	Field study	Jones 1983
Cebus apella	Field study	Janson 1984
Lagothrix lagotricha	Field study	Nishimura 1988
Old World monkeys		
Cercopithecus aethiops	Field study	Struhsaker 1967a
Erythrocebus patas**	Field study	Ohsawa *et al.* 1993
Miopithecus talapoin	Captive groups	Scruton and Herbert 1970
		Dixson and Herbert 1977
Macaca mulatta	Semi-free ranging group	Carpenter 1942; Manson 1992
M. mulatta	Field study	Lindburg 1983
M. fuscata	Field study	Tokuda 1961
M. fuscata	Captive groups	Hanby *et al.* 1971;
		Inoue *et al.* 1993
M. fascicularis	Field study	De Ruiter *et al.* 1992;
		De Ruiter and Van Hooff 1993
M sylvanus	Semi-free-ranging groups	Paul *et al.* 1993
M. maurus	Field study	Matsumura 1993
M. nigra	Captive group	Dixson 1977
M. arctoides	Captive group	Nieuwenhuijsen *et al.* 1987
M. arctoides	Semi-free-ranging group	Brereton 1994
M. thibetana	Field study	Zhao 1993
Mandrillus sphinx	Semi-free-ranging group	Dixson *et al.* 1993
Papio cynocephalus	Field study	Hausfater 1975
		Nöe and Sluijter 1990
P. ursinus	Field study	Seyfarth 1978
P. anubis	Field study	Packer 1979b
Great apes		
Pan troglodytes	Field study	Hasegawa and Hiraiwa-Hasegawa 1990
Pan paniscus	Field study	Kano 1992
Gorilla g. beringei	Field study	Harcourt *et al.* 1980

* A second male was present in this group of *Cebuella pygmaea*; the dominant male excluded this individual from proximity to the female during her peri-ovulatory period.
** Multiple influxes of males occurred into a one male unit of *E. patas*.
[+]The dominant male in Clarke's Study of *A. palliata* had priority of access to the female uring the probable time of ovulation, but did not copulate more frequently than subordinates at other times.

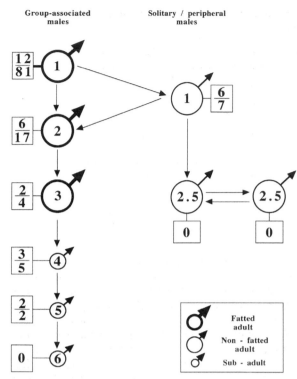

Fig. 4.1 Male dominance rank and mating success in semi-free-ranging mandrills (*Mandrillus sphinx*), during a single mating season in Gabon. Grouped-associated (fatted and sub-adult) males have a linear hierarchy in which male rank is positively correlated with mating success. Boxes indicate the numbers of females mated (upper) and numbers of observed ejaculatory mounts (lower) for each male. Solitary/peripheral, non-fatted, adult males mate opportunistically; only the highest-ranking peripheral male was observed mounting females (Author's unpublished observations.)

accounted for 70% of all ejaculations recorded during the annual mating season. Solitary males were not sexually inactive, however, so that the most dominant solitary male mated with six females and accounted for 7% of ejaculations. Sub-adult males within the group had very low mating success, in accord with their low ranks. This example illustrates how inclusion of sub-adult males within an analysis of male dominance and copulatory success might enhance the observed effect. Some authors have objected that positive correlations between male dominance and mating success are indeed biased in studies which include sub-adults in the analyses (Bercovitch 1986; McMillan 1989). However, in their comparative study Cowlishaw and Dunbar (1991) controlled for the presence of sub-adults and showed that a positive correlation between dominance and mating success still occurs when only adult male

hierarchies are considered. A second point which emerges from Fig. 4.1 concerns the ability of some solitary male mandrills, which exhibit lesser development of secondary sexual adornments and lower plasma testosterone levels than group-associated males (Wickings and Dixson 1992a), to copulate opportunistically with females. Here we see the possibility of alternative mating tactics in males which fail to attain the highest-ranking positions. There is the possibility that males pursue different patterns of social development (e.g., a solitary phase) and trajectories of reproductive growth (e.g., suppression of secondary sexual characteristics) within a given species and that such 'strategies' change over the course of the lifespan. I shall return to these questions of lifetime mating success and reproductive success in a moment.

The third example of effects of male dominance on mating success concerns the bonobo (*Pan paniscus*). Unlike chimpanzees, male bonobos have not been observed to form consortships with females or to exhibit 'possessiveness' towards particular females (i.e., intensive mate guarding by an alpha male). However, Kano (1992) has shown that although many males are sexually active in a free-ranging bonobo community, there is a positive correlation between male rank and mating success (Fig. 4.2A). Interestingly, this finding cannot be attributed to the inclusion of low-ranking sub-adult males in the analysis since the correlation between male rank and mating success achieves a still greater statistical significance ($P < 0.01$ as compared to $P < 0.05$) when only the 14 adult males in the community are considered.

A final example of relationships between male rank and sexual activity is provided by the study of stumptail macaques by Nieuwenhuijsen *et al.* (1987). Stumptails exhibit very high frequencies of both copulatory and masturbatory behaviour. In a captive group containing 18 adult males, the 4 highest-ranking individuals accounted for 87% of observed matings (Fig. 4.2B). Other studies have shown that the presence of such dominant males suppresses copulatory behaviour in lower-ranking stumptails (Estep *et al.* 1988). It might be imagined that higher frequencies of masturbation would occur in lower-ranking males as an alternative sexual outlet, since they have less access to female partners. However, this is not the case. Neither male rank nor copulatory frequency exhibited any significant correlation with masturbatory behaviour in the group studied by Nieuwenhuijsen and his colleagues. The significance of masturbatory behaviour in primates will be considered in the next chapter (see page 139).

Despite the positive correlations between male

Table 4.2 Relationships between male rank and copulatory frequencies in three captive groups of talapoin monkeys (*Miopithecus talapoin*)

Group number	Number of females in group	Males in order of rank	Males' hormonal status	Total mounts on all females
IV	4	1092	Intact	28
		1008	Intact	0
XIII	5	1017	C.T.	106
		1008	C.T.	0
		2044	Intact	0
XIV	5	2004	Intact	69
		1002	C.T.	0
		2043	Intact	0

C.T.=castrated and testosterone-treated. Data are from Dixson *et al.* (1975).

dominance rank and mating success in various primate species, it would be misleading to imply that such effects are universal. Various studies have failed to confirm that dominant males experience a mating advantage (e.g., in *Saimiri oerstedii*: Boinski 1987, *Cercopithecus mitis*: Cords *et al.* 1986; *Macaca mulatta*: Loy 1971, *M. fuscata*: Eaton 1978; *Pan troglodytes*: Tutin 1979). In the case of the rhesus monkey, Drickamer (1974a) demonstrated a most important relationship between male 'observability' and copulatory frequency. Dominant males of the island colony at La Parguera, in Puerto Rico, were more readily observed than others during field studies; when this variable was controlled for the correlation between male rank and mating success was no longer statistically significant (see also: Chapais 1983). Aside from observability and species differences in behaviour, a number of other variables affect measurements of male mating success. As the number of adult males in a social group increases, it might prove increasingly difficult for a high-ranking male to maintain priority of access to females. The problem increases if females are strictly seasonal in their reproductive biology, so that they ovulate during a restricted portion of the year. Under these conditions intermale competition may be intense,

A. Male Rank and Copulatory Frequency in Bonobos

B. Male Rank and Copulatory Frequency in Stumptail Macaques

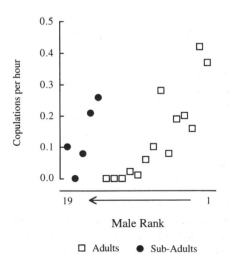

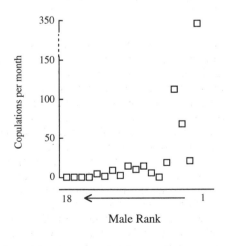

☐ Adults ● Sub-Adults

Fig. 4.2 Positive correlations between male rank and copulatory behaviour in A. Free-ranging bonobos (P < 0.05) and B. Captive stumptail macaques (P=0.001). Further discussion of these results is provided in the text. (Data on bonobos are from Kano 1992 and on stumptails from Nieuwenhuijsen *et al.* 1987.)

but dominance might not be the only, or the most effective, male tactic in securing matings and in maximising reproductive success (Paul 1997). In Boinski's (1987) studies of free-ranging squirrel monkeys, for example, no dominance hierarchy was measurable between adult males during the annual mating season. Males increased in weight at this time (the 'fatted' condition: Dumond and Hutchinson 1967) and females initiated copulations preferentially with the largest male in the group. In this case female choice, rather than male dominance *per se* influenced copulatory frequencies. In another seasonally breeding species, the ringtailed lemur (*Lemur catta*), a dominance hierarchy occurs among adult males, but it tends to break down during the annual mating season. In Koyama's (1988) field study of *L. catta* at Berenty, in Madagascar the second-ranking male in one group achieved highest mating success, but females mated with multiple partners during their brief oestrous periods. Koyama concluded that other factors, besides dominance, such as length of a male's residence in his group, and female choice, combined to influence mating success. Subsequent studies of semi-free-ranging *L. catta* by Pereira and Weiss (1991) have demonstrated a role for female choice in determining male mating success.

Where the number of females in breeding condition at any given time is not too large, such as in non-seasonal breeders or seasonal breeders in which females ovulate sequentially over a longer time period, then dominant males might be able to gain a mating advantage by concentrating copulations within the female's peri-ovulatory period. The best documented examples of this type concern Old World monkeys in which female sexual skin occurs. In baboons, for instance, the most likely period for ovulation to occur is during the last few days at the height of female sexual skin swelling, although there is considerable variability between females in this respect. Thus, ovulation might also take place as the sexual skin begins to deflate (*Breakdown* day) or on the day after breakdown (Wildt *et al.* 1977). In the yellow baboon (*Papio cynocephalus*) high-ranking males have a marked mating advantage, but their mating success is greatest during the last few days of maximal female swelling (Fig. 4.3). The same is true in the mandrill, where the alpha male in a semi-free-ranging group mate-guarded females and copulated most frequently during the last few days before sexual skin breakdown (Dixson *et al.* 1993). In this case it appeared that the alpha male terminated guarding episodes as soon as sexual skin breakdown had begun, indicating that females became less attractive sexually at this time. In species where no

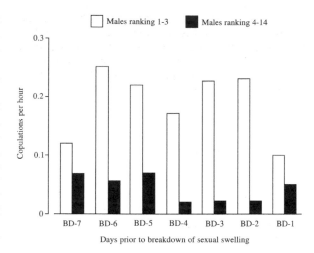

Fig. 4.3 Male rank and mating success in relation to the stage of the menstrual cycle in yellow baboons (*Papio cynocephalus*). The highest-ranking males in the troop have greatest mating success, and this effect is especially pronounced when ovulation is most likely, during the last few days before sexual skin breakdown (BD) in females. (Data from Hausfater 1975 and redrawn from Daly and Wilson 1983.)

sexual skin swelling occurs, it is more difficult for the (human) observer to decide whether dominant males concentrate sexual activity during the female's peri-ovulatory period. In the rhesus macaque Chapais (1983) used the termination of sexual consortships as an indicator of 'attractiveness breakdown' in the adult female and hypothesized that ovulation was most likely to occur during the four days preceding the breakdown day. Dominance rank correlated positively with copulatory frequency and time spent in consortship during the putative fertile period. However, it was not the case that dominance correlated in a simplistic manner with 'priority of access' to consorts (Altmann 1962), so that other factors such as sexual preference for particular partners (Herbert 1968; Loy 1971) and female choice were probably instrumental in producing the observed effects.

Among New World monkeys, it is unusual to find external indicators of female reproductive status. However, in the mantled howler (*Alouatta palliata*) the female's external genitalia become pinkish in colour and slightly swollen as ovulation approaches (Glander 1980). The dominant male in the group guards females and copulates with them during the phase of maximal genital enlargement (Jones 1985). In many other anthropoid species where female genital or behavioural cues of impending ovulation are either absent or not apparent to the human observer, it is still the case that males are more sexually active during the follicular phase of the female's cycle or during the peri-ovulatory period. It is incorrect, therefore, to assume that ovulation is necessarily *concealed* in species such as marmosets and tamarins, vervet monkeys, and langurs. (See Chapter 11, page 352 for further discussion of this subject).

Coalitions, alliances, and male mating success

The attainment and maintenance of high rank in some species depends upon an individual's ability to form successful alliances, or coalitions, with other group members (Harcourt and De Waal 1992). This is especially the case in baboons (Hall and DeVore 1965; Saayman 1971; Bercovitch 1988) and in chimpanzees (De Waal 1982, 1992; Uehara *et al.* 1994). In savannah baboons (*P. cynocephalus*) Bercovitch (1988) found that coalition formation is a 'low risk' tactic used by older males of medium or low rank; typically two males join forces in an attempt to oust another male from his consortship with a female. Most of the coalitions observed by Bercovitch targeted a younger, high-ranking newcomer to the group, and other authors have also found that younger, high-ranking males are most frequently targeted (Collins 1981; Smuts 1985). Coalitions were also aimed at those consortships where the female partner was close to ovulation (particularly on cycle days −2 and −1 before sexual skin breakdown, Fig. 4.4). Bercovitch also found that once two males had formed a coalition and had begun to threaten and harass a consorting male, a changeover typically occurred within a minute or so, although in one exceptional case harassment continued for 1.28 h. Coalitions usually involve males occupying adjacent ranks in the dominance hierarchy; but either the higher-ranking male (Noë and Sluijter 1990) or the lower-ranking individual (Bercovitch 1988) might obtain access to the consort female if the coalition succeeds.

In contrast to the rapidly forming and often transient coalitions of baboons, alliances among male chimpanzees are usually long-term affairs. Successful alliances secure mating opportunities for the males concerned, especially during periods of maximal sexual skin swelling in females (De Waal 1982, 1992). Changes in alliance structure can bring about loss of alpha male status and sometimes involve severe aggression and maiming of the deposed individual (De Waal 1986). Although close associations occur between particular adult males in the pigmy chimpanzee (*Pan paniscus*) and an aggressive rank

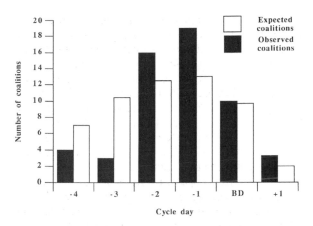

Fig. 4.4 The distribution of observed versus expected numbers of coalitions between male baboons (*Papio cynocephalus*) in relation to the stages of females' menstrual cycles. Observed coalitions increase in frequency as ovulation approaches and peak on days -2 and -1 before breakdown (BD) of female sexual skin swellings. (Redrawn from Bercovitch 1988.)

order exists among males in wild communities of this species, it is unclear whether alliances occur, or if they influence male rank and mating success.

Consortship, mate-guarding, and possessiveness

In a number of primates, temporary associations or male–female partnerships can occur. These consortships usually involve females in reproductive condition and are characterized by a 'high degree of reciprocally interactive behaviour and rapport' between the two sexes (e.g., in rhesus macaques: Carpenter 1942). From the male perspective, consortship is a mate-guarding tactic; the male attempts to associate and copulate with a female during the presumptive fertile period. In a few species, intensive guarding by a dominant male involves rather less interactive rapport with the female. This is the case in high-ranking male mandrills (Dixson *et al.* 1993) and also in alpha male chimpanzees which sometimes exhibit extreme possessiveness towards particular females at maximal sexual skin swelling (Tutin 1979; Hasegawa and Hiraiwa Hasegawa 1990).

Consortships have been described in many macaque species, in baboons, patas monkeys, howlers, ringtailed lemurs, orang-utans, and chimpanzees. (Hrdy and Whitten 1987). The durations of these associations and numbers of partners which a single female may consort with during her fertile period varies tremendously between species and between individuals (examples are provided in Table 4.3). In

Table 4.3 Consortship and mate-guarding in monkeys and apes

Species	Type of study	Male–female relationships	Source
New World monkeys			
Alouatta palliata	Field study	Females initiate a series of consortships with males during a 3–4 day, peri-ovulatory period.	Carpenter 1934
	Field study	Dominant male consorts with female and guards her during the 2–4 day fertile period.	Glander 1980; Jones 1985
Ateles paniscus	Field study	Female is sexually active for 8–10 days and actively solicits copulations. Consortships last 1–3 days.	Van Roosmalen and Klein 1988
Brachyteles arachnoides	Field study	Smaller, sexually mature, males associate with females and copulate prior to the presumptive peri-ovulatory period. Large males do not associate in this way, but mate opportunistically as members of 'mating aggregations' when ovulation is imminent.	Milton 1985a
Cebus apella	Field study	Females initiate consortships with the dominant male during the presumptive peri-ovulatory period and are guarded by the male during this phase. Subsequently, females mate with lower-ranking males.	Janson 1984
Old World monkeys			
Macaca mulatta	Semi-free-ranging groups	Females have 1–10 consorts per fertile period. High and medium-ranking females have more partners than low ranking females	Loy 1971
	Semi-free-ranging groups	Females have 1–9 consorts per fertile period and appear to follow a 'promiscuous' tactic.	Manson 1992
M. fuscata	Captive group	Some partners consort for several days; they travel and sit together as well as mating.	Hanby *et al.* 1971
	Semi-free-ranging groups	Mutual following and grooming between consorts. Adult males average 3.1 consortships, each lasting 5.7 days, during annual mating season.	Wolfe 1984b
M. fascicularis	Field study	High-ranking males have more consort partners. Male plays major role in maintenance of proximity between partners. Consortships last from 1 h–3 weeks. Partners may mate outside the consortship.	Van Noordwijk 1985
M. silenus	Field study	Females approached the single adult male in this group and associated with him prior to full sexual skin swelling. Consortships lasted several days. Consorts sometimes moved a little away from the main group.	Kumar and Kurup 1985
M. nigra	Captive group	Dominant male follows, sits with and copulates with a particular female at the height of swelling, but does not mate exclusively with her.	Dixson 1977
M. thibetana	Field study	Multiple consorts sometimes occur. A young adult male, plus two (rarely more) females maintain proximity, travel together and mate	Zhao 1993, 1994

Table 4.3—*Continued*

Species	Type of study	Male–female relationships	Source
M. sylvanus	Field study	Male often initiates consortship by maintaining close proximity to female ('proximity-possession principle'). However, females may solicit and copulate with 3–10 males in a single day ('promisicuous tactics'). Consociations rather than consort relations occur in this species.	Taub 1980
M. radiata	Captive group	Dominant males are involved in more consortships than others. Males play the major role in proximity maintenance and may copulate with other females as well as the consort partner.	Glick 1980
Papio ursinus	Field study	Male plays major role in maintaining proximity to female at the height of her swelling. Consortship declines as the sexual skin begins to deflate; subadult and juvenile males then begin to mate with the female.	Saayman 1970
P. cynocephalus	Field study	Male maintenance of proximity to his consort increases on day −3/−2 before breakdown of swelling. Male occasionally 'herds' female away from group. Female may have a number of consorts, however.	Hausfater 1975
P. cynocephalus	Field study	Young, dominant adult males employ 'solo' tactics to obtain consortships, older and lower-ranking males tend to employ 'combo' tactics (i.e. form coalitions) to oust other males.	Noë and Sluijter 1990
P. cynocephalus	Field study	Consortships last 1–8 days (mean: 4.5 days). Most females prefer the alpha male as a consort partner. Dominant females begin to consort earlier during sexual skin swelling and tend to have longer consortships than low-ranking females.	Seyfarth 1978
Mandrillus sphinx	Semi-free-ranging group	Alpha male follows, mate-guards and copulates with females at maximum swelling. Other males mate opportunistically.	Dixson *et al.* 1993
Great apes			
Pan troglodytes	Field study	Male initiates consortships lasting from 3 h–28 days. Male success depends upon female compliance and willingness to follow male away from group.	Tutin 1979
	Field study	High-ranking (usually alpha) male may exhibit 'possessiveness' towards a female at maximum sexual skin swelling. Prevents others from copulating.	Tutin 1979. Hasegawa and Hiraiwa-Hasegawa 1990
Pongo pygmaeus	Field study	Consortships last several days. Male follows and mates with female. Only fully developed adult males consort in this way; some consortships are initiated by younger females.	Galdikas 1981, 1985 Schurmann 1982

some instances the term consortship has been applied to situations where the sexes remain in proximity for a short time but where proof of reciprocal maintenance of proximity for sexual reasons is lacking. The initiation of a consortship may be due to the female partner (e.g. in *Alouatta palliata*: Carpenter 1934; *Cebus apella*: Janson 1984) or, more usually, the male seeks proximity to a female during the fertile phase of her ovarian cycle (e.g., in various macaque species, baboons, and chimpanzees as detailed in Table 4.3). In *Macaca fascicularis*, for example, once a consortship has begun, it is the female partner which usually initiates movements, and the male which follows her and maintains proximity (Table 4.3). However, Van Noordwijk (1985) points out that females are not passive in such relationships since they may 'wait and invite the consort partner, and they do not travel far when the male does not follow immediately'. Evidently, communi-

cation between sexual partners can be very subtle and for many species the precise roles played by males and females have yet to be analysed. Tutin (1979) made detailed observations on consortship behaviour in chimpanzees. A male seeks to initiate consortship by remaining close to a particular female, and grooming her, until the majority of other sub-group members have departed. The male then attempts to lead the female away from the remaining portion of the sub-group. The compliance of the female, in remaining silent and following the male is vital to the success of the male's tactic. Should the female vocalize, or respond to the distant calls of other males, then the interaction may be terminated prematurely (Fig. 4.5). Chimpanzee consortships can last for as little as a few hours or as long as 28 days.

The occurrence of consortships in various monkeys and chimpanzees does not mean that the partners mate exclusively with one another; a female

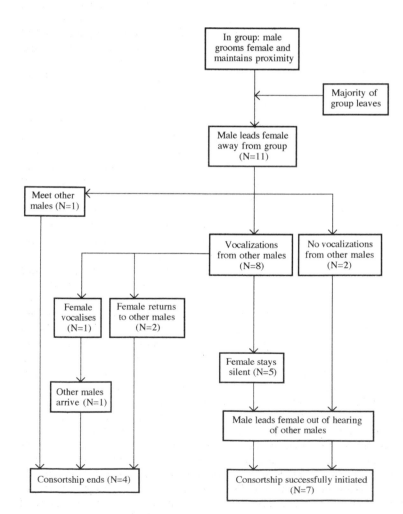

Fig. 4.5 Consortship initiation in free-ranging chimpanzees (*Pan troglodytes*). Analysis of 11 cases ending either in successful initiation of consortships by males (*N*=7) or failure to secure consortships (*N*=4). Females exert crucial influences upon male success in initiating consortships. (Redrawn and modified from Tutin 1979.)

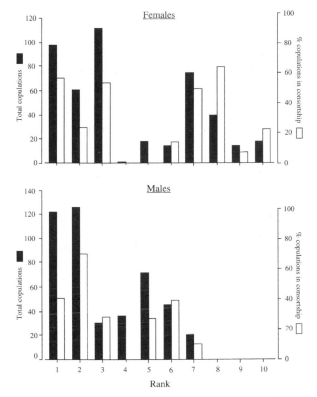

Fig. 4.6 Consortship and copulatory frequencies in free-ranging, long-tailed (cynomolgus) macaques (*Macaca fascicularis*). High-ranking animals of both sexes copulate most frequently and, although many copulations occur within consortships, both sexes frequently mate with other partners. (Data from Van Noordwijk 1985.)

might have a series of consorts during her fertile period (e.g. in *Macaca mulatta*: Manson 1992; *M. sylvanus*: Taub 1980; *Papio cynocephalus*: Hausfater 1975). Consortship does create a situation where copulations between partners are more likely, however. In *Macaca fascicularis* many or most copulations which occur during consortships are between the male and female in consort, yet either partner might mate with others (Fig. 4.6). In the mandrill, where dominant males follow and mate-guard females during the period of maximum genital swelling, there is a marked positive correlation between guarding and mating success, although it is not invariably the male guarding a female which succeeds in siring her offspring (Dixson *et al.* 1993). Possessiveness in chimpanzees is likewise a mate-guarding tactic employed to best effect by an alpha male. Tutin (1979) observed 39 incidents of possessive tactics by male chimpanzees and the most successful male, Figan, was also the highest-ranking in the community. During a 2-month period when four females developed fully swollen sexual skin, and

exhibited a total of seven menstrual cycles, Figan was possessive during five cycles and succeeded in interrupting copulations by lower-ranking males. The two cases where Figan did not monopolize matings occurred when two females were simultaneously at maximal swelling. In this circumstance, he showed a preference for the older female.

Relationships between consort behaviour and male reproductive success remain to be established. Tutin (1979) calculated that the vast majority of chimpanzee copulations (73%) result from opportunistic tactics, where a number of males mate with a single female at the height of her swelling. Twenty five percent of copulations occur when males exhibit possessive tactics and only two per cent result from consortships. Despite the low frequency of copulations attributable to consortships, Tutin calculated that seven out of fourteen pregnancies, for which there were sufficient data on the preceding sexual activity of females, resulted from consort tactics. Goodall (1986) has presented more detailed analyses of data from the same chimpanzee population at Gombe, indicating that consortship is much less effective in securing conception than was previously thought. Moreover, in their studies of sexual behaviour and sperm competition among chimpanzees of the Mahale Mountains, Hasegawa and Hiraiwa-Hasegawa (1990) calculated that most conceptions had probably resulted from multiple male matings, and opportunistic tactics, rather than from consortships. The DNA typing studies of Gagneux *et al.* (1997), already cited in Chapter 3, indicate that surreptitious matings between females and males in neighbouring communities account for more than half the offspring sired. Relationships between male sexual tactics and reproductive success in chimpanzees are still far from understood.

Correlations between male and female dominance rank and consortship behaviour have been studied in various macaque and baboon species. Figure 4.7 shows the positive correlation which exists between male rank and consortship success in the rhesus monkey. In savannah baboons, younger dominant males secure consortships more successfully than older individuals, which tend to be of medium or low rank (Noë and Sluijter 1990). Female rank also influences consortship quality; higher-ranking female baboons enter consortship significantly earlier during the follicular phase of the menstrual cycle than is the case in lower-ranking females (Fig. 4.8). Further, the consortships of dominant females also last longer, although this tendency is not statistically significant (Seyfarth 1978). In rhesus monkeys, female dominance rank correlates with numbers of consort partners; high and medium-ranking females

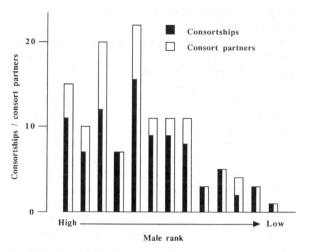

Fig. 4.7 Male rank and consortship success in a group of rhesus monkeys on Cayo Santiago. High-ranking males secure more consortships and have larger numbers of partners than subordinates. (Redrawn from Hill 1987.)

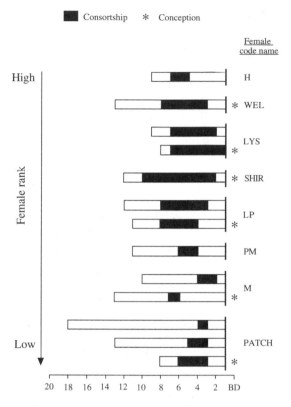

Fig. 4.8 The duration of sexual consortship, and its timing in relation to swelling and deturgescence of the female sexual skin in free-ranging baboons (*Papio cynocephalus*). Females are listed in order of descending rank. Horizontal bars indicate the duration of each female's sexual skin swelling and black bars indicate consortships. Consortships which resulted in conception are marked with an asterisk. (Redrawn from Seyfarth 1978.)

consort with more males during the fertile period than do low-ranking females.

Ultimately, the best test of the impact of consortship, or mate-guarding, upon reproductive success requires genetic studies to determine the parentage of offspring which are believed to result from such tactics. In the case of mate-guarding in mandrills, DNA typing has shown a positive correlation between male dominance, mate-guarding, and reproductive success (Dixson *et al.* 1993). For most species in which consortships occur, the necessary genetic and behavioural studies have yet to be reported.

Alternative mating tactics—a lifespan view

Copulation, like many other types of behaviour, is affected by age, so it should not surprise us to find that mating tactics in many primate species alter throughout the lifespan. The first challenge which must be faced by the pubescent male is to gain entry to the adult world of sexual relationships, and thus to impregnate females. In those species which have polygynous, multimale–multifemale, and dispersed mating systems, intermale competition is often pronounced. In captive stumptail macaques (*M. arctoides*) young males copulate openly in the group until testicular descent occurs, at approximately 3.3 years of age. Once their testes have descended, pubescent males come into increased conflict with the group's adult males; many copulations are interrupted by adults and younger males increasingly

engage in 'surreptitious' matings (Nieuwenhuijsen *et al.* 1988). Maturing male macaques and baboons usually emigrate from their social groups (Pusey and Packer 1987). In the Barbary Macaque (*M. sylvanus*), for example, pubescent males begin to copulate to ejaculation when they are aged between 3.5 and 4.5 years. The majority of emigrations from semi-free-ranging groups of Barbary Macaques occur between 3–5 years of age (66% of 71 emigrations observed by Paul and Kuester (1988) over an 8-year period: see Fig. 4.9). Some males begin to emigrate in the earliest stages of puberty, partly perhaps as a result of androgenic stimulation, at a time when they might be able to grow rapidly. They live as solitary males, or attempt to transfer directly into neighbouring groups. Emigration also lessens the risk that maturing males might copulate with

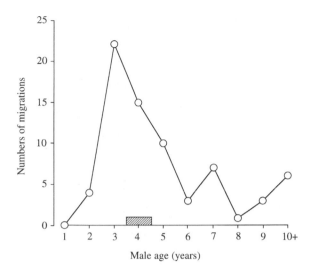

Fig. 4.9 Age-specific migration rates of male Barbary macaques (*Macaca sylvanus*) living as members of semi-free-ranging troops. Most migrations occur at between 3–5 years of age during the period when males first begin to ejaculate during copulation (as indicated by the horizontal bar). (Data refer to the years 1978–1985 and are taken from Paul and Kuester 1988.)

relatives in their natal groups and encourages genetic exchange between social units. In the case of the Barbary macaque, young males transfer directly to a new group and often succeed in copulating opportunistically with adult females as well as consorting with them, although such consortships usually occur outside the peri-ovulatory period (Kuester and Paul 1989). Multimale–multifemale groups are probably more permeable to transferring males than are polygynous, *one-male units*. Aggression does occur, however, and immigrants are not always readily accepted into the multimale–multifemale groups of baboons (Packer 1979a) or macaques (Dittus 1977). It is not uncommon, therefore, for emigrant males to live alone for a time (e.g., in pigtailed and long-tailed macaques, various forest guenons, and gorillas) or to band together, as all-male groups (e.g., in patas monkeys, geladas, and Hanuman langurs).

In nocturnal prosimians, such as bushbabies, pottos, and mouse lemurs, males also disperse at puberty, travelling some distance from their natal area (Charles-Dominique 1977; Bearder 1987). Among non-gregarious primates the challenge for such vagabond males is not to gain entrance to a social group, but to establish a ranging area with access to the home ranges of one, or more, females (Fig. 4.10). Again, males might not readily gain access to females for sexual purposes and some occupy peripheral ranges and are subordinate to the more successful 'A males' in the population (e.g., in *G.*

demidoff: Charles-Dominique 1977; *G. senegalensis*: Bearder 1987).

Figure 4.11 shows a schematic diagram of various alternative tactics employed by emigrant males in species with multimale–multifemale mating systems. Males emigrating from multimale–multifemale groups, either alone or in company with peers (e.g., in *M. sylvanus*: Paul and Kuester 1988), might transfer directly to a new group. Such males attempt to establish themselves within the male hierarchy and compete for copulations, as in *Papio cynocephalus* where young adult males tend to occupy the highest-ranking positions and achieve the majority of consortships (Noë and Sluijter 1990). In other cases, males remain alone after they emigrate, and it is not clear how long this phase may last. In Japanese macaques (*M. fuscata*) and pigtails (*M. nemestrina*) some males opt for long periods of solitary life. Solitary adult male Japanese macaques enter groups during the annual mating season and copulate with females, but emigrate again once the season ends (Nishida 1966). In longtailed macaques (*M. fascicularis*) a period of solitary life can occur at around 8–10 years of age, long after males have emigrated from their natal groups and joined a new group during puberty and early adulthood (Van Noordwijk and Van Schaik 1988). These older solitary males feed and grow rapidly before re-entering their groups; often such individuals acquire dominant positions in the male hierarchy after their solitary phase. Adult males which have fallen in dominance rank might ultimately emigrate from their groups, although they are unlikely to attain high status in any group they join subsequently (*M. fascicularis*: Van Noordwijk and Van Schaik 1985, 1988; *Papio cynocephalus*: Noë and Sluijter 1990). Solitary males might also linger at the fringes of groups during the annual mating season and attempt to mate in an opportunistic manner with female residents. This tactic is adopted by solitary male mandrills (Dixson *et al.* 1993 and author's unpublished observations). In this way, an emigrant male does not commit himself fully to joining a new group, with all the costs which surround his acceptance by resident males and females.

Figure 4.12 represents, in schematic form, the various tactics used by solitary males, or all-male groups, in those primates which have polygynous mating systems. A maturing male in a one-male unit might stay within his natal group, so that in the long-term he can 'inherit' the females. This tactic is employed by some male mountain gorillas (Harcourt and Stewart 1981). Males which emigrate, subsequently use various tactics to acquire female partners, and so establish the nucleus of a new one-male unit. Lone male gorillas, for instance, enter groups and 'take' one or more females; the gorilla is a non-female-bonded species and maturing females commonly transfer from their natal groups. Given that only one silverback is present and is likely to be the sire of such young females, emigration and transfer lessens the likelihood that inbreeding might occur (Harcourt *et al.* 1976). In the hamadryas baboon, a bachelor male sometimes takes a juvenile female from an established unit, guarding her and caring for her until her menstrual cycles commence (Kummer 1968, 1990). Alternatively, a bachelor hamadryas becomes the subordinate follower of a harem holder, eventually inheriting the unit's females from the ageing male. Essentially the same follower tactic has been described for male geladas (Dunbar and Dunbar 1975; Kawai 1979). Extra-group males may also attempt the takeover of a one-male unit by forceful, aggressive tactics. Group takeovers occur in geladas and have also been extensively studied in Hanuman langurs where the new alpha male sometimes kills small infants (Sugiyama 1967; Hrdy 1977; Sommer 1993). The issue of infanticide is complex and is considered separately, below.

Group takeovers in the gelada have been described in detail by Dunbar (1984). The length of tenure of a harem male is influenced by the size of his female unit; males with very large harems have shorter periods of tenure and are more at risk from takeovers. In terms of lifetime reproductive success, it seems likely that the follower tactic, described above, is as effective as an aggressive takeover tactic —an example of what might be an evolutionary stable strategy.

All-male groups do not necessarily attempt takeovers in order to gain access to females. In the patas monkey and in certain of the forest guenons (*Cercopithecus mitis*; *C. ascanius*), all-male groups infiltrate some established one-male units during the annual mating season so that a temporary multi-male–multifemale mating system is produced (Tsingalia and Rowell 1984; Cords 1987, 1988; Struhsaker 1988). Under these conditions, a number of males copulate with females, although the limited evidence available indicates that the original (tenured) male experiences some reproductive advantage (in the patas monkey: Ohsawa *et al.* 1993). Nor do the infiltrating males remain with the group

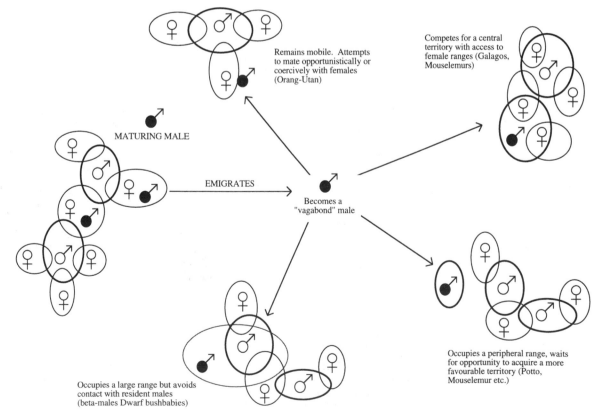

Fig. 4.10 Alternative tactics shown by maturing and adult males in primates which have dispersed (non-gregarious) mating systems.

or necessarily displace the resident male; additional males migrate between groups and many return to an extra-group existence once the mating season is over (Fig. 4.12).

In Chapter 7 various examples will be discussed of retarded development of secondary sexual characteristics in male primates, and the possible role of intrasexual selection in producing these effects (see page 197, Table 7.7). One set of tactics can involve the avoidance of intermale competition by developing males, and reduction of investment in secondary sexual adornments for at least a period of the lifespan. Retention of a juvenile coat colour by some maturing males in *Alouatta caraya* (Neville *et al.* 1988) or *Cercopithecus neglectus* (Kingdon 1980) might enable them to delay the timing of emigration and to become sexually active whilst still in their natal groups. Likewise, the failure of some uakaris (*Cacajao calvus*: Fontaine 1981) to develop muscular temporal swellings and other adult features, or the fact that some male squirrel monkeys become much

less 'fatted' than others during annual mating seasons (Boinski 1987) might represent alternative tactics and economy of investment in the face of intermale competition. Once maturing males emigrate from groups and opt for a solitary or all-male group existence it might still be adaptive in some circumstances to delay development of androgen-dependent secondary sexual characters and to reduce the likelihood of conflict. Almost certainly, the reduced development of sexual skin in solitary mandrills (Wickings and Dixson 1992a) and the slow growth of the proboscis in some members of *Nasalis* all-male groups (Bennett and Sebastian 1988) represents a transitory phase; growth proceeds if the male progresses to a more favourable reproductive situation. This is the case in the orang-utan, where growth of the cheek flanges, throat sac, and other features is retarded by the presence of a fully developed male, but proceeds rapidly if social suppression is removed (Kingsley 1982). Males which lack their full adornments and fail to hold territories are probably less

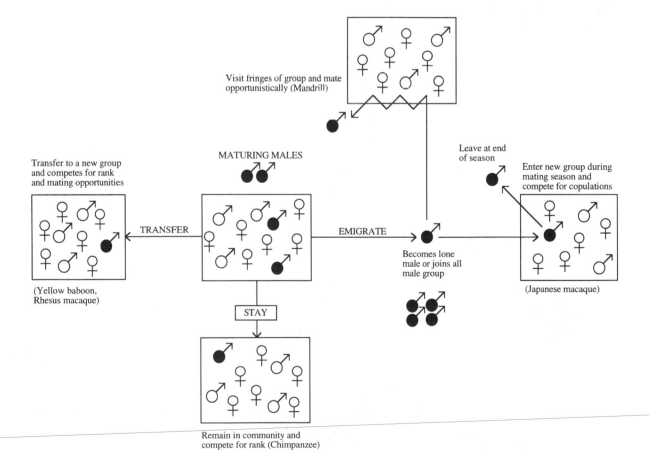

Fig. 4.11 Alternative tactics shown by maturing and adult males in primates which have multimale–multifemale mating systems.

attractive to females; there is some evidence for this in the orang-utan (e.g. Schurmann 1982: Galdikas 1985; Rodman and Mitani 1987).

Coercive matings

It is relevant at this point to mention another alternative tactic open to males; to coerce the female into copulation. Here I refer to *forced copulations* by males rather than to other possible forms of male coercion (see Clutton-Brock and Parker 1995). Such behaviour has been described in free-ranging male orang-utans, and usually involves non-flanged males. It represents an opportunistic mating tactic which is not thought to be particularly successful in securing conceptions (Rodman and Mitani 1987). However, a

final judgement must await the necessary genetic studies of paternity in free-ranging orang-utans.

Coercive matings have been reported in relatively few primate species (review by Smuts and Smuts 1993). Glick (1980), for example, studied a captive group of bonnet macaques (*M. radiata*) and noted that 'when a female refused the advances of an adult male, she was sometimes forced into copulation. Forced copulations were characterized by the aggressive pursuit and restraint of a screaming female, whose efforts to escape the male were unsuccessful'. By contrast, in free-ranging mantled howlers (*Alouatta palliata*) Jones (1985) observed only three coercive attempts at mating. 'In each case, the female was a young adult entering 'menarche' and the male's advances were successfully repelled, with an open-mouth, bared-teeth facial display accompanied by loud vocalizations'. Some degree of coercion may

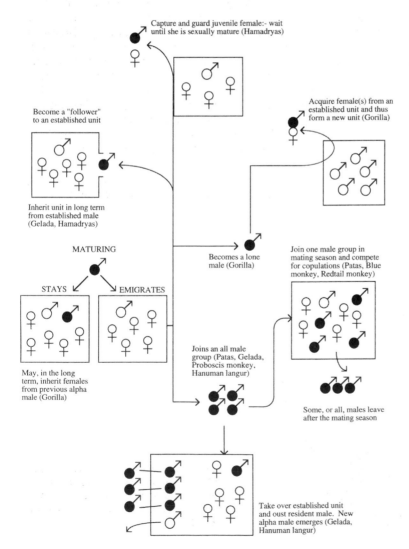

Fig. 4.12 Alternative tactics shown by maturing and adult males in primates which have polygynous mating systems.

occur in mating aggregations of the woolly spider monkey (*Brachyteles arachnoides*) since Milton (1985a) recorded that 'the female sometimes struggled or tried to move away from a new partner, though the male was usually successful in holding on to her and completing the copulation'. This example is by no means clear-cut, however, since the female also 'actively solicited copulation' when males had apparently 'lost sexual interest' in her.

Given the occurrence of marked sexual dimorphism in body weight and strength in many primates it is interesting to note how little evidence for successful coercive matings can be found; females are usually well able to refuse copulatory attempts in a wide variety of species (Dixson 1990a). Although coercive mating is an important tactic in some animals (as in the scorpion fly, *Panorpa*: Thornhill and Alcock 1983) it has probably not been significant, in evolutionary terms, among the primates. Human rape likewise represents an abnormal expression of male sexual behaviour, although not all rapes are truly paraphilic. The question of rape will be discussed in the next chapter, in the section dealing with paraphilias (see page 141).

Infanticide: the sexual selection hypothesis

Under natural conditions primates very rarely kill infants belonging to their own species. Adult males have been seen to kill infants, however, and the

significance of these infanticides has been much debated. Bartlett *et al.* (1993) have painstakingly analysed information on all known (and suspected) cases of infanticidal behaviour by male primates. Table 4.4, taken from their review, details 48 *observed* cases in 13 primate species, whilst Fig. 4.13 shows the percentages of infant deaths attributable to each species. It is notable that the vast majority of cases (*circa* 44%) refer to the Hanuman langur (*P. entellus*), especially the population at Jodhpur, with much smaller numbers in various monkeys (e.g., *Cebus olivaceus*, *Alouatta seniculus*, *Cercopithecus ascanius*, *Papio anubis*) and apes (*Pan troglodytes*, *Gorilla g. beringei*), all of which have either polygynous or multimale–multifemale mating systems. Sugiyama (1967) first drew attention to the occurrence of infanticides (and presumed infanticides) in Hanuman langurs at Dharwar, in western India. Deaths occurred after group takeovers by single males or by all-male bands, although in the majority of cases the infanticides were not observed directly. Sugiyama postulated an adaptive function for this behaviour. He suggested that lactating females which lost their infants might show earlier resumption of ovarian cycles and thus become available for impregnation by the new resident male. Subsequent observations of Hanuman langurs at Jodhpur (Mohnot 1971) and Mt. Abu in Rajasthan (Hrdy 1977) brought to light further examples of infant deaths after group takeovers.

Hrdy (1974, 1977) is best known for elaborating the sexual-selection hypothesis to account for male

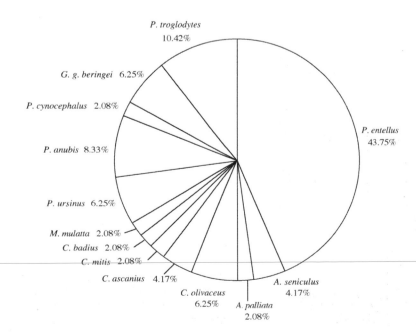

Fig. 4.13 Cases of observed infanticides in 13 species of non-human primates. (Redrawn from Bartlett, Sussman and Cheverud, 1993.)

Table 4.4 Summary of forty eight cases of attacks on infants by adult male primates, as reviewed by Struhsaker and Leland (1987) and by Bartlett *et al.* (1993)

| | | Infant | | Male Attacker | | | | |
| | | Sex | Age | Sire? | Days after takeover | Mates? | New infant? | |
Species	Site							Sources
1. *Presbytis entellus*	Dharwar	F	11.6 months	NO	21–28	VP	P	Sugiyama 1965, 1967
2. *Presbytis entellus*	Jodhpur	M	2.5 months	VU	15	YES	NO	Mohnot 1971
3. *Presbytis entellus*	Jodhpur	M	2.5 months	VU	19	–	NO	Mohnot 1971
4. *Presbytis entellus*	Jodhpur	F	2.0 months	VU	25	YES	NO	Mohnot 1971
5. *Presbytis entellus*	Mt. Abu	)						Hrdy 1974, 1977
6. *Presbytis entellus*	*Mt. Abu*	)						
7. *Presbytis entellus*	*Mt. Abu*	)						
				attacks by unidentified adult male, as observed by local people				
8. *Presbytis entellus*	*Mt. Abu*	)						
9. *Presbytis entellus*	Kanha	M	–	–	BT	–	–	Newton 1986
10. *Presbytis entellus*	Kanha	F	–	–	BT	–	–	Newton 1986
11. *Presbytis entellus*	Kanha	?	–	—	BT	–	–	Newton 1986
12. *Presbytis entellus*	Jodhpur	M	1.2 months	NO	?	YES	VP	Sommer 1987
13. *Presbytis entellus*	Jodhpur	M	3.7 months	–	?	YES	NO	Sommer 1987
14. *Presbytis entellus*	Jodhpur	M	1.5 months	NO	2 days	YES	NO	Sommer 1987
15. *Presbytis entellus*	Jodhpur	F	0.4 months	NO	2 months	–	–	Sommer 1987
16. *Presbytis entellus*	Jodhpur	M	1.9 months	NO	3–11 days	–	NO	Agoramoorthy and Mohnot 1988
17. *Presbytis entellus*	Jodhpur	F	3.4 months	NO	17 days	YES	VP	Agoramoorthy and Mohnot 1988
18. *Presbytis entellus*	Jodhpur	M	4.6 months	NO	18 days	YES	VP	Agoramoorthy and Mohnot 1988
19. *Presbytis entellus*	Jodhpur	F	1.0 months	–	3.5 months	YES	–	Agoramoorthy and Mohnot 1988
20. *Presbytis entellus*	Jodhpur	F	8.0 months	–	11 days	–	–	Agoramoorthy and Mohnot 1988
21. *Presbytis entellus*	Jodhpur	M	8.0 months	–	–	–	–	Agoramoorthy and Mohnot 1988
22. *Alouatta seniculus*	Hato Masaguaral	–	–	NO	–	–	NO	Rudran 1979
23. *Alouatta seniculus*	Hato Masaguaral	M	19.0 h	YES	Resident	YES	VP	Sekulic 1983
24. *A. palliata*	Hacienda La Pacifica	–	4.5 months	VU	8 months	YES	VP	Clarke 1981, 1983
25. *Cebus olivaceus*	Hato Masaguaral	M	<1.0 month	YES	Resident 5 weeks	YES	VP	Valderrama *et al.* 1990
26. *Cebus olivaceus*	Hato Masaguaral	M	9.0 months	VU	Resident	–	VP	Valderrama *et al.* 1990
27. *Cebus olivaceus*	Hato Masaguaral	F	9.0 months	VU	Resident	–	VU	Valderrama *et al.* 1990
28. *Cercopithecus ascanius*	Kibale	–	1.0 week	U	6–35 days	VP	P	Struhsaker 1977
29. *Cercopithecus ascanius*	Kibale	–	1 day	U	41–75 days	VP	P	Leland *et al.* 1984
30. *C. mitis*	Kibale	F	6 months	U	4–35 days	U	VU	Butynski 1982
31. *Colobus badius*	Kibale	M	1–1.5 months	NO	–	YES	–	Leland *et al.* 1984
32. *Macaca mulatta*	Jackoo	M	<1.0 year	?	–	NO	NO	Camperio-Ciani 1984
33. *Papio ursinus*	Moremi	M	4.0 days	NO	76 days	YES	–	Collins *et al.* 1984
34. *Papio ursinus*	Moremi	M	8.0 months	–	–	–	?	Collins *et al.* 1984
35. *Papio ursinus*	Moremi	F	6.0 months	–	Natal Male	–	–	Tarara 1987
36. *P. anubis*	Gombe	F	6.4 months	NO	9.5 months	YES	?	Collins *et al.* 1984
37. *P. anubis*	Gombe	F	8.5 months	NO	2 months	–	–	Collins *et al.* 1984
38. *P. anubis*	Gombe	F	7.2 months	NO	8 months	–	U	Collins *et al.* 1984
39. *P. anubis*	Gombe	F	8.4 months	U	Alpha male	–	U	Collins *et al.* 1984
40. *P. cynocephalus*	Amboseli	M	1.0 month	–	–	–	–	Shopland 1982
41. *G. g. beringei*	Karisoke	F	11.0 months	NO	–	VP	VP	Fossey 1984
42. *G. g. beringei*	Karisoke	F	3.0 months	NO	31 months	YES	NO	Fossey 1984
43. *G. g. beringei*	Karisoke	M	1.0 day	NO	–	–	VU	Fossey 1984
44. *Pan troglodytes*	Mahale	M	1.0 month	YES	Alpha male	–	?	Takahata 1985
45. *Pan troglodytes*	Mahale	M	6.0 months	YES	Alpha male	YES	U	Hamai *et al.* 1992
46. *Pan troglodytes*	Mahale	M	5.0 months	YES	Alpha male	–	–	Hamai *et al.* 1992
47. *Pan troglodytes*	Budongo	–	Newborn	–	–	–	–	Suzuki 1971
48. *Pan troglodytes*	Gombe	–	1.5–2.0 years	–	–	–	–	Bygott 1972

Sire?=attacker possible sire of victim.
Days after takeover=Days after takeover or immigration by attacker.
Mates?=did attacker mate with victim's mother?
New infant?=did victim's mother conceive by attacking male?
VP=very probably; P=Probably; U=unlikely; VU=very unlikely.

infanticide. Males which take over groups are deemed to benefit from killing (non-related) infants and subsequently impregnating the females. Various authors have criticized the sexual-selection hypothesis. The first problem is that many infanticides were not actually witnessed, so that the identities of the perpetrators are not known for certain (Schubert 1982). Sufficient attacks on infants have now been witnessed at Jodhpur to answer this criticism of the langur studies (Sommer 1993). Secondly, there is the possibility that infanticides only occur in Hanuman langur populations that are affected by overcrowding and human disturbance, resulting in higher than normal frequencies of intermale competition and group takeovers. This serious criticism was levelled by several experienced fieldworkers (Curtin and Dolhinow 1978; Boggess 1984). Newton (1986) conducted a quantitative study of population densities, occurrences of one male versus multimale–multifemale groups, and incidences of infanticide in Hanuman langurs. Interestingly, it transpired that infanticidal males are no more common in areas of high population density than in areas of low density. However, infanticide is absent in areas where the population consists primarily of multimale–multifemale groups and is highest where polygynous, one-male groups are the norm. The occurrence of polygyny, with the accompanying phenomena of all-male bands and aggressive takeovers of heterosexual groups, is a key factor in precipitating infanticide in Hanuman langurs.

The sexual-selection hypothesis predicts:

(1) that males should kill infants sired by other males, rather than attacking their own offspring;

(2) that females which lose infants in this way should then exhibit shorter interbirth intervals; and,

(3) the infanticidal male should sire the next offspring produced by a female whose infant he killed.

A fourth prediction relates to the occurrence of female proceptivity and copulation during pregnancy in Hanuman langurs and some other primate species. Hrdy (1979) views this as a female tactic to confuse paternity. By mating with immigrant males after a group takeover, pregnant females reduce the risk of infanticide once their offspring are born. Sommer (1993) analysed data from field studies of the Hanuman langur at Jodhpur in order to test these predictions, and was able to provide confirmatory evidence for the first three. Thus, in 95% of cases in which males killed infants (or were thought to have killed them) it seemed very unlikely, from field records of

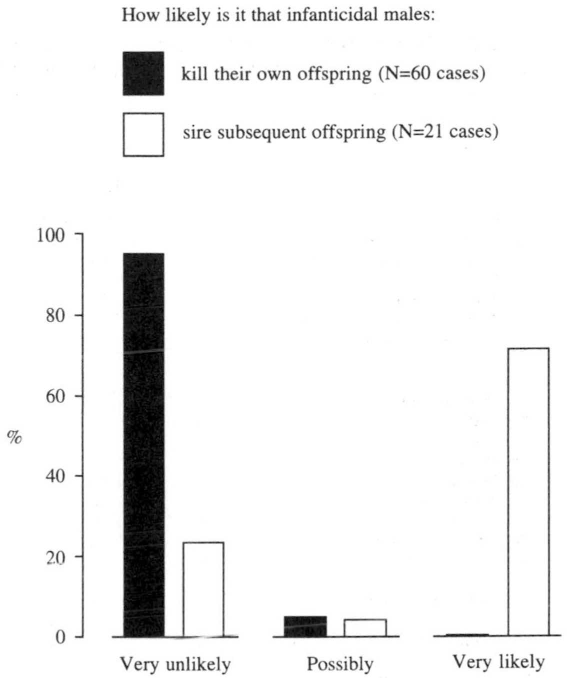

Fig. 4.14 Assumed genetic relationships between infanticidal male Hanuman langurs (*Presbytis entellus*), their victims, and subsequent offspring born to infant-deprived females. (Redrawn and adapted from Sommer 1993.)

group composition and sexual activity, that they could have sired those same offspring (Fig. 4.14). Following infanticides, the subsequent interbirth intervals of the affected females were shorter on average (mean = 12.8 months) than those of females which had raised infants in the normal way (mean = 16.7 months). Sommer also obtained positive evidence for the third prediction, namely that infanticidal males should be more likely to sire the next offspring born to an affected female. Infanticidal males were the *likely* sires of the next offspring in 76.2% of 21 cases for which there were sufficient data (Fig. 4.14). However, none of these estimates benefited from the use of DNA-typing procedures, so that the true biological relatedness of males and offspring is not known. Sommer did not find any evidence to support the 'paternity confusion' hypothesis, however. Pregnant females invited copulations (by head-shuddering, tail-lowering and presenting) and mated with males which were likely sires and males which had entered their groups subsequent to the conception period. Further, their behaviour towards both types of male was essentially the same throughout the gestation period (Fig. 4.15). There was no evidence of enhanced proceptivity and copulatory activity of pregnant females towards potentially infanticidal males. These observations are important, and we shall return to this subject in Chapters 11 and 12 in connection with the hormonal control of sexual behaviour in female primates.

In conclusion, the data on Hanuman langurs indicate a significant role for infanticidal behaviour in at least some populations of this colobine species. However, the same cannot be said for other primates; observations of infanticide are rare (Fig. 4.13) and circumstantial evidence is a poor substitute for hard data. The role of infanticide in the evolution of primate reproduction (e.g., Hrdy 1977; Hausfater and Hrdy 1984) and in relation to theories about the evolution of primate social organization (e.g., Dunbar 1988; Van Schaik and Dunbar 1990; Pereira and Weiss 1991) has been overestimated. Infanticides certainly occur in other mammals, such as lions, and it would be incorrect to deny that significant numbers occur in Hanuman langurs. However, fieldwork has not confirmed Struhsaker's (1977) prediction that 'we shall find this phenomenon to be the rule

rather than the exception' in primates. Accepting Sommer's (1993) estimate that infanticides have been described in 17 species, this represents approximately 7% of the extant species in the Order Primates and in most cases no more than a handful of incidents has been reported. I have taken the opportunity to make these points forcefully because, unpopular as these arguments might be in some quarters, it is necessary to place the problem of infanticide in perspective. Infanticide is neither *widespread* nor of *general* importance in the evolution of primate social or sexual behaviour.

Aside from the Hanuman langur, the only primate in which infanticide is known to be an issue, in evolutionary terms, is the human species. Child abuse and murder are significantly more likely to occur in families where a step-parent is present rather than in families containing both the natural parents. Moreover, it is stepfathers, rather than stepmothers, who are more likely to kill these children and especially young infants up to 2 years of age (Wilson *et al.* 1980; Daly and Wilson 1996).

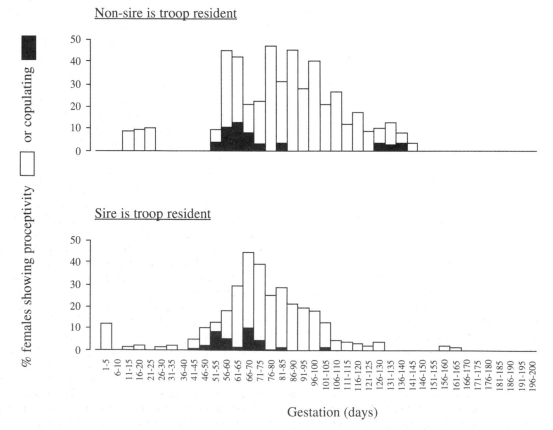

Fig. 4.15 Post-conception proceptivity (open bars) and copulatory behaviour (closed bars) in free-ranging female Hanuman langurs (*Presbytis entellus*) with males that are not likely to have sired their offspring (upper part of figure) and those that are the probable sires (lower part of figure). (Redrawn and modified from Sommer 1993.)

Analyses of Canadian and British data show that stepchildren are approximately 100 times more likely to be killed than are children living with both genetic parents (Creighton 1985; Daly and Wilson 1985). Daly and Wilson (1996) argue that 'evolutionary thinking suggests that step-parental affection will tend to be restrained: indulgence towards a mate's children, may have had some social utility for many millennia, but it must rarely have been the case that a stepchild's welfare was as valuable to one's expected fitness as one's own child's welfare'. These arguments apply to hunter-gatherer societies as well as to modern industrialized societies. Hill and Kaplan (1988) studied the Ache, a native people living in eastern Paraguay who have been hunter-gatherers until quite recently. Deaths (before 15 years of age), among children raised with a stepfather in the family were far more common (43.3% of 67 cases) than when the natural father was alive and still married to the mother (19.3% of 171 cases). The human situation is, of course, quite different to that described for polygynous groups of Hanuman langurs. In the human case we are dealing with small and relatively stable family units in which certainty of paternity is very high. Nonetheless, it appears that when a child is clearly not the offspring of the husband then its chances of survival are significantly diminished. Though we might feel reluctant to account for the deaths of young children by using such evolutionary arguments, it is hard to escape the conclusion that this category of human infanticide is affected by such mechanisms.

Harassment and interruption of copulation

Immature and adult members of over 30 primate species have been recorded harassing or interrupting the copulations of conspecifics (Neimeyer and Anderson 1983). Harassment has been used as a term to denote relatively mild interferences, such as approaches to the copulating pair and touching or slapping the participants. Harassment by immature individuals, and sometimes by adult females, has been observed in ringtailed lemurs, marmosets, langurs, patas monkeys, macaques, and many others (Table 4.5). Such behaviour does not necessarily lead to the termination of copulation; males usually continue to make pelvic thrusting movements and to ejaculate in these circumstances. Interruption of copulation is a more disruptive tactic employed by adult males, and especially by dominant males, to reduce the mating success of rivals. Interruption of copulation has been recorded in ringtailed lemurs,

squirrel monkeys, langurs, macaques, baboons, chimpanzees, bonobos, and many other species (Table 4.5).

Detailed studies of the harassment and interruption of copulation have been conducted on the stumptail macaque, since both types of behaviour occur frequently in groups of this macaque species (Gouzoules 1974; Neimeyer and Chamove 1983; Drukker *et al.* 1991; Bruce and Estep 1992; Brereton 1994). The question of interrupted matings will be considered first, since the functional significance of this behaviour is clearer than in the case of harassment. Alpha male stumptails rarely have their copulations interrupted by other males. Bruce and Estep (1992), for example, observed 829 copulations involving alpha males none of which was interrupted, although harassment by immature group members and adult females was commonplace. By contrast, 33% of copulations by beta males and 39% of those involving lower-ranking individuals were interrupted. Interestingly, during the course of their study Bruce and Estep observed that males which experienced rank reversals were only interrupted when they fell below alpha status. The alpha individual interrupted more copulations than any other male in the group and was the only male to interrupt matings involving the beta male. One clear effect of interruption was to shorten the duration of the post-ejaculatory 'pair-sit', which characterizes copulation in the stumptail. Pair-sits of alpha males lasted for 1.47 min on average and were not significantly different from those of lower-ranking males during non-interrupted matings (1.38 min). However, interruption of matings in lower-ranking males resulted in a shortening of the pair-sit, to on average duration of 0.88 min. Since one function of the stumptail's post-ejaculatory pair-sit might be to position the copulatory plug correctly within the female's ectocervix (Dixson 1987a), shortening of the pair-sit might have consequences for sperm competition and for the reproductive success of lower-ranking males.

As regards harassment of copulations; a number of theories has been proposed to account for such behaviour in stumptail macaques and in other primates. Male stumptails sometimes bite females during copulation. Gouzoules (1974) suggested that harassers attempt to divert the aggressiveness of the copulating male away from the female partner. Yet in many species harassments occur in the apparent absence of aggressive reactions by males. The *sexual-competition hypothesis* proposes that harassers reduce the mating success of copulating animals (Neimeyer and Chamove 1983) or attempt to gain future copulations at the expense of one of the mating individuals (Hall 1965; Loy and Loy 1977).

Table 4.5 Examples of sexual interference (harassment and interruption of copulation) in primates

Species	Type of study	Type of interference	Sources
Sexual interference by adult males			
Lemur catta	WG	Harassment and interruption of copulation	Jolly 1966; Koyama 1988
Callithrix jacchus	CG	Harassment and interruption of copulation	Rothe 1975; Abbott and Hearn 1978
Saimiri sciureus	SFG	Harassment by adult male sometimes causes disruption of copulation	Baldwin 1968; Dumond 1968
Macaca arctoides	CG SFG	Harassment and interruption, especially by higher-ranking males versus subordinates	Gouzoules 1974; Bruce and Estep 1992; Brereton 1994
M. mulatta	CG	Highest ranking males most often interrupt copulations	Wilson 1981
M. fuscata	WG	Highest-ranking males most often interrupt copulations	Stephenson 1975
M. fascicularis	WG	Dominant males sometimes harass copulations by subordinates	Wheatley 1982
Mandrillus sphinx	SFG	Alpha male interrupts copulations by others	AFD
Papio anubis	WG	Harassment and interruption of copulation	DeVore 1965; Smuts 1985
Presbytis entellus	WG	Harassment and interruption of copulation by other adults and sub-adults of both sexes	Yoshiba 1968
Colobus badius	WG	Interruption of copulation	Struhsaker 1975
Gorilla g. beringei	WG	Dominant male may interrupt copulations by others	Harcourt *et al.* 1980
Pan troglodytes	WG	Older males interrupt copulations by younger males	Tutin 1979
P. paniscus	WG	Dominant males interrupt subordinates	Kano 1992
Sexual interference by adult females			
Callithrix jacchus	CG	Harassment and interruption by dominant female in group	Rothe 1975; Abbott and Hearn 1978
Saimiri sciureus	SFG	Females harass copulating pair or attack male	Baldwin 1968 Dumond 1968
Macaca arctoides	CG SFG	Females and especially high-ranking non-pregnant females, harass copulations	Bruce and Estep 1992; Brereton 1994; Drukker *et al.* 1991; Gouzoules 1974
M. thibetana	SFG	Harassment most often involves higher-ranking females vs subordinate females.	Zhao 1993
M. mulatta	CG	Rarely by dominant females with most aggression directed at male	Wilson 1981
Erythrocebus patas	WG CG	Dominant females more frequently harass or interrupt lower-ranking females' copulations	Hall 1965; Hall and Mayer 1967
Presbytis entellus	WG	Most harassment or interruption of copulations involves higher-ranking females	Yoshiba 1968; Sommer 1989
Gorilla g. gorilla	CG	Most harassments involved a single adult female in the group	Hess 1973
Pan troglodytes	WG	Unusual. Dominant females with swellings sometimes intervene between mating partners and interrupt copulations	Goodall 1968; Nishida 1979

Table 4.5—*Continued*

Species	Type of study	Type of interference	Sources
Sexual interference by immature animals			
Lemur catta	WG	Harassment of copulations by juveniles	Jolly 1966
Callithrix jacchus	CG	Over 90% of copulations harassed by juvenile/subadult offspring	Rothe 1975
Saimiri sciureus	SFG	Juveniles may attempt to suckle as mother copulates	Dumond 1968
Ateles geoffroyi	CG	Playful harassment by juveniles of both sexes	Klein 1971
Macaca arctoides	CG	Harassment especially by high-ranking juveniles/sub adults of both sexes	Gouzoules 1974
M. mulatta	CG	Occasional harassment by juvenile males	Wilson 1981
M. fuscata	CG	Harassment involves approaches and mild threats by juvenile males	Hanby and Brown 1974
M. nigra	CG	Playful grappling by infants with copulating male	Dixson 1977
Papio anubis	WG	Juvenile males sometimes join older males in charging at copulating pair	DeVore 1965
Cercopithecus aethiops	WG	Juvenile males may grab or push copulating adult pair; adult male usually ignores but may threaten in response	Struhsaker 1967a
C. albogularis	CG	Two juvenile males harassed copulations and sometimes delayed occurrence of ejaculation	Rowell 1972
Erythrocebus patas	CG	Harassment by immature males sometimes resulted in interrupted copulations	Loy and Loy 1977
Presbytis entellus	WG	Harassment by infants and juveniles usually ignored by mating pair	Jay 1965; Hrdy 1977
Colobus badius	WG	Harassment by juveniles but copulation not interrupted	Struhsaker 1975
Pongo pygmaeus	FS	Infant or juvenile offspring sometimes bite or hit male during copulation with the mother. The male ignores harassment	Mackinnon 1974; Galdikas 1979
Pan troglodytes	WG	Non-aggressive or often fearful harassment by infants/juveniles of both sexes. Young females are more likely to harass copulations by their kin.	Tutin 1979 Goodall 1986
P. paniscus	WG	Juveniles often harass copulations by climbing on, or between, adults. Harassment often involves offspring of female partner	Kano 1992

WG=free-ranging group; SFG=semi-free-ranging group; CG=captive group; AFD=author's unpublished observations; FS=field study of orang-utans. The use of the term 'group' is not appropriate for orang-utans since the social system is non-gregarious in this species.

Tutin (1979) applied Trivers theories concerning parent/infant conflict to explain why infant chimpanzees sometimes harass their mothers during copulation. In theory, such harassment might reduce the likelihood of conception and hence delay the production of a sibling to compete for maternal attention. Neither the _sexual-competition_ nor the _parent–infant-conflict_ hypotheses have been adequately tested, however. A fourth hypothesis (_the sentinel hypothesis_: Neimeyer and Chamove 1983; Drukker _et al._ 1991) suggests that harassers draw attention to the mating pair, making it likely that other adult males will approach and interrupt the copulation. The alpha male is the main beneficiary of such behaviour (e.g., in the stumptail macaque) and passes on genes that influence harassment behaviour in his offspring. This far-fetched hypothesis receives little support from studies of stumptails, however, since it is the alpha male himself that _receives_ the majority of copulatory harassment (Drukker _et al._ 1991). A fifth theory proposes that harassment by adult females is a mechanism to 'test' the males' mating quality (Neimeyer and Chamove 1983). In the case of the stumptail one measurement of mating quality might be the ability of the male to maintain the pair-sit after ejaculation has occurred, even though he is harassed by other adult females. Interestingly, three separate studies have shown that pair-sit duration increases (by 28–83%, see Table 4.6) as a result of sexual harassment. Although consistent with the hypothesis of testing male quality, these observations do not provide convincing proof of its validity. It is of interest, however, that males pair-sit for longer when harassed; this is the reverse of what occurs when males interrupt copulations and is not consistent with some functions of harassment proposed above (sexual-competition and parent–infant-conflict hypotheses). Where harassment by infants and juveniles is concerned a sixth and less charismatic hypothesis is, of course, quite feasible. Young animals are naturally curious and playful; they respond to copulatory behaviour by investigating and attempting to play with the participants. Harassment in such contexts is an epiphenomenon and not a result of sexual selection.

Genetic assessment of male reproductive success

Behavioural measurements of mating success do not provide a certain guide to male reproductive success. Under natural conditions only a fraction of the total sexual activity in primate groups is ever observed or recorded. When females mate with multiple partners, the outcome cannot be gauged from behavioural data, since the female's reproductive tract provides the final arena for sperm competition. Some mating tactics, such as opportunistic copulations, are by their very nature difficult to observe and quantify. Genetic assignment of paternity is required to measure male reproductive success and to understand the relative success rates of the various tactics employed by males throughout their lifespan. The technique of DNA fingerprinting (sometimes called DNA-typing) has revolutionized this field since its discovery by Jeffreys _et al._ (1985). The Jeffreys technique, originally developed for work on human beings, has now been used to study the mating systems of numerous animals, including birds (Birkhead and Møller 1992), carnivores (Packer _et al._ 1991; Amos _et al._ 1993) and primates (Martin _et al._ 1992; De Ruiter and Inoue 1993). Although primate studies are still at a relatively early stage, important information has been obtained concerning relationships between dominance, mating success, and reproductive success for males of several species (Table 4.7).

Relationships between male rank and reproductive success are strongest in mammals where there are strong indications for effects of sexual selection upon masculine secondary sexual characters and behaviour (e.g., in the elephant seal: Le Boeuf 1974, and the red deer: Clutton-Brock and Albon 1979). Let us begin to view the genetic evidence pertaining to male dominance and reproductive success in primates by considering an extreme example; the mandrill (_Mandrillus sphinx_). The mating system of the mandrill has frequently, and incorrectly, been classified as polygynous, involving one-male units or harems, such as occur in geladas or hamadryas baboons (e.g., Stammbach 1987). However, studies of a large semi-free-ranging mandrill group have shown that adult males do not establish one-male units. Instead, a number of fully developed (_fatted_) adult males associate with the females all the year round

Table 4.6 Effects of harassment upon the duration of the post-ejaculatory 'pair-sit' in stumptail macaques

Type of study	Pair-sit duration (s)		Increase in duration (%)	Source
	No harassment	Harassment		
Semi-Free Ranging Group	47 (13)	73 (86)	55	Brereton 1994
Captive Group	60 (17)	110 (56)	83	Drukker _et al._ 1991
Captive Group*	71 (46)	91 (30)	28	Bruce and Estep 1992

Numbers of copulations are given in parentheses.
* Only copulations by the second-ranking male in the group were analysed.

Table 4.7 Genetic studies of paternity and reproductive success, as related to dominance rank in male primates

Species	Conditions of study	Comments	Sources
Lemur catta	Semi-free-ranging group. DNA typing	Low-ranking immigrant males introduced into the enclosure sired the majority of offspring over a 5-year period despite their low rank. Females exhibited mate choice for immigrant males	Pereira and Weiss 1991
Alouatta seniculus	Field study Blood proteins	In single and multimale troops the most dominant male sired all offspring during his period of tenure	Pope 1990
Macaca mulatta	Large captive group. Blood proteins+DNA typing	High-ranking males sire most offspring but some lower-ranking individuals are significantly successful	Smith 1993
Macaca mulatta	Semi-free-ranging group. DNA typing	Extra-group males succeed in siring offspring, as well as resident males	Berard *et al.* 1993
Macaca mulatta	Large captive group DNA typing	8 males out of 21 sired all offspring. Highest-ranking males most successful	Bercovitch and Nürnberg 1996
Macaca fascicularis	Field study Blood proteins+DNA typing	Alpha and beta-ranking males markedly more successful than others	De Ruiter *et al.* 1992
Macaca sylvanus	Semi-free-ranging groups DNA typing	Rank effects pronounced when younger adult (sub-adult) males included in analysis. Sub-adults have low reproductive success	Paul *et al.* 1993; Kuester *et al.* 1995
Macaca arctoides	Captive group DNA typing	3 consecutive alpha males sire almost all offspring over an 8-year period	Bauers and Hearn 1994
Macaca nemestrina	Captive group DNA typing	No correlation between male rank and paternity. Note, however, that the alpha of 5 adult males present was probably infertile	Gust *et al.* 1996
Erthyrocebus patas	Field study DNA typing	Resident males sire most offspring but intruder males also significantly successful	Ohsawa *et al.* 1993
Mandrillus sphinx	Semi free-ranging group DNA typing	Marked, positive correlation between male rank and reproductive success. Two highest-ranking males sired all offspring over a 5-year period	Dixson *et al.* 1993
Papio cynocephalus	Field study DNA typing	Dominant males have priority of access to fertile females and sire most offspring	Altmann *et al.* 1996
Pan troglodytes	Free-ranging community DNA typing	7 of 13 infants tested were sired by males in a neighbouring community and not by dominant males or as a result of consortships in the home community	Gagneux *et al.* 1997

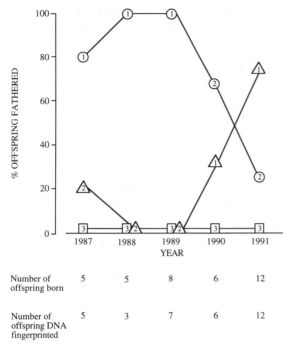

Fig. 4.16 Association between rank and paternity for three fatted adult male mandrills (numbers 7, 14 and 15) showing effects of rank changes upon percentages of infants sired over five years (1987–1991). Each male is represented by a separate symbol (male 7: {○}; male 14: {△}; male 15: {□}) and his rank (1, 2 or 3) is written inside the symbol. Note that male 15 remains at rank 3 and sires no offspring throughout the 5-year period. Male 7 commences at rank no. 1 and then falls to the No. 2 position; his fall in rank is associated with a gradual decrease in reproductive success during 1990 and 1991. The numbers of offspring born and numbers DNA fingerprinted are given for each year of the study. (Redrawn from Dixson *et al.* 1993.)

and compete for copulations during the annual mating season. Solitary (*non-fatted*) males engage in opportunistic matings (Wickings and Dixson 1992a; Wickings *et al.* 1993; Dixson *et al.* 1993). As described previously, male dominance rank and mating success exhibit a strong, positive correlation in mandrills, and this is reflected in the reproductive success of fatted (group-associated) males. Over a five-year period, two out of three resident fatted males sired all offspring for which it was possible to assign paternity by means of DNA-typing (Fig. 4.16). None of the three non-fatted adult males or three males which attained sub-adult status are known to have sired offspring, although a number of them copulated with the 12 adult females in the group. In Fig. 4.16 it can be seen that the original alpha male sired 80–100% of offspring born between 1987 and 1989,

but he was then deposed and occupied the second ranking position in the group. Loss of alpha status resulted in a fall in reproductive success, but the effect was gradual; the deposed male continued to father 67% and 25% of infants born during the next two years. To place these observations in context, one must keep in mind that this group occupied a 6-hectare rainforested enclosure and numbered 57 individuals by the end of the study. In the wild, where a multimale–multifemale group numbering 300–400 individuals has a home range of over 30 square kilometres, there might be considerably greater latitude for alternative mating tactics and greater reproductive success in subordinate or solitary males. Dominant male mandrills do not herd or guard a harem of females; instead each individual female is followed and guarded when she develops a sexual skin swelling. An analysis of guarding behaviour and mating success during conception cycles in a single mating season is shown in Fig. 4.17. The data refer to the last six days of maximal sexual skin swelling, and the day of sexual skin breakdown in 12 females, and thus include the likely day of ovulation. The dominant male (No. 14) mate-guarded and copulated with 11 of these females during at least part of this period and accounted for 24 (i.e. 77%) of the ejaculatory mounts observed. The second-ranking male (No. 7) mate-guarded three females, typically for shorter periods than the dominant male, and accounted for 13% of the ejaculatory mounts. Only two other males were observed to copulate during the presumptive peri-ovulatory period; a solitary/peripheral male (Male No. 13: twice) and a sub-adult group male (Male 5B: once). DNA-typing of offspring resulting from copulations during these cycles showed that the dominant male (No. 14) had sired nine infants and male No. 7 accounted for the remaining three (Fig. 4.17). Yet even in this situation, where a rigid male dominance hierarchy existed and females were guarded intensively during the presumptive fertile period, there were instances where behavioural observations failed to predict paternity. Whilst only male No. 14 was seen to guard and mate with females Nos. 5 and 6, male No. 7 sired their offspring. Likewise, male No. 7 copulated opportunistically with female No. 16 whilst male No. 14 succeeded in guarding her for 4 of the 6 days before sexual skin breakdown. In this instance male No. 7 sired the resulting infant.

Three points can be made on the basis of these studies. Firstly, no matter how thorough behavioural sampling might be, under natural conditions fertile matings can occur which are never observed. Paternity assignment on the basis of mating success is bound to be inaccurate. Secondly, the second-ranking male (No. 7) which succeeded in siring 25% of offspring had previously held alpha rank in the group. A previously dominant male still possesses the same genotype as during his period of alpha status. Indeed he has a proven 'track record' as an alpha male and should, in theory, still be attractive to females. This leads to the third point, namely that even in the mandrill where males invariably rank above all females on an individual basis, females probably exert significant mate choice. Females were observed to terminate guarding episodes by persistently avoiding and running from males. Likewise, females were observed to solicit copulations from certain males, whilst refusing mount attempts by others. Given the variety of alternative tactics employed by male primates and the additional variables of female choice, and preferences for particular partners (to be reviewed below), it should not surprise us to discover that male dominance does not correlate perfectly with reproductive success. Some further descriptions of DNA typing studies on various macaque species will be useful in order to amplify these points.

Smith (1993) measured long-term changes in male dominance rank and reproductive success in a large captive group of rhesus monkeys. He succeeded in assigning paternity for 226 (79%) of the offspring born in this group between the years 1978–1992. Average yearly reproductive success correlated with male rank; alpha males during this time-span had the highest success rates, whilst each low-ranking male sired approximately one third the number of infants achieved by top-ranking individuals (Fig. 4.18). Yet, one is impressed by the success of medium and low-ranking males revealed in Fig 4.18 and one suspects that in a free-ranging situation, where animals can easily avoid visual contact with one another, that variability in male success rates resulting from alternative tactics and female choice might be even more pronounced. This expectation is confirmed by the study of rhesus monkeys on Cayo Santiago by Berard *et al.* (1993) and by long-term research by Paul *et al.* (1993) on groups of semi-free ranging Barbary Macaques (*M. sylvanus*) at Affenberg Salem in Germany.

Berard *et al.* determined paternity for 11 of 15

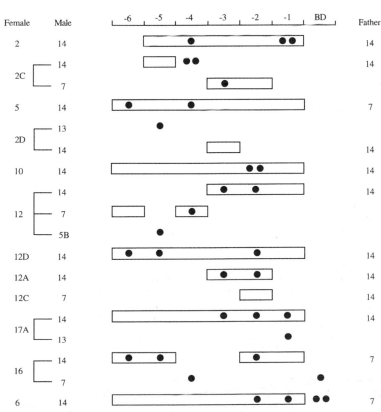

Fig. 4.17 Mating behaviour in relation to the females' conception cycles and reproductive success in male mandrills. Individual females are identified in the extreme left-hand column and those males which engaged in sexual activity are listed in the adjacent column. Sexual behaviour was scored during the seven days leading up to and including the day of sexual skin breakdown (BD). Male mate-guarding of females is indicated by the open boxes, and the occurrence of ejaculatory mounts by closed circles. The sire of each resulting offspring (as determined by DNA fingerprinting) is given in the right-hand column. (Redrawn from Dixson *et al.* 1993).

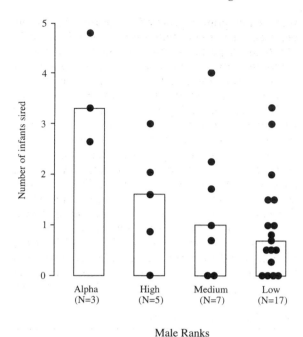

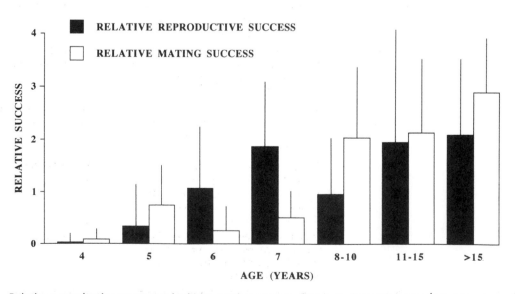

infants produced during a single birth season in a semi-free-ranging group of rhesus monkeys. This group contained 11 resident males. However, 9 additional males, from other groups, were also observed to copulate with the study group's females during the annual mating season. It transpired that the two highest-ranking resident males sired two offspring each, that males ranked at positions 6, 8 and 9 in the hierarchy each sired one offspring and that extra-group males accounted for four others. In a further four cases it was not possible to identify the male parents. This preliminary study indicated that, despite a positive correlation between male rank and consort frequency in this group, rank was not related in a simplistic fashion to reproductive success. The two highest-ranking males did account for 36% of offspring but, beyond these top-ranking positions in the group the dominance effect broke down; opportunistic copulations, effects of female choice, and sperm competition might all have influenced paternity distribution. Clearly, some extra-group males (including individuals of low rank in their home groups) enjoyed significant reproductive success (three such males sired four offspring between them). This study provides genetic evidence for the value of emigration and mating season mobility in macaque males (Lindburg 1969).

The study by Paul *et al.* (1993) of rank, mating effort, and reproductive success in male Barbary macaques also shows that male rank is not related in a simplistic way to reproductive success. Data were

Fig. 4.18 Male rank and average yearly reproductive success in a captive group of rhesus monkeys (*Macaca mulatta*). Data were collected over a period for 15 years. Note that although alpha and high-ranking males are most successful, some medium and low-ranking males sire as many, or more, offspring. Alternative male tactics, or female choice, are important determinants of male reproductive success in such cases. (Data are from Smith 1993.)

Fig. 4.19 Relative reproductive success and relative mating success of male Barbary macaques (*Macaca sylvanus*), living as members of a large semi-free-ranging group. The data are plotted in relation to male age. (Redrawn from Paul *et al.* 1993.)

collected for a large group of monkeys during four consecutive mating seasons and birth seasons. Several groups shared a 14.5 hectare enclosure and it was possible to determine parentage for 75 out of 83 offspring born into the focal group. Male rank was positively correlated with mating success in all 4 years, whilst rank also correlated with reproductive success in 3 out of 4 years (Fig. 4.19). However, the correlation between male rank and reproductive success disappeared if low-ranking sub-adult males were excluded from the analysis. Fig. 4.19 shows how mating success was low in males aged 4–7 years. Six- and seven-year-old males included some individuals with good reproductive success, however, indicating that by this age males have begun to contribute to the gene pool. Older males (aged 11 years and above) tended to show good concordance between mating success and reproductive success whilst males aged 8–10 years often had good mating

success but poor reproductive success. Paul *et al.* point out that their study supports the Bercovitch-McMillan hypothesis that inclusion of sub-adult males in statistical analyses can bias correlations between male rank and mating/reproductive success. The high reproductive success of some young adult males aged 6–7 years is most interesting and Paul *et al.* note that whilst dominant adults forced younger males to the periphery of the group during the day, these younger individuals sometimes 'acquired females at night or during dawn, or at dusk, when low visibility and the difficulty of manoeuvring in the sleeping trees reduced the effectiveness of the older males' coalitions'. Here then is a further male tactic, not considered so far, of guarding females and mating at night. In Table 4.8, I have gathered together the scattered evidence pertaining to nocturnal copulations or consortships in various monkeys and apes. It seems likely that much

Table 4.8 Occurrences of nocturnal copulation or mate-guarding in diurnal primate species

Species	Type of study	Nocturnal sexual activity	Source
Lemur catta	Field study	High-ranking male associates with sexually receptive females at dusk or emits loud vocalizations which may serve to attract such females to the sleeping site	Koyama 1988
Colobus badius	Field study	Female copulatory vocalizations ('quavers') heard at night, indicating occurrence of nocturnal copulation	Struhsaker 1975
Macaca sylvanus	Semi-free-ranging groups	Young adult males (aged 6–7 years) consort with females at dusk and during night in the sleeping trees	Paul *et al.* 1993
M. arctoides	Captive male-female pairs	24-h video recordings revealed that in 2 pairs males mounted on 2 of 58 nights and 9 of 55 nights respectively. No ejaculations occurred, however	Slob and Nieuwenhuijsen 1980
Mandrillus sphinx	Semi-free-ranging group	Alpha male continues to follow and mate-guard females at full sexual skin swelling as they enter the sleeping trees at dusk	Author's unpublished observations
Papio anubis	Field study	Consortships and copulations continue at night. Younger, high-ranking males are more likely to engage in consort takeovers during the night and to oust older, lower-ranking males	Packer 1979b; Smuts 1985
Pan troglodytes	Field study	Copulations seen twice on moonlit nights. However 'there have not been enough vigils at night to know how often this happens'	Goodall 1986
Pongo pygmaeus	Field study	Consortships are maintained at night and the male and female build their nests close together in the same tree	Mackinnon 1974

more remains to be discovered about this particular tactic. With regard to Barbary macaques, Paul *et al.* also remark that rank changes among males are fairly frequent and that females copulate with large numbers of partners in this species (see Table 3.5, page 38). Hence sperm competition, rather than dominance rank alone, might have important effects upon the outcome for paternity in this species and there is the distinct possibility that frequent copulations and depletion of sperm reserves might affect the fertility of males.

A final example of research which combines DNA-typing of paternity with patterns of sexual activity and male dominance rank concerns the longtailed macaques (*M. fascicularis*) of Ketambe in Sumatra (De Ruiter *et al.* 1992; De Ruiter and Van Hooff 1993). This remarkable study involved observing and trapping three groups of monkeys in their rainforest habitat. It was conducted under natural conditions, therefore, and previous work by Van Noordwijk and Van Schaik (1988) on this population had already established how a male's mating success alters throughout his lifetime. The results of De Ruiter *et al.* confirm a striking relationship between alpha status and male reproductive success in long-tailed macaque groups (Table 4.9). Second-ranking, beta males also sire more offspring than others and in this case, beta males are often individuals which previously occupied an alpha position in their groups. Other males have low reproductive success, but nonetheless succeed in siring occasional offspring. In longtailed macaques, the male tactic of emigrating at 8–10 years of age, to undergo a period of solitary life and rapid growth before challenging for the alpha position in the group, might be crucial for attainment of full reproductive potential (Van Noordwijk and Van Schaik 1988).

Female dominance, mating success, and reproductive success

Bateman's (1948) experiments with fruit flies (*Drosophila*) indicated that males experienced a reproductive advantage if they copulated with multiple partners, whereas females did not. Further, the variance in reproductive success of males was greater than for females. These fundamental observations have often been cited to explain the greater incidence of intermale competition and the eagerness of males to mate with multiple partners in a wide range of animals, including primates. A dominant male gorilla, gelada, or macaque can, presumably, sire a much larger number of offspring throughout

his lifetime than can be produced by any individual female, irrespective of her social rank. Moreover, variations in male reproductive success can be extreme, since subordinate males mate with fewer females or, in the case of polygynous species, may fail to secure a harem or gain only a limited period of tenure (e.g., Gelada; Dunbar 1984; Hanuman langur: Sommer and Rajpurohit 1989; Borries *et al.* 1991). Despite the general importance of *Bateman's principle*, it should not be taken to imply that females never gain an advantage from multipartner copulations or that intrasexual competition is confined to males. Among the primates there are numerous examples of species which have multimale–multifemale mating systems, where females mate with multiple partners during the fertile period (see Table 3.5, page 38). It has been estimated, for example, that an individual female chimpanzee receives as many as 135 ejaculations before conception (Hasegawa and Hiraiwa Hasegawa 1990) and might mate with most of the males in her community (Goodall 1986).

In his review of dominance and fertility among female primates, Harcourt (1987) pointed out that dominant female Japanese macaques mate more frequently than subordinates (Enomoto 1974; Stephenson 1975). However, 'subordinate females do not remain unmated, and no evidence exists to show that subordinate females mate too infrequently to conceive at the same rate as do dominant females'. Nonetheless, several hypotheses might be advanced to explain why females might gain a reproductive advantage from multipartner copulations and hence why interfemale competition could occur in sexual contexts. For instance, by copulating with a number of males, a female might reduce the likelihood of subsequent infanticide; i.e., multipartner matings represent a *paternity-confusion* tactic. This hypothesis might be discounted, given the weak evi-

Table 4.9 Male rank and reproductive success in free-ranging groups of Macaca fascicularis (after De Ruiter and Van Hooff 1993)

Group	Male ranks	Number of offspring/ total offspring	Percentage of total offspring
Large group	Alpha-male	11–13/21	52–62
	Beta-male	4–7/21	19–33
	8 other males	3–5/21	2–3
Small group 1	Alpha-male	8/11	73
	Beta-male	2/11	18
	0–2 other males	1/11	9
Small group 2	Alpha-male	12/13	92
	Beta-male	1/13	8

dence that infanticide has played any significant role in the evolution of primate reproduction (with the exception of the Hanuman langur: see pages 67–71 and Fig. 4.13). A second hypothesis relates to the importance of sperm competition in many primates which have multimale–multifemale or dispersed mating systems. Given that males differ in their fertility and semen quality, it is to the advantage of a female macaque, baboon, woolly spider monkey, or chimpanzee to become inseminated by the most able partner. Advantageous qualities of sperm, phallic anatomy, and copulatory patterns are important to females, as well as to males, since their offspring inherit the genes which control these traits. Hence there might be situations where females compete for multiple matings, rather than acting as passive objects of intermale rivalries. Competition can take the form of extravagant signals advertising female reproductive condition and sexual attractiveness, such as the sexual skin swellings of many Old World monkeys, bonobos, and chimpanzees (Harvey and May 1989). Competition may also take the form of dominance interactions, with higher-ranking females maximizing their access to multiple-partner copulations. It may be for this reason that high-ranking female savanna baboons begin to form consortships earlier during the follicular phase of the cycle than do lower-ranking females (Fig. 4.8). A particularly striking example of rank-order effects has also been described for female rhesus monkeys by Wallen (1990). In this case, a group of females was allowed to interact with a single adult male on a daily basis in a large, outdoor enclosure. Blood samples were collected from females in order to measure levels of 17β-oestradiol and to determine the timing of the pre-ovulatory oestradiol peak. Female rank and timing of onset of copulation were tightly coupled; high-ranking females began to mate some days before the oestradiol surge, whereas low-ranking animals only commenced mating once ovulation was imminent (Fig. 4.20). This remarkable finding emphasizes that, whilst all females might mate and probably all females achieve conception, high-ranking animals start earlier and receive more ejaculations. It is precisely this type of relationship between female dominance and copulatory behaviour which might be significant in terms of sperm competition.

A third hypothesis relating female rank to copulatory frequencies applies especially to species which have polygynous mating systems. Sperm competition is less important, since usually only a single adult male is present, and males tend to have relatively small testes in relation to their body weights (e.g., in the gorilla, gelada, or hamadryas baboon: Harcourt

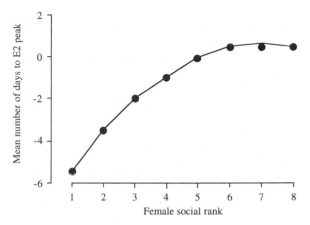

Fig. 4.20 The relationship between female social rank and the day of first mating, relative to the day of the female's oestradiol peak, in a captive group of rhesus monkeys (*Macaca mulatta*). Data refer to two ovarian cycles for each female. All data were collected during behavioural tests with a single male. (Redrawn from Wallen 1990.)

et al. 1981a). In these cases, sperm might be a 'limiting resource' for females within the one-male group and female–female competition for conceptions may occur. Zinner *et al.* (1994) have found evidence for such competition in a captive group of hamadryas baboons. Probability of conception in this group was negatively correlated with the number of females that showed synchronicity of sexual skin swelling and ovulation. Unfortunately, this report does not contain detailed information on sexual behaviour as related to female dominance rank. In the Hanuman langur, however, females have been observed harassing one another's copulations and low-ranking females are harassed by dominant individuals twice as often as vice versa (Sommer 1989). Sommer points out that since female langurs displace one another frequently over resources, such as food and water, harassment of copulations might serve to reduce competition for future environmental resources, by lessening the likelihood that the harassed female will reproduce. This view was also expressed by Hrdy (1977) as a result of her field work on Hanuman langurs. The alternative view, that females might compete for the limited sperm reserves available has been suggested by Small (1988). Interestingly, Sommer (1989) notes that female–female harassment decreased markedly during periods when extra-group males entered one-male units and that 'female competition for sperm is reduced during such periods, as numerous possible inseminators besiege the troop'.

Let us turn now to the question of whether female dominance rank is related to lifetime reproductive success rather than to mating frequencies. In reviewing this problem, Harcourt (1987) examined effects of nutritional constraints, socially 'stressful' stimuli, and differing rates of infant mortality upon the fertility and reproductive success of female primates. He cautioned that 'despite three decades of intensive field study of natural groups of primates, the number of reports of correlations between dominance rank and fertility remains small.... nevertheless, where correlations exist, they are nearly always positive'. Drickamer (1974b) reported that dominant female rhesus monkeys that begin to reproduce at a younger age, give birth earlier during the annual birth season and produce more offspring in the long-term than do subordinate females. Wilson *et al.* (1983) also reported an earlier age at first conception for daughters of dominant female rhesus monkeys. More recently, Bercovitch and Berard (1993) have analysed 30 years of field data on the Cayo Santiago rhesus monkey population in order to examine factors which affect female reproductive success. They found considerable variability between females in the numbers and survival rates of their offspring. Although high rank increased the likelihood that a female would have her first offspring at an earlier age, Bercovitch and Berard reported that dominance did not correlate with reproductive success across the lifespan. One cost of reproducing more rapidly might be that females do not live so long. Thus females which first gave birth at 3 years of age (*rapid reproducers*) tended to die younger (at a mean age of 9.0 years) than females which had their first offspring at 5 years or older. These *delayed reproducers* survived until 11.0 years, on average. The difference between the two groups was not statistically significant, however, and there was a trend in these data for rapid reproducers to have a larger number of surviving offspring than delayed reproducers for any given maternal life span (Fig. 4.21). Overall, high-ranking, rapid reproducers (N = 7) left 4.4 offspring whilst low-ranking, delayed reproducers (N = 2) left 1.5 offspring each. The results are not statistically significant (perhaps due to the small numbers of subjects involved) but they indicate a clear trend towards greater reproductive success in more dominant females.

Although female chimpanzees do not belong to female-bonded social groups and do not display marked dominance hierarchies, it has been possible to analyse effects of female rank upon long-term reproductive success in free-ranging chimpanzees (Pusey *et al.* 1997). Dominance relationships were analysed with respect to the direction of 'pant-grunt' vocalizations between females; this type of vocaliza-

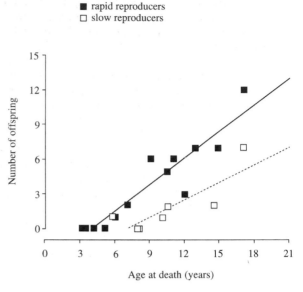

Fig. 4.21 The relationship between longevity and lifetime reproductive success in female rhesus monkeys (*Macaca mulatta*) on Cayo Santiago. The number of offspring refers to the number that survived to the age of sexual maturity. The solid line shows the regression between the two variables in *rapid reproducers* (females first give birth at 3 years of age: $y = 0.754x—3.04$; $r^2 = 0.88$) and the dashed line shows the relationship in *delayed reproducers* (females first give birth at 5 years or over: $y = 0.518x—3.65$; $r^2 = 0.71$). (Redrawn from Bercovitch and Berard 1993.)

tion signals submissiveness and correlates with receipt of aggression by females. With the benefit of a huge data set, spanning 35 years of fieldwork on the Gombe chimpanzees, Pusey *et al.* were able to demonstrate a positive relationship between females' ranks during their childbearing years and survival of their offspring to weaning age (5 years old: see Fig. 4.22). The daughters of higher-ranking chimpanzee females also reached sexual maturity earlier than daughters born to low-ranking mothers.

Table 4.10 summarizes information on relationships between female dominance, reproductive success and infant mortality in primates. Although examples are still limited, it is known that female rank, or receipt of aggression, can influence fertility in a number of species. As an extreme example, the dominant, breeding female in common marmoset groups suppresses gonadotrophin secretion in subordinate females (often her daughters) so that the latter are infertile. A similar type of female–female reproductive suppression has been documented in several other callitrichids (review: Abbott *et al.* 1993). Subordinate females in a variety of primate species may exhibit stress-related disruption of ovarian cyclicity. Low-ranking females in captive talapoin monkey groups may become hyperprolactinaemic, for example, and in this condition they fail to exhibit the oestrogen-induced surge of luteinizing hormone required for ovulation (Bowman et al. 1978). In captive baboons receipt of aggression can affect the menstrual cycle, particularly during the follicular phase when premature deflation of the sexual skin swelling can occur (Rowell 1970a). In the wild, female yellow baboons (*Papio cynocephalus*) have been observed to form attack coalitions (Wasser and Starling 1988). Two or more females form a coalition in order to attack a third animal, often a lower-ranking female in the follicular phase of the menstrual cycle. Females receiving such attacks exhibit a significantly increased number of cycles to conception and a longer interbirth interval. Attack coalitions also target pregnant females, resulting in abortion or premature parturition in some cases (Wasser and Starling 1988).

These observations encourage the view that intrasexual competition occurs among females in some primate species, and that one consequence of dominance relationships might be reproductive suppression (or disruption) in lower-ranking individuals. In macaque and baboon societies, which are female-bonded and strongly matrilineal in structure, maturing females tend to achieve similar ranks to those of their mothers. Dominance relationships exist between matrilines, therefore, and not just between individual females in different matrilines. Although relationships between matrilines are normally stable, aggressive overthrows have occasionally been witnessed in which a dominant matriline is deposed (e.g., in captive rhesus monkeys: Ehardt and Bernstein 1986; captive longtailed macaques: Chance et al. 1977; Japanese macaques: Gouzoules 1980; and olive baboons: Nash 1974). Severe fighting between females of two or more matrilines has been observed during such overthrows. Ehardt and Bernstein (1986) discuss the difficulties of explaining the causation and significance of matrilineal overthrows observed by various workers. In their own study, however, there were clear consequences for subsequent reproductive success. None of the deposed females produced offspring during the next birth season. Some spontaneous abortions occurred, as was also observed by Chance et al. (1977) in their study of longtailed macaques and by Nash (1974) for free-ranging olive baboons.

The differences in reproductive success between dominant and low-ranking females summarized in Table 4.10 might also be related to competition for resources, and especially for food. Given that females are at the core of many primate groups and that they band together exploit nutritional resources more effectively (Wrangham 1979, 1980), it is reasonable to suggest that competition for food might influence ovarian cyclicity, gestation, and lactation. There is no doubt that nutritional status has profound effects on reproductive physiology in female mammals (I'Anson et al. 1991). The proximate mechanisms by which nutrition affects fertility are complex and only partially understood. Nonetheless, it is of interest that fasting causes a slowing of the LHRH (luteinizing-hormone-releasing hormone) pulse generator in rhesus monkeys (Cameron et al. 1989) and that increases in cortisol secretion during low food intake also result in reduced LH (luteinizing hormone) secretion (Cameron 1988, 1989). In the human female, malnutrition is associated with amenorrhoea and reduced responsiveness of the pituitary gland to exogenous LHRH (Morimoto et al. 1980). One route to reduced fertility in subordinate female monkeys might operate via competition over food with resultant impairment of gonadotrophin secretion. Feeding competition has been documented in a few primate species and may be most intense where resources are 'clumped', either in nature or at the artificial feeding stations used in studies of captive groups (Harcourt 1987). Whether such competition influences fertility via nutritional mechanisms rather than directly via behavioural mechanisms is not known.

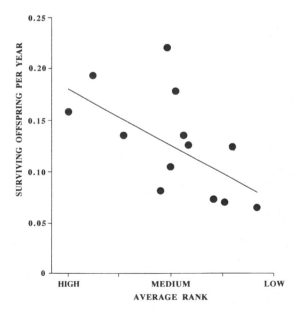

Fig. 4.22 Female rank and survival of offspring in free-ranging chimpanzees (*Pan troglodytes*). Annual production of infants surviving to age 5 years plotted against each mother's average rank during her childbearing years. (Redrawn and modified from Pusey *et al.* 1997.)

Female mate 'choice'

Darwin (1871) contended that sexual selection operates via choice of one sex by the other, (principally by female choice for the most attractive males) as well as via the agency of intrasexual competition. Despite a long-running controversy about the existence of female choice in animals, there is now ample evidence of its influence on reproductive success (Andersson 1994). Among avian species, for instance, experiments have shown that a variety of masculine secondary sexual adornments affect female choice (e.g., long tail feathers in the widow bird: Andersson 1982; and in the swallow: Møller 1988a; the coloured ocelli of the peacock's train:

Petrie *et al.* 1991; Petrie and Williams 1993). Traditionally, primatologists have concentrated upon intermale competition and dominance, as determinants of mating success. However, evidence has gradually accumulated that females of many primate species actively solicit copulations (i.e. they exhibit proceptivity: Beach 1976), and that they may refuse to copulate with certain males, whilst accepting others. The form and functions of proceptive displays will be described in the next chapter and need not be discussed here. In this section, the evidence for female choice is reviewed in relation to those masculine characteristics (whether structural or behavioural) which might influence female preferences. No experiments have been conducted on

Table 4.10 Dominance rank and fertility: comparisons of high-ranking and low-ranking female primates (high: low rank)

Species	W/F/C	Maturation	Birth rate	Season	Author
Cercopithecus aethiops	W	N	*N.S. 83%: 58% (4) (4)	−10: +20 (2) (2)	Whitten 1983
	W	N	N		Cheney *et al.* 1988
Macaca fuscata	W	N	N		Sugiyama and Ohsawa 1982
M. sinica	W		75%: 64% (15) (14)		Dittus 1979
	W		*98%: 56% (1) (1)		Dittus 1986
Papio cynocephalus	W	*5.5: 6.0 years (4) (4)	N	×	Altmann *et al.* 1988
Saguinus mystax	W		*100%: 0% (7) (7)		Garber *et al.* 1991
Theropithecus gelada	W		?N	×	Dunbar 1980, 1984
M. fuscata	F	4.8: 5.5 years (43) (38)	70%: 47% (43) (38)		Sugiyama and Ohsawa 1982
	F		?N		Takahata 1980
	F	@N.S. 0.3: 0.9 (9) (8)	N	*N.S. r_s=0.41 (17)	Wolfe 1984a
M. mulatta	F	*3.9: 4.4 years (19) (20)	83%: 73% (78) (70)		Drickamer 1974b
	F	*0.4: 0.1 ? (28) (26)	*61%: 57% ? (28) (80)		Meikle *et al.* 1984
C. aethiops	C		*122%: 95% (11) (19)		Fairbanks and McGuire 1984
M. fascicularis	C	N	N	N	Silk *et al.* 1981
M. fuscata	C	*N.S. 5.6: 6.1 years (17) (14)	*N.S. r_s=.49 (20)		Gouzoules *et al.* 1982
M. mulatta	C		*92%: 79% (64) (75)	?N	Wilson *et al.* 1978
	C	*0.4: 0.1 (27) (51)			Wilson *et al.* 1983
M. sylvanus	C	*0.6: 0.1 (7) (17)	N	N	Paul and Thommen 1984

Number of subjects in parentheses. W.F.C=wild (W), free-ranging and provisioned (F), captive (C) populations.
'Maturation'=age at first giving birth (years), or proportion of females giving birth at a given (young) age. 'Birth rate'= percentage of females giving birth each year. 'Season'=date of giving birth in breeding season, expressed as number of days compared to mean date. Data are presented only if a correlation is apparent. N=no correlation; x=correlation impossible (because no breeding season in that species); blank=correlation not examined. *=significant (P < 0.05, two-tailed) positive correlation (i.e., dominant females more fertile) in at least some groups or years; N.S.-=other groups, years, or measures indicated no significant correlation; @=significant negative correlation (dominant females less fertile); ?=correlation questionable. After Harcourt 1987.

primates to quantify masculine qualities which influence female choice, however. Thus, as far as the primates are concerned, we have little hard evidence about the importance of female choice in terms of maximizing genetic benefits to offspring (i.e. 'good genes' models of sexual selection), and whether masculine adornments or behaviour convey information concerning resistance to parasites (Hamilton and Zuk 1982), or other heritable qualities (Andersson 1994).

A number of studies has provided evidence that females choose dominant males as mating partners. A particularly striking example concerns the brown capuchin (*Cebus apella*) in which females in both free-ranging and captive groups persistently solicit copulations with the highest-ranking male (Janson 1984; Welker *et al.* 1990). Although females also mate with lower-ranking individuals during non-fertile phases of the ovarian cycle, the peak of copulations with the most dominant male results from a combination of female choice and male guarding behaviour. Keddy's (1986) field study of vervet monkeys (*Cercopithecus aethiops*) also indicates that females prefer high-ranking males. Further, dominant females are better able to exert choice, for instance by threatening subordinate males or refusing their mating attempts. That male dominance does not guarantee female preference is indicated by studies on Japanese macaques where old, high-ranking males might be refused by females. Perloe (1992) describes one such case, where females refused to copulate with two high-ranking males in their troop, whilst actively maintaining consortships with certain younger males of medium rank. Perloe speculates that females avoided forming consortships with old males which had previously mated with their mothers, so that mate choice might have been influenced by an *incest avoidance* mechanism. The same constraints might bias females to choose unfamiliar partners, such as newly immigrant males which are unlikely to be their relatives (e.g. in *Lemur catta*: Pereira and Weiss 1991; *Saimiri oerstedii*: Boinski 1987). In various species females are also more likely to mate with adult males rather than with sub-adults in the social group (e.g., *Cercocebus albigena*: Wallis 1983). In the orang-utan, for instance, females may be proceptive towards fully developed, territorial males (Schurmann 1982) but refuse the mating attempts of smaller, itinerant sub-adults (Mitani 1985; Galdikas 1985). It is unfortunate that none of these examples involves a species which has a monogamous mating system. The mechanisms which underlie mate choice in indris, owl monkeys, marmosets, and tamarins or the lesser apes are unknown. Female choice might play some

role in this process, as in 'monogamous' bird species (e.g., the barn swallow: Møller 1988a). Particularly intriguing is the case of the white-faced saki (*Pithecia pithecia*) where the male's white facial disc (not present in the female) may play an as yet unknown role in mate choice or intrasexual competition.

Sexual preferences—favouritism and friendship

Field and laboratory studies of primate sexual behaviour have revealed that sexual preferences for particular individuals can occur and that such preferences do not necessarily depend upon dominance rank. Yerkes (1939), for instance, recorded that the captive female chimpanzees he studied tended to prefer one of three males, 'Pan', during pair tests to observe sexual interactions. In his operant conditioning studies of sexual behaviour in pigtail macaques (*Macaca nemestrina*) Eaton (1973) found that females pressed bars to release some male partners into the test cage much more frequently than they released other males. Sexual preferences of males for particular females have also been recorded. When two or more females have full sexual skin swellings, for instance, males may consistently associate and copulate with one particular partner (e.g., in *Macaca nigra*: Dixson 1977; *M. maurus*: Matsumura 1993). In many of these cases the stimuli which affect sexual preference are not known. There are instances where males prefer older females as mating partners but it is hard to escape the notion that there are, as yet, undefined individual characteristics which affect sexual attractiveness in both sexes of some monkeys and apes. The most detailed studies available of sexual preference were carried out in the laboratory by Herbert (1968) and Everitt and Herbert (1969) using trios of rhesus monkeys. Each trio consisted of a male and two ovariectomized females which were either left untreated or given various dosages of oestradiol or progesterone. Under these conditions the male formed a close association with one female (the favourite), sitting with her and mating with her, whilst the second animal (the non-favourite) was largely excluded from their proximity. These preferences did not depend entirely upon the endocrine status of the females used in the tests. If both females were treated with oestradiol, the male consistently sat with, groomed with, and copulated with just one. When oestradiol was withdrawn from the favourite, the male now shifted greater attention to the hormone-treated non-favourite partner, but continued to mount, remain in proximity to and be groomed by the favourite partner for approximately 50% of observations (Fig. 4.23). Similar results occurred if the favourite partner was treated with progesterone, as well as oestradiol. Progesterone reduces female sexual attractiveness in the rhesus monkey (Baum *et al.* 1976). Manipulating the female's endocrine status in ways that influenced sexual attractiveness also had some influence on male sexual preference, but could not entirely explain the basis for such preferences (Everitt and Herbert 1969). These observations agree with those on captive rhesus monkey groups where males sometimes have favourite female partners and where such favouritism does not depend upon the stage of the female's menstrual cycle (Rowell 1963). Loy (1971) likewise records the occurrence of favouritism in semi-free-ranging rhesus monkeys on Cayo Santiago. Some of the cases quoted by Loy concerned female preferences for particular male partners. Neither field nor laboratory studies can explain the basis for such favouritism. We do not

know, for instance, whether some form of subtle facial or other communication occurs between the sexes in laboratory studies on trios of rhesus monkeys. Would altering the favourite's appearance, rather than her endocrine status, have a greater effect upon male sexual preference? Also it is unclear in these laboratory studies how far interactions and relationships between the females might have affected male sexual preference.

A most interesting study of special relationships between the sexes concerns the *friendships* which develop between some male and female olive baboons. Smuts (1985) found that some *Papio anubis* females form long-term associations with particular males, grooming and sitting with them and even sleeping huddled with them much more frequently than with other males (see Fig. 4.24). Males formed attachments to the infants of female friends (Fig. 4.25) and were observed to behave protectively towards them and their mothers more frequently than expected. It is possible, of course, that males might form friendships with females with whom they had consorted previously, and that the infants might be

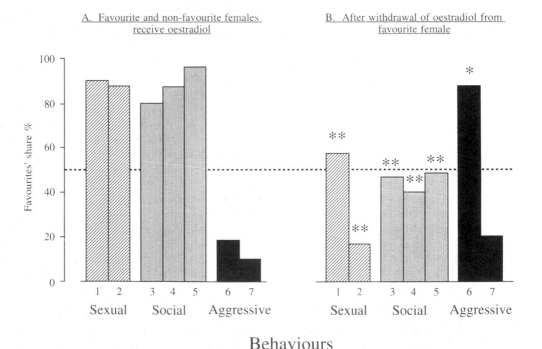

Behaviours

Fig. 4.23 Male sexual preference in six trios of captive rhesus monkeys, consisting of a single male and two ovariectomized, oestradiol-treated females. Effects of withdrawing oestradiol from the 'favourite' female are shown on the right hand side of the figure. The behavioural patterns are 1. numbers of males' mounts; 2. numbers of ejaculations; 3. time male and female spend in proximity; 4. male grooms female; 5. female grooms male; 6. male threatens female and 7. female threatened by other female. Values given are the mean percentage of each behaviour pattern directed towards favourite females. *P=0.05; **P=0.02 (Wilcoxon test). (Redrawn and modified from Everitt and Herbert 1969.)

Fig. 4.24 'Friendship' between a male and a female olive baboon (*Papio anubis*). Such long-term associations involve grooming, huddling, and sleeping together. (After Smuts 1985.)

Table 4.11 Comparison of probable fathers of infants and mothers' friendships with males during previous pregnancy and lactation in baboons (from Smuts 1985)

Female	Date of conception	Likely father[a]	Possible father[a]	Friends
AM	3/78	BZ	–	BZ, AO
LI	7/78	HD	CY	HD, VR, HM
HH	7/78	SK	BZ	SK, BZ
AM	11/78	AO	–	BZ, AO
CC	11/78	SK	BZ	CY
AT	11/78	CY	BZ	HC, AG, HS
JU	11/78	HD	VR	SK
CI	1/79	HD	TN	HC, VR

[a] Likely and possible fathers based on consortships during conception cycles and not upon DNA typing evidence. Friends of females were the likely fathers of their offspring in 50% of the cases listed above.

their own offspring. In the absence of genetic evidence, the behavioural data indicate that male friends *could* be the fathers of offspring in some cases, but that this is most unlikely to be the only mechanism underlying the formation of such relationships. Males might also benefit by gaining opportunities to mate with a female friend during her next period of sexual skin swelling. Again, the behavioural data indicate a poor correlation between friendship and subsequent presumed paternity of offspring (Table 4.11). Smuts (1985) points out, however, that males did experience some greater advantage in forming consortships with female friends than expected, although subsequent studies have not confirmed this view (Bercovitch 1991). Hopefully, some future DNA-typing analysis of paternity, coupled with detailed behavioural observations of sexual relationships might succeed in determining whether 'sex and friendship' influences reproductive success in baboons.

Long-term associations, equivalent to the friendships described in olive baboons, have also been recorded amongst rhesus and Japanese macaques (Chapais 1983; Takahata 1982a, b). In olive baboons, Smuts (1985) found that females preferred fully adult males which had resided in the troop for a long period. Males showed some slight preference for older and higher-ranking females as friends. In rhesus and Japanese macaques both sexes show some preference for high-ranking individuals when forming friendships. In all three species, males have larger numbers of these special relationships, on average, than do females (*P. anubis*: males 3.4; females 2.0; *Macaca fuscata*: males 7.3; females 1.0; *M. mulatta*: males 1.7; females 1.4; Smuts 1985). Thus far, it has not proven possible to identify a reproductive advantage to males as a result of special relationships (e.g., in the rhesus macaque: Manson 1994). Nor do rhesus or Japanese macaques form special affiliations with infants in the manner described for olive baboons. In the Barbary macaque (*M. sylvanus*), by contrast, adult males not infrequently associate with and pick up small infants. However, DNA-typing studies have shown that the infants concerned are *not the offspring of these males* (Paul, Kuester and Arnemann. 1992a) so that it would be incorrect to invoke kin-selection theory as

Fig. 4.25 A male olive baboon (*Papio anubis*) sitting with the offspring of a female 'friend'. (After Smuts 1985.)

an explanation of such behaviour. Given the extreme promiscuity of Barbary macaques, the females of which mate with multiple partners during the peri-ovulatory period of the cycle (Menard *et al.* 1992), it is hard to understand how males could identify their own offspring. Perhaps a more acceptable explanation for male–infant associations in Barbary macaques concerns the observation that males can avoid aggressive interactions with other males by picking up and transporting an infant (the *agonistic buffering hypothesis* proposed by Deag 1974). Deag's hypothesis has received support from detailed studies conducted by Paul *et al.* (1996) on male–infant interactions in semi-free-ranging groups of Barbary macaques.

Incestuous matings and incest avoidance

One important feature of mate choice concerns avoidance of mating with close relatives and its associated costs in terms of deleterious genetic effects upon offspring (in primates: Ralls and Ballou 1982). Since males often emigrate from their natal groups this provides one mechanism for *incest avoidance*. There is also a good deal of evidence that close relatives within social groups of monkeys and apes actively avoid mating with one another (e.g., in Japanese macaques: Enomoto 1974; rhesus monkeys: Missakian 1973; Barbary macaques: Kuester *et al.* 1994; olive baboons: Packer 1979a; and chimpanzees: Goodall 1986). Primates are notable for having extended periods of infant and juvenile development; mothers and offspring form long-term social relationships. This fact might reduce the likelihood of sexual attraction and copulation between mothers and their maturing sons, or between maternal siblings (Walters 1987; Pusey 1990). In family groups of marmosets, tamarins, or owl monkeys mature offspring have very rarely been observed to engage in any type of sexual activity with their parents or siblings. Kuester *et al.* (1994) describe the behavioural mechanism as follows: 'Since familiarity during infancy of one or both partners is often correlated with genetic relatedness, it can be used as a detector for kin'.

This explanation of Kuester *et al.* might be likened to the behavioural mechanism first proposed by Westermarck (1891) to account for the incest taboo in human societies. Westermarck argued that close association between human beings during childhood inhibits sexual attraction between such individuals later in life. Since close associations normally involve brothers and sisters within human families, the resulting lack of sexual attraction between siblings reduces the likelihood of incest and inbreeding. Although Westermarck's ideas were originally well received, they subsequently fell from scientific favour due to Freud's notions concerning incest. Freud held that human beings require strict laws and taboos concerning incest precisely because of a strong natural inclination to engage in such behaviour. That Westermarck was correct, however, is supported by at least two lines of evidence from studies of human beings, as well as by comparative observations on non-human primates. When children are raised as members of Kibbutzim in Israel, they very rarely marry members of their peer group. Shepher (1971) examined 2769 marriages in Kibbutzim and found that only 13 of these involved couples who had been raised in the same group. In most of these 13 cases the time spent together had been less than usual; eight couples had only joined the same peer group after 6 years of age. It seems that children raised in the Kibbutz environment play together and form friendships as expected, but that from puberty onwards their sexual interests lie outside the peer group. They regard other children with whom they were raised as siblings, and do not consider them to be appropriate sexual or marriage partners.

Parallel evidence supporting Westermarck's hypothesis comes from studies of effects of childhood association and marriage in China (Wolf 1995). Wolf gathered a huge body of data on the fate of marriages involving *sim-pua*, female infants who were adopted into families with the aim of marrying them to the son of the household. Wolf accumulated data on 14 200 women who were either *sim-pua* or who had married men with whom they had not been raised in the same family. Couples who had been raised together from infancy showed an aversion to marrying their designated partners; the result was unhappy and failed marriages with reduced rates of reproduction.

Among the non-human primates, long-term fieldwork on individually identified chimpanzees provides an excellent basis for examining occurrences of incestuous matings and mechanisms of incest avoidance (Goodall 1986). The best data concern relationships between mothers and sons and between maternal siblings. In these cases the genetic relatedness of individuals is well known from long-term observations of mothers and their successive offspring. The possibility of incestuous matings between fathers and daughters or between paternal siblings is much more difficult to analyse, however, since in the absence of DNA-typing to determine paternity the true identity of fathers is uncertain. Table 4.12 shows frequencies of attempted copulations and completed copulations between female chimpanzees and their mature sons. It is notable that four out of six males (aged between 16 and 22 years) never attempted to mate with their mothers, whilst in two further cases single copulations only were witnessed. Although females at full sexual swelling usually mate with multiple partners, and virtually all adult males in the community are involved in such copulations, the adult sons of such females rarely or never mate with them. Goodall's (1986) data on sexual interactions between maternal siblings are also impressive (Table 4.12). Copulations between siblings were rare, even during periods of frequent sexual activity involving other group members. For example, one female ('Fifi') was mated by all late-adolescent and mature males during her first full swelling *except* her two brothers, Figan and Faben.

Goodall (1986) also provides evidence that males which were the probable fathers of particular females were less likely to mate with them once they became adult. More informative in this respect, however, are the studies by Kuester *et al.* of kinship, familiarity, and mating avoidance in semi-free-ranging Barbary macaques. These authors were able to employ DNA-typing procedures and thus to ex-

amine the frequencies of sexual activity among paternal relatives. Sexual interactions between paternal kin were much more frequent than between maternal kin in Barbary macaques (Table 4.13). Father/daughter and brother/sister matings occurred in half of all possible dyads examined. However, only 2 out of 62 infants examined were found to be inbred, indicating that incestuous matings between paternal kin rarely led to production of offspring. Kuester *et al.*, thus point out that 'familiarity during early life was the key factor for establishing a mating inhibition'. Familiarity with maternal relatives led to reduced sexual interest, but it was also the case that familiarity with male 'caretakers' during infancy had a similar effect. As mentioned previously, male Barbary macaques often form affiliative relationships with infants and DNA-typing studies show that these infants are not necessarily close relatives. When female infants mature they are less likely to mate with previous caretakers. Incestuous matings do occur between fathers and daughters, however, since despite their close genetic relationship males have little opportunity to form close relationships with daughters. Indeed there is no evidence that males are able to recognize their own offspring in Barbary macaque groups, given that females mate with multiple partners around the time of conception.

Table 4.12 Frequencies of incestuous matings between mothers and sons, or between maternal siblings in free-ranging chimpanzees (from Goodall 1986)

Mother	Son	Age of son (years)	No. of cycles seen together	No. of copulations observed
Flo	Faben	16	2	0
		20	1	0
		22	2	0
Flo	Figan	14	1	0
		16	2	0
Olly	Evered	15	4	0
Sprout	Satan	18	2	1[+]
Melissa	Goblin	19	2	1[+]
		20	2	1
Nope	Mustard	16	2	0

		Numbers of copulations by females		
Sister	Brother	Total	With brother	With other mature males
Fifi	Figan (14)	250	1	249
Fifi	Faben (20)	250	2	248
Miff	Pepe (16)	100	1	99
Gremlin	Goblin (16)	128	26*	102

Data refer to sisters of late adolescent age; brothers' ages (in years) are given parentheses.
* Eight of these copulations were resisted by the female and four successfully so.
[+] Mother refused and pulled away before ejaculation.

Table 4.13 Incestuous interactions between familiar maternal relatives and between paternal relatives in groups of barbary macaques

Type of relative	Dyads N	Dyads with sex N	Dyads with sex %
Mother/Son	66	3	4.5
Sister/brother	123	3	2.4
Grandmother/grandson	13	0	0
Niece/uncle	50	3	6
Aunt/nephew	75	1	1.3
Cousins	33	3	9.1
Female/mother's cousin	11	2	18.2
Father/daughter	38	20	52.6
Brother/sister	85	44	51.7
Grandfather/granddaughter	2	0	0
Uncle/niece	4	2	50
Nephew/aunt	1	0	0
Cousins	3	1	33.3

Data from Kuester *et al* 1994.

The major histocompatibility complex and mate choice

The major histocompatability complex (MHC) is a cluster of highly variable genes, encoding for the production of glycoproteins (MHC class I and MHC class II molecules) which bind foreign antigens and present them to T-lymphocytes. The MHC plays a most important role in the recognition of infection and the triggering of appropriate immune responses. Can individuals recognize differences in the MHC of others and does this play any role in mate choice? Olfactory cues may be relevant in this context, since mice are able to distinguish between different human MHC-types on the basis of urinary cues (Ferstl *et al.* 1992) and, conversely, human beings can discriminate mouse strains differing only in their MHC (Gilbert *et al.* 1986). Under laboratory conditions, male mice may show mating preferences for females having a different MHC phenotype. Mice living under semi-natural conditions also prefer to mate with immunogenetically unrelated partners (Potts *et al.* 1991). One study has reported that women rate as more pleasant the odours of shirts worn by MHC-dissimilar men and also rate the odours of MHC-dissimilar men as being more reminiscent of their actual male partners (Wedekind *et al.* 1995).

Such information could be valuable for mate-choice in primates, whether on the basis of selection for partners with advantageous genetic traits affecting disease and parasite-resistance, or partners which are not too similar in immunogenetic qualities (as part of incest avoidance). In the latter regard *some studies* report that increased frequencies of shared MHC antigens are associated with infertility in human couples (Ober and Van der Ven 1996). In pigtailed macaques, parental sharing of MHC class I antigens is associated with an increased risk of pregnancy wastage (Knapp *et al.* 1996). The majority of prosimians and New World monkeys make use of specialized cutaneous glands and urine for communication in social and sexual contexts. Specialized glands also occur in a few Old World cercopithecoids, such as the mandrill, gibbon, chimpanzee, gorilla, and man (see Chapter 7 for information on this subject). Chemical cues might therefore convey important information concerning immunogenetic status of individuals in a population. In many nocturnal prosimians, which have dispersed mating systems, odour cues might thus convey important information affecting mate choice, given that direct contact between individuals is much less frequent than in diurnal, group-living primates. These possibilities have yet to be explored by experiment. Nor do we know whether the striking visual secondary sexual adornments of many diurnal primates, such as sexual skin, might vary according to the MHC genotype in different individuals. One study of birds has shown that MHC genotype is consistently associated with variation in a secondary sexual trait which influences female mate choice (spur length in male ring-necked pheasants: Schantz *et al.* 1996). The subject of MHC and mate choice in primates is a promising area for future research.

The problem of skewed birth sex ratios

Sex ratio biases have been reported in a number of primate species, including the greater galago, cebus and spider monkeys, various macaques and baboons, and the gorilla (Table 4.14). Numerous attempts have also been made to account for such biases in terms of maternal dominance rank, and to explain why dominant mothers should have either more sons, or more daughters, than expected (Hiraiwa-Hasegawa 1993). Trivers and Willard (1973) postulated that females in good reproductive condition should benefit by producing more sons than daughters. By investing more in sons and rearing them to peak physical condition, females maximize their own reproductive success, since successful males will produce more grandchildren than daughters can (Bateman 1948). Long-term studies of red deer (*Cervus elaphus*) by Clutton-Brock *et al.* (1988) support the validity of the *Trivers-Willard hypothesis*. Dominant hinds are heavier than low-ranking hinds; they produce more sons and their sons tend to experience a reproductive advantage when they mature. Unfortunately, it is not the case that dominant female monkeys necessarily invest in male, rather than female offspring. Such an effect has been recorded in some studies of spider monkeys and macaques, (Table 4.14), but in other cases no sex ratio bias could be identified (e.g., in an analysis of 645 conceptions in captive rhesus monkeys: Small and Hrdy 1986; and in a long-term study of captive stumptail macaques involving 75 offspring: Rhine 1994). Interestingly, one study of small captive rhesus monkey groups produced a result which ran counter to the Trivers-Willard hypothesis; high-ranking mothers produced more female offspring than did low-ranking mothers (Simpson and Simpson 1982). This result has since been confirmed by long-term studies of the same captive colony (299 offspring born over a 35-year period: Nevison *et al.* 1996; Fig. 4.26). Similar findings have also been reported for free-ranging yellow baboons by Altmann *et al.* (1988), but were not confirmed by Rhine *et al.* (1992).

An alternative argument, which seeks to explain sex ratios that are biased towards daughters rather than sons, concerns what is sometimes called the *advantaged daughter hypothesis*. Since female offspring in macaque and baboon groups tend to inherit their mothers' ranks, and to remain in their natal groups, it might be advantageous for high-ranking females to produce more daughters and to increase their matrilineal strength. Conversely, male offspring tend to emigrate as sexual maturity approaches and they derive little benefit from maternal rank (for exceptions see Manson (1993) on *M.*

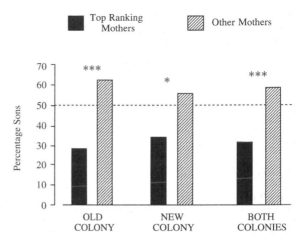

Fig. 4.26 Birth sex ratios and maternal rank in a captive colony of rhesus monkeys (Macaca mulatta) at Madingley, in Cambridge. Top-ranking mothers have significantly fewer sons (and more daughters) than other females. Data refer to 139 births between 1972 and 1981 (the 'old colony') and 160 births between 1982 and 1993 (the 'new colony'). *P < 0.05; ***P < 0.001. (Data are from Nevison *et al.* 1996.)

mulatta). On this basis, subordinate mothers might be expected to invest more heavily in male offspring, since these have the opportunity to emigrate and achieve higher rank in another group. There is also a third theory, the *local resource competition hypothesis*, which seeks to explain why lower-ranking females might bias the birth sex ratio towards production of sons, rather than daughters (Silk 1983, 1988). This hypothesis predicts that competition for resources (e.g. food) should benefit high-ranking females which harass the female offspring of

Table 4.14 Studies of birth sex ratio variations in primates: the possible effects of maternal rank

High-ranking mothers have a greater proportion of sons than low-ranking mothers	No biases in sex ratios	High-ranking mothers have a greater proportion of daughters than low-ranking mothers
Spider monkeys (WG) McFarland Symington 1987 Long-tailed Macaques (CG) Van Schaik *et al.* 1989 Barbary macaques (SFG) Paul and Kuester 1990	Yellow baboons (WG) Rhine *et al.* 1992	Bonnet macaques (CG) Silk 1988 Yellow baboons (WG) Altmann *et al.* 1988 Rhesus macaques (CG) Simpson and Simpson 1982 Nevison *et al.* 1996
Japanese macaques (CG) Aureli *et al.* 1990	Japanese macaques Fedigan *et al.* 1986 (SFG) Koyama *et al.* 1992 (WG)* Stumptail macaques (CG) Rhine 1994	

WG=free-ranging groups; SFG=semi-free-ranging groups; CG=captive groups.
*=provisioned free-ranging group.

lower-ranking mothers. In these circumstances subordinate females might produce more sons (which emigrate) rather than daughters, which remain in the group and exacerbate resource competition. Some studies have indeed shown that immature females receive more aggression from higher-ranking adult females than is the case for immature males (rhesus monkey: Simpson and Simpson 1985; bonnet macaque: Silk 1988).

Skewing of the birth sex ratio in favour of female offspring has also been documented for spotted hyenas (*Crocuta crocuta*) which live in complex, female-bonded social groups or *clans*. Holekamp and Smale (1995) describe an interesting shift in the sex ratio, favouring births of female offspring, when one clan gained access to a large tract of new territory. This circumstance arose when a neighbouring clan had been poisoned by local tribespeople. One third of Holekamp and Smale's study group split off and invaded the vacant territory. Births of female offspring outnumbered those of male offspring by two to one over the next three years; this effect was most pronounced where dominant mothers were concerned. Thus, as in the small groups of rhesus monkeys studied by Simpson and Simpson (1982) and Nevison *et al.* (1996), it might be advantageous

for dominant female hyenas to produce more daughters. This strengthens their matrilines within the clan, but it is only possible when sufficient resources are available to support growth of the social group.

Whilst skewed birth sex ratios certainly occur in some groups of various primate species (Table 4.14) no single hypothesis accounts for these occurrences. Setting aside arguments concerning the ultimate, evolutionary significance of skewed sex ratios, there is a pressing need to explain the proximate, physiological mechanisms which control these phenomena. Females might favour conceptions of one sex more than the other, or skewing might result from selective abortion of one sex (selective abortion or resorption of foetuses has been documented in the coypu, for example: Gosling 1986). Abortion is a costly tactic, however, and I know of no evidence that dominant female monkeys pursue this option. If one particular sex is favoured as regards occurrence of conception, then some remarkable mechanisms must exist within the female reproductive tract to enable recognition and selection of sperm which carry either an X or a Y sex chromosome. The limited evidence available on possible physiological mechanisms has been reviewed by Krackow (1995).

5   Sexual behaviour and sexual response

Successive generations of evolutionary biologists have based their contributions upon comparative studies of the anatomy and physiology of extant primates (Darwin 1859, 1871; Huxley 1863; Zuckerman 1932; Clark 1959; Schultz 1969; Hill 1972; Martin 1990). Diverse repertoires of sexual behaviour are to be found among prosimians, monkeys, and apes, so that comparative analyses of sexual patterns should also yield important information. Since there is no fossil record of sexual behaviour and no direct means of studying the origins of primate sexuality, the comparative approach might indeed provide the only means of shedding light upon these questions. This chapter will review patterns of precopulatory behaviour which serve to bring the sexes into proximity, as well as those postures, movements, and sexual responses which accompany copulation. Autoerotic (self-stimulatory) behaviour will also be considered, as well as abnormal forms of sexual activity. A discussion of sociosexual behaviour and homosexuality will be deferred until the next chapter.

Female sexuality: concepts and definitions
The question of oestrus

In non-primate mammals, female sexual receptivity (willingness to receive the male) is usually limited to the peri-ovulatory phase of the ovarian cycle and is strongly dependent upon ovarian secretion of oestrogen and progesterone. Heape (1900) defined this oestrous phase as a 'special period of sexual desire', noting that 'it is during oestrus, and only at that time, that the female is willing to receive the male'. In the literature on behavioural endocrinology it has become common parlance to equate 'oestrus' with the female's receptive period. The very word oestrus derives from the Greek for gadfly; the implication being that the female is goaded into a state of sexual heat at the approach of ovulation.

Among the prosimian primates, the concept of oestrus is as applicable as in the case of non-primate mammals. Female lemurs, lorises, and galagos, for example, exhibit limited periods of sexual receptivity

during the peri-ovulatory phase of the ovarian cycle. Ovariectomy abolishes receptivity in galagos, so that females vigorously refuse males' attempts to copulate and may attack them (Dixson 1978; Lipschitz 1994). However, among the anthropoids it has long been known that copulations occur at times other than when conception is likely. Heape was himself apparently aware of this fact (Nadler 1994). Recent reviews of the field and laboratory literature make it clear that copulations can occur throughout the ovarian cycle and, in a number of anthropoid species, after removal of the ovaries (e.g., Dixson 1983c, 1990a). For example, ovariectomy does not significantly reduce female receptivity in rhesus or stumptail macaques (Herbert 1970; Baum et al. 1978; Chambers and Phoenix 1987), although the *sexual attractiveness* of females might be reduced in such circumstances. In the common marmoset, males continue to ejaculate at high frequencies during laboratory pair-tests with ovariectomized, adrenalectomized females (Dixson 1990b). Ford and Beach (1952) noted that ovariectomized chimpanzees still occasionally engage in coitus and that ovarian hormones play only a minor role in the control of human sexual behaviour. They also noted the propensity of immature female monkeys and apes to exhibit sexual behaviour, whereas in subprimate mammals it is only at puberty 'with the first heat period' that 'females' sexually receptive responses appear'. Ford and Beach ascribed this relative independence of sexual receptivity from purely hormonal influences to the great enlargement of the neocortex of the brain which has occurred during anthropoid evolution.

Given the more flexible timing of receptivity in anthropoids, a number of students of primate sexual behaviour have argued that the oestrus concept is not applicable to monkeys and apes, any more than it is acceptable in descriptions of human sexuality (Keverne 1981; Dixson 1983c; Loy 1987; Wolfe 1991; Goy 1992). Despite attempts to redefine oestrus and to retain the term in descriptions of anthropoid sexuality (e.g., Hrdy and Whitten 1987, Nadler 1992), there are sound reasons to suggest that its usage should be dropped. When, as is often the case, field workers report that female monkeys or apes are 'in oestrus', no information is conveyed about physiological mechanisms beyond the misleading implication that sexual receptivity (or sexual initiating behaviour) is under rigid hormonal control. Further, continued usage of this terminology creates an artificial hiatus between studies of monkeys and apes on the one hand and anthropological and clinical studies of the human female on the other. Some anthropologists argue that since female monkeys or apes exhibit oestrus and yet women clearly do not, it stands to reason that oestrus must have been 'lost' during the course of human evolution (Lovejoy 1981; Hrdy 1988). In reality, the relative independence from hormonal control of human sexual receptivity is an extreme example of a phenomenon which exists in New World and Old World monkeys as well as among the great apes. It is, therefore, part of man's inheritance from pre-human ancestors.

Alternatives to the use of oestrous terminology have been available for over 20 years and are equally applicable to all female mammals. Beach (1976) proposed a schema which divides female sexuality into three components; *attractivity* (i.e., *sexual attractiveness*), *proceptivity*, and *receptivity*. Proceptivity refers to behavioural patterns displayed by females in order to initiate and maintain sexual interactions with males. An important distinction is made between female sexual invitational behaviour and sexual receptivity; the willingness of the female to accept the male and to permit copulation with intravaginal ejaculation. Sexual attractiveness refers primarily to non-behavioural cues, such as genital appearance or odours, which excite male sexual interest and arousal, thus increasing the likelihood that mounting will occur. Increased frequencies of copulation during the female's peri-ovulatory phase might result from non-behavioural effects of ovarian hormones (increased attractiveness) or from effects upon the brain, to increase either female proceptivity or receptivity (or both). In reality the distinction between female attractiveness and behavioural cues is not an absolute one. Proceptive behaviour might also accentuate non-behavioural signals that are sexually attractive to males (Fig. 5.1). Students of primate sexual behaviour have long been aware that ovarian hormones have powerful effects upon female attractiveness in certain species and that increases in the male's sexual activity might be attributable to the heightened attractiveness of the female (Herbert 1970). The following sections provide a comparative review and a discussion of the evolutionary significance of patterns of proceptivity, receptivity, and sexual attractiveness in female primates.

Proceptivity

Female primates use a variety of body postures, gestures, facial expressions and vocalizations as proceptive displays. Direct contact with the male might also be used to initiate or maintain sexual interactions. A comparative listing of proceptive displays is provided in Table 5.1. In some cases these motor patterns can occur in other social situations and they have not evolved specifically for communication in sexual contexts. 'Lipsmacking', for example,

Fig. 5.1 Proceptivity and sexual attractiveness in the chacma baboon (*Papio ursinus*). The female exhibits maximal sexual skin swelling and is in the peri-ovulatory phase of the menstrual cycle. She turns and begins to present to the male whilst engaging him in eye-contact. Here proceptive and attractive cues are combined in a complex fashion. (After Dixson 1990a.)

is an affiliative or distance-reducing display which occurs in both sexes of many catarrhine species, (Fig. 5.2) and in some platyrrhines also. Female catarrhines sometimes lipsmack as a proceptive behaviour, often combining this with eye-contact displays, sexual presentation to the male, or approach-retreat patterns (e.g., in talapoins, Celebes macaques, and baboons). As its name implies, lipsmacking sometimes has an audible component in addition to its role in visual communication. Rhythmic tongue movements often accompany the lipsmacking display but the tongue is not protruded to any great extent. However, in the New World family Callitrichidae and also in the howlers (genus *Alouatta*), rhythmic tongue protrusion ('tongue-flicking') plays a most important part in precopulatory invitational behaviour (Fig. 5.3). In the common marmoset, for instance, tongue-flicking forms part of a complex display in which the female becomes immobile, or crouches with her limbs flexed, stares at the male with her circum-aural tufts depressed, and (sometimes) lipsmacks. Proceptive tongue-flicking has been described in tamarins (*Saguinus midas*: Shadle *et al.* 1965; *S. oedipus*: Brand and Martin 1983) and in *Callimico goeldii* (Pryce, personal communication). It is likely that this pattern is common to the majority of callitrichids, therefore, and has been inherited from the common ancestors of this branch of the New World primates. One possibility is that the use of the tongue to bring material into contact with the nasopalatine ducts in the roof of the mouth, and thus to gain access to the vomeronasal organ (which is well developed and apparently functional in callitrichids), has become ritualized to provide a visual display of sexual interest. Callitrichids make extensive use of chemical communication, and both urine and cutaneous se-

Fig. 5.2 Lipsmacking facial expression in a rhesus monkey (*Macaca mulatta*). Lipsmacking is an affiliative facial expression used by both sexes in many catarrhine species and it sometimes functions as a sexual invitation. (Redrawn from Hinde and Rowell 1962.)

cretions play an important role in many aspects of their social communication (Epple *et al.* 1993). Tongue-flicking is not an exclusively sexual display, but it is fundamentally an affiliative behaviour, as is the case with lipsmacking among catarrhines. Rhythmic tongue movements also accompany allogrooming, so one might argue that tongue-protrusion could be derived from grooming behaviour, rather than from olfactory investigation of the scent-marks or genitalia of conspecifics. Tongue-flicks are sometimes exchanged during same-sex agonistic encounters, as in pair-tests between male marmosets (personal observation), but I do not believe that they reflect a purely aggressive

Fig. 5.3 Precopulatory tongue protrusion displays in New World monkeys. Left: 'Tongue-pumping' in the red howler (*Alouatta seniculus*). This display is used by both sexes in howler monkeys to invite copulation. Right: The proceptive tongue-flicking display of the female common marmoset (*Callithrix jacchus*). The female also crouches and stares at the male: The white circumaural tufts are often flattened, but not in this case. (Author's drawings from photographs.)

motivation. It is more likely that such behaviour represents a conflict between the tendency to approach a conspecific and the tendency to flee. It is interesting that lipsmacking can also occur in similarly ambivalent situations in certain Old World monkeys; the so-called 'aggressive lipsmacking' of patas monkeys (Hall *et al*. 1965), grey-cheeked mangabeys (Chalmers 1968c), talapoins (Dixson *et al*. 1975), and Celebes macaques (Dixson 1977).

Rhythmic tongue-protrusion or 'tongue-pumping' proceptive displays are also found in howler monkeys (see Fig. 5.3), having been reported for the mantled howler (*Alouatta palliata*: Carpenter 1934), red howler (*A. seniculus*: Neville 1972), Guatemalan howler (*A. pigra*: Horwich 1983), black howler (*A. caraya*: Jones 1983), and the brown howler (*A. fusca*: Mendes 1985). Interestingly, the female black howler also solicits the male by licking his face, hands, or

Table 5.1 Patterns of proceptive behaviour in female primates

Behaviour pattern	Species (examples)
Puffing out cheek pouches and drooling	Patas monkey
Rhythmic tongue protrusion	Common marmoset, Cottontop tamarin Goeldi's monkey, Howler monkey spp.
Eye-face (ears flattened, eyebrows raised)	Chacma baboon
Grimace and raised eyebrows	Brown capuchin
Lip retraction and open and closing mouth	Woolly monkey
Lip-smacking	Rhesus macaque, Celebes macaque, Baboon spp., Talapoin
Pout face and pursed lips	Proboscis monkey
Rapid, lateral head-shakes	Langur spp. Proboscis monkey
Head-bob and head-duck	Rhesus macaque
Puckered mouth expression	Redtail monkey
Engaging the male in eye-contact*	Common marmoset, Goeldi's monkey, squirrel monkey, Brown capuchin, Black spider monkey, Woolly spider monkey, Red colobus, Vervet, Talapoin, Rhesus macaque, Japanese macaque, Baboon spp. Mandrill, Grey-cheeked mangabey, Orang-utan, Gorilla, Chimpanzee, Bonobo, Man.
Sexual presentation postures* (quadrupedal stance)	Ringtailed lemur, Lesser galago, Tarsier, Black spider monkey, Squirrel monkey, Bald uakari, Proboscis monkey, Red colobus, Vervet, Talapoin, Rhesus macaque, Japanese macaque, Celebes macaque, Stumptail, Grey-cheeked mangabey, Baboon spp., Mandrill, Chimpanzee, Bonobo, Gorilla.
Sexual presentation postures (suspensory variants)	Slender loris (see text) Woolly spider monkey, Orang-utan.
Immobile crouching	Common marmoset
Tail curl	Patas monkey
Hand-slap	Rhesus macaque
Stiff-limbed strutting	Gorilla
Approach-retreat patterns	Bald uakari, Black spider monkey, Talapoin, patas monkey.
Parading	Goeldi's monkey, Celebes macaque, Mandrill
Seeking proximity to male	Golden lion tamarin, Rhesus macaque, Barbary macaque, Redtail monkey talapoin, Aye-aye
Backing-up to male and initiating genital contact	Black spider monkey, Orang-utan, Gorilla, Chimpanzee, Bonobo.
Licking male's face or hands	Black howler
Orogenital stimulation	Black howler, Orang-utan
Touch or pull male's hair or genitalia	Orang-utan, Chimpanzee
Smiling, embracing, kissing	Man
Grapple playfully with male	Slender loris, Potto, Black spider monkey, Aye-aye
Mount male (see text)	Japanese macaque
Vocalizations	Mouse lemur (trill),Squirrel monkey (purr), White-nosed saki (purr) Brown capuchin (hoarse whining sounds), Woolly monkey (tooth-chatter), Woolly spider monkey, Lion-tail macaque ('ho-ho' sounds), Patas monkey

*These behavioural patterns commonly precede copulation, so that numerous other examples might be cited.
Up-dated from Dixson 1983c, 1990a with additional references quoted in the text.

genitalia (Jones 1983). No other cebid genus has been reported to use tongue movements as a proceptive display, however, so their occurrence in *Alouatta* represents a parallel and separate evolutionary development from the tongue-flicking displays of the Callitrichidae.

Examples of proceptive facial displays in female primates are provided in Table 5.1, but the list is doubtless far from complete and it is difficult to make inferences about the evolution of these displays. It is noteworthy, however, that rapid side-to-side, head-shaking movements are used as sexual invitations by females of several of the Asiatic langur species (*Presbytis entellus*: Hrdy 1977; *P. cristata*: Bernstein 1968; and *P. senex*: Rudran 1973) as well as by the proboscis monkey (*Nasalis larvatus*: Yeager 1990b). Figure 5.4 shows the head-shaking proceptive display in *P. entellus*. This particular pattern has not been described in any of the African colobines (e.g., in *Colobus badius*, *C. polykomos*, or *C. guereza*). Head-shaking might thus represent a specialization of the Asiatic branch of the sub-family Colobinae. Further studies of the occurrence, or absence, of head-shaking displays in other Asiatic genera, such as *Pygathrix* and *Rhinopithecus*, would be helpful in resolving this question.

One subtle aspect of facial communication that is widespread among the anthropoid primates is establishment of eye-contact between the sexes as a prelude to copulation (e.g., in the gorilla: Hess 1973; talapoin: Dixson *et al.* 1975; black ape of Celebes: Dixson 1977; squirrel monkey: Baldwin and Baldwin 1981). Bielert (1986) was the first to define *eye-contact proceptivity* and to measure a peri-ovulatory peak in this behaviour during the menstrual cycle of the chacma baboon. Eye-contact with the male is often combined with a sexual presentation posture by females of this species (Fig. 5.1) and the same is true in many other species. In the marmoset, although the female does not present sexually to the male, she does engage in eye-contact displays (Fig. 5.3). These increase in frequency during the peri-ovulatory phase of the ovarian cycle and are stimulated by changes in oestradiol secretion and LHRH activity at this time (see Chapter 12 for a discussion of the neuroendocrine mechanisms involved).

Because eye-contact proceptivity occurs in New World and Old World monkeys, as well as in the great apes (Table 5.1) it almost certainly represents part of the phylogenetic inheritance common to all anthropoids. Facial communication is indeed complex in monkeys and apes; their facial musculature and facial expressions are more highly developed than in any other vertebrate order (Huber 1931; Van Hooff 1967). Subtle variants of eye-contact, and

Fig. 5.4 Proceptive side-to-side head-shaking display in a female Hanuman langur (*Presbytis entellus*). Similar proceptive displays have been described in other Asiatic colobines. (After Hrdy and Whitten 1987.)

not just gross facial movements, play a still poorly understood role in primate social communication. Prolonged 'peering' occurs in bonobos, for example, often being used by subordinates to initiate grooming, play, or food sharing with higher-ranking individuals (Idani 1995). In specifically sexual contexts, few observers familiar with monkeys and apes would dispute that communication between the sexes can be extremely subtle and that current scoring systems do not always adequately reflect the nature of these interactions. Eye-contact as a means of sexual initiation is probably much more important than currently realized. Homologous patterns also occur in man. Repeated eye-contact and raising of the eyebrows occurs as part of 'flirting' behaviour by women in Samoa, Europe, Africa, South America, and elsewhere (Fig. 5.5). It is a phylogenetically ancient trait, not determined by cultural factors (Eibl-Eibesfeldt 1970).

Turning to the use of body postures as proceptive visual displays, the most commonly described pattern is the so-called sexual presentation posture (Figs 5.1 and 5.6). Not all presentation postures fall within the definition of proceptivity, however. If a male monkey or ape attempts to mount and the female adopts a presentation posture in response to his initiative, this qualifies as a 'present to contact' and is indicative of receptivity rather than proceptivity (Baum *et al.* 1977). Presentations are not universally used among primates to initiate copulation. They are rare among prosimians, for example, having been described in tarsiers, the lesser galago, and the ringtailed lemur. Further examples might come to light, but the nocturnal way of life of many prosimians has doubtless limited the development of visual displays. Genital presentation is probably an adjunct to olfactory communication in such cases and this might represent one of the evolutionary origins of presentation behaviour among the diurnal anthropoids. Sexual presentations also occur among members of the same as well as the opposite sex, in various macaques, baboons, talapoins, chimpanzees, and many other anthropoids. These *sociosexual* presentations play an important role in affiliative and rank-related behaviour, as will be described in the next chapter. Likewise, the distinctive genital presentations of various callitrichid species are dominance displays, not proceptive behavioural patterns. Since female monkeys and apes sometimes present to males for sociosexual reasons, rather than to initiate copulation, it can be difficult to distinguish between the various types of display. In practice, given a thorough knowledge of the species concerned and of the social relationships between particular animals, it is usually possible to identify proceptive presentations. In the yellow baboon (*Papio cynocephalus*), for example, submissive females present with various degrees of leg flexure, or lateral curvature of the spine, to more dominant males or females. The proceptive female usually adopts a more upright stance rather than 'cringing' before the male during her presentation. Females sometimes present the rump for grooming, but it is usually clear, either from the context in which the display occurs, or from the exact posture adopted, that it is not a proceptive pattern (see Fig. 5.7, which compares the grooming and sexual presentations of the talapoin).

Several interesting variations of the sexual presentation posture are found in highly arboreal primate species. In the woolly spider monkey, for example, proceptive females employ the prehensile tail as well as all four extremities to 'hang-up' in front of males and to display their genitalia (Fig. 5.8). The female orang-utan uses highly variable quadrumanous postures to suspend herself in the branches and invite copulation. In the Lorisinae, copulation takes place with the female suspended below a branch. The female slender loris (*Loris tardigradus*) 'shows her readiness for copulation by moving suspended under horizontal branches for short distances' (Schulze and Meier 1995). Such an interaction, with the male closely following the female and sniffing her genitalia is depicted in Fig. 5.9. Further studies are required to determine whether female pottos, angwantibos, or slow lorises also invite copulation by moving in this fashion and whether female lorisines present in order to initiate copulation rather than in response to the males' contacts and mount attempts.

Proceptive presentations are often combined with other display elements (e.g., eye-contact with the male) and sometimes involve much more active patterns of solicitation. Approach-retreat locomotor patterns are not uncommon ('coy' behaviour involving approaches to and running from the male, as in the talapoin: Dixson *et al.* 1975; *Ateles* spp.: Van Roosmalen and Klein 1988; bald uakari: Fontaine 1981), as well as 'parading' before the male which might serve to draw attention to the female's sexual skin swelling (as in the Celebes macaque, liontailed macaque and mandrill). In the orang-utan, Nadler (1988) recorded that proceptivity 'consisted of the female approaching the male, combined with genital

Fig. 5.5 'Flirting' behaviour (equivalent to eye-contact proceptivity) in the human female. The display includes smiling, as well as head-lowering and breaking of eye-contact. (Photographs of a Turkana woman in Kenya, by H. Hass, after Eibl-Eibesfeldt 1970.)

presenting, rubbing the genitals on the male's body (including the head, back and hand), manual or oral manipulation of the male's penis or actually mounting the male'. Mountain gorilla females employ more subtle, but nonetheless direct tactics as Harcourt *et al.* (1981b) describe:

'Once near the male after her characteristically slow and hesitant soliciting approach, the female would stand facing him with her body slightly sideways on, as if waiting for a signal, when she would abruptly turn and back into the normal dorso–ventral copulatory position'.

The rhesus monkey can express her proceptivity merely by seeking close proximity to the male ('prox' behaviour: Czaja and Bielert 1975). As well as presenting sexually, the female also slaps her hand on the ground ('sporadic arm reflex': Carpenter 1942), 'head bobs' and 'head ducks' and makes threatening facial displays directed *away* from the male partner (both sexes use such 'threatening away' behaviour to solicit copulation: Ball and Hartman 1935). As we shall see in Chapter 11, peaks of proceptivity occur during the peri-ovulatory phase of the menstrual cycle in the rhesus monkey and there are some interesting differences in the timing of various display elements with respect to the pre-ovulatory oestradiol surge.

It is possible that hormonal changes during the peri-ovulatory period might stimulate increased activity as well as greater willingness to approach or remain close to prospective mates. Neuroendocrine mechanisms which control proceptivity are considered in detail in Chapter 12. It is sufficient at this point to note that increases in locomotor activity and 'restlessness' of females at mid-cycle have been noted in a number of field studies of primate sexual behaviour (e.g., in the chacma baboon: Saayman 1970; woolly spider monkey: Milton 1985a). In the mouse lemur, recent research has shown that captive females become more active (i.e., spend more time moving around their cages) on the day of oestrus and are also more proceptive at this time (Buesching *et al.* 1998). The mouse lemur is a nongregarious, nocturnal prosimian and increased activity during oestrus might be adaptive in bringing females into proximity with neighbouring males and also in facilitating a wider distribution of their scent-marks around the home range (Fig. 5.10).

Proceptive vocalizations have been described in primates, but their occurrence is not widespread (Table 5.1). In the mouse lemur, for example, the oestrus female emits a distinctive trilling call which is thought to be significant in sexual communication (Fig. 5.10). It would be of interest to determine whether females of other nocturnal primate species vocalize in this way during the peri-ovulatory phase of the ovarian cycle. Among the anthropoids, 'purring' sounds are given by female squirrel monkeys as a precopulatory display (Latta *et al.* 1967) as is also the case in the white-nosed saki (*Chiropotes albinasus*: however, only one captive female was observed: Van Roosmalen *et al.* 1981). The female woolly monkey retracts her lips, opens and closes her mouth rapidly, and emits 'tooth chattering' sounds

Fig. 5.6 Examples of proceptive sexual presentation postures in female primates. Upper: Female black ape of Celebes (*Macaca nigra*) presents to a male, looking back at him as she does so. The male grins, grasps the female's hips and examines her sexual skin swelling. Middle: Female Barbary macaque (*M. sylvanus*) presenting and making eye-contact with a male. Lower: Crouch-presentation by a female chimpanzee (*Pan troglodytes*) (Upper: After Dixson 1977; Middle: After Dixson 1983c; Lower: After Wickler 1967.)

(Eisenberg 1976; Nishimura 1988). Female woolly spider monkeys give a 'twittering' vocalization (Milton 1984, 1985a) and female capuchins (*Cebus apella*) also vocalize whilst soliciting the dominant male in the social group (Janson 1984). The sounds uttered by females of these New World monkey species are quite variable, as is also the case for the Old World monkeys listed in Table 5.1. It is not possible at this stage to identify homologies between the calls of different genera; nor is there any experimental evidence that these calls are arousing sexually to males or result in preferential approaches by males to the females concerned.

Tactile elements of female proceptive displays are also listed in Table 5.1. Elements derived from affiliative behaviour such as the playful grappling which occurs between male and female pottos and slender lorises (Epps 1976; Schulze and Meier 1995) and allogrooming between the sexes (e.g., in the rhesus monkey and chacma baboon) might serve as part of sexual initiation or maintenance of sexual activities. In those monkeys where the male makes a series of mounts before ejaculation it is not uncommon for the partners to engage in allogrooming during the intervals between mounts (e.g., in the rhesus monkey). Rarely, females have been observed to mount males in sexual contexts, raising the possibility that such behaviour might function to arouse the male and initiate copulation. Wolfe (1984b) describes such behaviour in female Japanese macaques (*Macaca*

fuscata). Half the females in the semi-free-ranging group studied by Wolfe either mounted males or sat on their backs, occasionally rubbing their genitalia against the males' bodies. Such behaviour occurred either before or during a series of mounts with intromissions, so that it could represent proceptive behaviour. Although sociosexual mounting is not unusual in monkeys and apes, the type of behaviour described by Wolfe is uncommon. Instances of receptive females mounting males have been described in some non-primate mammals, however, and Beach (1968a) suggested that these might be connected with attempts to arouse sexually sluggish males. Only twice have I seen female marmosets or owl monkeys mount males during laboratory pair-tests and in both cases the males concerned were hypoactive sexually. In such rare instances, sociosexual mounts by female monkeys upon males might represent proceptive behaviour.

The ability of female primates to use proceptive displays and hence to initiate copulation with preferred partners is related to the question of mate choice. In some cases females initiate the majority of matings (e.g., in *Gorilla*, *Ateles*, and *Alouatta*). In other cases it appears that non-behavioural cues are sufficient to attract males and to ensure that they initiate the majority of mounts. In laboratory studies of the greater galago and owl monkey, 100% of mounts were initiated by males (Eaton *et al*. 1973a; Dixson 1983a), in squirrel monkeys 89% (Wilson

Fig. 5.7 Contrast between presentation posture used to invite grooming (left) with that used to invite copulation (right) in the talapoin monkey (*Miopithecus talapoin*). (After Dixson *et al*. 1975.)

1977a), in free-ranging chimpanzees 82% (Van Lawick-Goodall 1969), in geladas 66% (Dunbar 1978), in vervets approximately 50% (Gartlan 1969), and in liontailed macaques 31% (Kumar and Kurup 1985). However, when sexual behaviour is observed in captivity it is important to consider the effect of experimental conditions upon the expression of proceptivity in females. In the rhesus monkey, for example, merely increasing the size of the cage used for pair-tests and allowing the female greater freedom to approach or avoid the male markedly enhances her proceptivity at mid-cycle (Wallen 1982).

Sexual receptivity

Sexual receptivity denotes the female's willingness to allow the male to copulate and to ejaculate intravaginally. In many subprimate mammals the receptive female adopts a reflexive posture during copulation. In rodents, such as the rat and hamster, *lordosis* involves concave arching of the spine with elevation of the rump and head (Fig. 5.11). It is a sexual reflex with a complex neuroendocrine basis (Pfaff 1980). Lordotic postures have been identified in certain prosimian species (e.g., in the ringtailed lemur: Evans and Goy 1968; ruffed lemur: Shideler *et al.* 1983). Given the rigid hormonal control of receptivity which occurs in prosimians, so that females may truly be said to exhibit oestrus in this suborder of the primates, it is not surprising that receptive posturing is more reflexive in nature. Careful observations of copulatory behaviour in lemurs, lorises, galagos, and other prosimians might bring to light further examples of lordotic postures in receptive females. Figure 5.12 shows a female ringtailed lemur exhibiting lordosis in response to manual stimulation by a human observer. In the rat, tactile stimulation by the male as he mounts and thrusts upon the receptive female activates her lordotic reflex (Pfaff 1980) and the same might also prove to be true in some prosimians.

Among the monkeys and apes, the absence of

Fig. 5.8 Female woolly spider monkey (*Brachyteles arachnoides*) presents sexually to a male by 'hanging-up' in front of him and displaying her genitalia. Note the prominent clitoris. (Author's drawing from a photograph by Professor Karen Strier.)

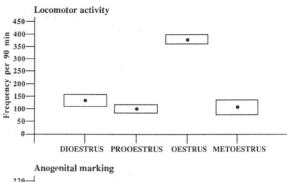

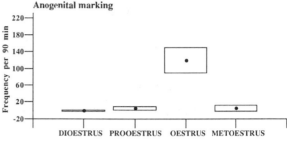

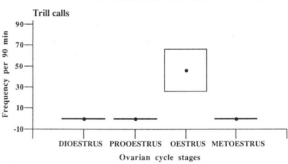

Fig. 5.9 A female slender loris (*Loris tardigradus*) is followed by a male. She moves into a distinctive precopulatory position, suspended below a branch, and the male attempts to mount. (After Schulze and Meier 1995.)

oestrus and of rigid hormonal effects upon sexual receptivity is associated with the lack of a reflexive copulatory posture. Female anthropoids do not exhibit lordosis, therefore, and receptivity is by no means obligatory or passive even during the peri-ovulatory phase of the cycle. Females may avoid a male's mounting attempts, actively refuse such attempts, or terminate the mount before intromission has occurred. Quantification of refusals, acceptances, and terminations of mounts thus provides more reliable information about female receptivity in anthropoids than measurements of the *lordosis quotient*, commonly used in studies of rodent sexual behaviour. Communication between the sexes continues during copulation and female monkeys and apes exhibit a variety of facial expressions and vocalizations during mating. In addition, physiological responses indicative of *orgasm* have been described in female stumptail macaques and some other monkey species. A full discussion of these phenomena is provided later in this chapter, when communication between the sexes during copulation is considered in detail.

Sexual attractiveness

Non-behavioural cues influencing female sexual attractiveness in primates include the oestrogen-dependent sexual skin swellings of certain Old World

Fig. 5.10 Effect of the ovarian cycle upon behaviour in the female mouse lemur (*Microcebus murinus*). During oestrus females show high frequencies of proceptive 'trill' calls, higher levels of locomotor activity and increased scent-marking behaviour. (Redrawn and modified from Buesching *et al*. 1998.)

monkeys and chimpanzees (Dixson 1983b). Oedema and pinkness of the swelling increase during the follicular phase of the menstrual cycle and are greatest during the peri-ovulatory period (Fig. 5.1). Experiments using models of sexual skin swellings, attached to the rumps of ovariectomized chacma baboons, have shown that the size and colour of the swelling are attractive and sexually arousing to males of this species (Girolami and Bielert 1987). The evolution of sexual skin swellings will be dealt with in detail in Chapter 7.

Olfactory stimuli, such as genital odours, have been implicated as signals of female attractiveness in numerous prosimians and New World primates. Scent-marking using cutaneous glands, urine, and genital secretions is widespread in both these groups.

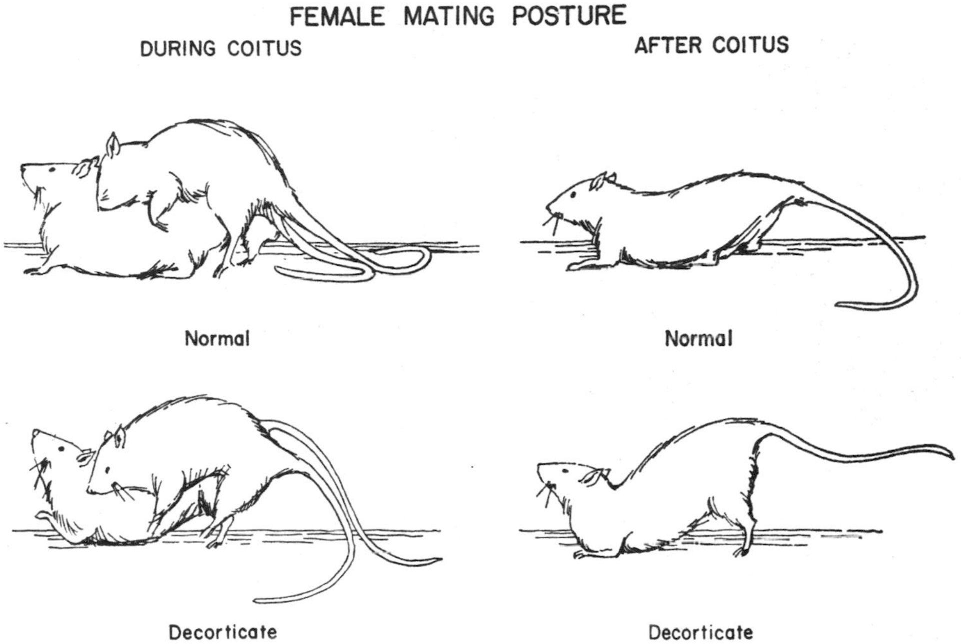

FEMALE MATING POSTURE

DURING COITUS   AFTER COITUS

Normal   Normal

Decorticate   Decorticate

Fig. 5.11 Lordotic postures in female rats. This reflexive posture occurs in the receptive female in response to tactile stimulation received from the male as he mounts, palpates the female's flanks, and makes pelvic thrusting movements. Removal of the female's cerebral cortex results in a more extreme and longer-lasting, lordotic response. (After Beach 1967.)

Males investigate these scent-marks or the genitalia of females before copulation, so it is likely that chemical cues (operating via the main olfactory system or via the vomeronasal organ and accessory olfactory pathways) signal female reproductive status (prosimians: Schilling 1979; Evans and Schilling 1995; New World primates: Klein 1971; Epple *et al.* 1993). Figure 5.10 shows the increased frequency of anogenital marking which occurs during the periovulatory period in female mouse lemurs. Increased marking activity, coupled with a propensity to roam more widely, might result in a greater distribution of chemical cues throughout the home range of the female. This could be one means by which females of non-gregarious nocturnal prosimians advertise their reproductive status to neighbouring males and hence attract a number of potential mating partners. The situation is analogous to certain diurnal catarrhines where the female's pink sexual skin swelling serves to attract numbers of males and to promote competition between them (e.g., in the chimpanzee, chacma baboon, or talapoin).

A more controversial issue concerns the existence of oestrogen-dependent vaginal odours as sexual attractants in Old World monkeys. Although scent-marking behaviour is rare among Old World monkeys, (e.g., sternal marking in the mandrill:

Jouventin 1975), males often sniff the genitalia of females before copulation. In some cases, human observers have commented upon the occurrence of strong-smelling vaginal secretions in females at mid menstrual cycle, as in *Macaca radiata* and *M. sinica* (Rahaman and Parthasarathy 1969a; Fooden 1979).

Fig. 5.12 Lordotic-type posture in a sexually receptive female ringtailed lemur (*Lemur catta*). The female adopted this position in response to manual pressure on the lumbar region by an experimenter. (Author's drawing, after Evans and Goy 1968.)

However, the most detailed experimental evidence relating to vaginal cues and sexual attractiveness has been obtained by studying the rhesus monkey. Despite reports that an oestrogen-dependent vaginal pheromone or *copulin* occurs in the rhesus monkey (e.g. Michael *et al.* 1977a) others have failed to confirm these findings (Goldfoot *et al.* 1976). Effects of hormones upon female sexual attractiveness will be dealt with in detail in Chapter 12, which includes a review and critique of research on olfactory cues and sexual attractiveness in the rhesus monkey. At this point it should be noted that it is doubtful whether vaginal odour plays a crucial role in attractiveness in this species (Goldfoot 1981), whilst its importance in other catarrhines has not been adequately studied.

In a few mammals, salivary secretions are important in sexual communication. In pigs, a primer pheromone contained in the boar's saliva stimulates the onset of oestrus in the sow (Short 1984); in the gerbil, males are able to distinguish between saliva samples collected at different stages of the female's ovarian cycle. Given that steroid hormones and other substances are secreted in saliva (Read 1993) the potential for saliva to act as a vehicle for chemical communication between the sexes should not be ignored. This is particularly so where some of the prosimians and callitrichid primates are concerned, since biting, licking, and muzzle rubbing on branches has been recorded in a number of species (e.g., *Microcebus murinus*: Buesching *et al.* 1998; *Saguinus mystax*: Heymann *et al.* 1989). In the case of the

mouse lemur, females muzzle rub most frequently when they are in oestrus. A most unusual and interesting example from the catarrhine primates concerns the patas monkey, in which proceptive females sometimes 'drool' saliva whilst displaying to males. Whether this is an epiphenomenon or significant in terms of olfactory communication is not known.

Male sexuality: concepts and definitions

Beach (1956) distinguished between a *sexual arousal mechanism* and an *intromission and ejaculatory mechanism* in the control of masculine sexual behaviour. Behavioural patterns which occur before copulation, such as the male following the female, investigating her genitalia, or engaging in invitational displays, are all indicative of sexual arousal. By contrast, those patterns which serve to bring about intravaginal ejaculation (mounting, with intromission, and pelvic thrusting) constitute the intromission and ejaculatory mechanism, referred to in this book as the *copulatory mechanism*. The distinction between these two mechanisms is useful because there are some significant differences in their neuroendocrine control. Thus, after castration adult male mammals (such as the rhesus monkey) may continue to show sexual interest in females long after they cease to intromit or ejaculate.

Once a male has ejaculated, a period of sexual inactivity usually follows, during which sexual arousal is reduced. This is referred to as the *post-ejaculatory*

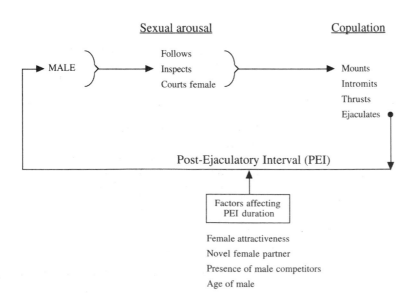

Fig. 5.13 Relationships between the behavioural components of male sexuality: the sexual arousal mechanism, copulatory mechanism, and post-ejaculatory interval.

refractory period. In the rat, for example, there is an absolute refractory period immediately after ejaculation, during which the male will not copulate. Thereafter, a more labile refractory phase occurs, so that favourable conditions (such as the presence of an attractive female) will arouse the male sexually. These components of male sexuality (sexual arousal, copulatory mechanism, and post-ejaculatory refractory period) are shown diagrammatically in Fig. 5.13, together with potential effects of feminine cues and behaviour upon the male's sexual activity. The functional significance of the post-ejaculatory refractory period is still debated and may differ between species; nor is its neurophysiological basis fully understood. It might be adaptive in allowing the male to replenish reserves of epididymal spermatozoa before the next copulation. The quiescent phase after mating might also be adaptive in lessening the likelihood that the male will interfere with a previously deposited copulatory plug.

Whilst there is plenty of evidence that males of various primate species are less likely to engage in copulatory behaviour immediately after ejaculation, little attempt has been made to measure the period

Table 5.2 Durations of post-ejaculatory intervals (PEI) and inter-ejaculatory intervals (IEI) in primates

Species	Conditions of study	Duration (min)	Sources
Post-ejaculatory intervals			
*Callithrix jacchus**	FAT+intact females	Median 12.2	Kendrick and Dixson 1984a
	FAT+OVX females	Median 20.8	Kendrick and Dixson 1984a
	FAT+proceptive intact females	Mean 12.8	Kendrick and Dixson 1983
	FAT+non-proceptive intact females	Mean 19.8	Kendrick and Dixson 1983
*Macaca nemestrina***	FAT: no sperm competition	Mean 42.6	Busse and Estep 1984
	FAT:+sperm competition	Mean 17.5	Busse and Estep 1984
M. nemestrina	FAT	5–38 (Median: 25.5)	Nadler and Rosenblum 1973
M. mulatta	FAT+intact females	Mean 26.8	Bielert and Goy 1973
M. arctoides	FAT: females at mid-cycle	Mean 1.1	Slob *et al.* 1978
M. thibetana	SFG	30.0–95.0 (Mean 55.3)	Zhao 1993
M. radiata	FAT	3.0–85.0 (median: 40.0)	Nadler and Rosenblum 1973
Papio ursinus	WG	Mean 56.0	Saayman 1970
Gorilla gorilla	FAT	14.0–30.0 (Median: 15.0)	Nadler 1976
Inter-ejaculatory intervals			
Lemur catta	WG	Minimum <4.0	Koyama 1988
Saimiri sciureus	FAT	Minimum 183.0	Clewe and DuVall 1966
*Macaca nemestrina***	FAT: no sperm competition	Approx 55.0	Busse and Estep 1984
	FAT+sperm competition	Approx. 30.0	Busse and Estep 1984
M. mulatta	WG	Minimum <6.0	Lindburg 1971
M. arctoides	LAB PAIRS: video study	3.1–139.4 (Mean 14.6)	Slob and Nieuwenhuijsen 1980
M. thibetana	SFG	40.0–160.0 (Mean 90.0)	Zhao 1993
M. sylvanus	SFG	Minimum 13.0	Kuester and Paul 1989
Papio ursinus	WG	Minimum 4.7	Bercovitch 1989
Mandrillus sphinx	SFG	Minimum <5.0	Dixson 1995b
Pan troglodytes	CG	Minimum <5.0	Allen 1981
Gorilla gorilla	FAT: intact females	14.0–30.0 (Median: 15.0)	Nadler 1976

FAT=Free access pair test conducted under laboratory conditions; CG=captive groups; WG=wild groups; SFG=semi-free-ranging groups; OVX=ovariectomized.

* Variability of the post ejaculatory interval in C. jacchus includes shortening of the PEI during pair-tests with highly proceptive females.

** In Macaca nemestrina the PEI and IEI are shortened if the male is removed from the pair test cage and allowed to observe a second male copulating with the female partner. This phenomenon is described in greater detail in Chapter 8 (see Fig. 8.21; page 243).

Table 5.3 Patterns of precopulatory behaviour in male primates

Behaviour pattern	Species (examples)	Sources
Sniffing or licking the female's genitalia*	Greater galago	Eaton *et al.* 1973a
	Slender loris	Schulze and Meier 1995
	Ringtailed lemur	Evans and Goy 1968
	Ruffed lemur	Shideler *et al.* 1983
	Fat-tailed dwarf lemur	Foerg 1982a
	Tarsier	Wright *et al.* 1986
	Squirrel monkey	Wilson 1977a
	Bald uakari	Fontaine 1981
	Spider monkeys	Klein 1971
	Owl monkey	Dixson 1983a
	Golden lion tamarin	Kleiman 1978
	Common marmoset	Kendrick and Dixson 1984a
	Goeldi's monkey	Heltne *et al.* 1981
	Pygmy marmoset	Soini 1988
	Howler monkeys	Glander 1980
	Woolly monkeys	Mack and Kafka 1978
		Nishimura 1988
	Red colobus	Struhsaker 1975
	Talapoin	Dixson *et al.* 1975
	Vervet monkey	Gartlan 1969
	Grey-cheeked mangabey	Chalmers and Rowell 1971; Wallis 1981
	Chacma baboon	Bielert 1986
	Mandrill	Author's observations
	Bonnet macaque	Rahaman and Parthasarathy 1969a
	Stumptail macaque[†]	Slob *et al.* 1978
	Black ape of Celebes	Dixson 1977
Visual inspection of the female's sexual skin*	Red colobus	Struhsaker 1975
	Black ape of Celebes	Dixson 1977
	Liontailed macaque	Kumar and Kurup 1985
	Grey-cheeked mangabey	Chalmers and Rowell 1971
	Chacma baboon	Bielert 1986
	Mandrill	Author's observations
	Chimpanzee	Goodall 1986
	Bonobo	Kano 1992
Rhythmic tongue protrusion	Pygmy marmoset	Soini 1988
	Common marmoset	Kendrick and Dixson 1984a
	Red-handed tamarin	Shadle *et al.* 1965
	Golden-lion tamarin	Kleiman *et al.* 1988
	Howler monkey	Carpenter 1934
Jaw quiver	Pygmy marmoset	Soini 1988
Protruded lips face	White-throated capuchin	Freese and Oppenheimer 1981
Grin-face	Talapoin	Dixson *et al.* 1975
Head-flag	Grey-cheeked mangabey	Wallis 1981
Eye-face	Chacma baboon	Saayman 1970
Grin and head-shake	Mandrill and Drill	Author's observations
Chin-thrust	Rhesus macaque	Hrdy and Whitten 1987
Silly grin	Rhesus macaque	Pomerantz and Goy 1983
Lip-smack	Rhesus macaque	Dixson 1977
	Black ape of Celebes	
Lip retraction, clonic jaw movements and tongue-flicks	Bonnet macaque	Kaufman and Rosenblum 1966
Jaw-thrust and pout	Pigtail macaque	Lindburg *et al.* 1985
	Liontail macaque	
Grimace, with bared teeth	Liontail macaque	Lindburg *et al.* 1985
High grin	Black ape of Celebes	Dixson 1977
Strutting	Pygmy marmoset	Soini 1988

Table 5.3—*Continued*

Behaviour pattern	Species (examples)	Sources
Pacing, jumping back and forth on branches Following females*	Squirrel monkey Red handed tamarin Cottontop tamarin Woolly monkey	Baldwin and Baldwin 1981 Oemedes and Carrol 1980 Mack and Kafka 1978; Nishimura 1988
	Talapoin monkey	Dixson *et al.* 1975
	Black ape of Celebes	Dixson 1977
	Mandrill	Author's observations
Oblique approach to female	White throated capuchin	Freese and Oppenheimer 1981
Reclining with penis erect	Orang-utan	Nadler 1988
Tree-leaping, branch-shaking, bipedal swagger, piloerection, seated posture with penis erect, stretching one or both arms towards female	Chimpanzee	Lawick and Goodall 1969; Goodall 1986
Seated posture with penis erect	Bonobo	Thompson-Handler *et al.* 1984
Raising arms, as if to embrace female	Mountain gorilla	Harcourt *et al.* 1981b
Tree-swaying	Orang-utan	Mackinnon 1974
Smiling, touching, holding, kissing	Man	
Grooming female	Common marmoset Red-handed tamarin Cottontop tamarin	Kendrick and Dixson 1984a Oemedes and Carrol, 1980
	Chacma baboon	Saayman 1970
	Rhesus macaque	Michael *et al.* 1966

* These behavioural patterns commonly precede copulation, so that numerous other examples might be cited.
† In stumptails, the male frequently inserts his finger into the female's vagina and sniffs it, prior to copulation.
Up-dated from Dixson 1983c.

of 'absolute' refractoriness in monkeys or apes. It is clear that a variety of factors can arouse males and reawaken their sexual activity. The sight of another male mating with a female shortens the post-ejaculatory interval in some cases (e.g., in the pigtailed macaque: Busse and Estep 1984). This might be an important factor in the dynamics of multipartner matings in relation to *sperm competition* as will be discussed more fully in Chapter 8. Indeed, the concept of a *post-ejaculatory interval* (PEI) may be more appropriate for many primates rather than some period of absolute refractoriness after ejaculation. Measurements of the PEI, though few in number, have been made in laboratory and field studies of several species and these are listed in Table 5.2.

Precopulatory behaviour in male primates

When animals are paired for brief periods in the laboratory in order to study sexual behaviour it is possible to use standard measurements of the male's sexual arousal, such as latencies to the first at-tempted mount or first intromission. These measurements are not possible when observing more permanent social groups. However, male primates exhibit various precopulatory behavioural patterns, including olfactory and visual inspection of the female genitalia, facial expressions, and body postures (Table 5.3). Some of these patterns indicate the male's attempts to invite the female to engage in copulation (the equivalent of female proceptivity) and some provide useful indices of the female's sexual attractiveness. As can be seen in Table 5.3, the propensity of males to investigate the genitalia of females before copulation is widespread among primates as in many other mammals. Sniffing and/or licking of the genitalia occurs in numerous prosimian and monkey species but is apparently rarer among the apes. As regards free-ranging mountain gorillas, for example, Harcourt *et al.* (1981b) comment that 'at no time did adults touch or sniff each other's genitalia.....although immatures of both sexes not infrequently sniffed oestrus females' rumps'. Chemical information concerning female attractiveness and reproductive status might derive from genital secretions and/or urine; such cues are probably of great-est importance among the prosimians and New World monkeys. It might be significant that males of several New World monkey genera pay particular attention to the urine of females when they enter breeding condition. Male spider monkeys sample the females' urine by licking or sniffing the places where micturition has occurred, as well as examin-ing the females' greatly enlarged clitorides (Klein 1971; Van Roosmalen and Klein 1988). Mantled howler males place the nose and mouth in female urine and raise their heads with lips slightly parted (Glander 1980). Such behaviour, like the *flehmen* of male ungulates, might serve to transport urine to the nasopalatine ducts and vomeronasal system, al-though direct proof of this is lacking. Likewise, in the bald uakari, urine sampling occurs and males 'catch urine in their hands...in order to smell and taste it' (Fontaine 1981). Visual inspection of the female genitalia also commonly precedes copulation and in those catarrhine species in which females have sexual skin swellings such inspections may be particularly frequent and obvious (see Figs 5.1, 5.6, 5.14, and Table 5.3, for examples).

As regards sexual initiating behaviour, males sometimes utilize the same displays as proceptive females, particularly where facial communication is concerned. Eye-contact is a mutually important form of precopulatory behaviour and males, as well as

Fig. 5.15 Adult male rhesus monkey (*Macaca mulatta*) engages a female in eye-contact and invites copulation with a 'jaw-thrust' display. (After Hrdy and Whitten 1987.)

females, may seek to establish eye-contact before copulation. Male common marmosets stare at fe-males and tongue-flick as an invitational display (Kendrick and Dixson 1984a). Male squirrel mon-keys look intently at the female, pace back and forth, and exhibit peculiar 'silent vocalizations' with their mouths open and abdominal muscles flexed, yet without any sound being produced (Baldwin and Baldwin 1981). Male chacma baboons sometimes return the distinctive 'eye–face' proceptive expres-sion made by females as a sexual invitation (Saayman 1970). Male chimpanzees gaze directly at females during their precopulatory displays (Goodall 1986) and male rhesus monkeys attempt to make eye-contact before inviting copulation with a distinc-tive 'jaw thrust' display (Fig. 5.15).

In a number of anthropoid species, males use different facial expressions to invite copulation from those which form part of the female's proceptive repertoire. It is interesting to consider how such sex differences in precopulatory displays might have evolved. In the mandrill, for example, adult males invite copulation by approaching and following the female, inspecting her sexual skin, and by 'grinning' at her (Fig. 5.16). The grin face of the adult male is complex; the lips are closed or slightly parted at the

Fig. 5.14 An adult male talapoin monkey (*Miopithecus talapoin*) looks at the sexual skin swelling of an adult female before mounting her. (After Dixson et al. 1975.)

Fig. 5.16 The distinctive 'grin-face' of the adult male mandrill (*Mandrillus sphinx*) is shown on the left. On the right, an adult male follows a female, grins, and looks at her sexual skin. These behavioural patterns commonly precede copulation in the mandrill.

centre of the mouth but raised at the mouth corners. The male looks intently at the female whilst grinning and may also shake his head from side to side, erect the nuchal crest, and make lipsmacking movements. Adult males also display towards one another in this fashion, the grin face and its associated display elements serving to 'reassure' the participants and to reduce the likelihood that aggression will occur between them. It is likely that its use in precopulatory contexts is a secondary specialization; the massive male reassures the female by performing a non-aggressive, affiliative facial display. Female mandrills, by contrast, do not use the grin face to invite copulation and rarely exhibit the more complicated elements of the males' displays, such as head-shakes or piloerection. Other less striking ex-

Fig. 5.17 Affiliative, distance-reducing, facial displays which are also used invite copulation. Left: 'Grin-face' of a male talapoin (*Miopithecus talapoin*). Right: 'High-grin' of a male black ape of Celebes (*Macaca nigra*). (After Dixson *et al*. 1975; Dixson 1977.)

amples of this phenomenon (for instance, the use of aggression-reducing displays by males in sexual contexts) include the 'high grin' of the black ape of Celebes and the 'grin face' of the talapoin (Fig. 5.17).

Copulatory behaviour in male primates
The evolution of copulatory postures

Copulation usually occurs in a dorso–ventral position in primate species, as is the case for other mammals. Since a hallmark of primate evolution was the development of grasping hands and feet, it is not surprising that males often hold females in various ways during copulation. The male often grasps the female's ankles or calves with his feet; the *double-foot clasp* mount of many Old World monkeys such as baboons, macaques, mangabeys,

and talapoins. (Fig. 5.18). In species where the male is much larger than the female, as in *Theropithecus* and *Mandrillus*, for example, the male's feet usually remain on the ground during copulation. Interestingly, sub-adult or juvenile male mandrills (*M. sphinx*) commonly use the typical double-foot clasp position, whereas adult males rarely do so. Likewise some species, such as *Erythrocebus patas* or *Presbytis entellus* may exhibit either foot clasp or non-foot clasp copulatory postures. The pattern is not obligatory, therefore, but is probably an ancient phenomenon; foot clasp responses may occasionally be observed in some prosimians (e.g., *Galago* spp.) as well as in some New World monkeys (e.g., *Cebuella pygmaea*: Soini 1988; *Callithrix jacchus*; Kendrick and Dixson 1984a; *Cebus capucinus*: Freese and Oppenheimer 1981; *Saimiri sciureus*: Baldwin and Baldwin 1981). It is common also for the male to

Fig. 5.18 Foot clasp mounting during copulation in some representative Old World monkeys. Upper left: Talapoin (*Miopithecus talapoin*: After Dixson *et al.* 1975). Upper right: Black ape of Celebes (*Macaca nigra*: After Dixson 1977). Lower left: Rhesus macaque (*Macaca mulatta*: After Mason 1960) Lower right: an isolation-reared male rhesus macaque fails to mount in the correct, double-foot clasp position. (After Mason 1960.)

Fig. 5.19 Copulatory posture of Garnett's galago (*Galago garnettii*). The male encircles the female's waist with his arms and bites her gently on the back.

grasp the female's hips with his hands during mounting or to encircle her waist with his arms. Male bushbabies, marmosets, tamarins, or owl monkeys often place their arms around the female's waist during mating. In the case of *Galago garnettii*, the male locks his fingers together and thereby exerts a strong hold over the female during prolonged intromissions (Fig. 5.19). Palpating movements, such as occur when the male rat rapidly stimulates the flanks of the receptive female with his forepaws during mounting (Pfaff 1980) have not been recorded in primates.

Although dorso–ventral mounting postures are relatively simple and stereotyped in primates, two variations of phylogenetic importance must be mentioned here. Firstly, among New World monkeys of the sub-family Atelinae a remarkable 'leg-lock' copulatory posture occurs, during which the male places his legs over the female's thighs (Fig. 5.20). This unusual position was first described for *Ateles belzebuth* and *A. geoffroyi* by Klein (1971). Subsequently, Eisenberg (1976) described the same copulatory posture for captive *A. fusciceps* and *Lagothrix lagotricha*. Field studies on *Brachyteles arachnoides* revealed that these extraordinary monkeys also utilize a leg-lock position during copulation (Milton 1985a). In all of these ateline species, both sexes also use the prehensile tail to anchor themselves to branches during copulations. The leg-lock posture has not evolved solely as an accompaniment to the prehensile tail, however; males in two other New World sub families (Alouattinae and Cebinae) also possess prehensile tails but do not exhibit leg-locking during copulation. Rather, it is the long duration of copulation in the sub-family Atelinae, where partners remain mounted for 6–35 min in *Ateles* (Klein 1971; Eisenberg 1976), up to 18 min in *Brachyteles* (Strier 1986), and 3–14 min in *Lagothrix* (Nishimura 1988),

Fig. 5.20 'Leg-lock' copulatory postures in free-ranging spider monkeys (*Ateles sp.*) . (After Klein 1971.)

which distinguishes this family. Intromissions are also of long duration in these monkeys, whereas in *Cebus* and *Alouatta* mounts and intromissions are usually quite brief. In the Atelinae, the evolution of the leg-lock allows the male to sit snugly behind the female, holding her in position and using his prehensile tail and feet to maintain a grip on the branches.

Turning now to the Old World, a second most interesting variant of the dorso–ventral copulatory position occurs in the prosimian subfamily Lorisinae, which comprises four genera (*Perodicticus* and *Arctocebus* in Africa, and *Loris* and *Nycticebus* in S.E. Asia). These slow-moving nocturnal prosimians are cryptic in their habits. In the potto (*Perodicticus potto*) copulations normally occur with the male mounted upon the female in the usual way but with both partners suspended, in an inverted position, beneath a branch. The same is the case in the slow loris (*Nycticebus coucang*) and slender loris (*Loris tardigradus*: Fig. 5.21), at least in captivity, and it is

Fig. 5.21 Inverted copulatory posture of the slender loris (*Loris tardigradus*). During copulation the male sometimes makes rapid, side-to-side head movements. (After Schulze and Meier 1995.)

Fig. 5.22 Aye-ayes (*Daubentonia madagascariensis*) copulating. This inverted posture has been observed occasionally in captive specimens; how frequently it may occur in the wild is not known. (After Winn 1994.)

probably also true of the angwantibo. The posture is probably phylogenetically ancient and was present in the stem forms from which the modern day Lorisinae are derived. Possibly, it is a protective behaviour which aids concealment of the mating pair, and allows them to drop to the ground if disturbed. Pottos, for instance, are known to react in this way when confronted by certain predators, although such behaviour has not been observed during copulation (Charles-Dominique 1977). The only other primate species which is known to copulate in an inverted position is the aye-aye (*Daubentonia madagascariensis*). Aye-ayes have occasionally been observed to adopt this position when mating in captivity (Fig. 5.22).

Whilst dorso–ventral copulatory postures are the norm among primates, ventro–ventral copulations have been observed among various hominoid species. A ventro–ventral copulatory posture has been described in free-ranging siamangs (*Symphalagus syndactylus*: Koyama 1971) although dorso–ventral copulations are probably usual for this species and for hylobatids in general. The orang-utan (*Pongo pygmaeus*) may copulate in either a dorso–ventral or ventro–ventral position in captivity (Nadler 1977) or in the wild, where a variety of suspensory positions occur in association with its highly versatile quadrumanous climbing behaviour (Fig. 5.23). Male orangs sometimes recline on their backs with the penis erect to invite copulation. The female then initiates intromission by sitting in the male's lap, facing towards him (Fig. 5.24). In the chimpanzee (*Pan troglodytes*) a common feature of precopulatory behaviour involves the male's penile display; females normally back towards the male, however, and the

copulatory posture is invariably a modified dorso–ventral one (Fig. 5.25), during which the male continues to sit upright with his thighs apart (Tutin 1979; Goodall 1986). The bonobo (*Pan paniscus*) is more versatile in its copulatory behaviour than the larger chimpanzee. During their field studies of bonobos in the Lomako forest, Zaire, Thompson-

Fig. 5.23 Sumatran orang-utans (*Pongo pygmaeus abelii*) copulating. The male (on the right) is reclining on his back as the female suspends herself above him. (After Rijksen 1978.)

Handler *et al.* (1984) observed 70 copulations, 26% of which were ventro–ventral and the remainder of the typical dorso–ventral orientation (Fig. 5.26). The ventro–ventral position was used more frequently by sub-adults rather than by adult males. This confirmed previous reports of frequent ventro–ventral copulations, often involving eye-contact between the partners, of a captive sub-adult male bonobo caged with two females at Yerkes Primate Centre (Savage and Bakeman 1978). The mountain gorilla (*Gorilla gorilla beringei*) copulates in a dorso–ventral position in the wild; all 69 copulations observed by Harcourt *et al.* (1981b) were of this type, 'although twice one pair briefly assumed the ventro–ventral position before turning to copulate normally'. When western lowland gorillas (*G. g. gorilla*) were observed during free-access pair-tests in captivity, several pairs exhibited ventro–ventral copulatory positions (Fig. 5.27), although 80% of copulations were of the dorso–ventral type (Nadler 1976, 1980). In discussing his findings, Ronald Nadler drew attention to the early work of Harold Bingham (1928) on 'sex development in apes'. Bingham had argued that the increased encephalization and intelligence of apes might be associated with less stereotyped patterns of behaviour and a closer resemblance to human patterns of sexual behaviour. In the light of modern evidence on the sexual behaviour of captive and free-ranging apes, Bingham's suggestion appears to be correct.

Although copulatory postures in man can be exceedingly variable, they are derived from basic dorso–ventral or ventro–ventral positions. There are valid comparisons to be made between human copulatory postures and those displayed by other hominoids. Ford and Beach (1952) drew attention to the occurrence of face-to-face copulatory postures in the majority of human societies. Variations of this posture include the typical *missionary position*, with the woman lying on her back and the man on top, employed by the great majority of American couples (Kinsey *et al.* 1948) as well as by the Alorese, Balinese, Chagga, Crow, Hopi, Lepcha, Marshallese, Pukapukans, Tikopia, Trobrianders, Trukese, and many others. However, variations occur in which the man squats or kneels before the woman 'and draws her towards him so that her legs straddle his thighs' (e.g., in the Balinese, Lepcha, Trobrianders, and Trukese). The man may then pull the woman up so that they embrace face to face. The woman may also sit upon the man as he reclines on his back. (Fig. 5.28) This pattern was the most common alternative to the missionary position among American couples recorded by Kinsey *et al.* (1948). It was noted as a secondary position for various societies (e.g., Crow, Hopi, Pukapukans, Trobrianders, Trukese, etc.) by Ford and Beach (1952) in their cross-cultural survey of human sexual behaviour. Ford and Beach pointed out that variants of the ventro–ventral copulatory posture might enhance the tactile stimulation of the clitoris required for female orgasm, whereas orgasmic responses are 'rare or absent in female apes or monkeys because their (dorso–ventral) method of mating deprives them of the very important source

Fig. 5.24 Penile display and copulation in the orang-utan (*Pongo pygmaeus*). Left: The male displays his erect penis whilst reclining on his back. Right: The female mounts the male in response to his display. (After Nadler 1988.)

of erotic stimulation'. This hypothesis, which it should be emphasized Ford and Beach regarded as 'highly speculative', must be abandoned in the light of a wealth of data on primate sexual behaviour

Fig. 5.25 Upper: Pre-copulatory penile display of a male chimpanzee. Middle: Copulation occurs after the female backs towards the male in response to his displays. Lower: Ejaculation has occurred and the female terminates the copulation by leaping away from the male. (After McGinnis 1973.)

obtained since their pioneering studies. We have seen that ventro–ventral postures occur in some apes in the wild as well as in captivity. Such postures probably also occurred in prehuman ancestors. Most importantly, 'orgasmic' responses are not confined to man and have now been described in a variety of monkeys and apes (see page 129). Nor are dorso–ventral copulatory postures absent in the human species. Ford and Beach recorded the occasional occurrence, in 12 societies they sampled, of intercourse with the couple lying side by side and the 'man entering the woman from the rear'. Moreover in eight societies (Crow, Dobuans, Hopi, Kurtatchi, Kuroma, Lepcha, Marshallese, and Wogeo) full dorso–ventral copulation occurred as an occasional position, 'with the man standing behind the woman while she bends over or rests on her hands and knees'. Although Ford and Beach state that this type of dorso–ventral position 'is apparently confined to brief and sudden encounters in the woods', such postures are frequently depicted in the literature on human sexual behaviour. Just how frequent dorso–ventral copulatory postures are in various western cultures is not known.

The evolution of intromission and ejaculatory patterns

Patterns of copulatory behaviour can be classified according to the presence or absence of particular behavioural components. Dewsbury (1972) has devised a schema in which mammalian copulatory patterns are divided into 16 possible types depending on the presence or absence of four components: genital locks, pelvic thrusting, multiple intromissions, and multiple ejaculations (Fig. 5.29) To apply this schema the first decision to be made is whether a species exhibits a mechanical tie or *lock* of the genitalia during intromission. Dewsbury (1972) described such a pattern in 17 out of 50 mammalian species (34%) for which data were available. It is not an uncommon pattern therefore, although 10 of 17 species in which it was identified belong to one mammalian order, the Carnivora. Among carnivores, dogs exhibit a pronounced genital lock during copulation, brought about by marked tumescence of the male's glans penis (Beach 1970; Hart and Kitchell 1966). Genital locking also occurs during copulation in some rodents, such as the Australian hopping mouse (*Notomys alexis*: Dewsbury and Hodges 1987), in at least one bat species (*Myotis nigricans*), and in the marsupial, *Antechinus flavipes* (Marlow 1961). The occurrence of this pattern is not simply a reflection of phylogenetic factors, therefore, and nor are the mechanisms which secure genital locking between the sexes the same in every case. In some rodents, for instance, the lock is correlated with the presence of a broad glans penis in the male, whereas in *Notomys alexis* the glans is elongated and is armed with very large penile spines which are responsible for anchoring the penis within the vagina (Breed 1981).

The second decision to be made in Dewsbury's schema is whether males make intravaginal, pelvic thrusting movements once intromission has occurred. In the rat and rabbit, males do not thrust during the very brief period of intromission (Larsson 1956; Rubin 1967) whereas in many mammals, including all primates so far studied, intromitted thrusting is the norm. The third decision concerns whether one or more (multiple) intromissions are required to attain ejaculation. In many mammals, a single intromission of variable duration is sufficient for the male to ejaculate. This is the case among the carnivores, perissodactyls, and artiodactyls listed by Dewsbury (1972) as well as for many rodents, primates, and marsupial species. However, multiple intromission patterns occur in a variety of mammals, such as in the rat where males typically intromit a number of times and ejaculate on the final mount of the series. Multiple intromission patterns of copulation also occur in some primates, including various macaque and baboon species.

The final decision used by Dewsbury (1972) in classifying mammalian copulatory patterns concerns the presence or absence of *multiple ejaculations*. Some species cease to copulate for an extended

Fig. 5.26 Upper: An adult male bonobo (*Pan paniscus*) makes a precopulatory penile display towards a female with a fully swollen sexual skin. Lower: The female mounts the male in a ventro–ventral position and copulation occurs. (Photographs by Dr C. Schmidt.)

period after ejaculation, whereas others are capable of a second ejaculation. It is this component of copulatory behaviour, resulting in the division of eight potential patterns into sixteen (Fig. 5.29), which is particularly difficult to define. As stated earlier, males of most species exhibit a post-ejaculatory interval or refractory period during which they are sexually quiescent. However, the duration of the post-ejaculatory interval can vary considerably, even within a single species (see Table 5.2). In the laboratory rat, the interval can be shortened by external stimuli that enhance male arousal (e.g., provision of a new female partner, handling the male, administration of electric shocks etc.: Sachs and Barfield 1976). Whilst the majority of primates studied fulfil the criteria for a multiple ejaculation pattern of copulation (Dixson 1983c; Dewsbury and Pierce

1989) there is reason to believe that this oversimplifies the very real interspecific differences in ejaculatory frequencies which occur under natural conditions. The following section details a modified version of Dewsbury's scheme which is more applicable to studies of copulatory behaviour in primates. I shall attempt to classify the patterns of all species for which sufficient data are currently available and then to make inferences concerning their functional and evolutionary significance.

Classifying primate copulatory patterns

The first problem in applying Dewsbury's schema to studies of primate copulatory behaviour concerns the single intromission copulatory pattern. Intromissions vary considerably in duration between species. In the majority of cases intromissions are of brief duration. For example, in free-ranging chimpanzees (*Pan troglodytes*), copulation involves an average of just 8.8 pelvic thrusts and lasts for approximately 7 s (Tutin and McGinnis 1981). A comparable pattern occurs in the bonobo (*P. paniscus*: (mean duration 12.2 s: Thompson-Handler et al. 1984), whilst in the gorilla it is slightly longer (*G. g. beringei*: mean duration 96 s and 27.5 thrusts: Harcourt *et al.* 1981b). In captive common marmosets, intromissions are exceedingly brief, averaging 5.2 s and with only a few pelvic thrusts (Dixson 1991a). These data, and others listed in Table 5.4, should be compared with those available for certain single intromission copulators in which durations of intromission are pro-

Fig. 5.27 Dorso–ventral and ventro–ventral copulatory postures in captive western lowland gorillas (*Gorilla g. gorilla*). (After Dixson 1981.)

Fig. 5.28 Female superior copulatory posture in *Homo sapiens*. (After Bancroft 1989.)

longed. Examples include the greater galago (*Galago garnettii*), in which intromissions lasting more than 2 h have been recorded (Eaton *et al.* 1973a), or various New World members of the sub-family Atelinae (e.g., *Brachyteles arachnoides*: intromissions range from 1 min 40 s to 8 min 18 s: Milton 1985a), or the orang-utan (*Pongo pygmaeus pygmaeus*: maximum copulatory duration of 46 min; median = 14 min: Nadler 1977). In all these examples, patterns of intromitted pelvic thrusting differ from those reported for brief, single-intromission copulations. For example, in *G. garnettii* and *P. pygmaeus* bouts of pelvic thrusting are interspersed with longer periods of rest, whilst in *B. arachnoides* the pair remain immobile once intromission occurs and then the male makes 5–10 hard thrusting movements during the final 'consummatory phase' of copulation (Milton 1985a). It seems likely that functional differences exist between species which share a single intromission pattern, but in which the duration of copulation is either relatively brief (typically <60 s) or prolonged (>3 min). For this reason I have separated the two types (Table 5.4) rather than adhering rigidly to Dewsbury's scheme.

A second problem concerns the possible occurrence of a genital lock during copulation in certain primate species. Dewsbury and Pierce (1989) conclude that 'functional locking does not occur in primates'. However, a number of authors have observed that male and female galagos become 'stuck

together' during copulation (*Galago demidoff*: Charles-Dominique 1977. *G. garnettii*: personal observation). Likewise, in the stumptail macaque (*Macaca arctoides*) the male can become 'tied' to the female during copulation and is dragged along if she moves off (Lemmon and Oakes 1967). Such effects do not always occur, however, and normally the pair sit quietly for a short period after ejaculation and can separate relatively easily during this time (Goldfoot *et al.* 1975). In my view such cases should be considered as potentially locking species to distinguish them from those primates where no evidence of a tie or lock exists.

A final problem, as mentioned previously, concerns the occurrence of single vs multiple ejaculatory patterns in primates. The vast majority of primates for which data exist are multiple ejaculators, in the sense that more than one ejaculation per hour is possible. Dewsbury and Pierce (1989) list only four out of 33 primate species for which they could locate adequate data which they considered to have a single ejaculation pattern; (*Galago crassicaudatus, Tarsius bancanus, Cebus nigrivittatus,* and *Cercopithecus aethiops*). Even these examples are of questionable validity. Under laboratory conditions male vervets (*C. aethiops pygerythrus*) can ejaculate more than once per hour (Girolami 1989). During prolonged intromissions male greater galagos (*G. crassicaudatus* and *G. garnettii*) exhibit periodic bouts of pelvic thrusting and it is possible that multiple

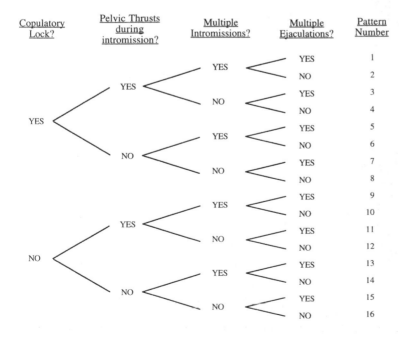

Copulatory Lock?	Pelvic Thrusts during intromission?	Multiple Intromissions?	Multiple Ejaculations?	Pattern Number
YES	YES	YES	YES	1
			NO	2
		NO	YES	3
			NO	4
	NO	YES	YES	5
			NO	6
		NO	YES	7
			NO	8
NO	YES	YES	YES	9
			NO	10
		NO	YES	11
			NO	12
	NO	YES	YES	13
			NO	14
		NO	YES	15
			NO	16

Fig. 5.29 Dewbury's schema for classifying masculine patterns of copulatory behaviour in mammals. (Redrawn from Dewsbury 1972.)

ejaculations occur (Dixson 1995a). The criterion for a multiple ejaculatory pattern as presently defined therefore has little value when applied to primate copulatory behaviour. More useful, from a functional standpoint, is knowledge of how frequently males ejaculate with fertile females under natural conditions. These data are useful for assessing possible effects of sperm competition. In Chapter 8 information on ejaculatory frequencies in primates is discussed with reference to mating systems and

relative testes sizes. The indications are that in species where females copulate with a single male (e.g., the monogamous siamang or polygynous gorilla) ejaculatory frequencies are lower than in multimale–multifemale systems (e.g. in chimpanzees and stumptail macaques) where females mate with multiple partners and males have large testes relative to their body weights (see Fig. 8.18, page 241).

With these caveats in mind, it is possible to modify Dewsbury's original schema to classify masculine

Table 5.4 Intromission durations in primates: distinguishing between single brief intromission (SBI) and single prolonged intromission (SPI) copulatory patterns

Species	Intromission duration (s)	Copulatory pattern	Sources
Tarsius bancanus	60–90	SBI	Wright *et al.* 1986
Cebuella pygmaea	4–10	SBI	Soini 1988
Callithrix jacchus	Mean 5 or 6	SBI	Dixson 1986a
Alouatta palliata	Mean 32	SBI	Carpenter 1934
Saimiri sciureus[a]	<30	SBI[a]	Baldwin and Baldwin 1981
Nasalis larvatus	Mean 41.5	SBI	Yeager 1990b
Miopithecus talapoin	<60	SBI	Dixson *et al.* 1975
Cercopithecus aethiops	<60	SBI	Gartlan 1969
Cercocebus albigena	Mean 13.9	SBI	Wallis 1983
Mandrillus sphinx	<60	SBI	AFD
Macaca radiata	<60	SBI	Glick 1980
	Mean 10		Shively *et al.* 1982
Pan troglodytes	Mean 7.0	SBI	Tutin and McGinnis 1981
	Mean 8.21		Hasegawa and Hiraiwa Hasegawa 1990
Pan paniscus	Mean 15.3	SBI	Kano 1992
Gorilla gorilla	med 96	SBI	Harcourt *et al.* 1981b
*Homo sapiens***	<120	SBI?	Kinsey *et al.* 1948
Microcebus murinus	>180	SPI	Martin 1973
Daubentonia madagascariensis	Mean 3720	SPI	Sterling and Richard 1995
Galago garnettii	780–15 600	SPI	Eaton *et al.* 1973a
G. moholi	420–720	SPI	Dixson 1989
Galagoides demidoff	3600	SPI	Charles-Dominique 1977
Arctocebus calabarensis	240[b]	SPI	Dixson 1989
Loris tardigradus	600–1020	SPI	Dixson 1989
Nycticebus coucang	180–420	SPI	Dixson 1989
Brachyteles arachnoides	100–496	SPI	Milton 1985a
Ateles belzebuth	480–1500	SPI	Klein 1971
A. paniscus	Mean 600	SPI	Van Roosmalen and Klein 1988
Lagothrix lagotricha	180–840	SPI	Nishimura 1988
Macaca arctoides	>180	SPI	Goldfoot *et al.* 1975
Pongo pygmaeus	med 840 max 2760	SPI	Nadler 1977

[a] Multiple intromissions also occur in *saimiri sciureus*.
[b] One observed copulation only.
med=median duration; max=maximum duration.
AFD=author's unpublished observations.
** The majority of men questioned by Kinsey *et al.* achieved ejaculation in less than 2 min during intercourse. This raises the possibility that a relatively brief (SBI) copulatory pattern is normal for human beings and that more prolonged copulation is a result of cultural evolution rather than of natural or sexual selection.

copulatory patterns on the basis of four criteria: 1, the presence or absence of a genital lock; 2, the presence or absence of pelvic thrusting during intromission; 3, the occurrence of single or multiple intromissions; and 4, whether or not there are prolonged intromission(s) (Fig. 5.30). Again, sixteen possible copulatory patterns emerge from this new classification, but no judgement is made about the presence or absence of multiple ejaculations. In practice, only four copulatory patterns are known to occur among primates; species that show these patterns are listed in Table 5.5. All species make pelvic thrusting movements during intromission. Among the few locking (or potentially locking species), a single prolonged intromission is the norm (pattern No. 3, for example, greater galago). Single prolonged intromissions do occur in a number of species which lack a genital lock (pattern No. 11: for example, woolly spider monkey). In the remaining two patterns, which also lack a genital lock, multiple brief intromissions (pattern No. 10: for example, rhesus macaque) or single brief intromissions (pattern No. 12: for example, chimpanzee) culminate in ejaculation. As can be seen in Table 5.5, the most widespread pattern among primates is No. 12. A single intromission with pelvic thrusting leads to ejaculation, the intromission is relatively brief (<3 min), and it is terminated promptly once ejaculation

has occurred. Species exhibiting this copulatory pattern include prosimians (e.g., *Lemur catta* and *Tarsius bancanus*), New World monkeys (e.g., *Alouatta palliata* and *Callithrix jacchus*), Old World monkeys (e.g., *Macaca radiata* and *Nasalis larvatus*), and apes (e.g., both species of the genus *Pan* as well as *Gorilla gorilla*). Phylogenetic factors alone do not limit the distribution of this pattern and it occurs in species with monogamous mating systems, polygynous one-male units and in some multimale–multifemale societies. This might represent a primitive pattern of copulation for *diurnal* primates.

More complex forms of copulation involving multiple, brief intromissions (MBI pattern) or a single intromission which is prolonged and maintained beyond the time of ejaculation (SPI pattern) have now been reliably documented for 28 primate species (Table 5.6). Again, the sample comprises prosimians, New and Old World monkeys, and one hominoid (*Pongo*), indicating that phylogenetic factors alone do not limit the distribution of MBI or SPI copulatory patterns. The question arises as to why these more complex forms of copulatory behaviour have developed during the course of evolution? One argument is that sexual selection has encouraged the development of these patterns in some primates with multimale–multifemale or dispersed (nongregarious) mating systems (Dixson 1987a, 1991b).

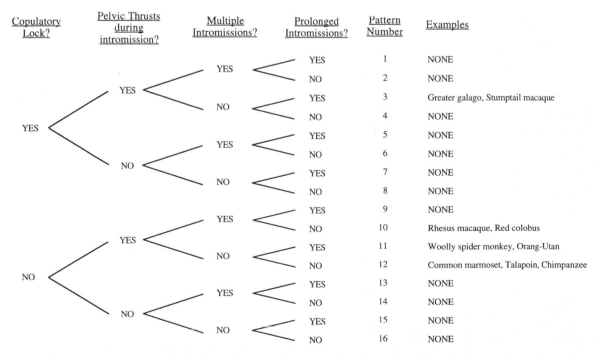

Fig. 5.30 A modified version of Dewsbury's (1972) schema which is applicable to the classification of masculine copulatory patterns in primates.

Under these conditions, in which females mate with more than one male, sexual selection has favoured the evolution of more complex penile morphologies (Dixson 1987a) and larger testes relative to body weight (due to sperm competition: Harcourt *et al.* 1981a). It is possible that complex copulatory patterns might also have evolved more readily in these circumstances. A second possibility is that natural selection might also restrict the evolution of copulatory patterns. Certain types of copulatory behaviour might be more common in diurnal versus nocturnal primates or terrestrial versus arboreal forms, due to predation pressures or other selective forces. Some of these possibilities are considered in Table 5.6 which provides information on mating systems, adult male body weights, activity patterns, and habitat preferences for the 28 primate species which are known to exhibit MBI or SPI copulatory patterns. Twenty six (93%) of these species have either a multimale–multifemale or a non-gregarious mating

Table 5.5 Classification of primate copulatory patterns, according to the criteria laid down in Fig. 5.30

Pattern no. 3	10	11	12
Possible genital lock: a single prolonged intromission and pelvic thrusts	No genital lock: multiple brief intromissions and pelvic thrusts	No genital lock: a single prolonged intromission and pelvic thrusts	No genital lock: a single brief intromission and pelvic thrusts
Prosimians			
Galago garnettii *G. crassicaudatus* *Galagoides demidoff* *Daubentonia madagascariensis*	None described	*Microcebus murinus* *Galago moholi* *Arctocebus calabarensis* *Loris tardigradus* *Nycticebus coucang*	*Lemur catta* *Varecia variegata* *Cheirogaleus major* *Tarsius bancanus*
New World monkeys			
None described	*Cacajao calvus* *Saimiri sciureus** *Leontopithecus rosalia*	*Lagothrix lagotricha* *Ateles belzebuth* *A. geoffroyi* *A. fusciceps* *Brachyteles arachnoides*	*Callithrix jacchus* *Saguinus oedipus* *Cebuella pygmaea* *Callimico goeldii* *Callicebus moloch* *Aotus griseimembra* *Cebus nigrivittatus* *Alouatta palliata*
Old world monkeys			
Macaca arctoides	*Colobus badius* *Macaca silenus* *M. nemestrina* *M. nigra* *M. maurus* *M. fascicularis** *M. mulatta* *M. fuscata* *M. thibetana* *Papio ursinus** *P. hamadryas*	None described	*Nasalis larvatus* *Miopithecus talapoin* *Erythrocebus patas* *Cercopithecus aethiops* *Mandrillus sphinx* *Cercocebus albigena* *Theropithecus gelada* *Papio anubis* *Macaca radiata* *M. sylvanus*
The apes and man			
None described	None described	*Pongo pygmaeus*	*Pan troglodytes* *P. paniscus* *Gorilla gorilla* *Homo sapiens***

*Pattern no. 12 also occurs in these species.
**Although human copulation may be prolonged (pattern no. 11) it is possible that the fundamental pattern involves a relatively brief single intromission of <2 min (See Table 5.4).
Data are from Dixson 1983c, 1991b; Dewsbury and Pierce 1989; and references quoted in the text.

system. The data therefore support the hypothesis that complex copulatory patterns occur more frequently in those primate species where females mate with a number of partners. An important caveat to this sexual selection hypothesis is that it *does not predict* that complex copulatory patterns (i.e., MBI and SPI patterns) will occur in *all primates* with a multimale–multifemale or a non-gregarious mating system. The occurrence of precopulatory competi-

tion between males, sexual preferences, and female choice before mating might all limit male access to females for copulatory purposes. In some cases, sexual selection might have acted primarily at the gonadal level or on penile morphology, rather than upon the intromission pattern. Therefore, it is not valid to cite the absence of complex copulatory patterns in certain multimale–multifemale primate species as evidence that sexual selection has not

Table 5.6 Occurrence of multiple brief intromission and single prolonged intromission copulatory patterns with respect to mating systems, adult male weight, activity patterns, and major habitat preferences in primates

Species	Mating system	Copulatory pattern	Male body weight (kg)	Activity pattern	Habitat preference	Sources
Prosimians						
Galago crassicaudatus	D	SPI	1.29	noct.	arb.	Eaton *et al.* 1973a
Galago garnettii	D	SPI	0.82	noct.	arb.	Eaton *et al.* 1973a; personal observation
Galago moholi	D	SPI	0.23	noct.	arb.	Dixson 1987a
Galagoides demidoff	D	SPI	0.07	noct.	arb.	Charles-Dominique 1977
Loris tardigradus	D	SPI	0.28	noct.	arb.	Dixson 1987a
Nycticebus coucang	D	SPI	0.73	noct.	arb.	Dixson 1987a
Arctocebus calabarensis	D	SPI	0.26	noct.	arb.	Dixson 1987a
New World monkeys						
Cacajao calvus	MM	MBI	4.10	diur.	arb.	Fontaine 1981
Saimiri sciureus	MM	MBI*	0.90	diur.	arb.	Latta *et al.* 1967; Wilson 1977a
Lagothrix lagotricha	Mm	SPI	6.67	diur.	arb.	Eisenberg 1976
Ateles belzebuth	MM	SPI	8.46	diur.	arb.	Klein 1971
Ateles geoffroyi	MM	SPI	7.48	diur.	arb.	Klein 1971
Ateles fusciceps	MM	SPI	8.89	diur.	arb.	Eisenberg 1976
Brachyteles arachnoides	MM	SPI	15.0	diur.	arb.	Milton 1985a
Leontopithecus rosalia	MP	MBI	0.56	diur.	arb.	Kleiman 1978
Old World monkeys						
Colobus badius	MM	MBI	8.38	diur.	arb.	Struhsaker and Leland 1979
Macaca silenus	MM	MBI	6.80	diur.	terr./arb.	Lindburg *et al.* 1985
Macaca nemestrina	MM	MBI	10.40	diur.	arb./terr.	Eaton and Resko 1974a
Macaca nigra	MM	MBI	10.40	diur.	terr./arb.	Dixson 1977
Macaca maurus	MM	MBI	9.50	diur.	terr./arb.	Matsumura 1993
Macaca fascicularis	MM	MBI*	5.90	diur.	arb./terr.	Shively *et al.* 1982
Macaca mulatta	MM	MBI	10.40	diur.	terr./arb.	Carpenter 1942
Macaca fuscata	MM	MBI	11.70	diur.	terr./arb.	Hanby *et al.* 1971
Macaca arctoides	MM	SPI	10.10	diur.	terr./arb.	Goldfoot *et al.* 1975
Macaca thibetana	MM	MBI	19.20	diur.	terr./arb.	Zhao 1993
Papio ursinus	MM	MBI*	28.26	diur.	terr.	Hall and DeVore 1965
Papio hamadryas	PG	MBI	18.30	diur.	terr.	Kummer 1968
Apes						
Pongo pygmaeus	D	SPI	73.40	diur.	arb.	Nadler 1977

Updated from Dixson 1991b.
D=Dispersed (non-gregarious); MM=multimale–multifemale mating system; PG=polygynous system; MP=monogamous pairs.
PI=maintenance of intromission for more than 3 min or during the post-ejaculatory phase.
* Single brief intromission patterns also occur in these species.
Body weights are from Harvey *et al.* (1987) and Martin and Dixson (in preparation).
noct.=nocturnal; diur.=diurnal.
arb.=arboreal; terr.=terrestrial

affected these behavioural patterns in others (Dewsbury and Pierce 1989). More crucial are observations of complex copulatory patterns in monogamous or polygynous species, since under these conditions females mate with a single partner and there should be little selection pressure for the development of complex copulatory behaviour. Examples of this type are few in number and of questionable accuracy. Dewsbury and Pierce (1989) quote only two: *Colobus guereza* (polygynous; MBI pattern: Hollihn 1973) and *Cercopithecus ascanius* (polygynous; MBI pattern: Struhsaker and Leland 1979). The problem in such cases is to assess the reliability of information on copulation, culled from reports which were not primarily intended to deal with sexual behaviour. In neither of the two examples quoted above was the occurrence of intromission recorded, so that the classification of an MBI pattern is dubious at best. The above paper by Hollihn (1973) has also been cited as evidence that an MBI pattern occurs in *Nasalis larvatus*, the proboscis monkey of Borneo (Dewsbury and Pierce 1989), yet in this case subsequent fieldwork has shown that this species, which lives in large troops consisting of polygynous one-male units (Yeager 1990a), has a single, brief intromission, copulatory pattern, as expected (Yeager 1990b).

Two exceptions to the sexual selection hypothesis do occur in Table 5.6; *Leontopithecus rosalia* has an MBI copulatory pattern and is thought to be monogamous (Kleiman 1978) whilst *Papio hamadryas* also exhibits an MBI pattern but is polygynous (Kummer 1968). Much recent information deals with the oc-

currence of facultative polyandry in callitrichids (Sussman and Garber 1987). However, *Leontopithecus* is the only callitrichid reported to have an MBI pattern of copulation, and at present there is no explanation for this difference. In the hamadryas baboon (*P. hamadryas*), one-male breeding units are banded together into very large troops. Certain specializations of the genitalia (female sexual skin and male genital morphology) indicate that polygyny in *P. hamadryas* has been secondarily derived from a multimale–multifemale mating system (Dixson 1983b; Kummer 1990). The multiple intromission copulatory pattern might also have originated in multimale–multifemale ancestral forms similar to those of some present day baboon species (e.g., *Papio ursinus* in Table 5.6).

Little attention has been given to the possible influence of natural selection rather than sexual selection upon the evolution of copulatory patterns in primates. In Table 5.6 it is evident that both arboreal and terrestrial primates exhibit complex copulatory patterns and that nocturnal and diurnal forms are represented in the sample. However, one useful correlation can be identified, concerning body weight, ecology, and SPI copulatory behaviour (Fig. 5.31). Prolonged intromissions occur in some nocturnal primates; small-bodied forms ranging from 67 g (*Galago demidoff*) to 1289 g (*G. crassicaudatus*). All are arboreal and cryptic in their habits. No small-bodied diurnal primate has evolved such a pattern. Multiple intromission patterns occur in some small, arboreal, diurnal monkeys (e.g., *L. rosalia*: 560 g, *S. sciureus*: 810 g). However, only certain large-bodied

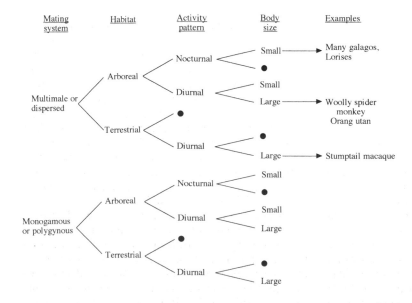

Fig. 5.31 Occurrence of single prolonged intromission (SPI) copulatory patterns in primates: interactions between the mating system and ecological factors. —• indicates that no forms exist in a given category, for example there are no nocturnal terrestrial primates. (After Dixson 1991b.)

arboreal monkeys are known to have SPI copulatory patterns and all have prehensile tails which provide additional anchorage in the trees (*Lagothrix* > 6 kg; *Ateles* > 7 kg; *Brachyteles* > 15 kg). The orang-utan, the only great ape which habitually mates in the trees, is a slow moving, quadrumanous climber. At over 70 kg in body weight, male orang-utans have little to fear from predators except man. For terrestrial monkeys, such risks are presumably much more significant. Indeed, only one primarily terrestrial monkey (*Macaca arctoides*) has an SPI pattern and it is noteworthy that intromission durations for this species are shorter than in the various arboreal monkeys which exhibit an SPI pattern. These observations lead to the hypothesis that risk of predation or some other selection process has limited the evolution of prolonged intromission patterns to cryptic nocturnal forms (e.g., *Galago*) or to large-bodied diurnal anthropoids, especially those which are arboreal.

The functional significance of complex copulatory patterns in primates has not been subjected to experimental analysis. Some insights are available from studies on other mammals. Prolonged single intromissions might serve as part of mate-guarding strategies. This suggestion applies particularly to certain of the nocturnal prosimians (e.g., *Galago* spp., *Loris* and *Arctocebus*) where the sexes are often dispersed throughout individual home ranges and copulations are limited to a circumscribed period of female receptivity or oestrus. Clearly, there are advantages for a male if he can maintain contact with a fertile female, rather than releasing her for further possible matings during the same night. Prolonged intromission might also assist sperm transport, the penis providing a mechanical means of keeping the ejaculate in contact with the os cervix; analogous to the functioning of a copulatory plug. Finally, as discussed elsewhere (see pages 318–19) prolonged intromissions might enhance neuroendocrine responses in the female partner which enhance fertility. This is an unexplored field as far as primates are concerned.

Turning now to the multiple intromission pattern, this might serve to dislodge copulatory plugs or coagulated semen from previous matings, and thus to reduce potential sperm competition. Plug removal is known to occur in laboratory rats (Wallach and Hart 1983) and has been reported in an anecdotal fashion in some primates (e.g., *Macaca nemestrina*: Busse and Estep 1984; *Lemur fulvus*: Brun et al. 1987). A second suggestion regarding the functional significance of the MBI pattern concerns the physical properties of cervical mucus and possible facilitation of sperm transport. Cervical mucous changes

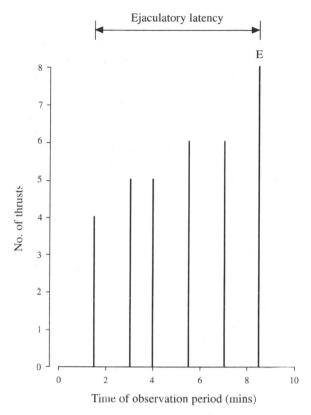

Fig. 5.32 Timing of a typical multiple mount/intromission copulatory series in the rhesus monkey (*Macaca mulatta*). The male makes a series of mounts with intromission and pelvic thrusting. The final, ejaculatory, mount of the series (E) is of longer duration and involves a greater number of thrusts. (Redrawn from Everitt and Herbert 1972.)

in physico-chemical composition during the periovulatory phase of the ovarian cycle and spermatozoa are able to penetrate the mucus at this time. Stretching of the mucus *in vitro* causes a realignment of its microstructure 'in such a way that spermatozoa swim in linear paths through the mucus interstices' (Overstreet *et al.* 1991). During intromission, the male's pelvic thrusting movements might assist this process, provided that the erect penis is long enough to contact the cervix. Multiple intromissions might be even more effective in 'priming' the cervix before deposition of semen at the cervical os. Figure 5.32 shows the timing of a typical multiple intromission series in the rhesus monkey. The final ejaculatory mount involves more intromitted thrusts and is of longer duration than the others. The total thrusting rate is much greater in the rhesus monkey than in species which exhibit a single, brief intromission to attain ejaculation. In the bonnet macaque (*Macaca radiata*) for instance, which is a single mount ejaculator, males thrust 17 times on average to attain ejaculation. The male rhesus, by contrast, mounts 8.5 times on average and delivers a total of 42 intromitted thrusts in order to ejaculate once. The cynomolgus macaque (*Macaca fascicularis*) can exhibit either pattern; 47% of copulations involve a single intromission with an average of 17 intromitted thrusts, whilst 53% of copulations involve an average of three intromissions (range 2–6) and 39 pelvic thrusts (Shively *et al.* 1982). Do such repeated intromissions evoke a neuroendocrine response from the female? Whilst such responses are important in some mammals (e.g., to facilitate prolactin secretion and support of the corpus luteum in the rat), there has been no demonstration of their importance in primates with MBI copulatory patterns. Do multiple intromissions result in the ejaculation of larger numbers of sperm? The tactile stimulation arising from multiple mounts and repeated bouts of pelvic thrusting might result in the release of larger numbers of spermatozoa from the reserves stored in the cauda epididymis (Bercovitch and Nürnberg 1996). This hypothesis has yet to be tested by experiment.

If a multiple intromission pattern provides some advantage in sperm competition between males in multimale–multifemale mating systems, why is the pattern not present in all species which have this mating system? A comparative approach to this problem was taken by Shively *et al.* (1982) in their studies of copulatory behaviour in the genus *Macaca*. Since both multiple and single intromission patterns occur among various macaque species, Shively *et al.* attempted to identify differences in social organization between macaque species which might account for differences in their copulatory patterns. They advanced the hypothesis that species which exhibit a single mount ejaculatory pattern (*Macaca radiata*; *M. sylvanus*; *M. sinica*; *M. arctoides*) tend to exhibit greater intermale tolerance and less competition than in MBI copulation species (e.g., *M. mulatta*; *M. fuscata*, and *M. nigra*). A detailed discussion of relationships between male rank and mating success in primates was included in Chapter 3. In respect of intermale relationships and copulatory patterns *in macaques*, the argument of Shively *et al.* has much to recommend it. In *Macaca sylvanus*, for instance, females may copulate repeatedly with large numbers of males during the peri-ovulatory period of the menstrual cycle when the sexual skin is fully tumescent. Taub (1980), for instance, records that 'a female might copulate with 2 to 10 (18 to 91%) of the group's sexually mature males on any one day'. Menard *et al.* (1992) reported very similar findings for two free-ranging troops and concluded that 'our data, indicating an essentially promiscuous pattern

of sexual behaviour, support the interpretation that a strong hierarchical social system is lacking in *M. sylvanus*'. Similar observations have been reported for free-ranging *Macaca radiata* by Simonds (1965) and Sugiyama (1971), whilst Barbara Glick (1980) reported some effects of male rank, such as greater access ,by high-ranking males to females during the peri-ovulatory period. Yet, one is also struck in her studies by the high frequencies of copulatory behaviour shown by free-ranging male *M. radiata* and the relative absence of aggressive competition for females. These findings contrast with the results of studies on macaques with MBI copulatory patterns, in which relationships between male rank and mating success are usually pronounced (e.g., free-ranging *Macaca mulatta*: Carpenter 1942; Kaufmann 1965; free-ranging *M. fuscata*: Tokuda 1961; captive *M. nigra*: Dixson 1977). *Macaca fascicularis* represents an intermediate case, since both single intromission and multiple intromission copulatory patterns occur in this species. Individual males may display either pattern, though the extent to which different populations in the wild might exhibit such behaviour is unknown. In their studies of captive *M. fascicularis*, Shively *et al.* (1982) found a positive correlation between male rank and copulatory behaviour. Field studies of this species at Ketambe in N. Sumatra reveal that 'a female may copulate with all group males in a single day' and yet 'at periods when females are particularly attractive, high-ranking males generally have better access and may keep other males away' (De Ruiter *et al.* 1992).

The hypothesis of Shively *et al.* presents a plausible case that among members of the genus *Macaca*, evolution of an MBI pattern of copulatory behaviour correlates with higher levels of intermale competition than occur in species where a single intromission pattern is the rule. These authors were able to study only two males of *M. arctoides*. They pointed out that in this species, which utilizes a single intromission copulatory pattern, field studies by Estrada and Sandoval (1977) and work on captive animals by Bertrand (1969) emphasized positive, affiliative behavioural interactions (grooming, huddling, etc.) among adult males. However, subsequent research by Estep *et al.* (1988) involving a captive group of 97 *M. arctoides* showed a clear positive correlation between male rank and copulatory frequency. Lower-ranking males were inhibited from displaying copulatory behaviour in the presence of dominant individuals; removal of dominant males led to greatly increased copulations by subordinates. At first sight this represents a contradiction of the hypothesis of Shively *et al.* However, in *M. arctoides* the copulatory pattern involves a much more prolonged single intromission (SPI pattern) which is maintained during the post-ejaculatory pair-sit. It is a specialized pattern and quite distinct from the brief single intromissions which occur, for example, in *M. radiata* or *M. sylvanus*. It is for that reason that *M. arctoides* appears in Table 5.6 as having a complex copulatory pattern, along with MBI species such as *M. mulatta* and *M. fuscata*. It is therefore possible that intermale competition has contributed to the evolution of a complex copulatory patterns in *M. arctoides* but that, for unknown reasons, sexual selection has favoured an SPI, rather than an MBI pattern.

Given that Shively *et al.*'s hypothesis finds support from studies of macaques, it is important to decide whether it is generally applicable to other genera of monkeys or apes. Table 5.7 presents relevant information on various New World monkeys, Old World monkeys, and chimpanzees. There is no doubt that Shiveley *et al.*'s hypothesis fails to account for many

Table 5.7 Primates (excluding macaques) in which a single brief intromission (SBI) copulatory pattern occurs. Do such species exhibit high intermale tolerance and low intermale competition, as predicted by Shively *et al.* 1982?

Species	Conditions of study	Intermale relationships	Is hypothesis of Shively *et al.* upheld?	Sources
Lemur catta	WG	Intense aggression and competition for mating opportunities occurs during the brief annual mating season.	No Koyama 1988	Jolly 1966;
Callithrix jacchus	CG	Dominant male in family or peer group has greatest opportunity to copulate with the single breeding female.	No Abbott and Hearn 1978	Epple 1975;
Cebuella pygmaea	WG	Dominant male observed mate-guarding and copulating with breeding female during post-partum phase. Second male excluded from mating.	No	Soini 1988
Alouatta palliata	WG	Female initiates copulations with a series of partners. Intermale tolerance is marked.	Yes	Carpenter 1965
Nasalis larvatus	WG	Males compete to associate with small units of females. All male groups are excluded from sexual interactions.	No	Yeager 1990b
Miopithecus talapoin	CG	Agonistic hierarchy is pronounced and only the highest-ranking male copulates.	No	Dixson *et al.* 197?
Erythrocebus patas	WG	Multimale influxes into one male units during annual mating season result in intense intermale aggression and copulatory competition.	No	Harding and Olse Ohsawa *et al.* 19?
Mandrillus sphinx	SFG	A pronounced agonistic hierarchy occurs. Intermale tolerance is low and the alpha male accounts for the majority of observed copulations.	No	Dixson *et al.* 199?
Theropithecus gelada	WG	Males compete to associate with units of females. Extra-group males are excluded until a unit leader is overthrown.	No	Dunbar 1984
Gorilla gorilla	WG	Some intermale tolerance occurs and several silverback males can co-exist in mountain gorilla groups, but only the dominant male copulates.	Possibly	Harcourt *et al.* 1?
Pan troglodytes	WG	Male rank affects mating success but intermale tolerance is high and females mate with most male non-kin.	Yes	Goodall 1986
Pan paniscus	WG	Rank correlates with male mating success, but tolerance is high and females mate with multiple partners.	Yes	Kano 1992

WG=free-ranging-groups; SFG=semi-free-ranging groups; CG=captive groups.

cases and thus cannot be accepted as a general explanation of the evolution of MBI versus SBI copulatory patterns in primates.

Communication during copulation

A simplistic account of primate copulatory behaviour might conclude after a discussion of the motor patterns required for intromission and ejaculation. Yet, in many primates one or both partners exhibit changes in facial expression and vocalization during copulation. In a few species the male may bite or grab the female's back or head (Fig. 5.19). Females of many species reach back to touch or grasp the male or look round to engage him in eye contact. Orgasmic responses may occur in females as well as behavioural indicators of the ejaculatory response in males (a pause in thrusting accompanied by body and leg tremor). At the conclusion of copulation females may either remain passive or run away from the male and thus terminate contact. It seems likely that selective forces have been at work to mould the evolution of some of these behavioural patterns. The alternative view, that they are epiphenomena which accompany the more essential copulatory movements, seems to be most unlikely. It is relevant to recall that females of prosimian species exhibit oestrus and may adopt reflexive lordotic postures during mating. This is the case, for instance, in *Lemur catta*, *Propithecus verreauxi*, and *Varecia variegata*. Some other prosimians probably exhibit lordotic responses also. The situation among anthropoid primates is quite different, as the female's mating posture is neither reflexive nor under rigid hormonal control. The greater complexity of communication between the sexes during copulation in anthropoids might be related to this fundamental difference in control of female receptivity. It is by no means obligatory or certain that female monkeys or apes will remain immobile and allow males to ejaculate intravaginally. In some cases, females have been observed to withdraw before ejaculation is complete (e.g., *Papio ursinus*: Saayman 1970; *Pan paniscus*: Thompson-Handler *et al.* 1984; *Pan troglodytes*: Goodall 1986). In those species with a multiple intromission copulatory pattern, the partners must remain in proximity over an extended period if the male is to achieve ejaculation. This does not always happen, as in *Macaca fuscata* where Hanby *et al.* (1971) found that only 34% of mount sequences ended in ejaculation. Although harassment by other group members accounted for some failures, in 23% of cases either the male or female

withdrew 'for no apparent reason'. Hence, communication between the sexes during copulation might play an important role in determining whether ejaculation occurs. Some of these behavioural patterns might also be important in cryptic female choice at the copulatory level (Eberhard 1996).

Facial communication during copulation

In many anthropoids, the female turns to look back at the male and establishes eye-contact with him during copulation. The importance of eye-contact as a component of sexual invitational behaviour has already been discussed (see pages 96–7). It seems likely that when females attempt to establish eye-contact with males during the mount this represents a continuation of precopulatory, eye-contact proceptivity. Such behaviour might facilitate the male's copulatory activity, enhance his arousal, and improve the chances that ejaculation will occur. There have been observations to suggest that males alter their pattern of pelvic thrusting in response to visual cues from females. In *Brachyteles arachnoides*, the male and female remain immobile for some minutes during the initial phase of intromission. Milton (1985a) recorded that 'this phase ended when the female turned her head to look at the male.....the female drew her lips back in a grimace and this expression was sometimes returned by her partner'. In Milton's view, the female's behaviour served to indicate her willingness for the male to commence pelvic thrusting and to ejaculate. Mutual eye-contact during ventro–ventral copulations in the pigmy chimpanzee (*Pan paniscus*) might also play some role in co-ordinating the male's thrusting movements. Savage and Bakeman (1978) observed that 'slow-motion study of filmed bouts revealed that often the speed and intensity of thrusting altered as a function of the partner's facial expression and/or vocalization'. In a few cases females contribute directly to thrusting movements. In the New World uakari, *Cacajao calvus*, the female sometimes 'complements the male's pelvic thrusts with visible pelvic thrusts of her own' (Fontaine 1981). Female mountain gorillas sometimes make thrusting movements during copulation and these tend to be slower and deeper than those of the male (Harcourt *et al.* 1981b). During copulations in which female orangutans adopt a superior position on the male it is usual for the female to make pelvic thrusting movements (Schurmann 1982; Nadler 1988).

Many examples can be cited of anthropoid species in which females look back at males during copulation. Such behaviour occurs, for instance, in New World monkeys, as in some callitrichids (*Callimico goeldii*: the female also makes tongue-flicking movements: Heltne *et al*. 1981; *Callithrix jacchus*: the female opens her mouth and may also tongue-flick as during proceptive displays: Kendrick and Dixson 1984a; *Cebuella Pygmaea*: rarely and with tongue-flicks: Soini 1988; *Saguinus fuscicollis*: briefly looking back and with a single recorded instance of tongue-flicking: Snowdon and Soini 1988; and *Leontopithecus rosalia*, in which the female looks back and also vocalizes coincident with the male's ejaculatory response: Kleiman *et al*. 1988). Several cebid species have also been observed to show such behaviour (*Ateles paniscus*: the female looks back and purses her lips: Van Roosmalen and Klein 1988; *Lagothrix lagothricha*: the female makes a 'click' or 'tooth chatter' display with the lips retracted, similar to her proceptive behaviour; the male may also do this: Nishimura 1988; and in *Brachyteles arachnoides*: Milton 1985a). Looking back during copulation has also been reported for females of many Old World monkeys, especially in the Cercopithecinae (e.g., in various *Macaca* species: *M. mulatta*: both sexes also lipsmack during eye-contact: Hinde and Rowell 1962; Zumpe and Michael 1968; *M. nigra*: both sexes lipsmack: Dixson 1977; *M. fuscata*: the female stares at the male's face: Wolfe 1984b; *M. arctoides*: as the female looks back both sexes may vocalize and give a distinctive facial expression with the mouth opened in an 'O' shape: Slob and Nieuwenhuijsen 1980; Slob *et al*. 1986; *M. silenus*: the female glances back repeatedly: Kumar and Kurup 1985 and also in *Cercocebus albigena*: Wallis 1983; *Papio ursinus*: sometimes in conjunction with a 'copulatory call': Saayman 1970; *Mandrillus sphinx*: author's observations and in *Miopithecus talapoin*: may be accompanied by a grimace and screeching vocalization at the end of the mount: Dixson *et al*. 1975). Among the apes, eye-contact between the sexes occurs during ventro–ventral copulations in *Pongo pygmaeus* (Schurmann 1982; Nadler 1988), *Pan paniscus* (Savage and Bakeman 1978; Thompson-Handler *et al*. 1984), and more rarely in captive *Gorilla g. gorilla* (Nadler 1976). During dorso–ventral copulations female chimpanzees (*Pan paniscus*) may look back and grimace whilst males occasionally lipsmack (Tutin and McGinnis 1981; Goodall 1986). In free-ranging mountain gorillas (*G. g. beringei*) the female sometimes looks back over her shoulder, with lips compressed, during dorso–ventral copulations (Harcourt *et al*. 1981b). In all these instances it is possible that the female's behaviour plays some role in co-ordinating and facilitating the male's copulatory thrusting movements and/or ejaculatory response. An evaluation of this hypothesis will require cross-species analyses of appropriate filmed records

of copulatory behaviour. Females are not obliged to seek eye-contact with males during their mounts and in some cases the absence of such behaviour might be significant. In his studies of captive orang-utans, Nadler (1977) noted that males were forceful in copulating with females at all stages of the menstrual cycle. Females could not escape from males during these 'free access pair-tests'. The males held the females down and frequently interrupted thrusting bouts to reposition their partners. Interestingly, Nadler notes that 'during copulation the female was relatively indifferent to the male. She sometimes examined the cage or looked off into space'. Yet during restricted access pair-tests, where females could control whether mating occurred, Nadler (1988) found that they frequently responded to male penile displays by mounting and thrusting in a ventro–ventral position. Eye-contact and female co-ordination of copulatory movements occurred under these conditions, as is also true for orangutans in the wild. Schurmann (1982) conducted field studies of *P. p. abelii* in the Gunung Leuser Reserve of Northern Sumatra, where he observed a similar male presentation display by an adult male ('Jon') towards a young female ('Yet'). After a prolonged courtship between this pair, initiated primarily by the female, she 'looked at Jon, touched him … positioned herself in a hanging posture … and lowered herself onto his penis'. Whilst facing the male she then became 'the active partner … performing all the pelvic thrusts'.

Vocal communication during copulation

Vocalizations occur during copulation in a variety of anthropoids including *Leontopithecus rosalia* (the female vocalizes during ejaculatory mounts: Kleiman *et al*. 1988), *Macaca fuscata* (a cackling vocalization is given by the female but she may cease to do so once the MBI series is underway: Hanby *et al*. 1971; Oda and Masataka 1992), *M. arctoides* (both sexes vocalize at 'orgasm': Slob *et al*. 1978; Goldfoot *et al*. 1980), *M. silenus* (females vocalize on 89% of mounts, emitting a series of 3–15 'ho-ho-ho' sounds: Kumar and Kurup 1985), *Papio ursinus* (the female's copulatory call is a series of staccato grunts increasing in frequency to panting barks as she runs off at the end of the mount; adult males give long low-frequency copulatory calls: Saayman 1970; Hamilton and Arrowood 1978), *M. nigra* (the male emits 'hoarse barks' while thrusting: Dixson 1977), *M. fascicularis* (the female vocalizes), *Miopithecus talapoin* (both sexes give high-pitched 'screeching' vocalizations as the male ejaculates: Dixson et al.

1975), *Colobus badius* (in *C. b. badius*, females emit 'quaver' vocalizations during copulation and males remain silent; in *C. b. preussi* they emit 'barks' or 'yelps': Struhsaker 1975), *Gorilla g. gorilla* ('cooing' sounds are given by both sexes during copulation, the female vocalizations being of longer duration: Nadler 1976), *Gorilla g. beringei* (females vocalize more frequently than males, emitting 'rapid, pulsating whimpers'; males give a 'harsher and more breathy' vocalization. Vocalizations increase at ejaculation and cease at the end of the mount: Harcourt *et al.* 1981b), *Pan troglodytes* (the female 'squeals' and the male gives 'panting' vocalizations at the climax of copulation: Goodall 1968, 1986), and *Pan paniscus* (females occasionally give a 'nasalized scream': Thompson-Handler *et al.* 1984).

What are the functions of these copulatory vocalizations? Although no certain answers can be given, several possibilities may be discussed. Vocalizations, like eye-contact and facial displays between the sexes, might assist the co-ordination and temporal patterning of copulatory movements. Secondly, where vocalizations coincide with the final stages of the mount, they might correlate with physiological changes associated with orgasm. Finally, it has been suggested that female copulatory vocalizations advertise the position of a copulating pair to other group members, and serve to increase intermale competition (Hamilton and Arrowood 1978; Hauser 1990). From an evolutionary standpoint, this last hypothesis is particularly interesting and deserves fuller consideration. Hamilton and Arrowood (1978) were the first to suggest, as a result of their field studies of chacma baboons (*Papio ursinus*), that female copulatory vocalizations might function to increase intermale competition. Adult females vocalized more frequently than younger females during the peri-ovulatory phase of the menstrual cycle. Similar findings had been reported some years earlier by Saayman (1970). Hamilton and Arrowood found that female baboons made copulatory sounds which were structurally more complex than those of female gibbons or human females. Calls were given more often during copulations with adult male baboons. Similar findings have been reported for chimpanzees (Hauser 1990) and Japanese macaques (Oda and Masataka 1992). In *Macaca silenus* adult females also vocalize more frequently (during 93% of mounts) than younger females (30% of mounts), as well as giving more intense calls earlier in the mounting series (Kumar and Kurup 1985). In this case, females do not necessarily vocalize more often during the ejaculatory phase of mounts, and nor do Japanese macaques (Hanby *et al.* 1971; Oda and Masataka 1992). This contrasts with the situation in the chacma baboon where the female's vocalization appears to be largely an 'involuntary' response which climaxes and continues after the male's ejaculation (Saayman 1970). In the chimpanzee also, the female's 'squeal' occurs during the ejaculatory phase, although in this case intromission is so brief that the degree of coordination between the sexes is difficult to assess (Goodall 1968). The female copulatory vocalizations of different species do not necessarily have exactly the same temporal patterning, therefore. Recent reviews of this topic indicate that females of multimale–multifemale primate species are more likely to emit complex sounds in association with mating than females of monogamous and polygynous species. Do these sounds affect the behaviour of conspecifics? Certainly the calls are loud enough to be perceived at a distance. Saayman (1970) noted that copulatory calls of female *Papio ursinus* were audible at 200–300 m and the sounds sometimes enabled him to locate animals in dense bushveld. Although female copulatory vocalizations were heard infrequently during observed matings in bonobos (*Pan paniscus*), 'similar vocalizations were frequently heard whilst searching for the animals'. (Thompson-Handler *et al.* 1984). The possibility that such sounds might arouse other group members to copulate was mooted by Struhsaker (1975) in his field studies of *Colobus badius badius*. Bouts of copulatory 'quaver' vocalizations by one female were sometimes 'immediately followed by similar bouts from other sources'. Struhsaker suggested that the female's vocalizations might induce 'a kind of social facilitation in mounting'. A similar phenomenon may occur in the talapoin monkey in which copulating animals may be located in dense forest by listening for their characteristic vocalizations (Rowell and Dixson 1975 and unpublished observations). However, there is little concrete evidence that female copulatory vocalizations attract other monkeys to the scene or that they provoke intermale competition. In their studies of captive Japanese macaques Oda and Masataka (1992) found that such vocalizations did not increase aggressive competition by other members of the group. Although dominant male Japanese macaques have been observed to disrupt copulations involving subordinate or juvenile males (Hanby *et al.* 1971; Wolfe 1978), Oda and Masataka were unable to relate the occurrence of female copulatory calls to any aspect of observed intermale competition for mating opportunities. In the circumstances their suggestion that the female's 'atonal' (copulatory) call 'evolved to incite intermale competition' is unconvincing. Similarly, in their field studies of chimpanzees in the Mahali mountains Hasegawa and Hiraiwa-Hasegawa (1990) found no consistent effect of female copulatory vocalizations upon the behaviour of other group members. Copulating pairs rarely evoked aggressive responses from adult males. Ten incidents of males hitting, threatening, and barking or charging at copulating animals were observed out of a total of 583 mounts with intromission. In 14 cases adult females responded to other animals which were mating but only once did this involve any physical intervention. These authors also record that younger females vocalized more frequently (during 68 of 88 copulations—79%) than adult females (240 out of 495 copulations—48%).

Whilst there is little evidence that female copulatory vocalizations encourage direct intervention or aggressive competition by other group members, the question remains open as to whether some more subtle form of communication is involved. The vocalizations emitted by various macaques, baboons, talapoins, chimpanzees, and other species might serve to alert other males in the group to the fact that mating is in progress and encourage them to approach or attempt mating at some *later time*. Direct and immediate intervention or competition need not occur, therefore. If this is so, however, the burden of experimental proof lies with those workers who have hypothesized that sexual selection has favoured the evolution of female copulatory calls in multimale–multifemale mating systems as a mechanism to incite intermale competition. One way forward might be to conduct 'playback' experiments using tape-recordings of female calls, in order to determine whether they attract males. Such an approach has proven valuable in studying the complex vocal repertoire of vervet monkeys (Cheney and Seyfarth 1990). Vocal communication in these monkeys is subtle and involves, for instance, the use of different alarm calls in response to different types of predator (e.g., leopards, eagles, or snakes: Struhsaker 1967b; Cheney and Seyfarth 1990). Complex usage of female vocal communication during copulation is hinted at by certain observations on captive stumptail macaques (*Macaca arctoides*) made by De Waal (1989). Cheney and Seyfarth (1990) quote De Waal to the effect that female stumptail macaques sometimes 'attempted to suppress their partners' copulatory grunts before they attracted the attention of the dominant male. Most of these attempts took the form of threats, but on one occasion a female placed her hand over her partner's mouth'. Such observations are thought-provoking, but wholly anecdotal. Experiments are needed to explore the possible functions of such behaviour; the temptation to imagine what the monkey is 'thinking' or 'feeling' in such circumstances is no substitute for experiment.

Tactile stimulation and the problem of orgasm

During ejaculation, males of many primate species exhibit myotonic reactions so that, as the male ceases to make pelvic thrusting movements (the 'ejaculatory pause') and semen is expelled, the body becomes rigid and muscular tremors occur in the legs, pelvis and trunk (Table 5.8). Such reactions have been observed in many New World and Old World monkeys as well as in the chimpanzee, gorilla and (more rarely) the orang-utan. They accord well with descriptions of myotonic responses during orgasm in the human male (Masters and Johnson 1966), and might be viewed as homologous with the same. Not all male primates show obvious ejaculatory responses and their absence might make it difficult for a human observer to identify whether ejaculation has occurred (e.g., in *Aotus trivirgatus*: Dixson 1983a; *Cercopithecus aethiops*: Gartlan 1969; *Macaca silenus*: Kumar and Kurup 1985). Male orang-utans sometimes exhibit a 'shuddering' of the body during presumptive ejaculations, whilst at other times the response is absent although emission of semen has occurred (Nadler 1977).

Although it is accepted that male primates exhibit orgasmic responses during ejaculation, the occurrence of orgasm in female non-human primates has been much disputed. Some authors have stressed that orgasm in the human female represents a specialized adaptation to reinforce emotional bonding between the sexes (Morris 1967; Beach 1974; Barash 1977). On this basis, human females are regarded as fundamentally different from female non-human primates since the latter are deemed not to experience orgasm during copulation (Ford and Beach 1952; Beach 1974). In his treatise 'The Evolution of Human Sexuality' Symons (1979) reviewed in some detail the evidence then available on copulatory responses in female non-human primates. He concluded that 'whilst the possibility that non-human female mammals experience orgasm during heterosexual copulation remains open, there is no compelling evidence that they do'. He also expressed concern that evidence pertaining to the occurrence of orgasm in female monkeys or apes 'has been obtained only among captive animals' and might be due to prolonged clitoral stimulation which does not occur under natural conditions.

Attempts to identify qualities which are absolutely unique to human beings are often doomed to failure. 'Man the Tool Maker' (Oakley 1949) is not alone in fabricating tools since chimpanzees possess this ability, albeit at a primitive level (Goodall 1964; McGrew 1992). Neither can self-awareness be denied to the chimpanzee (Gallup 1975), nor a primordial capacity for language acquisition (Gardner and Gardner 1969). As regards sexual behaviour and responsiveness, it is frequently the case that members of one sex are capable of exhibiting motor patterns and physiological responses which are more typical of the opposite sex, provided that conditions are favourable for their elicitation. Some female rats, for instance, perform a series of mounts with pelvic thrusting, culminating in an 'ejaculatory response' when treated with androgen and paired with other females. Among anthropoid primates, in which the hormonal control of sexual behaviour is less rigid than in many other mammals, sociosexual behaviour occurs in a variety of contexts and female–female mounting is relatively common (e.g., among macaques, baboons, talapoins: see Chapter 6). Female rhesus monkeys have occasionally been observed to mount male partners during laboratory pair tests; after performing a series of mounts with

pelvic thrusting one female exhibited 'orgasmic' responses involving tremor of muscles in the thighs and base of the tail (Michael *et al.* 1974). Subsequent experiments on stumptail macaques by Goldfoot *et al.* (1980) demonstrated that during female–female mounts the mounter exhibited uterine contractions similar to those which occur at orgasm in the human female. Chevalier-Skolnikoff (1974) had previously recorded that mounting and pelvic thrusting by female stumptail macaques sometimes culminated in the same kind of 'pause' and 'muscular body spasms' which occur during orgasm in males of this species.

Given that some female rhesus and stumptail macaques are capable of exhibiting orgasmic responses whilst mounting and thrusting, it seems probable that such responses might also occur during heterosexual copulation. This is indeed the case, not only in macaques but also in some other mon-

Table 5.8 Examples of behavioural responses observed at ejaculation during copulation in male primates

Species	Behavioural responses	Source
New World monkeys		
Callimico goeldii	thighs and pelvis quiver	Heltne *et al.* 1981
Cebuella Pygmaea	brief quiver of body	Soini 1988
Old World monkeys		
Macaca arctoides	Cessation of pelvic thrusting plus momentary full body inertia and rigidity	Slob and Nieuwenhuijsen 1980
	Vocalization and facial display	Slob *et al.* 1978
M. silenus	Cessation of pelvic thrusting, body tenses and facial grimace	Lindburg *et al.* 1985
M. mulatta	Cessation of pelvic thrusting, plus body and leg tremor	Carpenter 1942; Southwick *et al.* 1965
Papio ursinus	Cessation of pelvic thrusting and a 'rigid pause' of the body lasting 2–4 s	Hall and DeVore 1965; author's unpublished observations
Mandrillus sphinx	Cessation of thrusting, body tremor and rigidity	Author's unpublished observations
Cercocebus albigena	Tensing of the legs and pause in thrusting	Wallis 1983
Miopithecus talapoin	Cessation of thrusting, body and leg tremor, facial grimace and squealing vocalization	Dixson *et al.* 1975
Great apes		
Pongo pygmaeus	A shudder of the body at the end of final thrusting bout—in some cases only	Nadler 1977
Gorilla gorilla	Body rigid and with spasmodic muscle contractions and vocalization in some cases	Nadler 1976
Pan troglodytes	Cessation of thrusting, body tremor and climax of copulatory panting vocalizations	Goodall 1968; author's unpublished observations

keys and in the female chimpanzee. Figure 5.33, which is based upon the work of Slob *et al.* (1986), shows telemetric recordings of uterine contractions in a female stumptail macaque during three successive copulations. Behavioural measurements in both sexes are compared with the time-course of uterine changes. As the male intromits and thrusts, uterine movements occur as a mechanical consequence of the thrusting movements of the penis. Thrusting ceases as the male ejaculates, and yet in each case uterine contractions increase initially and then a gradual relaxation occurs over the next 15 s. The female exhibits her 'climax' facial expression (Fig. 5.34) during the phase of maximal uterine contractions, although Slob *et al.* (1986) state that the facial expression does not always accompany the uterine changes. The female's climax face closely resembles the 'ejaculation face' of the male and in some cases (e.g., mounts 2 and 3 in Fig. 5.33) both sexes exhibit the display synchronously during copulation. The responses measured by Slob *et al.* (1986) parallel the earlier work of Goldfoot *et al.* (1980) on isosexual mounting in female stumptail macaques. In that study, a female which exhibited the most intense behavioural signs of orgasm was implanted with 3 transmitters to monitor uterine contractions. When mounting a female partner this female showed 'an intense, sustained tonic uterine contraction', measured at 7.7×10^{-2} Newtons, which lasted for 50 s. Piloerection and changes in facial expression began 7–9 s after the onset of contraction and lasted for 30 s in total. Goldfoot *et al.* also recorded that 6 s after the onset of the major tonic uterine contractions, even clonic contractions occurred at intervals of 0.7 s. Ten seconds after the female ceased to exhibit the copulatory facial expression the uterus ceased to contract and returned to baseline tonus.

In human females orgasm is accompanied by a variety of genital, cardiovascular, respiratory, and muscular responses. The duration of orgasm varies among women but Levin and Wagner (1985) report that it lasts for 20 s on average. Heart rate and blood pressure increase during orgasm and contractions occur in muscles surrounding the outer third of the vagina (the *orgasmic platform*) as well as the anal sphincter and the uterus (Masters and Johnson 1966; Bohlen *et al.* 1982). The sensations reported by women during orgasm are variable but the experience of orgasmic sensations appears to be very similar in both sexes. A study by Vance and Wagner (1976) showed that male and female descriptions of orgasm were often indistinguishable, once the references to anatomical sex differences (e.g., concerning the genitalia) were removed. The following extracts from these studies are taken from Bancroft (1989).

'A heightened feeling of excitement with severe muscular tension, especially through the back and legs, rigid straightening of the body for about 5 seconds, and a strong and general relaxation and a very tired relieved feeling'.

'Combination of waves of very pleasurable sensations and mounting of tensions, culminating in a fantastic sensation and release of tension'.

'Begins with tensing and tingling in anticipation, rectal contractions starting series of chills up spine. Tingling and buzzing sensations grow suddenly to explosion in genital area, some sensations of dizzying and weakening-almost loss of conscious sensation, but not really. Explosion sort of flowers out to varying distance from genital area, depending on intensity'.

It appears that female monkeys of a number of species exhibit physiological responses during copulation that are similar, if not identical, to those which occur in women (Table 5.9). Increases in heart rate and uterine contractions occur during sexual climax in some female stumptail macaques, although the heart rate changes are less marked than those which accompany ejaculation in the male (Slob *et al.* 1986). Some female rhesus monkeys show increases in heart rate, reaching 'as high as 250 beats per minute during male ejaculation' (Phoebus 1982). Burton (1971) conducted experiments using three female rhesus macaques to determine the effects of clitoral and vaginal stimulation (using an artificial penis) upon genital responses. Two subjects had vaginal spasms in response to extended periods of genital stimulation. Wolfe (1984b) was able to observe female Japanese macaques as they self-stimulated the clitoris, usually by rubbing it back and forth between the index finger and the ischial callosity. This masturbatory behaviour resulted in rhythmic contractions of the vaginal and perianal musculature accompanied by vascular engorgement of the clitoris. Allen (1977) found that it is possible to evoke rhythmic vaginal contractions, clitoral tumescence, limb spasm, body tension, and other responses in a female chimpanzee by manual stimulation of the clitoris and vagina. Female chimpanzees masturbate occasionally. Temerlin (1975) observed such behaviour in a captive-raised female which 'laughed, looked happy and stopped suddenly' during daily bouts of self stimulation. Goodall (1986) recorded that free-ranging adolescent female chimpanzees at Gombe sometimes masturbate, 'laughing softly as they do so'.

Although female primates exhibit behavioural responses indicative of orgasm less frequently than males (the 'climax' face occurs once in every six copulations in female stumptail macaques, for instance: De Waal 1989), there is now plenty of evi-

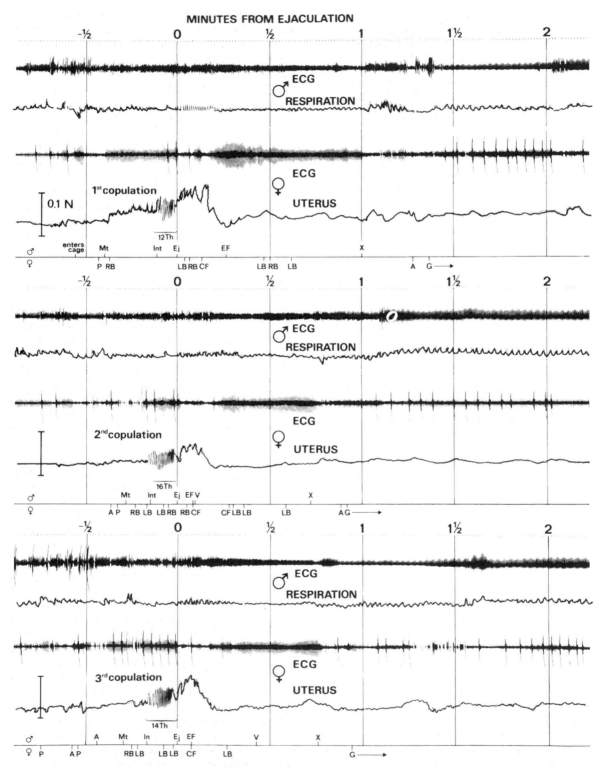

Fig. 5.33 Physiological changes and behavioural observations during three consecutive copulations in a pair of stumptail macaques (*Macaca arctoides*). Data shown are, for the male: ECG and respiration; for the female: ECG and uterine movements. Behavioural patterns indicated for the male are: A=approach; Mt=mount; Int=intromission; Th=number of pelvic thrusts; Ej=ejaculation; EF=ejaculation face; V=vocalizaton; X=end of intromission. Behavioural patterns for the female are: P=present; A=approach; RB=reach back with hand towards male; LB=look back towards male; CF=climax face; G=grooms male. N=Newton, contraction force. (After Slob *et al.* 1986.)

Fig. 5.34 The 'climax face' of the female stumptail macaque which occurs during orgasm. (After Chevalier-Skolnikoff 1974.)

dence that physiological responses similar to those which occur during orgasm in women also occur in some macaques and chimpanzees. In the absence of objective data on how female non-human primates perceive these events, there is no reason to deny the possibility that orgasmic sensations accompany the physiological changes, just as in human beings. Orgasm should therefore be viewed as a phylogenetically ancient phenomenon among anthropoid primates; the capacity to exhibit orgasm in the human female being an inheritance from ape-like ancestors. This view has been advanced by Allen and Lemmon (1981), Dixson (1990a), and Wolfe (1991) and it accords with the observed facts of female sexual physiology in monkeys, apes, and man. The argument that only abnormal amounts of clitoral stimulation produce such effects in non-human primates (Symons 1979) has also been challenged by Wolfe (1991). In monkeys and apes, the clitoris is situated at the base of the vagina and it is subject to direct stimulation during intromission and pelvic thrusting. Orgasmic responses occur as a result of normal bouts of pelvic thrusting during dorso–ventral copulations in the stumptail macaque (Slob *et al.* 1986). In the human female the clitoris is situated anterior to the base of the vagina and Wolfe (1991) argues that 'human females experience only indirect stimulation during heterosexual intercourse'. Studies by Fisher (1973) highlight the fact that many women require additional manual stimulation of the clitoris to achieve orgasm and do not always experience orgasm during intercourse in the absence of such manual stimulation by the partner. Kinsey *et al.* (1953) found that less than half the women interviewed attained orgasm during 90–100% of copula-

tions. In Fisher's (1973) study, which involved 300 married American women, the majority of subjects (>90%) reported that they had orgasms frequently during intercourse. However, in 95% of these women, orgasm was achieved by additional manual stimulation of the clitoris (before or after intercourse) and not solely by means of copulatory stimulation. Foreplay, or post-coital manipulation of the clitoris, might therefore play a substantial role in enhancing orgasmic responsiveness in women (Kinsey *et al.* 1953; Fisher 1973; Hunt 1974) and the relative absence of female orgasm in some cultures might reflect repressive attitudes towards such sexual expression (Mead 1967; Messenger 1971; Symons 1979).

The notion that the clitoris provides the sole sensate focus for stimuli which evoke sexual responses in female non-human primates might be incorrect. Very little experimental work has been reported and the tactile sensitivity of the vagina and cervix has rarely been considered. It is therefore of interest that intravaginal application of a local anaesthetic blocks some behavioural responses normally seen during copulation in female marmosets (Dixson 1986a). Yet application of local anaesthetic to the external genitalia, including the clitoris, is without effect upon the female's behaviour (Fig. 5.35). It is also noteworthy that some female chacma baboons (*Papio ursinus*) give copulatory vocalizations spontaneously at the height of sexual skin swelling, whilst urinating or defaecating (Hall 1962). In captivity, some ovariectomized talapoin monkeys also vocalize and grimace in the same fashion as during copulation when they urinate or defaecate, but only during maximal sexual skin swelling (Fig. 5.36). In such cases, one suspects that some deeper stimulation of the pelvic region during elimination might be responsible for triggering the reactions.

What, if any, are the adaptive functions of orgasm in the human female or in female non-human primates? Doubts have been expressed about whether orgasm fulfils any adaptive role in women (Symons 1979). Mead (1967) argues, from a cross-cultural perspective, that orgasm is not a fundamental and unlearned part of female sexual response and thus it contrasts with orgasm and ejaculation in the human male. The idea that orgasm evolved by sexual selection in the human female to facilitate bonding and long term relationships between the sexes seems far-fetched (Barash 1977). After all, females of other primate species, and particularly those with multimale–multifemale mating systems such as macaques and chimpanzees, exhibit orgasmic responses in the absence of such bonding or the formation of stable family units. By contrast gibbons, which are primarily monogamous, do not exhibit obvious signs of female orgasm. Indeed, to play the devil's advocate, one might argue that in stumptail macaques or chimpanzees the female's orgasm is rewarding, increases her willingness to copulate with a variety of males rather than one partner, and thus promotes sperm competition. Some attention has also been given to the possibility that female orgasm might play a role in sperm transport and enhance fertility. Morris (1967) advanced the peculiar argument that the human female might be likely to lose semen post-coitally if she walks upright, but that orgasm functions to exhaust her and ensure that she rests after intercourse has occurred. Such arguments lack comparative depth, however, since they do not apply to stumptail macaques and other monkeys in which

orgasms occur. In the human female, a negative pressure gradient develops between the vagina and uterus as a result of orgasm (Fox et al. 1970). Various workers have suggested that this might result in semen being transported rapidly into the cervical canal or the uterine cavity. Yet Masters and Johnson (1966) were unable to demonstrate such transfer of radio-opaque material across the cervix after (manually induced?) orgasms. To the best of my knowledge such rapid transfer has never been demonstrated in the human female. There is no evidence of any link between orgasmic potential and fertility in women (Bancroft 1989). In reviewing mechanisms of human sperm transport, Mortimer (1983) could not marshal convincing evidence for

Table 5.9 Examples of behavioural and physiological responses observed during copulation (C) or masturbation (M) in female primates

Species	Behavioural responses	Physiological responses	Source
Black spider monkey (C)	Head-shakes, look-back, pursed lips, touching mate and moving away—particularly as male ejaculates	No data	Van Roosmalen and Klein 1988
Common marmoset (C)	Look-back, open mouth, tongue-flick, struggling movements	Post-coital perivaginal contractions	Dixson 1986a
Talapoin (C)	Look-back and reach-back, keening vocalization and grimace	No data	Dixson *et al.* 1975
Rhesus macaque (C)	Reach-back and clutching reaction	Increased heart rate Contractions of vaginal, anal, and pelvic muscles	Zumpe and Michael 1968; Phoebus 1982
Rhesus macaque (M: by experimenter)	No data	Increased heart rate vaginal spasm, and anal contractions	Burton 1971
Stumptail macaque (C)	Look-back and reach-back. Climax face	Increased heart and respiratory rates Uterine contractions	Slob *et al.* 1986
Stumptail macaque (S)	Climax face	Uterine contractions	Goldfoot *et al.* 1980
Japanese macaque (M)	? none reported	Rhythmic vaginal and perianal contractions; clitoral tumescence	Wolfe 1984b, 1991
Chacma baboon (C)	Copulatory vocalization and withdrawal reaction	No data	Saayman 1970
Chimpanzee (M: by experimenter)	Limb spasm, body tension, grin-face and panting vocalization	Rhythmic vaginal contractions, clitoral tumescence, vaginal expansion and secretions	Allen 1977
Chimpanzee (M)	'laughing vocalizations'	No data	Temerlin 1975; Goodall 1986

Modified, after Dixson 1990a. S=during sociosexual mounts between females.

rapid transport as a result of orgasmic responses in the female. Indeed, he pointed out that this would involve bulk transfer of spermatozoa, including morphologically abnormal gametes, through the cervical canal. Mortimer reasoned that such a mechanism might be disadvantageous and that in any case there is no evidence for its occurrence.

Some students of primate sexual behaviour have suggested that vaginal and uterine contractions might serve to stimulate ejaculation and transport spermatozoa across the cervix in monkeys or apes (Allen and Lemmon 1981; Wolfe 1991). Rose and Michael (1978) showed that vaginal stimulation in the squirrel monkey results in the firing of neurons in the mesencephalic and pontine region of the brain. Neuronal responsiveness is increased during oestradiol treatment. They suggested that these phenomena might be associated with vaginal contractions during copulation, which in turn might influence ejaculatory latency. Such a role for orgasmic contractions of the vagina remains to be proven for any primate species and possible effects of uterine contractions upon sperm transport have not been studied. This is a neglected area of primate reproductive physiology.

From the evidence reviewed so far, it is clear that a variety of behavioural and physiological responses occur during copulation in female primates (Table 5.9). These differ between species and between individuals within a single species. It is now known that not all behavioural responses, such as facial expressions and vocalizations given by female primates, correlate with the occurrence of physiological events during orgasm. Thus, in their telemetric studies of the female stumptail macaque, Slob *et al.* (1986) found that the propensity of the female to look back and reach back towards the male partner was not associated with patterns of increased uterine contraction. The female's climax face, by contrast, was reliably associated with peaks of uterine activity. Previous research on the reaching back response or *clutching reaction* in female rhesus monkeys had suggested some link between this behaviour and female orgasm (Zumpe and Michael 1968). Female rhesus monkeys often look back and lipsmack at the male, and then reach back to grasp his leg or body during ejaculation. Such responses also occur in many other primate species (Table 5.10) but in some cases they are restricted to touching or stroking movements (e.g., in *Callimico goeldii*, *Cebus capucinus* and *Ateles Belzebuth*) rather than to actual clutching reactions. In *Brachyteles arachnoides*, for instance, the female reaches back to touch the male's genital region just before the end of copulation and this might assist him to dismount or signal him to do so (Milton 1985a). During isosexual mounting females of some species frequently reach back in exactly the same manner as during copulation (e.g., *Miopithecus talapoin*: Dixson *et al.* 1975: *Macaca nigra*: Dixson 1977). However, in the rhesus monkey and in some other macaques, such reactions are most commonly observed as the male ejaculates. Indeed, they probably represent a response to the male's myotonia and ejaculatory pause, rather than indicating the occurrence of orgasm in the female partner. Interestingly, if male rhesus monkeys fail to display myotonia and to ejaculate, as occurs for instance after transection of the dorsal penile nerves, then females also fail to show a clutching reaction even in response to prolonged bouts of intromitted pelvic thrusts (Herbert 1973). Expulsion of semen at ejaculation is not essential to trigger this reaction in the female rhesus monkey. Some castrated males continue to copulate and to exhibit an ejaculatory pause in the absence of expulsion of seminal fluid; females may still exhibit the clutching reaction under these conditions (Fig. 13.3, page 393).

As a final demonstration that the behavioural responses exhibited by female primates during copu-

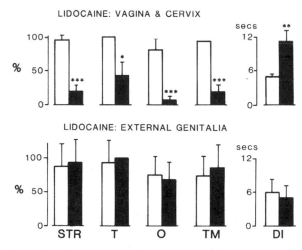

Fig. 5.35 Effects of local anaesthesia of the genitalia upon behavioural responses during copulation in the female common marmoset (*Callithrix jacchus*). Data are means (±SEM) percentages of control tests (open bars) or lidocaine-treated tests (solid bars) during which females showed struggling movements (STR), head-turning (T), opening the mouth (O), or termination of the mount after ejaculation (TM). Intromission durations (DI) are in seconds. Application of lidocaine intravaginally in eight females (upper histograms) had profound effects upon all these behavioural responses, whereas anaesthesia of the external genitalia, in the same subjects (lower histograms), had no effect. Sign test *P=0.03; ***P=0.004; Wilcoxon test **P=0.01. (After Dixson 1986a.)

lation can vary in their causation and functions, I shall describe some experiments on the common marmoset (*Callithrix jacchus*). Female marmosets show several responses during copulation—tongue-flicking, head-turning, mouth-opening, and struggling body movements. Head-turning, mouth-opening, and struggling body movements occur in response to penile stimulation of the female's internal genitalia and are abolished by intravaginal application of lidocaine (Fig. 5.35). Tongue-flicking, however, is not affected by this procedure; it probably represents a continuation of the female's precopulatory proceptive behaviour and is controlled by different neurological mechanisms. Hypothalamic lesions which abolish the female's proceptive tongue-flicking display also reduce tongue-flicking during copulation (Kendrick and Dixson 1986) but have no effect upon other responses shown by the female during intromission (Dixson 1990b). It is also noteworthy that some responses shown during intromission by female marmosets, such as struggling body movements, are responsible for terminating copulation. Intravaginal application of lidocaine, which renders the female relatively passive during copulation, also results in a doubling of the duration of intromission and a relative absence of termination by the female (Fig. 5.35). In some other primate species females may terminate the mount once ejaculation has occurred, or run away rapidly as the male partner dismounts. Some examples of such

behaviour, which occurs for instance in female chacma baboons and chimpanzees, are provided in Table 5.11. Active withdrawal by the female is uncommon among primates as a whole, but has been well-documented for particular species. It might indicate some nociceptive response to intromission and ejaculation rather than a 'pleasurable' sensation. The physiological basis of such responses and their possible significance remain a mystery.

Some conceptual issues in human sexuality
Gender identity and gender role

The terminology used to define sexual behaviour in non-human primates is also broadly applicable to human beings. Proceptivity, receptivity, and attractiveness occur in women, and men exhibit sexual arousal, copulatory behaviour, and reductions in sexual interest during the post-ejaculatory interval. Human males attempt to initiate sexual interactions (the equivalent of the female's proceptive behaviour) and vary in their sexual attractiveness. However, there are important conceptual problems which arise when describing sexuality in human beings rather than in other primate species. These differences relate to the greater cognitive development of human beings and to the concept of *self-awareness* or *consciousness* as it applies to human sexuality.

Firstly, there is the important question of *gender*

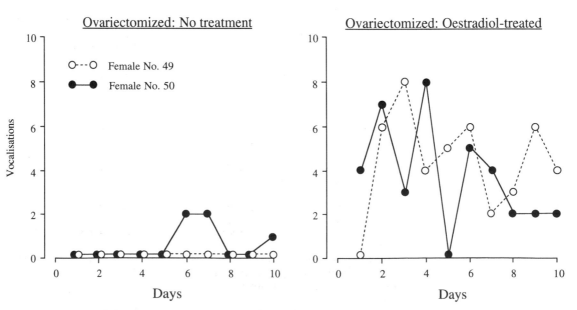

Fig. 5.36 Effects of oestradiol treatment upon 'spontaneous' copulatory vocalizations in two ovariectomized talapoin monkeys (*Miopithecus talapoin*). These females sometimes gave copulatory vocalizations whilst defaecating; treating them with oestradiol benzoate (17.5 mg/week) resulted in marked increases in vocalizations. Data are daily frequencies during observation periods which lasted for 160 min. (Author's unpublished observations.)

identity, defined by Money and Ehrhardt (1972) as 'the sameness, unity, and persistence of one's individuality as male or female (or ambivalent), in greater or lesser degree, especially as it is experienced in self-awareness and behaviour'. Self-awareness and individuality are important components of this definition. As human beings, we are able to reflect upon our individual sexual feelings (whether masculine, feminine or ambivalent) and to match these to our erotosexual preferences. It might be argued, and has often been stated, that gender identity is therefore an uniquely human quality (e.g., Ehrhardt and Meyer-Bahlburg 1981). However, if evidence for 'self-awareness' could be obtained for non-human primates, then it would be incorrect to deny that gender identity might also exist in such cases. The mirror experiments conducted by Gallup (1975) indicate that chimpanzees recognize themselves as individuals and hence might possess some degree of self-awareness. The same is apparently true of the gorilla (Denton 1993). In Gallup's experiments anaesthetized chimpanzees were marked on the upper surface of one ear and on the brows with an odourless red dye. Once they had recovered, their reactions to their own reflections were recorded and compared with responses before dye-marking had occurred. Chimpanzees showed a more than 25-fold increase in the frequencies with which they fingered the dye-marks, using the mirror to examine these otherwise visually inaccessible areas. As Denton (1993) puts it, 'The issue is that you cannot examine otherwise invisible portions of your body with the aid of a reflection unless you know who you are—that is, the animal is aware of its self'.

Self-awareness of one's image reflected in a mirror emerges during human development when infants are 24–30 months of age. Infants below this age are usually unable to respond to their reflections with signs of self-recognition. Monkey species so far tested fail to exhibit such self-recognition in response to a mirror, even in adulthood. In my experience, talapoin monkeys, marmosets, and rhesus monkeys fail to make this discrimination even after hundreds of hours of exposure to the one-way

Table 5.10 Occurrences of the reaching back or clutching reaction in female monkeys during copulations observed in captivity (C) or in the wild (W)

Species	Behaviour description and context	Source
New World monkeys		
Callimico goeldii	Sometimes touches male's shoulder or arm during pre-intromission phase	Heltne *et al.* 1981 (C)
Cebus capucinus	May touch male's thigh or anogenital region during copulation	Freese and Oppenheimer 1981 (W)
Ateles belzebuth	Sometimes rubs male's upper leg or back with one hand during copulation	Van Roosmalen and Klein 1988 (W)
Brachyteles arachnoides	Often puts her hand on male's genital region just prior to end of intromission	Milton 1985a (W)
Old World Monkeys		
Macaca nigra	Reaches back and grasps one of male's thighs during non-ejaculatory and ejaculatory mounts	Dixson 1977 (C)
M. arctoides	Reaches back and clutches male's leg most often during ejaculatory phase of mount	Slob *et al.* 1978 (C)
M. silenus	Females often reach back and 'press the male's leg against theirs during ejaculation'. Females clutch at male with one hand	Lindburg *et al.* 1985 (C) Kumar and Kurup 1985 (W)
M. mulatta	'Clutching reaction' with one hand or less frequently both hands alternately, during the ejaculatory phase of the mount	Zumpe and Michael 1968 (C)
Cercocebus albigena	Reaches back or grasps male's thigh briefly during copulation	Wallis 1983 (W)
Miopithecus talapoin	Usually reaches back with one hand and grasps male's thigh during copulation	Dixson *et al.* 1975 (C)

vision mirrors used for laboratory studies of social and sexual behaviour. Monkeys react to their reflections as if they are 'other' monkeys and either display aggressively towards them, or cease to respond at all after a sufficient period of habituation. Moreover, radical changes in facial appearance (shaving off the ear tufts of marmosets or shaving the scalp for the purposes of stereotaxic brain surgery) do not provoke increased self examination using a mirror once the monkeys have recovered and have been returned to the observation cage. It therefore appears that the African apes possess a greater extent of self-awareness, which is lacking in monkeys, and hence might also possess a degree of gender identity, homologous with that which characterizes human sexuality.

A second question concerns the concept of *gender role*, defined by Money and Ehrhardt (1972) as 'everything that a person says or does, to indicate to others or to the self, the degree in which one is male or female or ambivalent'. Included in this definition are habits of dress and speech that are usually, but not always, congruent with the individual's sexual preferences. A man with transsexual preferences in his private life, might adopt a heterosexual gender role in his professional life, dressing and speaking in a manner which is more easily tolerated in the work-place. Another man, equally transsexual in his preferences, might have 'come out of the closet' and make no attempt to disguise this in terms of his gender role in everyday life.

Money (1977a, 1991) employs the acronym gender-identity/role (G-I/R) to unite the two concepts. He points out that gender identity is 'the private expression of gender role', whilst gender role is 'the public expression of gender identity'. Exceptions occur, however, as in the case cited above of certain male → female transsexuals, where a man might continue to dress in masculine clothing in public whilst harbouring the desire to be a woman.

Do any of the non-human primates qualify as exhibiting gender role behaviour? The concept of 'social roles' has been applied to the behaviour of monkeys (Bernstein and Sharpe 1966). Setting aside the problem of gender identity and self-awareness for a moment, male and female monkeys or apes do display marked differences in social behaviour. For instance, large and powerful males may defend other

Table 5.11 Examples of behavioural patterns by which females terminate intromission or end the mount, once ejaculation has occurred

Species	Behavioural responses	Source
New World monkeys		
Callithrix jacchus	Female struggles and withdraws rapidly from male thus terminating intromission	Kendrick and Dixson 1984a (C); Dixson 1986a (C)
Ateles belzebuth	At the moment of ejaculation, the female may move forward with the male hanging on behind her	Van Roosmalen and Klein 1988 (W)
Brachyteles arachnoides	Female touches male's genital area as if to help him withdraw (or signal him to do so)	Milton 1985a (W)
Old World monkeys		
Papio ursinus	Female terminates the mount with a withdrawal reaction consisting of a forceful bounding away from the male.	Saayman 1970 (W)
Cercopithecus aethiops	Moves forward and causes male to dismount, then runs forward several feet.	Gartlan 1969 (W)
Miopithecus talapoin	As the male dismounts, the female often runs off rapidly	Dixson *et al.* 1975 (C)
Great apes		
Pan troglodytes	Female may dart forward at end of copulation or rush off after mount is terminated	Goodall 1968, 1986 (W)

C=observations of captive animals; W=observations made in the wild group members against potential predators (e.g., in the mountain gorilla: Schaller 1963). Female primates normally play the major role in caring for offspring. However, males of certain monogamous species also carry and care for their infants (e.g. marmosets or owl monkeys). These examples of sex differences in social behaviour in monkeys do not indicate that any self-awareness of gender-identity/role exists in the sense discussed by Money. However, granted that the African apes possess some form of self-awareness, it is possible that they might also experience some notions of the appropriateness of gender role as well as a sense of gender identity in their social and sexual lives.

Auto-erotism, including masturbation

In his 'Studies in the Psychology of Sex' Havelock Ellis (1902) defined human auto-erotism as 'spontaneous sexual emotion generated in the absence of an external stimulus proceeding, directly or indirectly, from another person'. Included within his definition were 'voluptuous day-dreams' and 'sexual orgasm during sleep' as well as acts of masturbation. 'Spontaneous' seminal emission occurs during sleep in the rat (Orbach 1961; Beach 1975a) and the rhesus monkey, although we have no notion of any relationship between such phenomena in animals and sexual fantasies or dreams. Havelock Ellis was well aware that behaviour similar to human masturbation occurs in animals and he cited bulls, goats, sheep, horses, stags, camels, and elephants as examples of mammals in which males in particular have been observed to masturbate. Male monkeys and apes were also known to masturbate, and in the case of apes it was noted that such behaviour took place 'even in freedom, according to the evidence of good observers'.

Self stimulation of the genitalia, equivalent to human masturbation, has been reported for a number of primates, especially among the Old World monkeys and apes (Table 5.12). Although such behaviour is probably more frequent among captive animals and might occur as a substitute for copulation, it is not 'abnormal' (e.g., as suggested by Hamilton 1914). Ransom (1981), for example, observed free-ranging adult male olive baboons masturbating on 26 occasions (with 11 ejaculations); sub-adult and juvenile males also masturbated, as did one young adult female 'by stroking her perineum and clitoris with the tip of her tail'. Carpenter (1942) and Lindburg (1971) saw occasional masturbation in free-ranging male rhesus monkeys; similar behaviour also occurs in male grey-cheeked mangabeys (Wallis 1983), male talapoins (author's

observations), and female Japanese macaques (Wolfe 1984b), living under normal conditions (Fig. 5.37). Schurmann (1982) records that a young female orang-utan invited copulation from an adult male by masturbating or by rubbing her genitalia against him. At the Gombe Stream Reserve, in Tanzania, adolescent female chimpanzees 'sometimes finger and tickle their genitals' (Goodall 1986). It is also the case that from earliest infancy manipulation of the external genitalia constitutes a normal part of behavioural development in many monkeys and apes, as well as in human infants. Free-ranging olive baboons may masturbate with the penis erect from as early as 15 weeks of age (Ransom and Rowell 1972). In the chimpanzee, erections occur from the age of 1 month (Plooij 1984). Baby male gorillas also explore and touch their genitalia from the first few weeks onwards; erections can occur and although Hess (1973) doubts whether this is equivalent to masturbation, the dividing line between such self-exploration and purposeful masturbation is hard to define. Few quantitative data exist for human infants, but Kinsey *et al.* (1948) reported that 50% boys aged between 3–4 years are probably capable of achieving orgasm. Self-stimulation of genitalia occurs in infants of both sexes, although it is subject to parental punishment and repression in many westernized societies.

The data assembled in Table 5.12 on masturbatory behaviour in male and female primates support the contention of Ford and Beach (1952) that human masturbation 'appears to have its evolutionary roots in the perfectly normal and adaptive biological tendency to examine, to manipulate, to clean, and incidentally to stimulate the external sexual organs'. This biological and anthropological viewpoint is in marked contrast to the old (but not so terribly old) literature on human sexuality which viewed masturbation as medically harmful and morally degenerate. Von Krafft-Ebing expressed the view that:

'Nothing is so prone to contaminate—under certain circumstances even to exhaust—the source of all noble and ideal sentiments, which arise of themselves from a normally developing sexual instinct, as the practice of masturbation in early years......If the youthful sinner at last comes to make an attempt at coitus, he is either disappointed because enjoyment is wanting, on account of defective sensual feeling, or he is lacking in the physical strength necessary to accomplish the act. This fiasco has a fatal effect, and leads to absolute psychical impotence'.

Far more damaging, one feels, was the effect of such moralizing by doctors, parents and educators, serving only to inculcate feelings of shame and sexual guilt in children and adolescents. The propensity to engage in genital self stimulation exists in so many of the Old World monkeys and apes that we may with confidence predict that it was also present in the early hominids. Male monkeys and apes sometimes masturbate even when females are available; the behaviour is not solely explicable as a substitute for copulation. In the stumptail macaque, for example, copulation and masturbation occurred at high frequencies in a captive group studied by Nieuwenhuijsen *et al.* (1987). However, whilst a positive correlation existed between male social rank and copulatory frequencies, there was no correlation between either rank or copulatory frequency and masturbatory activity (Fig. 5.38). Likewise, in the orang-utan, Nadler (1988) commented that captive males often masturbated to ejaculation, by thrusting against the cage bars or by hand and that this occurred 'nearly daily irrespective of mating'.

Examples of masturbatory behaviour in primates other than catarrhine species are few in number. I have found no reports of such behaviour for any of the prosimians. Amongst the New World monkeys, Maple (1977) states that male spider monkeys (*Ateles geoffroyi*) sometimes masturbate using the prehensile tail. Hershkovitz (1977) reports that captive female golden lion tamarins (*Leontopithecus rosalia*) 'often stroke the coiled tail across the perineal region for self stimulation'. The same author describes similar behaviour in a captive female Goeldi's monkey (*Callimico goeldii*), which often tail-coiled and rubbed the genitalia after basking in front of a heating lamp. However, tail-coiling constitutes a normal part of scent-marking displays in these callithrichids, so that it is difficult to be certain that the behaviour observed was masturbation in the strict sense. Hampton *et al.* (1966) report rhythmic pelvic thrusting against a substrate by a single captive specimen of the cottontop tamarin (*Saguinus oedipus*). Such isolated examples indicate that masturbation is neither common nor widespread in New World monkeys. Interestingly, mature sons in captive family groups of common marmosets (*Callithrix jacchus*) sometimes engage in orogenital self stimulation and also 'experience erections and spontaneous emissions during the mother's oestrus' (Rothe 1975). Since mature sons in these groups are unable to engage in copulatory behaviour, masturbation or spontaneous ejaculation may provide an alternative 'sexual outlet'.

The greater propensity for Old World monkeys and apes to engage in genital self-stimulation might relate to some cognitive difference between catarrhines and other primates. It might also be significant that only the apes and man have been reported to use objects to aid masturbation. Rijksen (1978)

observed rehabilitant orang-utans of both sexes modifying leaves or twigs and using them to stimulate the genitalia (Fig. 5.39). Nadler (1988) also observed similar behaviour in a captive female orang-utan, whilst Ford and Beach (1952) and Hediger (1965) reported the use of objects (a mango or pieces of wood) by captive female chimpanzees.

Although I have emphasized the occurrence of masturbation as a normal activity in non-human primates, there are examples where self-stimulation forms part of socially abnormal behaviour. It might be that its frequent occurrence in isolated captive animals is a consequence of boredom and deprivation. Raising laboratory-born rhesus monkeys in social isolation during the first year of life results in grossly abnormal social and sexual behaviour in adulthood (see Chapter 6 for a full treatment of this topic). Harlow (1965) found that when 12 such males, aged four and a half years, were paired with females in order to study sexual activity 'not a single male showed a normal mount, although three placed their hands on the presenting female's buttocks and masturbated with one hind foot....one male presented when presented to, and another male lay down, masturbated, and rolled around on a female presenting for grooming'. Hediger (1965) also observed masturbatory behaviour and absence of copulation in socially deprived male and female chimpanzees that had been taken from the wild at 1 or 2 years of age and raised to adulthood in a zoo environment.

It will be apparent from Table 5.12, as well as from descriptions provided in the text, that masturbation is, in general, more frequently observed in male primates than in females. This sex difference has also been reported for the human species (Kinsey *et al.* 1948, 1953).

The functional significance of masturbation remains obscure. It might provide a harmless adjunct to copulatory activity, particularly when opportunities for mating are absent or restricted. One recent suggestion is that masturbation in the human male might play some role in enhancing fertility (Baker and Bellis 1993a). This paradoxical conclusion is based upon the notion that if too many sperm are inseminated at copulation then fertility can be reduced (as in polyzoospermy: Wolf *et al.* 1984). However, if masturbation is significant as a reproductive tactic it is exceedingly hard to explain why more evidence of this is not available from studies of monkeys and apes. There is no relationship between copulatory activity and masturbation in stumptail macaques for example (Fig. 5.38). In the chimpanzee, which produces much higher sperm counts per ejaculate than does man, masturbation to ejaculation is exceedingly rare in the wild, despite the

Fig. 5.37 Masturbation in a female Japanese macaque (*Macaca fuscata*). The female manipulates her clitoris, by rubbing it against an ischial callosity. (Author's drawing from a photograph by Wolfe 1991.)

propensity of adult males to handle and stimulate their genitalia (Goodall 1986). It appears that if man is adapted to manipulate sperm numbers (or sperm quality) by masturbation, then he alone among the primates has evolved such tactics. Given the wide ranging effects of sperm competition upon the evolution of the genitalia and copulatory behaviour of non-human primates, such a conclusion is not justified.

Abnormal sexual preferences: the human paraphilias

The labelling of particular sexual preferences or practices as abnormal, perverted, or deviant, is a hazardous exercise. Down the ages, cultural and religious groups have varied in their views of what constitutes 'normality' in the sexual sphere. Masturbation, so severely stigmatized until recently, is a harmless sexual activity which also occurs in many monkeys and apes. Here, we are concerned with those sexual proclivities which are abnormal in the sense that they occur in a minority of people and adversely affect their ability to form, or to sustain, adequate sexual relationships. These are the *paraphilic* behavioural patterns, many of which are deemed to be deeply antisocial or unlawful. Paraphilias, as defined in Table 5.13, involve erotosexual preferences for inappropriate partners (as in paedophilia or zoophilia), parts of the body or objects (fetishism), or activities incompatible with normal relationships (e.g., sadism, rape, and lust-murder). Classification of the paraphilias is more complex than outlined in Table 5.13, however. Paedophilia, for example, includes individuals whose sexual preferences are exclusively homosexual, exclusively heterosexual, or bisexual. For some paedophiles aggression or sadism are important components of their

sexual fantasies or behaviour, whilst for others aggressive tendencies are not involved (Langevin *et al.* 1985a). It is also true that men who engage in rape belong to 'a mixture of types' (Langevin *et al.* 1985b), so that there is unlikely to be a single causative explanation for their behaviour.

The paraphilias are of theoretical importance because they pose questions about the development of normal human erotosexual preferences. We are also led to wonder (as with the question of masturbation) whether homologues of paraphilic behaviour might occur in the non-human primates. One route to the development of paraphilic behaviour might involve conditioning to inappropriate sexual stimuli during childhood or adolescence (Money 1977b, 1988). In fetishism, such conditioning results in erotic arousal in response to parts of the body (e.g., the feet or hair), items of clothing (such as shoes), or particular textures (e.g., rubber). Fetishism is more common in men; indeed the majority of paraphilias exhibit a male sex-bias (Table 5.13). Janus and Janus (1993), for example, reported that 11% of men and 6% of women in their North American sample had personally experienced fetishistic activities (Table 5.14). Money (1977b) points out that men are more prone to develop paraphilias which involve visual stimuli; and that this might reflect a fundamental sex difference in the importance of visual cues in human sexual arousal. We might hypothesize that, under normal conditions, men respond to a gestalt of sexually attractive stimuli. In heterosexual males these might include the female secondary sexual characteristics (shape and size of the breasts and buttocks, the waist-hip ratio and other features) as well as facial cues, feminine patterns of movement, and auditory or olfactory stimuli. Such preferences are not simply pre-programmed into males, however (the so-called *innate releasing mechanisms* of classical ethology); they are subject to conditioning during postnatal life. Precisely how and when this occurs is far from being fully understood. It is interesting to note, however, that even in adulthood the human male may exhibit classical conditioning of penile erection in response to unusual cues (Rachman and Hodgson 1968; Bancroft 1974). Similar findings have been obtained in experiments in adult rhesus monkeys, marmosets and chimpanzees (author's observations). Therefore, it is possible that conditioning of erectile and erotic responses to inappropriate stimuli in childhood or during adolescence might result in a fixation upon such stimuli in adult life. The case histories of abnormal sexual preferences recounted by Von Krafft-Ebing in 'Psychopathia Sexualis' include frequent references to the childhood or adolescent experiences of afflicted individuals. Two examples are cited below:

Case 96 (an example of events during childhood and puberty leading to fetishism in later life): 'Since his seventh year he had for a playmate a lame girl of the same age. It lies beyond doubt that [at puberty] the first sexual emotions towards the other sex were coincident with the sight of the lame girl. His fetish was a pretty lady who, like the companion of his childhood, *limped with the left foot*'.

Case 103 (an example of events during adolescence and subsequent fetishism): 'At the age of fourteen he was initiated into the pleasure of love by a young lady. This lady was a blond and wore her hair in ringlets; and, in order to avoid detection in sexual intercourse with her young lover, she always wore her usual clothing—gaiters, a corset, and a silk dress on such occasions. [In later years]....in order to awaken his desire, a woman *had to be* a blond, wear gaiters, a corset and a silk dress—in short she *had to be* dressed like the lady who had first awakened his sexual desire'.

Remarkable in such cases is the paramount im-

portance of the fetish object in sexual contexts; the female partner's role may be relegated to that of a vehicle for the fetish, or the fetish alone may suffice to arouse the man. A sexual relationship might be sustainable in these circumstances if the female partner is sympathetic and adaptable. Von Krafft-Ebing records the case of a woman who discovered early in her marriage that her husband was potent only if she wore an immense wig of his choosing. A new wig was required every three weeks, so that 'the result of this marriage was, after five years, two children, and a collection of seventy-two wigs'. In other cases, a relationship is not possible, as in the example described by Janus and Janus (1993) of a foot-fetishist who was unable to consummate his marriage. His wife recounted that the man repeatedly attempted to rub his erect penis against her feet but 'lost his erection entirely' if she attempted to initiate vaginal intercourse.

I have been unable to identify homologues of

Table 5.12 Occurrence of masturbation in captive (C), semi-free-ranging (SFR) and wild (W) Old World anthropoid primates

Species	Occurrences in males	Sources	Occurrences in females	Sources
Red colobus	W	Struhsaker 1975[1]	–	–
Talapoin	C	Dixson *et al.* 1975 Author's observations	C	Dixson *et al.* 1975
Rhesus macaque	C	Harlow 1965		
	SFR	Carpenter 1942		
	W	Lindburg 1967		
Japanese Macaque	C	Hanby *et al.* 1971	SFR	Wolfe 1984b; 1991
Black ape of Celebes	C	Dixson 1977	C	Dixson 1977
Tibetan macaque	SFR	Zhao 1993[2]	–	–
Stumptail macaque	C	Bertrand 1969	C	Chevalier-Skolnikoff 1974
	C	Nieuwenhuijsen *et al.* 1987		
Grey-cheeked mangabey	W	Wallis 1983	–	–
Sooty mangabey	C	Author's observations	–	–
Olive baboon	W	Ransom 1981	W	Ransom 1981
Chacma baboon	C	Bielert *et al.* 1980	–	–
Hamadryas baboon	C	Zuckerman 1932	–	–
Mandrill	SFR	Author's observations	–	–
Gelada	C	Schmidt (personal communication)	–	–
Orang-utan*	C	Nadler 1988	C	Nadler 1988
	SFR	Rijksen 1978*	SFR	Rijksen 1978*
	W		W	Schurmann 1982
Gorilla	C	Hess 1973	C	Hess 1973
Chimpanzee*	C	Hediger 1965 Author's observations	C	Hediger 1965*
	W	Goodall 1986	W	Goodall 1986
Man*		Kinsey *et al.* 1948* Ford and Beach 1952		Ford and Beach 1952 Kinsey *et al.* 1953*

* In these cases the use of objects to aid masturbation has been observed (see text).
[1] A single observation only, of one adult male. Ejaculation did not occur.
[2] Males aged 2–4 years. Ejaculations did not occur.

human paraphilic behaviour in monkeys or apes. This is not to say that abnormal sexual behaviour does not occur in non-human primates. Many examples of abnormal sexual behaviour may be cited among captive monkeys and apes, especially those animals reared in isolation and deprived of maternal comfort and peer contact during infancy (Harlow 1971). As discussed earlier, rhesus monkeys separated from their mothers and reared on inanimate 'surrogate' mothers during the first year of life are socially and sexually incompetent in adulthood. Males are most affected and may become aggressive during encounters with conspecifics, as well as making poorly orientated mounts on females or exhibiting social withdrawal and stereotypies. Isolate-reared rhesus monkeys have been shown to exhibit complete sexual responses when provided with an inanimate dummy as a partner, rather than a living member of the opposite sex. Males will mount the dummy and females will present to it during pair-

tests (Deutsch and Larsson 1974). However, there is no evidence of a 'fetishistic' preference for such a dummy; rather it represents a less threatening sex object for animals which are incompetent in their social contacts with living partners and hence abnormal in their sexual behaviour. Isolate-reared male monkeys do not become fixated sexually upon the feet or hands of females—at least there is no evidence of such fetish-like behaviour. Their aggression seems to be related to fearfulness in social encounters; it is not comparable to sadism in the human male. Male rhesus monkeys have been reared with dogs as maternal surrogates for the first 2.5 years of life (Mason and Kenney 1974). In adulthood, such males are not fixated sexually upon dogs; they approach and mount female rhesus monkeys readily during pair-tests, albeit their mounts are poorly orientated (Mendoza: personal communication). Early exposure to canine foster mothers might have been expected to engender a cross-species sexual prefer-

Table 5.13　Examples of human paraphilias, with brief descriptions of their characteristics (male > female indicates a higher incidence in men)

Paraphilia	Defining characteristics
Coprolagnia	The desire to defaecate upon others or to be defaecated upon as part of sexual activity (male > female).
Exhibitionism	Exposing the genitalia, e.g. 'flashing' in public (male > female)
Fetishism	Erotic preoccupation with a part of the body (e.g. the foot or the hair), articles of clothing (e.g. shoes), or textures of clothing (e.g. rubberware) rather than the whole partner (male > female).
Frottage	An erotic preference for rubbing the genitalia against the body of another—rather than engaging in intercourse (male > female).
Gerontophilia	Use of an elderly person as a sexual object (male > female).
Lust murder	Sexual pleasure associated with murder of the sexual object (male > female).
Masochism	Obtaining sexual gratification by being hurt, humiliated, bound, or degraded (male similar to female).
Necrophilia	Use of a dead body as a sexual object (male > female).
Paedophilia	Use of an immature person as a sexual object (male > female).
Rape	Coercive sexual activity without consent of the partner (male > female)
Self strangulation	Induction of hypoxia (e.g. by hanging oneself) to heighten orgasmic pleasure during masturbation (male > female).
Sadism	Obtaining sexual pleasure from acts of cruelty (male > female).
Scoptophilia	Voyeurism or Peeping Tomism-clandestine viewing of others who are naked and/or engaged in sexual activity (male > female).
Urophilia	The desire to urinate upon others, or to be urinated upon, as part of sexual activity (male > female).
Zoophilia	Use of an animal as a sexual object (male > female)

This listing is not exhaustive. Information from Allen 1961; Money 1977b; and Bancroft 1989.

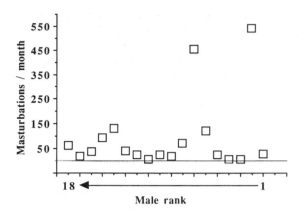

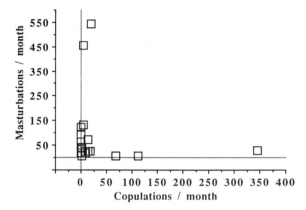

Table 5.14 Percentages of men and women who report having experienced certain paraphilic activities

Type of Paraphilia	% Men	% Women	Sample sizes Men	Sample sizes Women
Sado-masochism	14	11	1336	1406
Dominance/bondage	11	11	1331	1403
Fetishism	11	6	1329	1399
Verbal humiliation	5	7	1342	1411
Urophilia	6	4	1291	1381
Coprolagnia	1	0	1241	1310
Necrophilia	0	0	1291	1367
Paedophilia	2	0	1340	1413

Data compiled from Janus and Janus (1993).

Fig. 5.38 Masturbatory frequencies, male rank and copulatory frequencies in a captive group of stumptail macaques (*Macaca arctoides*). Male rank shows no correlation with frequencies of masturbation (upper graph). Nor is masturbatory frequency significantly correlated with frequencies of copulatory behaviour in these males (lower graph). (Data are from Nieuwenhuijsen *et al.* 1987.)

ence (an equivalent of zoophilia in human beings), but such is not the case. Cross-species matings have been recorded, but they are either 'facultative' in the sense that they occur in the absence of conspecific partners, or they are cases of 'mistaken' mate-choice. Mitchell (1977), for example, recounts the case of copulation between a male rhesus monkey and a female baboon which occurred after the animals had been caged together for a long period in the absence of other monkeys. Closely-related monkey species sometimes hybridize in the wild (e.g., certain *Cercopithecus* species: see Chapter 9 for examples), but consistent cross-species sexual preferences have not been found and it seems highly unlikely that they would occur.

If true paraphilic behaviour is absent in monkeys,

why does it occur in man and are the paraphilias universally present in different human cultures? It is certainly true that children can be emotionally damaged if deprived of love during the first few years of life. Bowlby's (1969, 1973, 1980) reports underline the harm that befalls children raised in orphanages and deprived of maternal affection. Such individuals might have problems in cementing close relationships if they are placed with foster parents; they often do badly at school, lack self esteem, and are stated to be precocious sexually in some cases. Whether they are more likely to exhibit paraphilic behaviour at puberty or in adulthood is unknown, but the notion that maternal deprivation creates such sexual abnormalities is a gross oversimplification of a complex problem. Whilst the unnatural and sometimes harsh conditions of a childrens' home might place some individuals at greater risk, it must be stated that many men who exhibit paraphilic behaviour were not orphaned or reared under deprived conditions. It is also of interest that, from generation to generation, categories of paraphilic activity remain relatively constant. Fetishists in Von Krafft-Ebing's time were perhaps more likely to be fixated upon hands, or handkerchiefs and velvet items than nowadays, but feet, shoes, and hair functioned as fetishes over a century ago and continue to do so in modern times. Bizarre erotosexual activities, as in the use of excretory products (coprolagnia and urophilia), dead bodies (necrophilia), and animals (zoophilia) have been recorded since ancient times. In considering what might underlie conditioning to such inappropriate stimuli it is relevant to mention two factors which affect human social development but which are absent (or virtually so) in non-human primates. Firstly, there is the phenomenon of self-awareness, expressed sexually in an individual's gender identity and gender role. As discussed previously, the concepts of gender identity

Fig. 5.39 A female rehabilitant orang-utan (*Pongo pygmaeus abelii*) using a piece of vine to stimulate her genitalia. The use of objects for masturbation has been recorded in both sexes of the orang-utan. (Author's drawing from a photograph in Rijksen 1978.)

and gender role are not applicable to monkeys, although homologues might exist in the great apes. Higher cognitive abilities, coupled with the emergence of self-awareness during the third year of life in children, might create additional risks for sexual development which are not present in monkeys. Secondly, and related to this last point, is the sheer complexity of human social learning, and the many cultural and religious taboos with resultant feelings of guilt or shame which attach to sexual activities. In societies which cloth the human body from childhood onwards and which regard childhood expressions of sexuality as totally unacceptable, is it surprising that paraphilic behaviour sometimes occurs? A child exposed to harsh physical punishment might,

in rare cases, become prone to sado-masochistic activities in later life. A child who is made to feel guilty and shameful with respect to genital and excretory functions might confuse them in sexual contexts and in some way become subject to coprolagnic or urophilic fantasies. Given the reluctance of many societies to allow children to play together naked, to see adults naked, or to gain any practical experience of sociosexual contacts, is it surprising that they sometimes fixate upon what *is available and visible* (i.e., the feet, shoes, or hair) rather than upon biologically appropriate stimuli?

Of the greatest interest from a comparative viewpoint are those few anthropological studies which provide detailed information on sexual attitudes and behaviour in societies where sexual guilt or chastisement of sexual activity in children is rare. The study by Money *et al.* (1977) of the Yolngu, an aboriginal tribe living in Northern Australia, is valuable in this respect. Money *et al.* comment that 'the straightforward attitude of the Yolngu towards nudity and sex play in young children allows these children to grow up with a straightforward attitude towards sex differences, towards the proper meaning and eventual significance of the sex organs and towards their own reproductive destiny'. The Yolngu certainly have cultural taboos concerning sex; marriages are arranged and polygyny is the rule. At age 12 or 13 a girl might be betrothed to a man who is 40 or 50 years old. In these circumstances sexual problems do occur and intercourse for some women is a perfunctory affair, lacking in orgasm. However, a remarkable feature is 'the relative freedom in Yolngu society, from sexual behaviour disorders'. Paraphilic behaviour is virtually absent; Money *et al.* report a single possible case of adolescent scoptophilia and another of zoophilia in a mentally defective man. The key difference between these people and westernized societies appears to lie in their tolerance of childhood sexuality, nudity, and lack of guilt concerning sexual matters. In view of the degree of human misery caused by the paraphilias listed in Table 5.13, the observations of Money *et al.* are especially important. Paraphilias are not universally present in human societies, their incidence could be greatly reduced if tolerance and education in sexual matters were more widespread. This is a most important but socially sensitive area of sex research.

6   Sociosexual behaviour and homosexuality

Sociosexual patterns

In Zuckerman's (1932) text 'The Social Life of Monkeys and Apes' he recognized that some monkeys, such as the hamadryas baboons (*Papio hamadryas*) which he had studied on 'Monkey Hill' at London Zoo, do not confine the use of sexual patterns to copulatory contexts. Male monkeys sometimes present their hindquarters to other males, a fact that was well known to Darwin (1876). Male monkeys may also mount one another and similar behaviour occurs between females of certain species. Often, the animals engaged in such *sociosexual* activities use motor patterns normally associated with sexual behaviour as a social greeting, to deter aggression, or to reinforce rank-related interactions. The form and functions of sociosexual patterns vary between species, but the important point is that motor patterns normally associated with sex are sometimes incorporated into the non-sexual sphere of social communication.

Examples of sociosexual mounting, presentation, and genital inspection are shown in Figs 6.1, 6.2, and 6.3. Isosexual mounts and presentations are commonly observed in many Old World anthropoids such as macaques, baboons, mangabeys, mandrills, talapoins, patas monkeys, the red colobus, the proboscis monkey, various langur species, and in the great apes. As we shall see, such behaviour also occurs in some New World genera, but the sociosexual activities of squirrel monkeys, howlers, marmosets, and tamarins show some important differences from those of their Old World relatives. Mutual embracing, accompanied by genital inspection, has also been recorded in some Old World monkeys (e.g., black apes of Celebes, stumptail macaques, and mangabeys—see Fig. 6.3). Among the apes, female bonobos (*Pan paniscus*) show an unusual form of sociosexual behaviour, during which they rub their genitals together; this pattern is referred to as 'G–G rubbing' (Fig. 6.4). Male bonobos also rub their erect penes together (so-called 'penis fencing') but this is much rarer than the G–G rubbing displays of females (Kano 1992). G–G rubbing

serves a tension-reducing function in bonobos as, for example, when these animals are feeding upon some prized item such as sugar cane. Mutual handling of the genitalia occurs in chimpanzees, and often among adult males it acts as a reassurance behaviour, for instance when sub-groups meet and mingle (Fig. 6.5). In the mandrill (*Mandrillus sphinx*) it is usually the higher-ranking males that touch or manipulate the penes of subordinates, often in response to sociosexual presentations by lower-ranking males.

Does sociosexual behaviour reflect dominance rank?

Hanby (1977) has aptly stated that 'most commonly, sociosexual gestures appear to function as tension-reducing, comforting, or appeasing mechanisms, rather than as sexual or dominance expressions'. This view runs counter to an older literature which ascribes a dominant role to the mounter and a submissive role to the mountee or presenting partner during sociosexual interactions (Zuckerman 1932; Maslow 1936–1937; Wickler 1967). Hanby, by contrast, concludes that 'equating the mounter's role with maleness and dominance and the recipient's with femaleness and submission is an outmoded and outworn concept'. As a generalization I believe this to be correct. Attempts to find simple correlations between agonistic rank and rankings based upon sociosexual mounts or presentations have usually proven fruitless. Bernstein (1970), for instance, found no correlations between dominance hierarchies based upon agonistic behaviour and those constructed by measuring mounting behaviour in captive groups of *Cercopithecus aethiops*, *Cercocebus atys*, *Theropithecus gelada*, *Macaca nemestrina*, *M. fascicularis*, or *M. nigra*. Yet, in some catarrhines there are examples where presentation behaviour indicates submissiveness and mounts are made primarily by dominant individuals. In the talapoin monkey (*Miopithecus talapoin*), for instance, presentations and mounts which occur during, or immediately after, an aggressive encounter between two monkeys are affected by the rank-relationship between the animals. In one captive group, 80% of such presentations were made by lower-ranking monkeys to dominant aggressors whilst 87% of mounts were by dominant individuals on monkeys of lower-rank that they had just threatened or attacked (Dixson *et al*. 1975). It should be mentioned that only 36 presentations and 63 mounts following aggressive encounters occurred during 53 h of observations on this group of monkeys (three males and five females). Neither females' presentations to males nor males' mounts on females were included in the analysis since these might have been sexual rather than sociosexual behaviour. The aggression-induced presentations and mounts constituted only 15% and 19% respectively of the total presentations and mounts in this talapoin group during the same period. In other situations, higher-ranking monkeys sometimes presented to, and were mounted by,

lower-ranking animals. Females in talapoin groups seem to prefer to mount or present to a particular female partner, often the same animal they sit with or groom most often (Dixson *et al*. 1975). Such behaviour can form part of social relationships between females and even occurs occasionally during play between them (Gautier-Hion 1971).

Presentation can function as a social, submissive, or greeting gesture towards more dominant animals in wild and captive rhesus monkeys (Carpenter 1942; Altmann 1962), wild Nilgiri langurs (*Presbytis johnii*: Poirier 1970), wild grey-cheeked mangabeys (*Cercocebus albigena*: Chalmers 1968c), and wild and captive baboons (Zuckerman 1932; Bolwig 1959; Hall and DeVore 1965). Sometimes the submissive nature of a presentation may be identified by observing the exact form of the posture employed. Not all presentations are the same, as was shown by Hausfater and Takacs (1987) in their studies of free-ranging yellow baboons (*Papio cynocephalus*).

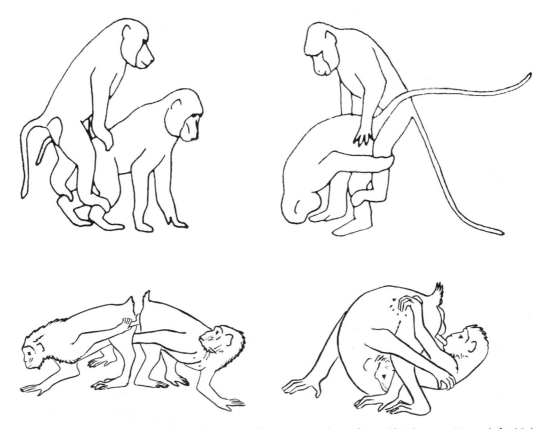

Fig. 6.1 Male–male sociosexual mounts and presentation postures in anthropoid primates. Upper left: Male–male double-foot clasp mounting in guinea baboons (*Papio papio*). (After Hanby 1977). Upper right: Male–male double-foot clasp mounting in bonnet macaques (*Macaca radiata*). (After Hanby 1977). Lower left: Mutual presentation and genital manipulation between male stumptail macaques (*Macaca arctoides*). (After Chevalier-Skolnikoff 1974). Lower right: Mounting and oro-genital contacts between male stumptail macaques (*M. arctoides*). (After Chevalier-Skolnikoff 1974.)

Fig. 6.2 Female–female sociosexual mounting postures in anthropoid primates. Left: Double-foot clasp mounting and reaching back response in female black apes of Celebes (*Macaca nigra*). (After Dixson 1977.) Right: Female stumptail macaque (*M. arctoides*) mounted 'piggy back' fashion upon another female. (After Chevalier-Skolnikoff 1974.)

These authors were able to define six categories of presentation posture and to show that submissive animals often postured with their legs flexed in a distinctive manner (Fig. 6.6). In the mandrill (*Mandrillus sphinx*), sociosexual presentations are usually given by monkeys to those which rank above them. 'Rank' was determined by quantifying agonistic encounters and displacement—avoidance behaviour in the manner applied to baboons by Rowell (1967a). Mandrills often show leg-flexure and pronounced curvature of the spine to look back at a dominant animal when presenting sociosexually (Frei and Dixson, unpublished observations). This type of cowering during presentation is especially obvious when a young male presents to an adult male in the group or when a female presents to a much higher-ranking individual of either sex. The form of the

Fig. 6.3 Mutual lateral embracing and genital investigation between two adult female Celebes macaques (*M. nigra*). (After Dixson 1977.)

Fig. 6.4 Two adult female bonobos (*Pan paniscus*) mutually embrace and engage in 'G–G rubbing' behaviour. (After Kano 1992.)

presentation contrasts markedly with that used to solicit copulation and, as in the talapoin, submissive mandrills may present in this way as a result of receiving aggression. Indeed, this type of sociosexual behaviour might serve to terminate aggressive interactions and forms part of what de Waal has called *reconciliation* behaviour. An examination of Fig. 6.6 also calls to mind Darwin's (1872) views on *antithesis* in the form and function of displays which are used for visual communication. The facial expressions or postures of a confident or aggressive animal often contain components which are the opposite of those used by a submissive monkey. The sociosexual presentation of a submissive female mandrill or yellow baboon, with legs bent, hindquarters lowered and spine flexed laterally, is in many ways the antithesis of the upright, four-square, sexual presentation given by a confident female towards a prospective mate. Further studies of sociosexual communication in primates would benefit from fine-grained quantification of the various displays employed together with the exact contexts in which they occur.

Evidence of a positive relationship between social rank and patterns of sociosexual behaviour has also been provided by Srivastava *et al.* (1991) for free-ranging Hanuman langurs (*Presbytis entellus*). The group studied by Srivastava *et al.* contained a single resident male and a maximum of 15 adult females. Their observations centred upon mounting interactions between adult females. Of the 524 mounts observed by the authors, 84% were by females on others of lower rank. Figure 6.7 shows a typical female−female mount involving a double-foot clasp position and pelvic thrusting movements by the mounter. Figure 6.7 also shows relationships between female rank and such mounting activity. Whilst females did mount higher-ranking partners, the frequency of this behaviour was much less than predicted by chance if the differences in rank between partners exceeded two places in the hierarchy. By contrast, higher-ranking females mounted lower-ranking ones at greater than expected frequencies for rank differences of 1, 2 or 3 positions. Thus, whilst it is an oversimplification to interpret mounting solely as a dominance-related activity in Hanuman langurs, or in other primates, there are contexts where mounts and presentations relate to social rank. Srivastava *et al.* (1991) observed that female Hanuman langurs which were mounting sometimes behaved aggressively, head-bobbing and teeth grinding at mountees, or displacing the latter. Other displays which were associated with sociosexual mounts included jumping and mutual embracing; these were thought to serve a reassurance function. In many other species it has also been observed that presentations or mounts might serve some form of distance-reducing or reassurance behaviour between animals before they sit together or groom one another (Dixson *et al.* 1975; Dixson 1977; Hanby 1977). This does not imply that it is necessarily the mounter which always receives grooming or the presenting animal which initiates a grooming

Fig. 6.5 Genital manipulation as an affiliative behaviour in male chimpanzees (*Pan troglodytes*). In this case the behaviour occurred in response to calls from a neighbouring community. (After Wrangham and Smuts 1980.)

bout. Indeed, in the study by Srivastava *et al.* 79.2% of grooming bouts which followed mounting interactions involved mounters grooming mountees.

Penile erection and sociosexual communication

In males belonging to a wide variety of primate species erections occur in contexts other than during copulation. In some cases these erections can be viewed as epiphenomena, or side effects of the autonomic activity which occurs when males are aroused during play or aggression. In other cases erections have become *ritualized* and incorporated into sexual or sociosexual displays. Examples of both types of erectile response are considered in the following sections.

Erection in response to the sight or proximity of a female

Male monkeys sometimes exhibit penile erection when observing a female or following her and inspecting her genitalia (*Miopithecus talapoin*: Dixson *et al.* 1975; *Macaca nigra*: Dixson 1977; *Macaca mulatta*: Pomerantz 1990; *Mandrillus sphinx*: author's observations). Male common marmosets (*Callithrix jacchus*) develop such *anticipatory erections* during laboratory pair-tests, before the female has been released into the test cage and before the animals can see one another. Males are apparently conditioned to the testing procedures and erections occur in anticipation of sexual contact (Lloyd and Dixson 1988). In none of these species do males employ erections as sexual invitations to females, however. It is easy to imagine that precopulatory erections, indicating the male's sexual arousal, might become incorporated into invitational displays. Examples of such ritualized behaviour are provided by the great apes (*Pan troglodytes*: Yerkes 1939; Goodall 1986; *P. paniscus*: Kano 1992; *Pongo pygmaeus*: Nadler 1988). Chimpanzees and pigmy chimpanzees both use erections as part of elaborate visual displays to solicit copulation (see Chapter 5). In the orang-utan, in which the penis is obscured by long abdominal hair, males may none the less recline on their backs and exhibit the erected penis prior to copulation. This display also acts as a sexual invitational posture (see Chapter 5). Occasional attempts have been made to

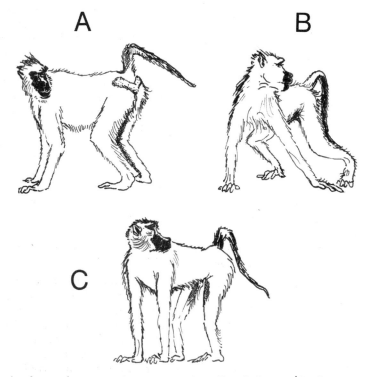

Fig. 6.6 Variations in the form of presentation postures in yellow baboons (*Papio cynocephalus*). A. Submissive presentation with legs partially flexed (this sub-adult male is presenting to a more dominant adult male). B. An adult female presents submissively to a high-ranking male. Note the pronounced curvature of the spine and leg flexure. C. Sexual presentation posture by an adult female. There is no leg flexure or curvature of the spine. (Author's drawings, after Hausfater and Takacs 1987.)

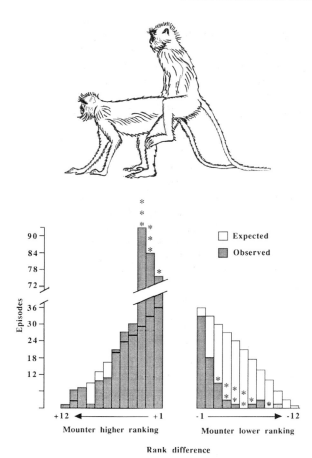

Fig. 6.7 Relationships between female rank (based upon measurements of displacement behaviour) and sociosexual mounting in free-ranging Hanuman langurs (*Presbytis entellus*). Mounters are higher-ranking than mountees in the majority of cases, especially when the females concerned are close to one another in rank. *P < 0.05; **P < 0.01; ***P < 0.001. (Redrawn from Srivastava *et al*. 1991.)

implicate erection as a precopulatory display during human evolution, and the prominent erection of the chimpanzee has been cited as a parallel case of sexual selection for visual display (Short 1980). However, the elongated filiform penis of the chimpanzee has probably evolved in tandem with the female's large sexual swelling and is adapted for its role during intromission (Dixson and Mundy 1994). There is little evidence that large penes evolved for precopulatory display purposes in any of the primates.

Erection during sleep

In man, nocturnal penile tumescence (NPT) typically accompanies periods of rapid eye movement (REM) sleep. Indeed, these responses have proven valuable to clinicians studying the causes of erectile dysfunction (Bancroft 1995). NPT is an epiphe-

nomenon and probably results from decreased tone in the sympathetic innervation of the penis. An inhibitory pathway from the thoracic/lumbar level which innervates the genitalia is perhaps 'disinhibited' during REM sleep (see Chapter 13, pages 430–4 for a discussion of the neural control of penile erection). Unfortunately, there have been no studies of NPT in non-human primates.

Erection during play, grooming or when seeking 'reassurance'

Penile erections have been observed during play in some primates (e.g., in the Sumatran orang-utan *Pongo pygmaeus abelii*: Rijksen 1978) or whilst males are being groomed (*M. talapoin*: Dixson *et al*. 1975; *M. nigra*: Dixson 1977; *M. mulatta*: Herbert 1973; *M. sphinx*: author's observation). Infant male Japanese macaques sometimes have penile erections whilst being groomed by their mothers (Hanby and Brown 1974). The same is true of infant rhesus monkeys (author's observation) and probably of many other species. Baby boys exhibit penile erection on occasion, whilst being dried after a bath, for example, (Kinsey *et al*. 1948). Erections in situations where males are 'distressed' or 'seeking reassurance' have also been noted by some authors. Rijksen (1978) has described infant and juvenile male Sumatran orang-utans having erections and appearing to be distressed when their mothers moved away and they were unable to catch up. Adult male chimpanzees also exhibit erection, often accompanied by mutual genital touching, as a form of reassurance behaviour (Fig. 6.5). All these cases demonstrate the wide variety of arousing stimuli which might elicit erections, but there is no implication that the erections themselves serve any function in visual communication. In the following section examples of ritualized penile displays are reviewed but it is as well to remember that such ritualization could be based upon genital responses in non-sexual contexts.

Erections during aggression and ritualized penile displays

Males in one-male units of proboscis monkeys (*Nasalis larvatus*) sometimes display aggressively to males outside their units by branch-shaking, vocalizing, and exhibiting erections (Yeager 1990b). Marler (1969) observed erections in a similar context in resident males of the one-male units of *Colobus guereza* and I have observed these displays between males in two captive groups of *C. polykomos*. It is possible, but unproven, that such erections form part of the overall aggressive display. However, erections might simply occur because the aggressive male is highly aroused; in Hanuman langurs (*Presbytis entellus*), for example, males tooth-grind and exhibit erections during some aggressive episodes but they do not actively display erections in a stereotyped manner. Wickler (1967) drew attention to the fact that male baboons or vervets resting in the fringes of their social groups, sometimes sit with legs apart and with their genitalia exposed to view. Erections may occur in these circumstances and Wickler argued that such males might serve as 'guards', signalling their presence to adjacent groups. He also compared human phallic symbolism with the hypothesized genital displays of male non-human primates. The evidence presented in support of his ideas was limited and circumstantial. Little has been added to our knowledge that would support his view. If a male primate has brightly coloured sexual skin and a large phallus it is always tempting to assume that some genital display is associated with this. Morris (1967), for instance, noted that the facial colours of the mandrill (*M. sphinx*) are similar to the red and blue sexual skin that covers its genitalia. He suggested that the face 'mimicked' the genitalia for display purposes. In reality, mandrills do not employ penile displays or genital displays, except for sociosexual presentation postures which are given much more frequently by females and immature animals than by adult males. Moreover, the facial displays of male mandrills are identical to those of its sister species the drill (*M. leucophaeus*) which lacks the brilliant blue and red face mask.

The best documented case of ritualized genital displays in primates concerns the squirrel monkeys (genus *Saimiri*) which do not possess large or brightly coloured external genitalia. Squirrel monkeys of both sexes exhibit three genital display postures which are often accompanied by penile or clitoral erections (Ploog 1967). In the *open position* one animal displays at some distance from the recipient whilst in the *closed position* the displaying monkey is almost touching the recipient and appears 'to jab the partner frontally with its penis'. Two monkeys may also display simultaneously in a *counter position*. These display types are shown in Fig. 6.8 which also summarizes the direction of displays both within and between the sexes. Ploog (1967) pointed out that he had never observed females displaying to males by using a closed position and that males never answered a female display by adopting a counter posture. Why this should be so is not known. Unlike the penile displays of chimpanzees, the genital displays of squirrel monkeys are not sexual invitations. On the contrary, they appear to serve a variety of communicatory functions related to dominance rank and affiliative behaviour. However, from Ploog's (1967)

PARTNER RELATIONS IN DISPLAY BEHAVIOUR

Open position	Closed position	Counter-position
♂ → ♂	♂ → ♂	♂ ↔ ♂
♂ → ♀	♂ → ♀	♂ ↔ ♀
♀ → ♂	—	—
♀ → ♀	♀ → ♀	♀ ↔ ♀

Fig. 6.8 Genital displays of the squirrel monkey (*Saimiri sciureus*). Upper: The animal on the left displays in the open position, whilst looking at the recipient. Lower: The closed counter-position in which the partners touch, but do not look at each other. The table summarizes the direction of display types observed within and between the sexes in captive squirrel monkey groups. (After Ploog 1967.)

account it is far from clear what information might be transmitted by such displays and we still know far less about this type of sociosexual behaviour than about presentation-mounting interactions in catarrhines. Genital displays might not be limited to the genus *Saimiri* among the South American primates. Penile erections are said to occur as ritualized displays in captive black howler monkeys (*Alouatta caraya*: Jones 1983) but no function has been described for this behaviour. Likewise, capuchins (genus *Cebus*) might employ some form of ritualized penile display but there are insufficient data to decide upon its significance (Freese and Oppenheimer 1981). Contributions to knowledge in these areas could be made by studying behaviour in captive groups of monkeys.

Sociosexual presentation in New World primates

Among the New World primates some species of marmosets and tamarins (family *Callitrichidae*) employ sociosexual presentations in ways that are quite different from those described for Old World monkeys and apes. Female callitrichids, such as the common marmoset (*Callithrix jacchus*), do not exhibit hindquarter presentation as a proceptive behaviour. Yet, both sexes do employ a stereotyped genital presentation with the limbs held stiffly and the tail arched over the back (Fig 6.9). The display occurs in agonistic contexts and it is usually the dominant partner that presents the genitalia to an animal of lower-rank (Epple 1975). Chemical stimuli play a crucial role in social communication among callitrichids, including the communication of social rank (Epple 1974, 1980; Epple *et al*. 1993). In some marmosets and tamarins only the dominant female in a family group breeds. Olfactory cues contribute to the suppression of ovarian cyclicity in subordinate females (Abbott *et al*. 1993). The genitalia of both sexes are richly endowed with cutaneous scent-marking glands and both sexes mark substrates with urine as well as with cutaneous secretions (Epple *et al*. 1993). It is possible, therefore, that the stereotyped presentation postures of these monkeys, which display the male's scrotum or the female's pudendal pad so conspicuously (Fig. 6.9), serve to reinforce chemical cues indicative of high social rank and reproductive potential. There are indications that presentation displays have non-submissive functions in some other New World primates, but they are poorly documented. In the black howler (*A. caraya*) adult females sometimes present the vulva and males present the scrotum during stereotyped displays. Jones (1983) comments that the vulval display appears to mean 'Stop, slow down', whilst the male's scrotal display, in which the testes are descended during the quadrupedal presentation, appears to reduce aggression by signalling other males to 'back off'.

The development of sociosexual behaviour

Sociosexual patterns of mounting, presentation, mutual embracing, and genital inspection or manipula-

tion begin to develop during infancy in many monkeys and in the great apes (Fig. 6.10). There are frequent reports that male monkeys less than one year of age are capable of mounting other group members, including adult females. Such mounts include pelvic thrusting movements and penile erections, and in some cases they result in vaginal intromission (*Macaca radiata*: Rosenblum and Nadler 1971; *M. arctoides*: Trollope and Blurton Jones 1972; *Papio anubis*: Owens 1973; *Macaca mulatta*: Goy *et al*. 1974; *M. fuscata*: Hanby and Brown 1974; *M. nigra*: Dixson 1977; *Mandrillus sphinx*: author's observations). In the black ape of Celebes (*M. nigra*), males less than one year old are unable to grasp the adult female's calves with their feet (the double-foot clasp position). Instead they employ a modified posture, standing on top of the female's rump. These early rehearsals of copulatory patterns find an outlet in sociosexual interactions with members of both sexes. Infant male monkeys may mount one another, as well as infant females, either spontaneously (as during play) or in response to a presentation posture. The development of sociosexual behaviour has been studied most extensively in the rhesus monkey (Fig. 6.11). Early studies revealed that quantitative differences occur between the sexes in frequencies of a variety of behavioural patterns. Infant males showed higher frequencies of mounting, rough-and-tumble play, play initiation, and aggression than

Fig. 6.9 Genital presentation posture of a male common marmoset (*Callithrix jacchus*). (Author's drawing, after Epple 1967.)

females did, whilst female infants showed higher frequencies of sociosexual presentations (Goy 1968, 1978; Goy and Resko 1972; Phoenix 1974a). Male rhesus monkeys aged only 2–4 months are capable of mounting and making pelvic thrusting movements and by 6 months of age 50% of males can attain a double-foot clasp position (Goy *et al.* 1974). The foot clasp position is not ubiquitous among primates, however, and it is lacking during infant development in some species (e.g., *Saimiri sciureus*: Baldwin and Baldwin 1973; *Pan troglodytes*: Goodall 1968; and probably in *Gorilla gorilla*: Hess 1973; Nadler 1986). Other sociosexual patterns also appear early in development; in the squirrel monkey, for instance, infant males begin to use penile display postures at 49 days of age and will display in response to their own mirror image (Fig. 6.10).

The normal development of sociosexual behaviour depends upon at least two factors: exposure of the developing male brain to an appropriate hormonal environment and the effects of early social experience. Effects of neuroendocrine stimuli during gestation and early in postnatal development will be dealt with in Chapter 10. We are concerned at this point with how the infant's relationships with its mother and with other group members might influence sociosexual development.

Social stimuli and sociosexual development

The mother–infant relationship makes an important contribution to the development of sociosexual pat-

Fig. 6.10 Development of sociosexual patterns of behaviour in infant anthropoids. Upper left: Presenting and mutual genital manipulation by infant male stumptail macaques (*Macaca arctoides*). (Redrawn from Bertrand 1969). Upper right: An infant female stumptail mounts a juvenile female. (Redrawn from Bertrand 1969). Lower left: Double-foot clasp mounting in infant male rhesus macaques (*Macaca mulatta*). (After Keverne 1985). Lower right: A 49-day-old male squirrel monkey (*Saimiri sciureus*) displays its genitalia in the open position to a human observer. (Redrawn from Ploog 1967.)

terns in monkeys such as macaques and in chimpanzees. Hanby and Brown (1974) observed that infant Japanese macaques (*Macaca fuscata*) of both sexes made thrusting movements against the mother's body during the first postnatal week. By week two infants had begun to 'get off' their mothers and re-establishment of contact sometimes involved a type of presentation by the mother, encouraging the infant to climb on to her body (Fig. 6.12). There is no implication that these patterns are sexually motivated, merely that the infant learns from its earliest days to rehearse movements and make contacts that are valuable for sociosexual development. In rhesus macaques, infants that get off their moth-

ers are frequently 'beckoned' by other group members and in particular by juvenile females. Such beckoning involves facial displays, eye-contact, and often a type of presentation posture (Fig. 6.13) as an invitation to the infant to approach and establish contact (Hinde and Simpson 1975). Communication and contact between the infant and other group members, as well as with the mother, might be expected to influence sociosexual development.

The importance of social stimuli has been studied extensively in the rhesus monkey, by isolating infants and allowing only controlled amounts of contact with conspecifics. The most extreme experiments, by Harlow and his colleagues (Harlow 1971),

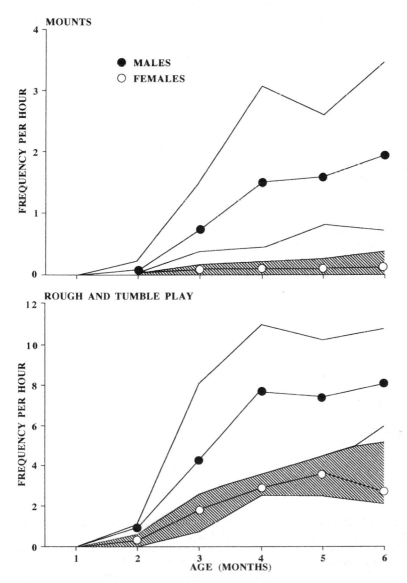

Fig. 6.11 Development of sex differences in frequencies of sociosexual mounting and rough-and-tumble play in rhesus monkeys (*Macaca mulatta*) living as members of captive breeding groups. Data are median hourly frequencies (±interquartile intervals) for 14 males and 20 females during the first 6 months of life. (Brown, Nevison and Dixson: unpublished data.)

involved rearing infants in complete isolation from their mothers and with no physical contact with other monkeys. Isolated infants were provided with various types of *surrogate mothers* which provided the opportunity to cling or receive milk from an artificial nipple. These experiments demonstrated the qualities that facilitate the infant's attachment to its mother. For instance, when given a choice between a wire surrogate mother which also provided milk, and a surrogate covered by towelling, infants chose the softer, towel surrogate. The tactile qualities of the mother were of paramount importance to the infants (Fig. 6.14). Harlow's experiments also produced fundamental information about the consequences of social isolation for longer-term behavioural development in rhesus monkeys. Isolation during the first year of life has profound effects upon the development of sexual and social behaviour, and these effects are more serious and long-lasting in males than in females. As adults, isolate-reared males either fail to mount females or

mount in a poorly orientated fashion, without attaining the double-foot clasp posture or intromission (Harlow and Harlow 1966; Missakian 1969; Goldfoot 1977). Nor do they rehabilitate to any marked degree. Some isolate-reared females succeed in copulating, although maternal behaviour is poor in such individuals and they are likely to neglect or abuse their offspring (Harlow 1971). Isolate-reared rhesus monkeys are socially withdrawn; they are aggressive towards conspecifics and prone to stereotypies such as rocking and self-biting. Their ineptness in sexual and sociosexual encounters might well reflect a broader problem: an inability to function normally in a social group or to form social relationships. Deutsch and Larsson (1974) provided male and female rhesus monkeys which were reared on surrogate mothers with a dummy instead of a live partner. The males mounted the dummy and made pelvic thrusting movements with the penis erect, whilst the females presented sexually to the dummy (Fig. 6.15). Hence the *copulatory*

Mount   Gestures

Boarding   Mounts

Dorsal   (mount)
riding

Dismount   Gestures

Fig. 6.12 Early mounting patterns of infant Japanese macaques (*Macaca fuscata*) on their mothers. (After Hanby and Brown 1974.)

mechanism is operative in such deprived males, and proceptivity is possible in females, given the provision of a suitably non-threatening and inanimate stimulus to elicit their sexual responses.

The mother is not the only source of stimulation required for the normal development of sociosexual patterns. Infant rhesus monkeys reared with their mothers, but with only limited access to peers, show marked deficits in sociosexual behaviour. Goldfoot (1977) compared the behavioural development of male infants reared with their mothers and given either 30 min per day or 24 h per day of access to peers. He found that six males that were allowed to interact freely with peers throughout their first 12 months of life had all developed the normal, double-foot clasp mounting pattern when observed in year 2. This accords with the observation that over 90% of rhesus males in captive social groups of this species will start to double foot clasp mount by 1-year of age. By contrast, none of the males restricted to 30 min daily access to peers exhibited the double-foot clasp posture when observed during their second year (Goldfoot 1977). Harlow *et al.* (1966) also reported that male rhesus monkeys reared for the first year with their mothers only, subsequently failed to exhibit double-foot clasp mounting when given access to peers. If, however, infants are reared without their mothers but with access to peers, then sociosexual mounting develops in some individuals. Peer grouping of male rhesus monkeys, isolated from their mothers at 3–6 months of age, results in 30% of subjects developing the foot clasp mount by 18–24 months, which is much later than normal males (Goldfoot 1977).

The experiments described above, and many others conducted at Wisconsin Primate Centre and Oregon Primate Centre during the 1970s and 1980s, have shown that if male rhesus monkeys have adequate access to their mothers and to peers during the first year then they develop normal sexual responses and are able to copulate to ejaculation in adulthood. By contrast, males deprived of normal social contacts in infancy fail to develop foot clasp responses and to copulate normally as adults (Goy

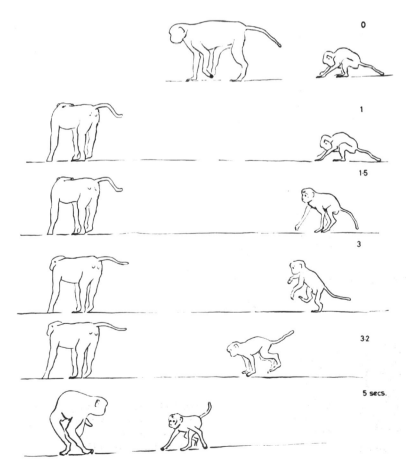

Fig. 6.13 A rhesus monkey mother leaves her infant and then 'beckons' it by establishing eye-contact and presenting. Drawings were made from a film and the approximate times of the frames are shown on the right. (After Hinde and Simpson 1975.)

and Goldfoot 1974; Goldfoot 1977). These findings have been interpreted as indications of purely social effects upon behavioural development, rather than effects of the social milieu upon the infant male's neuroendocrine system which, in turn, influences behaviour. Whilst social effects are undoubtedly of paramount importance, it is now known that increased testosterone secretion occurs in infant male rhesus monkeys (and males of some other primates) during the first 5 or 6 postnatal months (Plant 1982). These postnatal changes in testosterone secretion (as well as the more marked increases in testosterone which occur during foetal life) are now believed to influence the later maturation of the pituitary–testicular axis during puberty (Mann *et al.* 1989) and possibly to have subtle influences upon sexual behaviour (Eisler *et al.* 1993). These questions will be considered in detail in Chapter 10. It is sufficient at this point to note that experimental manipulations, such as isolation of the infant from its mother or introducing it to strange peers, might suppress the pituitary–testicular axis, but that no experiments have been conducted to examine this possibility.

In addition to the detailed studies of rhesus monkeys, several earlier workers had described the effects of social deprivation upon behavioural development in the chimpanzee (Nissen 1954; Davenport and Menzel 1963; Rogers and Davenport 1969). Nissen studied a group of captive chimpanzees which had been reared by human caretakers. These animals failed to exhibit copulatory behaviour, even when females were at maximal sexual skin swelling. However, when paired with sexually experienced chimpanzees, two out of four females eventually permitted copulation, whereas none of the six human-reared males mated. Davenport and Menzel (1963) and Rogers and Davenport (1969) studied chimpanzees which had been removed from their mothers within 12 h of birth and 'reared in closed boxes in isolation from humans and other chimpanzees'. From 3 years of age, these isolate-reared subjects were caged in pairs or groups in order to observe social and sexual behaviour. One might reflect that it is unlikely that such draconian experiments could be repeated nowadays, given current awareness of the intellectual capacities and rarity of these apes. Some interesting results emerged from this work. At 6–8 years of age the restriction-reared chimpanzees failed to intromit during pair-tests with restriction-reared females. However, when housed with sexually experienced chimpanzees in a social group, five restriction-reared females copulated (two became pregnant) whilst three out of five isolate-

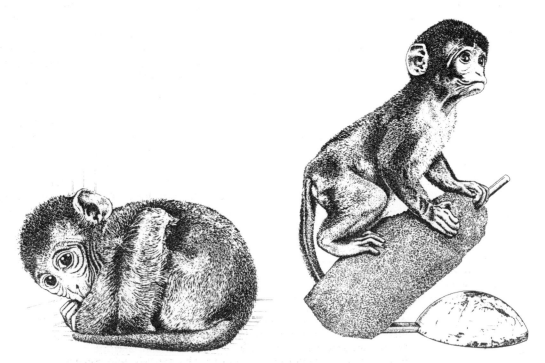

Fig. 6.14 Left: An infant rhesus monkey, reared in social isolation, huddles and sucks its fingers. Right: An infant rhesus reared in social isolation, but with access to a cloth 'surrogate mother' is less withdrawn and more advanced in its behavioural development. (From Keverne 1985; After Harlow and Harlow.)

Fig. 6.15 An adult male, isolation-reared, rhesus monkey mounts a dummy female. Note the double-foot clasp posture. Some isolation-reared males are capable of making a complete series of mounts, with pelvic thrusting movements and ejaculation. (Author's drawing, after Deutsch and Larsson 1974.)

reared males gained intromissions and two ejaculated. Although some improvements occurred in the males' behaviour with the course of time, intromissions were frequently interrupted by 'the performance of stereotyped repetitive behaviour [posturing, gross body-swaying and eye-poking] which disrupted social processes'.

Hence, isolation-rearing causes deficits in sexual behaviour in chimpanzees which are more pronounced in males than in females. Some chimpanzees of both sexes exhibit improvements of sexual function, however, and the capacity to recover seems to be greater than that of rhesus monkeys reared in isolation from their mothers. Moreover, Nissen's chimpanzees, which were reared by human caretakers, experienced more drastic effects upon sexual behaviour in later life than chimpanzees reared in isolation from their mothers and human beings (Rogers and Davenport 1969). This finding, if correct, has obvious implications for captive breeding programmes in zoological gardens, where it is still not unusual for infant chimpanzees to be hand-raised if their parents neglect them.

Is sociosexual behaviour equivalent to homosexual behaviour?

Despite the fact that some primatologists refer to isosexual mounts or presentations in monkeys and apes as *homosexual behaviour* (Maple 1977; Nadler 1990; Wolfe 1991; Vascy 1995), I do not think sociosexual patterns are equivalent to homosexuality in the human sense (see also Wallen 1997). Human homosexuality is not defined with reference to motor patterns (e.g. male mounts male, or female mounts female) but in relation to an erotic preference for members of the same sex, rather than the opposite sex, as partners in sexual activities. The usual practice is to refer to men with sexual preferences for male partners as homosexuals, whilst women who prefer female partners are called lesbians. Bisexuals have varying degrees of erotic preference for partners of both sexes. Kinsey recognized that as far as humans are concerned a spectrum of sexual preferences can exist, ranging from the exclusively homosexual (or lesbian) at one extreme, to exclusively heterosexual at the other. In their pioneering report on 'Sexual Behaviour in the Human Male' Kinsey *et al.* (1948) adopted a seven-point scale, ranging from O (exclusively heterosexual) to 6 (exclusively homosexual) to classify erotosexual preferences. The *Kinsey Scale* (reproduced in abbreviated form in Table 6.1) has been extensively used in studies of human sexuality and is still widely employed (McWhirter and Reinisch 1988).

Given that homosexuality is not necessarily an absolute quality, but exists as a spectrum of possibilities throughout the population, are there incidences of homosexual orientations in non-human primates? The answer to this question is 'possibly', but examples are rare and tend to occur under unusual circumstances. For instance, Rijksen (1978) in his field studies of Sumatran orang-utans, noted a few

Table 6.1 The Kinsey *et al.* (1948) rating scale for heterosexual-homosexual orientation

Scale point	Definition
0	Exclusively heterosexual
1	Predominantly heterosexual, only incidentally homosexual
2	Predominantly heterosexual, but more than incidentally homosexual
3	Equally heterosexual and homosexual
4	Predominantly homosexual, but more than incidentally heterosexual
5	Predominantly homosexual, but incidentally heterosexual
6	Exclusively homosexual

examples of homosexual contacts among young males. These involved 'rehabilitant' orang-utans, animals which had been kept in captivity for varying periods and were being reintroduced to a natural way of life in the rainforest. One rehabilitant male (a juvenile) was observed to repeatedly fellate another youngster. When this male finally achieved reintroduction to the wild and departed from the area, the second male began to engage in oral sex with a new partner. He had apparently learned to engage in this behaviour and continued to practise it. Rijksen also observed mounts involving ejaculation between young male rehabilitants. He noted that: 'the sub-adult male Sibujong playfully forced the adolescent male Bobo into a crouching position to have ano-genital contact. Afterwards, when both animals were roaming free, a second homosexual interaction between them was observed. In both cases the sub-adult ended the contact after ejaculation'.

It should be noted at this point that despite the high frequency of male–male mounts in monkeys, such as macaques, baboons, talapoins, mandrills, and many others, the occurrence of penile erection, anal intromission or ejaculation in these circumstances is exceedingly rare. From my own observations of talapoins, mandrills, and macaques I should say that isosexual mounts between males are characterized by a general lack of erectile responses. I have never observed anal intromission in these species. Although some authors assume that anogenital contacts occur during male–male mounts, most reports lack explicit information on this subject. Chevalier-Skolnikoff (1974) observed anal intromission between male stumptail macaques and Erwin and Maple (1976) reported such behaviour in a pair of male rhesus monkeys. This latter study is of particular interest because the males concerned exhibited a preference for each other when given the choice of associating with a female. These males had been reared in captivity and allowed access only to their mothers during the first 8 months. We have seen in previous sections that denying male rhesus monkeys access to peers during the first year of infancy adversely affects social and sexual behaviour in later life. This pair of males was caged together, after removal from their mothers, until approximately 30–33 months of age. For the next 2–3 years, as they passed through puberty and entered adulthood, each male was tested for sexual and social responses when paired with 1 year old males, and with adult strangers of both sexes. It is noteworthy that both of these males, despite early social deprivation, mounted female partners and ejaculated during pair-tests. Yet, when the two males were re-

united at 63 months of age, 'they displayed multiple mutual mounting with thrusting and intromission'. No ejaculations were observed during these male–male mounts, despite anaesthetizing the animals to check for the presence of semen in the rectum (Erwin and Maple 1976). When the pair was caged with a female, the males contacted one another, sat together, groomed, and mounted each other. By contrast, their interactions with the female were aggressive and she had to be removed as a result of attacks and bites by both males. Hence, in this exceedingly rare case, we have an example of male rhesus monkeys exhibiting some degree of sexual and social preference for each other involving anal intromission (but not ejaculation) during isosexual mounts. Yet, both males copulated and ejaculated with females. Perhaps it might be accurate to equate their behaviour with that of bisexual human males, but it is exceedingly difficult to say how far their interactions were socially motivated rather than reflecting an 'erotic' or sexual preference. It is necessary to recall that some human males who are primarily heterosexual in orientation will, nonetheless, exhibit *facultative homosexuality* under extreme circumstances (e.g. when imprisoned for long periods: Kinsey *et al.* 1948).

If we turn now to the question of female–female sociosexual behaviour in non-human primates, is there any indication that females develop erotic preferences for one another in a lesbian fashion? Here, I think the answer must be 'no', but with the proviso that some females of certain species can show orgasmic responses during isosexual interactions as well as during copulation. The best documented case concerns the stumptail macaque (*M. arctoides*) in which some females produce a 'copulation face' and undergo orgasmic uterine contractions when mounting other females (Goldfoot *et al.* 1980). The physiology of orgasm was discussed in Chapter 5 (see pages 129–36), where it was noted that the uterine responses of female stumptails during isosexual mounts resemble those of the human female during orgasm. Whether female stumptails find these experiences 'pleasurable' in the human sense is unknown, but it seems reasonable to suggest that they may do so. Whether females might then deliberately mount other females in order to experience such responses, or whether preferences for particular partners might develop, are unresolved questions. Vocalizations and genital responses during female–female mounts have been recorded in some other primate species. In talapoins, for instance, both partners may retract the lips and emit a 'keening' vocalization similar to that given by females during copulation (Dixson *et al.* 1975). Again there is no evidence that such responses are anything more than epiphenomena when they occur during sociosexual activity.

Patterns of sociosexual behaviour, such as presentations, mounts, and mutual embraces are much more commonly observed in primate species which live in larger social groups with multimale–multi-female mating systems (e.g., macaques, baboons, mandrills, mangabeys, and some langurs) or polygynous systems (e.g., hamadryas baboons, geladas, and proboscis monkeys) rather than in monogamous species. This makes perfect sense if one assumes that the need for complex social communication is greater in larger groups. As far as homosexuality is concerned, it is relevant to ask whether homosexual pairings or preferences have been reported in any primate species which has monogamy as the principal mating system. I know of no such case in marmosets, tamarins, owl monkeys, titi monkeys, or in the lesser apes. Homosexual pair-bonding would be biologically maladaptive in such species and one can see little possibility for any form of positive selection pressure. In groups of marmosets or tamarins a number of non-reproducing adults may be present, including mature offspring of the dominant (reproductive) female, which assist in carrying her twin offspring. Yet there is no evidence for any form of facultative homosexual activity in such circumstances. I mention this because of an argument advanced by E. O. Wilson (1978) concerning a possible adaptive function for human homosexuality. Wilson suggested that homosexuals, whilst not contributing directly to the gene pool, might assist their relatives in some way and thus contribute to future generations via kin-selection. There is no evidence, however, that homosexuals act as 'helpers' to their heterosexual relatives or assist the offspring of relatives in any additional way. As Bancroft (1989) has commented: 'such an explanation must sound ironic to modern-day homosexuals who are so often alienated from their families'.

Does sociosexual behaviour occur in man?

Rather than asking whether the sociosexual activities of monkeys and apes show any homologies with human homosexuality, it might be more fruitful to consider the possibility that sociosexual behaviour occurs in man, just as in many non-human primates. Given that sociosexual patterns are found in all of the great apes and are evident during infancy and juvenile life in orang-utans and gorillas, though much less commonly in adulthood, it is reasonable to

suggest that sexual patterns might occur during play and other activities in human children. In our own westernized societies, where so much guilt and opprobrium surrounds the subject of sex, especially in relation to children, it is no easy task to acquire meaningful data on childhood sexuality. With the publication of Freud's (1905) 'Three Essays on the Theory of Sexuality', the notion that man possesses a *sex-drive* or *libido* which commences its development in infancy was introduced to the world. It is not my purpose here to discuss Freud's theories regarding phases of infantile sexual development (oral phase, anal phase, phallic phase, etc.), merely to point out that the concept of infantile sexuality has existed for over 90 years.

Self stimulation of the genitals has been observed in baby boys and girls during the first year of life. Girls might begin to display such behaviour a little later than boys but, in both sexes, orgasm has been observed at as early as 6 months of age (Bakwin 1973; Galenson and Roiphe 1974). There is no suggestion that genital stimulation is accompanied by erotic imagery or sexual thoughts such as occur during pubescent or adult masturbation. Infants perceive genital stimulation as pleasurable and so may persist in the behaviour if not punished.

In older infants there is good evidence that 'sexual play' as well as self-directed behaviour is perfectly normal for our species, given appropriate opportunities and a climate of sexual tolerance. Two anthropological studies in particular can be cited to support this view. Firstly, there is Malinowski's (1932) report on 'The Sexual Life of Savages', which deals with sexual and social behaviour in the Trobriand Islands. Malinowski comments that 'children initiate each other into the mysteries of sexual life in a directly practical manner at a very early age'. Trobriand children 'indulge in plays and pastimes in which they satisfy their curiosity concerning the appearance and function of the organs of generation and incidentally receive, it would seem, a certain amount of positive pleasure'. Genital manipulation and oral stimulation occur during such games as well as childish simulation of intercourse, which the Trobrianders refer to as 'katya'. Adults find it perfectly acceptable that small children behave in this way during their daily games in the bush. The natives told Malinowski that it was part of the childrens' play to have such intercourse 'They give each other a coconut, a small piece of betel-nut, a few beads or some fruits from the bush, and then they go and hide and katya'. Provided that this behaviour occurred playfully in the bush and away from the family hut it was not discouraged in any way.

A second example of infantile sex play in the

human species is provided by Money *et al.* (1977) in their studies of the Yolngu aborigines of Arnhem Land, in the Northern Territory of Australia. For the Yolngu it is acceptable and amusing for small children to simulate sexual intercourse as part of their play behaviour. The behaviour is publicly tolerated up to the age of approximately 8 or 9 years. Although the Yolngu do not keep exact records of their childrens' ages, Money et al. point out that it is at this age when boys undergo an initiation ceremony of circumcision ('Dhapi'). Whilst the Yolngu do not actively educate their children in sexual matters and do not make love in their presence, there are opportunities from infancy to practise and rehearse sexual patterns during play. Money *et al.* comment that 'Little children......learn by what they see. What they see may be the play of other children scarcely older than themselves, or the play and sexual activity of older children and teenagers on whom they are spying in the bush'.

Kinsey *et al.* (1948, 1953) also obtained information concerning 'sex play' in North American children and some of these data are reproduced in Fig. 6.16. These activities included oral–genital and mutual genital contacts, but rarely (in 3 out of 52 girls interviewed) was vaginal penetration involved. Oral–genital contacts occurred between boys and between girls as well as between partners of opposite sex. Moreover, some boys attempted femoral or anal intercourse with each other. Langfeldt (1981) has also described mounting behaviour by Norwegian boys and girls aged 4–5 years in playgroup settings.

These examples serve as reminders that not all

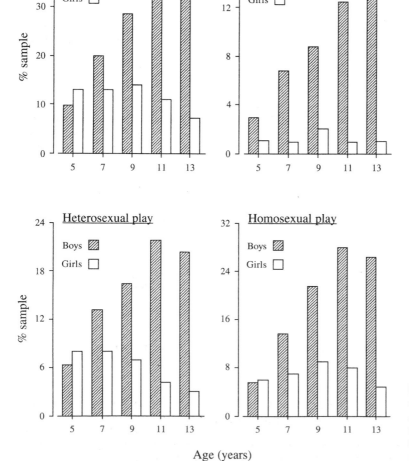

Fig. 6.16 Sociosexual play in human children. Percentages of boys and girls, aged between 5 and 13 years who reported engaging in various types of sexually-related play. (Data from Kinsey *et al.* 1948, 1953).

'sex-play' is heterosexual, just as not all sociosexual activities in monkeys and apes involve oppositely-sexed infants or juveniles. Money (1991) refers to the sociosexual behaviour of monkeys as 'erotosexual rehearsal play' and compares this directly with 'erotosexual play' in human children. I prefer the term 'sociosexual'. The mounts and presentation postures of infant monkeys and apes have a far wider significance in social communication; they are not solely 'erotosexual' activities. It might be helpful to view some of the kindred activities of small children, such as the mounting behaviour of 4 and 5 year olds described by Langfeldt (1981), as part of learning to communicate socially and to establish relationships and not just as sexual rehearsals. Interestingly, Kinsey *et al.* (1953) used the term 'sociosexual development' in their description of childhood

sex play.

It is well known that depriving human infants or very young children of maternal love and contact has profound effects upon their mental health and social development. This is particularly so when infants experience severe deprivation (e.g., when reared in orphanages) during the first 3 years of life (Bowlby 1969, 1973, 1980). We have seen that maternal deprivation and peer deprivation both influence the development of social behaviour in monkeys such as the rhesus macaque. Deprived animals are sexually as well as socially inept as adults; the effects are particularly profound and long-lasting in the case of males. What effects, if any, are produced in human children by restricting their interactions with peers and preventing infantile sociosexual contacts remains unknown. Money (1991) has com-

Table 6.2 The incidence of homosexuality in various human societies

Country, reference, and question asked	Age of respondents	Sample size	Homosexual	
			% male	% female
USA (Diamond 1993)				
Q. Do you usually consider yourself heterosexual, usually heterosexual, bisexual, usually homosexual or homosexual?	>18 years	1024 male 990 female	1.7 (K6) 1.9 (K5–6)	0.2 (K6) 0.3 (K5–6)
USA (Harris Poll 1988)				
Q. Any same sex partner 1. In previous 5 years 2. Last year 3. Last month	15–50 years	739 male 409 female	1. 4.4 (K1–6) 2. 3.8 (K1–6) 3. 1.8 (K1–6)	3.6 (K1–6) 2.8 (K1–6) 2.1 (K1–6)
Philippines (Hart, D. 1968)	Post-pubertal	729		
Q. ?		(total population)	1.6 (K1–6)	0
Japan (Jap. Assoc. Sex. Ed. 1981)**				
Q. Ever any homosexual body contact?	10–22 years	2505 male 2485 female	5.8 (KI–6)	5.0 (K1–6)
Thailand (Sittitrai *et al.* 1992)				
Q. Any same sex experience during adulthood? Hetero/ usually hetero/bisexual/ usually homosexual/homo-sexual?	Post-pubertal	983 male 1285 female	0.3 (K6) 3.4 (K1–6)	0.9 (K6) 1.2 (K1–6)
Great Britain (Wellings *et al.* 1990)	16–59 years	1000	9.0 (K1–6)	4.0 (K1–6)
Q. Any same sex experience?				
Holland (Van Zessen and Sandfort 1991)				
Q. Sexual behaviour in last 12 months? Q. Ever in life same sex experience?	18–50 years	421 male 580 female	3.3 (K5–6)* 8.3 (K1–5)* 12.0 (K1–6)	0.4 (K5–6)* 6.9 (K1–5)* 4.0 (K1–6)

Kinsey (K) scale ratings are given in parentheses, where applicable (see Table 6.1 for Kinsey Scale definitions).
* Indicates sexual preferences regardless of behaviour during preceding 12 months. Modified from Diamond 1993.
** Personal communication to Diamond 1993.

mented on these issues as follows:

'The great majority of the world's children are raised under a strict religious taboo on sex. Their infantile erotosexual rehearsal play, whether alone or with playmates, is subject to severe, often brutal, reprisals if it is discovered. Yet the evidence from rhesus monkeys confirms that copulatory rehearsal in infancy is an age-specific precursor and absolute prerequisite of successful copulation and breeding in adulthood'.

Clearly there is a need for further scientific research in this area of human sexuality. Yet, given society's attitudes to these matters, it is difficult to envisage how effective research might be designed and carried out.

Human homosexual behaviour

Homosexuality is an ancient phenomenon and has occurred in human societies throughout recorded history (Boswell 1980). In modern times Kinsey *et al.* (1948, 1953) reported that in North America the incidence of homosexuality was 4% in men and 2% in women. These percentages referred to people who were exclusively homosexual from adolescence (Kinsey Scale: level 6). A larger proportion of respondents had some degree of homosexual experience; thus 37% of white males sampled had experienced orgasm at least once in their lives as a result of homosexual activities. The sample studied by Kinsey *et al.* might not have been truly representative of the population as a whole; it contained a substantial number of prison inmates, for example. This is not intended as a carping criticism of the original work but merely as an explanation of why further research on the incidence of homosexuality in the USA and elsewhere has been valuable. A most useful review of current data has been provided by Diamond (1993). These studies, conducted in six countries, show that frequencies of male homosexuality and lesbianism are less than those proposed by Kinsey *et al.* In the USA, for example, less than 2% of men questioned by Diamond considered themselves to be exclusively or almost exclusively homosexual (Kinsey Scale 5–6), whilst the incidence of lesbianism was between 0.2% (exclusively so) and 0.3% (almost exclusively lesbian). These data also indicate a finding common to other studies; the incidence of homosexuality is consistently higher in males than in females (Table 6.2). If we consider those studies which questioned whether any type of homosexual activity had occurred (i.e. a possible Kinsey 1–6 rating) then the percentages are higher;

in Great Britain up to 9% of men (aged 16–59 years) and 4% of women had some same-sex experience (Wellings *et al.* 1994). In Japan 5.8% of males (aged 10–22 years) and 5.0% of females reported having had homosexual body contact. These contacts and experiences included single instances of a variety of homosexual activities but, even so, the frequencies fall far short of those obtained by Kinsey *et al.*

Despite some improvements in attitudes towards homosexuals in North America and Europe and the tendency for 'gays' to be more open about their sexuality ('coming out of the closet') there is no doubt that homosexuals are in a minority in modern societies and are often the subject of bigotry and discrimination (Bancroft 1989). Greater tolerance for homosexual behaviour has been noted in a few countries, as in the Philippines (Hart, D. 1968; Whitam and Mathy 1986). Diamond (1993) also comments upon 'the comparatively liberal attitudes toward homosexual or bisexual activity in Denmark, the Netherlands, Philippines, and Thailand'. However, the most remarkable examples of homosexual integration occur in a number of Melanesian societies (Herdt 1984). A particularly interesting and well documented case concerns the Sambia, a tribe which lives in the mountains of Papua New Guinea. When Sambian boys are 7–10 years old, they begin an initiation period which lasts until they are 19 years of age and which involves ritualized homosexual activities (Herdt 1987; Baldwin and Baldwin 1989). For the first 7–10 years, boys are raised by their mothers, in company with other women and female children. However, once the age of initiation is reached, boys are taken by their fathers to live in the mens' house. Here they are initiated into homosexual activities and learn to perform fellatio upon older (pubertal and post-pubertal) members of the tribe. The Sambia believe that ingestion of semen by young boys strengthens them and assists their development towards manhood. Such behaviour is viewed as beneficial and a normal rite of passage, preparing the boy for a heterosexual way of life when he reaches maturity. Once boys reach puberty, they play the active (insertee) role during oral sex with younger boys. This behaviour continues until approximately 19 years of age when young men are considered eligible for marriage. Once a man has reached this stage, homosexual relations are regarded as taboo, although they continue in a few cases (Baldwin and Baldwin 1989; Herdt and Stoller 1989). In some other Melanesian societies (e.g., Malekula, Marindanim, and East Bay) it is culturally acceptable for married men to have homosexual relationships with younger males (Herdt 1984).

The example provided by the Sambia and some other Melanesian tribes is a valuable one, since it reminds us that age-structured homosexual activities can be perfectly acceptable in some cultures. The ritualized homosexual activities engaged in by young male Sambians do not detract from their ability to switch to heterosexual contacts once manhood is attained. No stigma attaches to their earlier homosexual activities, indeed 'for the Sambia, ritualized homosexuality is seen as a logical and necessary precursor to heterosexuality' (Baldwin and Baldwin 1989) which enables boys 'to grow big and live a long life' (Herdt 1981). However, it would be incorrect to say that the heterosexual relationships of adult Sambians are affectionate and well-adjusted emotionally or physically, at least from a western cultural perspective. Sambian women apparently fail to reach orgasm during intercourse, for instance (Herdt 1981, 1984) whilst marital arguments and violence towards women are commonplace (Herdt 1982). The issue is complicated because Sambian men marry women from outside their own villages. Marriages are arranged, rather than resulting from personal choice and 'men and their kin feel that they cannot trust these in-marrying women'. Men also believe that 'women can pollute them (via menstrual blood) and deplete them of their semen' (Herdt and Stoller 1989). Sambian culture is far from sexually tolerant. Without knowing more about sexual behaviour and close relationships in married Sambians it is best to reserve judgement on the question of whether ritualized homosexuality during boyhood might have lasting effects upon their behaviour as adults.

Homosexual patterns and relationships

Homosexual males engage in a variety of sexual activities, including kissing, mutual masturbation, fellatio, and anal intercourse. Depending upon the preferences and relationships between the participants, these activities tend to be reciprocal ones, for the mutual pleasure of both partners. The stereotype of the exclusively 'active' homosexual male or exclusively 'passive' partner (during oral or anal sex) is incorrect (Marmor and Green 1977; Masters and Johnson 1979). The relative frequency of these behavioural patterns can be listed in the order: fellatio (active or recipient), being masturbated, masturbating the partner, anal intercourse (insertor role), anal intercourse (insertee role), and mutual body contact. The most preferred activities are fellatio (either giving or receiving) and the active role in anal intercourse (Bell and Weinberg 1978). Bell and Weinberg also recorded that among lesbians the most frequent sexual patterns are mutual masturbation, oral sex, and body-rubbing. Lesbians also expressed a preference for oral sex as the most pleasurable technique. Whilst lesbians frequently kiss and stimulate one anothers breasts or genitals, they rarely use artificial aids (dildos) for intravaginal stimulation (Marmor and Green 1977; Bancroft 1989).

Homosexual activities can occur within long-term, stable ('close-coupled') relationships, as part of more 'open' relationships in which sex occurs with partners outside the established couple, or during transient sexual contacts. Bell and Weinberg (1978) found that lesbians are more likely to form *close-coupled* relationships (28% of their sample) than are male homosexuals (10% of their sample). By contrast, male homosexuals are more likely to engage in transient and frequent sexual contacts than are lesbians. This applied particularly to younger men and these 'functionals', as Bell and Weinberg referred to them, constituted 15% of the sample as compared to only 10% of lesbians questioned. Some male homosexuals can be exceedingly active sexually and their life-time frequencies of sex partners greatly exceed those of lesbians. Symons 1979 has advanced an interesting evolutionary argument to account for this sex difference. He proposes that male homosexuals are less constrained in their behaviour than heterosexual males, who must relate to women in order to achieve sexual gratification. Women are much less likely to seek multiple partners than men; women are the 'limiting resource' for which males must *compete* (see Chapter 4 for a discussion of male and female sexual tactics in primates). No such constraints apply between male homosexuals, however. Among lesbians, the greater female propensity to form lasting relationships is still apparent. Lesbians are also less likely to be unhappy in relationships or to experience sexual dysfunctions than is the case amongst male homosexuals. Bell and Weinberg categorized 12% of their male homosexual sample as 'dysfunctionals' as compared to 5% of the lesbians questioned.

The biological basis of homosexuality

Given that homosexuality is widespread throughout human societies, and that it occurs in a small but consistent percentage of the population, what are the causes of the phenomenon? Considerable debate has centred upon three possible causative factors: genetic mechanisms, disruption of normal neuroendocrine processes during foetal development, and experiential factors during infancy and childhood. A full discussion of the neuroendocrine mechanisms which affect sexual differentiation of the brain and behaviour is provided in Chapter 10. Studies which purport to show differences in pituitary luteinizing hormone secretion between homosexual and heterosexual males are described on pages 299–300, and Swaab's observations on the sexually dimorphic nucleus of the anterior hypothalamus in the human brain and Le Vay's studies of the anterior hypothalamus in homosexual men are dealt with on pages 306–9. Suffice it to say that none of these researches has succeeded in pinning down a causative basis for human homosexuality. However, there is good evidence that an individual's genetic make up can affect his/her predisposition towards homosexuality and that experiential factors during childhood are crucial in shaping gender identity and later erotic preferences. I shall consider the genetic evidence first and then deal with experiential factors.

Genetic contributions to homosexual development

The most compelling evidence for genetic effects derives from studies of identical (monozygotic) and non-identical (dizygotic) twins. The question has been framed as follows: 'given that one twin is known to be homosexual, what likelihood is there that the second twin will show agreement (*concordance*) in his/her erotosexual orientation?' In an early report on this problem (Kallmann (1952) claimed a remarkable 100% concordance for homosexuality among monozygotic (MZ) twins, whilst concordance was only 10% for dizygotic (DZ) twins. Subsequent work has not confirmed this extreme degree of concordance for MZ twin pairs. However, a study by Whitam *et al.* (1993) found that 22 out of 34 sets of MZ male twins (64.7%) were concordant for homosexuality, 2 sets (5.9%) exhibited partial

concordance, whilst only 4 out of 14 sets of DZ male twins (28.6%) exhibited such concordance. Only four pairs of female MZ twins and nine pairs of DZ (female–male) twin pairs were available for study. Whitam *et al.* point out that lesbianism is less common than male homosexuality. Further, the authors located their subjects through articles and advertisements in 'gay' publications which are read more widely by male homosexuals. Nonetheless, Whitam *et al.* found that three out of four sets of MZ female twins were concordant for homosexuality compared with three out of nine sets of DZ twins. Their results are summarized in Table 6.3.

Monozygotic twins not only look identical but are reported to share remarkable similarities of personality and behaviour. This apparently holds true for many (but not all) cases where a homosexual orientation exists. Consider the following example cited by Whitam *et al.*:

'A pair of homosexual MZ twins, aged 35, reported that they did not reveal their sexual orientation to each other until fairly late. After mutual revelation, they discovered that both, completely independently and unknown to each other, simultaneously for about a year while living in different cities, had photographed shirtless construction workers over the age of 40 and later at home masturbated to the photographs. This reported behaviour is of course difficult to understand in terms of any contemporary explanatory system other than coincidence'.

In a number of studies of MZ twins, one of whom is known to be homosexual, concordance for sexual orientation with the second twin has consistently ranged from 50% to 65%. (Bailey and Pillard 1991; Buhrich *et al.* 1991; Bailey *et al.* 1993; Whitam *et al.* 1993). These studies support the idea that human sexuality is influenced in part by the genome. This is not to say that human erotosexual orientation is simplistically determined by the individual's genetic endowment or that postnatal, experiential factors are unimportant in this respect.

It is important to remember that a significant proportion of identical twins are highly *discordant*

Table 6.3 Concordance and discordance for homosexuality in identical (monozygotic) and non-identical (dizygotic) human twins

Type of twinning	Number of twin pairs	% Concordance for homosexuality		
		Concordance	Partial concordance	Discordance
Monozygotic (males)	34	64.7	5.9	29.4
Monozygotic (females)	4	75	0	25
Dizygotic (males)	14	28.6	0	71.4
Dizygotic (male–female)	9	33.3	22.2	44.4

Modified from Whitam *et al.* 1993.

for sexual orientation. Thus, Whitam *et al.* (1993) found that 10 out of 34 pairs of MZ male twins contained one homosexual and one heterosexual individual. In nine of the 10 discordant pairs, one twin scored 6 on the Kinsey scale (exclusively homosexual), whilst his brother scored zero (exclusively heterosexual). In the tenth case the scores were 6 and 1, respectively. In the single case of discordance noted for the four pairs of female MZ twins, the difference was again extreme, one twin scored 6 and her sister scored zero on the Kinsey Scale.

In a thoughtful discussion of the MZ twin-studies Turner (1994) points out that 'twinning occurs only after several cell divisions of the embryo and the outcome is known to be highly variable. The range includes, but is not limited to, one twin forming a lump in the chest of the survivor, through a conjoint pair to two or more perfectly concordant individuals'. Turner cites the case of female conjoint twins (so-called 'Siamese twins') reported by Cuny (1956) in which marked differences in personality were apparent despite the fact that the girls shared the same torso and a single pair of legs. Discordance between MZ twin girls might occur, for instance, because of skewed X chromosomal inactivation (Richards *et al.* 1990). A number of X-linked traits might exhibit discordance in MZ twins, including Kallmann Syndrome, Duchenne muscular dystrophy, and colour blindness (Nance 1993). Turner (1994) cautions that 'such events must raise concern over the validity of published reports on the concordance rates of MZ twins'.

Despite Turner's (1994) reservations concerning the studies of concordance rates in MZ twins, they nonetheless provide strong support for the existence of a genetic predisposition towards homosexuality. The nature of this genetic input has been much debated and remains controversial. The observation that male homosexuals are more likely to occur on the female side of certain family trees led Hamer and his colleagues to search for genetic markers of homosexuality on the X chromosome (Hamer *et al.* 1993; Hu *et al.* 1995). Hamer reported that 33 out of 40 pairs of homosexual brothers shared markers in the region known as Xq28, which lies at the tip of the long arm of the X chromosome. It is incorrect, however, to regard Xq28 as a 'gay gene'. Firstly, this region probably contains several hundred genes, whose functions remain to be determined. Secondly, although the findings of Hamer *et al.* have been supported in some studies (Turner 1995), others have not replicated them (D'Alessio 1996). Sexual preferences are so complex, especially so in large-brained animals such as the higher primates, that it is very difficult to conceptualize how a gene, or small cluster of genes, might so totally

control their development. It might be relevant in this regard to consider the possible effects of genomic imprinting upon brain development. Although the developing foetus inherits half its genes from each parent, maternal and paternal alleles are not expressed equally in all tissues. Recently it has been shown, for instance, that maternal genes make a greater contribution to development of the cortex of the mouse brain, whilst paternal genes are preferentially expressed in various limbic areas (Keverne *et al.* 1996). If, indeed, the propensity to develop as a homosexual in human males is inherited via female progenitors then might this be mediated via genomic imprinting with its widespread effects upon the developing cortex and limbic system? Some 'alteration' in the imprinting process might be expected to produce a spectrum of changes and perhaps to affect behaviour and personality in other ways besides erotosexual preferences. This is pure speculation, of course, but it might prove to be more useful than speculation based upon the relative sizes of tiny hypothalamic nuclei in gay and heterosexual men (Le Vay 1993).

Homosexuality and birth order

The likelihood that a man will develop as a homosexual is correlated with the number of older brothers in his family. The larger the number of older brothers, the greater the likelihood of a homosexual orientation becomes; the numbers of older sisters has no effect upon sexual orientation, however (Blanchard and Bogaert 1996a, b). This observation also applies to paedophilic men—homosexual/bisexual paedophiles tend to have a larger number of older brothers than do heterosexual paedophiles (Bogaert *et al.* 1997). Figure 6.17, reproduced from the work of Blanchard and Bogaert (1996a), shows data for 302 homosexual men and for an equal number of heterosexual individuals. As the numbers of older male siblings in this group of 604 men increased, so also did the representation of homosexuals within the sample. It will be noted, for example, that 363 of the men questioned had no older brothers and that 45% of them were homosexuals. Whilst only seven men in the sample had four or more older brothers, five of these subjects (71%) were homosexuals.

It is possible that social factors connected with having large numbers of older brothers might affect the development of sexual orientation in men (Blanchard et al. 1995). However, an alternative immunologically-based hypothesis to account for birth-order effects upon homosexuality has been

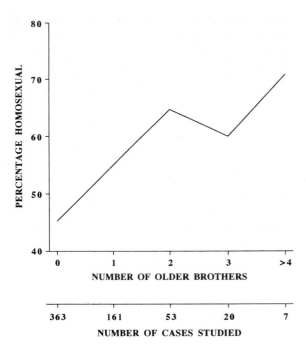

Fig. 6.17 Relationship between the number of older brothers a man has and the likelihood that he is a homosexual. Data are for 302 homosexual men and 302 matched heterosexual men. (Redrawn and modified from Blanchard and Bogaert 1996a.)

proposed by Blanchard and Klassen (1997). They suggest that maternal immune reactions to male foetuses might have some cumulative effect, increasing the likelihood that males born later in the sequence might develop homosexual propensities. The mechanism proposed by Blanchard and Klassen is that mothers who produce sons are progressively immunized against minor histocompatibility antigens (H–Y antigens) which are controlled by genes on the Y chromosome (and thus produced only in male foetuses). Given that H–Y antigen is strongly expressed in the brain, as well as in the epidermis and sperm cells, it might influence the development of neural substrates which affect sexual orientation. How this might occur is not understood. Blanchard and Klassen also caution that, whatever role birth order and H–Y antigen might play in the causation of human homosexuality, this is only one of a constellation of factors which might affect sexual orientation.

Experiential contributions to homosexual development

As far as male homosexuality is concerned, considerable efforts have been made to link childhood experiences and problems of the parent–child relationship to the later emergence of same-sex erotic preferences. Much less is known about how experiential factors might shape lesbian sexuality. Bieber *et al.* (1962) interviewed 100 homosexual and 100 heterosexual males who were undergoing psychoanalysis. They concluded that certain qualities of parent–child relationships, such as an absent or emotionally distant father and a 'close binding' or 'intimate' mother–son relationship were more typical of homosexuals than of heterosexuals. However, more recent studies of men not receiving medical treatment indicate that it is the more 'neurotic' homosexuals who also report having had disturbed relationships with their parents (Siegelman 1974, 1978).

A relationship is, of course, a two-way process. The personality and behaviour of a boy destined to become homosexual might influence parental attitudes and responses. It is not solely a question of parents influencing their children and affecting their sexual preferences in later life. Bieber *et al.* (1962) also found that homosexual men were more likely to recall being treated as 'sissies' during childhood and having avoided the more boisterous, athletic pursuits of other boys. In this regard, Green (1985; 1987) has conducted extensive, longitudinal studies on behavioural development of effeminate boys ('sissy boys'). The effeminate boys in Green's sample preferred playing with girls' toys, were interested in cross-dressing, and avoided the rough games of their male peers. A key factor in the background of these children seemed to be parental acceptance, and sometimes active encouragement, of effeminate behaviour during their early years. Encouragement was sometimes given by a mother who had wanted a daughter, rather than a baby boy; 15% of mothers dressed their little boys in girls' clothes. Mothers also tended to be over-protective and attentive towards their effeminate sons, whereas the fathers tended to be emotionally distant or were absent. In some cases 'fathers had made considerable efforts to engage their sons in mutually enjoyable activities but, receiving no positive reinforcement for the effort, they retreated from the relationships' (Marmor and Green 1977). Some fathers regarded their sons as 'mamma's boys'. Many of them (about one third of those studied) were notable for their physical beauty in childhood, although non-effeminate sons in the same families did not suffer from gender identity problems. Green (1987) interviewed two-thirds of the sixty-six males in his 'feminine boy' group when they were adolescents or young adults, and used the Kinsey scale to quantify their sexual orientation. Three quarters of these subjects were rated as being homosexual or bisexual, by comparison with just one of the 'masculine boys' in the control group. Only one of the effeminate boys was transsexual as an adolescent and wanted to be surgically reassigned as a woman.

Possible effects of experiential factors upon the development of a lesbian sexual orientation have been examined by studying tomboyish behaviour in girls. Girls who are 'tomboys' tend to prefer rough and tumble games and are more athletic than their female peers. Doll play is less preferred than expected and such girls tend to be more interested in career prospects than in domestic or family pursuits. These comments concerning tomboyish girls are only generalizations and are not intended to stereotype their personalities. Girls who have been exposed to high concentrations of androgens during foetal life, due to congenital adrenal hyperplasia (see page 285) also exhibit tomboyish behaviour (Money and Ehrhardt 1972) and, as we shall see in Chapter 10, there is some evidence that such women are also more likely to report bisexual or lesbian tendencies. Even in these cases, where both prenatal (androgenic) and postnatal (tomboyish) influences occur, effects upon sexuality are not invariably present or pronounced. Tomboyish behaviour, of itself, is not associated with lesbianism later in life. Thus, 'in contrast to the effeminate boy who usually carries into adulthood considerable problems relating to his gender identity, the tomboyish girl seldom has any difficulty in adapting to an adult (heterosexual) female role' (Bancroft 1989).

7 Sexual selection and sexually dimorphic traits

Among the primates, numerous examples can be found of morphological differences between adult males and females within a single species. An obvious example is body size. Males of polygynous species, such as the gorilla and the proboscis monkey, are much larger and heavier than females. Sexual dimorphism in body size also occurs in many multimale–multifemale societies, such as those of baboons and macaques, but it is virtually absent in monogamous primates such as the lesser apes and many of the callitrichids. Similar sex differences can occur in the size of the canine teeth, which are especially large in polygynous species (Fig. 7.1). However, as discussed below, relationships between the mating system, body size, and canine size are not equally represented throughout the Order Primates. The kinds of relationship depicted in Fig. 7.1 are much more applicable to the Old World monkeys and apes (catarrhines) than to the New World monkeys or the prosimians. It is also among the catarrhines that the most striking examples of masculine secondary sexual adornments are to be found. These include capes of hair in male geladas and hamadryas baboons, bony paranasal swellings in mandrills and drills, and the impressive cheek flanges of the adult male orang-utan. Red sexual skin occurs on the perineum and genitalia of various macaques and hamadryas baboons, whilst in the gelada the male possesses a brilliant red chest patch. The skin of the scrotum is blue in many of guenons, the patas monkey, and the mandrill; in this last species both red and blue sexual skin occurs on the face as well as on the genitalia. Nor is sexual skin limited to males among catarrhine species; females of many macaque species exhibit pink or reddish perineal sexual skin, whilst in baboons, mangabeys, red colobus monkeys, talapoins, and chimpanzees the sexual skin becomes swollen and oedematous during the follicular and peri-ovulatory phases of the menstrual cycle. By contrast, sex differences in the size and histological composition of cutaneous scent-marking glands are much more widespread among the prosimians and New World primates than among the catarrhines. Male ringtailed lemurs possess particularly large and complex glands on their shoulders and wrists; the female homologues of these structures are much less well-developed. Sternal cutaneous glands are often larger in adult males than in adult females; this is the case in greater and lesser galagos, for instance, as well as in most New World monkey genera. Finally, there are numerous examples among the primates of loud and sexually dimorphic vocal

Body size dimorphism

Relative canine size

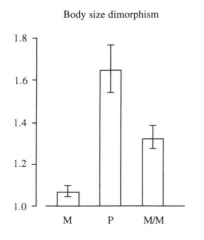
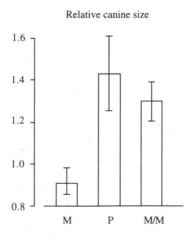

Fig. 7.1 Body size dimorphism (adult male weight divided by adult female weight) and relative canine size (in males) in primate genera with different breeding systems. Sample sizes for M (monogamous), P (polygynous), and M/M (multimale–multifemale) groups respectively are: body size dimorphism (12, 9, 14) and relative canine size (4, 7, 9). Data are shown as means ± S.E.M. (Redrawn from Clutton-Brock 1991).

displays. Sex differences in vocalizations, such as in the calls of howler monkeys, the great calls of male orang-utan, and booming vocalizations of male guenons, are associated with sexual dimorphisms in the vocal tract, especially where the larynx, laryngeal sacs, and hyoid apparatus are concerned. Since all these sexually dimorphic traits might affect relationships between and within the sexes, it is of importance to examine their evolutionary background in the various primate lineages and their possible relevance to the functioning of mating systems.

Sexual dimorphism in body weight

Throughout the animal kingdom, there is a tendency for adult females to be larger than adult males of their own species. This observation applies to the invertebrates and lower vertebrates, but among mammals and birds the male is usually the larger sex (Darwin 1871; Andersson 1994). Notable exceptions occur, however, so that in blue whales, leopard seals, certain bats, and some other mammals, females are larger than males. In such cases Ralls (1976) suggests that large size is advantageous to females for reasons of fecundity and that 'a big mother is often a better mother'. Where males are markedly larger than females this is usually attributed to the effects of intrasexual competition for access to mating partners, such as occurs in highly polygynous mammals like the elephant seal. An alternative argument suggests that larger male size is also selected for because it acts as a defence against predators. Studies of the primates have consistently shown that sexual dimorphism in body weight is greatest in polygynous forms, less so in multimale–multifemale societies and minimal in monogamous species (Clutton-Brock *et al.* 1977; Clutton-Brock and Albon 1979). That the situation is more complex than this has been shown by Martin *et al.* (1994). Striking examples of sexual dimorphism in body weight are found among the Old World monkeys and apes; few of the New World monkeys exhibit such pronounced dimorphism and among the prosimians males and females tend to be very similar in size (Fig. 7.2). The Old World catarrhines contain by far the largest members of the Order Primates, and include massive species such as the great apes, baboons and macaques. Since sexual dimorphism in body size tends to increase with increased body size (*Rensch's rule*), it is possible that the differences between catarrhines and smaller-bodied New World primates and prosimians shown in Fig. 7.2 might reflect this. However, Rensch's rule

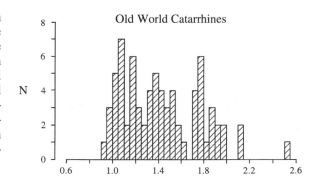

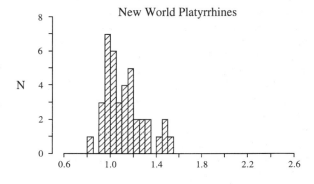

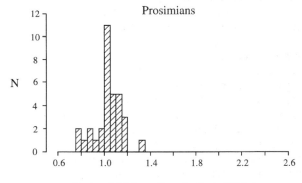

Ratio of Male to Female Body Weight

Fig. 7.2 Extent of sexual dimorphism in body weight, based on ratios for adult males and females in Old World anthropoids (N=73 species), New World anthropoids (N=39 species) and prosimians (N=33 species). Marked sexual dimorphism is much more common among the Old World monkeys and apes, than in prosimians or New World monkeys. (Redrawn from Martin *et al.* 1994.)

only accounts for a small proportion of the observed differences in sexual dimorphism among primates (Martin *et al.* 1994). Two additional factors might be involved. Firstly, the catarrhines contain many terrestrial and semi-terrestrial representatives, whereas all New World monkeys are arboreal as are the vast majority of prosimians, especially the smaller-bodied nocturnal forms. The long history of terrestrial life among catarrhines, such as the baboons, macaques, and the African apes, might have contributed to the development of larger body size as a means of defence against predators. In polygynous catarrhines, such as the gorilla and gelada, massive males might indeed be more successful in defending their groups against predators, as well as in competing among themselves for access to females. Secondly, there is the possibility that intermale competition is more intense in many of the Old World species, such as macaques and baboons, than in New World primates such as the squirrel monkeys, woolly monkeys, or howlers. An objective test of this hypothesis is difficult but might be secured by comparing the socionomic sex ratio (numbers of adult females ÷ numbers of adult males in the social group) with the degree of body weight dimorphism in primates. Despite attempts to conduct such an analysis, the inclusion of large numbers of monogamous primates has biased the results. Martin *et al.* (1994) conclude that 'there has, as yet, been no convincing demonstration of a link between sexual dimorphism in body weight and the level of competition among males for access to females'.

Sexual dimorphism in body weight arises during postnatal development either because males grow more rapidly than females or because they continue to grow for a longer time and attain their maximum weight at a greater age than females. Figure 7.3 compares the growth in body weight of a highly sexually dimorphic catarrhine species, the mandrill, with that which occurs in a monogamous New World primate, the common marmoset. In the mandrill, the two sexes weigh the same at birth but by two years of age males are significantly heavier than females. Males continue to grow more rapidly than females during adolescence and reach adult weight by approximately 8 years of age. Between 5 and 7 years of age, when males exhibit an adolescent 'growth spurt', they gain weight at a rate of 5.1 kg. per year. Females by contrast, gain approximately 1.85 kg in weight per year and complete somatic development by 5 years of age. Matching this bimaturism in somatic growth is a marked sex difference in age at sexual maturity. Female mandrills begin to exhibit sexual swelling and to undergo menstrual cycles at 2.75–4.5 years of age; the average age at first conception is 3.95 years. Yet, male mandrills are still pubescent at 5–7 years of age and do not reach full sexual maturity until approximately 8 years (Fig. 7.3). In monogamous species, the situation is markedly different. Both sexes grow at a similar rate and attain sexual maturity at similar ages. In the marmoset, for example, males and females attain an adult body weight of 300 g by 12–15 months, puberty

begins at between 8–12 months and sexual maturity is reached by 18 months of age (Fig. 7.3). Later attainment of puberty and sexual maturity in males of sexually dimorphic species, and concordance of reproductive maturity in males and females of monogamous primates, is a general phenomenon, as can be seen from the examples listed in Table 7.1. This has obvious repercussions for sexual behaviour and reproductive success in males of sexually dimorphic species. Such males invest more in body growth

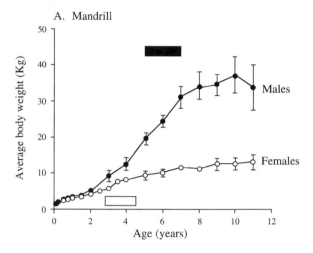

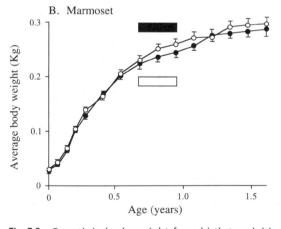

Fig. 7.3 Growth in body weight from birth to adulthood in two primate species. A. The Mandrill (*Mandrillus sphinx*), a species which has a multimale–multifemale mating system and which exhibits marked sexual dimorphism in body size. (Data are from Wickings and Dixson 1992b). B. The Common marmoset (*Callithrix jacchus*) which has a monogamous mating system in which males and females are very similar in body size. (Redrawn from Abbott and Hearn 1978). Data are means (±S.E.M.) for males (filled circles) and females (open circles). Horizontal bars indicate the approximate age-span for attainment of puberty (filled bars: males; open bars; females.)

than females and reproductive maturity is delayed. In many catarrhines it is the maturing males which must emigrate and compete for mating opportunities outside their natal groups, whilst females are philopatric. Larger size might be adaptive for survival, as well as for intermale competition, when males emigrate from their groups.

Special mention must be made of the prosimians, since several authors have commented upon the relative absence of sexual dimorphism in body weight in this suborder of the primates (Crook 1972; Clutton-Brock and Harvey 1977; Martin 1990; Jolly 1984; Martin *et al.* 1994). Kappeler (1991) has been able to analyse long-term data on the body weights of 26 prosimian species housed at the Duke University Primate Centre. Whilst no sex differences in body weight were found in 19 species, males were significantly heavier than females in all the galagos examined (*Galago crassicaudatus*; *G. garnettii*, *G. moholi*, and *G. demidoff*) as well as in the slow loris (*Nycticebus coucang*) and tarsier (*Tarsius syrichta*). In mouse lemurs (*Microcebus murinus*) females were the larger sex; adult males weighed 83% of the female values, on average. A resumé of body-weight data for the eight prosimian families examined by Kappeler is shown in Fig. 7.4. Whilst it is true that prosimians do not exhibit the extremes of sexual dimorphism seen in anthropoids, larger males do occur in some nocturnal forms which have dispersed mating systems. Intermale competition for access to

central ranging areas, and hence to females, is intense in some of these species (e.g., *G. demidoff*; Charles-Dominique 1977; *G. moholi*: Bearder and Martin 1979). Dominant males are often larger and heavier than others in such mating systems and it is possible that sexual dimorphism might be more pronounced if only body weights for dominant males, rather than all adult males, were included in statistical analyses. Nocturnal prosimians are probably constrained by their ecological niches to remain small and cryptic in their habits. These constraints probably limit selection for increased male body weight. The fact that the sexes are of equal size in many diurnal lemurs has been correlated with the occurrence of female dominance in certain of these species. In *lemur catta*, which is the best documented case, females outrank males, particularly in feeding contexts, even though males are extremely aggressive towards each other during the annual mating season and use their larger canines for fighting purposes. It would be incorrect to generalize that 'female dominance' is the rule in diurnal prosimians, however, and there is no evidence that female body weight is significant in this context. It seems possible that fundamental differences do exist between prosimians and anthropoids in the expression of sexual dimorphism in body weight due to sexual selection. Kappeler (1991) points out that the differences are unlikely to be explained by differences in body size between the two suborders (i.e.

Table 7.1 Age of onset of puberty in male and female monkeys and apes, as related to the degree of sexual dimorphism in adult body weight

Species	Age at puberty (years)		Adult body weight	Sources
	Male	Female	Male/female	
Callithrix jacchus	0.66–1.0	0.66–1.0	1.0	Abbott and Hearn 1978
Aotus griseimembra	0.75–1.0	0.75–1.0	1.0	Dixson 1983a and unpublished observations
Saimiri sciureus	2.0–3.0	2.5	1.29	Kaplan *et al.* 1981
Miopithecus talapoin	5.0–6.0	3.5	1.27	Rowell 1977a, b
Erythrocebus patas	3.0–4.0	2.1	1.78	Rowell 1977a
Macaca mulatta	3.0–3.5	2.0–3.0	1.44	Wilen and Naftolin 1976; Bercovitch and Goy 1990; Dixson and Nevison 1997
Papio cynocephalus	5.5–6.0	4.5	1.33	Altmann *et al.* 1977
Mandrillus sphinx	5.0–7.0	2.75–4.5	3.0	Wickings and Dixson 1992b
Pongo pygmaeus	6.0–8.0	5.1–7.1	1.86	Dixson *et al.* 1982; Markham 1995
Pan troglodytes	7.0–11.0	7.3–10.2	1.34	Young and Yerkes 1943; Winter *et al.* 1980

Age of pubertal onset, rather than age at 'sexual maturity' has been used here, since in practice it is often very difficult to pinpoint the age of sexual maturity in primates. Animals may be sexually mature physiologically but fail to reproduce for social reasons. Pubertal onset may be measured more objectively with respect to testicular enlargement and increased testosterone secretion (in males) and secretion of ovarian steroids, changes in sexual skin morphology or occurrence of menarche (in females).

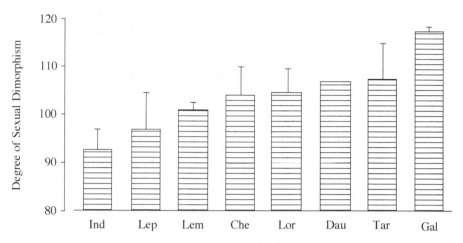

Fig. 7.4 Mean degree of sexual dimorphism in body weight (male weight ÷ female weight × 100), based on the respective species means, along with the standard error of the mean for each of eight prosimian families (and subfamilies). Ind = Indriidae, Lep = Lepilemuridae, Lem = Lemuridae, Che = Cheirogaleidae, Lor = Lorisinae, Dau = Daubentoniidae, Tar = Tarsiidae and Gal = Galaginae. (Redrawn from Kappeler 1991.)

because many prosimians are quite small and dimorphism is less pronounced in smaller species). Thus, examination of extinct large-bodied subfossil lemur species has also failed to reveal sex differences in body size (Godfrey 1988; Jungers 1990).

Sexual dimorphism in canine tooth size

Field studies have provided ample evidence that

aggression between adult males can lead to severe injuries, particularly in certain anthropoids which have multimale–multifemale or polygynous mating systems. Examples are shown in Fig. 7.5 of canine slash wounds and scarring inflicted during such fights in howler monkeys and macaques. Aside from the possible effects of sexual selection upon canine dimorphism, many other variables must be considered in studies of the evolution of canine size. These include taxonomic differences, possible effects of

Fig. 7.5 Canine slash wounds resulting from intermale aggression in primates. Left: Adult male red howler (*Alouatta seniculus*). (After Crockett and Eisenberg 1987). Right: A young adult immigrant male *Macaca thibetana* severely injured by resident males in the troop. (After Zhao 1994.)

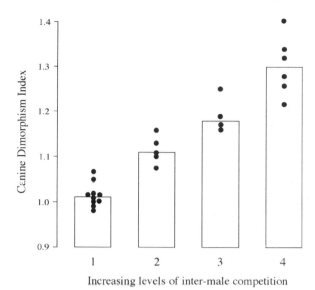

Fig. 7.6 Relationship between levels of intermale competition and degree of canine sexual dimorphism in a sample of New World Monkey Species. Each species is represented by a filled circle; histograms are means for each of the four levels of inter-male competition. (Data from Kay *et al.* 1988.)

increasing body size, variations in feeding ecology and terrestrial versus arboreal modes of life. Although the exact roles played by each of these variables have been much disputed, the majority of studies agree that intrasexual selection has accentuated male-biased sexual dimorphisms in canine size in polygynous primates and in many species which have multimale–multifemale mating systems (Fig. 7.1). Monogamous primates, by contrast, usually exhibit little or no sexual dimorphism in canine size (Leutenneger and Kelly 1977; Harvey *et al.* 1978; Kay *et al.* 1988; Beauchamp 1989; Martin *et al.* 1994). Comparisons at the mating system level are relatively crude, however, since they preclude a more detailed analysis of the degree of intermale competition in relation to sexual dimorphism in canine size. To address this question Kay *et al.* (1988) used a four-point scale with which to rate levels of intermale competition in New World monkey species; level 1 indicated minimal competition and level 4 maximal competition between the adult males in any particular social system. They found a clear-cut, positive relationship between levels of male competition and the degree of sexual dimorphism in canine size among New World monkeys (Fig. 7.6). This approach is particularly useful, as it explains some apparent exceptions to the rule that males in multimale–multifemale societies have larger canines than females. In the woolly spider monkey (*Brachyteles arachnoides*), for example, the canines are of similar

length in both sexes (Milton 1985b), at least as far as the more northerly populations are concerned. The southern populations do exhibit some sexual dimorphism in canine length (Desa *et al.* 1993). Adult male woolly spider monkeys are not overtly aggressive to one another in sexual contexts; instead they form mating aggregations around a female when she is approaching ovulation. Many males copulate with each female under these conditions (Milton 1985a). Kay *et al.* (1988) rated Brachyteles males at level 2 on their four-point scale of intermale competition, despite the occurrence of a multimale–multifemale mating system in this species (Fig. 7.6).

A more complicated model of competition intensity and frequency, developed by Plavcan and Van Schaik (1992) has been used to measure the possible effects of intrasexual competition (among females, as well as among males) upon canine sizes in prosimians and anthropoid primates (Plavcan *et al.* 1995). The important contribution made by these studies is that they have provided evidence that the development of larger canines 'is correlated with the intensity of intrasexual contest competition in both sexes'. Plavcan *et al.* (1995) note that 'relative canine size among females is almost as variable as that among males, so that the forces that shape female canine size must play a substantial role in the evolution of canine dimorphism in primates'. The forces shaping canine size might differ between the sexes, so that among males competition for access to mates is most important (i.e. sexual selection) whereas among females, intrasexual contests for access to food and other resources (natural selection) provides the probable mechanism. The formation of coalitions between females, as in macaques, baboons, vervets, and mandrills in which females form matrilineal dominance hierarchies, is associated with a reduction in relative canine size relative to those species where females compete intensely but not in coalitions. This coalition effect among female macaques and baboons accentuates the degree of canine sexual dimorphism, since males have long canines in these species whereas females have shorter teeth than expected.

At the taxonomic level, an important observation concerns the prosimian primates, since they exhibit relatively little sexual dimorphism in canine size, or body size, despite the differences in their mating systems. Plavcan *et al.* (1995) argue that competition intensity is associated with variations in canine size in both sexes among many prosimians, and that the selective forces involved probably resemble those which occur among anthropoids. Sexual dimorphism in canine size is not totally lacking in prosimians and there are some important differences between the families which constitute this suborder (Kappeler

1993). Adult male ringtailed lemurs, brown lemurs, and galagos have larger canines than females, for example, whereas sexual dimorphism is lacking in lorisids and females of *Propithecus diadema* have larger canines than males. Various authors have put forward hypotheses to account for lack of consistent relationships between sexual dimorphism in canine size and mating systems among the prosimians (reviewed by Kappeler 1996). Perhaps female choice plays a greater role than intermale competition in many of these species (Kappeler 1993) or female oestrous synchrony during restricted mating seasons limits the ability of males to monopolize access to mating partners (Martin *et al.* 1994; Plavcan *et al.* 1995). Plavcan *et al.* even suggest that 'pair-bonding' is fundamentally important in many gregarious prosimians and that the canines are not sexually dimorphic for this reason. None of these arguments is convincing.

Sexual dimorphism in vocal anatomy and display

Long distance vocalizations often play an important role in social communication, particularly in those primates which inhabit rainforests where visual contacts between groups are limited. In some cases both sexes utter loud calls, whereas in others it is only adult males which vocalize in this way. Examples include the loud, dog-like barks of the Madagascan indris (Pollock 1979), the roaring and barking calls of South American howler monkeys (Carpenter 1934), the loud calls of various forest-living guenons in Africa (Gautier 1971), the distinctive cries of bushbabies (Bearder *et al.* 1995), and the complex duets of male and female siamangs (Chivers 1974). Such vocalizations are thought to serve as intergroup location and spacing signals. In many cases, groups in a given area synchronize their displays and call most frequently during the early morning, soon after dawn. In the howler monkey, however, bouts of loud calling also occur throughout the day (Sekulic 1982). Various observers have noted age and sex differences in vocal characteristics (e.g., in *Varecia variegata*: Pereira *et al.* 1988; in *Hylobates* and *Symphalangus*: Haimoff 1984), so that choruses might transmit information concerning group composition as well as about the spatial distribution of animals in the rainforest. It is also the case that marked sexual dimorphism of the vocal apparatus occurs in some species, whereas in others the larynx is essentially the same size in both sexes.

Loud calls are given by both sexes in many primates which have monogamous mating systems; all these species are arboreal and territorial in their behaviour. Perhaps the best-studied examples are the lesser apes, for which Haimoff (1984) has produced a valuable comparative analysis of song structure. Haimoff distinguishes four sequences in the structure of song bouts: an introductory sequence; an organizing sequence; the great call; and male song (Fig. 7.7). During the introductory or 'warm-up' sequence, either the female alone produces a series of single notes (e.g., in *Hylobates klossii*, *H. moloch*, *H. muelleri*, and *H. agilis*) or both sexes commence a simple duet (in *H. Pileatus* and *Symphalangus syndactylus*). In *H. lar*, the situation is more variable, so that either sex may commence the duet. With the exception of *H. klossii* and *H. pileatus*, the next phase (organizing sequence) of the display involves both sexes co-ordinating their locomotor and vocal activity in preparation for uttering the great call. Great calls may be given by the female partner alone (as in *H. klossii*, *H. moloch*, and *H. muelleri*) or both sexes may sing together for at least part of the sequence (as in *H. hoolock*, *H. pileatus*, and *S. syndactylus*). As the female reaches the climax of her great call, both partners brachiate rapidly or shake the branches. An exceptional display sequence occurs in *H. concolor*, however (Fig. 7.7). In this species, there is no introductory sequence and instead the male leads off, with a distinctive male song, whilst the female interjects her great call at intervals throughout his display.

Haimoff (1984) stresses that time is required for pairs of gibbons or siamangs to perfect their complex vocal duets. Perhaps the duets themselves might play some role in cementing the relationship between the sexes and in signalling the status of that relationship to neighbouring family groups.

Despite the sex differences in vocal display which occur in various gibbons (Fig. 7.7), anatomical studies of the larynx show that its morphology is not sexually dimorphic except in the case of the *H. concolor* where a small throat sac is present in the adult male (Miller 1933). This might be connected with the exceptional degree of sexual dimorphism in song structure in this species. Intra-laryngeal saccules occur in other *Hylobates* species, but in *Symphalangus* they protrude from the larynx in both sexes to form a large sac beneath the throat. This laryngeal sac is inflated by males and females as they perform their duets. Laryngeal sacs act as reservoirs to prolong vocalizations and to increase their volume. They have evolved independently a number of times in various primate groups (Starck and Schneider 1960; Hill 1972). Table 7.2 provides details of the distribution of laryngeal sacs and other vocal specializations in primates, as well as sex differences in their morphology and functions during vocal displays. I have not encountered a case of marked sexual dimorphism in the larynx among monogamous primates, although there is a need for more thorough studies to examine this question. In the indris, for example, both sexes possess a large laryngo-tracheal sac, situated dorsal to the trachea with its opening just below the cricoid cartilage. Males and females utter the same dog-like howls during intergroup displays and the laryngeal sac is presumably important in amplifying these sounds (Fig. 7.8). Only two other lemuroids are known to possess these dorsal laryngeal sacs; the mouse lemur (*Microcebus murinus*), in which they have been reported to play no special role in vocal displays (Hill 1953), and the ruffed lemur (*Varecia variegata*). The mating system of ruffed lemurs is by no means certain; some authors believe them to be monogamous (e.g., White 1991) whilst Pereira *et al.* (1988) suggest that they have a fusion–fission type of social organization. Both sexes employ loud calls, consisting of roars and shrieks; females roar whilst males utter shrieks most frequently during group choruses (Pereira *et al.* 1988). These authors comment that roar-shriek choruses in ruffed lemurs are equivalent

to the inter-group spacing or intra-group rallying calls of other primates, since they often occur during the early morning and are 'contagious', being taken up by other animals in a group and by neighbouring social groups. The sex differences in vocal display and monomorphism of the laryngeal apparatus in *Varecia* are reminiscent of the situation in the monogamous lesser apes. However, it is intriguing that *Varecia* has large testes relative to its body weight, indicative of sperm competition, and a multi-male–multifemale mating system rather than strict monogamy as in the lesser apes (see Fig. 8.3, page 220).

Among the monogamous New World monkeys, enlargement of the larynx and occurrence of laryngeal sacs has been described in a number of genera (Table 7.2). In both male and female owl monkeys (*Aotus*) the hyoid and thyroid cartilages are dilated and a large ventral air sac communicates with the larynx via paired openings in the thyro-hyoid membrane. Expansion of the hyoid and thyroid also occurs in both sexes of the titi monkey (*Callicebus*) and in the white-faced saki (*Pithecia pithecia*).

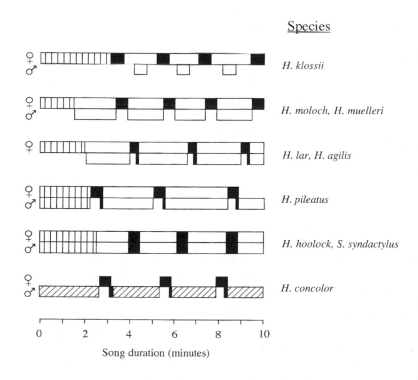

Species

H. klossii

H. moloch, H. muelleri

H. lar, H. agilis

H. pileatus

H. hoolock, S. syndactylus

H. concolor

0   2   4   6   8   10

Song duration (minutes)

Song Sequences:

▭ Introductory   ■ Great-call

▭ Organizing   ▨ Male-song

Fig. 7.7 Stylized diagrams illustrating the sequential structure of the duet song bouts produced by all gibbon taxa, and which sex participates vocally in each of the sequences. (Redrawn from Haimoff 1984.)

Selection pressures have apparently operated on both sexes in many monogamous primates to favour the development of laryngeal specializations which enhance their loud calls. Competition between females as well as between males occurs in territorial contexts. In this regard, it is of particular interest that vocal displays exhibit sex differences in some cases, as in the lesser apes. It is possible that these vocalizations also play some role in the initial courtship phases of mate selection, as well as in cementing the social and sexual relationship between partners. Mate choice, as well as intrasexual competition, may have played some role in the evolution of vocal duets.

Sexual dimorphism of laryngeal structure, coupled with a propensity for males alone to give loud calls or to give more pronounced calls than females, has evolved in various forest-living primates which have polygynous or multimale–multifemale mating systems. In the Old World monkeys, there is a median ventral laryngeal sac in both sexes and it is larger in colobines than in cercopithecines (Hill 1966). The functions of this vocal sac have been studied in forest guenons, many of which are polygynous (Gautier 1971, 1988). Gautier records eight such species in which males emit distinctive 'booms' as part of their loud call displays (*C. mona*; *C. campbelli*, *C. pogonias*, *C. wolfii*, *C. hamlyni*, *C. neglectus*, *C. nictitans*, and *C. mitis*). Females do not vocalize in this way and their laryngeal sacs are much smaller

Table 7.2 Specializations of vocal anatomy and sex differences in the structure of the vocal organs of primates

Species	Mating system	Specializations of vocal anatomy	Sexual dimorphism in vocal anatomy
Indri indri	Monogamy	Large median dorsal laryngo-tracheal sac	No
Microcebus murinus	Dispersed	Dorsal laryngo-tracheal sac extends along sides of neck and under the skin of the back. Its function is uncertain	No
Varecia variegata	?	Dorsal laryngo-tracheal sac	No
Aotus spp.	Monogamy	Dilated hyoid bone and thyroid cartilage. A large, median ventral air sac opens into the larynx via paired openings in thyro-hyoid membrane	No
Callicebus	Monogamy	Dilated hyoid bone and thyroid cartilage	No
Pithecia pithecia	Monogamy	Dilated hyoid bone and thyroid cartilage	No
Alouatta spp.	Multimale–multifemale and polygyny	Greatly expanded, hollow, hyoid bone and laryngeal saccules	Male > female especially in *A. seniculus*
Colobus guereza *C. polykomos*	Polygyny	Large laryngeal sac absent, but a sub-hyoid sac and laryngeal saccules do occur	Male > female
Cercopithecus spp. e.g. *C. mona*, *C. pogonias*, *C. mitis*	Polygyny	Very large ventral laryngeal sac	Male > female
Hylobates spp.	Monogamy	Intralaryngeal saccule	No
H. concolor	Monogamy	Small laryneal sac present in male	Yes
S. syndactylus	Monogamy	Very large distensible laryngeal sac	No
Pongo pygmaeus	Dispersed	Huge laryngeal sacs extend below the throat, into the axillae	Yes Present in male
Gorilla gorilla	Polygyny	Paired, lateral, laryngeal sacs extending under skin of neck, axillae and chest	Yes Much larger in male
Pan troglodytes	Multimale–multifemale	Paired, lateral, laryngeal sacs extending under skin of neck, axillae and chest	No. Large in both sexes
Homo sapiens	Polygyny/ monogamy	Enlarged thyroid cartilage and thickened vocal cords	Yes Yes Characteristic of males and correlates with deeper voice

Principal sources: Hill 1953–1974.

than those of adult males. *C. pogonias* males produce a deep and impressive boom (personal observation) and possess a large laryngeal sac, whilst in *C. cephus* the booming vocalization is absent and the laryngeal sac is greatly reduced in size. The importance of the laryngeal sac in production of loud calls by male guenons has been demonstrated by Gautier (1971, 1988). In *C. neglectus* the sac is inflated during just that portion of the loud call when the booming sound is emitted (Fig. 7.9). Production of a fistula in the vocal sac markedly diminishes sound production (Gautier 1971).

Loud calls are not given by all *Cercopithecus* species. They are absent in *Cercopithecus aethiops*, *C. l'hoesti*, and *C. solatus* as well as in two closely related genera: *Miopithecus* and *Erythrocebus*. All these monkeys, with the exception of *Miopithecus*, are more terrestrial in habit than other *Cercopithecus* species and, with the exception of *Erythrocebus*, tend to form multimale–multifemale groups rather than one-male, polygynous units. Loud calls might not be adaptive in *C. aethiops* or *E. patas*, since they frequently live in more open, savannah conditions, rather than in rainforest, and are less territorial in their behaviour.

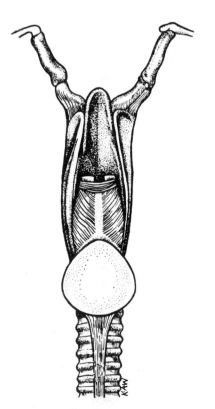

Fig. 7.8 The dorsal median air-sac of the indris (*Indri indri*) . (After Hill 1972 and redrawn from Starck and Milne-Edwards.)

Among the colobine monkeys, a sub-hyoid sac and laryngeal saccules are present in *Presbytis*, *Colobus polykomos*, and *C. guereza* (Napier and Napier 1967). In *C. guereza*, adult males in one-male units emit a loud 'roaring' call as an intergroup display (Marler 1972). In *Presbytis johnii* and *P. entellus*, adult males also emit loud 'whoops', often in association with jumping displays. Whoops are heard during the early morning or in association with intergroup encounters during the day. In *P. johnii* only the leader male of a one-male unit gives such calls, whereas in *P. entellus* several males in a multimale–multifemale unit may vocalize together, as may members of an all-male band (Hohmann 1989). It appears that in the colobines also, vocal sacs play an important role during whooping displays. Hohmann (1989) observed male *P. entellus* inflating the vocal sac during calling. In one case, a male had been bitten in the throat and the vocal sac was perforated. This impaired the male's ability to produce loud calls.

Since only adult males produce whooping or roaring calls in many colobine species and the laryngeal sacs are implicated in sound production, it seems likely that sex differences in tracheal structure occur. Beyond the statement that vocal sacs are larger in adult male than adult female colobines (Hill 1966), I have been unable to determine whether they are larger in males of polygynous forms (e.g., *P. johnii*) than in colobines which habitually form multimale–multifemale groups (e.g., *Colobus verus*; *C. badius*) and which lack loud calls.

Among the New World primates, an extreme example of sexual dimorphism in the vocal apparatus is provided by the howler monkeys (genus *Alouatta*). Howler groups may contain a single sexually active male, several adult males, or an age-graded series of males (Crockett and Eisenberg 1987). In adulthood both sexes have an egg-shaped, hollow hyoid bone and various laryngeal sacs which are inflated during deep roaring calls. These specializations are more pronounced in males, and their calls are deeper and more powerful than those of female group members. In the most extreme case the adult male red howler (*A. seniculus*) has a hyoid bone which is five times larger than that of the adult female. The bone causes a swelling beneath the male's chin and is so large that the breast bone is also sexually dimorphic, being deeply notched in the adult male to provide space for the vocal apparatus (Fig. 7.10).

Both sexes in howler troops vocalize in chorus, especially during the early morning hours. These calls are believed to play some role in intertroop location and spacing, although their precise functions have been much debated (Sekulic 1982; Crockett and Eisenberg 1987; Schön 1971). Sekulic

(1982) made a detailed study of call frequencies in four troops of *H. seniculus* with overlapping home ranges. She found that roaring occurred throughout the day and was not restricted to a 'dawn chorus'. More than 94% of roaring vocalizations which occurred during the day were directed at monkeys which were visible and not members of the troop; that is at solitary individuals as well as at neighbouring troops. During intertroop encounters, most calls were given by those males in troops with the smallest number of resident males. An adult male having no competitors within his troop vocalized more frequently than males which had competitors. Sekulic suggests that 'the loud roars of red howler monkeys are used in assessment of their opponents, as an alternative to energetically expensive chases and fights'. Such calls might 'repel solitary males and subordinate males in neighbouring troops who may attempt to replace them'.

The hyoid bone varies in size between the various *Alouatta* species, being largest and most sexually dimorphic in *A. seniculus* and least so in the mantled howler (*A. palliata*). Crockett and Eisenberg (1987) point out that species differences in hyoid volumes and in socionomic sex ratios are not correlated in the expected fashion in howler monkeys. The mantled howler has a greater socionomic sex ratio (i.e., ratio of adult females per adult male) than either the red howler or black howler, yet sexual dimorphism in hyoid volume and body weight is not so pronounced in the mantled howler (Table 7.3). At the present time there is no complete explanation of these findings. However, it is interesting that in the red howler 40% of troops contain a single male whilst most of the remainder contain just 2 males (Rudran 1979). Perhaps the hyoid is largest in red howlers, in association with a greater degree of intermale competition and female defence polygyny in this species.

Finally, the great apes require special mention, since in the orang-utan, gorilla, and chimpanzee, both sexes possess paired air sacs, emerging from each side of the larynx and forming an extensive system of pouches in the neck, chest, and axillae. As expected, sex differences in the size of these sacs are pronounced in the polygynous species, the gorilla and the orang-utan. In the gorilla, the male inflates the laryngeal sacs during his chest-beating display, beating the bare skin of his chest with cupped hands to produce loud 'pock-pocking' sounds which carry for up to a kilometre in the forest (Fig. 7.11). By comparison, the chest-beats of females and juveniles result in dull, thumping sounds (Schaller 1963). Gorillas chest-beat in a variety of social contexts, but the adult male's display is important both for intergroup communication and to intimidate intruders (Schaller 1963; Dixson 1981). In the orang-utan, fully developed territorial adult males advertise their

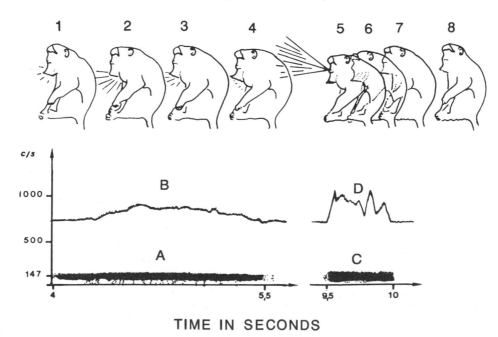

TIME IN SECONDS

Fig. 7.9 Use of the laryngeal sac during 'booming' vocalizations in the adult male De Brazza's monkey (*Cercopithecus neglectus*). 1–4: inflation of vocal sac accompanied by sound (spectra A, B); 5–7: emission of booming call (spectra C, D), 8: conclusion of call. (Modified from Gautier 1971.)

presence by means of long calls. This extraordinary vocal display, which lasts for 1–2 minutes, is accompanied by inflation of the male's extensive system of laryngeal sacs (Fig. 7.12). Females do not vocalize in this way and their laryngeal sacs are much smaller than those of adult males. It is thought that the males' long calls function for intermale spacing in the forest; whether they have any significance in attracting females is not known (Rodman and Mitani 1987).

In the chimpanzee, both males and females utter

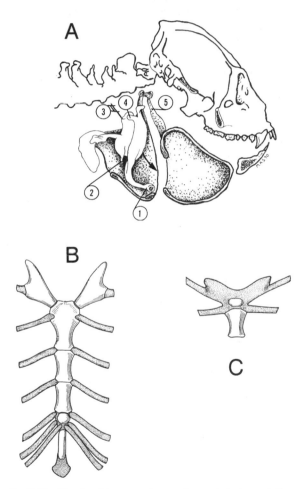

Fig. 7.10 The hyoid apparatus and associated specializations in the howler monkey (*Alouatta seniculus*). A. Adult male. Skeleton of the hyolaryngeal apparatus (after Schön 1971). 1. Processus furcatus. 2. Lingula of cuneiform cartilage. 3. Arrow through thyrocuneiform space of left side. 4. Lingula of epiglottic cartilage. 5. Cornu branchiale I. B. Sternum of adult male to show the pronounced suprasternal notch which accommodates the enlarged vocal apparatus (after Klima 1972). C. Anterior portion of the sternum in an adult female: the suprasternal notch is lacking. (After Klima 1972.)

a distinctive loud call, referred to as 'pant-hooting', and laryngeal sacs are similarly developed in both sexes. Pant-hoots serve to maintain contact between sub-groups in the fusion-fission society of the chimpanzee, and they might also enable sub-group members to locate one another as they travel through the forest. Food availability affects call frequency and males often pant-hoot when a large food source is located (Wrangham 1977). Individual and population differences in call structure have been described by Marler and Hobbett (1975) and Mitani *et al.* (1992). Mitani and Nishida (1993) have also found that pant-hooting frequencies in adult male chimpanzees are influenced by a number of social factors. They studied seven of the ten adult males in the 'M' Unit group of chimpanzees, some 90 individuals in total, living in the Mahale Mountains of Tanzania. The pant-hoot frequencies of these seven males are shown in Fig. 7.13A, together with their social ranks. Rank and call frequency were found to be positively correlated. Moreover, the frequency with which males pant-hooted was affected by the proximity of their 'alliance partners'. As described in Chapter 4, male chimpanzees form social alliances with one another and success in forming alliances affects male rank and mating success. Mitani and Nishida found that a male's pant-hooting frequency increased when alliance partners were nearby, compared with when such partners were absent. By contrast, the proximity or absence of females in reproductive condition (i.e. with large sexual skin swellings), did not affect call frequencies (Fig. 7.13C). However, some marked individual variations in male pant-hooting did occur when swollen females were nearby. Two high-ranking males (SU and NS in Fig. 7.13A) called more often at such times whilst the low-ranking males MU, AJ, and BE called less often. The alpha male (NT) also pant-hooted less frequently at such times. Mitani and Nishida point out that male rank has marked effects upon mating frequency and mating tactics in the chimpanzee. An alpha male can control sexual access to a fully swollen female chimpanzee. There might be little benefit to an alpha male if he pant-hoots when an attractive female is nearby, since he is already in a strong position to monopolize copulations. Lower-ranking males might not benefit from pant-hooting either, since they 'are neither preferred partners nor able to compete effectively with other males'. Higher-ranking individuals might succeed in attracting swollen females by pant-hooting, however, so that the vocalization might be significant in intersexual communication for a sub-set of male chimpanzees. Further information is required to test this proposition: we have no information on whether

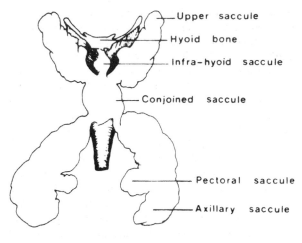

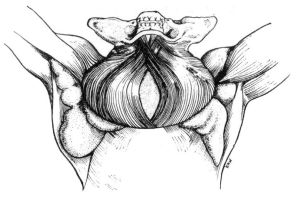

Fig. 7.12 Inflated air-sacs of an adult male orang-utan (*Pongo pygmaeus*), showing the extent of the sacs, extending to beyond the armpits and partly covered by the platysma myoides decussating in the median line. The sacs are inflated during the male's distinctive 'long-call' displays. (After Hill 1972.)

Fig. 7.11 Chest-beating display of an adult male western lowland gorilla (*Gorilla g. gorilla*) together with a dissection of the laryngeal sacs, which are inflated, to act as resonators, during the display. (After Dixson 1981.)

swollen females respond differently to the pant-hoots of particular males, for instance.

This brief review of loud calls and vocal anatomy in primates indicates that sexual dimorphism of laryngeal structure is greatest in those forest-living, arboreal species which have polygynous mating systems. Sexual selection has favoured the development of large laryngeal sacs in male guenons, howlers, and orang-utans, for instance, and in the howler monkey the male's hyoid bone is also markedly larger than that of the female. Although females also produce loud calls in some species (e.g. howlers) it is usual for the adult male alone to vocalize in polygynous species (e.g. the 'booms' of male guenons and 'whoops' of male langurs). In those cases where multimale–multifemale social groups also occur (e.g. in certain langurs and howlers), the suspicion is that they have a strong tendency to form functionally one-male, polygynous units and that this circumstance has favoured the evolution of larger vocal structures in the male sex. This contrasts with the situation in various monogamous primates, where both sexes vocalize (albeit with a different contribution to the duet, as in gibbons) and where the structure of the larynx and

Table 7.3 Hyoid volumes and socionomic sex ratios in adult howler monkeys

Species	Socionomic sex ratio*	Hyoid volume (mL)		
		Male	Female	Female/male (%)
Alouatta palliata	2.45	4.5	2.0	44
A. seniculus	1.68	69.5	12.5	18
A. belzebul	–	56.5	12.5	22
A. fusca	–	39.0	8.5	22
A. caraya	1.95	23.0	8.0	35

*Socionomic sex ratio=nos. of adult females / nos. of adult males.
Data from Crockett and Eisenberg 1987.

A. Male rank and pant-hoot frequencies

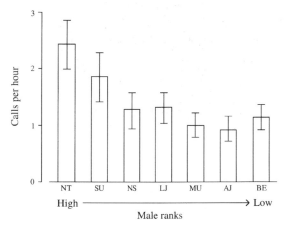

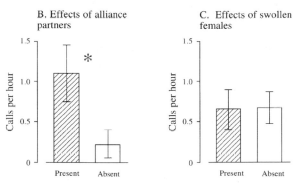

Fig. 7.13 Social factors which influence pant-hooting by free-ranging male chimpanzees (*Pan troglodytes*). A. High-ranking males vocalize more frequently than lower-ranking individuals. Data are means per hour (±SEM) based on 25 h of observations on each male. B. Males pant-hoot more often when their male alliance partners are in the vicinity. (*P < 0.05). C. The presence or absence of females having sexual skin swellings has no consistent effect upon males' pant-hooting. The significance of these findings is discussed in the text. (Redrawn and modified from Mitani and Nishida 1993.)

larygeal sacs is essentially the same in both sexes (Table 7.2).

Although the enlargement of the vocal apparatus and evolution of loud calls in various primates might be ascribed to intrasexual selection, there is always the possibility that intersexual selection has played some ancillary role in this process. The vocal duets of male and female gibbons, for instance, might play some as yet poorly understood role in pair formation and maintenance of a long-term monogamous relationship. The long call of the male orang-utan might function to attract prospective mates, as well as to repel rival males. There are insufficient data from field studies to assess these possibilities. In the

spider monkey (*Ateles*), however, it has been noted that females in reproductive condition will alter their movements in response to the distant calls of adult males (Van Roosmalen and Klein 1988). In the mouse lemur (*Microcebus murinus*) the advertisement calls of adult males exhibit individual variations in acoustic structure; females are capable of recognizing individual males on the basis of their calls and females play an active role in mate choice (Zimmermann and Lerch 1993; Buesching *et al.* 1998). In some galago species, also (e.g. *G. moholi* and *G. crassicaudatus*) females respond to the loud calls of certain males by approaching and following them as dawn approaches; these animals sometimes sleep together in nests during the daylight hours (Bearder, personal communication). Much remains to be learned about the role played by loud calls in inter-sexual communication in primates.

For those primate species in which the vocal apparatus is sexually dimorphic, it is interesting to speculate whether any sex differences might occur in those parts of the central nervous system which innervate the vocal apparatus. Such sex differences exist in other vertebrates, as for example in the sonic motor control nucleus of the oyster toad-fish (Fine *et al.* 1984) and in the vocal control nuclei of songbirds, such as the canary (Nottebohm and Arnold 1976). To my knowledge, no studies have examined the laryngeal control nuclei in guenons, howler monkeys, or orang-utans to determine whether neuronal sizes or numbers are greater in males than females.

Sexual dimorphism in cutaneous glands and scent-marking displays

Specialized cutaneous glands are found in the majority of prosimian species and New World monkeys, being located in the circumgenital region or on the ventral surface of the thorax in most cases. All New World monkey genera examined by Epple and Lorenz (1967) were found to possess glands in the sternal or epigastric regions. It is likely that the possession of such glands is a phylogenetically ancient feature of New World primates. By contrast, very few Old World monkeys have specialized cutaneous glands. They are present in the sternal region of both *Mandrillus* species (Hill 1970) and there is limited evidence for their occurrence in several of the guenons (e.g., *C. neglectus*: Loireau and Gautier Hion 1988). Interestingly, all the hominoid genera have been found to possess cutaneous glandular fields on the sternum (*Hylobates*, *Symphalangus*, and *Pongo*) or in the axillae (*Pan*, *Gorilla*, and *Homo*).

Table 7.4 Occurrence of specialized cutaneous glands and of scent-marking displays in primates

Primate group	Number of genera studied	Genera with specialized cutaneous glands	Genera which mark using glands	Genera which mark using urine
Prosimians	17	16 (94%)	16 (94%)	13 (76%)
New World monkeys	16	15 (94%)	14 (87.5%)	11 (69%)
Old World monkeys	14	2 (14%)	2 (14%)	0 (0%)
Apes and man	5	5 (100%)	0 (0%)	0 (0%)

Data Sources: Napier and Napier 1967; Epple and Lorenz 1967; Hershkovitz 1977; Schilling 1979; Coimbra Filho and Mittermeier 1981; Geissmann 1987; Mittermeier *et al.* 1988; Loireau and Gautier-Hion 1988; and author's observations.

However, none of the apes or man engages in scent-marking behaviour whereas the majority of prosimians and New World monkeys mark their environments with cutaneous glandular products and with urine (Table 7.4). In certain galagos (e.g. the greater galago and Allen's galago) and lorises stereotyped 'urine-washing' displays occur in both sexes. The animal balances on one hand and foot whilst cupping the contralateral hand and foot below the genital region. Urine is deposited on the palmar and plantar surfaces; the display may be repeated using the hand and foot on the opposite side of the animal's body. Interestingly, this highly stereotyped display is also found in several New World genera (including *Aotus*: Fig. 7.14) and was probably present in the common ancestors of prosimian and platyrrhine primates. It is highly effective in distributing urine throughout a complex, three-dimensional home range (as in various noctur-nal, forest-living prosimians: Charles-Dominique 1977).

In association with the greater development of chemical communication in prosimians and New World primates, the olfactory system is also more complex in these primates than among the catarrhines. This is especially the case in the prosimians, which possess larger olfactory bulbs than in other primates. An accessory olfactory system, involving a vomeronasal organ and accessory olfactory bulbs, is likewise highly developed in prosimians (e.g., in the mouse lemur: Evans and Schilling 1995) but present also in various New World primates (e.g. in *Aotus* and in callitrichids: Hunter *et al.* 1984; Abbott *et al.* 1993). Accessory olfactory pathways project to hypothalamic areas which regulate neuroendocrine functions and sexual behaviour; these pathways are especially important in mediating the effects of olfactory cues upon reproduction (Johns 1979;

Fig. 7.14 Homologous urine-washing displays in (left) a prosimian (*Galago crassicaudatus* and (right) a New World monkey (*Aotus lemurinus*). (After Dixson 1983b).

Fig. 7.15 An adult male sifaka (*Propithecus verreauxi*) marking a vertical trunk with the throat gland. (After Richard 1976.)

Keverne 1979a; Bronson 1989). Such pheromonal effects have recently been documented in a prosimian species (*Microcebus murinus*: Perret 1992) and a New World monkey (*Callithrix jacchus*: Abbott *et al.* 1993), as described below. It is probable that the stimulatory and inhibitory effects of pheromones upon pituitary-gonadal function are widespread among prosimians and New World primates, whereas among the catarrhines it is noteworthy that the vomeronasal system is usually lacking except during foetal development.

These introductory remarks are necessary, because in any assessment of the effects of sexual selection upon the evolution of scent-marking glands and chemical cues in male and female primates, the importance of phylogenetic differences must be recognized at the outset. Scent glands and urinary chemical cues serve a variety of communicatory functions in prosimians and New World primates, as in other mammals. A role in territorial marking has been proposed, as in *Propithecus verreauxi* (Fig. 7.15), in which males sternal and neck-rub most frequently at the territorial margins (Mertl-Millhollen 1979). Females do not scent-mark in this way, although they do exhibit genital-marking behaviour. In *Lemur catta*, males engage in intergroup 'stink fights' during their annual mating season and also tail-mark, spur-mark, and genital-mark during intragroup aggressive encounters (See Figs 3.9 and 7.16). 'Urine-

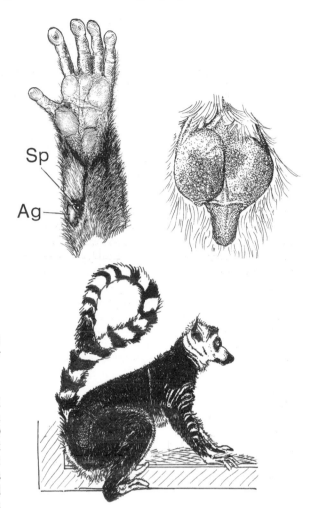

Fig. 7.16 Scent glands and marking behaviour in male ringtailed lemurs (*Lemur catta*). The antebrachial gland (Ag) and its associated spur (Sp) is shown in the upper part of the figure, as well as the male's genitalia, viewed from the perineal aspect. The scrotum is richly supplied with cutaneous glands, which are used in genital marking displays (lower figure). (Author's drawings, after Evans and Goy 1968.)

washing' displays are exhibited by both sexes in various galago species, and in *G. alleni* Charles-Dominique (1977) has demonstrated that adult males most frequently mark the boundaries of their territories.

Olfactory stimuli might transmit information about age, sex, and social rank of individuals, as well as providing lasting signals concerning their presence in a particular area. Olfactory discrimination experiments have shown that brown lemurs (*L. fulvus*) and greater galagos (*G. crassicaudatus*) can distinguish an individual's identity, age, and sex by means of chemical cues alone (Harrington 1977; Clark 1982). Epple and her co-workers have reported similar findings in callitrichids, such as *Saguinus fuscicollis* (Epple *et al.* 1993). The dominant, breeding female and male in a captive social group also mark more frequently than subordinate animals. The scent-marks of dominant individuals evoke a measurably greater response, in terms of olfactory inspections and marking responses, during olfactory discrimination tests. Epple refers to each individual tamarin or marmoset as possessing a 'chemical fingerprint', denoting its species identity, individual identity, age, sex, and dominance status. Changes in reproductive status might also be signalled by chemical cues. Rendering male marmosets (*Callithrix jacchus*) anosmic has measurable effects upon their copulatory behaviour with females during the postpartum phase (Dixson 1992, 1993a). In *Saguinus oedipus*, transfer of urine and circumgenital cutaneous secretions from a peri-ovulatory female to the genitalia of reproductively inactive females activates mounting by their male partners (Ziegler *et al.* 1993). The precise nature of the chemical cues which influence female attractiveness in these species is unknown.

Sex differences in the size and histological structure of scent-marking glands have been noted in a variety of primates. Examples are provided in Table 7.5; this list is by no means exhaustive. Fig. 7.17 shows sex differences in the histological structure of the sternal glands in *Galago garnettii*. A sparsely-haired sternal field occurs in both sexes but its area is much larger in adult males. The dermis in the sternal region contains two types of specialized gland. Large multiacinar sebaceous glands open into hair follicles via well-defined ducts. These glands are largest in adult males. Beneath this sebaceous layer, deep in the dermis, is a further layer of large, coiled apocrine glands, each of which is supplied with a narrow duct opening to the surface of the skin. Apocrine glands are of similar size in both sexes and large, active glands occur in juveniles as well as in adult specimens. There is evidence that the sebaceous glands in particular are androgen-dependent in *G. garnettii* (Dixson 1976, 1980). Castration leads to atrophy of the sebaceous layer in the sternal field, whilst testosterone treatment reverses this affect (Fig. 7.17). The brachial (sebaceous) and antebrachial (apocrine) glands of male *Lemur catta* are also stimulated by testicular androgens

Table 7.5 Examples of sexual dimorphism in the cutaneous glands of primates

Species	Cutaneous area	Direction of sex difference	Type of evidence	Primary mating system
Prosimians				
Galago garnettii	Sternal	M > F	E, H	D
Galago moholi	Sternal	M > F	E	D
Lemur catta	Brachial Antebrachial	M > F	E, H	MM
Hapalemur griseus	Brachial Antebrachial	M > F	E	?
New World monkeys				
Pithecia pithecia	Throat	M > F	E	MG?
Saguinus oedipus	Circumgenital	F > M	E, H	MG/PA
S. fuscicollis	Circumgenital	F > M	E, H	MG/PA
Callithrix jacchus	Sternal	M > F	E	MG/PA
Old World monkeys				
Mandrillus sphinx	Sternal	M > F	E, H	MM
M. leucophaeus	Sternal	M > F	E, H	MM
Apes and man				
Pongo pygmaeus	Sternal	M > F	E, H	D/PG?
Pan troglodytes	Axillary	M > F	H	MM
Gorilla gorilla	Axillary	M > F	H	PG
Homo sapiens	Axillary	M > F	H	PG/MG

Type of evidence: E=gross external morphology of gland; H=histological studies.
Mating system: MG=monogamous; PA=polyandrous; PG=polygynous; MM=multimale–multifemale; D=dispersed.
Data sources: Dixson 1976; Andrimiandra and Rumpler 1968; Napier and Napier 1967; Epple *et al.* 1993; Hill 1970; Dixson 1981; Geissmann 1987; and author's observations.

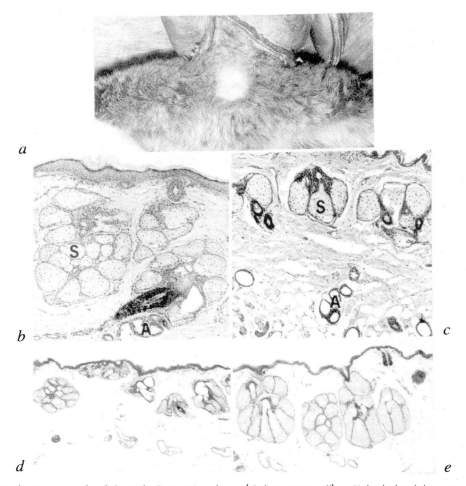

Fig. 7.17 Sternal cutaneous glands in male Garnett's galagos (*Galago garnettii*). a. Naked glandular area on the chest of an adult male. b. Section of sternal gland of an intact adult male (X 63). Large sebaceous (S) and apocrine (A) glands occur in the dermis. c. Section of sternal skin of a juvenile male (X 63). Sebaceous glands are smaller than in adults. d. Sternal gland of an adult castrate, sectioned to show atrophic sebaceous glands (X 32). e. Same male as d., after treatment with testosterone propionate, showing marked growth of the sebaceous glands (X 32). (After Dixson 1976.)

(Andrimiandra and Rumpler 1968), as is the subcaudal glandular complex in *Aotus lemurinus* (Dixson 1983a, 1994). Glandular fields in other primates likewise consist of sebaceous and apocrine elements, either in combination or alone. Much enlarged apocrine glands characterize the axillary organ which is found in the African apes and man (Fig. 7.18). The axillary organ is larger in adult males than in females; male gorillas, for example, have a distinctive pungent odour which is different from that of the female. It should not be imagined, however, that cutaneous glands are always larger in males than in females. Examples of much enlarged glandular elements in females occur in various tamarin species. In *Saguinus fuscicollis*, for example, sebaceous and apocrine glands occur in the circumgenital skin of both sexes but the apocrine glands of adult females are very large (especially so in the labia majora) and among the most complex of their type in mammals (Fig. 7.19). Sex differences in marking activity also occur in tamarins. *S. fuscicollis* females consistently mark more frequently than males by rubbing the circumgenital and supra-pubic regions against substrates, although the sex different is quantitatively small (Epple 1980). In *S. oedipus*, however, this sex difference in marking frequency is much more pronounced (Epple *et al.* 1988). Greater frequencies of sternal-marking (as opposed to genital marking) patterns have been noted in males of a variety of primate species. Examples include *Propithecus verreauxi*, in which only males sternal-rub (Mertl-Millhollen 1979), *Daubentonia madagascariensis*, in which histological evidence of a sternal gland is lacking (Winn 1994), and *Cebuella pygmaea* in which only males have been observed (rarely) to chest-rub (Soini 1988). In the Old World monkeys,

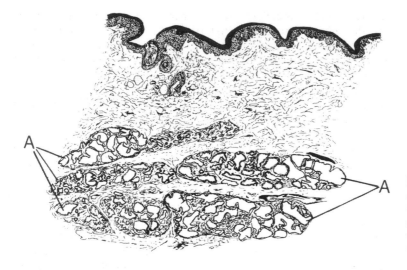

Fig. 7.18 Section of skin from the axilla of an adult male western lowland gorilla (*Gorilla g. gorilla*) to show the layers of apocrine glands (A) which constitute the 'axillary organ'. (After Dixson 1981.)

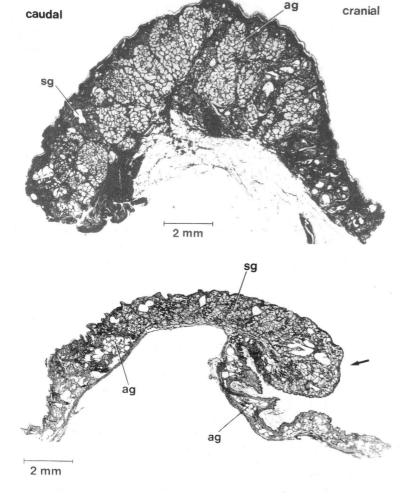

Fig. 7.19 Sexual dimorphism in the circumgenital glands of the saddle-back tamarin (*Saguinus fuscicollis*. Upper: Section through the circumgenital skin of an adult female. Lower: section through the circumgenital skin of an adult male. sg=sebaceous glands; ag=aprocrine glands. (After Epple *et al.* 1993).

although both sexes of the Mandrill rub their chests against branches and tree-trunks, it is the adult, breeding males which have the largest sternal glands and which mark most frequently (personal observation). As a final example, in *Lemur catta* which possesses highly specialized glands on the shoulder (*brachial*) and wrist (*antebrachial*), adult males mark most frequently by rubbing the tail between the shoulder and wrist complexes and then 'flicking' it to disperse the scents. Males also use the horny spur on the wrist gland to gouge substrates prior to deposition of antebrachial secretions.

From what has been said so far, it is clear that phylogenetic factors, rather than mating systems, are of the greatest explanatory value in accounting for the distribution of cutaneous glands and scent-marking displays in primates. Olfactory cues probably assist nocturnal prosimians to navigate throughout their home ranges as well as being involved in social communication. The New World monkeys, which have a long and unbroken history of arboreal life in tropical rainforest, have retained the propensity to scent-mark using cutaneous glandular secretions and urine. It is probably significant, however, that males in a variety of primate species (and mating systems) mark in territorial contexts using glands on the neck, sternum or epigastric surfaces, and that this behaviour is less pronounced in females. Male greater galagos and lesser galagos occupy larger territories than females, for instance, and the larger size of the male's sternal gland might be significant in this context. It is also intriguing that in the few cases where female scent-marking glands are superior to those of males in size and histological complexity monogamous mating systems are the norm. This is the case in *S. fuscicolis* and *S. oedipus* where females have such well-developed circumgenital glands. Competition between females occurs in many of the monogamous callitrichids, such that only a single dominant female breeds in each family group. Intrasexual competition might thus have favoured the evolution of complex scent glands in female callitrichids.

Pheromones can be classified under two broad headings—signalling pheromones and primer pheromones (Bronson 1968). Signalling pheromones evoke behavioural responses. Since this book deals with sexual behaviour, the most relevant chemical cues are those which might stimulate sexual responses, such as mounting by males in the presence of an attractive female. This problem will be discussed fully in Chapter 12, when we shall consider effects of hormones upon sexual attractiveness. At this point I wish to focus upon examples of primer pheromones; these are distinguished because they evoke a cascade of physiological changes in the recipient, rather than a prompt behavioural response. Primer pheromones might operate via the accessory olfactory system to either stimulate, or inhibit, the reproductive endocrinology and fertility of the recipient individual. A well-documented example of such effects is inhibition of ovarian cyclicity in subordinate female marmosets (Abbott *et al.* 1993). In family groups of common marmosets and some other callitrichids (e.g., *S. fuscicollis* and *S. oedipus*) the dominant, reproductive female inhibits, or partially inhibits, the ovarian cycle of her mature daughters or other subordinates. In the marmoset, Abbott and his colleagues have shown that this effect is brought about by both behavioural and olfactory stimuli, which combine to cause suppression of the GnRH pulse generator in subordinate females and hence to inhibit gonadotrophin secretion and ovarian cyclicity. Removal of a subordinate from her group dis-inhibits the ovarian cycle, but this effect is delayed if she is exposed to the scent marks of the dominant female (Barrett *et al.* 1990). The vomeronasal system plays a significant role in mediating the pheromonal effect. Since subordinates assist the breeding female in carrying infants in the group (e.g. Ingram 1977), this mechanism of reproductive suppression is advantageous to the dominant female, enabling her to recruit potential helpers to rear her twin offspring.

A second well-documented example of primer pheromones and reproductive competition concerns the mouse lemur (*Microcebus murinus*). In this case, the chemical cues are contained in urine, with which both male and female mouse lemurs mark their home ranges. Dominant males occupy ranges which overlap those of a number of females. Dominant males also exhibit higher frequencies of urine-marking than subordinates, which occupy peripheral ranges and are presumed to have less contact with females. Mouse lemurs are seasonal breeders; changes in photoperiod (increasing day length) trigger the onset of the mating season and induce increases in spermatogenesis and testosterone secretion in males. Perret and her colleagues have shown that social and urinary cues also play a crucial role in co-ordinating endocrine changes and behaviour during the mating season (Perret 1992). Thus, in the wild, urine marking by dominant mouse lemurs might play a most important role in intermale competition. Primer pheromones in female urine also affect male reproductive endocrinology and behaviour in this species. Captive male mouse lemurs are relatively unaggressive to one another if group-housed, unless females are introduced. Under these conditions males become more aggressive, scent-marking frequencies increase, and dominance is established between males with consequent effects upon their

Table 7.6 The secondary sexual adornments of adult males and occurrences of sexual dichromatism in primates

Species	Adult male features	Primary mating system	Sources
Prosimians			
Lemur macaco	Black coat colour (female: reddish brown)	MM	Napier and Napier 1967
L. mongoz	Cheek hair reddish (female: white cheeks)	MG	Napier and Napier 1967
New World monkeys			
Saimiri sciureus	'Fatting' of the shoulders, arms and head	MM	Dumond and Hutchinson 1967 Baldwin and Baldwin 1981
Alouatta caraya	Black coat colour (female usually brown)	MM/PG	Crockett and Eisenberg 1987
A. fusca clamitans	Black coat colour (female olive/yellowish)	MM/PG	Crockett and Eisenberg 1987
Cacajao calvus	Head hairless and red with muscular temporal swellings	MM	Fontaine 1981
Pithecia pithecia	White face mask and black pelage (female brownish)	MG?	Napier and Napier 1967
Old World monkeys			
Nasalis larvatus	Nose large and elongated	PG	Yeager 1990a
Rhinopithecus roxellana	Fleshy flaps at corners of mouth	?	Kavanagh 1983
Cercopithecus aethiops	Blue scrotum, red perineum and prepuce	MM	Wickler 1967
C. neglectus	Blue scrotum	PG (MG?)	Kingdon 1980
C. diana	Blue scrotum	PG	Kingdon 1980
C. hamlyni	Blue scrotum	PG	Kingdon 1980
Miopithecus talapoin	Blue scrotum and perineum	MM	Dixson and Herbert 1974
Allenopithecus nigroviridis	Red anal field	MM?	Hill 1966
Erythrocebus patas	Blue scrotum, red perineum	PG	Dixson 1983b
Mandrillus sphinx	Red nasal strip and cobalt blue paranasal ridges. Bony maxillary swellings. Red anal field and penile shaft. Blue and violet tints on scrotum and rump. Longer hair on nuchal and shoulder regions, orange beard. 'Fatted' appearance of rump and flanks in dominant males	MM?	Hill 1970 Wickings and Dixson 1992a
M. leucophaeus	Lacks the red and blue face mask of *M. sphinx*. Face is blackish. Lower lip red. Hair on neck and shoulders moderately long. Rump and scrotum blue and red tints as in *M. sphinx*. Beard is white	MM?	Hill 1970 Zuckerman and Parkes 1939
Theropithecus gelada	Long cape of hair on shoulders, back and head. Red sexual skin on chest	PG	Matthews 1956 Crook 1966
Papio hamadryas	Red sexual skin on rump and face. Swelling of rump. Long cape of white hair on shoulders, back and sides of head	PG	Kummer 1968 Hill 1970

Table 7.6—*Continued*

Species	Adult male features	Primary mating system	Sources
Papio ursinus	Mane of hair on shoulders and back longer than in female, but not so pronounced as in P. hamadryas	MM	Napier and Napier 1967 Hill 1970
Macaca mulatta *M. fuscata* *M. assamensis*	Red sexual skin on perineum and scrotum. Face reddish or flecked with red	MM	Hill 1974 Vandenbergh 1965
Apes and man			
Hylobates concolor	Black pelage with white cheek patches (female: fawn with black patch on crown)	MG	Napier and Napier 1967
H. hoolock	Black pelage with white hair above eyes and a 'preputial tuft' (female is brown)	MG	Napier and Napier 1967
H. pileatus	Black pelage (Female is black and white with longer, pale hairs on side of crown)	MG	Napier and Napier 1967
Symphalangus syndactylus	Long preputial tuft, resembling a short tail	MG	Napier and Napier 1967
Pongo pygmaeus	Pronounced fibrous/fatty cheek 'flanges': plus (particularly in the Bornean sub-species) a hump on top of the head. Long hair on shoulders, back and arms. Beard: orange/whitish	PG/D?	Napier and Napier 1967
Gorilla gorilla	White saddle of short hair on back. Pad of fibrous/fatty tissue on top of head: giving head a 'mitre' like appearance, especially in eastern races	PG	Schaller 1963
Homo sapiens	Beard, body hair, pubic hair. Loss of scalp hair results in baldness in some males	PG/MG	Tanner 1962

Mating System: MG=monogamous; PG=polygynous; MM=multimale–multifemale; D=dispersed.
?=the primary mating system is in doubt, or is disputed by some authorities.

plasma testosterone (Perret 1992). Air which has been odorized with urine from a female in pro-oestrus induces an increase in plasma testosterone in the male mouse lemur, whereas anoestrous female urine is without effect. In this case, experimental evidence has been obtained for effects of olfactory cues from females upon male reproductive function in a nocturnal prosimian which has a dispersed mating system.

It is likely that primer pheromonal effects occur in other prosimians and evidence pertaining to this question has been reviewed by Izard (1990). Synchronicity of oestrus occurs in free-ranging *Lemur catta* (Jolly 1966; Pereira 1991) and *Propithecus verreauxi* (Jolly 1966; Richard 1985) as well as in captive *Varecia variegata* (Foerg 1982b), *Lemur fulvus*

(Boskoff 1978), and *Galago crassicaudatus* (Izard 1990). However, the precise role of olfactory cues in producing these effects in prosimians remains to be demonstrated. The possibility, for instance, that the complex scent-marking displays of male *L. catta* might have neuroendocrine effects upon females in the social group was suggested long ago by Evans and Goy (1968). In captive *Varecia variegata* the male's testicular volume increases synchronously with the seasonal onset of oestrus in the female. Izard (1990) comments that 'exchange of olfactory information [might] precipitate physiological changes that result in coordination of reproductive activities'. Izard has also demonstrated that the onset of oestrous cycles at puberty is hastened in *G. garnettii* and *G. senegalensis* by housing the prepubertal female with an adult male partner. This might be equivalent to the stimulatory pheromonal effect demonstrated in mice by Vandenbergh (1983). The significance of the masculine influence upon sexual maturation in these galagos might relate to their dispersed mating system. Males regularly visit and court females in neighbouring home ranges; similar behaviour occurs in pottos and has been studied in detail by Charles-Dominique (1977). Perhaps, frequent contact with a male, as well as exposure to his scent-marks, hastens the onset of oestrus in mature females or the commencement of ovarian cyclicity in immature females, thus providing a mating advantage for the resident male. No studies of nocturnal prosimians have gathered precise data on the timing of oestrus among females in a particular area, or on the dynamics of intermale competition if synchronicity of oestrus occurs among females. Nor is the identity of any potential primer pheromone in prosimians known. In the case of the mouse lemur, the active agent in the urine of dominant males is ether-soluble (i.e., lipophilic). It is puzzling that the substance can be volatilized and that pheromonal effects do not depend upon direct contact with the recipient, such as might be expected if the effect was mediated via the vomeronasal organ (Perret 1992).

Sexual skin and other secondary sexual traits in adult males
Blue and red sexual skin

Adult males of a number of Old World monkeys possess a coloured sexual skin, covering the perineum, genitalia and adjacent areas. The sexual skin can be blue, as in some *Cercopithecus* monkeys, *Miopithecus talapoin* and *Erythrocebus patas*, or red, as in *Macaca mulatta* and *Papio hamadryas*, or a mixture of the two colours, as in the genus *Mandrillus*. Table 7.6 lists the secondary sexual adornments which occur in adult male primates, including the distribution of sexual skin among male catarrhines. Blue sexual skin is due to the presence of melanin in the dermis (e.g., in *C. aethiops*: Machida and Giacometti 1967 and in *M. talapoin*: Dixson and Herbert 1974). Light which impinges on the melanic layer is reflected back through the epidermis and, by a process of 'Tyndall scattering', a bluish hue is attained. The scattering effect depends upon collagen in the skin and is altered if the water content of the epidermis is reduced (in *C. aethiops*: Price *et al.* 1976). It appears that a reduction in the blue coloration of the scrotum is possible, but it does not result from hormonal changes as might be expected. Indeed, castration or testosterone treatment has no effect upon the blue perineal or scrotal colours of

the adult male talapoin (Dixson and Herbert 1974), and seasonal changes in testosterone secretion have no effect upon the blue scrotum of the adult patas monkey (Bercovitch 1996). Hill (1955b) observed changes in the blue sexual skin, as well as the red skin, on the face and rump of the male mandrill, as a result of perfusing cadavers with a red dye. He suggested that vascular mechanisms contributed to the brilliant blue tints of mandrills as well as to the scarlet coloration of the nasal strip and perineal field. This is unlikely, however. If the red sexual skin is pinched or squeezed it becomes pale, temporarily, until the blood flows back into the compressed area. The blue paranasal skin is not affected by this treatment and results, apparently, from structural pigmentation and the presence of the layer of dermal melanin.

Red or pink sexual skin is affected by testicular hormones. Castration results in a fading of the red sexual skin in the male rhesus monkey, for example, and the coloration is restored by testosterone treatment (Vandenbergh 1965). Castration of the male hamadryas baboon results in the swollen red rump diminishing in size and colour (Zuckerman and Parkes 1939), whilst castration of the adult patas monkey causes the red perineal field to fade, but has no effect upon the blue scrotum (Dixson 1983b). It seems, therefore, that red and blue sexual skin are affected by entirely different physiological mechanisms and, in consequence, we might expect that they sometimes fulfil different roles in visual communication.

There are strong correlations between the occurrence of sexual skin and mating systems in Old World monkeys. Sexual skin is far more common in species which have polygynous or multimale–multifemale mating systems (Table 7.6). However, this is by no means an absolute rule, so that, as in the female, there are cases where a multimale–multifemale mating system occurs but there is no sexual skin (e.g., in various *Presbytis* species and in some macaques). Some polygynous forms such as *Nasalis* or *Gorilla* lack masculine sexual skin whereas others (e.g., *Cercopithecus hamlyni*, *C. aethiops*, and *E. patas*) have brightly coloured genitalia.

As regards the evolution of blue sexual skin, it is found in just four genera of Old World monkeys; *Cercopithecus*, *Miopithecus*, *Erythrocebus*, and *Mandrillus* (Table 7.6). Since *Cercopithecus*, *Miopithecus*, and *Erythrocebus* are phylogenetically very closely related (Gautier-Hion *et al.* 1988), it is likely that their common ancestors also possessed blue, melanin-based sexual skin on the scrotum and perineum. Kingdon (1980) has noted that these colours are best developed in those *Cercopithecus* species which have a more terrestrial, or semi-terrestrial, life-style. The same is true of *Erythrocebus*. The mandrill is also a semi-terrestrial species and inhabits dense rainforest where visibility at lower levels is extremely poor. The bright blue and contrasting red hues of the mandrill's sexual skin, on the face as well as the rump, might be adaptive in signalling the presence and positions of males as the troop progresses through the forest. There is little evidence that male monkeys which possess blue sexual skin make use of their coloration in specific visual displays. Although a 'red, white, and blue display' (exhibiting the blue scrotum, red prepuce and white rump pelage) occurs in West African vervet males in aggressive contexts (Durham 1969), no particular genital displays are found in other *Cercopithecus* species (Kingdon 1980), in *Erythrocebus* (Bercovitch 1996), or in *Mandrillus* (personal observation).

Red sexual skin in male hamadryas baboons and in some macaque species, resembles the female structures, since it occurs in the genital and perineal regions. In *P. hamadryas*, especially, the resemblance between the two sexes is heightened because the male's rump is swollen, in the manner of the female's sexual skin (Fig. 7.20). Wickler (1967) ascribes the evolution of this phenomenon to a process of *socio sexual mimicry*. In his view, the male's coloration and swelling mimic the female structures and are of value during sociosexual communication, since males sometimes present their hindquarters to higher-ranking individuals as an affiliative display or to signal submission. Presentation of a 'female' signal is adaptive in this context in reducing the likelihood that aggression will ensue. Wickler extends his argument to include other species, such as the rhe-

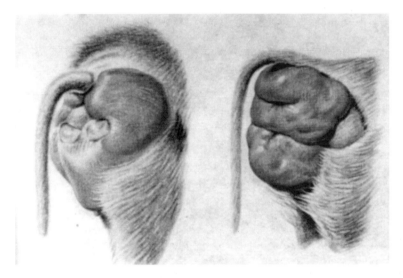

Fig. 7.20 Upper: Rump of (left) an adult male hamadryas baboon (*Papio hamadryas*) compared to (right) the swollen sexual skin of an adult female. Lower: Sexual skin and other secondary sexual adornments of the gelada (*Theropithecus gelada*) to show (left and centre) the adult female and (right) the adult male. (After Wickler 1967.)

sus macaque, where both sexes possess red sexual skin on the rump, to the gelada where the male has a red patch of sexual skin on the chest homologous with that of the adult female (Fig. 7.20), and to the mandrill where the male's facial colours have been suggested to mimic the hues of its genitalia (Morris 1967). Seductive as these arguments might be, there is no experimental evidence to support them, and much of the circumstantial evidence on sociosexual behaviour is contradictory. In various anthropoids, males present to and mount one another for a variety of reasons (see Chapter 6). Yet, in many cases, sexual skin is lacking in males or does not correspond to the females' sexual skin in coloration. Male talapoins present to one another, for instance, yet the sexual skin is blue, whilst the female has a pinkish swelling. In various baboons, macaques, and mangabeys, females have pink swellings yet males lack any correspondence of perineal or scrotal coloration. Although the male gelada has a red patch of sexual skin on the chest, there is no specialized display by which males utilize this feature during sociosexual encounters with other males. The coloration of the chest-patch does alter, however, dependent upon aggressive behaviour in this species, as described below. The brilliant facial colours of the male mandrill are not employed for overt intermale displays. A specialized grinning facial expression is often used by mandrills and adult males in particular; they combine this display with lateral head movements and erection of the nuchal crest. The mandrill's grin is an affiliative display (Van Hooff 1967) and is frequently used as a precopulatory invitation by the male (author's observations). Exactly the same facial expression occurs in the closely-related drill (*M. leucophaeus*) which lacks the striking facial colours of the male mandrill. As regards the assumed 'mimicry' of the genitalia by the male mandrill's face, Grubb (1973), has advanced an interesting argument to counter this hypothesis. He suggests that the physiological mechanisms available for production of cutaneous hues in anthropoids are limited in their scope. Vascular changes and dermal melanin will cause reddish or bluish tints whenever they occur. Grubb proposes that, in the mandrill, 'not fortuitously, but because the physiological mechanisms for production of sematic pigments are limited, the face has convergently come to acquire bright colours similar to those of the genitalia'. The resemblance between facial and genital colours is not due to sociosexual mimicry, therefore.

Despite criticisms of Wickler's (1967) hypothesis of sociosexual mimicry in primates, it is intriguing that the rump of the male hamadryas baboon is so large, and similar to the female's swelling, as well as

being brightly coloured (Fig. 7.20). In this connection, Kummer (1968) described a particular type of non-agonistic display, frequently used between harem males as a greeting. This *notifying behaviour*, as Kummer termed it, has been studied in greater detail by Colmenares (1990, 1991) and includes (often reciprocal) 'lipsmacking, ear-flattening, grimacing, hindquarter presentation, touching or grasping the partner's hindquarters and/or the genitalia and mounting'. In Colmenares view these displays

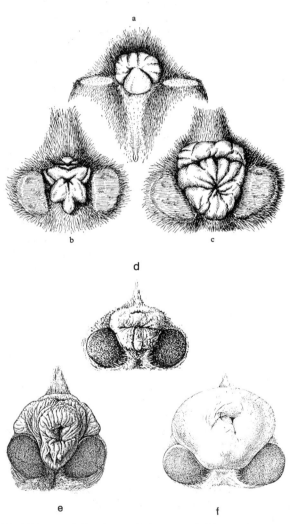

Fig. 7.21 Specializations of the sexual skin in the olive colobus (*Colobus verus*) and red colobus (*C. badius*), including the peculiar 'pseudoswellings' found in immature males of these species. a. pseudoswelling of an immature male olive colobus. b. partly swollen sexual skin of an adult female olive colobus. c. fully swollen sexual skin of an adult female olive colobus. d. pseudoswelling of an immature male red colobus. e. detumescent sexual skin of an adult female red colobus. f. fully swollen sexual skin of an adult female red colobus. (a, b, c: After Hill 1952; d, e, f: After Kuhn 1967.)

play an important role in resolving possible conflicts between harem holders without the use of overt aggression. Interestingly, other *Papio* species, in which one-male units do not occur, also lack notifying displays between adult males. Perhaps, therefore, Wickler's hypothesis is valid for the hamadryas baboon, but should not be extended to include catarrhines in general. There are two further examples, however, where male monkeys develop 'pseudo-swellings' suggestive of the sexual skin swellings of female conspecifics. But, in these cases, it is juvenile males which possess pseudo-swellings. Immature males of the red colobus (*Colobus badius*) and olive colobus (*C. verus*) are distinguished by swellings of the circumanal and perineal tissues, similar to the female's sexual skin (Hill 1952; Kuhn 1967: see Fig. 7.21). In the olive colobus, little sign of this structure can be seen once males reach adulthood (Booth 1957), whereas, in the red colobus the swelling may persist to some extent (Kuhn 1967). Wickler (1967) has also ascribed a sociosexual function to these juvenile male swellings. He suggests that young males present to adults as an appeasement gesture and that the swelling might be adaptive in this context. Since the time Wickler advanced this argument, field studies by Struhsaker (1975) on the red colobus have confirmed that juvenile males do present frequently to adult males in this species. A 'type one' present, involving flexure of all four limbs and raising the posterior, was observed 123 times, 86 of which involved juveniles. Adult males were the usual recipients of these presentations, but they did not mount the young animals. During adolescence, as the young males' external genitalia enlarge and the swelling becomes less prominent, they are often harassed by adults and may emigrate from the natal troop (Clutton-Brock and Harvey 1976; Struhsaker 1975). Unfortunately, very little is known about the social organization or behaviour of the olive colobus monkey (Booth 1957). Nor do we have any information about possible hormonal mechanisms which influence the juvenile male's swelling or its reduction in the adult animal. Sadly, both the red colobus and olive colobus do not thrive in captivity and have rarely been exhibited in zoological gardens.

It is possible that sexual skin plays some role in intermale display and competition. The blue scrotal coloration of the male vervet may fade in subordinate males, for instance, presumably as a result of epidermal changes and loss of water content (Price *et al.* 1976). At the anecdotal level Hall (1967) noted increased aggression by an adult male patas monkey towards a younger male in which the blue scrotal colour had recently developed at puberty. The red chest patch of male geladas also alters as a result of intermale agonistic encounters during harem takeovers in the wild. Typically, a harem male has brightly developed sexual skin whilst others, which live in all-male groups, tend to have pale chest patches. When a resident male is deposed in a takeover, the red sexual skin on the chest fades in colour, a change which occurs rapidly, whilst a newly dominant male exhibits red flushing of the chest-patch (Dunbar 1984). It has been claimed that the nasal strip of the adult male mandrill can change in colour, and that its red hue becomes more pronounced in aggressive males (Darwin 1872; Kingdon 1992). I have been unable to confirm this by long-term observations on semi-free-ranging mandrills. However, lower-ranking and peripheral or solitary males tend to have paler sexual skin (and lower plasma testosterone) than dominant, group-associated males (Wickings and Dixson 1992a). Loss of social rank might also be associated with reduction in red coloration in male mandrills, but this is a gradual process, not a rapid one, as in the gelada (author's unpublished observations). We might ask, since human observers are aware of these changes in male vervets, geladas, or mandrills, whether females and other conspecifics in the social group also monitor these events. It is entirely possible that a male's ability to develop and maintain these brightly coloured structures signals his health and vitality, his ability to tolerate parasites (Hamilton and Zuk 1982), or to advertise his 'good genes' by producing a costly advertisement (such as one which might increase risk of predation: Zahavi 1975). No experiment or comparative study has been conducted which would allow us to judge whether these hypotheses have any force in explaining the evolution of primate sexual skin. Whilst it is known that female swellings increase masculine sexual arousal (in the chacma baboon: Girolami and Bielert 1987), there is no experimental evidence that sexual skin in male primates is attractive to female conspecifics. It would be most interesting to conduct the required experiments using brightly coloured males such as hamadryas baboons or mandrills.

Capes of hair and facial adornments

Table 7.6 lists the occurrence of sexual dichromatism in the pelage among primates, together with those species in which males possess manes, capes of long hair, beards, and other adornments, such as the long and fleshy nose of the proboscis monkey (*Nasalis larvatus*), cheek flanges of the orang-utan, or bony paranasal swellings of the mandrill. In a few cases it is known that secondary sexual characters are androgen-dependent. The best-documented example concerns the long white cape of hair in the adult male hamadryas baboon, which is lost gradually after castration and is restored by replacing testosterone (Zuckerman and Parkes 1939). Interestingly, loss of the cape has been observed in male hamadryas baboons in the wild after receipt of aggression and loss of rank (Kummer 1990). The most pronounced development of secondary sexual adornments occurs in catarrhines which have polygynous, one-male unit mating systems, such as the hamadryas, gelada, proboscis monkey, and gorilla, but multimale−multifemale mating systems are also represented in Table 7.6. In the New World red uakari (*Cacajao calvus*), for example, both sexes lose the hair on the head at maturity, and the skin of head becomes reddish in hue. However, the adult males in multimale−multifemale groups of this species are readily identified, because of the pro-

nounced muscular bulges on the temporal region of the skull. Examples of facial adornments in males are shown in Fig. 7.22, including the distribution of facial hair in *Homo sapiens*. These various specializations are thought to be due to effects of sexual selection, either by intermale competition or by intersexual selection, or both. In truth, we know very little about the evolution or functions of any of the masculine adornments shown in Fig. 7.22. In some cases, such as the peculiar flaps at the corners of the upper lip in the golden snub-nosed monkey (*Rhinopithecus roxellana*) we do not even know the mating system of the species. However, there are some indications, in certain polygynous and multimale−multifemale anthropoid mating systems, that suppression of male secondary sexual development can occur, and that it might result from intermale competition. In *C. calvus*, for instance, some males in a semi-free-ranging troop develop rapidly at puberty,

Fig. 7.22 Specialized facial adornments in adult male primates. Upper row: left. Red uakari (*Cacajao calvus*), Centre: Proboscis monkey (*Nasalas larvatus*), right. Golden snub-nosed monkey (*Rhinopithecus roxellana*). Lower row: left: Mandrill (*Mandrillus sphinx*), centre. orang-utan (*Pongo pygmaeus*), right. Man (*Homo sapiens*). (Author's drawings, from photographs.)

exhibiting testicular enlargement and growth of the distinctive temporal bulges, whereas others in the same group might take several years to exhibit these changes in genital and secondary sexual traits (Fontaine 1981). In *Alouatta caraya*, some maturing males retain, for a time, the brown coat colour found in juveniles and adult females instead of changing to a black pelage (Neville *et al*. 1988). Such males are sometimes sexually active, despite their immature coloration. In the mandrill, peripheral or solitary

Table 7.7 Examples of suppression of secondary sexual development in male primates: possible effects of intermale competition

Species	Traits affected	Primary mating system	Sources
Prosimians			
Galagoides demidoff	Body weight of subordinate males is 75% of that in dominants.	D	Charles-Dominique 1977
G. moholi	Body weight of 'B' males is 93% of that in dominant, territorial 'A' males. 'B' males have smaller sternal glands.	D	Bearder 1987
New World monkeys			
Cacajao calvus	Muscular temporal 'bulges' on the head of the adult male develop more slowly and testicular growth is retarded.	MM	Fontaine 1981
Alouatta caraya	Retention of the (brown) juvenile coat colour, rather than transition to an adult (black) pelage.	MM/PG?	Neville *et al*. 1988
Old World monkeys			
Nasalis larvatus	Delayed growth of the adult male's large nose in males in all-male groups, or in maturing offspring in a captive group.	PG	Hollihn 1973; Bennett and Sebastian 1988
Cercopithecus neglectus	Retention of juvenile (russet/grey) pelage in presence of dominant male.	MG/PG?	Kingdon 1980
Macaca fascicularis	Attainment of full body weight is delayed until males emigrate and spend a period as solitary males at around 9 years.	MM	Van Noordwijk and Van Schaik 1985
Mandrillus sphinx	Muted secondary sexual colours, smaller testes and lower plasma testosterone in subordinate, peripheral and solitary males.	MM/PG?	Wickings and Dixson 1992a
Apes and man			
Pongo pygmaeus	Delayed cheek flange, vocal sac and hair growth in maturing males, when fully developed male is present.	PG/D?	Kingsley 1982
Homo sapiens	Pubertal development: growth of external genitalia, pubic hair, larynx and muscularity show great individual variations. Are social factors involved?	PG/MG	Tanner 1992

Mating systems: MG=monogamous; PG=polygynous; MM=multimale–multifemale; D=dispersed.
From Dixson 1997a.

males have muted secondary sexual characteristics and their mating and reproductive success is greatly exceeded by more colourful, dominant males which associate with the females (Wickings and Dixson 1992a; Dixson *et al*. 1993). In the orang-utan, suppression of male cheek flanges, throat sac, and hair growth can occur in maturing males in captivity when a fully developed male is present (Kingsley 1982). This delay in full secondary sexual growth does not mean that such males are infertile or sexually inactive, however. Spermatogenesis and increased testosterone secretion occurs at 6–7 years of age in captivity (Dixson *et al*. 1982) and in the wild such non-flanged males follow females and attempt to copulate with them (Galdikas 1985).

All these examples, and a number of others summarized in Table 7.7, indicate that some of these striking secondary sexual adornments are probably connected with intermale competition and subject to suppression by dominant individuals within the mating system. An advantage might accrue to the suppressed individual, however, in lessening intermale competition for a certain period of time and thus allowing him to mate opportunistically with females. Less brightly coloured or well-adorned males might lack sexual attractiveness to females, although very little is known about this subject. In the orang-utan, for instance, there is evidence that females tend to avoid the advances of non-flanged males, whilst preferring to approach fully-developed, territorial males (Schurmann 1982; Rodman and Mitani 1987).

Table 7.7 also includes several examples of reproductive suppression among male nocturnal prosimians. In these cases (such as *Galagoides demidoff*) it is not brightly coloured characters that are suppressed, since these nocturnal animals do not possess visual adornments of that kind. Rather, it is body weight and behaviour which are affected, so that dominant 'A' males are larger and have more ready access to female home ranges whereas subordinate males are smaller and attempt to avoid competition with A males or to occupy peripheral ranges (Charles-Dominique 1977). In these cases, as in those described above for various anthropoid species, it might be assumed that suppression is likely to represent a transient phase in the lifetime reproductive strategy of the male. If an opportunity to acquire higher rank or to occupy a vacant territory arises, full development of secondary sexual characters might then proceed. Among the anthropoids a most interesting example of suppression among maturing males has been described for free-ranging long-tailed macaques (*Macaca fascicularis*) by Van Noordwijk and Van Schaik (1985). These authors observed that 'the only males growing to maximum body size were those who were semi-solitary for several months at

the age of about 9 years'. It appeared that it was 'impossible for a male to achieve rapid growth while a full-time member of a social group'. It would be of particular interest to determine whether such a phenomenon occurs in other primate species where maturing males emigrate from their social groups either to live alone for some time (as in some macaques, howler monkeys, and gorillas) or as members of all-male groups (e.g., in patas monkeys, geladas, Hanuman langurs, and proboscis monkeys). In the case of the mandrill, semi-solitary or solitary males do continue to grow and ultimately exceed the group-associated 'fatted' males in head–body length (Wickings and Dixson 1992a).

The problem of fluctuating asymmetry

Van Valen (1962) described three kinds of morphological asymmetry in animals—antisymmetry, directional asymmetry, and fluctuating asymmetry. Antisymmetry involves enlargement of a bilateral feature on one side of the body, but the side enlarged varies randomly. The enlargement of one claw in fiddler crabs provides an example of this type of asymmetry. Directional asymmetry, by contrast, involves consistent enlargement on one side and with no variation of the side involved. The left testis is consistently larger than the right one in avian species, for example, whereas in man it is the right testis which is usually larger than the left. Both antisymmetry and directional asymmetry of morphological characters are thought to be strongly determined by genetic mechanisms. Fluctuating asymmetry, on the other hand, involves small, random variations of bilateral characters due in large part to environmental factors (e.g., nutritional constraints, parasitic infections), which impinge upon an organism throughout its development. All bilateral structures are prone to exhibit tiny differences between the two sides of the body. Although an organism is genetically programmed to develop symmetrically, with its forelimbs of equal length or its ears identical in size, tiny differences emerge during growth of the two sides of its body. Evolutionary biologists have been attracted to this subject because measurements of fluctuating asymmetry might provide some indication of an organism's ability to maintain developmental homeostasis in the face of environmental stressors (Parsons 1992). Fluctuating asymmetry might be greater in sexually selected characters (e.g., in avian feather adornments: Møller and Hoglund 1991; Manning and Hartley 1991; in primate canine teeth: Manning and Chamberlain 1993), although not all studies confirm this view (tail length in long-tailed birds: Balmford *et al.* 1993; forceps length in earwigs: Tomkins and Simmons 1995). Methods of measurement also vary between studies, so the reader should not assume that fluctuating asymmetry has been assessed in the same way by different authors.

The possibility that fluctuating asymmetry in male secondary sexual adornments might be affected by environmental stressors has been examined by Møller (1992, 1993) using the barn swallow (*Hirundo rustica*). Experimental infestation of barn swallows with fowl mites produced a higher level of fluctuating asymmetry in tail length among males, but not among females. Likewise, nuclear contamination at Chernobyl, in the former USSR, led to an increase in fluctuating asymmetry of the tail feathers of male barn swallows. The length of the male barn swallow's forked tail is important for female mate choice and male reproductive success in this species (Møller 1988a). Moreover, experimental production of asymmetric tails in male barn swallows adversely effects their attractiveness to females (Møller 1992). Whether these manipulations influence female choice directly, or via some secondary cue (e.g., reduced aerodynamic ability in asymmetric males) is a moot point.

Evidence for the direct effects of male asymmetry upon female choice has been obtained by Swaddle and Cuthill (1994) in experiments on female mate choice in zebra finches (*Taeniopygia guttata*). If male zebra finches are marked using coloured leg bands, then females show significant preferences for those males with matching bands on the two legs. Symmetrically colour-banded males also exhibit greater reproductive success; they produce more offspring which survive to the fledgling stage than do asymmetrically marked males (Swaddle 1996). In this case, since all males wear the same number of bands on each leg and only the colours of the bands vary, it appears that females must be responding directly to bilateral symmetry of coloured cues. This occurs despite the fact that the cues are artificial, rather than representing natural secondary sexual traits. An especially clear demonstration of how symmetry in a naturally occurring masculine secondary sexual trait affects female choice concerns the brush-legged wolf spider (*Schizocosa ochreata*). Males of this species have tufts on their forelegs and the tufts play an important role in courtship and agonistic displays. However, approximately 20% of male spiders lose one of their forelegs due to injuries during development. Although the leg is regenerated, its decorative tuft is usually not replaced. Such males, which lack a tuft on one of their forelegs, are less attractive to females (Uetz *et al.* 1996). Females are also less receptive if tested with a male having one of the leg tufts shaved off, or if shown video record-

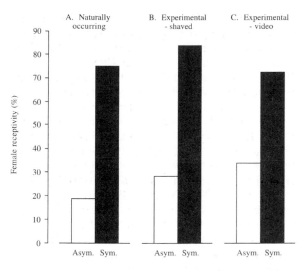

Fig. 7.23 Receptivity of females to symmetric and asymmetric males in the brush-legged wolf spider (*Schizocosa ochreata*). Decorative tufts on the forelegs of the male spider play an important role in courtship and agonistic displays. Asymmetry in male tufts was either A. Naturally occurring (a result of loss and regeneration of one forelimb). B. Experimentally induced (one tuft shaved), or. C. A digitized and manipulated video image of a courting male was displayed on a microscreen television. In each case, females were preferentially receptive to symmetric males. (Redrawn and modified from Uetz *et al.* 1996.)

ings of such a male's courtship displays (Fig. 7.23). This is a striking example of how traumatic alteration of a symmetrical male adornment can markedly affect female mate choice. It is not, of course, an example of how fluctuating asymmetry in such adornments might affect sexual interactions. Careful measurements of the leg tufts of intact males might reveal correlations between symmetry and the reproductive success of certain individuals. As they stand, Uetz *et al.*'s experiments might indicate that females prefer symmetrical males because of their past (or future) fighting potential, given that the leg tufts are also used during intrasexual agonistic displays. A male which has avoided damage to his forelegs, or which has managed to regenerate a damaged limb in a more complete fashion, might possess superior genetic qualities.

Few studies have measured fluctuating asymmetry in sexually dimorphic traits in primates. Effects of sexual selection upon enlargement of the canine teeth in male primates were considered earlier in this chapter. In polygynous primates and those living in multimale–multifemale groups, intrasexual selection has, in many cases, contributed to the evolution of larger canines in adult males. Fluctuating asymmetry in canine length in such species, is greater among males than females (Manning and Chamberlain 1993). In the gorilla there is some evidence that environmental stressors (habitat destruction) might have contributed to higher levels of fluctuating asymmetry in sexually selected features (the male's canines) but not in non-sexually selected traits (premolar size: Manning and Chamberlain 1994). Whether fluctuating asymmetry in male secondary sexual adornments might influence female choice in primates has not been examined. It would be of interest, for example, to measure asymmetries in facial adornments such as the large cheek flanges of the male orang-utan or the bony paranasal swellings of the male mandrill.

Thornhill *et al.* (1996) have reported effects of male fluctuating asymmetry upon female sexual response in human beings. Orgasm during copulation was more frequent for those women whose male partners exhibited lower levels of fluctuating asymmetry. However, the measurement of fluctuating asymmetry employed was a composite one, based upon absolute left-right differences in widths of the feet, ankles, hands, wrists, elbows, and ears as well as measurements of ear length. None of these traits is sexually selected and none is likely to be important for male sexual attractiveness or female mate choice. It was hypothesized that male fluctuating asymmetry might correlate positively with facial attractiveness, but this was not the case in the study by

Thornhill *et al.* (1995). Mens' fluctuating asymmetry did predict their body weights, however. Thornhill *et al.* also attempted to relate the 'types' of orgasms women experienced to the fluctuating asymmetries of their male partners. 'High sperm retention' orgasms (i.e. those which occurred during/after the male's ejaculation: Baker and Bellis 1993b) were more likely with men whose fluctuating asymmetry indices were lowest. These bizarre findings take no account of the absence of robust evidence for the occurrence of 'sperm retention' orgasms in women (see page 272 for a discussion of this subject). Moreover, it is difficult to understand why women should not decide much earlier in sexual contexts whether male fluctuating asymmetry is important (e.g., in relation to attractiveness and mate choice) rather than delaying the process until the point of orgasm.

That perfect symmetry is not always advantageous in sexual contexts has been shown in a most interesting study of facial attractiveness by Swaddle and Cuthill (1995). These authors used computerized modifications of photographs of human faces to produce images which were perfectly symmetrical, non-symmetric (25° transformation from normal), or mirror images of the norm. Both male and female subjects rated symmetrical faces (of either sex) as least attractive; normal and mirror images of normal faces were most preferred and intermediate scores of attractiveness were assigned to non-symmetric

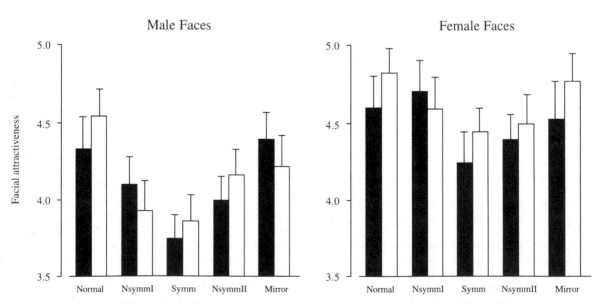

Fig. 7.24 Photographs of normal (slightly asymmetric) human faces are rated as more attractive than manipulated (more symmetric) versions of those same faces. Shown are mean ($\pm$S.E.M.) facial attractiveness scores for (left) male and (right) female sitters, as rated by male (filled bars) and female (open bars) subjects. Normal=unmanipulated photographs; Mirror=Mirror image version of normal photograph; Symm=perfectly symmetric face; NsymmI and NsymmII=25° transformations from normal. There is an overall effect of these manipulations ($F_{4,120}=4.44$, $P=0.003$). (Redrawn and modified from Swaddle and Cuthill 1995.)

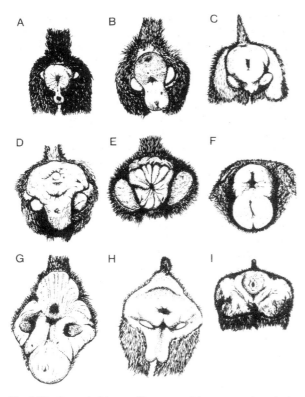

Fig. 7.25 Sexual skin swellings at mid menstrual cycle in female catarrhines. A *Cercocebus albigena*; B. *Cercocebus atys*; C. *Mandrillus sphinx*; D. *Papio ursinus*; E. *Colobus verus*; F. *Pan troglodytes*; G. *Miopithecus talapoin*; H. *Macaca nemestrina*; I. *Macaca nigra*. (After Dixson 1983b.)

faces (Fig. 7.24). The asymmetries which occur in normal human faces might arise during development for a number of reasons, including differences in muscle tone on each side of the face during voluntary facial expression (Ekman *et al*. 1981). Computerized images of human faces which are perfectly symmetrical might be less attractive because they appear abnormal and lacking in emotional expression (Swaddle and Cuthill 1995).

Secondary sexual characters in the adult female
The problem of female sexual skin

In many catarrhine species, adult females exhibit cyclical changes in coloration and swelling of skin in the anogenital region and rump. This aptly named sexual skin is typically pink or reddish in colour; in the gelada (*Theropithecus gelada*) it also occurs on the chest, so the monkey is sometimes called the 'bleeding heart baboon'. The pink or reddish hue of sexual skin results from its vascular specializations and is under hormonal control. In the rhesus monkey there is a plexus of large, thin-walled blood

vessels just below the epidermis (Collings 1926). Ovariectomy causes the red colour to fade and application of an oestrogen-containing cream directly to the sexual skin restores its colour (Herbert 1966). Swelling of the sexual skin is due primarily to an oestrogen-dependent oedema (Zuckerman and Parkes 1939; Krohn and Zuckerman 1937). In species which possess pronounced swellings (see Fig. 7.25), ovariectomy therefore causes the swelling to decrease, an effect which is reversed by oestradiol treatment (e.g., in *Papio ursinus*: Saayman 1970; *Miopithecus talapoin*: Dixson and Herbert 1977). Water retention is mostly extracellular (Krohn and Zuckerman 1937), although some intracellular retention occurs within the connective tissues of the sexual skin (Aykroyd and Zuckerman 1938). Progesterone antagonizes the effects of oestrogen upon the sexual skin. If progesterone is administered to a female in which the sexual skin is fully swollen, then deturgescence occurs (e.g., in the baboon: Gillman 1940; Gillman and Stein 1941).

Changes in coloration and swelling of the adult female's sexual skin therefore reflect changes in oestrogen and progesterone secretion during the menstrual cycle or in pregnancy. Increased reddening and swelling of sexual skin during follicular development, followed by detumescence during the luteal phase of the menstrual cycle, has been documented for many species (Dixson 1983b). Although the temporal relationship between sexual skin changes and the event of ovulation is not rigid, in the few species which have been adequately studied, the tendency is for ovulation to coincide with maximum sexual skin swelling or with the onset of detumescence. Hendrickx and Kraemer (1969) found that the optimum time for mating to result in conception was three days before the onset of detumescence in *Papio anubis* and *P. cynocephalus*. They therefore suggested that ovulation can precede detumescence by 3 days in these species, as previously suggested for *P. ursinus* by Gillman and Gilbert (1946). However, extensive laparoscopic studies (Wildt *et al*. 1977) have shown that only 17.8% of ovulations occur 2–5 days before detumescence, whereas 38.5% of follicles ovulate on the last day of maximal swelling and 26.9% on the first day of detumescence. In the chimpanzee, ovulation likewise coincides with the onset of sexual skin detumescence (Graham *et al*. 1973).

Table 7.8 lists the various genera and species of Old World monkeys and apes, and indicates whether a sexual skin is present, along with a description of its morphology at mid-menstrual cycle. Sexual skin has a peculiar and discontinuous distribution, being found in all or some members of eleven genera and absent in eight others. All species of *Cercocebus*,

Table 7.8 Sexual skin morphology in female Old World monkeys and apes: descriptions of sexual skin morphology at mid-menstrual cycle in adult females

Species	Sexual skin morphology
Colobus badius	Pronounced, pink swelling involves the vulval and clitoral areas, and sometimes includes the circumanal field. Sub specific variations in swelling size occur.
Procolobus verus	Pronounced pink swelling includes circumanal field as well as the vulval and clitoral areas.
Simias concolor	Pink swelling of the vulval and clitoral areas has been described; a complete morphological description is lacking for the adult female.
Macaca	
silenus–sylvanus group:	Swelling is large and pinkish; it involves skin in the sub caudal, circumanal, paracallosal, vulval and clitoral areas.
fascicularis group:	Little or no oedema occurs. Red/pink sexual skin occurs in sub caudal, circumanal, paracallosal, vulval and clitoral areas.
sinica group:	Swelling is usually absent. Some individuals show puffiness and pink coloration of the circumanal and vulval skin.
arctoides group:	No oedema occurs and colour change is absent or confined to slight pinkness of the vulval and clitoral skin.
Papio hamadryas	Very prominent pink swelling involves a dorsal circumanal field and a ventral lobe, consisting of vulval and clitoral areas.
P. ursinus	Swellings are pronounced, becoming largest in *P. ursinus*.
P. cynocephalus *P. anubis* *P. papio*	The tail root is not included in the swelling which consists of a dorsal (circumanal) lobe and a ventral or pubic lobe (vulval and clitoral areas).
Mandrillus sphinx *M. leucophaeus*	A prominent pink swelling of the circumanal, vulval and clitoral areas. In M. sphinx some females show red and blue coloration of facial sexual skin but this is more muted than in adult males.
Cercocebus atys *C. torquatus* *C. galeritus*	A prominent pink swelling of the circumanal, vulval and clitoral areas, similar to Mandrillus.
Cercocebus (=Lophocebus) *albigena* *C. aterrimus*	Pink swelling includes vulval and clitoral areas but only slight involvement of circumanal field.
Theropithecus gelada	Red, non-oedematous sexual skin covers the circumanal, paracallosal, vulval and clitoral regions. This is fringed by oedematous white vesicles. An '8'-shaped area of red skin, fringed by white vesicles also occurs on the chest and a small patch of sexual skin is situated on the lower abdomen.
Erythrocebus patas	Pink/reddish circumanal and vulval skin only. No oedema occurs.
Cercopithecus aethiops	Anal field and vulva may show pink/reddish coloration. No oedema occurs.
Miopithecus talapoin	Pronounced pale pink swelling includes a subcaudal lobe, as well as circumanal, paracallosal and pubic swelling.
Allenopithecus nigroviridis	Large swelling-probably similar morphologically to *M. talapoin*, but a complete description is lacking for this species.
Hylobates lar	A small swelling of the circumvulval area includes the labia minora, urethral eminence and dorsal vaginal wall.
Pan troglodytes *P. paniscus*	Large pink swelling of the circumanal, vulval and clitoral areas. The swelling is longer-lasting in *P. paniscus*.

Up-dated from Dixson 1983b using data from Tenaza 1989; Dahl and Nadler 1992; and references cited in the text.

Mandrillus, *Theropithecus*, *Papio*, and *Pan* have a well-developed sexual skin. Within the genus *Cerco-pithecus* females of the subgenera *Miopithecus* and *Allenopithecus* have prominent swellings which are absent in the remaining species. In *Cercopithecus aethiops*, however, there is a bright red sexual skin covering the perianal field and some pink coloration also surrounding the vulva and clitoris. A similar arrangement occurs in female patas monkeys (*Erythrocebus patas*). *Macaca* species exhibit the greatest variations of sexual skin to be found in any genus. Fooden (1976) recognizes four species group-ings of macaques (Table 7.9). Well-developed swellings occur in the *silenus–sylvanus* group, red sexual skin with little swelling is found in the *fascic-ularis* group whilst in the remaining two groups (*sinica* and *arctoides*) sexual skin activity is much reduced. Among the African colobines swellings are found in the subgenera *Procolobus* (*Colobus verus*) and *Piliocolobus* (*Colobus badius*), in which there are important subspecific variations in the size of the swelling (Struhsaker 1975) but not in the sub-genus *Colobus* (*C. guereza* and *C. polykomos*). No sexual skin had been described for any of the Asiatic colobine species, however, until 1989 when Tenaza recorded its occurrence in a rare Langur species (*Simias concolor*) on South Pagai island off the west coast of Sumatra. A definite swelling, which varies in size among females, occurs in this species (Fig. 7.26A). Tenaza recorded one individual on which a large pink swelling 15 cm wide and 8 cm deep was clearly visible.

Among the lesser apes, Carpenter (1940, 1941)

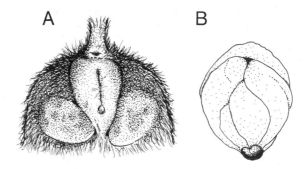

Fig. 7.26 Female sexual skin swellings of A. *Simias con-color*. (Author's drawing, after Tenaza 1989) and B. *Hylo-bates lar*. (Author's drawing, after Dahl and Nadler 1992.)

described changes in labial colour and turgescence during the menstrual cycle in *Hylobates lar* but a clear description of a definite sexual skin swelling in this species has appeared only recently (Dahl and Nadler 1992). The swelling is modest in size, pinkish in colour and involves 'convoluted lobes and bands of labia minora and the fleshy lobes of the urethral eminence' as well as some slight eversion of the dorsal vaginal wall (Fig. 7.26B). In the siamang (*Symphalangus syndactylus*) some small degree of swelling of the female genitalia has been observed in the wild (Chivers 1974), but no anatomical studies have been reported and it is unclear whether true sexual skin occurs in this species.

The vast majority of the details summarized above and in Table 7.8 were unknown to Charles Darwin in 1876 when he contributed an article to 'Nature' entitled 'sexual selection in relation to monkeys'.

Table 7.9 Genital morphology in the four species of macaques defined by Fooden 1976

Species group	Adult female		Adult male	
	Sexual skin	Vagina and cervix	Penile morphology	Baculum
silenus–sylvanus; *silenus;* *sylvanus, nemestrina, hecki nigrescens, nigra, tonkeana, ochreata, maurus, brunnescens*	Pronounced oedema and colour changes	Uterine cervix and cervical colliculi moderately large	Glans broad and blunt	Moderate length
fascicularis: *fascicularis, mulatta, fuscata, cyclopis,*	Little oedema, coloration pronounced, e.g. in *mulatta* but not in fascicularis	Uterine cervix and cervical colliculi moderately large	Glans narrow and bluntly bilobed	Moderate length
sinica: *sinica, radiata, assamensis, thibetana*	Oedema minimal or absent, little evidence of colour change except, e.g. during pregnancy	Vaginal lining spiny, uterine cervix and cervical colliculi greatly hypertrophied	Glans sagittate	Large
arctoides: *arctoides*	Oedema absent, colour absent except at mid-cycle in some individuals	Unique dorsal vesticular colliculus	Glans greatly elongated and pointed	Very large

After Dixson 1983b.

None the less, Darwin was of the opinion that such brightly coloured skin must in some way serve 'as a sexual ornament and attraction'. Since that time a wealth of circumstantial evidence has accumulated to indicate that changes in colour and swelling of the female's sexual skin might influence her attractiveness to males. Thus, in a variety of species, sexual interactions (and in particular ejaculations) are most frequent during the follicular phase or at mid-ovarian cycle. In those species where females possess a swelling, therefore, copulations are most frequent when the skin is fully tumescent during the peri-ovulatory phase of the cycle (e.g., *Miopithecus talapoin*, in captivity: Scruton and Herbert 1970; in the wild: Rowell and Dixson 1975; *Cercocebus albigena*, in captivity: Rowell and Chalmers 1970; in the wild: Wallis 1983; *Colobus badius*, in the wild: Struhsaker 1975). Other species can copulate throughout the cycle, but the majority of ejaculations still occur during the follicular or peri-ovulatory stage (e.g., in captive *Macaca nemestrina*: Eaton and Resko 1974a; *M. nigra*: Dixson 1977; and *Pan troglodytes*: Young and Orbison 1944; or in free-ranging *Macaca silenus*: Kumar and Kurup 1985; *Papio ursinus*: Saayman 1970; *Pan troglodytes*: Tutin 1980; Goodall 1986; *P. paniscus*: Furuichi 1987). In the talapoin monkey (*Miopithecus talapoin*), males normally remain on the fringes of the troop, except during the annual mating season (Rowell 1973; Rowell and Dixson 1975). Females develop their prominent sexual skin swellings during the annual mating season and males move into the central part of the troop to compete for copulations. The frequency of association between the sexes increases in direct proportion to the number of females with fully swollen sexual skins (Fig. 7.27).

While the observations cited above are consistent with the view that the female's sexual skin acts as a visual attractant to males, they in no way prove it, since changes in female proceptivity or receptivity and odour cues might also affect frequencies of copulatory behaviour. In the gorilla, for instance, copulations are often restricted to mid-cycle, yet the female lacks a sexual skin and shows only limited changes in labial swelling (Nadler 1975; Harcourt et al. 1981b). Some indication that sexual skin swellings are attractive to males comes from studies in which the male's visual inspections of the female's rump have been quantified. Male talapoins, Celebes macaques, and mandrills all look at the sexual skins of females much more frequently when they are swollen, and such inspections frequently precede attempts at copulation (Dixson and Herbert 1977; Dixson 1977).

A variety of alternative functions have been pro-

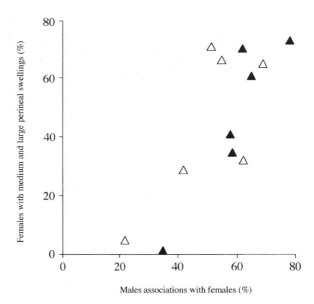

Fig. 7.27 Correlation between the percentage of females having medium/large sexual skin swellings and percentages of males' associations with females observed in two free-ranging troops of talapoins (*Miopithecus talapoin*) during the annual mating season in Cameroon. (Redrawn from Rowell and Dixson 1975.)

posed to explain the evolution of sexual skin in female catarrhines, but none is convincing. It has been suggested that sexual skin might play some role in thermoregulation (Keverne 1970), water balance (Rowell 1967b), or in volatilization of vaginal secretions to release sexually attractive odours (Michael *et al.* 1976). The tactile feedback provided by a tumescent sexual skin swelling might stimulate female proceptivity or sexual receptivity, a suggestion originally made by Zuckerman (1932) which has never been tested by experiment. Alternatively, in some species the swelling creates a funnel leading to the vaginal introitus which might facilitate intromission by the male (e.g., in *Macaca nigra*: Dixson 1977). All these possibilities have been discussed in greater detail elsewhere (Dixson 1983b).

The most convincing evidence that sexual swellings are visually arousing and attractive to males has been obtained by Bielert and his colleagues in a series of experiments on captive chacma baboons (Bielert and Girolami 1986; Bielert and Anderson 1985; Bielert and van der Walt 1982; Girolami and Bielert 1987; Bielert *et al.* 1989). When single-caged male chacma baboons are allowed to view an ovariectomized female they rarely masturbate or have seminal emissions, but once the female has been treated with oestrogen, emissions occur more frequently. Since these responses do not occur if the

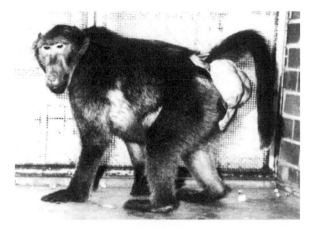

Fig. 7.28 An ovariectomized adult female chacma baboon (*Papio ursinus*) fitted with an artificial sexual skin swelling, in order to measure its possible effects upon sexual arousal in males. (After Girolami and Bielert 1987.)

female is screened off from the males, so that they can hear and smell her but have no visual contact, the sight of the female is required to produce the observed effects. However, oestrogen treatment might influence the female's behaviour (e.g., her proceptive displays) as well as the degree of sexual skin tumescence. Girolami and Bielert (1987) therefore performed further experiments in which ovariectomized females were fitted with an artificial swelling (Fig. 7.28). Under these conditions, which did not influence female invitational behaviour, eight out of nine males showed an almost 20-fold increase

in frequencies of seminal emissions (coagulated emissions resulting primarily from masturbation accumulated beneath the males' cages and were counted daily). Results of these experiments are summarized in Fig. 7.29. Further studies by Bielert *et al.* (1989) sought to determine the importance of the colour of female sexual skin in stimulating the males' sexual arousal. Positive evidence was obtained for the importance of red coloration, as well as degree of swelling in stimulating the males' masturbatory responses. However, unnatural coloration, such as a black model of the female swelling, also elicited some response. Experiments of this nature are exceedingly difficult to control, particularly where primates are concerned. Effects of novelty might interfere with the quest to determine the physiological mechanisms which govern female attractiveness. In his experiments on sexually attractive odour cues in female rhesus monkeys, Goldfoot (1981) records, for instance, that novel artificial odours applied to the rump of ovariectomized females sometimes activated males' mounting responses. Studies of zebra finches by Nancy Burley (cited by Trivers 1985), in which subjects were marked with coloured leg bands, produced the fortuitous finding that males banded red and females banded black were more attractive to the opposite sex and experienced a significant increase in reproductive success. It is thought that red leg bands enhanced the effect of male's red beak and improved his attractiveness.

Experiments by Bielert *et al.* with model swellings

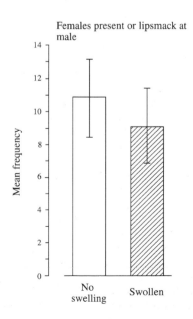

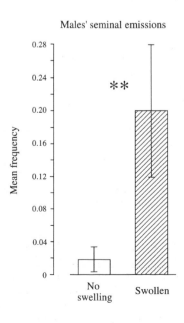

Fig. 7.29 Effects of an artificial sexual swelling fitted to an ovariectomized baboon upon female proceptive responses towards males, and males' seminal emissions. The sexes were caged separately (visual, olfactory, and vocal contacts were possible) throughout this experiment. Males showed significant increases in seminal emissions when the female was fitted with an artifical swelling (**P < 0.005). The presence of an artificial swelling did not affect female proceptivity. (Data are from Girolami and Bielert 1987.)

have at long last provided quantitative evidence for Darwin's contention that sexual skin acts as a sexual attractant. Girolami and Bielert (1987) refer to the swelling of *Papio ursinus* as a *sexual releaser* but stress that they are not suggesting that it represents an *innate releasing mechanism* in the sense defined by classical ethology (Lorenz 1950; Tinbergen 1951). I believe that they are correct to stress this reservation because there is no evidence that male primates are innately predisposed to exhibit sexual arousal in response to the sight of female swellings. Learning and experiential factors must play a key role in this regard. Indeed, captive adult chimpanzees, which normally respond to the sight of a swollen female by showing erection, will exhibit exactly the same reaction to the sight of a disposable surgical glove once they have become conditioned to associate this with donation of semen samples by masturbation (author's unpublished observations). It seems likely that, under natural conditions, males also become conditioned to associate female swellings with the pleasurable sensations of copulation and ejaculation, and thus respond to the sight of a swollen female by showing sexual arousal.

If the female's coloured sexual skin or sexual skin swelling is so attractive to male baboons, talapoins, and chimpanzees, why do these structures not occur in all monkeys and apes? None of the New World primates possesses a sexual skin, despite the strong parallel evolution of social organization and mating systems between New World and Old World monkeys. However, New World monkeys make much more extensive use of olfactory communication in reproductive contexts than catarrhines. We have seen that specialized cutaneous scent-marking glands and urine-marking displays are more prevalent in New World monkeys, and in prosimians, than in Old World monkeys and apes (Table 7.4). In connection with this, it is probable that New World primates have evolved in a more stable arboreal environment than Old World forms, many of which are terrestrial or are derived from previously more terrestrial types. It has been argued elsewhere, and in more detail, that sexual skin evolution might have been favoured among catarrhines because of their more terrestrial ancestry, and greater reliance upon visual cues rather than odour as a means of communicating reproductive status (Dixson 1983b). There is also some evidence that colour vision is more acute in catarrhines than in some New World primates (e.g., *Saimiri* or *Cebus*: De Valois and Jacobs 1971). If this is correct then the visual system of New world primates might not be so well adapted to perceive the pink or reddish hues which characterize female sexual skin. However, more information is required on this ques-

tion. In one New World species, the red uakari (*Cacajao calvus*), the head reddens and becomes hairless in adulthood. It is hard to imagine why these remarkable changes occur unless they have some significance in social communication. Returning to the catarrhines, the vast majority of species in which females have a coloured sexual skin and/or pronounced swelling of the sexual skin during the peri-ovulatory phase of the menstrual cycle, also have a multimale–multifemale mating system (Table 7.8). Swellings might function to attract several males in a multimale group, thus 'enabling the female to choose (actively or passively) between them' (Clutton-Brock and Harvey 1976). The female's swelling might be viewed as a product of sexual selection to enhance attractiveness under conditions where females mate with multiple partners. Indeed, swellings might encourage sperm competition, if the female mates with a series of partners during her peri-ovulatory period (Harvey and May 1989).

Swellings occur in females of certain catarrhine species which do not have a multimale–multifemale mating system. Some of these apparent exceptions to the sexual selection hypothesis of Clutton-Brock and Harvey (1976) are explicable, however, whereas others remain enigmatic. In *Papio hamadryas* and *Theropithecus gelada*, for instance, females possess a sexual skin yet the mating-system revolves around one-male groups (a single large adult male associated with a group of females plus their offspring). However, the social systems of both species are complex. As described in Chapter 3, the hamadryas troop consists of bands, clans, and one-male units (Kummer 1990) and in the gelada each one-male unit is similarly embedded within a much larger 'herd' system (Dunbar and Dunbar 1975). As Struhsaker and Leland (1979) have pointed out, these one-male units are not comparable with the more spatially isolated one-male units of many *Cercopithecus* species. Indeed it is entirely possible that the ancestors of hamadryas baboons and geladas lived in multimale social groups, just as macaques and baboons do today. Sexual skin might have arisen in such ancestral forms and has been retained to the present day (Dixson 1983b). A similar argument has been applied to *Mandrillus*, since there was some evidence that the large groups of mandrills or drills consist of one-male units (*M. sphinx*: Jouventin 1975; *M. leucophaeus*: Gartlan 1970). However, more recent studies, particularly of the mandrill, indicate that multimale–multifemale groups probably represent the mating system (Dixson *et al*. 1993) and that males are unable to monopolize a 'harem' of females in the same way as in *Papio hamadryas* or *Theropithecus gelada* (see also pages 74–7). The prominent swellings of female mandrills and drills probably arose as a result of sexual selection in a multimale–multifemale mating system, in the same way as hypothesized for macaques or chimpanzees.

More baffling from the standpoint of such evolutionary arguments is the occurrence of sexual swelling in females of a monogamous primate species (the white handed gibbon, *Hylobates lar*: Dahl and Nadler 1992) and in the rare Asiatic langur, *Simias concolor*, which lives in spatially separate, polygynous one-male units (Tenaza 1989). Tenaza points out that there should be very little selection pressure for females of *S. concolor* to attract multiple mating partners. However, he also raises the interesting possibility that females in one-male units might compete among themselves for the attention of the single male present. Swellings might be selected for on the basis of intrasexual competition among females, rather than solely on the basis of intersexual selection, as an attractant to the male. In Chapter 4, relationships between female social rank and reproductive success were discussed more fully (see pages 80–4). It is important to note here that the social environment can profoundly effect ovarian cycles in certain primates and that, in those species with sexual swellings, the size and colour of the sexual skin provide striking visual evidence of such phenomena. In her studies of captive baboons (*Papio anubis*) Rowell (1970a) showed that the early follicular phase of the cycle was greatly lengthened in animals which had received prolonged aggression from other females, or which had been removed from the social group and caged alone (in tactile isolation). Field studies of two troops of yellow baboons (*Papio cynocephalus*) in Tanzania (Wasser and Starling 1988) have further shown that females may form *attack coalitions* consisting primarily of 'menstrual and premenstrual females, midpregnant females and females about to resume cycling'. These coalitions are aggressive mainly towards cycling females, females in the follicular stage of the menstrual cycle being attacked more often than those in the luteal phase. Although Wasser and Starling (1988) did not comment specifically on effects of aggression upon the size of the female's swelling, it is clear that menstrual cycles were disrupted. Recipients of aggression required a greater number of cycles to become pregnant and interbirth intervals were lengthened. Attacks also involved some pregnant females and these were more likely to have their offspring late or to have abortions. The authors conclude that attack coalition behaviour in female baboons 'is a form of reproductive competition whereby females attempt to suppress the reproduction of others'.

These examples, drawn from research on captive and free-ranging baboons, have been cited to expand upon Tanaza's (1989) suggestion that competition between females in one-male units of *Simias concolor* might in some way account for the evolution of sexual skin swellings. Unfortunately we know very little about the social and sexual behaviour of these monkeys. Moreover, no other Asiatic colobine has a sexual skin and previously I argued (incorrectly) that sexual skin had not arisen in any of the ancestors of the Asiatic colobines, perhaps because of their greater propensity to form one-male groups rather than multimale–multifemale assemblages (Dixson 1983b). Nonetheless, we may hypothesize that in species with swellings, such as baboons, mangabeys, mandrills, and chimpanzees, the female advertises her 'fitness' and ability to overcome social or other environmental obstacles to ovulation by producing a fully tumescent sexual skin. Males which mate with such females are attempting to fertilize the 'best' females available. In multimale–multifemale groups there might be greater competition between females in reproductive contexts, and swellings have arisen in some species as result of intrasexual selection as well as intersexual selection to encourage sperm competition. Indeed, both processes might have operated in tandem and there is no requirement to argue for a unitary causation of sexual skin evolution.

Two further possible mechanisms which might have affected evolution of female sexual skin, were mentioned briefly in relation to sexual skin in male catarrhines: the Hamilton and Zuk hypothesis (1982) and the Zahavi handicap hypothesis (1975). Hamilton and Zuk (1982) proposed that the colourful adornments and vigorous courtship displays of many male passerine birds evolved, in part, as indicators of successful resistance to parasitic infection. Males better able to resist or tolerate parasites can maintain brighter secondary sexual adornments or more vigorous courtship displays, these being subject to sexual selection by female choice. This hypothesis has been tested with variable results in a number of vertebrate taxa (Zuk 1992), but never in non-human primates. Further, all tests of the hypothesis have attempted to account for the evolution of masculine advertisements. Yet there is no theoretical reason to exclude the secondary sexual adornments in females from consideration; the sexual skin swellings of female catarrhines being a prime example. Swelling cycles are disrupted, for instance, in female chimpanzees suffering from gastric illnesses (Yerkes 1943) or in chacma baboons (*Papio ursinus*) on a poor diet (Gillman and Gilbert 1946). However, whilst the female's health and condition might affect swelling size and thus influence a valuable visual cue regarding her reproductive status, there is no evidence concerning the specific effects of parasites in this context. A comparative analysis, for example, of numbers of parasite species affecting primates in which females possess or lack swellings, might be useful in this regard. However, such analyses are fraught with problems (Zuk 1992). Further, given the peculiar distribution of sexual skin among catarrhine species (Table 7.8) it is hard to imagine why parasites should have played an important role in the genesis of such signals indicating fitness in talapoins, red colobus monkeys, or chimpanzees, but have failed to exert such effects in kindred forms living in quite similar habitats (e.g., guenons, black and white colobus monkeys, or gorillas). A vast, if scattered, literature exists on the endo and ectoparasites of non-human primates, so a comparative analysis should be possible. I fear, however, that it would not produce a decisive result.

Zahavi's (1975) handicap model proposes that certain exaggerated traits might have evolved, despite being 'costly' or disadvantageous to the individuals which possess them, precisely because these individuals are displaying superior survival ability. The handicap thus becomes subject to sexual selection by female choice. This controversial theory might explain, for example, the evolution of exaggerated features, such as the huge tail of the peacock or plumage of certain male birds of paradise, which impede the owner's movements and might attract predators as well as potential mates. It remains an open question as to whether the sexual skin swellings of female primates are costly in an evolutionary sense. Conspicuous pink swellings, which advertise the female's reproductive status at a distance might also make her more conspicuous to carnivores or birds of prey. Whether such females are more prone to predation is unknown. The formation of swellings

requires the sequestering of water during the follicular phase of the cycle and its greater elimination during the luteal phase (Krohn and Zuckerman 1937). There is no reason to believe that this imposes a significant physiological handicap upon the female, since water is readily available to most species or is obtained via a diet of leaves and fruits. Exceptions might occur in species inhabiting hot and dry environments (e.g., *Papio hamadryas* in Ethiopia or Saudi Arabia) but are unlikely to furnish an explanation for sexual skin evolution among catarrhines as a whole. Swellings are large structures (Fig. 7.25) and may be scratched or cut occasionally (e.g. by thorny vegetation). Moreover, in some forms (e.g., baboons, mandrills, mangabeys, and chimpanzees) the swelling involves the anal field and the animal may become fouled by its own faeces. Yet there is no evidence that any of these factors represent significant (life threatening) handicaps to the animals concerned. Rather, the swellings appear to be the result of runaway sexual selection (Fisher 1930) to maximize their visual impact as sexually attractive advertisements to males. Added to this mechanism, inter-female competition might also constrain the individual's ability to produce and maintain a swelling, and thus to attract males and to encourage sperm competition. These arguments are more convincing as explanations of how sexual skin swellings arose among catarrhines than either the Hamilton and Zuk hypothesis or the Zahavi hypothesis.

The evolution of sexual skin morphology in various catarrhine lineages

Moderate swelling and pinkness of the external genitalia can be observed during the peri-ovulatory phase of the ovarian cycle in many female mammals

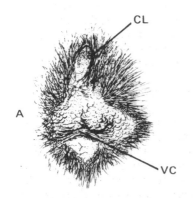
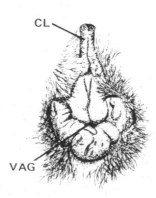

Fig. 7.30 External genitalia of an ovariectomized female Garnett's galago (*Galago garnettii*); A. Untreated. B. After treatment with oestradiol. VC=Vaginal closure membrane; VAG=Vaginal opening; CL= Clitoris. (After Dixson 1983b.)

(e.g., in rats and ferrets) including prosimian primates which lack a true sexual skin. In females of *Lemur catta* the external genitalia increase from 1.5 to 3.0 cm in length and develop a pinkish vulval coloration during oestrus (Jolly 1966). Swelling and pinkness of the vulva during oestrus has been recorded in many other prosimians (e.g., *Microcebus murinus*: Van Horn and Eaton 1979; *Perodicticus potto*: Manley 1966; *Galago crassicaudatus* and *G. garnettii*: Dixson and Van Horn 1977; *Varecia variegata* and *L. macaco*: Bogaert *et al*. 1977). In *Tarsius*, these changes are particularly striking and resemble those found in catarrhine sexual swellings (Catchpole and Fulton 1943). These changes are also oestrogen-dependent in prosimians and may be induced by administration of 17β-oestradiol to ovarectomized females (e.g. in *Galago garnettii*: Fig. 7.30).

The majority of extant prosimians are nocturnal, and diurnal forms such as *L. catta* are derived from nocturnal ancestors and still possess adaptations (such as a reflecting layer or *tapetum lucidum* in the retina) which arose in their nocturnal ancestors (Martin 1972; Charles-Dominique 1977). Colour vision, and particularly the perception of reddish hues in the long-wavelength portion of the light spectrum, might be lacking or very poorly developed in these animals (De Valois and Jacobs 1971; Pariente 1979). It is unlikely, therefore, that reddening of the female genitalia during oestrus acts as a visual signal in prosimians, although the extent of swelling could have some significance. It is far more likely that olfactory cues from the female's genitalia and urine play the major role in communicating her sexual status (Evans and Goy 1968; Schilling 1979; Perret 1992). However, if we make the not unreasonable assumption that similar changes in the female external genitalia occurred during oestrus in ancestral prosimians, then they could have provided the starting point for evolution of the sexual skin. The development of colour vision in the diurnal ancestors of modern monkeys and apes was adaptive both in terms of feeding ecology (identifying fruits, etc.) and communicatory biology (use of brightly coloured pelage or of sexual skin). In this context, the origin of sexual skin could lie in elaborations of the slight swelling and pinkness of the vulval and clitoral area to produce advertisements visible at a greater distance. Once established, runaway sexual selection would have favoured increased swelling and incorporation of skin in adjacent areas (i.e. the circumanal and paracallosal regions) to produce the highly complex swellings found in modern catarrhines.

The development of sexual skin at puberty in female catarrhines is instructive, since the skin un-

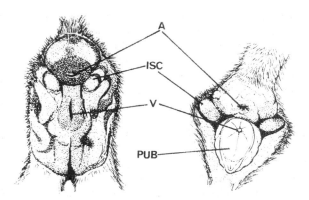

Fig. 7.31 Pubescent sexual skin of (left) a female rhesus monkey (*Macaca mulatta*) after treatment with oestradiol. (From Zuckerman *et al*. 1938, after Dixson 1983b) and (right) a female talapoin (*Miopithecus talapoin*) aged 3.5 years during the annual mating season. (From Rowell 1977a, after Dixson 1983b). A=Anus; ISC=Ischial callosity; V=Vaginal opening, PUB=Pubic lobe of sexual skin.

dergoes a number of ontogenetic changes which might recapitulate stages in its evolutionary past. Observations on a variety of species (e.g., *Macaca mulatta*; *M. fascicularis*; *Miopithecus talapoin*; *Mandrillus sphinx*, and *Pan troglodytes*: Dixson 1983b), confirm that it is the vulval, pubic, and anal regions of the sexual skin which swell and redden first in response to oestrogenic stimulation during puberty. In the talapoin (*M. talapoin*), for instance, Rowell (1977a) observed that females aged two and a half years exhibit only slight puffiness of the vulval and clitoral area during the annual mating season in Cameroon. At three and a half years these changes are more pronounced. The pink pubic lobe reaches 1.5 cm in diameter and a small degree of oedema extends into the circumanal skin (Fig. 7.31). Finally, in fully mature females, the swelling includes the region lateral to the ischial callosities as well as the pubic lobe, circumanal region, and ventral tail base. These regional differences in sexual skin sensitivity are due to qualitative differences in the skin itself, rather than to variations in its innervation or vascular supply. The crucial factor might relate to the regional distribution of oestrogen receptors; receptors for both oestrogen and progesterone occur in sexual skin. Transplantation of the pubic area of the prepubertal sexual skin in female rhesus monkeys does not alter its characteristic swelling in response to exogenous oestrogen (Zuckerman *et al*. 1938). Administration of 17β-oestradiol to ovariectomized adult talapoin monkeys results in swelling first of the pubic lobe, then of the circumanal and paracallosal regions, mimicking in a telescoped fashion the series of changes observed in adolescent females during consecutive mating seasons. The evolution of sexual skin has mainly involved changes in target organ sensitivity, rather than changes in concentrations of circulating oestrogens. Indeed, many female New World monkeys exhibit very high levels of circulating oestrogen (e.g., oestrone: Bonney *et al.* 1979, 1980) in the absence of any changes in their external genitalia.

If the series of changes observed in sexual skin swelling and coloration during puberty in female catarrhines reflects the evolutionary history of sexual skin, then certain species (e.g., among the *fascicularis*, *sinica* and *arctoides* species groups of the genus *Macaca* and in *Erythrocebus patas* or *Theropithecus gelada*) provide intriguing information. In these examples, sexual skin swelling and coloration is greater during puberty than in adulthood and declines during successive menstrual cycles throughout adolescence. This is the case in the rhesus monkey (*Macaca mulatta*) for instance; the changes in sexual skin development throughout puberty were documented in detail by Zuckerman *et al.* (1938). Swelling and slight redness of the vulval lips, pubic area, and circumanal field occurs first and a 'blister swelling' may develop in the central zone (Fig. 7.31). However, the oedema then gradually reduces in the central area and extends laterally along with increased reddening of the sexual skin. These changes occur progressively throughout adolescence and can take up to two years. Swelling and reddening of other areas, such as the inner margin of the hind legs, the flanks, back, or face may occur in certain individuals. Ultimately, oedema becomes much reduced and the red sexual skin achieves its adult distribution. Sexual skin activity in some other macaques (e.g. *M. radiata*, *sinica*, *assamensis*, and *arctoides*) is also greater during puberty than in adulthood, when little or no oedema occurs and colour changes are limited. Vulval swellings occur during adolescence in *Erythrocebus patas*, whereas the adult female lacks any swelling. In the gelada (*Theropithecus gelada*) the female's chest-patch and rump have a pinkish coloration fringed by white oedematous vesicles (Fig. 7.20). Yet, the adolescent female exhibits greater swelling of entire chest-patch area and a deeper purplish coloration on the rump (Dunbar 1977). My interpretation of these phenomena is that ancestral forms of these species might have possessed more pronounced sexual skin swellings. Present day representatives exhibit much reduced oedema or coloration of the skin in adulthood whilst the transitory changes during adolescence recapitulate the evolutionary history of the sexual skin. In those macaque species in which females have least sexual skin development, such as

M. arctoides or *M. sinica*, other genital features (e.g., the structure of the female's cervix and the glans penis in the male) are exceedingly complex and specialized. I have argued elsewhere (Dixson 1983b) that possession of swellings is a primitive feature for macaques—thus, swellings are well-developed in the *silenus–sylvanus* species group which has the widest and most discontinuous geographical distribution range and is thought to have dispersed earlier than other species groups of macaques (Fooden 1976, 1980). The *silenus–sylvanus* group is also characterized by possession of simpler penile morphologies than the other three species groups of macaques (i.e. the *fascicularis*, *sinica* and *arctoides* groupings of Fooden 1976). The reduced sexual skin of adult females in these groups, with limited colour change and little or no oedema, represents a derived condition rather than a primitive stage in sexual skin evolution. This point is important because it raises the possibility that the discontinuous distribution of sexual skin among catarrhines might result from secondary reduction or loss of the structure in some cases, as well as from its separate evolution in different lineages.

An alternative explanation for increased sexual skin activity during adolescence in female catarrhines has been proposed by Anderson and Bielert (1994). Pubescent females pass through a period of *adolescent sterility* (e.g., rhesus: Hartman 1931; chimpanzee: Young and Yerkes 1943). Given this period of infertility in younger, cycling, females it would be adaptive for males to copulate preferentially with adult partners. There is some evidence for such preferences, for example, in *Macaca fascicularis* (Van Noordwijk 1985) *M. arctoides* (Kalkstein 1991), and *Papio anubis* (Scott 1984; Smuts 1985). However, in all these cases, and others reviewed by Anderson and Bielert (1994), adolescents with the largest sexual swellings had the greatest success in eliciting mounts by males. Anderson and Bielert therefore suggest that enlarged swellings in adolescent monkeys represent supernormal stimuli which compensate to some degree for their lower sexual attractiveness to adult males. In some cases adolescent females also display heightened proceptive behaviour by comparison with adults (e.g., *Papio cynocephalus*: Rasmussen 1983; *Pan troglodytes*: Goodall 1986). Bielert and Anderson (1985) have previously argued that the larger swellings which result from repeated non-conception cycles in isolated adult female baboons (e.g., as reported in *P. anubis* by Rowell 1970a) are more sexually arousing to adult males. The proposition that larger adolescent swellings in baboons, macaques, and other species represent supernormal sexual advertisements is in-

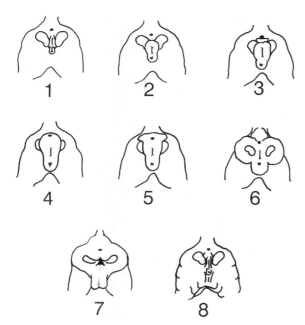

Fig. 7.32 Diagrammatic representation of possible stages in sexual skin evolution. Diagrams represent appearance of the adult female's rump at mid-ovarian cycle. (1) No sexual skin—slight oedema in the vulval and clitoral area. (2) Obvious swelling involving vulval and clitoral region only. (3) Vulval and clitoral sexual skin plus slight involvement of perianal skin. (4) Vulval, clitoral, and full circumanal swelling. (5) Slight lateral extension of the sexual skin from the perianal area has occurred. (6) Sexual skin involves the entire region lateral to the ischial callosites in addition to the vulval, clitoral, circumanal, and subcaudal area. (7) Pronounced swelling is present in the circumanal, subcaudal, and paracallosal regions, but vulval swelling is reduced. (8) A coloured sexual skin covers the rump as in (7) but oedema is limited or absent. (After Dixson 1983b.)

triguing and cannot be discounted. It remains my view, however, that this possibility should not obscure the arguments advanced for reduction of sexual skin swelling during the evolution of various macaques, guenons, and geladas.

Figure 7.32 depicts possible theoretical stages in sexual skin evolution among catarrhine primates. Among extant species, there is a series of sexual skin types which might be equivalent to evolutionary stages. The simplest condition, involving the vulval and clitoral areas only, is exemplified by *Simias concolor* or *Hylobates lar*. Next, are swellings which involve part of the circumanal field as well as the vulval and clitoral skin; such swelling occurs in *Cercocebus albigena*. More complex swellings including the entire anal field occur in *C. atys* as well as in *Mandrillus*, *Papio*, and *Pan*. The most complex swellings, which also include the paracallosal region

and ventral tail root are found in *Miopithecus* and in the *silenus–sylvanus* macaque group. The final stage in sexual skin evolution, represented by Fig. 7.32 (8), involves loss of oedema but retention of varying amounts of coloration in the sexual skin. Reduction of the sexual skin is hypothesized to have occurred in *Macaca*, *Cercopithecus* and, to a lesser extent, in *Theropithecus*. The suggested mechanism is reduction of distance-visual advertisements in females due to selection pressure to prevent hybridization (Dixson 1983b). Thus, in the *fascicularis*, *sinica*, and *arctoides* macaque groups swelling is greatly reduced or absent in the adult female and a variable degree of red coloration persists on the rump. Since all these species have a multimale–multifemale mating system, and sexual selection favours development of female swellings in these circumstances (Clutton-Brock and Harvey 1976), it is difficult to see why these macaques should have lost these specializations. However, red sexual swellings are visible at a great distance and are highly attractive to males. Lone males occur in many macaque species and they might enter troops and sire offspring. Interchange of males between groups also occurs. In conditions where hybridization between closely related forms is likely, but is not favoured by natural selection, one might predict that sexually attractive distance cues such as the female's swelling could undergo reduction or replacement by proximal cues. It is interesting that in *M. radiata* and *M. sinica* the female produces copious cervical secretions, the odour of which is believed to attract males (Rahaman and Parthasarathy 1969a; Fooden 1979).

Retention of sexual skin swellings in members of the *silenus–sylvanus* macaque group indicates that the factors influencing their speciation might have differed from those in the other species groups of macaques. The *silenus–sylvanus* group has a wide and discontinuous distribution and is thought to have dispersed earliest during macaque evolution. Isolation of various populations by climatic and geographical changes has played a major role in speciation (Fooden 1969, 1975, 1991). The seven Celebesian (Sulawesian) macaques, for instance, developed in partial isolation from each other on a group of islands which later became united to form a single land mass.

Secondary loss of sexual skin might also have occurred in the African guenons (genus *Cercopithecus*). This genus, along with patas monkeys (*Erythrocebus*), talapoins (*Miopithecus*), and Allen's swamp monkey (*Allenopithecus*) comprise a single tribe (the Cercopithecini) within the subfamily Cercopithecinae. The four African genera are very closely related; indeed *Miopithecus* and *Allenopithecus* have traditionally been regarded as subgenera of *Cercopithecus*. Some authors (e.g. Lernould 1988) also include patas monkeys (*Erythrocebus*) as members of the genus *Cercopithecus*. Prominent sexual skin swellings occur in *Miopithecus* and *Allenopithecus*, whereas in *Erythrocebus patas* and *Cercopithecus aethiops* sexual skin is limited to some redness (without oedema) around the vulva and peri-anal field. Females of other *Cercopithecus* species lack sexual skin. One hypothesis is that ancestral Cercopithecini possessed sexual skin and that it has undergone reduction in lineages which gave rise to modern guenons and patas monkeys (Dixson 1983b). The ancestors of *Cercopithecus* in all probability formed part of an early Miocene radiation of semi-terrestrial or terrestrial Cercopithecinae. However, it is still disputed as to whether the Miocene ancestors of cercopithecine monkeys were more terrestrial than colobines during this period, or whether both groups were primarily ground-living (Pickford and Genut 1988). Cheek pouches developed in ancestral ceropithecines, whereas they are lacking in the more arboreal colobines. Some authors have argued that the retention of cheek pouches in *Cercopithecus* monkeys might provide one indication of their previously more terrestrial way of life (Napier and Davis 1959; Napier and Napier 1967; Napier 1970). The majority of terrestrial or semi-terrestrial monkeys live in multimale–multifemale groups, so that sexual skin might have evolved in this context in ancestors of modern-day Cercopithecini. However, today's *Cercopithecus* species are predominantly arboreal and have primarily one-male (polygynous) mating systems. Influxes of additional males into one-male groups occur from time to time (Struhsaker 1988; Cords 1987, 1988). One-male units of forest guenons are spatially isolated, and adult males employ loud vocalizations as a spacing mechanism (Gautier 1971; Gautier and Gautier 1977). At least 25 *Cercopithecus* species have developed to fill various arboreal ecological niches and three species occupy semi-terrestrial niches. Many *Cercopithecus* species are sympatric and are very similar in size and general body proportions. However, the facial and body hair is so distinctly patterned and coloured in the various types, that species and races are readily distinguishable by their pelage alone (Kingdon 1980, 1988). This is important because, although troops of the same species tend to avoid each other, troops of different *Cercopithecus* species commonly form polyspecific associations (Gautier and Gautier-Hion 1969; Gautier-Hion 1988). Mechanisms must exist to reduce the likelihood that cross-species matings will occur in such associations. Hybridization does occur under such circumstances (a list of wild and captive hybrids is provided by Lernould 1988), but it is comparatively rare and hybrids might be reproductively disadvantaged (e.g., reduced sexual attractiveness or fertility in *C. ascanius/C. mitis* hybrids: Struhsaker *et al.* 1988). A partial isolating mechanism might be provided by temporal separation of the mating seasons, but many forest guenons have similar annual reproductive cycles. Indeed, Butynski (1988) comments upon the 'considerable interspecific synchrony of mating and birth seasons among forest guenons living in the same area'. Instead, the distinctive facial patterns and pelage coloration of forest guenons probably play a vital role in visual signalling and species isolation (Kingdon 1988). In association with the evolution of visual cues which enable immediate species recognition, selection might have favoured the reduction of distance cues that might cause sexual attraction between the various species. Red sexual skin swellings are highly attractive to males, are visible at a considerable distance, and are similar in appearance in different species. Reduction of sexual skin in females might therefore have occurred during the evolution of *Cercopithecus* monkeys. Cues which operate over a shorter distance (e.g., genital odour) or are controllable by the signaller (e.g., subtle proceptive displays) offer more species-specificity.

The gelada (*Theropithecus gelada*) offers a final, and peculiar, example of sexual skin evolution accompanied by reduction of oedema. A patch of pinkish sexual skin occurs on the female's chest and lower abdomen, as well as on her rump (Fig. 7.20). Changes in coloration of the skin provide a poor indicator of the stage of the menstrual cycle but a fringe of pinkish vesicles does show changes in oedema (Matthews 1956; Alvarez 1973; Smith and Credland 1977). Copulations are most frequent at mid-cycle when vesicles on the chest and rump are swollen (Dunbar 1978). While it is feeding the gelada spends much of its time sitting and shuffling forward in a seated position. This leaves its hands free to search for the grasses, seeds, and roots that form the bulk of its diet (Crook and Aldrich-Blake 1968; Jolly 1970b). This method of feeding is probably an ancient characteristic of the gelada and is associated with adaptations of its jaw muscles and cheek teeth for masticating small tough objects, such as seeds (Jolly 1970a). In male geladas, there are prominent fat-filled, cushion-like structures on the rump which might have evolved in association with its unusual feeding posture. The reduction of sexual skin oedema on the female's rump to a fringe of vesicles might also be attributable to these causes. The development of sexual skin on the chest could have occurred because this area is much more visible as the female sits and feeds (Wickler 1967). Jolly (1970b) has suggested that the divergence between *Theropithecus* and *Papio* might have occurred during the Pliocene and that the ancestral form would have been a more generalized, smaller, monkey, perhaps resembling *Cercocebus*. It is also logical to suggest that the ancestral form possessed a sexual skin involving the vulval and clitoral areas and possibly part of the anal field as well (Fig. 7.32 (3 and 4)). It seems unlikely that the paracallosal region was affected, since this area is not involved in modern day *Cercocebus*, *Papio*, or *Mandrillus*. Paracallosal sexual skin therefore developed in the line leading to *Theropithecus* after it had diverged from the line which gave rise to *Papio*.

The evolutionary sequence depicted in Fig. 7.32 must have occurred a number of times, with slight variations, in different catarrhine lineages. Hence, the swellings of female mandrills and chimpanzees resemble one another, but it is not implied that their remote common ancestors possessed such a structure. Given recent evidence on the occurrence of swellings in female *Simias concolor* (Tenaza 1989) and *Hylobates lar* (Dahl and Nadler 1992) it is necessary to revise the previous schema (Dixson

1983b) and to propose that female sexual skin has developed independently at least five times during catarrhine evolution:

(i) in the common ancestors of the sub-family Cercopithecinae;

(ii) in the African branch of the sub-family Colobinae among the common ancestors of *Colobus badius*, *C. kirkii* and *C. verus*;

(iii) in the Asiatic branch of the Colobinae in the genus *Simias*;

(iv) among the lesser apes in at least some species of *Hylobates*; and

(v) among the Pongidae in the common ancestor of *Pan troglodytes* and *P. paniscus*.

Secondary reduction and modifications of sexual skin might also have occurred in members of several cercopithecine genera (*Macaca*, *Cercopithecus*, *Erythrocebus*, and *Theropithecus*) as described above. These two processes have resulted in the peculiar discontinuous distribution of sexual skin among the extant forms. The evolutionary history of female sexual skin is summarized in Fig. 7.33; the schema is inevitably speculative, since sexual skin is a relatively labile structure which can probably undergo

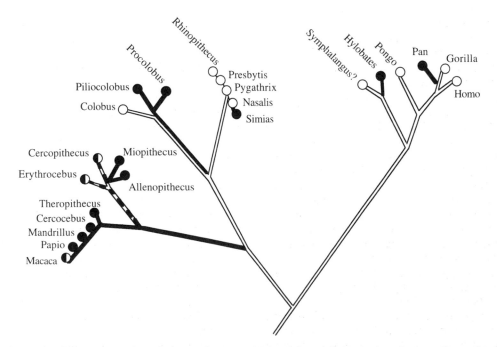

Fig. 7.33 Diagrammatic representation of the evolutionary history of sexual skin in female catarrhine primates. Black circle=a prominent sexual skin present in all members of the genus; black semicircle=sexual skin undergoing a process of secondary reduction within the genus open circle=sexual skin has never occurred within the genus. The lengths of the branches of the tree are arbitrary and do not represent a time scale. The black branches denote the occurrence of sexual skin in ancestral forms. (Redrawn and modified from Dixson 1983b.)

development quite rapidly when driven by sexual selection. The arguments presented here have changed since they were originally formulated, and further revisions might be necessary in the future.

Secondary sexual features of the human female

The information reviewed in this chapter has some relevance for our understanding of the evolution of secondary sexual features in the human female. Women have enlarged breasts, buttocks, and thighs but no trace of sexual skin activity. Modest cyclical changes in breast size occur during the human menstrual cycle, but breast volume is greatest during the luteal phase rather than during the late follicular phase, as is the case with sexual skin swellings. The absence of sexual skin is predictable, given that man does not have a multimale–multifemale mating system in the manner of chimpanzees or macaques. Polygyny, rather than rigid monogamy, is the most common form of human mating system; in 84% of the 185 human societies considered by Ford and Beach (1952) 'men are permitted by custom to have more than one mate at a time if they can arrange to do so'. These findings accord with the view that the physical characteristics of adult human beings evolved in the context of sexual selection within a partially polygynous mating system. Sexual dimorphism in body size and muscular development in human beings, as well as the occurrence of a beard and relatively small testes in relation to body weight, are indicative of a polygynous/monogamous ancestry. The enlarged buttocks and breasts of the human female also arose in this type of mating system,

therefore, and a variety of mechanisms have been proposed to explain this. Secondary sexual changes begin at puberty and are caused by hypertrophy of adipose tissue around the buttocks and thighs, as well as hypertrophy of adipose and stromal tissues in the breasts. Indeed, the human female is unique among the primates, in developing prominent breasts before her first pregnancy. However, it is incorrect to say that breast enlargement does not occur in other primates during adulthood; among the great apes some parous females have pendulous breasts. Whilst breast development is greater in the human female and begins earlier than in other hominoids, it is not a unique characteristic therefore. Very large buttocks and breasts adorn neolithic statuettes excavated in various parts of Europe and dating back some 25 000 years (Fig. 7.34). Among modern human races greatly exaggerated buttocks (Steatopygia) are found in Kalahari bushwomen (Schultz 1928) and are considered a sexually attractive feature (Fig. 7.34). The fat deposits might be adaptive as energy reserves, since bushmen live in a harsh environment in which nutritional constraints during pregnancy and lactation are considerable. Among the !Kung hunter-gatherers, for example, there is a period of adolescent sterility (averaging 17 months) after menstrual cycles commence at puberty (16.6 years); the first infant is produced at 18.8 years and interbirth intervals average 4.1 years (Howell 1979). The argument has been advanced that the fat-padded buttocks and breasts of the human female evolved in ancestors living in harsh savannah environments as signals to indicate the young female's nutritional status and potential for successful reproduction (Cant 1981; Reynolds 1991). Males selected

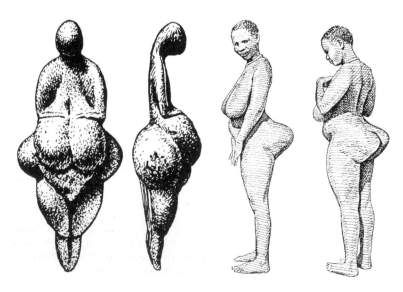

Fig. 7.34 Left: Two views of a stone-age figurine from Laspugue, Haute Garonne. Right: Hottentot woman from Bersabe. (After Wickler 1967.)

younger females with marked secondary sexual development for these reasons; hence the origin of male sexual preferences for women with shapely breasts or buttocks. Bielert and Anderson (1985) regard the early onset of breast enlargement in pubescent females as an example of a supernormal stimulus to attract males and to enable the female to experiment with sexual relationships during the phase of adolescent sterility. Cant's (1981) view is that fat deposition has become localized due to selection for more effective visual signals. Fattening of the wrists or ankles might impede movement, for instance, and fattening of the abdomen could provide an ambiguous signal. It has also been argued that the breasts mimic the appearance of the buttocks, thus providing a more efficient signal during upright walking. This argument is analogous to that

advanced to explain the evolution of sexual skin on the chest and rump of female gelada baboons (Wickler 1967; Morris 1967). It is also pertinent to mention that the African precursors of man lived in relatively hot environments where a general thickening of the layer of sub-cutaneous fat might have been disadvantageous for thermoregulation. Hence concentration of fat reserves in the buttocks and breasts might have been more efficient physiologically, as well as providing visual cues regarding female reproductive potential (Reynolds 1991).

The question of how body fat distribution might affect sexual attractiveness in women has been investigated by Singh (1993) and Singh and Luis (1995). By measuring waist circumference and circumference of the hips, at the point where the buttocks are largest, it is then possible to calculate a ratio be-

Choice of Marriage Partner

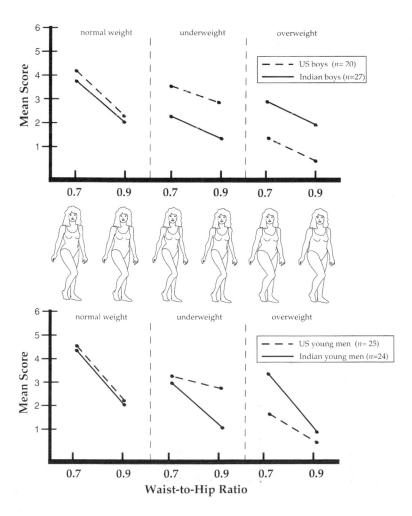

Fig. 7.35 Correlation between female waist-to-hip ratio (WHR) and sexual attractiveness in women. Boys and men of North American (USA) and Indian ethnic origin, consistently rate line drawings of women having a WHR of 0.7 as more attractive marriage partners. (Figure by kind permission of Dr. Devendra Singh.)

tween these two dimensions (the waist-to-hip ratio or WHR). This simple ratio provides an indication of how fat is distributed between the upper and lower body, and the amounts of intra-abdominal versus subcutaneous fat which are present. The WHR has been found to correlate with computer tomographic measurements of the intra-abdominal/subcutaneous fat ratio (Ashwell *et al.* 1985) and deep abdominal fat (Després *et al.* 1991). Men and women differ significantly in their waist-to-hip ratios, and this sexual dimorphism emerges during puberty. As the maturing female's hips and buttocks enlarge the WHR decreases, so that in adulthood it ranges from 0.67–0.80 in healthy subjects whereas in adult males the range is from 0.85–0.95 (Singh 1993). This sex difference plays a significant role in sexual attractiveness. When men were asked to rate line drawings of female body shapes for sexual attractiveness, they consistently judged drawings representing normal body weight and a low WHR (0.7) as most attractive (Fig. 7.35). This combination was preferred over drawings representing underweight or overweight women and those with higher WHR values (0.8, 0.9 or 1.0). Since men belonging to different ethnic groups (Afro-Americans, American caucasians, and Indians) exhibit the same preferences the phenomenon might be applicable to human males in general rather than being culture-specific (Singh and Luis 1995).

Singh argues that the evolutionary importance of the WHR in relation to female attractiveness might reside in its usefulness as an indicator of reproductive condition and health. Obese women with high WHR values have been reported to be more likely to suffer from menstrual irregularities and hirsutism (Hartz *et al.* 1984) as well as reduced fertility (Kaye *et al.* 1990; Zastra *et al.* 1993). In pubescent girls

lower WHR values are associated with higher levels of oestrogens and greater gonadotrophic activity (De Ridder *et al.* 1990). Singh, therefore, proposes that women with lower WHR values (i.e. values of approximately 0.7) are more attractive because the WHR 'is used as a reliable and honest signal of a woman's reproductive potential and health'. There is also evidence that female figures in which a low WHR is combined with large breast size are rated as being particularly attractive, feminine, and desirable by men (Singh and Young 1995).

Like the waist-to-hip ratio, sex differences in human facial characteristics and in facial attractiveness might have been selected for on the basis that they are indicators of reproductive condition and fertility. Beautiful female faces have been found to differ from average female faces in consistent ways (Johnston and Franklin 1993; Perrett *et al.* 1994). Especially attractive are combinations of facial cues which include a smaller lower jaw, full lips, and large eyes. Fullness of the lips might be influenced by oestrogens during puberty, and increases in parallel with fat deposition elsewhere up to the age of 14 years (Fargas 1987). Jaw enlargement, by contrast, is a masculine characteristic, and increases during the growth spurt in pubertal boys. Johnston proposes that attractive facial cues in women, such as lip fullness, a smaller jawline and proportionately larger eyes, may be a product of sexual selection favouring features of the face and body that are reliable cues of female fertility. The idea is intriguing and intuitively more convincing than the notion that attractive female faces represent the average of faces in a population (Langlois and Roggman 1990; Langlois *et al.* 1994) or that a child-like (neotenous) quality might underlie feminine facial beauty (Jones 1995).

8 Sperm competition

In 1970, Geoffrey Parker wrote a classic paper on sperm competition in insects. Parker suggested that sexual selection due to competition between spermatozoa of rival males would occur when females mated with a number of partners rather than a single male. Studies on artificial insemination in several vertebrate species have shown that when equal volumes of semen, or equal numbers of spermatozoa, from two males are mixed prior to insemination, then one male might sire a disproportionate number of the resulting offspring (e.g. rabbit: Beatty 1960; mouse: Edwards 1955; cattle: Beatty et al. 1969). In experiments using poultry, Martin et al. (1974) showed that inseminating equal numbers of spermatozoa from a Colombian and a Leghorn cockerel resulted in the Colombian male siring only 34% of offspring. This result occurred irrespective of the male's age, season of the year, total sperm number, breed of hen inseminated, or interval from insemination until the time of egg laying over a 15-day period. However, when the proportions of spermatozoa in the mixture were progressively increased in a stepwise fashion to favour the Colombian cockerel, then the proportion of offspring sired by this male increased, in line with mathematical predictions. Hence, it is reasonable to suggest two things, firstly that individual differences might exist between males in sperm or semen quality and, secondly, that a male might improve his chances of siring offspring if he can ejaculate a larger number of spermatozoa within the female's reproductive tract at the appropriate time to ensure conception. *Sperm competition* might therefore have favoured production and ejaculation of larger numbers of spermatozoa in species where females mate with more than one partner during the fertile period. Equally, sexual selection might have operated to increase the fertilizing efficiency of spermatozoa, or to improve the qualities of other portions of the ejaculate in which spermatozoa are suspended. This chapter reviews the evidence for these kinds of sperm competition in primates.

Relative testes size and mating systems in primates

Enormous differences in the weight of the testes, relative to body weight, among primates were first documented by Schultz (1938). However, a functional explanation for these differences was lacking until Roger Short (1979) applied sperm competition theory to explain differences in relative testis size among the great apes. The chimpanzee, which lives in multimale–multifemale groups, has very large testes. Indeed the combined weight of the testes had been measured at 118.8 g, as compared with 35.3 g for the much larger male orang-utan (Fig. 8.1) or 29 g for the massive gorilla. Short (1979) was able to relate these differences to variations in the mating systems of the great apes. In the multimale–multifemale mating system of the chimpanzee, many males might copulate with a single female during the phase of maximal sexual skin tumescence. Goodall (1986) recorded that five adult and two subadult male *Pan troglodytes schweinfurthii* copulated 84 times with a single female during an 8-day period. Under these conditions, sexual selection has led to the evolution of large testes, capable of supplying greater numbers of spermatozoa per ejaculate. The pygmy chimpanzee or bonobo (*Pan paniscus*) also lives in large social groups with a similar mating system to that described for *P. troglodytes* (Thompson-Handler et al. 1984; Furuichi 1992). The testes are very large in bonobos and, although quantitative measurements are not available, it is reasonable to suggest that sexual selection has had similar effects upon the evolution of testicular size in both chimpanzee species. In the gorilla, by contrast, a single silverback male in the social group accounts for the vast majority of copulations with the resident females (Harcourt et al. 1980). In this more strictly polygynous mating system the opportunity for sperm competition is minimal, and the adult male's testes are exceedingly small relative to body weight. Male orang-utans are relatively non-gregarious and mate with females which enter their home ranges (Mackinnon 1974; Schurmann 1982; Galdikas 1985). Consortships can develop and this, coupled with the arboreal habits of orang-utans, their slow locomotion, and low population density, might limit the number of males with which a female mates during her fertile period. Nonetheless, more than one male might attempt copulation, since younger males also pursue females and attempt to force matings (Mackinnon 1974). Relative testis size is somewhat greater in the orang-utan than in the gorilla.

In human males the testes are not large in relation to body weight; they are much smaller both absolutely and relative to body weight than those of the chimpanzee and resemble the condition found in the orang-utan (Table 8.1). Relative testes weight in man is modest and not indicative of strong selection via sperm competition. This accords with reports that the majority of human societies have polygynous mating systems or are monogamous (Ford and Beach 1952). However, since women might mate with more than one partner or conceive with a male other than the husband, sperm competition is possible in *Homo sapiens*, though whether it has played a significant role during human evolution remains highly debatable (Smith 1984).

The more general applicability of findings concerning relative testis size and mating systems in the Hominoidea was demonstrated by the broader comparative studies of Harcourt *et al.* (1981a) on over 30 species of anthropoid primates (Fig 8.2). Monkeys living in multimale–multifemale social groups, such as various species of *Macaca*, certain baboons (e.g. *Papio ursinus* and *P. cynocephalus*), or squirrel monkeys (*Saimiri* spp.) have larger testes relative to their body weights than polygynous forms such as certain langurs (e.g. *Presbytis obscura* and *P. cristata*) or black and white colobus (*Colobus polykomos*).

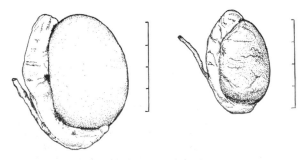

Fig. 8.1 Testes of (left) an adult chimpanzee (*Pan troglodytes*) and (right) an adult orang-utan (*Pongo pygmaeus*). The scales are in 1 cm divisions. (Author's drawings.)

Pair-living, monogamous primates, such as the owl monkeys (*Aotus* spp.) or gibbons (*Hylobates* spp.) also have relatively small testes. This is not to deny that extra-pair copulations occur in some pair-living primates (Palombit 1994; Reichard 1995), as certainly occurs in many 'monogamous' bird species (Birkhead and Møller 1992). However, the opportunities for sperm competition are much less in pair-living primates, such as owl monkeys, than in many multimale forms such as the chimpanzee. Hence, sexual selection has favoured the evolution of larger testes in species where females commonly mate with a number of partners rather than remaining with, or being monopolized by, a single male.

Further tests and applications of relative testis size in primates
The prosimian primates

The study by Harcourt *et al.* (1981a) did not include any of the prosimian primates. However, a later report (Dixson 1987a) showed that large testicular volumes occur in lemurs which live in multimale–multifemale social groups (e.g. *Lemur fulvus* and *L. macaco*). Moreover, nocturnal prosimians which have non-gregarious or dispersed social systems (e.g. *Microcebus murinus*, *Cheirogaleus major*, *Galago senegalensis*, and *G. demidoff*) were also found to have very large testes relative to body weight (Fig 8.3). These findings have been confirmed and extended in subsequent studies of testes size in prosimians (Dixson 1995a; Kappeler 1997b). In many nocturnal prosimians, males occupy separate home ranges and are intolerant towards each other (e.g. *Perodicticus*, *Arctocebus*, *Galago alleni*, *G. senegalensis*, *G. demidoff*, *Daubentonia*: Charles-Dominique 1977; Charles-Dominique and Bearder 1979; Sterling 1993). The home ranges of males tend to be larger than those of females and, since they overlap the ranges of a number of females, this might suggest that a type of nocturnal polygynous system is in operation (Martin 1981; Bearder 1987). Although there is no evidence that females mate with a single male under these conditions, this seems unlikely, given the dispersed nature of the system and difficulties for males in guarding individual females at the height of oestrus. Mating aggregations, in which a number of males cluster around or follow an oestrous female, have indeed been recorded for several species (*Galago crassicaudatus*: Clark 1985; *G. senegalensis*: Bearder, personal communication; *Nycticebus coucang*: Barrett 1985; *Daubentonia madagascariensis*: Sterling and Richard 1995). Hence

Table 8.1 Adult testes weights and body weights in the apes and man

Species	Numbers and origin	Mean combined testes weights (g)	Mean body weight (kg)	Sources
Hylobates lar	?	5.5	5.5	Kennard and Willner 1941
H. cinereus	7 (W)	4.6	5.54	Schultz 1938
Pongo pygmaeus	2 (W+C)	35.3	74.64	Schultz 1938
P.p. pygmaeus	3 (C)	36.5	101.3	Dixson *et al.* 1982
P.p. abelii	10 (C)	31.28	96.2	Dahl *et al.* 1993
Pan troglodytes	3 (C)	118.8	44.34	Schultz 1938
P.t. troglodytes	10 (C)	157.9	55.89	Dixson and Mundy 1994
Gorilla gorilla heringei	3 (W)	28.96	164.66	Hall-Craggs 1962; Dahl, personal commun.
G.g. gorilla	9 (C)	15.01	171.69	Dahl, personal commun.
Homo sapiens	3 (N.America)	50.2	63.54	Schultz 1938
H. sapiens	5 (India)	35.175	?	Jit and Sanjeev 1988
H. sapiens	1743 (Switzerland)	39.06	?	Zachmann *et al.* 1974
H. sapiens	? (Sweden)*	49.56	?	Taranger *et al.* 1976

* At 17 years of age. W=wild; C=captive.

sperm competition might occur in non-gregarious nocturnal prosimians and sexual selection might have favoured the evolution of larger testes for this reason. Likewise, in many lemurs which live in multimale–multifemale groups, larger testes relative to body weight might maximize reproductive success. However, an important caveat to this hypothesis concerns the occurrence of strict patterns of seasonal breeding in many prosimians. Hence in groups of ringtailed lemurs (*L. catta*) mating occurs during a period of two weeks when individual females are

in oestrus for periods of 24 hours or less (Jolly 1966; Pereira 1991). A restricted mating season could have led to selection for larger testis size in some primates, irrespective of the mating system. That this is not the case has been demonstrated by more recent studies of relative testes sizes in nocturnal prosimians and anthropoid primates (Dixson 1995a; Harcourt *et al.* 1995). In the absence of a restricted mating season, sexual selection still favours the evolution of large testes. This is the case in the chimpanzee, for instance, and also in the potto (*Per-*

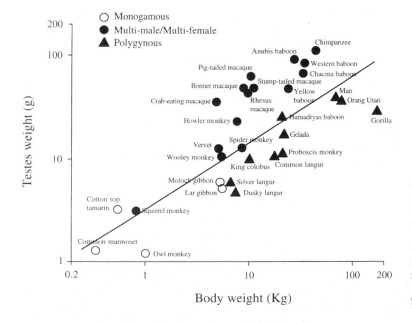

Fig. 8.2 Relative testes weights and mating systems in anthropoid primates. A logarithmic plot of combined testes weights versus body weight for anthropoids having monogamous, polygynous or multimale–multifemale mating systems. (Based on Harcourt *et al.* 1981a; redrawn and modified from Short 1985.)

odicticus potto), a non-seasonally breeding prosimian (Fig. 8.3). The same is likely to be true for the Aye-aye (*Daubentonia*) which has 'voluminous' testes (Hill 1953), a non-gregarious social system, and a non-seasonal pattern of reproduction (Sterling 1994).

The marmosets and tamarins (Callitrichidae)

Although marmosets and tamarins have traditionally been classified as monogamous primates, living in extended family units (e.g. Kleiman 1977), field studies of some species have shown that a single reproductive female might mate with more than one partner (e.g. *Callithrix humeralifer*: Rylands 1982, 1986; *Saguinus fuscicollis*: Goldizen 1987; *Saguinus mystax*: Garber *et al.* 1996). This has led to suggestions that facultative polyandry might occur, so that two or more males copulate with a single female and co-operate in rearing her offspring (Goldizen 1987; Sussman and Garber 1987). It is not known whether facultative polyandry occurs throughout this primate family or if it is confined to particular species or populations. However, in the context of intermale competition, it would be of interest to know whether sexual selection has favoured the evolution of larger relative testis size in any of these monkeys. In a comparative study Harcourt et al. (1981a) noted that relative testis weight in the cottontop tamarin (*Saguinus oedipus*) was greater than expected for a monogamous primate, whilst in a second callitrichid, the common marmoset (*Callithrix jacchus*), the testes were small in relation to body weight, consistent with a monogamous mating system. In a subsequent study (Dixson 1987a) testicular volumes were also found to be larger than expected in several callitrichid species, but the numbers of specimens measured were very small (*Cebuella pygmaea*, N = 1; *Callithrix argentata*, N = 1 and *Saguinus nigricollis*, N = 2). Moreover, in none of these cases were testicular volumes and body weights obtained from the same specimens. This point is crucial since adult body weights can vary enormously within callitrichid species and there are differences between captive and free-ranging individuals, or even between separate populations in the wild (e.g. in *Saguinus mystax* and *S. fuscicollis*; Garber and Teaford 1986). That intraspecific variations in testicular weight occur will be appreciated by considering Fig. 8.4. Individual data on body weights and testicular volumes in *Cebuella pygmaea*, *Callithrix jacchus*, and *Saguinus oedipus* have been plotted, as well as mean values for each species. In all cases except one (the highest value for *Cebuella*) testicular volume estimates and body weights were obtained from the same individuals (Dixson 1993a). Whilst a few male *C. jacchus* and *S. oedipus* do have relatively large testes, the mean values for these species are low, as are those for *Cebuella*, and do not indicate effects of sexual selection due to sperm competition. Why testicular volumes vary so much between males remains unexplained, however, and could reflect some type of intermale competition and suppression of gonadal activity. Garber *et al.* (1996) also raise this possibility with respect to marked individual differences in

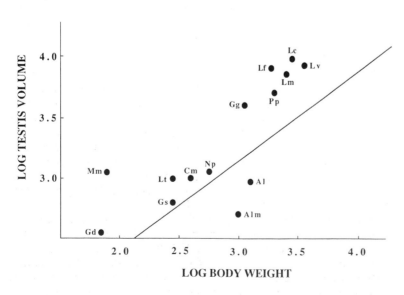

Fig. 8.3 Logarithmic plot of mean testis volume versus body weight in 14 species of nocturnal primates. Np. *Nycticebus pygmaeus*; Lt. *Loris tardigradus*; Pp. *Perodicticus potto*; Gs. *Galago senegalensis*; Gg. *Galago garnettii*; Gd. *Galagoides demidoff*; Mm. *Microcebus murinus*; Cm. *Cheirogaleus major*; Al. *Avahi laniger*; Lc. *Lemur catta*; Lf. *lemur fulvus*; Lm. *lemur macaco*; Lv. *Varecia variegata*; Alm. *Aotus lemurinus*. In the case of *L. tardigradus*, mean testis volume was calculated by dividing testicular weight by 1.05 (the specific gravity of the tissue). The principal axis line (slope 0.718) is derived from a previous study of testis volume versus body weight in a larger sample of primate species (Dixson 1987a). (Redrawn from Dixson 1995a).

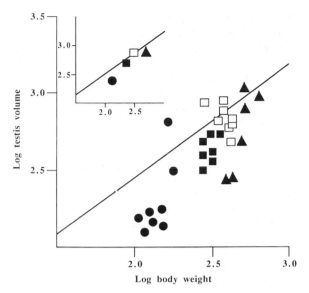

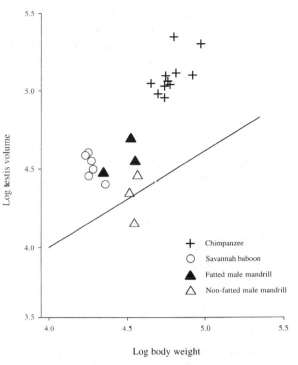

Fig. 8.4 Logarithmic plot of mean testis volume versus body weight in three callitrichid species. The lower graph shows values for individual specimens of *Cebuella pygmaea* (●); *Callithrix jacchus* (captive individuals) (□):, free-ranging (■) and *Saguinus oedipus* (▲). The upper graph shows mean values for each species. On both graphs the principal axis line (slope 0.718) is derived from studies of testis volume versus body weight in a larger sample of primate species (Dixson 1987a). (Redrawn from Dixson 1993a.)

testes volumes of free-ranging moustached tamarins (*S. mystax*). It is also apparent that testicular volumes for captive *C. jacchus* exceed those for wild trapped specimens (Fig. 8.4) which raises questions about seasonal and nutritional effects upon testicular function. Whilst a broader analysis, involving larger numbers of callitrichid species, would certainly be valuable, the available evidence does not indicate that sperm competition has played a significant role in affecting relative testis size in this family of New World primates (Dixson 1993a).

The mandrill and drill

Returning to the Old World, a particularly interesting test of testis size/body weight relationships in relation to mating systems is provided by the West African genus *Mandrillus*. Figure 8.5 shows estimates of testicular volume relative to body weight in adult male mandrills compared with data on two known multimale–multifemale species—the chimpanzee (*Pan troglodytes*) and the savannah baboon (*Papio cynocephalus*). The group-associated (fatted) male mandrills which have the most colourful sec-

Fig. 8.5 Logarithmic plot of mean testis volume versus body weight for individual adult specimens of the mandrill (*Mandrillus sphinx*), Savannah baboon (*Papio cynocephalus)* and chimpanzee (*Pan troglodytes*). The principal axis line, as in Figs 8.3 and 8.4, is taken from a larger scale study of testis volume versus body weight in primates (Dixson 1987a). 'Fatted' male mandrills, which are the most sexually active and reproductively successful males, have very large testes in relation to their body weight, indicative of possible effects of sexual selection via sperm competition. (Baboon data are from Bercovitch 1989; Mandrill and chimpanzee data are from the author's measurements.)

ondary sexual adornments also have the largest testes relative to body weight. Relative testis size in these males, which include the most sexually active and reproductively successful individuals, is similar to that measured for the multimale–multifemale mating system of savannah baboons. Simple measurements of testicular size in these mandrills, therefore, provide a strong indication that sperm competition occurs in this species. The more solitary (non-fatted) mandrill males have smaller testes relative to body weight, however, and although they copulate opportunistically with females during the mating season, DNA fingerprinting studies have shown that no offspring were sired by these animals (Dixson *et al.* 1993). As discussed in the previous chapter, it is possible that these solitary or peripheral males exemplify an alternative reproductive strategy, delaying full development of the secondary sexual characteristics and testes but nonetheless growing to great size and awaiting opportunities to acquire tenure within a multimale–multifemale group. Hopefully, field studies of the mandrill will eventually resolve these questions. Limited data on testes size in the drill (*M. leucophaeus*) show that these monkeys also have large testes relative to body weight, indicating the occurrence of sperm competition (Dixson 1987a; Harcourt *et al.* 1995).

Relative testis size and mating systems in non-primates

Kenagy and Trombulak (1986) have shown that the principles distilled from studies of primates are applicable to eutherian mammals in general (whilst recent studies of marsupials also indicate that relative testes sizes vary in a consistent fashion depending upon their mating systems: Rose *et al.* 1997). Kenagy and Trombulak compiled data on combined testes weight and body weight for 113 mammalian species. Gross testicular size increases with body size but smaller mammals must devote a greater amount of body mass to testicular tissue than larger species do. On allometric principles, a 10-g mammal should have testes representing 1.8% of body weight whereas in a whale weighing 10 000 kg the testes should be equal to 0.04% of body weight. In practice, deviations from these allometric predictions frequently occur. Kenagy and Trombulak concluded that testis size is not affected by factors such as whether the gonads are retained intra-abdominally (the testicondid condition which occurs in cetaceans, elephants, and rhinoceroses) rather than being contained in a scrotum. Nor are there obvious relationships between terrestrial, aquatic, or aerial modes of locomotion and relative testis size. They confirmed, however, that where females tend to mate with a number of males (as, for example, in rats or sheep) sexual selection has favoured the evolution of larger testes. Relationships between testis size and body size remain enigmatic for certain species of mammals. Lions, for example, have relatively small testes despite the fact that more than one male in the pride is sexually active and copulatory frequencies are exceedingly high (Bertram 1975). The DNA fingerprinting studies of Packer *et al.* (1991) are relevant here, since they show that prides are usually controlled by coalitions consisting of closely related males. Kin-selection might therefore operate in these

circumstances, where there is less intrasexual competition between males within the pride. As a final example, it is interesting to note that the largest testes, relative to body weight, of all mammals occur among the toothed whales (*Odontoceti*). In the porpoise (*Phocoena phocoena*), for instance, the testes weigh 2.32 kg, representing a staggering 4% of body mass. Evidence from field studies is limited, because of the great difficulties of observing mating behaviour of cetaceans in the wild. However, many dolphins and porpoises occur in multimale–multifemale schools and there is some evidence that their mating systems are promiscuous. This might, therefore, represent an extreme example of sexual selection by sperm competition. The alternative view, that aquatic copulation somehow requires larger testes relative to body weight, is not supportable, since in the whalebone whales (*Mysticeti*), for instance, the testes are not larger than expected.

Testis size, sperm production, and ejaculate quality
The compartments of the testis

The testis contains both sperm-producing tissue (seminiferous tubules) and non-sperm-producing (intertubular) elements, including blood vessels, nerves, connective tissue, and the interstitial cells of Leydig, which secrete sex steroids such as testosterone. It is necessary, therefore, to consider the proportions of tubular and intertubular tissues within the testis since, for instance, larger relative testis size might be due to proliferation of non-gamete-producing intertubular tissue. That this is not so is shown by Fig. 8.6, which provides data on intertubular tissue areas in histological sections from 13 primate species, including monogamous forms such as the gibbon and owl monkey, as well as multimale–multifemale species such as the macaques and the chimpanzee. Among Old World primates, the testes of multimale–multifemale forms such as *Macaca* contain relatively more seminiferous tissue than polygynous langurs (*Presbytis*) or monogamous gibbons (*Hylobates*). Schultz (1938) demonstrated this by examining histological sections of testes, drawing them on to heavy paper, cutting out the tubular and intertubular areas, and comparing their weights. He calculated that a macaque possesses 4.9 g seminiferous tissue and 2.3 g intertubular material per kilogram body weight. A langur, by contrast, possesses 0.29 g seminiferous tissue and 0.13 g intertubular material per kilogram body weight. Accordingly, he suggested that macaques have seventeen times more gamete-producing tissue than langurs per unit body weight. Sperm competition theory was unknown to Schultz, of course, but his results are reassuring in terms of modern thinking on this subject. However, they do not reflect the whole picture because, when a larger sample of primate species is considered, some discrepancies occur (Fig. 8.6). Among the New World primates, areas of intertubular tissue are very small indeed in some monogamous or polyandrous callitrichids, averaging 7–10% in histological sections from *Callithrix jacchus*, *Saguinus fuscicollis*, or *Leontopithecus rosalia*. Over 90% of the cross-sectional area of the testis in these forms consists of seminiferous tubules; there is no suggestion that the relative volume of gamete-producing tissue is reduced in these monkeys. Indeed it is larger than in certain multimale–multifemale Cebids (*Cebus apella* and *Lagothrix lagotricha* in Fig. 8.6). It seems likely that phylogenetic factors, and mating systems, might influence the relative proportions of tissues in compartments of the testis.

Among New World primates, individual seminiferous tubules tend to have larger cross-sectional areas, on average, among callitrichid species than among cebids, as well as occupying a proportionately larger volume within the testis (Dixson 1983a). Few species and limited numbers of specimens have been examined, however. Further comparative studies would be valuable and might reveal, for instance,

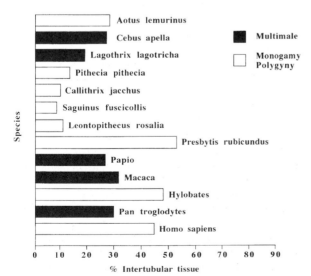

Fig. 8.6 Percentages of intertubular tissue (as opposed to seminiferous tubules) in the testes of various primate species having monogamous/polygynous or multimale–multifemale mating systems. (Data from Schultz 1938 and Dixson 1983a.)

that some of the differences between macaques and langurs identified by Schultz represent variations at the phylogenetic level, rather than being due solely to effects of sexual selection.

Rates of sperm production and storage

Sperm production rates are not known for the vast majority of primate species, but it is evident that some major differences occur between species and that these might be relevant to sperm competition. To cite an extreme example, the owl monkey, which is the only nocturnal anthropoid primate and has a monogamous mating system (Moynihan 1964; Wright 1981), also exhibits a very low rate of spermatogenesis. Very few seminiferous tubules in owl monkey testes contain spermatozoa and it was originally suggested that captive *Aotus* suffer from testicular atrophy due to an irreversible deficiency of vitamin E (Hunt *et al.* 1975). However, extensive studies of captive male owl monkeys showed that they were completely fertile and, moreover, that treatment of a prepubertal male with large doses of vitamin E did not alter the histological appearance of the seminiferous epithelium in adulthood (Dixson 1983a). It is unknown whether free-ranging owl monkeys exhibit this histological picture, but it seems reasonable to assume that the condition is normal for *Aotus* and is commensurate with its monogamous mating system. How fertility is ensured by a monkey which produces few spermatozoa and copulates infrequently remains unknown (Dixson 1994).

Data on sperm production rates, sperm reserves, and transit times through the epididymis in nine mammalian species, including the rhesus monkey and man, have been analysed by Møller (1989). This information was sufficient to show that sperm production rates and reserves have been moulded by sexual selection. When the effect of body size is controlled for, it becomes evident that sperm production rates and sperm reserves are positively correlated with testes size. These differences in sperm production and storage are in accord with differences between the mating systems of the various species studied, including the rhesus macaque (multimale–multifemale, high sperm competition pressure) and man (polygyny/monogamy, low sperm competition pressure).

Problems of ejaculate quality

Information on ejaculate volumes, sperm densities, and motilities is available for a number of primate species and derives mostly from studies concerned with captive animals and research on fertility regulation and artificial insemination. As such, the data are variable and less than perfect, given that semen samples are usually obtained from anaesthetized subjects by electroejaculation, using transrectal or penile stimulation. Ejaculate volumes, sperm densities, and motilities might differ in these circumstances from those which result from coitus. Møller (1988b) was able to demonstrate that primate species with relatively large testes produce ejaculates which contain a greater percentage of motile sperm and larger absolute numbers of motile sperm. Primates with multimale–multifemale mating systems, such as the chimpanzee or various macaques and baboons, produced the highest quality ejaculates, particularly in terms of sperm motility.

The studies reviewed so far are consistent with the notion that sperm competition theory (Parker 1970) is applicable to primate mating systems. Testis weight, in relation to body weight, is greater in those mating systems (multimale–multifemale and dispersed or non-gregarious systems) in which individual females mate with a number of males during the periods when conception is likely to occur. The testes are smallest in relation to body weight in those primates with monogamous and polygynous mating systems, since each female usually mates with a single male under these conditions. Information on sperm production and ejaculate quality also strengthens the argument that sexual selection has enhanced the ability of males in multimale–multifemale mating systems to maximize production of spermatozoa and numbers of fully motile spermatozoa.

Does sexual selection influence sperm morphology?

If kin-selection theory can be used to explain many instances of 'altruistic' behaviour involving close relatives, it is logical to suggest that a male's gametes might also assist each other in sperm competition vis à vis other males (Trivers 1985). Spermatozoa might be divisible into 'egg-getter' sperm, which attempt to fertilize ova, and 'helper' sperm which might function to speed up transport of the egg-getter morphs, to provide them with nutrients or to block the spermatozoa of rival males within the female's reproductive tract.

Trivers (1985) has cited several fascinating examples of morphological variations among spermatozoa in various invertebrates. For example, some species of Lepidoptera and pentatomid bugs produce two

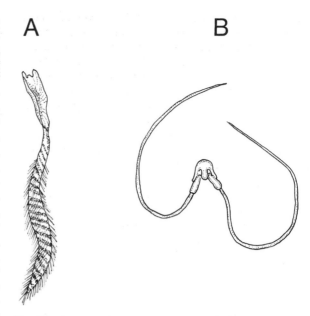

Fig. 8.7 Co-operative sperm. A. Giant helper sperm, or spermatozeugma, produced by some molluscs, which transports hundreds of fertilizing sperm attached to it. B. Binary sperm of the Virginia opossum (*Didelphis virginiana*). (Redrawn from Trivers 1985, after Hyman and Fretter.)

types of male gamete; normal, nucleated cells and anucleate spermatozoa. In the case of pentatomid bugs a large, specialized, lobe of the testis, the so-called 'Harlequin lobe', produces only anucleate spermatozoa. In the moth *Manduia sexta* 96% of spermatozoa are of the anucleate type. In some lepidoptera it has been established that these anucleate spermatozoa are absorbed by the female, but whether they have some nutrient function or actually interfere with spermatozoa from other matings is unknown.

A clear case of sperm morphs which improve transport of other spermatozoa in the same ejaculate is provided by molluscs, some of which produce a type of giant, non-fertilizing spermatozoan called a spermatozeugma (Fig. 8.7A). The spermatozeugma carries hundreds of normal spermatozoa attached to its tail, and these are released as the giant sperm swims through the female's reproductive tract. Among mammals, the most striking examples of co-operation among spermatozoa are found in South American marsupials, such as the Virginia opossum (*Didelphis virginiana*). In these marsupials (but not in the Australian representatives of the Metatheria) most spermatozoa are linked in pairs (Fig. 8.7B). The pairing occurs at the sperm heads and involves the plasmalemma overlying the acrosomal region of each gamete. This arrangement seals the acrosomal surfaces of both gametes within a compartment. Indeed, it has been suggested that the major function of this peculiar system is to protect the acrosome region (Phillips 1970). More convincing, however, is the notion that these tandem spermatozoa swim more rapidly and hence assist one another in the race to reach the female's ova (Taggart *et al.* 1993). In *D. virginiana*, pairing of the spermatozoa occurs in the lower segment of the corpus epididymis (Temple-Smith and Bedford 1980) and they separate only when they reach the female's oviduct (Rodger and Bedford 1982). Experiments conducted *in vitro* have shown that tandem sperm swim much more rapidly through viscous media than do single gametes. This adaptation is likely to be advantageous in situations where competition occurs between the ejaculates of two or more males. Field studies of *D. virginiana* indicate that such competition probably occurs in the wild. These opossums are relatively non-gregarious animals, but during mating seasons (there are two seasons each year) males double the area of their home ranges, to overlap the ranges of 5–9 females. Males are intensely aggressive towards each other in their attempts to monopolize oestrous females (Ryser 1992). Unfortunately it is not known how frequently females copulate with more than one male or whether multiple paternity of litters occurs in this species. Sperm competition would seem likely, however, as an explanation for the evolution of tandem spermatozoa.

Among the mammals in general, morphological specializations, such as the tandem spermatozoa of South American marsupials are most unusual. No such specializations have been described among the primates and anucleate spermatozoa or giant morphs, such as the spermatozeugma, are unknown. Examples of spermatozoa from various species are shown in Fig. 8.8. In their review of sperm dimensions in 284 mammalian species Cummins and Woodall (1985) found some differences in the lengths of the midpiece and principal piece of spermatozoa in various groups, whilst head lengths were relatively uniform. Nonetheless, Baker and Bellis (1988, 1989a) have proposed that functional differences might occur between spermatozoa in some mammals. Baker and Bellis suggest that some sperm might be part of a relatively small population of 'egg-getter' gametes in the ejaculate, adapted to ascend the female's reproductive tract and effect fertilization. Others might be 'kamikaze' sperm, which occupy the female tract at strategic points and inhibit 'egg-getter' spermatozoa in ejaculates from other males. Baker and Bellis propose that kamikaze sperm might appear morphologically normal in some

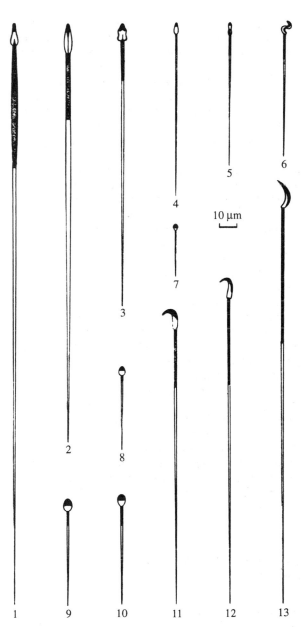

Fig. 8.8 Examples of spermatozoa of marsupials (Nos. 1–6) and eutherian mammals (Nos. 7–13). The acrosome and mid-piece are shaded. 1. Honey possum *Tarsipes rostratus*; 2. Marsupial rat *Dasyuroides byrnei*;. 3. Short-nosed brindled bandicoot *Isoodon macrourus*; 4. Tammar wallaby *Macropus eugenii*; 5. Brush-tailed possum *Trichosurus vulpecula*; 6. Koala *Phascolarctos cinereus*; 7. Hippopotamus *Hippopotamus amphibius*; 8. Man *Homo sapiens*; 9. Rabbit *Oryctolagus cuniculus*; 10. Ram *Ovis aries*; 11. Golden hamster *Mesocricetus auratus* ; 12. Laboratory rat *Rattus norvegicus*; 13. Chinese hamster *Cricetulus griseus* . (After Setchell 1982.)

cases, but that their physiology and behaviour would

distinguish them from 'egg-getter' spermatozoa. A second possibility is that morphologically abnormal sperm, which do occur as a proportion of gametes ejaculated in a variety of mammals, might be kamikaze sperm. This latter suggestion can be evaluated in primates, with reference to work on man and the great apes. Table 8.2 shows data collected by Seuánez (1980) on sperm morphology in human males and in male gorillas, chimpanzees, bonobos, and orang-utans. Twenty-seven percent of spermatozoa in human ejaculates and 29% of spermatozoa in gorilla ejaculates were structurally abnormal. Abnormalities included spermatozoa with irregularly-shaped, large or vacuolated heads, for instance. This high degree of pleiomorphism in human and gorilla sperm was not matched in the chimpanzee, bonobo, or orang-utan, in which >95% of gametes were morphologically normal (Table 8.2). Given the small relative testes sizes of man and the gorilla, coupled with mating systems in which sperm competition is low, it is hard to imagine why sexual selection would have favoured the evolution of morphologically abnormal or 'kamikaze sperm' in these species. Conversely, if pleiomorphism had occurred to produce kamikaze sperm, one might expect to see higher frequencies of such gametes in the chimpanzee or bonobo, since sperm competition is important in their multimale−multifemale mating systems. For these and other reasons, Harcourt (1989, 1991) is critical of the kamikaze sperm hypothesis and his scepticism seems to be soundly based. This begs the question of why so many abnormal spermatozoa occur in the ejaculates of gorillas and man. It should be born in mind that, despite the fact that subjects studied by Seuánez (1980) were fertile, many gorillas exhibit poor fertility or actual atrophy of the testes in captivity (Dixson *et al.* 1980). Is it possible that the high frequencies of abnormal gametes in gorillas represent partial testicular or epididymal dysfunction in captivity? Likewise are ejaculates of many human males affected by environmentally abnormal conditions, such as the wearing of tight clothing (which raises the temperature of the scrotum and might affect spermatogenesis: Bedford 1974) or low-level environmental toxicity effects? Recent studies suggest that sperm counts in western societies have fallen rapidly over the past 50 years (Carlsen *et al.* 1992). Parallel studies of sperm morphology in wild male gorillas and in non-industrialized human societies are required to tackle these questions.

An alternative suggestion made by Baker and Bellis (1989a), that kamikaze sperm 'might be morphs only in a physiological or behavioural sense' is exceedingly difficult to evaluate. It should not be dismissed out of hand until appropriate experiments have been conducted. The suggestion that many spermatozoa in the ejaculate which appear perfectly normal might function to block gametes of rival males might seem far-fetched. However, very little is known about potential interactions between spermatozoa, or other constituents of the ejaculates of rival males, within the female reproductive tract. In the chimpanzee, for instance, as many as five males have been observed copulating with a single female over a 2-min period (Goodall 1965). The potential for interactions between the ejaculates of such males exists, but we have no evidence as to whether

Table 8.2 Comparison of the morphology of spermatozoa of man and the great apes

	Homo sapiens	*Pan troglodytes*	*Pan paniscus*	*Gorilla gorilla*	*Pongo pygmaeus*
No. of individuals	4	3	1	2	2
Normal, i.e. modal	73.0	95.5	98.0	71.0	98.5
Abnormal					
Large head	2.0	0.3	–	2.3	–
Small head	0.5	–	0.5	0.8	–
Tapered head	0.3	0.2	–	1.0	–
Densely staining	0.6	0.2	0.5	4.5	–
Vacuolated head	2.4	3.0	1.0	1.3	0.3
Irregularly shaped	18.4	0.2	–	16.0	0.8
Multiple heads	0.5	–	–	0.5	–
Abnormal midpiece	–	0.3	–	0.5	–
Cytoplasmic droplets	–	–	–	0.3	–
Immature cells	2.3	0.3	–	2.0	0.5
Total:	100.0	100.0	100.0	100.2	100.1

For each individual 200 cells were scored and all results are expressed as percentages.
After Seuánez (1980).

kamikaze strategies occur.

Has sexual selection influenced the evolution of sperm length?

Cummins and Woodall (1985) assembled and analysed data on sperm dimensions in 284 mammalian species. They found that, whereas headlength is quite similar throughout the mammals, the length of the sperm mid-piece and principal piece (i.e. the tail) is much more variable. Cummins and Woodall also pointed out that a tendency exists for overall sperm length to be negatively correlated with body weight; i.e. small mammals tend to have longer gametes than large mammals. However, the correlation is weak and there are exceptions, such as among the Chiroptera in which sperm length is positively correlated with body weight. More recent studies by Gage (1998) have shown that there is no correlation between body weight and sperm length for a sample of 445 mammalian species. Moreover, among the primates, Cummins and Woodall (1985) observed no correlation between body weight and sperm length for 16 species, a finding since confirmed independently by two further studies (Harcourt 1991; Gomendio and Roldan 1991). That exceptionally long sperm can occur in relatively small animals is exemplified by the case of the fly *Drosophila bifurca*, which produces sperm measuring 58.29 mm, or 20 times the length of the fly itself (Pitnik *et al.* 1995).

Given that body size has no consistent effect upon sperm length, we might examine other variables as possible determinants of sperm length. One hypothesis explored by Gomendio and Roldan (1991) is that sperm length varies depending upon the mating system and that sperm are longer in primate species in which females mate with a number of males. Under these conditions, which promote sperm competition, selection might favour longer gametes which swim more rapidly and hence have an advantage in securing fertilization. Gomendio and Roldan (1991) showed that sperm length and maximum velocity are, in fact, positively correlated; the data considered were on just five species (man, domestic cattle, golden hamster, house mouse, and rabbit). Since longer sperm might indeed swim more rapidly, it is, therefore, interesting that sperm length is greater in 'polyandrous' primate species, where females mate with a number of males, than in 'monandrous' primates, where females typically have one partner (Gomendio and Roldan 1991). Similar observations on eupyrene sperm length in relation to mating systems have been reported for butterflies by Gage (1994).

The data set used by Gomendio and Roldan comprised 28 primate species (8 monandrous and 20 polyandrous species). Some of the criteria used by these authors to classify the mating system are highly debatable, however, such as the inclusion of *Erythrocebus patas* as a multimale rather than a single male species and the classification of *Daubentonia* as monandrous. Moreover, *Macaca speciosa* and *M. Arctoides* appear as separate species in the analysis, when they are in fact one species (*M. speciosa* being the older Latin name for *arctoides*). These problems might explain why Harcourt (1991), who also analysed sperm length and mating systems among primates, came to a different conclusion, namely that 'there is no evidence that polyandrous genera (N = 8) differ from monandrous genera (N = 7) in the overall length of their sperm.'

Decisions concerning the mating systems of primate species and their choice for inclusion in statistical analyses of sperm length vs mating system might seriously bias the results obtained. Such a bias might be avoided, however, by comparing sperm lengths, not to mating systems directly, but to measurements of relative testis weight in the same species. This approach is more objective and is free from *a priori* judgements about the polyandrous or monandrous nature of mating systems (Dixson 1993b). Since relative testis size is greatest in primate species which face sperm competition sperm length should also be greater under these conditions if the Gomendio-Roldan hypothesis is to be supported.

Table 8.3 lists 15 primate species for which there are reliable data on both relative testis weight and sperm lengths. Species numbers 1–9, which include various macaques, baboons, and the chimpanzee, all have relatively large testes; data points for these species in log–log plots of testes weight vs body weight fall above the regression line. All these species have multimale mating systems and their sperm lengths range from 69.88–90.40 μm (x ± S.D.: 80.06 ± 7.37 μm). Species numbers 10–15 in Table 8.3, which include the common marmoset, gelada, gorilla, and man, have smaller testes relative to their body weights and have either monogamous or polygynous mating systems. Also included in this second group, for which data points in log–log plots of testes weight vs body weight fall below the regression line, is the orang-utan. Despite having a non-gregarious social organization (Mackinnon 1974; Galdikas 1985) orang-utans are such slow moving, arboreal apes that it is unlikely that females mate with many adult males during each peri-ovulatory period (Schurmann 1982). However, for the pur-

Table 8.3 Sperm lengths in primate species with larger relative testes weights vs species with smaller relative testes weights

Species with larger relative testes weights	Primary mating system	Sperm length (μm)
1. *Macaca nemestrina*	MM	90.40
2. *Macaca fascicularis*	MM	77.06
3. *Papio anubis*	MM	72.95
4. *Macaca arctoides*	MM	89.41
5. *Macaca mulatta*	MM	79.35
6. *Pan troglodytes*	MM	69.88
7. *Papio cynocephalus*	MM	74.30
8. *Cercopithecus aethiops*	MM	86.64.
9. *Saimiri sciureus*	MM	80.60
Species with smaller relative testes weights		
10. *Homo sapiens*	PG	58.80
11. *Callithrix jacchus*	MP	50.0
12. *Theropithecus gelada*	PG	87.30
13. *Pongo pygmaeus*	D	66.58
14. *Hylobates lar*	MP	63.50
15. *Gorilla gorilla*	PG	61.11

Mating systems: MM = Multimale–Multifemale; PG = Polygynous; MP = Monogamous; D = Dispersed.
From Dixson (1993b).

poses of the present analysis judgements concerning the mating system are not essential, since species have been selected solely on the basis of differences in testis weight, corrected for body weight. There is a statistically significant difference between the two groups; genera with larger testes relative to body weight have the longest sperm (genera with larger testes = 5, genera with smaller testes = 6; U = 5, P < 0.05). This effect is not without exceptions, however, since *Theropithecus gelada*, which has relatively small testes and a polygynous mating system, nonetheless has large spermatozoa (Table 8.3). Statistical analyses of the data on relative testes size and sperm length lend support to the Gomendio-Roldan hypothesis. Primate genera with larger testes in relation to their body weights also have, on average, longer spermatozoa. However, the results of this study (Dixson 1993b) must be treated with appropriate caution, as is also the case with previous reports (Gomendio and Roldan 1991; Harcourt 1991). There are more than 200 extant primate species, less than 10% of which have been adequately studied to determine both sperm lengths and relative testis weights. Indeed re-analysis of this problem (Howes and Dixson, unpublished data) using a slightly larger data set has failed to confirm the original result (genera with larger testes = 11;

sperm length averages 77.6 μm. Genera with small testes = 8; sperm length averages 70.07 μm. U = 58, not significant). In future, therefore, it would be valuable to construct log–log plots of testis weights versus body weights for a much larger sample of primates than is currently available. Prosimians should also be included in such analyses. More accurate measurements of sperm length will also be required to expand the existing data base. Measurements of sperm length vary considerably in different studies of a single species. Replication of studies, to include measurements of an adequate number of spermatozoa from several healthy adult males of each species, is required to establish confidence regarding sperm lengths. It will be appreciated that this represents a difficult task, particularly where rare species are involved.

The male accessory reproductive organs and sperm competition

The accessory reproductive organs of male primates consist of the prostate and bulbo-urethral glands, which are derived from the urogenital sinus during embryonic development, and the ductuli efferentes of the testis, the epididymis, vas deferens, and seminal vesicles, all of which are derivatives of the mesonephric (Wolffian) duct system (Fig. 8.9). During copulation, sperm are transported from the cauda epididymides, via the vasa deferentia, to the urethra and are mixed with secretions of the seminal vesicles, prostrate, and bulbo-urethral glands before ejaculation. It is appropriate, therefore, to include some discussion of the functions of the male accessory reproductive organs in this chapter on sperm competition. For instance, it is possible that the anatomy and physiology of the accessory organs in primates which copulate frequently (e.g. chimpanzees or macaques) might differ from species which are monogamous and less affected by sexual selection and sperm competition (e.g. gibbons or owl monkeys).

Structure and functions of the vasa deferentia

The distance which sperm must travel from the cauda epididymis to reach the urethra is considerable; the vas deferens is approximately 19 cm long in macaques, such as the stumptail and rhesus monkey (Ramos 1979), and 38 cm long in man and the great apes (Martin and Gould 1981). Given the relatively brief duration of intromission in many primate species, coupled with the requirement to transport many millions of spermatozoa from the cauda epididymis before ejaculation, it is not surprising that

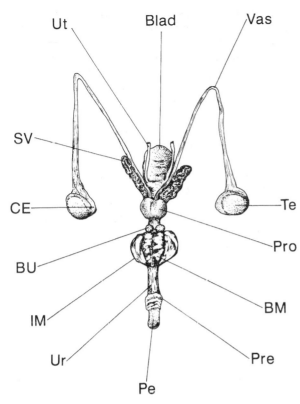

Fig. 8.9 Dissection of the reproductive tract of an adult male orang-utan (*Pongo pygmaeus*) to show the major organs. (Author's drawing). Blad = bladder; BM = bulbocavernosus muscles; Bu = bulbo-urethral gland; CE = cauda epididymis, IM = ischiocavernosus muscle; Pe = penis; Pre = prepuce; Pro = prostate; Sv = seminal vesicle; Te = testis; Ur = urethra; Ut = ureter; Vas = vas deferens.

the urethral end of the vas; stereocilia are shorter and more irregular, and epithelial cells contain many cytoplasmic granules which are lacking at the epididymal end of the organ. It is also the case that the circular muscle layer becomes progressively thicker, whilst the inner longitudinal muscle layer decreases in thickness, towards the urethral end of the vas deferens (Batra 1974). In some species, the terminal portion of the vas deferens is expanded to form the ampulla (*Presbytis entellus*: David and Ramaswami 1971; *Macaca arctoides* and *M. mulatta*: Ramos 1979; *Homo sapiens*: Vendrely 1985). Enormous ampullae occur in the stallion and jackass, for example, but their functions are not known. Various authorities agree, however, that the vasa deferentia must have other functions besides transporting spermatozoa to the urethra. Phagocytosis of defective sperm occurs in the terminal portions of the vas deferens (rat: Cooper and Hamilton 1977; macaque: Ramos 1979). Moreover, sperm transport in the vas is under complex neural control and involves movement of luminal contents towards the epididymis as well as in a urethral direction. Prins and Zaneveld (1980) demonstrated this in their radiographic studies of fluid transport in the vas deferens of the rabbit. They injected small volumes of ethiodiol (iodinated poppyseed oil) into the vas of adult rabbits at its junction with the epididymis. The injections were made in the urethral direction so that the ethiodiol travelled down the vas, away from the cauda epididymis. However, within 24 h ethiodiol had been transported proximally into the cauda epididymis. These observations were made using sexually inactive buck rabbits. When a male was allowed to mount a female (without intromission), ethiodiol was transported from the cauda epididymis, whilst ejaculation resulted in rapid emptying of the contents of the vasa deferentia (Fig. 8.11). Once sexual activity had ceased, transport of ethiodiol from the vas deferens back into the cauda epididymis was again observed. This process began within 10 min of ejaculation and all ethiodiol gradually returned to the epididymis over the next 24 h. Prins and Zaneveld were also able to demonstrate that several ejaculations were sometimes required to remove all the dye from the epididymis and vas deferens. These observations show that spermatozoa might be transported in both directions by the vas deferens. Perhaps, one function of the secretary epithelium lining the lumen of the vas is to produce substances which aid in protecting and maintaining spermatozoa which are not ejaculated, but which are gradually returned for re-storage in the cauda epididymis.

Given that some primate species copulate frequently, particularly species such as the chimpanzee,

the vas deferens is the most muscular of all the tubular organs in the body. Three muscle layers can be distinguished—an outer layer of longitudinal muscles, a middle layer of circular muscles, and an innermost longitudinal muscle layer. These features can be appreciated by examining Fig. 8.10, which shows transverse sections of vasa deferentia from the great apes and man. By contrast with its thick and muscular wall, the central lumen of the vas deferens is very small and is lined by columnar secretory epithelium. A complicating factor is the occurrence of regional differences in the muscle coat and epithelial lining of the vas deferens throughout its length. These differences have been studied in relatively few mammalian species (e.g. rat: Hamilton and Cooper 1978; rhesus and stumptail macaque: Ramos 1979) Major differences in epithelial structure occur between the proximal and distal regions of the vas deferens. Longitudinal folding of the epithelium is much more complex towards various macaques, and woolly spider monkeys in which sperm competition is pronounced, it is possible that sexual selection might have favoured adaptations of the vasa deferentia to cope with increased sexual activity. Such mechanisms might involve thickening of one or more of the muscle layers in the wall of the vas deferens, to facilitate more rapid sperm transport during sexual stimulation. This possibility, first suggested by Smith (1984), is supported by comparisons of the structure of the vas deferens in hominoids (Fig. 8.10). The cross-sectional area of the vas deferens is greatest in the chimpanzee, mainly because of thickening of its muscular coat rather than enlargement of the central lumen. This observation correlates with the larger relative testis size and more frequent copulatory behaviour of chimpanzees, as compared to orang-utans, gorillas, or human males. The vas deferens of the orang-utan is markedly smaller in cross-sectional area, and this accords with the smaller relative testis size of this species. However, it should be noted in Fig. 8.10 that the gorilla, which has very small testes and

which copulates infrequently, nonetheless has a thicker and more muscular vas than either the orang-utan or man. Similar structural variations occur in other primate genera but there has, as yet, been no systematic study of vas deferens structure in relation to mating systems in primates or other mammals. Such an endeavour would be worthwhile, but it would require sampling of the vas along its entire length to take account of regional variations in its structure. Such anatomical comparisons are relatively crude and it might be equally important to consider whether sexual selection has influenced the neural control of sperm transport by the vas deferens. Mechanisms such as those described by Prins and Zaneveld (1980) in the rabbit might be most highly developed in species where males ejaculate frequently, and where improved proximal transport of non-ejaculated sperm for re-storage in the cauda epididymis would be advantageous. Sperm stores are not inexhaustible, so that conservation of non-ejaculated sperm could occasionally be important. For example, in those monkeys where males mount

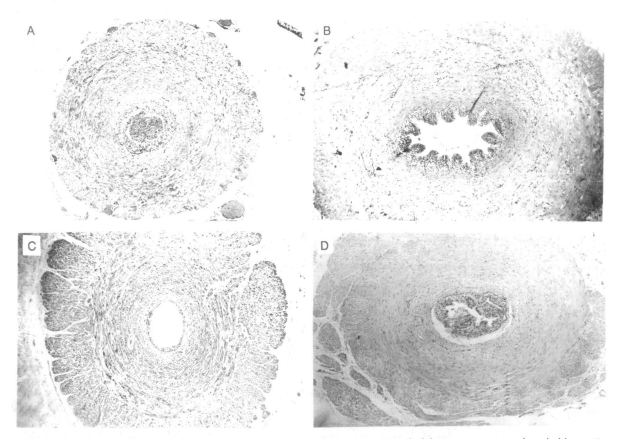

Fig. 8.10 Transverse sections of the vas deferens in (a) *Pongo pygmaeus* (×44), (b) *Pan troglodytes* (×56), (c) *Gorilla gorilla* (×44) and (d) *Homo sapiens* (×56). All sections were stained using haematoxylin and eosin. (After Dixson 1987a.)

and intromit a number of times before ejaculation (as in certain macaques and baboons) it might be

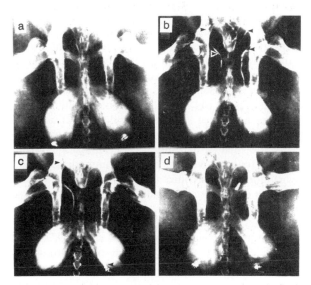

Fig. 8.11 Radiographs showing distal transport of epididymal dye into the vasa deferentia of a buck rabbit after sexual stimulation, and the proximal movement back into the epididymides during subsequent sexual rest. Arrows indicate the dye. a. Before sexual stimulation dye is in the cauda epididymis. b. Immediately after sexual stimulation dye is in the vasa deferentia. c. 10 min after stimulation proximal movement is apparent. d. 7 h after stimulation almost all dye is back in the epididymis. (After Prins and Zaneveld 1980.)

important to protect and recover sperm from the vas if the male is forced to break off a mount series. Such interruptions in the mounting series are by no means uncommon (e.g., in the Japanese macaque: Hanby *et al*. 1971). The vas deferens is innervated by neurons of the sympathetic nervous system (Sjöstrand 1965) and release of norepinephrine has been postulated to control the powerful muscular contractions which facilitate sperm transport (in man: Ventura *et al*. 1973). Denervation of the muscles might be achieved by stripping the mesentery along the vas (Birmingham 1970) and the sympathetic supply is also damaged by vasectomy. At present it is not known whether the sympathetic innervation differs between species in relation to patterns of sexual behaviour and frequencies of ejaculation.

The seminal vesicles and prostrate

A prostate gland is present in all mammals (Price 1963; Hamilton 1990) whereas seminal vesicles are lacking in marsupials, monotremes, and carnivores, and have become reduced or lost in some species of insectivores, rodents, lagomorphs, bats, and primates (Price and Williams-Ashman 1961). Interestingly, the seminal vesicles differ greatly in size among mammals and the primates are no exception to this (Figs. 8.9 and 8.12). Historically, the seminal vesicles were so named because it was believed that they acted as stores for spermatozoa prior to copulation. This view has been discounted by most authorities,

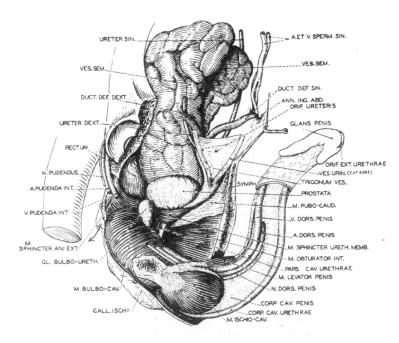

Fig. 8.12 Dissection of the reproductive tract of an adult male rhesus monkey (*Macaca mulatta*). Note the very large size of the seminal vesicles (ves. sem.) in this multimale–multifemale species. (After Wislocki 1933.)

although there are still occasional reports that spermatozoa occur in the seminal vesicles. Ponig and Roberts (1978), for example, identified spermatozoa in *post-mortem* samples of seminal vesicles from *Saimiri sciureus*, *Erythrocebus patas*, *Macaca mulatta*, *Pan troglodytes*, and *Homo sapiens*. They revived the suggestion that the vesicles might act as storage reservoirs for sperm. However, earlier workers who had given careful consideration to this possibility failed to find significant numbers of sperm in the seminal vesicles of various mammals (hedgehog: Marshall 1911; rabbit, bull, and ram, Walton 1962). Reports of occasional occurrences of spermatozoa in the human seminal vesicles might be due to abnormal *post-mortem* conditions and consequent movement of gametes into the glands from the vas deferens (Beams and King 1933). The same phenomenon might explain occasional occurrences of sperm in the seminal vesicles of some non-human primates.

Far more important than any possible role in gamete storage are the secretory functions of seminal vesicles. For this reason some physiologists prefer to use the term 'vesicular glands' rather than seminal vesicles to describe these organs. The vesicles are tubular and consist of an outer fibrous coat, a thick muscle layer (of longitudinal and circular muscles), and an innermost layer of secretory epithelium. The epithelium contains tall columnar cells and basal cells, the whole being thrown into elaborate folds (David and Ramaswami 1971; Harrison and Lewis 1986; Hamilton 1990). Each seminal vesicle might be branched, or it might consist of a single coiled tube. In New World monkeys, for instance, each vesicle consists of a single, unbranched tube in *Saimiri*, *Cebus*, *Lagothrix*, and *Ateles* (Hill 1960, 1962). The tube is folded upon itself to form a lobulated mass but, when straightened out it can be of considerable length (e.g. 7.0 cm in *Saimiri*: Hill 1960). Some Old World monkeys have much larger seminal vesicles, forming complex, lobulated masses. In *Cercocebus aterrimus* these masses are 5 cm long and consist of a single unbranched tube, whilst in *Macaca* the massive vesicles are branched, so that 'as many as 15 secondary tubules and a lesser number of tertiary branchings are recognized' (*M. mulatta*: Hill 1974). In man the seminal vesicles are approximately 5 cm long in situ but, upon straightening the seminal vesicle, it is found to be approximately 10 cm long and to possess 5 or 6 diverticulae (Lowsley *et al.* 1942).

The seminal vesicles are chiefly responsible for producing the bulk of the fluid portion of the ejaculate. This is the case, for example, in the human male, where the vesicles provide 60% of the fluid and 30% is produced by the prostate gland (Harper

1994). Therefore, one function of the seminal vesicles might be purely mechanical; to provide a transport medium for spermatozoa when they are transferred to the female tract. However, the secretions of the seminal vesicles and prostate contain a bewildering variety of chemical constituents, which vary qualitatively and quantitatively among mammalian species (Mann and Lutwak-Mann 1981; Harrison and Lewis 1986; Brooks 1990). Human seminal plasma has a pH of 7.2–7.8, due mainly to the alkaline nature of seminal vesicular secretion. Prostatic secretion by contrast is more acidic (pH 6.5). This low pH is harmful to spermatozoa (Harper 1994). However, vaginal pH rises immediately after ejaculation and becomes alkaline for at least several minutes in the human female. Effects of intercourse upon vaginal pH in women were recorded originally by Masters and Johnson (1966) in their classic studies of human sexual response. Fox *et al.* (1973) used radiotelemetry to record vaginal pH continuously during human coitus. They found that vaginal pH rose from 4.3 to 7.2 within 8 s of ejaculation, and that pH gradually fell to pre-ejaculatory levels within 4 min. Masters and Johnson, however, recorded more long-lasting elevations of pH after intercourse and noted that in some cases 'the active neutralizing action of the ejaculate has been recorded for as long as 16 hours'. A second possible function of the seminal vesicles, therefore, might be to buffer vaginal pH after ejaculation and assist survival of spermatozoa during the period when they are transferred to the cervix and thence to higher portions of the female tract.

Seminal plasma contains a variety of chemicals, being rich in fructose and prostaglandins (19-hydroxylated prostaglandins in the case of the human male). Prostatic fluid, by contrast, is rich in citric acid, acid phosphatase, zinc, and magnesium (Table 8.4). However, interspecific variations have been recorded among various primates, including macaques and baboons (Table 8.4) as well as in the human male (for review: see Harrison and Lewis 1986). It has long been known that some of these constituents (e.g., fructose) have some nutrient or protective value for spermatozoa. Spermatozoa are capable of glycolytic fructolysis in the presence of fructose; this is an anaerobic process which results in production of lactic acid. The metabolic pathway for conversion of fructose to lactic acid by spermatozoa was elucidated by Mann (1964). A further function of seminal vesicular secretion might be a nutrient one, to provide fructose as a metabolic substrate for maintenance of sperm motility. A more speculative suggestion concerns the possible role of accessory sexual glandular secretions as immunosuppressants. It is possible that components of the seminal plasma might bind to sperm and mask sperm-specific antigens, or act directly to suppress immunological responses by the female towards the 'foreign' antigens carried by the male's sperm (Hancock 1981).

Finally, a well-established function of the seminal vesicles in many species is to produce protein-like substances which coagulate when mixed with the enzyme vesiculase from the cranial lobe of the prostate. Such coagulation of seminal vesicular secretion when mixed with secretions of the cranial prostate was originally demonstrated by Van Wagenen (1936) in studies of the rhesus monkey. Coagulation produces a soft, white gelatinous material, but in some primates a more solid, rubbery *copulatory plug* is formed. The form and functions of copulatory plugs will be considered in more detail below. For the present, it is sufficient to note that a coagulum or plug might enhance a male's fertility; for instance by maintaining a sperm-rich fraction of the ejaculate in close contact with the os cervix and its cervical mucus (e.g. in the rhesus monkey: Settlage and Hendrickx 1974). Thus, another function of the seminal vesicles might be to assist coagulation of ejaculated semen and enhance male fertility.

Having considered the possible functions of the seminal vesicles in aiding fertility in mammals, we can now examine why the seminal vesicles vary in size among primates. It is impossible to conduct a broad, cross-species comparison of the biochemical composition of secretions of the seminal vesicles in primates, since relatively few species have been studied. Moreover, even at the gross anatomical level few anatomists have weighed the seminal vesicles of primates or conducted detailed studies to examine possible regional variations in histology. Nonetheless, linear measurements of the vesicles and anatomical drawings of their gross appearance are available for a range of anthropoids. The seminal vesicles are small in the Callitrichidae, inconspicuous in *Aotus*, and vestigial in specimens of *Pithecia* and *Callicebus* examined by Hill (1960). All the genera listed so far are monogamous, or possibly polyandrous in the case of certain callitrichids. Moreover, in the monogamous lesser apes the seminal vesicles are small, approximately 15 mm long in *Symphalangus*, and of moderate size in *Hylobates*. By contrast, the seminal vesicles of certain species are very large indeed. Examples include various macaques, such as the rhesus monkey (Fig. 8.12),

Table 8.4 Some biochemical constituents of prostate (P) and seminal vesicle (SV) tissue in macaques and baboons (after Harrison and Lewis 1986)

Species	Prostatic acid phosphatase	Total lactate dehydrogenase isoenzymes	Citric acid	Protein	Fructose
Macaca mulatta					
P—cranial	11.3 ± 1.4	139.4 ± 36	11.3 ± 1.9	70.0 ± 13	0.85 ± 0.27
P—caudal	743.3 ± 147	244.9 ± 101	12.0 ± 1.8	64.0 ± 6	0.60 ± 0.14
SV	4.5 ± 1.2	51.0 ± 11	4.0 ± 2.3	58.3 ± 9	2.36 ± 1.2
M. fascicularis					
P—cranial	120.0 ± 97	144.4 ± 15	6.9 ± 1.2	72.2 ± 8	0.53 ± 0.18
P—caudal	1585.3 ± 598	213.0 ± 41	9.5 ± 1.8	60.5 ± 4	0.74 ± 0.11
SV	54.9 ± 31	99.5 ± 20	2.3 ± 1.7	52.7 ± 7	0.67 ± 0.15
M. nemestrina					
P—cranial	37.1 ± 18	131.8 ± 16	9.3 ± 1.3	74.1 ± 5	0.61 ± 0.12
P—caudal	1499.9 ± 399	174.4 ± 20	10.3 ± 2.5	66.7 ± 5	0.60 ± 0.10
SV	4.0 ± 0.9	68.6 ± 12	2.5 ± 0.8	65.2 ± 5	1.81 ± 0.30
M. arctoides					
P—cranial	145.0 ± 87	175.9 ± 36	8.8 ± 2.4	69.0 ± 4	2.29 ± 0.71
P—caudal	1872.0 ± 795	313.1 ± 76	17.3 ± 1.6	73.1 ± 6	2.53 ± 1.10
SV	33.9 ± 25	118.3 ± 78	6.8 ± 1.3	64.5 ± 5	10.57 ± 6.00
Papio cynocephalus					
P—cranial	68.9 ± 24	237.5 ± 83	5.7 ± 1.5	67.1 ± 8	0.21 ± 0.07
P—caudal	1031.0 ± 505	275.7 ± 62	7.4 ± 2.1	71.4 ± 7	0.38 ± 0.10
SV	4.1 ± 1	71.5 ± 8	2.6 ± 1.3	66.0 ± 7	0.51 ± 0.11

Units of measurement: Prostatic acid and phosphatase $= \mu$g/mg protein min^{-1}; total lactate dehydrogenase enzymes $=$ BB units, μg protein^{-1}; mg, g^{-1}, wet weight; fructose $=$ mg, g^{-1} wet weight.

stumptail, and the black ape of Celebes (*M. nigra*), all of which have very large seminal vesicles. Seminal vesicles are of great size in the mandrill, in which breeding males sustain high rates of ejaculation in a multimale–multifemale mating system. In the polygynous system of geladas, by contrast, the seminal vesicles are of moderate size and smaller than those of most baboon species. Among the apes, chimpanzees have much larger seminal vesicles than those of gorillas or orang-utans. The human seminal vesicles are somewhat larger than those of a gorilla, smaller than those of the orang-utan, and much smaller than those of the chimpanzee. These relationships are in line with observations on the mating systems of the apes and man. Species in which adult males mate frequently and in which sperm competition is greatest tend to have larger seminal vesicles than those which are monogamous or live in polygynous one-male units. Seminal vesicle size would be expected to scale with body size, of course. Therefore, as species increase in size, larger vesicles would be required to provide sufficient secretions to cope with dilution factors and the increased volume of the female's reproductive tract. As with relative testis size, however, species differences in the size of the seminal vesicles cannot be accounted for on the basis of body weight alone. Nor are they simply the result of taxonomic effects. Figure 8.13 presents data on seminal vesicle sizes and mating systems for 27 primate genera. Seminal vesicles are rated on a four-point scale—1. vestigial, 2. small, 3. medium sized, or 4. large. Ratings of relative seminal vesicle sizes for polyandrous genera (multimale–multifemale and dispersed mating systems in Fig. 8.13) average 3.7 by comparison with 2.0 in monandrous genera (monogamous and polygynous mating systems in Fig. 8.13). This difference is statistically significant (Mann Whitney U-test; P < 0.002; Dixson 1997b).

The general conclusion which emerges from this preliminary analysis of seminal vesicle sizes in primates is, that in mating systems where females mate with more than one partner during the fertile period and in which sperm competition is more likely to occur, the vesicles are larger than in monandrous forms. Not surprisingly, therefore, a correlation also exists between relative testis sizes and seminal vesicle sizes; residuals of testes weights in genera having large seminal vesicles are significantly greater than in genera having small/medium-sized vesicles. Ejaculatory frequencies are also greatest in those genera having the largest seminal vesicles (Fig. 8.14). The dynamics of sperm competition, which occurs when ejaculates of two or more males are mixed within the female's reproductive tract, remain virtu-

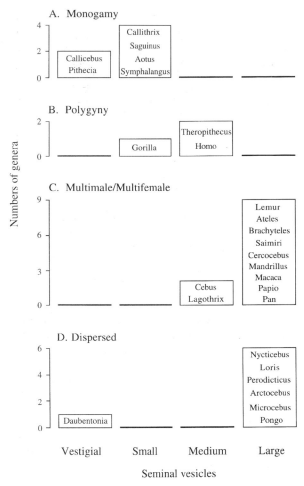

Fig. 8.13 Seminal vesicles sizes and mating systems in 27 primate genera. Seminal vesicles are rated on a four-point scale as 1. Vestigial, 2. Small, 3. Medium-sized, or 4. Large. Mating systems are classified as A. Monogamous, B. Polygynous, C. Multimale–multifemale, or D. Dispersed. Numbers of genera in each category are shown as open bars, with the genus names included within each bar. (After Dixson 1997b.)

ally unknown in primates. It is possible that sexual selection might have influenced the biochemical composition of accessory sexual glandular secretions, and not merely the relative sizes of the seminal vesicles. In fruitflies (*Drosophila melanogaster*), for example, secretions of the male's accessory sexual glands play an important role in stimulating females to lay more eggs. However, increased exposure to these products also increases female death rate (Chapman *et al.* 1995).

Given that the seminal vesicles are energetically expensive organs, producing secretions rich in proteins, fructose, and other constituents, it is reasonable to propose that natural selection might have

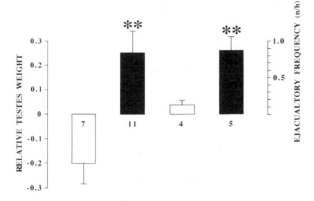

Fig. 8.14 Relative testis weights and ejaculatory frequencies in primate genera with large seminal vesicles (rating =4; closed bars) compared with genera with smaller seminal vesicles (ratings, 1, 2 or 3; open bars). Data are means (+S.E.M.) for the numbers of genera indicated at the foot of each bar. **P=0.002. (Redrawn from Dixson 1997b.)

favoured reduction in size of the vesicles under conditions where copulation is relatively infrequent and the need for large ejaculate volume and coagulum formation is reduced. This might explain the very small size of the vesicles in primarily monogamous forms such as *Aotus* and various *Callithrix* and *Saguinus* species of the New World, as well as among the lesser apes of the Old World. Loss of the seminal vesicles in *Callicebus* and *Pithecia pithecia* might represent the most extreme expression of this trend. Sufficient fluid for transport of the spermatozoa is, presumably, produced by other accessory sexual glands (prostate and bulbo-urethral glands). Prostates are present in both genera, as in all other mammals. In the case of *Callicebus*, the prostate is long and narrow, being situated on the dorsal surface of the urethra, whilst in *Pithecia pithecia* it is confined to the dorsal and lateral aspects of the prostatic urethra (Hill 1960). Reduction of the seminal vesicles does not preclude provision of nutrients such as fructose, which is also produced by the prostate (Table 8.4) and also by the ampullary region of the vas deferens.

A most intriguing case of atrophy of the seminal vesicles among primates is provided by the aye-aye (*Daubentonia madagascariensis*). The vesicles are very small and diffuse in this species; the prostate is very large, however, and encircles the urethra (Petter Rousseaux 1964). Aye-ayes are non-gregarious and have a dispersed mating system. Fieldwork on the island of Nosy Mangabe has shown that males have larger home ranges which overlap those of a number of females (Sterling 1993). Copulations are prolonged in this species. The occurrence of prolonged

intromission was predicted for the aye-aye on the basis of its peculiar penile morphology. Thus, the baculum is very long (Dixson 1987b) and the glans penis exhibits morphological specializations which recall those found in the Canidae (Hill 1953). Hill also cites Kollmann to the effect that the testes of *Daubentonia* are 'voluminous'. All these specializations are indicators of a mating system in which sexual selection by sperm competition has favoured the evolution of large testes, a complex penile morphology and copulatory behaviour. If Hill is correct that the glans penis of *Daubentonia* bears some functional resemblance to that of canids, it is possible that a true copulatory lock occurs in this primate, similar to that which occurs in dogs. Carnivores, moreover, lack seminal vesicles, as does *Daubentonia*, and reduction or loss of seminal vesicles has been correlated also with copulatory behaviour involving a genital lock in various muroid rodents (Hartung and Dewsbury 1978) and in the Australian hopping mouse (*Notomys alexis*: Dewsbury and Hodges 1987; Breed 1990). Breed has also noted that, in captivity at least, female *Notomys* sometimes copulate and form a genital lock with more than one male during oestrus. A soft coagulum is produced, rather than a hard copulatory plug, and this might serve to reduce backflow of spermatozoa within the female's vagina. If the hypothesis concerning *Daubentonia* is correct, field studies, or research on the several captive colonies of these prosimians which now exist, might reveal that prolonged copulations involve a genital lock and that the male produces some form of soft coagulum, but not a hard copulatory plug. The penis itself might function to prevent backflow of semen during lengthy intromissions. This might provide a partial explanation for the reduction of seminal vesicles in a variety of mammals (such as carnivores and certain muroid rodents) where a true genital lock occurs, rather than a less binding form of anchorage between penis and vagina.

The structure and functions of primate copulatory plugs

Two fractions occur in the ejaculates of monkeys and apes; a liquid fraction and a coagulum (Wildt 1986). As previously stated, coagulation is usually due to mixing of protein components of the seminal vesicular secretion with the enzyme vesiculase produced by the cranial lobe of the prostate gland. In rodents, such as the rat and guinea pig, the clotting enzymes (which include a transglutaminase) are produced by the coagulating glands (Notides and Williams-Ashman 1967; Williams-Ashman *et al.*

1977) and mixing of fluid from the rat coagulating gland with rhesus monkey seminal vesicular secretion also results in coagulum formation (Van Wagenen 1936). If the cranial lobe of the prostate is removed surgically in the rhesus monkey, coagulation of semen is prevented (Greer *et al.* 1968)

The coagulated portion of the semen, which comprises 55–68% of the ejaculate (Wildt 1986) gradually contracts, releasing spermatozoa as it does so (Van Pelt and Keyser 1970; Settlage and Hendrickx 1974). Indeed, in the rhesus monkey, 30%–70% of spermatozoa are contained in the coagulum (Hoskins and Patterson 1967; Roussel and Austin 1967). Likewise, in the chimpanzee, semen coagulates rapidly when ejaculated, but if it is incubated *in vitro* for 1 h at 37°C then a liquified fraction is obtained which contains 51% of the spermatozoa, although it represents only 26.5% of total ejaculate volume (Marson *et al.* 1989). Liquefaction of coagulated semen is brought about by proteolytic enzymes contained in the prostatic secretion. In man, liquefaction of semen occurs in only 20 min after ejaculation, but the process takes longer in macaques and some other primates, so that some experimenters use enzymes to digest the coagulum to release spermatozoa more rapidly (for review see Wildt 1986).

Some primates produce a more substantial, rubbery copulatory plug, rather than a soft coagulum (Table 8.5). There have been no systematic studies of the occurrence and functions of copulatory plugs among primates. However, a plug is produced by the chimpanzee (Tinklepaugh 1930; Dixson and Mundy 1994), *Lemur fulvus* (Brun *et al.* 1987), *Lemur catta* (personal observation), *Microcebus murinus* (Martin 1973), *Macaca arctoides* (personal observation), *Arctocebus calabarensis*, *Nycticebus coucang*, and *Loris*

tardigradus (Manley 1967). Moreover, prompt coagulation of semen and formation of plug-like deposits in the female's vagina also occurs in many macaques and baboons, as well as in the mandrill (personal observation). It is clearly a question of degree as to whether one refers to these deposits as coagula or plugs. However, whereas coagulated semen often remains gelatinous (e.g. in man and in the common marmoset), in some cases it hardens to a firm consistency (e.g. in the ring-tailed lemur and chimpanzee). Examples of copulatory plugs from various primate species are shown in Fig. 8.15. In general these plugs are deposited against the female's cervix and take on the contours of the os cervix, fornices, and vaginal walls as the semen coagulates rapidly after ejaculation. A similar effect has been described for the copulatory plugs of muroid rodents (Baumgardner *et al.* 1982).

In man, if the ejaculate is collected in fractions as it issues from the urethra, the first portion (there are 6 in all) contains a high concentration of spermatozoa suspended mostly in prostatic secretion, which brings about subsequent liquefaction of coagulated semen (Tauber *et al.* 1980; Mortimer 1983). If the same applies to other primates, then it is possible that partitioning of the ejaculate results in release of a sperm rich fraction close to, or in contact with, the os cervix and in the most advantageous position for transport into the cervical mucus. In those species with a well-formed coagulum or plug this might assist in preventing loss of spermatozoa by backflow within the vagina of the female. An alternative, but not mutually exclusive, hypothesis is that a copulatory plug might obstruct semen deposition and sperm transport by a second male. If this is the case, then one might expect to find copulatory

Table 8.5 Primate species in which copulatory plugs are known to occur, together with descriptions of their mating systems

Species	Primary mating system	Sources
Microcebus murinus	Dispersed	Martin 1973
Lemur fulvus	Multimale–multifemale	Brun *et al.* 1987
L. catta	Multimale–multifemale	Author's unpublished data
Arctocebus calabarensis	Dispersed	Manley 1967
Loris tardigradus	Dispersed	Manley 1967
Tarsius spp	Dispersed–monogamous*	Hill 1955a
Ateles geoffroyi	Multimale–multifemale	Goodman and Wislocki 1935
Brachyteles arachnoides	Multimale–multifemale	Milton 1985a
		Strier, personal communication
Macaca arctoides	Multimale–multifemale	Author's unpublished data
Pan troglodytes	Multimale–multifemale	Tinklepaugh 1930
		Dixson and Mundy 1994

*Monogamy has been mooted in the case of T. spectrum.

PRIMATE COPULATORY PLUGS

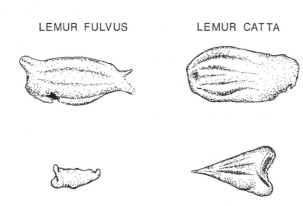

LEMUR FULVUS   LEMUR CATTA

LORIS TARDIGRADUS   PAN TROGLODYTES

Fig. 8.15 Examples of copulatory plugs in various primate species. The cervical end of the plug is on the right hand side in each case. (Author's drawings.)

plugs in those species where females mate with more than one partner during the fertile period. The fact that substantial copulatory plugs are found in certain lemurs, stumptail macaques, and chimpanzees but are lacking in *Aotus* or *Callithrix jacchus* accords with this hypothesis (Table 8.5). However, we lack comparative data on a wide enough range of primate species to evaluate this hypothesis. Copulatory plugs which act as physical barriers to subsequent matings have been described in insects: Parker 1970; spiders: Austad 1984, snakes: Devine 1984, and some other animal groups. Voss (1979) also proposed a similar 'chastity belt' function for copulatory plugs in rodents, but experimental studies have failed to support this notion (Dewsbury 1988). The suggestion that copulatory plugs assist in minimizing loss of spermatozoa from the female's tract and enhance sperm release and transport mechanisms is supported by experiments on rats (Blandau 1945; Matthews and Adler 1977, 1978). Unfortunately, no comparable experimental evidence exists for any primate species.

Possible effects of repeated ejaculations

A common argument advanced by evolutionary biologists is that males often invest very little in their offspring and are therefore able to produce millions of mobile spermatozoa, whilst females invest heavily in offspring and produce much smaller numbers of larger egg cells. Dawkins (1976) commented that

'since each sperm is so tiny, a male can afford to make many millions of them every day .... he is potentially able to beget a very large number of children in a very short time, using different females. This is only possible because each new embryo is endowed with adequate food by the mother in each case'. Such arguments should not be interpreted to mean that the male capacity to produce ejaculates is infinite or without 'costs' in physiological terms. There is now evidence that the ability of males to copulate repeatedly and to fertilize females is constrained, varies tremendously among vertebrates, and has been moulded by sexual selection in certain cases (Dewsbury 1982, 1988; Nakatsuru and Kramer 1982).

As an example of the costliness of producing large numbers of ejaculates, we might consider some research on sheep. During oestrus, Soay sheep are inseminated about once an hour (Grubb and Jewell 1973). In the bighorn sheep (*Ovis canadensis*) dominant males which succeed in guarding oestrous females ('tending' behaviour) also copulate, on average, 0.90 times per hour (Hogg 1984). Sperm competition might be intense under these conditions. Among domestic sheep females can invite and mate with more than one ram during the periovulatory phase of the ovarian cycle (Lindsay and Fletcher 1972). Rams are affected by their frequent copulatory behaviour; exposure to ewes in oestrus results in a 15–42% reduction in testicular weight and 21–36% reduction in numbers of spermatozoa in the testes over the course of 7–8 weeks (Knight *et al.* 1987). Synott *et al.* (1981) used an intravaginal device fitted to ewes in order to collect spermatozoa during natural matings. This system was employed by Fulkerson *et al.* (1982) to estimate numbers of spermatozoa ejaculated before, or after, rams mated with a female in natural oestrus. By averaging sperm counts before and after the natural mating, an estimate of the numbers of spermatozoa ejaculated with the female in natural oestrus was obtained. Approximately 60 million spermatozoa are required per ejaculate to secure fertility, and if the count falls below approximately 50 million spermatozoa, then fertility is significantly decreased. A ram which is 'rested' (i.e. not sexually active) can easily supply sperm counts far in excess of these values. However, rams which are sexually active within a flock of ewes exhibit rapid decreases in sperm numbers per ejaculate. Experiments by Synott *et al.*, summarized in Fig. 8.16, show that 14 rams, initially capable of producing between 1721–10 949 million sperm per ejaculate after natural matings, were only producing 3–557 million sperm per ejaculate after 6 days of sexual activity. During this period, the daily copulatory rate dropped from an average of 18.4 to 11.5, although some males continued to ejaculate at similar or even higher frequencies to those measured on the first day of the experiment. It is evident from the data of Synott *et al.* that striking decreases in sperm counts occur in rams as a result of frequent copulations. In 8 of 14 males, sperm numbers in the ejaculate after 6 days had fallen well below the critical level required to ensure fertility.

In reviewing this work on sheep Lindsay (1984) points out that semen samples are normally obtained from rams which are sexually rested. The possibility that rams mating under unrestricted conditions might fail to produce sufficient spermatozoa in a single ejaculate to ensure fertility has, he says, 'received little attention'. In this context it is relevant to point out that the vast majority of data available on sperm counts in non-human primates refer to samples collected by electroejaculation from males whose mating activity was not recorded. This being the case, we have little notion of the real effects of repeated ejaculations upon semen quality in monkeys or apes. One notable exception is an experiment performed by Marson *et al.* (1989) on chimpanzees. Six semen samples were obtained (by masturbation) at hourly intervals from 2 adult chimpanzees. The experiment was repeated after some days, giving a total of 3 or 4 samples per time point. Results are shown in Fig. 8.17. Ejaculate volume increased over the 5-h experiment (volume at time 0: $x \pm SD$ 2.6 $\pm$ 0.7 mL; and at 5 h: 4.7 $\pm$ 0.6 mL) mainly because of increased secretion by the seminal vesicles (as indicated by the increased volume of the coagulum and by progressively higher concentrations of fructose in the ejaculate: Fig. 8.17). Total

sperm counts also changed significantly and decreased from an initial value of 1278 $\pm$ 872 million to 587 $\pm$ 329 million after 5 h. A male chimpanzee that ejaculates six times over the course of 5 h experiences a 54% drop in sperm numbers in the ejaculate. These data refer to effects of masturbation; it is possible that still larger numbers of spermatozoa are ejaculated during normal coitus. In captivity, male chimpanzees are capable of very high ejaculatory frequencies; Allen (1981) records one male which mated three times in succession with an inter-ejaculatory interval of 5 min. In nature, where groups of males follow and copulate repeatedly with a female when her sexual skin is swollen, ejaculatory frequencies, particularly in younger males, are high (maximum = 0.72 per hour recorded by Tutin 1979). Under such conditions, it is possible that even the impressive sperm reserves of male chimpanzees might become depleted. Even if such depletion does not compromise fertility, it could place a male at a potential disadvantage in sperm competition. However, male social rank also influences copulatory behaviour and more dominant males can exclude others to some extent. Goodall (1986) records the case of a high-ranking male ('Evered') that was the only male present and copulating with a female during the 9 days of her sexual skin swelling. In this case, a total of 21 copulations (2.33 per day) was recorded; it is well within the capacity of a male chimpanzee to maintain maximal sperm counts per ejaculate under these conditions.

Information on ejaculatory frequencies in male primates is very limited, particularly under natural conditions (Dixson 1995b). The available data, however, indicate that copulations are more frequent in

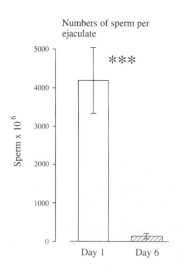

Numbers of sperm per ejaculate

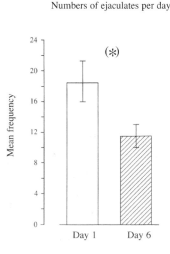

Numbers of ejaculates per day

Fig. 8.16 Effects of sustained sexual activity upon numbers of spermatozoa per ejaculate and ejaculatory frequencies in rams. Data refer to 14 rams allowed to mate naturally with a flock of ewes. By Day 6, a highly significant decrease in numbers of spermatozoa per ejaculate has occurred (***$P = 0.001$) whilst ejaculatory frequencies are slightly reduced (*$P = 0.079$). (Data are from Synott *et al.* 1981.)

multimale–multifemale mating systems than in pair-living (monogamous) or one-male unit (polygynous) systems (Table 8.6 and Fig. 8.18). In a videotaping study of stumptail macaques, for instance, it was noted that one male isolated with a single female in a laboratory cage ejaculated 38 times in one day. Ejaculatory frequencies in free-ranging multimale–multifemale stumptail groups are unknown but copulations occur at very high frequencies in captive groups (Table 8.6). It appears that, like chimpanzees, male stumptails are capable of frequent ejaculations and that intense sperm competition occurs in these macaques. In mandrills, living under semi-natural conditions in Gabon, dominant males have been observed to copulate and ejaculate 2 or 3 times per hour during the annual mating season. Since observations were only possible for 2–3 h out of 12 hours of daylight it seems likely that these males are copulating at least 10 times a day. Relative testis size is large in mandrills, but sperm reserves might be depleted by frequent copulations in this species, given that the mating season spans 5 months of the year, during which time females enter reproductive condition and develop sexual skin swellings (Dixson *et al.* 1993). Field studies of a variety of primates with multimale–multifemale mating systems indicate that their copulatory frequencies are high (e.g. *Brachyteles arachnoides*: Milton 1985a; *Pan paniscus*: Furuichi 1987; *Macaca sylvanus*: Taub 1980; *Macaca radiata*: Glick 1980; *Papio cynocephalus*: Bercovitch 1989; *Alouatta palliata*: Carpenter 1934). By comparison, copulatory frequencies in monogamous primates are typically low (Kleiman 1977). In a videotape recording study of captive pairs of marmosets (*Callithrix jacchus*) all copulations were recorded during the 30 days after parturition (post-partum conception is the norm in this species). Under these conditions, males typically ejaculated 1–3 times at most each day during the peri-ovulatory phase (Dixson and Lunn 1987). In the monogamous owl monkey only 19 attempted copulations were observed during 278 h of observations on five captive family groups, giving an hourly frequency of mating attempts of 0.07 for males of this species (Dixson 1983a).

Does mating order influence male reproductive success?

Differential effects of mating order upon male reproductive success have been convincingly demonstrated in insects and in birds (Thornhill and Alcock 1983; Birkhead and Møller 1992). Sperm storage occurs within the female reproductive tract of both these groups; female insects store male gametes in spermatothecae; in birds numerous tiny sperm storage tubules are located at the utero-vaginal junction. Among insects the last male to mate usually enjoys a reproductive advantage (although this is not inevitably so for all arthropods: Zeh and Zeh 1994). Specializations of the penis might function either to scoop out spermatozoa deposited by previous males (as in the damsel fly *Calopteryx maculata*: Waage

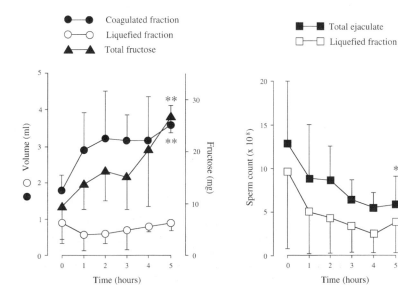

Fig. 8.17 Effect of repeated (hourly) ejaculations (by masturbation) upon semen volume and sperm counts in chimpanzees (*Pan troglodytes*). Data are means ($\pm$S.D.). *P < 0.05; **P < 0.001 compared with values at time 0 h. (Redrawn from Marson *et al.* 1989.)

1979) or to displace such gametes and pack them deeper into the female's spermatotheca (the 'last in, first out' advantage which is believed to occur in certain dragonflies: Thornhill and Alcock 1983). As an example of how effective the 'last male advantage' principle can be, data from experiments conducted by Smith (1979) using the waterbug (*Abedus herbertii*) are shown in Fig 8.19. In this species, the male carries the eggs laid by a number of females on his back. Paternal investment is high, therefore, and there has been heavy selection pressure to avoid possible cuckoldry. Males invariably copulate with each female before allowing her to oviposit several eggs, and repeat copulations at intervals throughout the process of egg deposition. This bug includes individuals with a dorsal stripe (a genetically dominant trait) as well as unstriped individuals. Smith (1979) cleverly demonstrated a last-male advantage by mating females sequentially with striped or unstriped partners and counting the numbers of each morph among the resulting offspring.

Last-male sperm precedence in birds has been demonstrated in a number of species after natural matings (chicken: Warren and Kilpatrick 1929; zebra finch: Birkhead *et al.* 1989) or artificial insemi-nations (chicken: Bonnier and Trulsson 1939; duck: Cheng *et al.* 1982). An example of the advantage conferred upon the last male to mate is shown in Fig. 8.20 from a study by Birkhead (1988) of domes-tic fowl. Sperm removal by copulation is unlikely to explain such effects, given that the majority of avian species lack a penis. 'Stratification' of sperm, due to the second male's gametes packing the spermatozoa of previous males deeper into the sperm storage tubules to produce a last in, first out advantage has been proposed (e.g. in chickens: Van Krey *et al.* 1981). However, the evidence for such effects is far from complete and other explanations, such as phys-iological suppression of rival ejaculates and the occurrence of 'kamikaze' sperm are still debated (Birkhead and Møller 1992).

These studies of insects and birds provide an interesting contrast with research on mating order and sperm precedence in mammals. With rare ex-ceptions (such as in *Chiroptera*: Racey 1979) female mammals do not store spermatozoa for extended periods before the time of fertilization. The life span of spermatozoa within the female's reproductive tract is limited and ova are only viable for a short period. Survival times for viable spermatozoa in the

Table 8.6 Hourly frequencies of ejaculation in primates with multimale–multifemale (MM), polygynous (PG), or monogamous (MG) mating systems

Species	Study type	No. of males	Ejaculation		Mating system
			Range	Mean	
*Lemur catta**	F	5	1–1.75	1.35	MM
Callithrix jacchus	CG	6	0.095–0.202	0.155	MG
Aotus lemurinus	CG	5	–	0.07	MG
Papio ursinus	F	–	–	0.75	MM
P. cynocephalus	F	9	0.41–1.46	0.83	MM
*Mandrillus sphinx**	SF	6	0–3	0.85	MM
*Macaca mulatta**	CG	–	0–0.69	0.38	MM
M. fascicularis	CG	–	0–1.33	0.69	MM
*M. radiata**	CG	–	0.44–2.44	0.97	MM
M. arctoides	CG	18	0–11	1.91	MM
*M. sylvanus**	F	21	0–4.14	2.28	MM
M. nigra	CG	1	–	0.63	MM
*M. thibetana**	F	7	0–0.9	0.45	MM
*Erythrocebus patas**	F	5	0.19–0.74	0.43	MM**
Theropithecus gelada	F	–	–	0.18	PG
Symphalangus syndactylus	F	–	–	0.005	MG
Pan paniscus	F	5	0.19–0.42	0.27	MM
P. troglodytes	F	–	0.03–1.14	0.52	MM
Gorilla gorilla	F	3	0.32–0.38	0.35	PG
Homo sapiens	NA	3342	0–0.119	0.025	PG/MG

F=Field study; CG=captive group(s); SF=semi-free-ranging group; NA=North American population.
*=a species with pronounced mating seasonality. **E. patas* is classified as a MM species because in this study influxes of additional males during the mating season created a multimale–multifemale group. Data on *Mandrillus sphinx* include the author's unpublished observations and the mean frequency relates to the highest-ranking male in the group. Data are from Dixson (1995b).

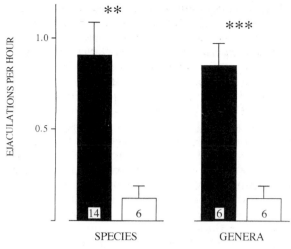

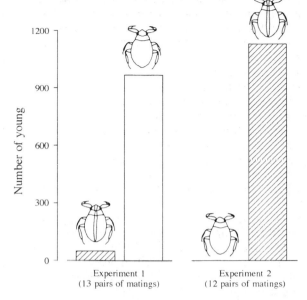

Fig. 8.18 Mean (+S.E.M.) hourly frequencies of ejaculation in 'Monandrous' primate species and genera (monogamous and polygynous mating systems as indicated by open bars) compared with 'polyandrous' species and genera (multimale–multifemale mating systems as indicated by black bars) **P < 0.002; ***P < 0.001 (Mann–Whitney U test). At the foot of each histogram, numbers of species /genera are indicated, derived from Table 8.6. (Redrawn from Dixson 1995b.)

Fig. 8.19 Last male sperm precedence in the Waterbug (*Abedus herbertii*). The striped morph is caused by a single dominant allele. Females which mate first with a homozygous striped male and then with an unstriped male produce a vast majority of unstriped offspring (experiment 1). Reversal of the mating order, so that striped males mate last, produces mainly striped offspring (experiment 2). (Redrawn from Smith 1979.)

female genital tract have been estimated as 6–12 h (mouse), 30–36 h (rabbit), 30–48 h (cow, sheep), and 30–45 h in man (Johnson and Everitt 1988). There is some evidence, however, that human spermatozoa might remain motile (but not necessarily viable) for longer periods (5–7 days: Mortimer 1983). Sperm viability is not of similar duration in all mammals; in the horse for example, spermatozoa remain viable in the female tract for up to 6 days (Johnson and Everitt 1988). These authors list survival times for ova in various mammals and, in all cases, one day is

the maximum period (e.g. in cow, sheep, horse, and man) whilst some species have significantly shorter periods (e.g. 6–8 h in the rabbit; 6–15 h in the mouse). We have very little information on survival times for spermatozoa or ova in any of the non-human primates. Some surprises might await us, particularly as regards viability of spermatozoa within the female reproductive tract.

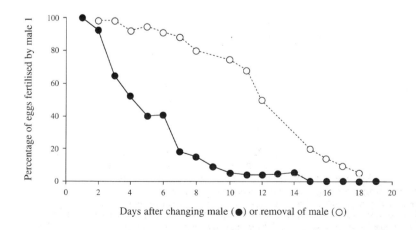

Fig. 8.20 Sperm competition in the domestic fowl. Upper curve (o---o) shows the proportion of eggs that are fertile following the cessation of insemination by a male. Lower curve (●—●) shows the proportion of eggs fertilized by the first male after sequential inseminations by two males. The difference between the two curves is the result of sperm precedence by the second male. (Redrawn from Birkhead and Møller 1992; after Birkhead 1988.)

Given that spermatozoa are not stored long-term in the female tract in most mammals and that ova have a limited viability also, some authors have argued that overt sperm precedence mechanisms do not exist in mammals (Ginsberg and Huck 1989; Birkhead and Hunter 1990). Instead, paternity might depend upon the timing of copulation, sperm transport, and capacitation with respect to ovulation. Only a relatively brief time window is available for spermatozoa which have undergone capacitation in the female tract to complete the transition to rapid swimming (the activation process) in response to the presence of the female gamete in the oviduct. This line of reasoning might be employed to explain the fact that first-male advantages have been recorded in some experiments (e.g., guinea pigs: Martan and Shepherd 1976), and second male precedence (e.g., prairie voles: Dewsbury and Baumgardner 1981) or no effect in others. In the golden hamster a 'lottery' or 'raffle' effect holds; males allowed to ejaculate more frequently with a female sire more offspring, but there is a complex relationship also between mating order and differences in reproductive success depending upon the male's genetic strain (Ginsburg and Huck 1989).

At the present time not a single experimental study has examined effects of mating order and sperm competition upon reproductive success in primates. It might be assumed that in those species where females mate with more than one male during the fertile period, sperm competition will follow the lottery or raffle principle (Parker 1990). However, in the absence of experimental data some caution should be exercised. In anthropoid primates copulations are not necessarily restricted to the peri-ovulatory period, since there is no rigidly controlled period of oestrus, such as occurs in most sub-primate mammals (Ford and Beach 1952; Keverne 1981; Dixson 1983c; Loy 1987). As we shall discuss in Chapter 9, the cervix in some species is remarkably complex and spermatozoa commonly lodge in the cervical crypts before entering the higher portions of the female reproductive tract. Given that we do not know how long spermatozoa remain viable within the female tract in macaques, baboons, spider monkeys, and chimpanzees, for example, it would be premature to dismiss the possibility of mating order effects, or the importance of subtle competitive tactics among males some days before ovulation.

As regards the raffle principle of sperm competition, it is relevant to note that social facilitation of repeated copulations has been observed in a number of primate species with a multimale–multifemale type of mating system. In the woolly spider monkey (*Brachyteles arachnoides*), for instance, Mil-

ton (1985a) notes that 'the female was closely followed by many males and as soon as one copulated, another often took his place'. Coagulated seminal fluid was expelled from the female's vagina 'indicating that considerable sperm displacement occurred with successive copulations'. In the chimpanzee, Goodall (1986) reports that 'frequency of copulation can be correlated with social excitement and generally high levels of arousal in the adult males'. As an example she reports: 'I watched one party as it arrived at a new food source: the attractive (swollen) female climbed into the tree along with eight bristling males, each of whom copulated with her in quick succession'. In the Barbary macaque (*Macaca sylvanus*) females copulate frequently with a variety of partners during the fertile period (Taub 1980; Small 1990). 'A female might copulate with 2–10 of the group's sexually mature males on any one day of oestrus. Females appear to copulate with the greatest number of different males when ovulation is most likely, i.e. when perineal swellings are largest' (Taub 1980). These examples should be considered alongside other data, summarized earlier (Chapter 3, pp. 37–9), concerning multiple copulations by females of various primate species.

The social stimulation which occurs when a number of males are in proximity to a single attractive female, or when several such females are present in a group of monkeys, might also result in a shortening of the refractory period which normally follows ejaculation. The Coolidge effect might operate under such circumstances—i.e. a male which has copulated to exhaustion might show resumption of mounting if a second, attractive female becomes available. The sight of another male mounting can also act as a trigger to resume copulation. Busse and Estep (1984) conducted experiments using captive pigtail macaques (*Macaca nemestrina*) which showed that a 60% reduction in duration of the post-ejaculatory interval occurred if the male observed a second male copulating with his partner. The effect occurred irrespective of whether the second male had attained ejaculation (Fig. 8.21). This finding almost certainly has parallels in social groups of various monkey species and in the chimpanzee. I have observed male Celebes macaques (*Macaca nigra*), talapoins (*Miopithecus talapoin*), and mandrills (*Mandrillus sphinx*) to remount a female within a few minutes of a previous ejaculation in response to sexual interest shown by a second male. Similar observations have been reported by Ohsawa *et al.* (1993) for free-ranging patas monkeys. Females are, of course, not necessarily passive in these situations; increased female sexual initiating behaviour (proceptivity) and attractiveness (e.g. genital swelling and/or odour) during the peri-ovulatory period might play a crucial role in promoting sperm competition. On current evidence from other mammalian species (Ginsberg and Huck 1989; Birkhead and Hunter 1990) it seems most likely that a raffle effect occurs in these circumstances rather than sperm precedence as in many birds of insects.

Do social or sexual stimuli affect sperm counts?

Males in multimale–multifemale primate groups tend to have larger testes relative to bodyweight and to be capable of repeated ejaculations with the same female, or with several partners, during a single day. Can a male vary the sperm content of the ejaculate to maximize his potential for success in sperm competition? Baker and Bellis (1989b) addressed this question in a study of human sexual behaviour. Ten couples contributed semen samples, obtained (by condom) as a result of intercourse or by masturbation. The subjects were asked to record the time they had spent together since their previous copulation, as well as the times at which all the various samples were collected. The hypothesis advanced by Baker and Bellis was that sperm counts should increase with increasing risk of possible sperm competition. This prediction was borne out by the results of the study. A negative correlation was found between sperm counts resulting from copulation and time spent together since previous intercourse. No such correlation was found for masturbation and, in general, sperm counts produced by masturbation were lower than those resulting from sexual intercourse. Time since last intercourse was not a significant predictor of sperm numbers produced at copulation, however, and the results are interpreted by Baker and Bellis as confirmation of the view that possible sperm competition (during the absence of the female partner) results in a boosted sperm count once the pair are reunited. These results are certainly intriguing, particularly because man is not a primate species in which intense sperm competition has moulded the evolution of the reproductive system. Whether such effects occur in macaques or chimpanzees is unknown and would be exceedingly difficult to verify. It might be possible to conduct the necessary experiments in sheep, however, since intra-vaginal collection of semen samples is possible (Synott *et al.* 1981) and sperm competition is known to be intense in these animals.

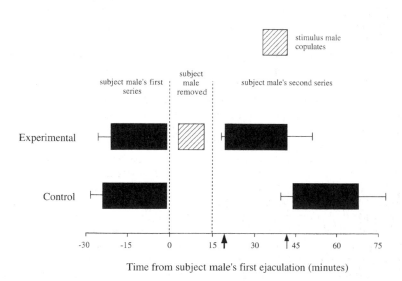

Time from subject male's first ejaculation (minutes)

Fig. 8.21 Intermale competition and reduction of the post-ejaculatory interval in pigtail macaques (*Macaca nemestrina*). Black bars show means (+S.E.M.) for the timing of successive copulatory series by subject males, paired with females during 12 experimental and control tests. (Bars are aligned to the subject male's first ejaculation). After the subject male has ejaculated, he is removed and allowed to observe the female interact with a second stimulus male (experimental series) or no second male (control series). Subject males which have observed a second male copulating, show a marked reduction of their PEIs when reunited with the female. The large arrow on the left of the time axis indicates the mean PEI in experimental tests. The smaller arrow, on the right, indicates the mean PEI in control tests. (Redrawn and modified from Busse and Estep 1984.)

9 Sexual selection and genitalic evolution

Sexual selection and the evolution of male genitalia

'It is very remarkable, considering that the organs have the same rather limited functions to perform, how varied the male genitalia of primates are in their morphology' Hill (1972)

The variations in genital morphology referred to by Osman Hill include differences in penile length, or complexity of distal morphology, including the size and shape of keratinized penile spines, the presence or absence of a *baculum* (*os penis*), and other features. Species within a single genus present remarkable differences in their penile structure and this circumstance has been of considerable value in taxonomic studies of primates, as in many other animal groups (Hill 1953–1970; Fooden 1976; Hershkovitz 1977).

Given that the penis functions as a conduit for urine or for semen, why should its structure be subject to such variability? Before the publication, in 1985, of William Eberhard's treatise 'Sexual Selection and Animal Genitalia', four hypotheses had been advanced to account for the evolution of genitalic complexity in male animals: 1, the lock-and-key hypothesis; 2, genitalic recognition; 3, pleiotropism; and 4, mechanical conflict of interest. It will be useful to summarize these hypotheses, as they might apply to primates as well as to other animals, before discussing Eberhard's intriguing theory regarding the effects of sexual selection via female choice on the evolution of penile morphology.

The lock-and-key hypothesis

This states that males of certain species have developed complex genitalia in order to mesh with equally complex female structures which 'recognize' only the correct key provided by males of the same species. Complementary meshing of the male and female genitalia is viewed as a mechanism to reduce the risk of interspecific copulations and hence to act as a species-isolating mechanism (Mayr 1963). A powerful objection to the lock-and-key hypothesis is that selection should favour recognition mechanisms which operate before genital contact between the sexes (Alexander 1964; McGill 1977). Among the anthropoid primates individuals are members of social groups and sexual recognition of conspecifics does not depend upon genital contact. Indeed, although penile morphologies might be highly variable among primates, there are very few instances where females exhibit complementary adaptations of vaginal morphology. In two species of the greater bushbaby, for instance (*Galago crassicaudatus* and *G. garnettii*), males differ in the shape of the glans penis, degree of projection of the distal end of the baculum, and size and morphology of the penile spines (Fig. 9.1). None of these differences, however, is matched by anatomical specializations of the female genitalia—the vagina is quite similar in both species (author's unpublished observations). Moreover, in captivity male greater galagos will copulate with females of the opposite species, although such pairings have rarely resulted in hybridization (Dixson and Van Horn 1977). Figure 9.2 shows diagrams of the female reproductive tract and male genital morphologies in several species of the genus *Macaca*. Whilst the cervix shows considerable interspecific variability among Macaque species, the vagina and introitus are not modified to intermesh with the male's penis in a lock-and-key fashion, except possibly in the case of *Macaca arctoides* (Fooden 1967). Copulations should be possible between Macaque species; indeed hybridization has occurred among certain species in captivity (Bernstein and Gordon 1980).

Under natural conditions, species-isolating mechanisms doubtless occur in primates, but they involve ecological and geographic factors, or behavioural mechanisms distinct from copulation. Charles-Dominique (1977) conducted an ecological study of five sympatric prosimian species in Northern Gabon. Two species of lorisines, the potto (*Perodicticus potto*) and the angwantibo (*Arctocebus calabarensis*) occur in the rain forest along with the galagines, the needle claw bushbaby (*Euoticus elegantulus*), Allen's bushbaby (*Galago alleni*), and Demidoff's bushbaby (*G. demidoff*). Penile morphologies differ markedly among these prosimian species (Fig. 9.1) but this is not because of the requirement for a lock-and-key fit with the female genitalia. Species-isolation is

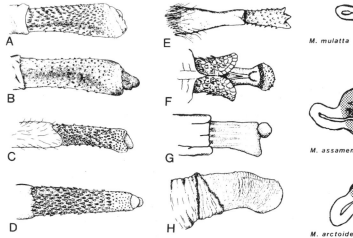

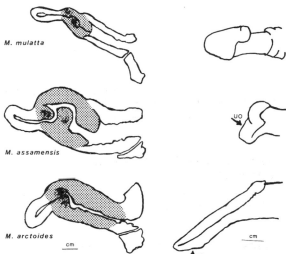

Fig. 9.1 Examples of penile morphology in primate species which have a non-gregarious ('dispersed') mating system. (a) *Galago crassicaudatus*; (b) *G. garnettii*; (c) *G. alleni* (d) *G. demidoff* (e) *Euoticus elegantulus*; (f) *Arctocebus calabarensis*; (g) *Nycticebus coucang*; (h) *Pongo pygmaeus*. Right lateral views of penes except (d) (dorsal) and (f) (perineal). (After Dixson 1987a; e–g redrawn from Hill 1953, 1958.)

Fig. 9.2 Comparative morphology of male and female genitalia in macaques. Shaded areas indicate the extent of the uterine cervix; uo=opening of urethra. (Redrawn and modified from Fooden 1991.)

secured by ecological and behavioural mechanisms. The two lorisines differ in their use of vertical strata in the rain forest and in their feeding ecology. The potto occurs mainly at canopy level and feeds on fruits and gums whilst the smaller angwantibo occurs in the shrub layer in tree fall zones and feeds principally upon insects. Differences in ecological niche also occur between the three galagines (Table 9.1). Species recognition and sexual recognition probably involves olfactory communication and vocalizations (particularly in the bushbabies) in these nocturnal prosimians. These factors, rather than differences in penile morphology, secure species-isolation under natural conditions.

Among monkey species there are several circumstances in which a significant risk of interspecific matings might occur. Firstly, we have seen that males commonly emigrate from their natal groups and enter other groups during annual mating seasons (e.g., in *Macaca*, *Cercopithecus*, and *Presbytis*). It might therefore be possible for such males to attempt entry into a group of a closely related species; Fooden (1967), in discussing the complementary genital specializations of male and female stumptail macaques (Fig. 9.2), suggested that they originated in the context of species-isolation, to avoid interspecific matings between the ancestors of *M. arctoides* and *M. mulatta*. The argument is persuasive because the pronounced dorsal colliculus at the opening of the female's vagina matches well with the

Table 9.1 Ecological niche separation in five sympatric prosimian species in the rainforest of Gabon

Species	Body weight	Vertical	Preferred supports	Diet
Lorisinae				
Perodicticus potto	1100	Forest canopy	Large branches and lianes	Fruits
Arctocebus calabarensis	210	Undergrowth	Foliage and lianes	Insects
Galaginae				
Euoticus elegantulus	300	Forest canopy	Large branches	Gums
Galago alleni	260	Undergrowth	Thin trunks and liane bases	Fruits
Galagoides demidoff	61	Forest canopy	Dense foliage and lianes	Insects

Information compiled from Charles-Dominique (1977).

peculiar sagittate glans penis of the male stumptail. However, this is probably the sole example of its kind among primates.

A second circumstance, and a more common one, where hybridization might occur concerns zones of overlap between the geographic ranges of closely related species. This is the case, for instance, among various baboon (*Papio*) species in Africa, as well as among *Macaca* species (e.g., on the island of Sulawesi). Re-establishment of contact between previously allopatric species might explain occurrences of hybridization in the wild (e.g., between *Papio hamadryas* and *P anubis*: Nagel 1973). On Sulawesi, interbreeding occurs between *Macaca hecki*, which occupies the Northern peninsula, and *M. tonkeana* which is found in the central region of the island. M. *tonkeana* does not interbreed with *M. maura*, however, in the zone of overlap between their ranges in the S.W. peninsula of Sulawesi (Groves 1980). All three species are very similar morphologically and the differences between their genitalia are not remarkable. The lack of a hybrid zone between *M. tonkeana* and *M. maura* is presumably due to behavioural/ecological or physiological factors. The ancestors of these two species were separated for some time by lowering of the intervening land bridge and a marine incursion. Groves (1980) hypothesizes that by the time contact was re-established 'the differential adaptations of the two had already proceeded beyond the reproductive isolation threshold'.

The third circumstance which might increase the risk of interspecific matings is the occurrence of polyspecific associations between groups of monkeys. Striking examples of such associations occur in certain forest-living guenons (Gautier-Hion 1988). As an example, in Gabon *Cercopithecus nictitans* has often been observed in association with *C. pogonias* or *C. cephus*, and all three species might associate together for varying time periods. It is believed that monkeys in such associations benefit in terms of increased foraging efficiency and perhaps also in the avoidance of predators (Gauter-Hion 1988). *Cercopithecus* monkeys are distinguished by their striking facial patterns and pelage coloration; these features differ markedly between species and might be exceedingly important as species-recognition mechanisms (Kingdon 1980). There are also marked differences in the vocal repertoires of the various species. Adult males give species-specific long calls (Gautier 1971; Gautier and Gautier-Hion 1983). By contrast, differences in penile morphology between *Cercopithecus* species, including those which form polyspecific associations, are quite modest (Hill 1966) and not sufficient to prevent interspecific copulations. Indeed such matings do occur occasionally and hy-

brids have been identified in the wild (e.g., between *Cercopithecus ascanius* and *C. mitis* in East Africa: Aldrich-Blake 1968; Struhsaker *et al.* 1988), as well as in captivity. Hybridization between *C. nictitans* and *C. cephus* has not been reported in the wild. However, when both species were kept in a naturally rainforested enclosure in Gabon hybridization occurred between a male *C. nictitans* and a female *C. cephus*. The female hybrids were fertile (author's unpublished observations). In their 8-year studies of *C. ascanius* and *C. mitis* in the Kibale forest of W. Uganda, Struhsaker *et al.* (1988) observed three hybrid crosses between male *mitis* and female *ascanius*, together with their six back crosses. The female hybrids (N = 2) were fertile and well integrated within the *C. ascanius* troops. One of these females was observed mating with both *C. ascanius* and *C. mitis* males, but her offspring were sired only by a *C. mitis* male which stayed with the *C. ascanius* troop for six years. Hybrid females had low reproductive success, however, and the single hybrid male observed might have been infertile. Despite these costs of hybridization, Struhsaker *et al.* (1988) advance the intriguing argument that it might sometimes be advantageous for *C. mitis* males to hybridize with *C. ascanius* females when competition for females of their own species is intense. The hybrids are larger than pure bred *C. ascanius* and tend to achieve more dominant positions in the social group. Since dominant individuals have priority of access to food, there might be advantages to hybridization which offset, to some extent, the disadvantages of reduced attractiveness and reproductive success.

The information reviewed in this section provides little support for the hypothesis that diversity of penile morphologies among primates can be accounted for by the lock-and-key hypothesis. Complex meshing of genital morphologies between male and female primates is exceedingly rare; this phenomenon occurs much more frequently among insects and other invertebrates (Eberhard 1985, 1996). Species-isolating mechanisms in primates depend upon geographic barriers, ecological niche separation, and behavioural mechanisms which precede the requirement for genital contact and copulation. Eberhard (1985) rejected the lock-and-key hypothesis as a general explanation of the evolution of male genitalic complexity, basing his arguments primarily upon studies of a wide range of invertebrates. He pointed out, however, that interspecific pairings do occasionally occur, despite differences in courtship signals (e.g. in insects, Berlese 1925; Brown 1981; Leslie and Dingle 1983). Genitalic differences might help to prevent such occurrences under natural conditions, although the necessary field data are lacking for most animal groups. Field data on primates are more complete, however. The field studies on *Cercopithecus*, *Macaca*, and *Papio species* indicate that hybridization does occur in nature, but rarely. Once the point of copulation has been reached, interspecific differences in genital morphology are probably of little importance in preventing intravaginal ejaculation.

The genitalic recognition hypothesis

This is a variant of the lock-and-key hypothesis which argues that females 'recognize' conspecific males by virtue of the *stimulation* received during copulation. Fertilization occurs in response to appropriate copulatory stimuli. In the female rat, for instance, multiple intromissions by the male trigger neuroendocrine changes which are required for support of the corpus luteum during pregnancy (Adler 1969; Diamond 1970). In those mammals which are induced ovulators, the correct pattern of copulatory stimuli is required to release the pre-ovulatory surge of luteinizing hormone (e.g. in mustelids, cats, and some other carnivores: Ewer 1973; and in the rabbit: Harris 1965). No such parallels are known among primates. Female monkeys and apes are typically spontaneous ovulators and coital stimuli are not required for support of the corpus luteum during the ovarian cycle or in the initial stages of gestation. There is no evidence, therefore, that female primates identify males as members of the same species on the basis of genital stimulation during copulation.

The pleiotropism hypothesis

Ernst Mayr (1963) advanced the view that changes in penile morphology during evolution have occurred as a result of pleiotropic, or incidental, genetic effects. He viewed diversity in male genitalia as a side effect of genetic changes in other systems which were favoured by natural selection. On this basis, provided that the penis continues to be functional during copulation, pleiotropic effects are tolerated and not subject to negative selection. Eberhard (1985) has pointed out that this hypothesis fails to account for the lack of pleiotropic effects in other organs of the body. Why should the penis be subject to such bizarre morphological side effects of genes whilst other organs are not? Given the crucial importance of the penis in placing spermatozoa advantageously within the female tract, surely its morphology 'would be especially likely to be subject to selection' (Eberhard 1985). A powerful argument against the pleiotropism hypothesis concerns those

species in which the male transfers spermatozoa to the female by using secondary structures, rather than the primary genitalia. In such cases it is these secondary structures, such as the pedipalps of male spiders or the heterocotylus of male cephalopods, which have undergone species-specific morphological changes. If pleiotropic effects explain bizarre morphologies of primary genitalia, why are such effects absent in these cases?

Mechanical conflict of interest hypothesis

This hypothesis develops from the idea that males and females might have conflicting interests during copulation; males of some species might attempt to restrain the female, for instance, or force her to accept their gametes rather than allowing her a degree of discrimination or choice. In some cases male genitalia might evolve to act as 'snippers' or 'syringes' to bypass barriers in the female tract and gain access to ova (Lloyd 1979) or to damage the female genitalia and thus prevent subsequent matings. Eberhard (1985) discusses possible examples of this type among invertebrates and the theoretical objections to accepting such an explanation for their evolution. I have been unable to identify a single case among the primates where the mechanical conflict of interest hypothesis might be applicable. Objections to this hypothesis and to the other possible explanations of evolution of complex penile morphologies reviewed so far are summarized in Table 9.2.

Eberhard's hypothesis: sexual selection by female choice

The failure of previous theories to account for the evolution of complex genital morphologies in male animals led Eberhard (1985) to propose a new explanation for such phenomena. Comparative studies of many invertebrate groups led him to conclude that in species where females mate with a number of partners, rather than with a single male, the penis can function as an *internal courtship device* encouraging successful sperm transport and fertilization. In species that employ internal fertilization, male gametes are rarely shed directly on to the female's ova. In insects, for example, females might store spermatozoa for extended periods in spermatothecac (Thornhill and Alcock 1983). Among mammals, including the primates, spermatozoa must migrate through the female's reproductive tract and traverse anatomical and physiological barriers in the cervix, uterus and uterotubal junction before encountering ova within the fallopian tubes. Numerous opportunities exist, therefore, for the female to exert her influence (often called 'choice', but with no conscious decision-making being implied by the use of this term) upon the fate of spermatozoa. Eberhard surmised that small adaptive features of male genital morphology, or of copulatory and associated behaviour, might become subject to rapid evolutionary change. The same process of runaway sexual selection which has affected some masculine secondary sexual characters (e.g. the plumage of peacocks, argus pheasants, or birds of paradise) might also apply, under appropriate conditions, to the primary genitalia of male animals. The sexual selection by female choice hypothesis predicts that in species where females mate with a single male there will be little selection pressure for the development of complex male genital morphologies. Among insects this is so for termites, which have a monogamous mating system. The simplified genitalia of male termites are probably derived characters; these insects are descended from cockroach or mantid-like ancestors and species-specific complex male genitalia occur in these groups.

Intermale competition occurs if more than one male mates with a single female during the fertile period. Such competition occurs between ejaculates of different males (sperm competition: Chapter 8) Equally, it is possible to argue that intermale competition, rather than female choice, could explain the evolution of complex male genitalia. In practice, neither possibility is mutually exclusive. There is a parallel here with effects of sexual selection upon the evolution of masculine secondary sexual adornments. Some adornments might play a dual role, being used in intermale competition, and as signals to females indicating the degree of male prowess and fitness. The important point about Eberhard's hypothesis is that it recognizes the impact of female genital structures and physiology as selective filters affecting male reproductive success and hence the evolution of male genital morphology.

Penile morphology and mating systems in primates

Figure 9.3 shows the results of a study on penile morphology and mating systems in 130 primate species (Dixson 1987a). Four characters were noted for each species: length of the flaccid *pars libera*, length of the baculum (*os penis*), size of the keratinized spines on the distal portion of the penis, and complexity of the distal region in terms of its overall shape. Features such as overall length or complexity of distal penile morphology are very difficult to quantify in practice. Therefore, each character was rated on a 5 point scale (1 = least development and 5 = maximum development of each character). Although open to subjective errors, this approach has

Table 9.2 A summary of arguments against various hypotheses to explain evolution of complex penile morphologies in animals

1. Lock and key
   1. Selective context is probably relatively uncommon.
   2. Selection favours earlier species discrimination by females.
   3. Female structures often do not evolve in step with those of males.
   4. Female structures often cannot exclude males of other species.
   5. Rapid genitalic divergence occurs in species isolated from all close relatives.
   6. Correlation exists between spermatophore complexity and male–female contact.

2. Genitalic recognition

   Points, 1, 2, 5 and 6 above also apply to this hypothesis.

3. Pleiotropism
   1. Selective neutrality in genitalia is not likely.
   2. Reason why pleiotropism acts in genitalia rather than other organs is not explained.
   3. Rapid divergent evolution does not occur in primary genitalia when other structures transfer sperm.
   4. Rapid divergent genitalic evolution does not occur in species with external fertilization.

4. Mechanical conflict of interest
   1. Sexual selection often favours females not overcoming male manipulations.
   2. Female structures often do not evolve in step with those of males.
   3. Obvious anticlasper devices are lacking in females.
   4. Some male genitalia are not manipulative.

Adapted from Eberhard (1985).

the virtue that it enables numerical comparisons to be made between large numbers of species.

The various species were placed in four categories dependent on their mating systems (monogamous, polygynous, multimale–multifemale and dispersed). Eberhard's theory predicts that complex penile morphologies will occur more often in mating systems where females mate with a number of males (multimale–multifemale and dispersed systems) rather than with a single partner (monogamous and polygynous systems). These predictions are born out by the data (Fig. 9.3). Elongation of the penis (scale points 4 and 5) was present in 51% of multimale species and in 67% of dispersed primates sampled. Only 8%

of monogamous and 14% of polygynous species rated 4 or 5 for this trait. Multimale species also had higher frequencies of 4 and 5 scale ratings for complexity of distal penile morphology, size of penile spines and baculum length. These effects were far more pronounced, however, among the dispersed species. Maximum scores (scale 5) occurred in 73% of dispersed species for complexity of distal morphology, 37% for size of the penile spines and 86% for baculum length. None of the pair-living or polygynous species studied rated a maximum score for these traits. Subsequent studies have largely confirmed these findings. However, the most questionable correlation concerns the penile spines. Large

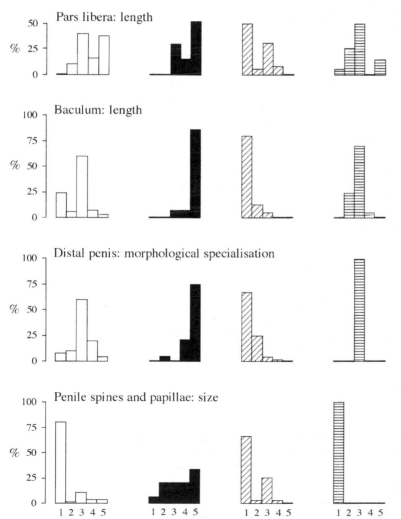

Fig. 9.3 Relationships between penile morphology and mating systems in primates. Data shown for each mating system (multimale–multifemale, non-gregarious/dispersed, paired/monogamous, polygynous) are the percentage of species which display particular specializations of penile morphology (graded on a scale of 1–5 for increasing complexity). (Redrawn from Dixson 1987a.)

spines are far more common in prosimians than in anthropoids, so that phylogenetic constraints complicate interpretations of the role played by sexual selection in the evolution of these interesting structures (Harcourt and Gardiner 1994).

Descriptions of penile morphologies in dispersed and multimale–multifemale species

Examples of penile morphologies in primates with non-gregarious (dispersed) mating systems are shown in Fig. 9.1. Penile morphology is exceedingly complex among non-gregarious nocturnal prosimians. The baculum is elongated in all species which have been examined and it might protrude from the distal pole of the penis above the urethral opening. A horny pad covers the distal end of the baculum in *Cheirogaleus* and *Microcebus* and the glans penis is bifid in the latter genus. Three lappets occur at the distal end of the penis in *Euoticus*, surrounding the urethral opening, and smaller frills and lappets also adorn the apex of the penis in other genera (e.g., *Perodicticus*, *Loris*, and *Nycticebus*). Penile spines are much enlarged in *Galago*, *Euoticus*, and *Arctocebus*. They vary greatly in shape, being tridentate in *G. garnettii* but single-pointed in *G. crassicaudatus*, *G. alleni*, and *G. demidoff*. In *Phaner* there are two enormous spines on either side of the glans. This little-known species is thought to be relatively solitary in its social organization (Pollock 1979). Also living a non-gregarious life is the rare aye-aye, (*Daubentonia*). Recent field work confirms that males live alone, within large home-ranges which overlap the smaller ranges of several females (Sterling 1993; Sterling and Richard 1995). The penis of *Daubentonia* is a morphologically complex, elongated organ, clothed in recurved spines distally. There is a specialized arrangement of erectile tissue which is reminiscent of that found in the *Canidae* (Hill 1958).

Complex penile morphologies are also found in many primates with multimale–multifemale mating systems. Interspecific variability between members of a single genus is a frequent condition and accords with the prediction of the female choice hypothesis. Among the lemurs three enlarged spines occur on either side of the base of the glans in *Lemur fulvus* and the baculum projects above the urethral opening (Fig. 9.4). In *L. catta* the spines are smaller whereas in *Varecia variegata* spines are lacking and the glans penis is filiform (Hill 1953). In the New World spider monkeys (*Ateles*) the penis is elongated with a relatively simple morphology but bears deeply pigmented barbs in some species (*A. geof-*

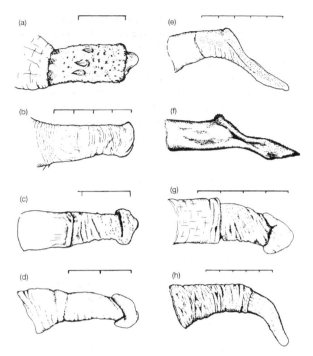

Fig. 9.4 Examples of penile morphology in primate species which have a multimale–multifemale mating system: (a) *Lemur fulvus*; (b) *Ateles belzebuth*; (c) *Saimiri boliviensis*; (d) *Macaca fascicularis*; (e) *M. Arctoides*; (f) *M. arctoides* (penis erect); (g) *Papio cynocephalus*; (h) *Pan troglodytes*. Right lateral views of formalin-fixed specimens, except (f), which is of a living animal. Scales are in 1-cm divisions. (After Dixson 1987a.)

froyi: Hill 1962). In *Brachyteles* the penis is elongated and mushroom-shaped at its distal end, as is the glans penis in *Lagothrix* (Hill 1962). Small spines occur on the glans in at least some *Lagothrix* species, whereas in *Saimiri* the glans is cone-shaped with quite large spines on its surface and on the distal penile shaft (Fig. 9.4). In studies of the New World sub-family *Pithecinae* Hershkovitz (unpublished data) has described several examples of specialized penile morphologies. Three genera comprise this sub-family: *Cacajao*, *Chiropotes*, and *Pithecia*. These genera are characterized by reduction of the baculum and by differences in the shape of the glans penis and size of the penile spines. The penile spines are minute in *Pithecia pithecia*, whilst in *Chiropotes* and *Cacajao*, which live in multimale–multifemale groups, spines are large and hooked.

Turning to the anthropoids of the Old World, the penis is more uniform in shape than in New World monkeys and typically consists of a thin shaft surmounted by a rounded or helmet-shaped glans. Once again, species-specific variations are frequently encountered and complex penile morphologies are more likely to occur in multimale–multifemale mating systems. In the genus *Macaca*, for instance, all 19 species live in multimale–multifemale groups and the mating system is believed to be of the same type —females might mate with more than one male during the peri-ovulatory period. The glans penis is remarkably variable in macaques. It is simple and broad in the *silenus–sylvanus* species group, narrow and bluntly bilobed in the *fascicularis* group, sagittate in the *sinica* group, and greatly elongated with large penile spines in *Macaca arctoides*. In *Pan* there is also an interspecific difference in penile morphology, for in *P. troglodytes* the external urinary meatus is a simple slit-shaped aperture, whereas in the bonobo (*P. paniscus*) it is 'Y' shaped and surrounded by four lappets (Izor et al. 1981). In both species the penis is greatly elongated and this might represent an adaptation to secure adequate depth of intromission with females during maximum sexual skin swelling (Dixson and Mundy 1994).

Descriptions of penile morphologies in monogamous and polygynous forms

In primates which live in family groups consisting of an adult pair plus offspring the male usually has a small and relatively unspecialized penis. This is so for many of the marmosets and tamarins (family Callitrichidae) of the New World, such as *Callithrix*, *Saguinus*, and *Cebuella* (Fig. 9.5). Debates about the prevalence of polyandry in Callitrichids have yet to be resolved as this will require comparative studies of sufficient numbers of species and populations in the wild (Sussman and Garber 1987; Goldizen 1987; Ferrari and Lopes Ferrari 1989; Dixson 1993a). It is possible that certain specializations, such as moderate enlargement of the penile spines in *Callithrix* or the relatively longer penis of *Callimico* (Goeldi's monkey) and *Leontopithecus*, have been influenced by a propensity of females to mate with more than one partner. At the moment this is purely speculative, since there is little evidence that multimale matings have influenced evolution of larger relative testes size in callitrichids (i.e. by sperm competition: see Chapter 8, pages 220–1). In the various *Aotus* and *Callicebus* species of the New World, the penis is very short and morphologically unspecialized. The length of the flaccid pars libera in one specimen of *Aotus* examined represented only 2.5% of head-body length. In *Callithrix jacchus* (N = 3) it averaged 6.2%. These values in pair-living primates are much lower than in the non-gregarious *Galago garnettii* (11%) or the multimale–multifemale chimpanzee (18%).

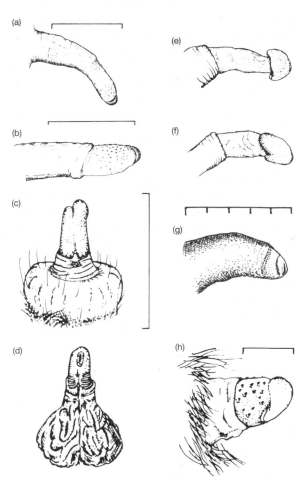

Fig. 9.5 Examples of penile morphology in primate species with a monogamous (pair forming) (a–d, h) or polygynous (e–g) mating system: (a) *Saguinus oedipus*; (b) *Callithrix jacchus*; (c) *Cebuella pygmaea*; (d) *Pithecia pithecia*; (e) *Colobus guereza*; (f) *Erythrocebus patas*; (g) *Gorilla gorilla*; (h) *Hylobates agilis*. Right lateral views of formalin-fixed specimens except (c) (Perineal view redrawn from Hershkovitz 1977), (d) (Perineal view redrawn from Hill 1960) and (h) after Matthews (1946). Scales are in 1-cm divisions. (After Dixson 1987a.)

Among the lesser apes, in which a primarily monogamous mating system is the rule, the penis is short and unspecialized. This observation applies to both *Hylobates* and *Symphalangus*. Quite large spines were described on the glans penis of *H. agilis* by Matthews (1946) although it is unclear whether other species in the genus exhibit this trait (Fig. 9.5).

In Chapter 3 it was noted that polygynous mating systems are widespread among Old World monkeys; one-male units occur in many *Cercopithecus* species and in *Erythrocebus*, as well as in various Asiatic colobines (e.g., *Nasalis*) and in the gorilla. It is unwise to adopt rigid definitions. As discussed earlier, mating systems can be more flexible than previously recognized in primates. I refer here to the primary mating system of each species. One-male units of *Erythrocebus patas* and of *Cercopithecus* might be subject to influxes of extra-group males from time to time. Females might mate with a number of males in such circumstances (Cords 1987, 1988) but the primary mating system is a polygynous one-male unit. Penile morphologies are mildly divergent among species of genera such as *Cercopithecus* and *Presbytis* (Fig. 9.6), but these differences are much less than those encountered in non-gregarious forms (e.g. *Galago*), or in certain multimale–multifemale species (e.g. *Macaca*). Likewise, among the great apes, the polygynous gorilla has a short and relatively unspecialized penis, whereas the organ of the chimpanzee is elongated and filiform. In the orang-utan, which has an unusual type of non-gregarious mating system, the penis is moderately specialized. It is 6.5–10.0 cm. long (Dahl 1988) and the glans is sculpted by peculiar keratinized ridges (Fig. 9.1). Also, as we shall see in a moment, the baculum is larger in *Pongo* than in either *Pan* or *Gorilla*.

The information reviewed so far indicates that penile morphologies in primates conform to Eberhard's (1985) hypothesis of sexual selection by female choice. Morphologies tend to be more complex among primates in which females mate with more

than one male during the fertile period. Numerical comparisons of morphological complexity (Fig. 9.3) confirm this impression. Although the analysis used is simplistic, and does not make any allowance for phylogenetic factors, re-analysis of the same data shows that the same trends are present if comparisons are made at the intergeneric or interfamilial level (Eberhard, personal communication; Verrell 1992). How has female choice (or intermale competition) wrought these effects upon the evolution of penile morphologies in primates? The following sections deal in turn with various structural characters, (baculum, penile spines, etc.) and speculate upon their functions in copulatory behaviour and the possible effects of sexual selection in each case.

The evolution of the baculum

An os penis or baculum is present in five mammalian orders: *Primates, Rodentia, Insectivora, Carnivora,* and *Chiroptera* (the first letters of these five orders provide an apt mnemonic). The baculum, which is the most diverse of all bones in its morphology (Romer 1962), forms by ossification of the distal region of the corpora cavernosa and extends forwards into the glans penis. Radiographs of the penes of various monkeys and apes, to show their bacula, are shown in Fig. 9.7. The development of the baculum is affected to some extent by androgens and it enlarges during puberty, as does the overall size of the penis (e.g. in *Macaca*: Fooden 1975; Dixson and Nevison 1997). The length and shape of the baculum varies considerably among mammals, and there has been no agreement as to why this should be so. In bats (*Chiroptera*), for instance, the baculum is typically small but in *Scotoecus* (the African house bat) and some members of the genus *Pteropus* it is greatly elongated. Loss of the baculum has also occurred in certain bats (e.g. in the *Phyllostomids*: Smith and Madkour 1980). Functional explanations for these variations are lacking. Among primates, a similar situation prevails; some species have greatly elongated bacula (particularly among non-gregarious nocturnal prosimians such as *Galago, Daubentonia,* and *Perodicticus*) whereas in others it is reduced or absent. Absence of a baculum has been noted in some New World monkeys, such as members of the Atelinae (*Ateles, Alouatta,* and *Lagothrix*). In *Aotus, Callicebus,* and *Pithecia* the bone is greatly reduced, whereas in *Chiropotes* and *Cacajao* it is absent in specimens so far examined (Dixson 1987b). The woolly spider monkey (*Brachyteles*) of the New World has been reported to possess a baculum (Napier and Napier 1985) in contrast to the closely-related spider monkey (*Ateles*). This is

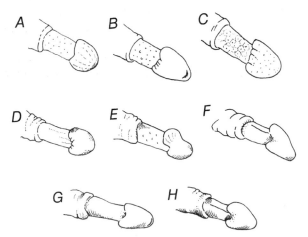

Fig. 9.6 The majority of langurs (*Presbytis*), guenons (*Cercopithecus*), and patas monkeys (*Erythrocebus*) have primarily polygynous mating systems. Their penile morphologies are less divergent than those of primates with multimale–multifemale or dispersed mating systems. A. *Presbytis senex*; B. *P. thomasi*; C. *P. entellus*; D. *Cercopithecus mona*; E. *C. petaurista*; F. *C. albogularis*; G. *C. Neglectus*; H. *Erythrocebus patas*. (Redrawn from Hill 1958 (A–C) and Hill 1966 (D–H.)

a surprising observation which requires confirmation. The tarsiers are notable for the absence of a baculum, as is man (Schultz 1969; Hill 1972). Ossification of the distal septum of the corpora cavenosa has been reported, as a rare aberration, in human males and in one case a distinct boney rod was present in the penis. As far as is known all other primate species possess a baculum (Table 9.3).

Figure 9.8 shows a logarithmic plot of baculum length vs. body weight in ten families and subfamilies of prosimians, monkeys, and apes. Some points of phylogenetic interest emerge from consideration of this figure and of Table 9.3. Among the prosimians elongated bacula are found in the Lorisidae (e.g. in *Perodicticus*, *Loris*, and *Galago*), and in the genera *Daubentonia*, *Microcebus*, and *Cheirogaleus*. However, the three Lemur species examined and *Propithecus verreauxi* all have much shorter bacula, comparable with those of monkeys of similar size. Compared with prosimians or Old World monkeys the majority of New World monkeys have relatively small bacula. In callitrichids the bone varies in length from 1.5 mm. in *Saguinus inustus* to 3.5 mm in *Leontopithecus rosalia* (Hershkovitz 1977). These data on marmosets and tamarins, which weigh 300–700 g, might be compared with those on certain small prosimians, such as *G. demidoff*, which at 63 g has a baculum measuring 12.1–14.2 mm, or *G. senegalensis* (237 g), in which the bone is 16.7 mm long. Indeed, no New World monkey has an exceptionally large baculum and in some genera the bone is greatly reduced in size (e.g. *Aotus* and *Pithecia*), whereas in others it has been lost altogether (*Cacajao*, *Chiropotes*, *Ateles*, *Lagothrix*, and *Alouatta*). By contrast, all the Old World monkeys and apes so far examined possess bacula. However,

as can be seen from Fig. 9.8 members of the subfamily Colobinae (such as *Presbytis*, *Nasalis*, and *Colobus*) tend to have shorter bacula relative to body weight than genera of the sub-family Cercopithecinae. In the proboscis monkey (*Nasalis larvatus*), for instance, massive males weighing 20 kg have bacula 7.9 mm in length, whereas in a cercopithecine monkey of similar weight, such as a baboon, the baculum might be more than 20 mm long.

Despite their large size, the great apes have smaller bacula than many Old World monkeys. The orang-utan has a longer baculum (14.6–15.0 mm) than the gorilla (12.5 mm), chimpanzee (6.0–7.8 mm), or bonobo (8.6 mm). There is a tendency towards reduction of the baculum in hominoids and this process might have progressed further in the African apes than in the Asiatic genus *Pongo*.

The remarkable differences between the relative lengths of primate bacula summarized above cannot be accounted for solely on a phylogenetic basis. In addition, there is a marked correlation between the occurrence of an elongated baculum and copulatory patterns which involve prolonged intromission and/or maintenance of intromission after ejaculation has occurred (Dixson 1987b). Figure 9.9 summarizes information on prosimians and monkeys in which intromission is either prolonged, or in which intromissions are brief and in which copulation terminates once ejaculation has occurred. Logarithmic plots of baculum length vs body weight were analysed in the two groups by the principal axis technique, in order to produce lines of best fit. The slopes of these lines were 0.252 (prolonged-intromission group) and 0.677 (brief-intromission group). The principal axes are widely separated and no overlap occurs in the scatter of points plotted for

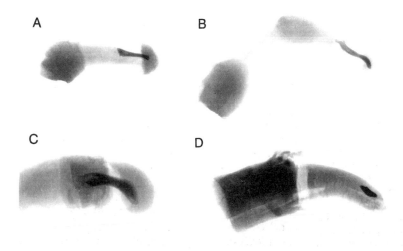

A   B

C   D

Fig. 9.7 Radiographs of penes to show the position and shape of the baculum. A. *Colobus guereza*; B. *Ceropithecus aethiops*; C. *Mandrillus sphinx*; D. *Pan troglodytes*.

Table 9.3 Baculum length, glans penis length, and length of the erect penis in adult male primates

Species: (female mating system)	Baculum length (mm)	Number measured	Sources	Glans penis length (mm)	Number measured	Sources	Pars libera length (mm)	Number measured	Sources
Lemur catta (P)	11.5	1	Dixson 1987a	15.75	2	AFD	43.0	1	AFD
L. fulvus (P)	9.0	?	Petter; Rousseaux 1964	15.1	2	Petter; Rousseaux 1964	ND	–	–
Varecia Variegata (P)	12.3	1	Dixson 1987b	ND	–		ND	–	–
Hapalemur griseus (?)	11.0	?	Petter; Rousseaux 1964	ND	–		ND	–	–
Cheirogaleus medius (P)	14.0	?	"	ND	–	Petter; Rousseaux 1964	ND	–	–
C. major (P)	ND	–	–	11.0	1	"	ND	–	–
Microcebus murinus (P)	11.0	?	"	6.5	1	"	ND	–	–
Avahi laniger (M)	ND	–	–	12.0	1	"	ND	–	–
Propithecus verreauxi (P)	7.6	2	Dixson 1987b	ND	–		ND	–	–
Daubentonia madagascariensis (P)	28.0	?	Petter; Rousseaux 1964	ND	–		ND	–	–
Loris tardigradus (P)	14.2	1	Dixson 1987b	ND	–		ND	–	–
Perodicticus potto (P)	21.0	1	Schultz 1969	ND	–		ND	–	–
Galago senegalensis (P)	16.7	1	Dixson 1987b	ND	–		ND	–	–
Galagoides demidoff (P)	13.1	2	Dixson 1987b	ND	–		ND	–	–
G. crassicaudatus (P)	22.4	4	Dixson 1987b	ND	–		ND	–	–
G. garnettii (P)	ND	–		20.0	1	AFD	33.2	1	AFD
Callithrix jacchus (M)	2.0	1	Hershkovitz 1977	5.8	3	AFD	22.4	5	Dixson 1986a
C. argentata (M)	2.1	2	Hershkovitz 1977	ND	–		ND	–	–
C. humeralifer (M)	2.1	3	Hershkovitz 1977	ND	–		ND	–	–
Cebuella pygmaea (M)	2.2	2	Dixson 1987b	ND	–		ND	–	–
Saguinus oedipus (M)	1.7	9	Hershkovitz 1977; Dixson 1987b	ND	–		ND	–	–
S. fuscicollis (M)	1.9	5	Hershkovitz 1977	ND	–		ND	–	–
S. labiatus (M)	1.7	1	Hershkovitz 1977	ND	–		ND	–	–
S. midas (M)	2.2	1	Hershkovitz 1977	ND	–		ND	–	–
S. mystax (M)	1.6	1	Hershkovitz 1977	ND	–		ND	–	–
S. nigricollis (M)	1.6	1	Hershkovitz 1977	ND	–		ND	–	–
S. bicolor (M)	2.4	1	Hershkovitz 1977	ND	–		ND	–	–
Leontopithecus rosalia (M)	3.0	1	Hershkovitz 1977	ND	–		ND	–	–
Callimico goeldii (?)	1.8	1	Hershkovitz 1977	ND	–		ND	–	–
Aotus trivirgatus (M)	2.2	1	Dixson 1987b	3.6	1	AFD	10.0	1	AFD
Cebus apella (P)	8.5	1	Dixson 1987b	ND	–		ND	–	–
C. capucinus (?)	10.1	1	Dixson 1987b	ND	–		ND	–	–
Saimiri boliviensis (P)	9.0	1	Dixson 1987b	ND	–		ND	–	–
S. sciureus (P)	ND	–		10.0	1	Hill 1960	ND	–	–
Presbytis comata (M)	6.7	1	Dixson 1987b	ND	–		ND	–	–
P. vetulus (M)	12.5	1	Hill 1958	ND	–		ND	–	–
Nasalis larvatus (M)	7.9	3	Dixson 1987b	ND	–		ND	–	–
Pygathrix nemaeus (?)	15.3	2	Dixson 1987b	ND	–		ND	–	–
Colobus guereza (M)	12.4	2	Dixson 1987b	9.5	1	AFD	ND	–	–

Table 9.3—*Continued*

Species: (female mating system)	Baculum Length (mm)	Number measured	Sources	Glans penis length (mm)	Number measured	Sources	Pars libera length (mm)	Number measured	Sources
C. polykomos (M)	14.2	2	Dixson 1987b	ND	—	—	ND	—	—
C. badius (P)	13.3	3	Dixson 1987b	ND	—	—	ND	—	—
Procolobus verus (P)	9.8	1	Dixson 1987b	ND	—	—	ND	—	—
Cercopithecus aethiops (P)	16.5	5	Dixson 1987b	12.0	2	AFD	ND	—	—
C. mitis (M)	16.3	2	Hill 1966	ND	—	—	ND	—	—
C. mona (?)	23.7	1	Dixson 1987b	ND	—	—	ND	—	—
C. neglectus	17.7	1	Hill 1966	ND	—	—	ND	—	—
Miopithecus talapoin (P)	9.5	2	Hill 1966	8.0	4	AFD	ND	—	—
Erythrocebus patas (M)	15.8	2	Dixson 1987b	ND	—	—	ND	—	—
Allenopithecus nigroviridis (P)	ND	—	Dixson 1987b	10.0	1	Hill 1966	ND	—	—
Cercocebus albigena (P)	21.2	2	Dixson 1987b	ND	—	—	ND	—	—
C. torquatus (?)	14.3	2	Dixson 1987b	ND	—	—	ND	—	—
C. aterrimus (?)	ND	—	Dixson 1987b	ND	—	—	125.0	1	AFD
Mandrillus sphinx (P)	24.8	2	Dixson 1987b	23.0	3	AFD	121.5	2	AFD
M. leucophaeus (P)	21.0	1	Hill 1970	ND	—	—	ND	—	—
Papio anubis (P)	30.0	1	Hill 1970	ND	—	—	ND	—	—
P. cynocephalus (P)	22.5	2	Dixson 1987b	14.5	1	AFD	ND	—	—
P. hamadryas (M)	21.6	1	Hill 1970	ND	—	—	140.0	1	AFD
P. papio (P)	22.0	1	Hill 1970	ND	—	—	ND	—	—
P. ursinus (P)	26.2	1	Dixson 1987b	ND	—	—	ND	—	—
Theropithecus gelada (M)	26.0	1	Hill 1970	ND	—	—	ND	—	—
Macaca nemestrina (P)	21.7	5	Hill 1974	17.85	6	Fooden 1975	ND	—	—
M. nigra (P)	23.8	3	Fooden 1969; Hill 1974	13.0	1	Hill 1974	ND	—	—
M. fascicularis (P)	13.1	5	Fooden 1969; Dixson 1987b	16.55	4	Fooden 1969	ND	—	—
M. mulatta (P)	17.0	1	Fooden 1974	18.7	1	Fooden 1975	93.0	4	AFD
M. fuscata (P)	19.0	4	Fooden 1972	20.0	2	Fooden 1975	ND	—	—
M. assamensis (P)	25.8	3	Dixson 1987b	ND	—	—	ND	—	—
M. sinica (P)	20.0	1	Hill 1974	16.0	1	Fooden 1980	ND	—	—
M. thibetana (P)	24.4	3	Fooden 1966	ND	—	—	ND	—	—
M. arctoides (P)	53.1	6	Dixson 1987b	54.1	6	AFD	82.6	6	AFD
Hylobates lar (M)	8.5	4	Hill 1958	ND	—	—	ND	—	—
Pongo pygmaeus (P)	13.55	2	Dixson 1987b	38.1	5	Dahl 1988	85.0	5	Dahl 1988
Pan troglodytes (P)	6.9	2	Hill 1958	X	—	—	144.0	11	Dixson and Mundy 1994
P. paniscus (P)	8.5	1	Izor et al. 1981	X	—	AFD	170.0	1	Dahl 1988
Gorilla gorilla (M)	12.6	2	Hill 1958	8.0	1	AFD	65.0	1	Short 1980
Homo sapiens (M)	X	—	Dixson 1987b	ND	—	—	165.0	2,310	Gebhard and Johnson 1979

ND=no available data; X=structure does not occur; AFD=author's previously unpublished data; M=females usually mate monandrously; P=females usually mate polyandrously.
From Dixson and Purvis, in press.

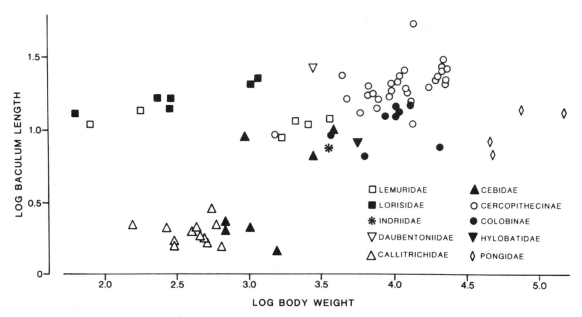

Fig. 9.8 Logarithmic plot of baculum length versus body weight for 10 families of primates. (After Dixson 1987b.)

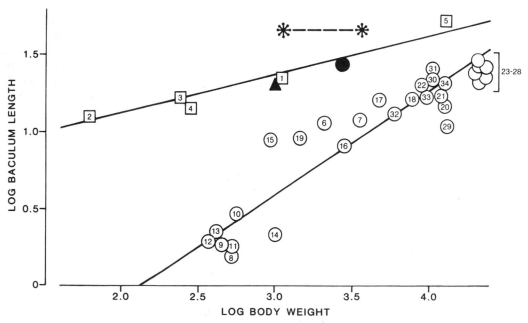

Fig. 9.9 Logarithmic plot of baculum length in relation to patterns of copulatory behaviour in 34 species. □ =Species with long intromission times and/or post-ejaculatory maintenance of intromission. ○=Species in which copulation is usually brief and intromission is terminated promptly after ejaculation. Principal axes are plotted separately for the two groups. Two additional species are plotted for which only anatomical data are available: {▲} *Perodicticus potto;* ● *Daubentonia.* Species studied are as follows: 1. *Galago garnettii,* 2. *G. demidoff* 3. *G. Senegalensis,* 4. *Loris tardigradus.* 5. *Macaca arctoides,* 6. *Lemur catta,* 7. *V. variegata,* 8. *Saguinus oedipus,* 9. *S. fuscicollis,* 10. *Leontopithecus rosalia,* 11. *Callimico goeldii,* 12. *Callithrix jacchus,* 13. *C. argentata,* 14. *Aotus lemurinus,* 15. *Saimiri sciureus,* 16. *Cebus apella,* 17. *Cercopithecus aethiops,* 18. *C. mitis,* 19. *Miopithecus talapoin,* 20. *Erythrocebus patas,* 21. *Cercocebus atys,* 22. *C. albigena,* 23. *Mandrillus sphinx,* 24. *M. Leucophaeus,* 25. *Papio ursinus,* 26. *P. anubis,* 27. *P. hamadryas,* 28. *Theropithecus gelada,* 29. *Macaca sylvanus,* 30. *M. nemestrina,* 31. *M. nigra,* 32. *M. fascicularis,* 33. *M. mulatta,* 34. *M. fuscata.* *∗ — ∗* =fossil adapid described by Von Koenigswald (1979). (After Dixson 1987b.)

the two groups. Further analyses of residual baculum length in primates show that, once phylogenetic factors have been controlled for, species with single prolonged intromission (SPI) copulatory patterns have significantly longer bacula than those which employ either single brief (SBI) or multiple brief intromission (MBI) patterns (Dixson and Purvis, in press; see Fig. 9.10).

Primates with elongated bacula tend also to be species in which intromission is maintained into the post-ejaculatory period. All these species have either a dispersed type of mating system (the nocturnal prosimians in Fig. 9.9) or a multimale–multifemale system (*Macaca arctoides*). These are mating systems in which females mate with more than one male, and in which it is proposed that sexual selection has had the greatest impact upon the evolution of complex penile morphologies and copulatory patterns. Elongated bacula appear to be an example of this type of specialization. Precise data on copulatory behaviour are lacking for certain prosimians in which males possess an elongated baculum (e.g. *Perodicticus*, *Euoticus*, and *Daubentonia*). Data points for baculum length vs body weight for these species fall close to the principal axis for PI species, leading to the prediction that they too should exhibit prolonged intromission patterns with maintenance of intromission once ejaculation has occurred. For *Daubentonia*, at least, this prediction seems to be correct (Winn 1994; Sterling and Richard 1995).

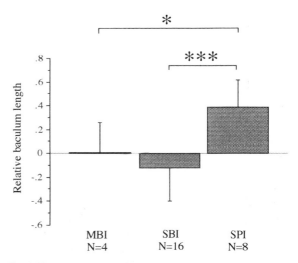

Fig. 9.10 Relative baculum length and intromission patterns in primates. MBI=multiple brief intromission pattern; SBI=single brief intromission pattern; SPI=single prolonged intromission pattern. N=number of genera studied. *P < 0.05; ***P < 0.001. (Data are from Dixson and Purvis, in press.)

The only fossil baculum attributable to a primate species has been described by Von Koenigswald (1979). Post-cranial remains of a lemur-like creature were discovered in Eocene oil shale deposits at Messel, close to the German city of Darmstadt. The foot demonstrates a broad and opposable great toe

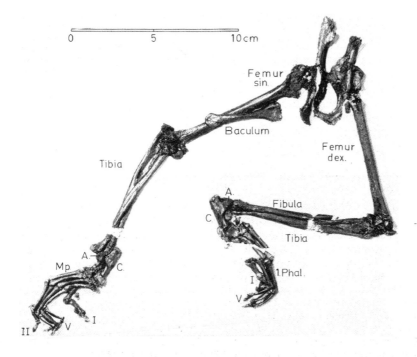

Fig. 9.11 The only known fossil baculum from a primate species belongs to an adapid which lived during the Eocene. It possessed a baculum of remarkable length (46 mm) and we surmise that it had an SPI copulatory pattern. (After Von Koenigswald 1979.)

and the second toe is armed with a toilet claw as in modern prosimians. There is also a remarkable baculum, 46 mm in length (Fig. 9.11). The fossil baculum described by Von Koenigswald in 1979 and a second specimen from the same site belonged to an adapid of moderate size, approximating perhaps to a modern greater galago. The conclusion is that copulation with prolonged intromission occurred in this extinct form and that the mating system was either of a dispersed type, as in many modern nocturnal prosimians, or (less likely) that it lived in multimale–multifemale groups.

It is important to emphasize that, whilst possession of an elongated baculum correlates with an SPI copulatory pattern in primates, prolonged intromissions can occur in species which lack a penile bone. This is true of certain marsupials (e.g., *Antechinus flavipes*: Marlow 1961) because a baculum is lacking in the Metatheria. Among the primates, prolonged intromissions occur in *Ateles* and *Lagothrix* (Klein 1971; Nishimura 1988), as well as in *Homo*, yet the baculum has been lost in all three genera. One possibility is that the present day copulatory patterns of these genera developed after the baculum had become greatly reduced, or had been lost, in ancestral forms (Dixson 1987b).

The hypothesis that baculum length might have increased in some primates due to sexual selection in association with a prolonged intromission pattern is testable by comparative studies of other mammalian orders which possess a penile bone. The carnivores and pinnipeds (Grand Order Ferae: Eisenberg 1981) make excellent subjects for such studies. Like the primates, the Grand Order Ferae contains over 200 species which vary in weight from about 70 g (the least weasel: *Mustela rixola*) to over 2500 kg (elephant seals: *Mirounga leonina*). Moreover, all the pinnipeds and carnivores, with the exception of the Hyaenidae, possess a baculum. Figure 9.12 shows a logarithmic plot of baculum length vs body weight in 66 carnivore species together with a plot for those species for which quantitative data on intromission durations have been adequately documented (Dixson 1995c). Elongated bacula characterize many carnivores in the families Ursidae, Canidae, Mustelidae, and Procyonidae as well as in all pinnipeds studied. SPI copulatory patterns are well represented in all these groups and, if data ever become available on all the species shown in these families in Fig. 9.12, the prediction is that most of them will be found to exhibit this type of mating pattern. An exception in Fig. 9.12 is provided by the cat family (Felidae). The baculum is shorter in felids than in other carnivores studied and, interestingly, intromission durations are much shorter in those

species for which data are available (e.g., the domestic cat, tiger, and lion).

Given that baculum length has been subject to modification by sexual selection, what functions does it serve during prolonged intromissions in primates or carnivores? A variety of possible functions has been proposed (Long and Frank 1968; Patterson and Thaeler 1982; Dixson 1987b). Indeed, there is no reason to expect that any single hypothesis can account for the evolution of diversity in bacular structure. Long and Frank (1968) proposed that in some mammals, such as mustelids, the complex shape of the distal end of the baculum might assist the male in gaining intromission. Since males in some of these carnivore species are much larger than females, the baculum might help in overcoming the 'problem of friction and vaginal resistance for the enlarged penis'. In *Mustela vison*, for instance, the distal end of the baculum is hook-shaped and grooved, whilst in the Racoon (*Procyon*) there is a bifid tip to the bone. The American badger (*Taxidea taxus*) has a wedge-shaped, screw-like tip to the baculum. A similar hypothesis has been suggested by Bonner for the elephant seal, in which 'it seems that the penis is inserted before it is fully erect and this would be possible only with a baculum'.

It seems unlikely that an elongated baculum evolved in primates solely to assist males in gaining intromission. In the majority of cases in which elongation has occurred, (e.g., in *Galago*, *Microcebus*, and *Loris*), males are very similar to females in body weight and there is no indication that the long but relatively thin penis encounters undue resistance as the male attempts penetration. In the stumptail macaque, however, males are larger than females. The penis is fully erect before intromission yet the female's vaginal introitus is very small in this species and guarded by a large dorsal colliculus (Fig. 9.2). In this case, the baculum might assist the male to attain intromission. Indeed, although the baculum is relatively small in all other Old World monkeys and apes, it might, once the corpora cavenosa are erect, impart additional stiffness to the glans.

It has long been accepted that the baculum might act as a supporting rod for the penis during copulation (Romer 1962). It is situated dorsal to the urethra and, in some cases, is grooved on its perineal surface and partly protects the urethra. In species with prolonged intromission, this arrangement might be important because of the considerable pressure exerted upon the penis during copulation. An extreme case is provided by the *Canidae* in which a genital lock occurs during copulation owing to engorgement of the distal portion of the penis within the female's vagina. Such locking postures can last for 15–30 min in the red fox (*Vulpes vulpes*: Lloyd 1980) and for over 60 min in the domestic dog (Beach 1968b). The bone might strengthen the penis and facilitate the flow of semen which occurs, periodically, throughout the prolonged intromission (Hart and Kitchell 1966). In mustelids, prolonged intromissions are common (e.g., 1–2 h in weasels and martens: Ewer 1973; >14 min in the marine otter: Kenyon 1969; and 24 min in the Canadian otter: Liers 1951). Fractures of the baculum have been recorded in the European otter (*Lutra lutra*) and it has been suggested that vigorous and prolonged copulation might cause these injuries (Laidler 1982).

Among primates, there is certainly a case to be made for a protective or supporting role for elongated bacula. In galagos, such as *G. garnettii*, intromissions can last for up to 2 h and since bouts of

pelvic thrusting occur at intervals throughout this time, it is possible that multiple ejaculations occur. The baculum might assist in protecting the urethra during this process. Moreover, it is important to remember that prolonged intromission might itself represent a type of mate-guarding behaviour by the male.

The baculum plays an important role in supporting the elongated glans penis of the stumptail macaque (*Macaca arctoides*) during copulation. In this species the male maintains intromission for a short period after ejaculation, the so-called post-ejaculatory pair-sit. This is not a complete genital lock, as the partners can usually separate with a minimum of difficulty (Goldfoot *et al.* 1975). However, a remarkable, crescent-shaped copulatory plug is moulded by the expanded portion of the urethra

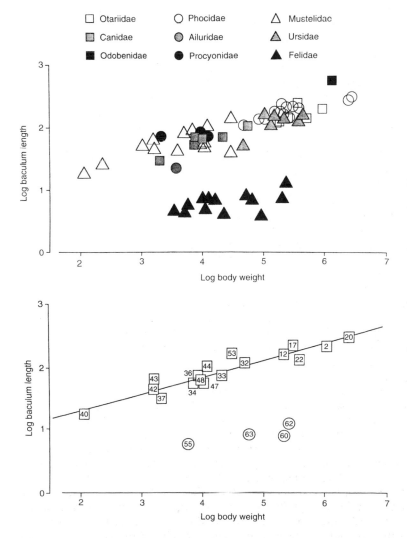

Fig. 9.12 Upper: Logarithmic plot of baculum length versus body weight for 66 species of carnivores and pinnipeds. Lower: Logarithmic plot of baculum length versus body weight for carnivores and pinnipeds which are known to have prolonged intromission (□) or short intromission (○) durations. Of the 66 species plotted in the upper graph, 21 are represented in the lower part of the figure. Long intromission species: 2. *Eumetopias jubata*; 12. *Halichoerus grypus*; 17. *Leptonychotes weddelli*; 20. *Mirounga angustirostris*; 22. *Ursus arctos horribilis*; 32. *Canis lupus*; 33. *Canis familiaris*; 34. *Vulpes vulpes*; 36. *Nyctereutes procyonides*; 37. *Fennecus zerda*; 40. *Mustela nivalis*; 42. *Martes martes*; 43. *Martes foina*; 44. *Meles meles*; 47. *Lutra lutra*; 48. *Lutra canadensis*; 53. *Enhydra lutris*. Short intromission species: 55. *Felis sylvestris catus*; 60. *Panthera leo*; 62. *Panthera tigris*; 63. *Panthera onca*. (After Dixson 1995c.)

within the glans penis. The post-ejaculatory pair-sit might ensure that the plug is positioned correctly within the enlarged ectocervix of the female before the male dismounts.

A third hypothesis is that the baculum might assist in stimulating the female during copulation and facilitate certain neuroendocrine responses which are important for fertility. Patterson and Thaeler (1982) present this view as a modification of the lock-and-key hypothesis (i.e. equivalent to the genitalic recognition hypothesis outlined at the beginning of this chapter). They suggest that the baculum and penis might both 'serve in the reproductive isolation of the species'. Whilst this view is probably incorrect, it is known that phallic stimulation causes neuroendocrine responses in females of various mammalian species. Examples include release of prolactin and support of the corpus luteum in the rat, reflexive release of the luteinizing hormone surge required for ovulation (e.g., in certain carnivores), and release of oxytocin from the posterior pituitary gland. The specific contribution made by the baculum in such cases is unknown, although various authors have suggested that it might play a role in induced ovulation (e.g., in weasels: King 1989 and badgers: Neal 1986). The complex morphology of the projecting distal pole of the baculum might be relevant in such cases. Induced ovulation is unknown among primates, nor is there any evidence of effects of penile stimulation upon release of prolactin in the female. Further, various felid species are induced ovulators and yet the baculum is relatively small and unspecialized in members of this carnivore family (Dixson 1995c). Postcoital release of oxytocin has been reported in women (Carmichael *et al.* 1987), but human males lack a baculum. Despite this negative view, I know of no studies which have sought to determine whether neuroendocrine changes occur during copulation in females of those primate species which possess elongated bacula. Is it possible, for instance, that penile stimulation might *hasten the onset* of the LH surge in such spontaneously ovulating females, rather than being an essential trigger as in induced ovulators?

Finally, it is necessary to discuss the potential role of the baculum in facilitating sperm transport within the female's reproductive tract. It has been mentioned that in most primates with elongated bacula the tip of the bone projects from the distal pole of the penis with the urinary meatus on its perineal surface (see Fig. 9.1). This arrangement, coupled with the fact that the penis is typically very long in these animals, might bring the tip of the baculum into contact with the os cervix during copulation. This might assist transfer of semen into the cervical

canal. Frills and lappets on the distal glans (e.g. in *Euoticus* and *Perodicticus*) might fulfil a similar function. In this case, the female's cervix is viewed as a potential barrier to spermatozoa and sexual selection by cryptic female choice has encouraged the evolution of phallic morphologies which attempt to overcome this obstacle to sperm transport.

The evolution of penile spines

Like the baculum, the occurrence of the keratinized penile spines on the glans penis of many prosimians, monkeys, and apes represents a primitive character of the Order Primates. Similar structures occur on the glans penis of some other mammals, such as rodents, where they have been extensively studied as an aid to taxonomic studies (Hooper and Musser 1964) or in an attempt to analyse their functions in sexual behaviour (Beach and Levinson 1950; Phoenix *et al.* 1976).

In primates, as in rodents, penile spines exhibit marked interspecific variations in size and shape. Are these variations merely pleiotropic effects of genes which regulate other more important systems, or do they have functional significance? Firstly, I shall review information on spine structure and then discuss the possible functions of penile spines in reproduction.

Penile spines are made up of many, overlapping layers of keratinized material. They occur either as single-pointed conical or elongated structures, or as multi-pointed complexes. As we have seen, penile spines differ considerably in size in prosimians, monkeys, and apes. They can be classified into three categories on the basis of size and shape:

1. simple spines: small single-pointed structures of moderate length (Fig. 9.13A: common marmoset);

2. robust simple spines: single-pointed structures but much enlarged and often thickened at the base (e.g., Fig. 9.13B: *Galagoides demidoff*); and

3. complex spines: multi-pointed structures typically of large size. The number of points can vary from two or three per spine (e.g., *Galago garnettii*) up to six in (*Microcebus murinus*: Fig. 9.13C and D).

Type 1, simple spines, occur widely among the primates. Type 2, robust simple spines, are found in a few anthropoids (e.g., *Macaca arctoides*; *Ateles belzebuth*) but more commonly in certain prosimians (e.g., in *Euoticus*, *Galago crassicaudatus*, *G. alleni*, *Avahi*, and *Phaner*). Complex penile spines are restricted to the prosimians, being found, for example, in *Lemur fulvus*, *L. catta*, and *Microcebus murinus*). It has been reported that certain lorisines possess low hexagonal plates upon the glans penis, rather than the usual, pointed structures (in *Loris*, *Nycticebus*, and *Perodicticus*: Hill 1958). However, in specimens I have examined by scanning electron microscopy, simple penile spines of typical shape were identifiable in *Nycticebus coucang* and remnants of the bases of such spines were visible in a single skin sample from *Loris tardigradus*. I am uncertain whether spines occur in *Perodicticus potto*, but the morphology of penile spines in other lorisines appears to conform to the pattern established for primates in general.

In evolutionary terms, Type 1 spines appear to represent the primitive and unspecialized condition. Complexity has been achieved either by increased size of the individual spines or by fusion of two or more spines to produce complex structures. Spines are androgen-dependent and grow at puberty as a result of androgenic stimulation (e.g. in the rat: Beach and Holz 1946; and in the talapoin monkey: Dixson and Herbert 1974). Fusion of the dermal primordia which give rise to single spines would produce multi-pointed structures. This process might occur on different areas of the glans penis in a single species. In *G. garnettii* (Fig. 9.1) and *L. fulvus* (Fig. 9.4), for example, simpler spines occur on the distal glans whilst those in the mid-region and base are more complex. This observation raises the question of whether the various spine morphotypes might

fulfil different functional roles within a single species.

Five functions have been proposed for penile spines in rodents and some of them might also be applicable to primates (Dixson 1995a). They can be summarized as follows.

1. Spines are often situated above tactile receptors in the dermis of the glans penis; their deflection during pelvic thrusting movements might enhance tactile signals required to achieve intromission and/or ejaculation. First suggested for the rat (Beach and Levinson 1950; Hart 1978) this hypothesis finds support from work on the common marmoset, *Callithrix jacchus* (Dixson 1991a). Selective removal of the Type 1 spines which clothe the glans penis of the marmoset can be achieved using a thioglycollate cream which weakens the keratin. Males treated in this way still intromit and ejaculate but the time required to attain intromission is significantly lengthened, as is the duration of intravaginal thrusting. These subtle, but statistically significant, changes in copulatory behaviour indicate that simple penile spines enhance tactile feedback from the glans penis. I suggest that this is the basic function for this kind of Type 1 penile spine, which is the most widely distributed and primitive form found in primates. In those cases where more than one

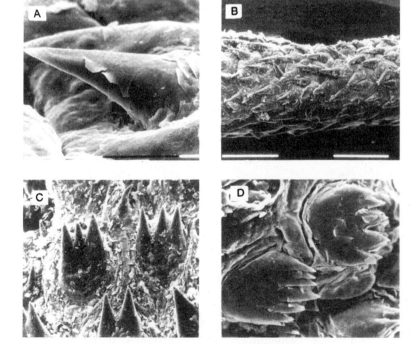

Fig. 9.13 Scanning electron micro-graphs of penile spines in various primates. A. Type 1, simple spine of *Callithrix jacchus*. B. Type 2, robust spines on the glans penis of *Galagoides demidoff*. C. Complex spines of *Galago garnettii*. D. Complex spines of *Microcebus murinus*. (After Dixson 1995a.)

morphotype of spine occurs on the glans, as described above, simple spines occur on the distal glans where they are ideally positioned to assist in sensory feedback during the pre-intromission phase of pelvic thrusting.

2. Spines might assist in imparting tactile stimulation to the female and thus facilitate neuroendocrine responses affecting sperm transport. This hypothesis has been described above when considering possible functions of the baculum. Whilst copulatory stimuli are relevant to these processes, for example in the rat or in induced ovulators, there is little experimental evidence for primates, and no definitive evidence for any species, that penile spines are responsible for such effects. Given the large size and rasping points of Type 2 or Type 3 spines, it seems likely that females could perceive their presence during copulation. However, females often appear relatively passive during intromission (e.g. in the greater galago) and there is no evidence, for instance, that orgasmic reactions are more likely in females of primate species where males possess such structures.

3. Spines might provide tactile stimulation that enhances the female's sexually receptive posture during oestrus. This hypothesis again derives from work on rats. Tactile stimuli from the male, as he mounts, palpates the female's flanks, and thrusts vigorously against her perineal area, facilitate the lordotic reflex of the receptive female (Pfaff 1980). Male rats make a series of mounts with intromission before ejaculation occurs. Penile stimulation of the female's vagina and cervix during the mount series also enhances the expression of lordosis (Rodriquez-Sierra *et al.* 1975). It is possible that similar responses occur in prosimians, since females exhibit oestrus and, in some cases, adopt a lordotic posture during copulation (Dixson 1983c). No experiments comparable to those performed on rodents have examined these questions in prosimian primates. This argument is less applicable to female anthropoid primates since, as far as is known, none adopt sexually reflexive, lordotic postures during copulation (Dixson 1983c, 1990a). However, given the lengthy duration of intromission in some nocturnal prosimians, such as *Galago garnettii* (120 min), *G. demidoff* (40 min), and *Loris tardigradus* (10–17 min), it is possible that sexual selection by female choice has encouraged the evolution of complex male genitalia which stimulate the receptive female and induce her to remain immobile.

4. Penile spines might assist in removal of coagulated semen or copulatory plugs deposited by previous matings and are thus relevant to reducing potential sperm competition. The repeated intromissions made by male rats before ejaculation dislodge plugs left by previous matings (Wallach and Hart 1983). Since penile spines are situated on the distal end of the glans penis as well as on the sides of the glans in the rat, they might assist in dislodging plugs. Although spines have been described in the majority of primates, relatively few species employ a multiple intromission pattern of copulation before ejaculation. In Table 5.6 on page 121 only 14 examples are listed, and almost all these monkeys have multimale–multifemale mating systems. Multiple intromissions before ejaculation would certainly be expected to dislodge or remove previous ejaculates. Bizarre specializations for this purpose have arisen in some insects such as damselflies in which there is a scoop-shaped appendage, armed with spines, on the male intromittent organ. This is used to remove sperm from the female's spermatotheca before the male deposits his own gametes (Waage 1979). However, among those primates which make a number of intromissions before ejaculation (with the exception of *Cacajao*) penile spines are of the small, simple, Type 1 variety. It is not impossible that these structures might assist in dislodging copulatory plugs, but one might expect them to be larger in this case. In *Cacajao*, however, spines are large, nearly 1 mm long according to Hershkovitz, who suggests they are important in breaking up copulatory plugs.

5. Penile spines might fulfil a mechanical function by gripping the walls of the vagina during copulation, thus producing a type of genital lock. In the Australian hopping mouse, *Notomys alexis*, the glans penis is elongated and armed with impressive Type 2 spines (Breed 1981). A true genital lock occurs during copulation in the hopping mouse (Dewsbury and Hodges 1987). The long glandes and large penile spines of various prosimian species, and especially those in the genera *Galago*, *Euoticus*, and *Arctocebus*, indicate the possibility of a similar copulatory mechanism. Copulation involving prolonged intromissions, with intermittent bouts of pelvic thrusting, has been reported for various galagos and for *Arctocebus*. During mating tests conducted in the laboratory it is often difficult to separate copulating *G. garnettii*; the penis becomes firmly lodged in the vagina during prolonged intromissions (author's unpublished observations). Charles-Dominique (1977) describes a similar phenomenon in *G. demidoff*. The older anatomists were aware of the possibility that penile spines in

some prosimians might function in this way during copulation, referring to them as 'grappling spurs' (Clark 1959; Hill 1972).

The evolution of distal penile complexity

The evolution of long penes and complex distal penile morphologies in members of some non-gregarious or multimale–multifemale primate mating systems might be attributed to two principal causes. Firstly, males might attempt to displace the sperm of competitors. Secondly, males might attempt to 'encourage' the female to make use of their spermatozoa, by placing it in the most advantageous position, or by stimulating uptake and transport of spermatozoa within the female's tract. In this latter case the penis functions as an 'internal courtship device' (Eberhard 1985). In the absence of experimental data, it is only possible to speculate, briefly, upon the ways in which various specializations of penile morphology in primates might be related to these two functions. For instance, the elongated filiform penis of the chimpanzee probably acts as a probe, to force its way through the copulatory plugs deposited by previous males within the female's vagina. The mushroom-shaped glans of *Brachyteles*, or plunger-like arrangement found in *Lagothrix*, might function to lodge semen firmly against the female's os cervix, in the manner described for the rat (Sachs 1982). As previously described, in prosimians such as *Galago*, *Nycticebus*, and *Perodicticus* a projecting baculum might be important in facilitating sperm deposition at the cervical os. The frills and lappets surrounding the urethral opening might also contact the cervix during copulation and encourage sperm uptake (e.g. in the bonobo, *Pan paniscus*). In the New World Pithecines, *Cacajao* and *Chiropotes*, Hershkovitz has described a small, bluntly pointed, but labile glans. The penis is long in these animals and the urinary meatus is surrounded by erectile lappets. Hershkovitz suggests that these structures contact the female's os cervix during copulation to enable direct delivery of sperm into the uterus. Whether intra-uterine insemination occurs in these, or in any other, primate species remains to be determined, however.

The striated penile muscles and morphological changes in the glans penis during copulation

Mention must be made of the striated penile muscles, situated at the base of the penis, since they are responsible for expulsion of semen during ejaculation and for reflexive penile movements which accompany intromission or ejaculation. These muscles

might therefore play a crucial role in the process of 'internal courtship' of the female and they might be concerned also in intermale competition to displace rival ejaculates. Several of these muscles have been described; of especial interest as regards sexual reflexes are the ischiocavernosus and bulbocavernosus (sometimes called bulbospongiosus) muscles. The functions of both these muscles have been extensively studied in the rat; a diagrammatic representation of their positions and effects upon penile reflexes is shown in Fig. 9.14. The paired ischiocavernosus muscles arise from ischia of the pelvic girdle and insert along the sides of the vascular erectile bodies (corpora carvernosa) of the penile shaft. Movements of these muscles are important in causing 'flip'-like reflexive movements which assist the male in elevating the penis and attaining intromission. Indeed, if the ischiocavernosus muscles of

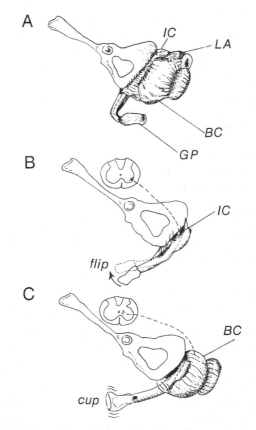

Fig. 9.14 Functions of the striated penile muscles in the rat. A. penis in non-erect state. B. penis erect. Ischiocavernosus muscles (IC) cause 'flip' responses. C. penis erect. Bulbocavernosus muscles (BC) cause 'cup' responses of the glans penis (GP). LA=Levator ani muscles. Motor neurons originating in the lumbar region of the spinal cord innervate the IC and BC muscles and control reflexive penile movements, as indicated in B and C. (Redrawn from Hart and Melese-d'Hospital 1983.)

the rat are removed, males are 'virtually incapable of achieving penile insertion' (Sachs 1983). By contrast, the large, paired bulbocavernosus muscles surround the corpus spongiosum at the base of the penis. At its distal end the corpus spongiosum provides the erectile component of the glans penis; hence when the bulbocavernosus muscles contract, they bring about a marked flaring or 'cupping' response in the glans penis of the rat (Fig. 9.14). This reflexive response occurs during intromission in the rat and serves two functions. Firstly, during non-ejaculatory intromissions the penile cup assists in dislodging plugs from previous matings. Secondly, during the final and ejaculatory intromission, the penile cup functions to pack and seal the male's own copulatory plug against the female's cervix. After removal of the bulbocavernosus muscles from rats Sachs (1983) found that their female partners failed to become pregnant, although copulatory stimuli were sufficient to induce the neuroendocrine state of pseudopregnancy in some cases. Electromyographic studies have shown that the proximal division of the bulbocavenosus musculature is especially important in producing these reflexive changes in the glans penis of the rat during copulation (Holmes *et al.* 1991).

The striated penile muscles are not essential for penile erection in the rat, as this is controlled by vascular mechanisms; this is also the case in man (Bancroft 1989) and probably other primates also. In some mammals, however, the striated penile muscles play an important role in the mechanism of penile erection (bull and ram: Watson 1964) and maintenance of a genital lock during copulation (dog: Hart 1972a). In the male elephant the muscles are especially well developed and function to move the penis in order to locate the female's vaginal orifice, and then to make thrusting movements once intromission is attained.

The above examples serve to show that the striated penile muscles play important and variable functions in mammalian sexual behaviour and fertility. It is a unfortunate that little or no experimental work has sought to define the functions of these muscles in prosimians, monkeys, or apes. Interesting anatomical variations occur among various primate genera and examples of these are shown in Fig. 9.15. In primates, as in the rat, the ischiocavenosus muscles usually arise from the pelvic ischia and insert along the sides of the penis in the tough tunic which surrounds the corpora cavernosa (e.g. in *Callithrix jacchus*: Fig. 9.15A; *Pongo pygmaeus*: Fig. 9.15B). However, in some genera these muscles form a more compact mass, surrounding the perineal as well as the lateral aspect of the penile base (e.g. in *Cercopithecus*: Fig. 9.15C). In the rhesus macaque,

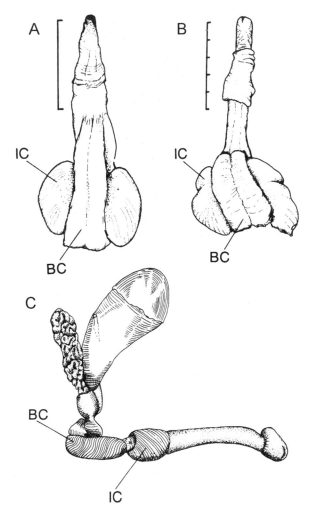

Fig. 9.15 Variations in the striated penile muscles of primates IC=ischiocavernosus muscles; BC=bulbocavernosus muscles. A. *Callithrix jacchus*. (After Dixson 1987a); B. *Pongo pygmaeus* (After Dixson 1987a); C. *Cercopithecus ascanius* (Redrawn from Hill 1966). Scales in A and B are in 1-cm divisions.

Wislocki (1933) described a separate, smaller, pair of muscles which arise medially to the ischiocavenosus muscles and insert along the cranial border of the penis (the levator penis muscles: see Fig. 8.12, page 231). These muscles, he suggested, function 'to elevate and straighten the pendulous end of the penis'. The bulbocavenosus muscles insert along the perineal surface of the penis, medial to the ischiocavenosus muscles, in some primate species (e.g. *C. jacchus* and *P. pygmaeus*). However, in some monkeys they are situated more proximally and surround the penile bulb (e.g. in *Macaca* and *Cercopithecus*). In this position, the muscles are especially well situated not only to control the series of contractions which result in ejaculation, but also to cause

Fig. 9.16 Male woolly monkey (*Lagothrix lagotricha*) showing the large size of the penis and plunger-shaped glans during full erection. (Author's drawing, after Nishimura 1988.)

morphological changes in the glans penis during this process. Marked changes in distal penile morphology certainly occur during erection and ejaculation in primates, but they are poorly described in anatomical studies which almost invariably concern the flaccid penis. Studies of erection in the dog brought to light certain specializations of penile morphology (the urethral process, *corona glandis*, and *collum glandis*) which were previously unknown (Hart and Kitchell 1966). Marked changes in penile morphology occur during erection in some primates. In *Cebus*, for instance, the glans sometimes expands rapidly and repeatedly during penile erection, in a similar way to the reflexive cupping response seen in the rat. In *Lagothrix lagotricha*, the glans is particularly prominent and plunger-shaped in erection (Fig. 9.16). Penile spines, which normally point proximally on the glans, become elevated and prominent during erections (e.g. in *C. jacchus*: Dixson 1991a). The glans pulsates rapidly during ejaculation in some species, presumably as a result of contractions of the bulbocavernosus muscles (e.g., in *Mandrillus sphinx* and *Miopithecus talapoin*). A quantitative study of the sizes and functions of the striated penile muscles in primates would be very worthwhile. It is my impression that the bulbocavernosus muscles are more massive in species such as macaques, where females mate with a number of males. Males of such species ejaculate more frequently than monogamous and polygynous forms and produce larger volumes of semen. The bulbocavernosus muscles might be larger for this reason. One additional hypothesis,

however, is that these muscles control reflexive penile movements during copulation and that sexual selection by female choice has resulted in the evolution of larger penile muscles.

Sexual selection and the evolution of female genitalia

Some observations on evolution of the clitoris

The clitoris, the homologue of the male's penis, is present in all primate species, but it varies considerably in size and structure among the various genera. In some prosimians (in *Lorisines* and *Hapalemur*: Hill 1972) the urethra tunnels the clitoris and opens at its tip, but this arrangement is exceptional and does not occur in any of the anthropoids. A small bone, the homologue of the baculum in the male's penis, occurs in the clitoris of some prosimians (Hill 1953, 1972). Erection of the clitoris has been observed in a number of monkey species (e.g. in *Cercopithecus* species during sexual presentation: Hill 1966; in *Saimiri* during genital displays: Ploog 1967; in *Macaca fuscata* during masturbation: Wolfe 1984b; and in *Aotus lemurinus* during stimulation of the external genitalia of the anaesthetized female: author's unpublished observations. Hill (1974) observed that that the clitoris of macaques 'recalls the penis, in miniature, being composed of a pair of crura...surmounted apically by the glans clitoridis. Crura and glans are composed of erectile tissue'. A rich plexus of free sensory nerve endings occurs in the epithelium of the glans and more complex tactile receptors (Meissners and Pacinian corpuscles) also occur in the glans or shaft of the clitoris. Therefore, it seems that the clitoris of certain monkeys and apes is capable of the same erectile responses and sensory feedback as the clitoris of the human female and might be equally important as a sensate focus during copulation. The precise role of the clitoris in female monkeys or apes is unknown and no experiments have examined the effects of clitorectomy upon sexual behaviour in non-human primates. Removal of the clitoris and portions of the labia is routinely practised as a form of female circumcision in parts of Africa (e.g. Somalia and Sudan). This barbaric operation is said to reduce the pleasure of sexual intercourse for the woman and to render her more faithful to her spouse (Lightfoot-Klein 1989).

The clitoris is relatively small in the majority of anthropoid species (exceptions are shown in Fig. 9.17) and in certain cases (e.g. *Cercocebus*, *Mandrillus*, and Pan) it is embedded within the pubic lobe of the sexual skin when it is swollen. Perhaps, even in these cases, the pressure of the male's pubic region against the female sexual skin during copula-

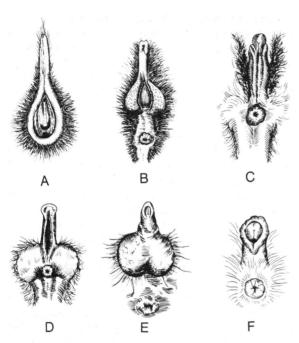

Fig. 9.17 Female external genitalia of various primates to show the position and size of the clitoris. A. *Microcebus murinus*; B. *Galago alleni*; C. *Lagothrix lagotricha*; D. *Ateles paniscus* (female); E. *Ateles paniscus* (male, for comparison with D); F. *Cercopithecus neglectus*. (Redrawn from Hill 1953, 1962, 1966.)

tion results in stimulation of the clitoris. Enlargement of the clitoris has occurred in members of four New World primate cebid genera, *Ateles*, *Brachyteles*, and *Lagothrix* (comprising the sub-family Atelinae) and to a lesser extent in *Cebus* (sub-family Cebinae). The clitoris in *Cebus capucinus* has been measured at 18.0 mm; in *Ateles belzebuth* it measures 47.0 mm. Indeed, whilst clitoral hypertrophy is pronounced in all species of the Atelinae which have been examined (descriptions of the clitoris in *Lagothrix* and *Brachyteles* are provided by Hill 1962), spider monkeys (*Ateles*) have the largest clitorides of any primate species, and the organ superficially resembles the male's penis (Fig. 9.17). However, unlike the penis, the clitoris in *Ateles* is grooved along its perineal surface, and the epithelium which lines this broad, shallow groove 'is smooth and more like a mucous membrane' (Hill 1962). There is no evidence that the clitoris plays a more significant role in sensory feedback during copulation in *Ateles* than in other primates, nor is it erected or used in any form of visual display (Klein 1971; Klein and Klein 1971; Eisenberg 1976; Eisenberg and Kuehn 1966; Van Roosmalen and Klein 1988). A more likely functional explanation for the enlarged clitoris of *Ateles* has been provided by Klein (1971). As female spider monkeys travel through the trees they

frequently deposit urine on the branches and foliage. Males lick and sniff areas where females have urinated, as well as handling and inspecting the female's genitalia before copulation. Klein (1971) observed that the peculiar, elongated clitoris retains and distributes urine droplets as the female moves around. The urine is voided at the base of the clitoris, flows down the shallow groove on its perineal surface, and is held by the skin folds on each side of the groove. Field studies of other Atelines also indicate that chemical cues are important in communicating the female's reproductive status (*Brachyteles*: Milton 1985a; Strier 1986; Nishimura *et al.* 1988), although this must remain an inference since no experimental studies have examined the role of olfactory cues in sexual attractiveness in this group of monkeys. All New World monkeys possess specialized cutaneous glands and many employ urine for scent-marking purposes (Epple and Lorenz 1967; Epple *et al.* 1993; Coimbra-Filho and Mittermeier 1981; Mittermeier *et al.* 1988). Why, then, has selection favoured clitoral hypertrophy for olfactory communication in *Ateles*, *Brachyteles*, and *Lagothrix*, but not in *Saguinus*, *Saimiri*, *Cacajao*, or a host of other New World monkeys? The key to this problem lies in a consideration of the social organizations and mating systems of the Atelinae. The social organization of these monkeys has strong parallels with the fusion-fission societies of chimpanzees and bonobos. *Ateles* live in communities of up to 30 individuals, but the animals rarely move in a single, cohesive group. Instead, the group fragments into sub-groups, averaging 3–8 individuals in the case of *A. geoffroyi* (Carpenter 1935; Eisenberg and Kuehn 1966; Freese 1976). Sub-groups are flexible in composition and can contain only adult females, only adult males, or a combination of both sexes as well as juveniles and infants. Interestingly, adult females usually outnumber adult males in the population (e.g., in *A. geoffroyi* the ratio of adult males to adult females is 1:1.8 (Cant 1978) or 1:2.2 (Coelho *et al.* 1976)). Similar field data exist for other species in this genus (Van Roosmalen and Klein 1988). When female spider monkeys enter mating condition they become intensely sexually active for approximately 8–10 days (e.g. in *A. paniscus*: Van Roosmalen and Klein 1988). During this period females roam widely, and they respond to the long-calls of distant males by approaching them and copulating with a variety of partners. The efficient distribution of attractive urinary cues by the pendulous clitoris might be adaptive, therefore, in broadcasting olfactory signals as widely as possible and thus attracting males from the various sub-groups. This mechanism, coupled with the heightened proceptivity of the female, encourages multimale matings and sperm competition.

It is noteworthy that sub-grouping and a fusion–fission type of social organization have also been reported in *Brachyteles* (Milton 1984, 1985a) and in *Lagothrix* (*L. Lagotricha*: Ramirez 1988). In all these species a multimale–multifemale mating system prevails; males have large testes relative to body weight and sperm competition is presumed to occur. Milton (1985a) records that as female *B. arachnoides* enter reproductive condition, they range widely, vocalize, and scent-mark. Males congregate in order to copulate repeatedly with individual females during this period. The thickened and elongated clitoris of *Brachyteles*, as in *Ateles*, probably represents an extreme case of sexual selection to enhance the efficiency of urine-marking and to attract males. The situation is analogous to the evolution of sexual skin as a visual advertisement to attract males, and to promote sperm competition, in chimpanzees and various catarrhine monkeys. New World monkeys have a long and unbroken history of arboreal evolution, however, whereas the Miocene ancestors of Old World monkeys included terrestrial forms. One extreme arboreal specialization found in *Ateles*, *Brachyteles*, and *Lagothrix*, the prehensile tail, finds no counterpart in the Old World. Further, many New World monkeys retain a well-developed vomeronasal organ and accessory olfactory system, an inheritance from their prosimian ancestors, which is lacking in catarrhine primates, except during foetal life, or as a rare atavism in adult animals. Under these circumstances, it is understandable that sexual selection might favour specializations to enhance olfactory communication, rather than the production of sexual skin swellings.

Among non-primate mammals, the most extreme example of clitoral hypertrophy occurs in the spotted hyena (*Crocuta crocuta*), in which the female's clitoris is the same size as the male's penis. Moreover, the clitoris is capable of erection and a false scrotum is present (Matthews 1939; Kruuk 1972). Female spotted hyenas are larger than males and occupy dominant positions in the social group or clan. Kruuk (1972) discovered that hyenas within the clan 'greet' one another by means of a complex display involving mutual visual inspection, sniffing, and licking of the genitalia. He suggested that the enlarged clitoris and false scrotum of the female has evolved to facilitate recognition and acceptance between conspecifics during these greeting displays. Gould (1983), however, points to an alternative route for evolution of the enlarged external genitalia in female spotted hyenas. In females of this hyena species circulating concentrations of androgens are higher than in adult males (Racey and Skinner 1979). Gould connects this to the large body size and socially dominant nature of females in the clan

system. Androgenic stimulation to produce larger and more aggressive females has also caused the androgen-sensitive external genitalia to hypertrophy. The enlarged clitoris and false scrotum are by-products of this evolutionary process. Gould (1983) suggests that, once genital enlargement had occurred in female hyenas, the clitoris and false scrotum were available for use in greeting displays but that they did not evolve *a priori* for such purposes. Gould's hypothesis has been criticized, however, and is not accepted by all workers who study spotted hyenas (e.g. East *et al.* 1993).

As regards clitoral hypertrophy in spider monkeys and other atelines it is unlikely that the evolutionary mechanisms involved are the same as those proposed by Gould (1983). Firstly, females in the family *Atelinae* are not larger than males; sexual dimorphism in body weight shows a male bias in this group. Secondly, females are not strikingly aggressive or dominant over males, although female spider monkeys do show some propensity to lead sub-groups during daily movements through the forest (Van Roosmalen and Klein 1988). There is no evidence that adult females are exposed to high levels of circulating androgens during development or in adulthood. Certainly, there is a need for endocrinological studies to verify this point. However, I suggest that in the case of *Ateles*, *Brachyteles*, and *Lagothrix*, sexual selection has operated directly upon the clitoris to favour its enlargement for scent-marking and sexual communication, and not as a side effect of androgenic stimulation to enhance aggressiveness and social dominance. With regard to the physiological mechanisms involved, several possibilities might be suggested. The clitoris might be more sensitive to circulating androgens during foetal development due to possession of more receptors for the steroids or a greater enzymatic (5-alpha-reductase) capacity to convert testosterone to dihydrotestosterone (DHT). DHT is responsible for stimulating penile growth during development in males and would likewise be expected to promote clitoral development. It is also possible that other hormones (e.g., growth hormone) might promote clitoral hypertrophy. This would result in specific effects upon the clitoris, rather than global effects involving heightened levels of circulating androgens during foetal development or in adulthood.

The vagina

It is possible that the anatomy and physiology of the vagina have been affected by sexual selection, so that, as for the penis, consistent differences might be identifiable between species belonging to different mating systems. Vaginal length could provide a useful criterion, as this measurement is highly variable between species and between genera (Table 9.4). However, once the effects of body size and phylogenetic factors have been controlled for, there seems to be little correlation between the mating system and residual vaginal length (Dixson and Purvis, in press). Relatively few species have been measured, however, and even on the basis of the small sample listed in Table 9.4 one exception emerges. Genera in which adult females exhibit

marked swelling of the sexual skin at mid-menstrual cycle have long vaginae. We have seen that sexual skin swellings are found in catarrhines, such as some of the macaques, baboons, the talapoin, red colobus monkey, and chimpanzee, all of which have multimale–multifemale mating systems. Swellings are visually sexually arousing to males (Girolami and Bielert 1987), serving to attract multiple partners and thus to encourage the likelihood of sperm competition (Harvey and May 1989). Swellings also cause marked elongation of the entrance to the vagina during the period when conception is most likely to

Table 9.4 Vaginal and uterine lengths in adult female primates

Species	Vaginal length (mm)	Number measured	Sources	Uterus and cervix length (mm)	Number measured	Sources
Lemur catta (P)	46.0	1	Hill 1972	ND	–	–
Cheirogaleus major (P)	15.0	1	Petter Rousseaux 1964	ND	–	–
Propithecus verreauxi (P)	44.4	1	Eckstein 1958	ND	–	–
Callithrix jacchus (M)	22.5	3	Dixson 1986a	11.2	1	Eckstein 1958
Alouatta seniculus (M?)	35.0	1	Hill 1962	26.0	1	Hill 1962
Lagothrix lagotricha (P)	27.0	?	Hill 1960	ND	–	–
Cacajao melanocephalus (?)	47.0	1	Hill 1960	ND	–	–
Saimiri sciureus (P)	12.0	?	Hill 1960	19.75	?	Hill 1960
Procolobus verus (P)	ND	–	–	35.0	1	Eckstein 1958
Cercopithecus mitis (M)	44.0	1	Hill 1966	44.0	1	Hill 1966
C. neglectus (?)	45.0	1	Hill 1966	31.0	1	Hill 1966
C. cephus (M)	53.0	1	AFD	45.0	1	Hill 1966
C. ascanius (M)	40.0	1	Hill 1966	32.0	1	Hill 1966
Miopithecus talapoin (P)	25.0	1	Hill 1966	10.0	1	Hill 1966
Erythrocebus patas (M)	38.0	1	Hill 1966	28.0	1	Hill 1966
Cercocebus albigena (P)	73.0 (SW)	1	AFD	ND	–	–
C. torquatus (?)	32.0	1	Hill 1974	42.0	1	Hill 1974
C. aterrimus (?)	ND	–	–	22.0	1	Hill 1974
Mandrillus sphinx (P)	92.0 130.0 (SW)	2 1	AFD	ND	–	–
M. leucophaeus (P)	80.0	1	Hill 1970	41.0	1	Hill 1970
Papio hamadryas (M)	ND	–	–	54.5	1	Hill 1970
P. ursinus (P)	ND	–	–	73.0	1	Hill 1970
Theropithecus gelada (M)	40.0	1	Hill 1970	42.5	1	Hill 1970
Macaca nemestrina (P)	24.0	1	Hill 1974	40.0	1	Hill 1974
M. nigra (P)	39.0	1	Hill 1974	21.0	1	Hill 1974
M. fascicularis (P)	24.5	1	Hill 1974	40.0	1	Hill 1974
M. mulatta (P)	63.0	12	AFD	35.0	2	Eckstein 1958
M. arctoides (P)	55.0	1	AFD	49.5	1	Hill 1974
Hylobates hoolock (M)	33.0	1	Matthews 1946	26.0	1	Matthews 1946
H. Lar (M)	ND	–	–	26.0	1	Eckstein 1958
Pan troglodytes (P)	126.0 169.0 (SW)	19 15	Dixson and Mundy 1994	63.5	4	Eckstein 1958 Dixson and Mundy 1994
Gorilla gorilla (M)	90.8	1	AFD	101.0	1	Eckstein 1958
Homo sapiens (M)	110.0	?	Bancroft 1989	71.0	2	Eckstein 1958

ND=no available data; AFD=author's previously unpublished data; SW=measurement made when sexual skin fully swollen; M=females usually mate monandrously; P=females usually mate polyandrously.
From Dixson in Purvis, in press.

occur (Dixson and Mundy 1994). Although the data are limited, it seems that elongated penes are also found in those species in which females possess swellings. Hill (1974) commented upon this correlation between male and female genital morphologies with respect to the mangabeys (genus *Cercocebus*). Measurements of penile and vaginal length in the chimpanzee confirm that the penis is elongated and that males must vary in their ability to contact the female's os cervix during intromissions at mid-menstrual cycle. Vaginal depth increases by 50% in some females at the height of swelling (Fig. 9.18). The elongated and filiform penis of the chimpanzee might thus have co-evolved with the female's swelling, its function being to facilitate placement of the copulatory plug as close as possible to the female's os cervix. Males with shorter penes would suffer a reproductive disadvantage.

An additional function of sexual skin swellings in female catarrhines might therefore be a mechanical one; to extend the operating depth of the vagina and to impose sexual selection pressure upon males via cryptic female choice (Eberhard 1985, 1996). Penile elongation and distal morphologic specializations might result from these female-imposed selective

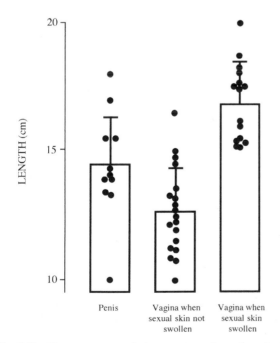

Fig. 9.18 Measurements of the erect penis and vaginal length (when the female's sexual skin is not swollen and when it is at full swelling) in chimpanzees (*Pan troglodytes*). Data are individual values with the overall mean and standard deviation indicated by the histogram and bar. (Redrawn and modified from Dixson and Mundy 1994.)

forces. Eberhard's concept of cryptic female choice at the genitalic level serves as a valuable reminder that the anatomy and physiology of the female's reproductive tract might influence male reproductive success and that sperm competition is not solely an intrasexual contest.

Once a male has succeeded in copulating and depositing spermatozoa at the cervical os, then the gametes must survive and migrate into the cervical mucus and thence to higher regions of the female reproductive tract. Yet the vagina constitutes a potentially hostile environment for spermatozoa, its low pH being detrimental to sperm motility and survival. Vaginal pH in the human female ranges from 3.5–4.0 before coitus (Masters and Johnson 1966). However, immediately after ejaculation, vaginal pH rises to approximately 7.0 (Fig. 9.19) due to the buffering effect of seminal fluid. This buffering assists sperm survival and motility in the vagina; the effect lasts for some hours as can be seen in Fig. 9.19. It is of interest, therefore, that if normal (i.e. fertile) men experience repeated ejaculations then the ability of their semen to buffer vaginal pH is reduced. A similar phenomenon has been observed in some subfertile men. Masters and Johnson (1966) have documented cases where spermatozoa were rendered immobile soon after ejaculation despite adequate buffering of vaginal pH. They note that such effects provide 'proof positive of an obvious functioning role for the vagina in conceptive and contraceptive physiology'. Cryptic female choice might operate at the vaginal level to 'test' the ejaculates of rival males and to discourage the onward migration of poor quality gametes. Almost nothing is known concerning this possible avenue of sexual selection in non-human primates.

The cervix

The structure of the cervix and changes in cervical mucus composition during the ovarian cycle both have important effects upon sperm migration and survival. In the genus *Macaca*, all species of which are considered to have multimale–multifemale mating systems, the cervix is especially complex in structure. In *Macaca fascicularis* and *M. nemestrina*, for example, the cervical canal is tortuous and lined with thick mucosa containing deep crypts. *Macaca arctoides* possesses a relatively straight cervical canal, but a greatly hypertrophied ectocervix occurs in this species (Fig. 9.2). Histological studies of the reproductive tract in female *M. fascicularis* have revealed how spermatozoa migrate from the vagina, through

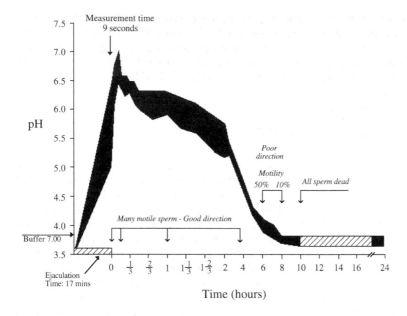

Fig. 9.19 Buffering action of human seminal fluid upon vaginal pH and its relationship with sperm motility after intercourse. (Redrawn from Masters and Johnson 1966.)

the cervix and into the uterus after copulation (Jasczak and Hafez 1972). For the first six hours after ejaculation, most spermatozoa are found in the upper vagina, in the external os of the cervix, and in the lower portion of the cervical canal. By 24 h after ejaculation sperm numbers are much reduced; most gametes have migrated through the cervix, although many are also retained in the cervical crypts and others have been removed by phagocytosis (Fig. 9.20).

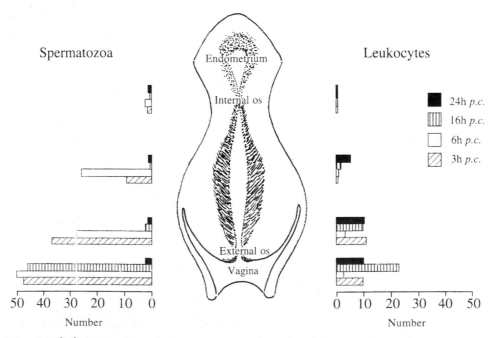

Fig. 9.20 Post-coital (PC) distribution of spermatozoa and leucocytes in the vagina and uterine cervix of female long-tailed macaques (*Macaca fascicularis*) at 3, 6, 16 and 24 h after copulation. (Redrawn from Jasczak and Hafez 1972.)

The cervix might therefore function as a reservoir for spermatozoa, releasing them into the uterus gradually after ejaculation. Whether the cervix provides an arena for sperm competition or cryptic female choice remains a matter for speculation. Not all primates with multimale–multifemale mating systems possess anatomically complex cervices—the macaques are exceptional in this regard. Other species in which females mate with multiple partners and in which sperm competition is important can, nonetheless, have a simple cervix with a straight canal. The female chimpanzee has a simple cervix, for example. However, the structure and chemical properties of cervical mucus might be more important in this regard than the shape of the cervical canal and its crypts. In macaques, Jasczak and Hafez (1972) noted that 'although spermatozoa appeared to migrate at random, they were often densely crowded in strands of mucus. Some strands of mucus containing spermatozoa disappeared into the crypts of cervical mucosa, while others continued through the internal os to the uterus'. In the human female, the properties of cervical mucus alter during the menstrual cycle and the mucus becomes more permeable to spermatozoa for several days around the time of ovulation (Tsibris 1987). Abnormal sperm motility is associated with failure to gain entry to the cervical mucus (Feneux *et al.* 1985), and morphologically abnormal gametes are often excluded (Hanson and Overstreet 1981). Those spermatozoa which are successful in migrating into the mucus might remain there for varying periods of time. Sperm numbers remain relatively constant for four hours and then decline gradually (Tredway *et al.* 1975). In exceptional cases motile sperm have been found in the cervical mucus for up to five days (Gould et al. 1984), although whether such gametes are still capable of effecting fertilization is debatable. Nonetheless, the evidence pertaining to human beings and macaque females indicates that the cervix acts both as a filtering mechanism and as a temporary reservoir for spermatozoa during their migration into the uterus (Overstreet *et al.* 1991).

The uterus

Contractions of the uterus and vagina occur during the post-ejaculatory period and contribute to rapid transport of spermatozoa to the oviducts. Such rapid transport has been described in a number of mammals, including the mouse, rabbit, cow, ewe, and human female (Table 9.5). By 5 min after coitus in the rabbit, for example, spermatozoa are present in the ampullae of the oviducts. It is often stated that spermatozoa transported in this fashion are unlikely to fertilize ova. In the rabbit, many gametes are damaged during rapid transport (Overstreet *et al.* 1991). In species which copulate before ovulation, rapidly transported spermatozoa might pass through the oviducts and into the peritoneal cavity before ova are shed. However, as Harper (1994) has pointed out 'in species such as the human, in which intercourse might occur before, after, or at the same time as ovulation, it cannot be definitively stated that the first spermatozoa do not fertilize the ovum'. It is generally held that the second, slower phase of sperm transport, involving gradual release of gametes from the cervical reservoir, is responsible for fertilization. Yet the cautionary note sounded by Harper regarding human beings also applies to other primates, and especially to many anthropoid species in which copulation is not limited to a restricted period of oestrus. We should consider the possibility that the rapid transport mechanism might even favour a male chimpanzee, stumptail macaque, or baboon which succeeds in mating last and coincident with ovulation. Whether such events influence sperm competition in primates is not known.

A number of physiological mechanisms has been proposed to account for rapid sperm transport in mammals. Contractions of the uterine myometrium might be stimulated by coitus-induced release of oxytocin (in women: Carmichael *et al.* 1987), by postaglandins in semen (e.g., in man: Templeton *et al.* 1978), or by platelet-activating factor (PAF) which is present in spermatozoa (Kumar *et al.* 1988) as well as in the uterus (e.g., in the human endometrium: Alecozay *et al.* 1989). However, the non-pregnant uterus does not contract in response to oxytocin (Kumar 1967) and it is uncertain whether either postaglandins or PAF could account for the

Table 9.5 Time taken by spermatozoa to reach the oviduct after copulation or artificial insemination

Species	Time	Region of tube
Mouse	15 min	Ampulla
Rat	15–20 min	Ampulla
Hamster	2–60 min	Ampulla
Rabbit	Several min	Ampulla
Guinea pig	15 min	Ampulla
Bitch	2 min–several hours	Oviduct
Sow	15 min	Ampulla
Cow	2–13 min	Ampulla
Ewe	6 min–5 h	Ampulla
Woman	5–68 min	Oviduct

Data are from Harper 1982, 1994.

required contractile responses. Nor is it known how spermatozoa in the uterus are transported in the correct direction to locate the uterotubal junction (reviewed by Harper 1994).

The question of orgasm in female primates was discussed in Chapter 5 (see pages 129–36). Although some authors have ascribed a role for female orgasm in sperm transport, the evidence for such effects is far from convincing (Wolfe 1991; Baker and Bellis 1993b). Nor is the rapid sperm transport mechanism referred to above dependent upon orgasmic responses. Overt signs of orgasm are lacking in various female mammals in which rapid sperm transport occurs after copulation. In the human female, small numbers of gametes are rapidly transported to the oviducts after artificial insemination (Settlage *et al.* 1973). Masters and Johnson's (1966) pioneering experiments on orgasm in the human female were also cited in Chapter 5. These experiments involved six women. A radio-opaque fluid similar in viscosity to semen was placed in the vagina and maintained in contact with the cervix by means of a contraceptive diaphragm. Occurrence of orgasm (whether as a result of intercourse or masturbation is not stated in Masters and Johnson's (1966) account) failed to cause transfer of radio-opaque material into the cervical canal or uterine cavity (Fig. 9.21). Masters and Johnson did observe a slight dilatation of the os cervix, lasting for 20–30 min, during the post-orgasmic resolution phase of the sexual response cycle. They suggest that this phenomenon, which occurs mainly in women who have not previously been pregnant, 'might passively improve spermatozoal access to the cervical canal'. Experimental studies of orgasm in relation to sperm transport have rarely been attempted in primates and there is a clear need for further work in this area.

Spermatozoa have been recovered from the uterus for up to 24 h after coitus in both the human female (Rubinstein *et al.* 1951) and in *M. fascicularis* (Vandevoort *et al.* 1989). It has been concluded that most spermatozoa have reached the oviducts by this time, although the fate of gametes which remain in the cervix is by no means certain (Harper 1994). Leucocyte numbers increase in the uterus and cervix after insemination and it seems likely that many spermatozoa are removed by phagocytosis (Pandya and Cohen 1985; Barratt *et al.* 1990). Profet (1993) has advanced the theory that menstruation serves to remove bacteria and other pathogens introduced into the female tract at copulation. This idea is based upon the notion that sperm are 'vectors of disease' and that bacteria 'regularly cling to sperm tails and are transported to the uterus'. Menstrua-

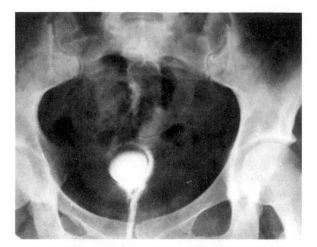

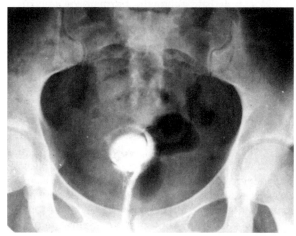

Fig. 9.21 Evidence that radio-opaque material is not transported across the cervix at orgasm in the human female. Radiographs were taken before (upper) and after (lower) occurrence of orgasm whilst radio-opaque fluid (white area) was maintained in contact with the os cervix by means of a contraceptive diaphragm. (After Masters and Johnson 1966.)

tion results in the shedding of endometrial tissue, along with potentially infectious material, and is accompanied by a massive infiltration of leucocytes throughout the uterine cavity. Profet therefore proposes that menstruation evolved as a mechanism to rid the reproductive tract of pathogens, and that menstruation whether 'overt or covert is either universal or nearly so among mammalian species'. It is argued, for example, that menstruation occurs covertly in prosimians and New World monkeys, by contrast with the more copious externally visible bleeding which typifies catarrhine species. There are some major problems with this *pathogen hypothesis*, however. Despite assertions to the contrary, menstruation of any significant volume occurs only in the Old World monkeys, apes and man. The shedding of a few erythrocytes is trivial by comparison. Prosimians show little or no signs of such activity and among New World primates only a few species exhibit slight blood loss (e.g. *Ateles geoffroyi*: Goodman and Wislocki 1935; *Cebus apella* : Nagle and Denari 1983). Given that some female prosimians and New World monkeys engage in multiple copulations involving copious amounts of semen (*Lemur catta*: Koyama 1988: *Brachyteles arachnoides*: Milton 1985a), it is surprising that they show no evidence of menstruation, whereas even monogamous and polygynous catarrhines, which copulate much less frequently, display obvious blood loss. There are numerous cases of anthropoid species copulating throughout pregnancy, when menstruations cease. As an example, the reader is referred to Fig. 4.15 (page 70) showing data, collected under field conditions, on copulatory behaviour during pregnancy in the Hanuman langur. Moreover, in the non-pregnant female, sperm which traverse the uterotubal junction and enter the oviduct, whether by initial rapid transport or by more gradual recruitment after copulation, are not subject to removal by menstruation.

It might be imagined, if the pathogen hypothesis is correct, that menstruation would be heaviest in those Old World monkeys or apes in which sperm competition is most prevalent. Yet under normal conditions a female great ape or Old World monkey is either pregnant or lactating for much of the time. Menstrual cycles are few and menstruations are comparatively light in such circumstances. There is little difference between the various great ape species in the degree of menstrual flow, despite the marked differences in their mating systems and importance of sperm competition. Many observations cited on menstruation in apes and in other Old World catarrhines derive from studies of captive animals. Females which undergo multiple non-conception cycles in captivity have heavier menstruations, just as some women have heavier periods after repeated cycles. Such observations reflect abnormal physiology and are not relevant to the evolutionary question.

The uterotubal junction and fallopian tubes

Spermatozoa do not pass freely from the uterine cavity into the fallopian tubes; the uterotubal junction acts as a filtering mechanism, so that relatively small numbers of gametes are recoverable from the distal oviduct (ampulla) in various mammals (Table 9.6). How the uterotubal junction operates as a filter is not fully understood. This region varies tremendously in its anatomical complexity in different mammals, but the notion that it acts as a simple muscular sphincter has usually been rejected (Woodruff and Pauerstein 1969). The human uterotubal junction is convoluted and measures 1.0–2.0 cm in length. Woodruff and Pauerstein reported that small inert particles could pass in either direction across this region throughout the ovarian cycle. More detailed ultrastructural studies in the pig have shown that a muscular ligament (the infundibular-cornual ligament) undergoes rhythmic contractions during the peri-ovulatory period of the ovarian cycle. This ligament is situated in the uterotubal junction and its contractions might serve to move spermatozoa into the oviduct (Persson and Rodriguez-Martinez 1990; Harper 1994). Despite the putative importance of the uterotubal junction in sperm selection and transport, I know of no detailed comparative study of its histology in primates. It would be inter-

Table 9.6 Numbers of spermatozoa ejaculated, sites of deposition, and numbers of sperm arriving in the oviduct in various mammals

Species	No. of spermatozoa (in millions) per ejaculate	Site of sperm deposition	Number of sperm in ampulla of oviduct
Mouse	50	Uterus	<100
Rat	58	Vagina	500
Rabbit	280	Vagina	250–500
Ferret	–	Uterus	18–1600
Guinea pig	80	Vagina and uterus	25–50
Bull	3000	Vagina	A few
Ram	1000	Vagina	600–700
Boar	8000	Uterus	1000
Man	280	Vagina	200

Data are from Harper 1982, 1994.

esting, for example, to compare primates which engage in sperm competition (such as the chimpanzee) with those in which matings are primarily monandrous (as in the gorilla or gibbon). A careful study, which takes account of phylogenetic factors as well as differences in sexual behaviour, might demonstrate some correlations between the structure of the uterotubal junction and mating systems. However, as in the case of the cervix, biochemical mechanisms might be more important than anatomical complexity in determining sperm transport through this region of the female reproductive tract.

Spermatozoa traversing the uterotubal junction during the slower, second phase of gamete transport do not ascend the oviduct in a steady stream. Instead, they are retained in the lower portion of the oviduct (the isthmus) and remain there before ovulation. This is the case in the cow and sheep, for example (Hunter 1988), and might also be true of mammals in general. Information on isthmic sperm storage in primates is very limited, however. Sper-

matozoa in the isthmus are closely associated with the oviductal epithelium (e.g., hamster: Smith and Yanagimachi 1991; bull: Hunter *et al.* 1991). In some marsupials, the spermatozoa are retained in crypts in the oviductal epithelium (e.g. in *Didelphis virginiana*: Rodger and Bedford 1982). These close associations between spermatozoa and the oviductal epithelium are thought to be of importance, not only for temporary storage of the gametes, but for maintenance of their viability and fertilizing capacity. *In vitro* experiments, using bull spermatozoa cultured with oviductal cells, have shown that association with oviductal cells renders sperm capable of hyperactivated locomotion, as well as maintaining their fertilizing capacity (Pollard *et al.* 1991). Little or nothing is known about potential interactions between the oviducts and male gametes in primates. Nor do we know how sperm from a number of males, stored in the isthmus of the oviduct, might interact and whether physiological conditions there might favour certain gametes over others, so provid-

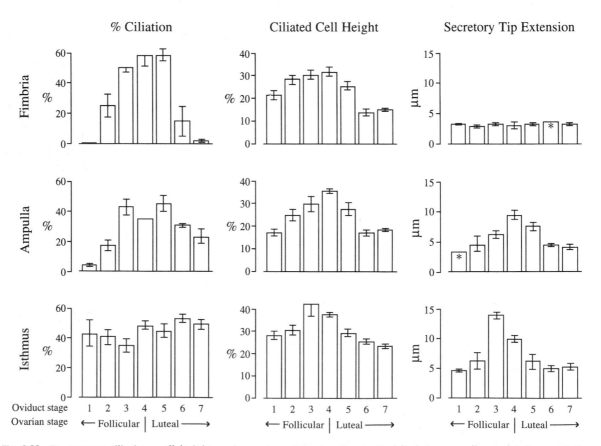

Fig. 9.22 Percentage ciliation, cell height and secretory tip extension in the fimbria, ampulla, and isthmus of the oviduct of long-tailed macaques (*Macaca fascicularis*) at different times during the menstrual cycle. Oviduct stages are; 1. preciliogenic–ciliogenic; 2. ciliogenic–ciliated; 3. ciliated–ciliogenic; 4. ciliated–secretory; 5. early regression; 6. late regresson; 7. full regression. Data are means ±S.E.M. (Redrawn from Brenner and Slayden 1994.)

ing a further avenue for cryptic female choice.

Onward migration of spermatozoa from the isthmus into the ampulla of the oviduct occurs during the peri-ovulatory period. It has been suggested that the ova, or chemical cues associated with their presence in the oviduct, might stimulate such migration of spermatozoa. However, this hypothesis is not accepted by some authorities (see Yanagimachi 1994). The oviduct contracts rhythmically during the peri-ovulatory period and these contractions propel spermatozoa towards its distal (ovarian) end (Battalia and Yanagimachi 1980). Oviductal contractions are stimulated by hormonal changes and it is likewise the case that the ciliation and secretory activity of the oviduct alters markedly depending upon the stages of the ovarian cycle. In *Macaca fascicularis*, ciliation of the oviductal epithelium increases under oestrogenic stimulation during the follicular phase of the menstrual cycle and then regresses during the luteal phase (Brenner and Slayden 1994). This cycle is most pronounced in the distal portions of the oviduct (fimbria and ampulla) and less so in the isthmus. Conversely, cycles of epithelial secretory activity are greatest in the isthmus, where sperm are stored, and least in the fimbria (Fig. 9.22). The cilia beat towards the uterine end of the oviduct and transport the ovum in this direction. Spermatozoa move in the opposite direction to the ciliary currents, both as a result of oviductal contractions and by swimming actively.

Although the events described above might bring spermatozoa into proximity with the ovum in the ampulla of the oviduct, they are incapable of penetrating the egg vestments and of entering the ovum itself unless three processes have been accomplished during their sojourn in the female's tract. These three processes, shown in diagrammatic form in Fig. 9.23, are *capacitation*, *hyperactivation*, and the *acrosome reaction*. Although these events concern the sperm itself, they must be described briefly here because all are affected by the female's reproductive physiology and hence might be subject to selection pressure via cryptic female choice.

Capacitation involves removal of a coating from the sperm plasma membrane and molecular changes which are not fully understood (Yanagimachi 1994). The phenomenon was originally described independently by Chang (1951) and Austin (1951), who showed that a period of exposure to the female tract and its fluids was necessary to enable spermatozoa to acquire fertilizing ability. In species such as the primates, in which spermatozoa are deposited in the vagina at ejaculation, capacitation might begin in the cervix (Overstreet *et al.* 1991), but the process also continues at higher levels within the female's

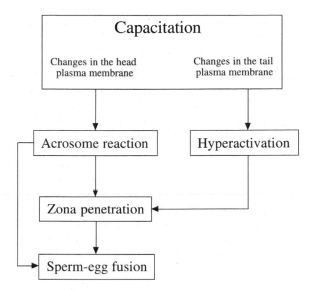

Fig. 9.23 Possible relationships between sperm capacitation, the acrosome reaction and hyperactivation. (Redrawn from Yanagimachi 1994.)

tract and might be completed in the isthmus of the oviduct (Yaganimachi 1994).

As spermatozoa begin their onward migration from the isthmus, they exhibit rapid, figure-of-eight, movements, which Yaganimachi has referred to as hyperactivation. Figure 9.24 shows hyperactivated locomotion in human spermatozoa. These vigorous movements might assist gametes to break away from the reservoir of sperm in the isthmus and migrate further up the oviduct (Demott and Suarez 1992).

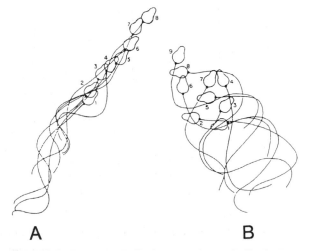

A   **B**

Fig. 9.24 Diagrams showing patterns of movement in human spermatozoa before (A) and during (B) hyperactivation. (From Yanagimachi 1994, after Morales *et al.* 1988.)

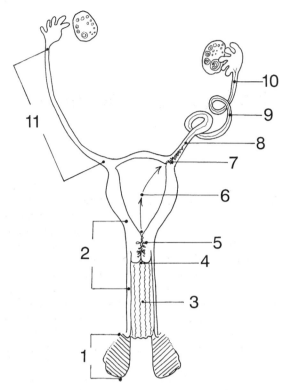

11

10

9

8

7

6

5

4

3

2

1

Fig. 9.25 Some of the factors discussed in this chapter which influence the transport, or storage, and viability of spermatozoa within the female reproductive tract. 1. Sexual skin swelling and vaginal elongation affects the male's ability to achieve maximal insertion before ejaculation. 2. Myometrial and vaginal contractions affect rapid sperm transport. 3. Low vaginal pH reduces sperm survival; semen buffers vaginal pH. 4. The physico-chemical properties of cervical mucus alter during the cycle and affect sperm transport through the cervix. 5. Cervical crypts act as reservoirs for storage and gradual release of spermatozoa. 6. Sperm are transported through the uterine cavity towards the uterotubal junction. 7. Filtering and further selection of sperm can occur at the uterotubal junction. 8. Temporary storage of spermatozoa occurs in the isthmus of the oviduct. 9. Sperm capacitation, the hyperactivation of swimming movements and the acrosome reaction occur in response to oviductal stimuli. 10. The ovum and its vestments exert selective effects upon sperm and may release chemical cues which attract spermatozoa. 11. Oviductal length may be greatest in species in which sperm competition is most intense.

Hyperactivated sperm have been found in the ampulla of the oviduct during the peri-ovulatory period in several mammals (e.g., in the rat, rabbit, sheep, and pig). However, there is little information con-

cerning the hyperactivation of sperm in any primate species, with the exception of human beings (Morales *et al.* 1988).

The third process which is essential to prepare spermatozoa for fertilization of ova is the acrosome reaction. The acrosome forms a cap covering the anterior end of the sperm head and it varies greatly in morphology among different species. In man, for instance, it forms a simple rounded structure, whereas in the rat it is crescent-shaped and pointed. The acrosome reaction involves multiple fusions between the inner and outer acrosome membranes, which lead to escape of the acrosomal contents. These changes in the sperm head render it capable of penetrating the zona pellucida surrounding the ovum and then fusing with the egg plasma membrane.

This brief account has attempted to describe some of the many ways in which the female's reproductive anatomy or physiology imposes constraints upon males during the processes of sperm deposition, storage, and transport. Sperm competition should not be viewed as a 'sprint-race' between the gametes of rival males, but rather as a race over hurdles. The hurdles are the anatomical and physiological barriers provided by the female's vagina, cervix, uterotubal junction, and oviduct, as well as by the ovum and its vestments. At all these levels, the possibility of sexual selection by cryptic female choice exists in female primates (Fig. 9.25) as in females of other mammalian species, in reptiles (Olsson *et al.* 1996) and in the numerous invertebrate groups discussed by Eberhard (1996). In order to investigate these problems effectively, a marriage is required between modern techniques of reproductive physiology and studies of evolutionary biology. Recently, Gomendio and Roldan (1993) examined oviductal length in a small sample of mammalian species and found that the occurrence of long oviducts (relative to body size) correlates with the production of high sperm numbers by males. This finding might be related to greater losses of male gametes and importance of sperm competition in such species. The sample examined was, by necessity, very small and only 2 of the 11 species included belonged to the Order Primates. Measurements of oviductal length in primates are scarce. This is just one area where relatively simple techniques could be used to generate data sufficient for the testing of evolutionary hypotheses.

10　Sexual differentiation of the brain and behaviour

Sexual behaviour is *sexually dimorphic*, in the sense that certain activities are typical of one sex, rather than the other (e.g., female monkeys present their hindquarters; males mount and make pelvic thrusting movements). However, use of the term *sexually dimorphic behaviour* does not imply that a pattern occurs exclusively in one sex. We have seen, for example, that in monkeys and apes both sexes may employ hindquarter presentation and mounting postures as part of sociosexual communication (see Chapter 6). Such *heterotypical* behaviour has been described in many vertebrates (Haug *et al*. 1991). Sexual dimorphism in behaviour is a question of degree, therefore, rather than of absolute differences between males and females (Goy and McEwen 1980). Nor are behavioural sex differences limited to sexual activities; the sexes may also differ in the frequencies or contextual aspects of aggression, parental care, play, and locomotor activity. In the human sphere, sex differences have been described in visuo-spatial perception, language development, mathematical abilities, and other cognitive functions (Maccoby and Jacklin 1994; Halpern 1987; Kimura 1992). The origin of dimorphic patterns of sexual behaviour represents part of a much larger problem, therefore. During development genetic, hormonal, and environmental factors can influence many behavioural processes. This chapter explores the role played by hormones, secreted during foetal and early postnatal life, in shaping behavioural development in male and female primates.

The organization hypothesis

The concept that testosterone (or its metabolites), secreted by the testes, in some way *organizes* neural mechanisms which control sexual behaviour has its roots in research on guinea-pigs and rats, but it is applicable also to other mammals, including the primates (Phoenix *et al*. 1959; Young *et al*. 1964; Money and Ehrhardt 1972; Baum 1979; Goy and McEwen 1980). The hypothesis has been advanced that testosterone both *masculinizes* and *defeminizes*

the developing brain (Beach 1975b), and that such effects occur (although not necessarily concurrently) during *critical periods*, the timing of which varies in different species. In the guinea pig, a long-gestation rodent, the critical period occurs *in utero* (Dantchakoff 1938; Phoenix *et al*. 1959). Treatment of the pregnant female with testosterone propionate (TP) masculinizes the genitalia of her female offspring. As adults such females show increased frequencies of mounting (indicating behavioural masculinization) as well as decreased ability to exhibit lordosis in response to oestrogen and progesterone treatment (indicating defeminization of behaviour). The guinea pig is a *precocial* species, in which the offspring are born well advanced in neurological development. The eyes of the neonate are open, for instance, and for the few species of mammals studied so far, the critical period for neural organization by testosterone occurs before the eyes open (Goy and McEwen 1980). The rat, by contrast, is an *altricial* species; the offspring are born hairless and unable to see. In the rat, the critical period extends into early postnatal life. This discovery made it possible to castrate neonatal males in order to examine their behaviour and responsiveness to hormone treatments in adulthood (Grady *et al*. 1965). Neonatal castration, especially if performed before Day 5, was effective in reducing defeminization of the brain. As adults such males displayed lordosis much more readily in response to oestrogen and progesterone treatment than was the case in normal males (Grady *et al*. 1965; Clayton *et al*. 1970). Neonatally castrated rats will mount females when treated with TP in adulthood, but deficits in intromission and ejaculation occur (Beach and Holz 1946). This might be due, in part, to impaired phallic development, since the penis does not grow normally if deprived of neonatal androgen. However, Hart (1972b) has shown that exposure of the brain to androgen during the neonatal period is necessary for the normal development of the ejaculatory response in male rats. Effects of androgen upon the neural substrates which control mounting occur before birth in the rat, however (Booth 1979). Gestation lasts for 22 days in this rodent species and the testis first acquires its ability to secrete testosterone on day 16 or 17 (Feldman and Bloch 1977). Elevated levels of testosterone are then measurable until postnatal day 10 (Resko *et al.* 1968). Prenatal androgenization of female rat foetuses (beginning on day 14 or 16 of gestation) masculinizes the genitalia, and mounting activity may then be elicited (in response to TP) from such females in adulthood. Neonatal administration of TP alone does not masculinize females' behaviour, however, whilst combined pre- and postnatal androgen treatment produces complete masculinization of sexual behaviour (including the ultrasonic vocalizations which occur after ejaculation in male rats: Sachs *et al.* 1973). Masculine patterns are not absolutely dependent upon the stimulation of phallic growth in androgenized females, since, under exceptional circumstances (oestradiol benzoate treatment accompanied by painful stimuli during behavioural testing), ovariectomized rats exhibit mounting and intromission/ejaculatory type responses in the absence of TP treatment (Barfield and Krieger 1977). The capacity to exhibit masculine patterns of behaviour therefore exists in rudimentary form in the brain of the genetic female rat. Exposure of the female brain to minute amounts of testosterone *in utero* may occur normally, due to the close juxtaposition of female and male foetuses in the uterus; this could explain some degree of latent masculinization (Baum 1979).

Research on other mammalian species has shown that the critical period for sexual differentiation of the brain and subsequent behaviour occurs prenatally in the sheep (Clarke *et al.* 1976), whilst in beagles it takes place before birth and in the neonatal period (Beach *et al.* 1972). In the beagle, exposure of five female foetuses to TP between days 24–33 post conception produced a greater defeminizing effect than in five foetuses treated between days 28–37. The degree of behavioural masculinization was the same in both groups. Such observations confirm that the masculinizing and defeminizing actions of testosterone upon behaviour do not necessarily occur concurrently and 'are not merely opposite sides of the same coin' (Beach *et al.* 1972). In the genetic male, the presence of testosterone during appropriate developmental periods masculinizes the brain and, in certain species, defeminization can also occur. In the genetic female, however, ovarian steroids are not generally thought to be involved in organization of the brain; rather it is the absence of androgen which allows brain development to progress in a fundamentally feminine direction (Baum 1979; Goy and McEwen 1980).

Sexual differentiation at the genital level

The effects of gonadal hormones upon the development of the male and female genitalia were studied and understood before hypotheses concerning their organizing effects upon the brain were formulated. Since appropriate genital development plays such an important role in sexual behaviour, and in our own species crucially affects sex assignment at birth, it is important to review the basic mechanisms at this point.

Sexual differentiation of the genitalia begins when the embryonic genital ridge develops into either an ovary or a testis. In mammals, differentiation of the testis is controlled by a testis-determining gene, situated on the Y chromosome of the male foetus. In man, this event occurs during the seventh week of gestation. In the genetic female, ovaries differentiate from the medullary tissues of the genital ridge, provided two XX chromosomes are present, rather than an XY chromosome pair as in the genetic male. Primordial genital ducts also develop during the period when the gonadal differentiation is in progress; initially both masculine (Wolffian duct) and feminine (Mullerian duct) primordia are present (Fig. 10.1). In the genetic female, the Mullerian ducts give rise to the fallopian tubes and uterus as well as the upper portion of the vagina. Proliferation of the Mullerian ducts occurs in the third month of gestation, whilst the Wolffian structures atrophy. In the genetic male the reverse is true; the Wolffian ducts differentiate to form the vasa deferentia, prostate, and seminal vesicles, whilst the Mullerian ducts undergo regression (Fig. 10.1). Secretions of the foetal testes play a key role in causing these changes to the genital duct system. In experiments on rabbits Jost (1961, 1970) demonstrated that the testes produce a substance which inhibits Mullerian duct development in the genetic male (Mullerian-duct-inhibiting substance), whilst androgenic secretions of the testes promote growth of the Wolffian derivatives. In man, various forms of hermaphroditism have been described, resulting from abnormalities of gonadal and genital duct development. In true hermaphrodites, possessing an ovary on one side and a testis on the opposite side of the body, it is possible for Mullerian duct development to predominate in association with the ovary, whilst on the contralateral side Wolffian derivatives develop due to a local effect of testicular secretions (Money and Ehrhardt 1972).

In contrast to the internal genital ducts, the external genitalia of male and female derive from the same primordia during foetal development (Fig. 10.2). Jost's experiments also demonstrated that testicular androgens are responsible for masculinization of the external genitalia, whilst, in the absence of androgenic stimulation, the genitalia differentiate in the feminine direction. In the male, androgen stimulates growth of the genital tubercle to form the corpora cavernosa and glans penis; in the female the much smaller homologue of the male organ develops as the clitoris. Fusion of the labial folds produces a scrotum in the male, whilst, in the female, the folds do not fuse and form instead the labia surrounding the entrance to the vagina. These peripheral effects of testicular androgen upon the genitalia require the metabolism of testosterone to form dihydrotestosterone (DHT); DHT is the active androgen affecting growth of the male external geni-

talia. A rare congenital anomaly results in a deficiency of the enzyme (5-alpha-reductase) required for metabolic conversion of testosterone to DHT. Human male infants born with this condition present a superficially feminine appearance of their external genitalia (Imperato-McGinley *et al.* 1974). The subsequent behaviour and gender identity of these individuals has been the subject of much controversy, as discussed later in this chapter.

The organization hypothesis and primate behavioural development

Primates typically have long gestation periods and

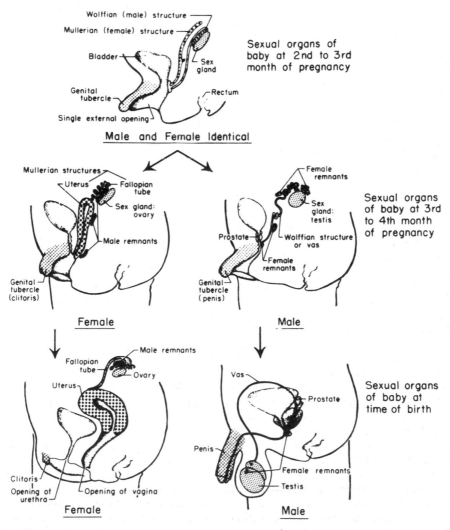

Fig. 10.1   Sexual differentiation of the genitalia in the human foetus. (After Money and Ehrhardt 1972.)

their offspring are born with the eyes open, hair present, and grasping reflexes well developed. The expectation is that the presumptive critical period for the action of androgen upon sexual differentiation of the brain should occur *in utero* in primates. However, measurements of circulating testosterone during gestation and postnatal development in macaques and human beings have shown that a postnatal surge of testosterone also occurs, peaking at about 3 months and ending by the 5th or 6th postnatal month (Fig. 10.3). The significance of this *postnatal surge* is poorly understood, whereas the evidence for prenatal organizing effects of androgen is better established, particularly in the case of the rhesus monkey (Goy 1968, 1978; Goy *et al.* 1978;

Goy *et al.* 1988; Phoenix 1974a; Pomerantz *et al.* 1986). I shall review this evidence first, and then consider the question of the postnatal testosterone surge in monkeys and human infants.

In their experiments on the rhesus monkey, Goy, Phoenix, and their co-workers adopted the same strategy as previously applied to the guinea pig; pregnant females were treated with testosterone propionate (TP) in an attempt to masculinize (or defeminize) the developing female offspring. The resulting *pseudohermaphroditic* female offspring were found to exhibit greater than normal frequencies of certain sexually dimorphic behavioural patterns, including rough-and-tumble play, play initiation, threat, and double-foot clasp mounting (Fig.

EXTERNAL GENITAL DIFFERENTIATION IN THE HUMAN FETUS

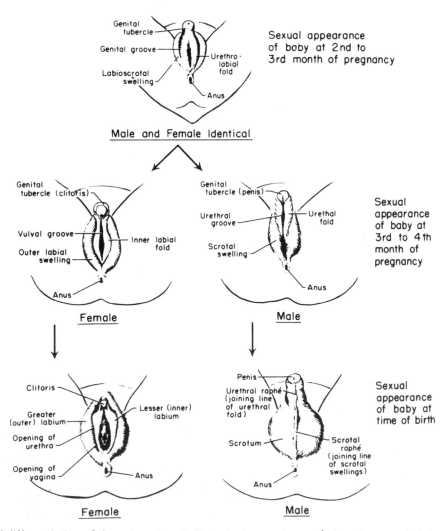

Fig. 10.2 Sexual differentiation of the external genitalia in the human foetus. (After Money and Ehrhardt 1972.)

10.4). Their performances were, on average, intermediate between those of normal genetic females and normal males, but considerable individual variability was noted as regards the behavioural outcome of prenatal androgenization. Masculinization of the external genitalia also resulted from prenatal TP treatments, so that a well-formed penis and scrotum were present at birth in pseudohermaphroditic females (Fig. 10.5). Prolonged exposure to TP *in utero* produced the most pronounced effects upon behavioural and genital masculiniza-

tion. Frequencies of rough-and-tumble play or mounting (either with peers or with mothers) were highest in females which had received TP prenatally for 35 days or more, whereas effects were minimal, or absent, in females treated for only 15 days (Goy 1981). Goy accordingly proposed that the developing brain of the rhesus monkey is sensitive to androgen for an extended period and that the hormone concerned is probably testosterone. Testosterone is the major androgen present in the foetal circulation of the rhesus monkey and occurs at higher levels in

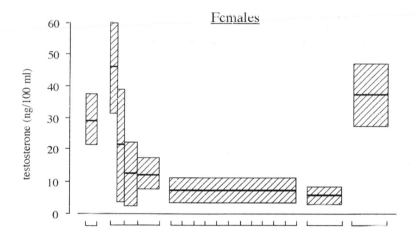

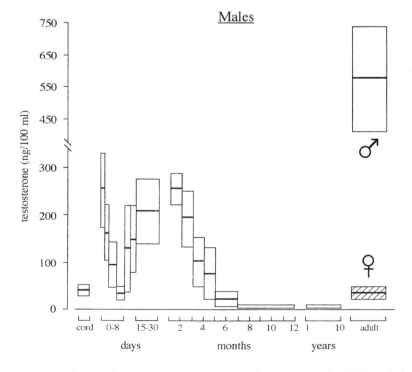

Fig. 10.3 Plasma testosterone levels in human males (open bars) and females (hatched bars) during postnatal development. Note the presence of a postnatal testosterone surge in infant males. Data are means (±S.D.). Redrawn from Bancroft 1989; after Forest *et al*. 1976.)

males than females from day 100 of gestation, the earliest time at which blood samples can be obtained (Resko 1970a, 1974; Goy and Resko 1972).

Pseudohermaphroditic female rhesus monkeys vary as regards the degree to which their external genitalia are masculinized by prenatal TP treatment (Goy *et al.* 1978). Closure of the urogenital groove occurs, as in a male foetus, so that no vaginal opening remains and a well developed penis and scrotum are present in some cases. Goy (1981) concluded that '15 mg of TP per day for 55 or more days beginning between day 40 and day 45 of gestation induced penises and scrotums in females that were indistinguishable from corresponding structures in the intact males at birth'. These peripheral

effects of TP create problems of interpretation when the behaviour of pseudohermaphroditic females is considered. Firstly, several differences have been documented in the behaviour of rhesus monkey mothers towards male infants, as opposed to females. Mothers inspect the genitalia of male offspring more frequently, for example, and are sometimes less restrictive towards male babies than towards females. It is possible that pseudohermaphroditic females, with fully masculinized external genitalia, might evoke maternal responses more appropriate for the nurturing of male offspring. This is unlikely to effect the masculinization of behaviour in pseudohermaphroditic females, however. Thus, pronounced behavioural masculin-

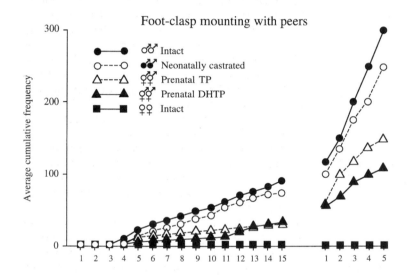

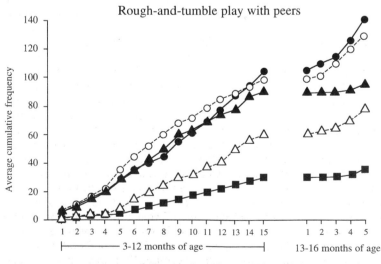

Fig. 10.4 Cumulative average frequencies of foot clasp mounting (upper) and rough-and-tumble play (lower) during the first 16 postnatal months in infant rhesus monkeys. Subjects are: normal males, neonatally castrated males, normal females, and female pseudohermaphrodites, exposed prenatally to either testosterone propionate (TP) or dihydrotestosterone propionate (DHTP). (Redrawn and modified from Goy 1978.)

Fig. 10.5 A pseudohermaphroditic infant female rhesus monkey (*Macaca mulatta*) showing masculinization of the external genitalia as a result of exposure to testosterone propionate during foetal development. (Author's drawing from a photograph by Dr Robert Goy and Dr Charles Phoenix.)

ization occurred after exposure to dihydrotestosterone propionate (DHTP) *in utero* (Fig. 10.4) even though the dosage of DHTP had less effect upon genital masculinization than did TP (Goy *et al.* 1978; Goy 1981). The results of these and more recent studies indicate that behavioural masculinization is independent of genital masculinization in pseudohermaphrodites (Goy *et al.* 1988). A second, more important, problem concerns the possible effects of genital masculinization upon the behavioural potential of pseudohermaphroditic females in adulthood. Because a vaginal opening is lacking in these animals, normal copulation and expression of female sexual receptivity is impossible. Further, the anomalous genitalia of these females might affect the responses shown towards them by male and female conspecifics in social and sexual contexts. Some adult pseudohermaphroditic females display masculine copulatory patterns in response to TP treatment (Eaton *et al.* 1973b). They mount other females more frequently during laboratory pair-tests; one of the seven pseudohermaphrodites studied by Eaton

et al. (1973b) also intromitted, made shallow pelvic thrusting movements, and ejaculated (body tremor and emission occurred). This female, as well as two other pseudohermaphrodites, 'frequently masturbated to ejaculation in their home cages'. To what extent such results reflect hormonal effects upon the brain and upon masculine sexual preferences, as distinct from effects upon the genitalia and genital sensitivity, is hard to assess. The problem of *gender-identity* (an individual's perception of his/her own masculinity or femininity) is also raised by such studies. This concept was discussed in Chapter 5 and related to the broader problem of self awareness in man and other primates. Since great apes are known to exhibit self awareness in the context of examining their own images in a mirror, there is no reason to deny that they may also be aware of their gender. Gender identity is probably not unique to human beings, therefore. Monkeys, however, do not display self awareness during mirror tests and probably also lack a homologue of human gender identity. If this is so, then the behaviour of pseudohermaphroditic female rhesus monkeys is probably not due to altered gender identity, resulting from the presence of anomalous (masculinized) genitalia.

Whilst good evidence exists for organizational effects of testosterone, resulting in masculinization of behaviour in rhesus monkeys, there is reason to believe that behavioural defeminization does not occur in the way described for rodents or beagles. Oestrogen-treated, adult pseudohermaphroditic female rhesus monkeys display patterns of proceptivity and receptivity which do not differ significantly from those exhibited by normal females (Phoenix *et al.* 1983). Intromission is impossible in the absence of a vagina, but females permit male partners to mount and thrust during pair-tests. They also present to males, and although some of these displays may be motivated by sociosexual factors (e.g., submission in agonistic contexts), Phoenix *et al.* comment that 'their present to contact ratio, prox rate, and low rate of aborting male invitations to copulate argue for retention of the capacity to respond as a female to oestrogen stimulation'. In order to study the sexual receptivity of the female pseudohermaphroditic rhesus monkey in greater detail, Phoenix *et al.* (1984) carried out an ambitious experiment to construct an artificial vagina in one subject and thus render normal copulation possible. The female selected was 17 years old and had previously been found to exhibit greatest masculinization of behaviour in response to TP in adulthood (this pseudohermaphrodite exhibited intromission and ejaculation during oppositely-sexed pair tests, as described above). Figure 10.6 shows the genital ap-

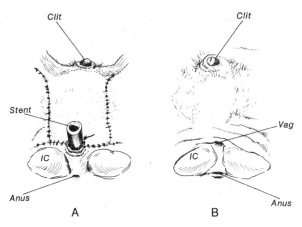

Fig. 10.6 Surgical construction of a vagina in an adult pseudohermaphroditic female rhesus monkey (*Macaca mulatta*). A. Just after the operation. B. Once healing has occurred. Vag=vaginal opening; IC=ischial callosity; Clit= enlarged clitoris (not removed). (Author's drawings, after Phoenix *et al*. 1984.)

pearance of this female subsequent to vagino-plasty (a plug or stent was placed in the vaginal opening in order to maintain patency). The penis was not removed and so the female still displayed an anomalous genital appearance, which might have influenced the behaviour of males during pair-tests. This female was oestrogen-treated and tested with nine male partners at various times after surgery. Receptivity was apparently normal, but only one male succeeded in gaining intromission (on two occasions) perhaps due to anatomical problems rather than to adverse responses (e.g., refusals) by the female partner. Phoenix *et al*. (1984) therefore concluded that 'rhesus females, treated prenatally with testosterone, have the capacity to display behaviour characteristic of either sex. Whether any defeminization occurs needs to be explored more fully'. These authors stressed the differences between the rhesus monkey and some other mammals, since psychosexual differentiation can occur without defeminization in these monkeys, a finding which also 'bears important implications for sexuality in human beings and needs further study'.

Before considering the problem of psychosexual differentiation in man, two further points must be emphasized regarding the experiments on pseudo-hermaphroditism in the rhesus monkey. Firstly, although behaviour is masculinized in female rhesus monkeys exposed prenatally to androgen, the secretion of luteinizing hormone remains normal in such animals. Menarche is delayed (occurring at 36.8 months on average in pseudohermaphrodites as compared with 29.2 months in normal females).

However, as adults they exhibit ovulatory menstrual cycles. In the rat, by contrast, treatment of the developing female with testosterone propionate during the critical period results in a tonic rather than a cyclical pattern of LH secretion in adulthood, and ovulation is abolished (Barraclough 1966). Such androgenized female rats do not show an LH surge in response to an oestrogen challenge, indicating a failure of the *positive feedback* mechanism which controls ovulation in normal females (Finn and Booth 1977; Booth 1979). In the rhesus monkey, prenatal androgenization does not result in anovulatory sterility and nor does a single large dose of testosterone administered to the newborn female disrupt ovarian cyclicity in adulthood, in the manner described for rats (Treolar *et al*. 1972). In contrast to rodents, adult male rhesus monkeys are capable of displaying an LH surge in response to an oestrogen challenge. Treatment of castrated male rhesus monkeys with oestradiol causes an LH surge (Karsch *et al*. 1973), whilst implantation of ovarian tissue into such males results in cyclic changes in steroid secretion. In the rat, ovarian transplants fail to induce cyclical activity unless castration is performed soon after birth, in order to prevent masculinization or defeminization of LH-control mechanisms (Harris 1965). In rhesus monkeys, it seems that neither behaviour nor gonadotrophin regulation is defeminized by androgen during development. A second major difference between the rhesus monkey and some sub-primate mammals concerns the fact that a *non-aromatizable* androgen (dihydrotestosterone) causes behavioural masculinization in the female in the same manner as TP (Goy 1981; Goy and Robinson 1982). In the rat, however, testosterone acts as a *prohormone* and is converted (by the enzyme aromatase) to oestrogen in the brain before exerting its effects upon masculinization and defeminization of behaviour (Baum 1979). The relevance of the *aromatization hypothesis* to the hormonal control of sexual behaviour in adult male primates will be assessed in greater detail in Chapter 13 (see pages 397–8).

The organization hypothesis and psychosexual differentiation in man

Studies of people whose mothers were treated with steroids during pregnancy and studies of foetal endocrinological abnormalities (such as congenital adrenal hyperplasia, testicular feminization, and Imperato-McGinley syndrome), have produced information about the effects of prenatal hormones upon gender-related behaviour in man. A spectrum of gender-related behavioural and cognitive differences has been described. Men perform better than women on tasks requiring spatial skills (Maccoby and Jacklin 1974; Halpern 1987). Studies of rats indicate that such a sex difference also occurs in regard to visual orientation during the performance of maze tasks (Williams and Meck 1991). Castration of the newborn male rat interferes with the development of such behaviour, whilst treatment of female rat pups with oestradiol benzoate masculinizes their spatial abilities (Williams *et al.* 1990). By contrast, women tend to be better than men at tests which require verbal fluency, such as finding words which all begin with the same letter, or at tasks involving perceptual speed, such as sorting items which match one another (Halpern 1987; Kimura 1992). Men are, on average, better at aimed-throwing of objects and at certain types of mathematical reasoning (Kimura 1992). There is also some evidence for greater asymmetry of cerebral function in the control of speech in men, whilst language functions are more equally represented in both hemispheres in the female brain (McGlone 1980). This latter hypothesis has been disputed, however (Fairweather 1982). Nor is it clear how far the development of these various cognitive sex differences might be influenced by prenatal (or postnatal) hormones. Some gender differences in behaviour, such as preferences for athletic pursuits, rough-and-tumble activities, playing with toys such as trucks rather than dolls, probably are influenced by prenatal testicular hormone secretion. The problem of controlling for effects of social environment upon the development of such sexually dimorphic behaviours is especially difficult. Moreover, where human sexuality is concerned, problems of interpretation are immense and different methods of study must be used than in work on rodents or monkeys. Human beings clearly do possess a concept of gender identity and it is well advanced by the time a boy or girl reaches three years of age (Money and Ehrhardt 1972). Children have self awareness of their individuality as male or female (or ambivalent) and this fundamental sense of gender identity should also fit with their 'gender role' in society. Gender role may be defined as 'everything that a person says and does, to indicate to others or to self, the degree to which one is male or female or ambivalent' (Money 1991). As discussed in Chapter 5, the gender role adopted by an adult in society is not always concordant with erotosexual preferences, however. For example, some male homosexuals adopt the dress and social role typical of heterosexual men whilst others may dress in public as 'flaming queens' (Money 1991). In what follows, an attempt is made

to review current knowledge of how the prenatal hormonal environment influences subsequent gender-related behaviour and gender identity/role in man. The specific question of homosexuality is then discussed at greater length in a separate section.

Congenital adrenal hyperplasia

Congenital adrenal hyperplasia (CAH) is due, in most cases, to 21-hydroxylase deficiency which affects steroidogenesis by the adrenal glands. In the most serious (*salt wasting*) form of the condition, cortisol secretion is greatly reduced and aldosterone secretion is also affected. In the *simple virilizing* form of CAH the adrenocortical deficiency is less pronounced. Both types of CAH are characterized by hypersecretion of adrenocorticotrophin (ACTH) which, in the absence of sufficient negative feedback by cortisol, stimulates the adrenal cortex to secrete androgens in abnormally large amounts. Depending upon the timing and extent of androgen secretion during foetal development, the external genitalia of the female infant may then become masculinized, so that an enlarged clitoris and empty scrotum are present in the newborn (Fig. 10.7).

Genital incongruencies require surgical correction and it is also necessary to treat the affected individuals with glucocorticoids to combat cortisol deficiency. Salt-wasting babies are especially at risk and require mineralocorticoid replacement in addition to glucocorticoids. There have been many reports concerning behavioural development in girls affected by congenital adrenal hyperplasia. Table 10.1 shows data from Money and Ehrhardt (1972) comparing tomboyish behaviour, energy expenditure in recreation, aggression, and clothing preferences in CAH girls—girls exposed to masculinizing effects of exogenous synthetic progestagens *in utero* and girls

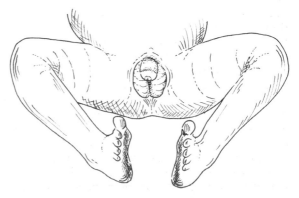

Fig. 10.7 Human female infant with congenital adrenal hyperplasia, showing enlarged clitoris and labia due to androgen exposure *in utero*. (Author's drawing.)

Table 10.1 Tomboyishness, energy expenditure, and clothing preferences in girls with congenital adrenal hyperplasia (CAH), Turner's Syndrome (TS), and progestin-induced hermaphroditism (PIH), compared with matched controls (MC)

Behavioural signs	PIH vs MC	CAH vs MC	TS vs MC
Evidence of tomboyism			
1. Known to self and mother as tomboy	$P \leq 0.05$	$P \leq 0.01$	NS
2. Lack of satisfaction with female sex role	NS	$P \leq 0.05$	NS
Expenditure of energy in recreation and aggression			
3. Athletic interests and skills	$P \leq 0.05$	$P \leq 0.10$	$P \leq 0.05$
4. Preference of male versus female playmates	$P \leq 0.05$	$P \leq 0.01$	NS
5. Behaviour in childhood fights	NS	NS	$P \leq 0.05$
Preferred clothing and adornment			
6. Clothing preference, slacks versus dresses	$P \leq 0.05$	$P \leq 0.05$	NS
7. Lacking interest in jewellery, perfume and hair styling	NS	NS	$P \leq 0.05$

Data from Money and Ehrhardt 1972.
NS = no statistically significant difference.

exhibiting Turner's syndrome. Girls with CAH tend to perceive themselves, or are recognized by their mothers, as being *tomboyish*. A significant number express dissatisfaction with a female gender role, express a preference for male playmates, and some preference for athletic pursuits involving higher energy expenditure. This latter finding might equate in some way to increased rough-and-tumble activities in androgenized female monkeys.

Girls whose mothers were treated with synthetic progestagens during pregnancy also exhibit some masculinization of energy expenditure and clothing preferences as well as tomboyish behaviour (Table 10.1). In Turner's syndrome, which is characterized by a 45 X karyotype in many individuals, the gonads fail to differentiate and remain as primitive streaks.

The developing foetus is not exposed to gonadal hormones and the resulting baby appears genitally female but fails to develop normal feminine characteristics because of the absence of ovarian steroids at puberty. Table 10.1 shows that girls with Turner's syndrome are less athletic than controls, less aggressive, and less interested in jewellery or perfumes.

As far as expectations and imagery of adult gender role in CAH girls are concerned, Money and Ehrhardt also found marked effects (Table 10.2). Girls with CAH were, on average, more interested in career prospects and less concerned about marriage than controls; they also showed less interest in caretaking of younger children and entertained fewer fantasies about having their own children in the future. No significant effects were noted in girls with

Table 10.2 Anticipation and imagery of maternalism, marriage and romance in girls with congenital adrenal hyperplasia (CAH), Turner's syndrome (TS) and progestin-induced hermaphroditism (PIH), compared to matched controls (MC)

Anticipation and imagery	PIH vs MC	CAH vs MC	TS vs MC
Maternalism			
1. Toy cars, guns, etc. preferred to dolls	$P \leq 0.05$	$P \leq 0.05$	NS
2. Juvenile interest in infant care-taking	NS	$P \leq 0.001$	NS
3. No daydreams or fantasies of pregnancy and motherhood	NS	$P \leq 0.05$	NS
Marriage			
4. Wedding and marriage not anticipated in play and daydreams	NS	$P \leq 0.05$	NS
5. Priority of career versus marriage	$P \leq 0.05$	$P \leq 0.05$	NS
Romance			
6. Lack of heterosexual romanticism in juvenile play and daydreams	NS	NS	NS
7. Lack of adolescent (age 13–16) daydreams of boyfriend and lack of dating relationships	NS	NS	NS
8. Lack of homosexual fantasies reported	NS	NS	NS

Data from Money and Ehrhardt 1972.
NS = no statistically significant difference.

Turner's syndrome, by comparison with controls (Table 10.2). They showed an essentially feminine gender identity in the absence of foetal gonadal hormones and in the absence of the second X chromosome (Money and Ehrhardt 1972).

Although these results are consistent with masculinizing effects of prenatal androgen in CAH girls, some reservations and criticisms may also be expressed. Exposure to androgens during foetal life does not necessarily reduce parental interest in primates, as in the case of certain monogamous non-human primates where males as well as females actively carry and care for their offspring (e.g., marmosets and owl monkeys: Dixson 1983a). Slipjer (1984) has also pointed out that many children who have the condition of congenital adrenal hyperplasia are unwell and that chronic illness (due to diabetes for example) can result in masculinization of behavioural development in some girls. Girls with the chronic salt wasting form of CAH show greater behavioural masculinization than simple virilizers; this might be linked to the fact that chronic salt wasting girls are subject to illness due to dehydration. In their studies of CAH girls Hines and Kaufman (1994) failed to demonstrate greater frequencies of rough-and-tumble play in their subjects compared with a control group. The expected sex difference in rough-and-tumble activities was found, boys showing higher frequencies of these activities than normal girls (Fig. 10.8). Moreover, CAH girls did show different toy preferences (for trucks rather than dolls) when compared with the control group. Hines and Kaufman observed behaviour in a laboratory setting rather than asking the subjects to express their preferences. The authors point out that social constraints might have reduced the performances of CAH girls in rough-and-tumble play and suppressed their initiation of play with boys rather than girls in the experimental set-up. This seems entirely possible and underscores the important effect of socialization (e.g., parental influences) upon behavioural development in CAH subjects, as well as any effect of prenatal androgenization.

As adults, women with CAH experience normal menstrual cycles; cyclical secretion of gonadotrophins and ovarian steroids occurs, just as previously described for pseudohermaphroditic fe-

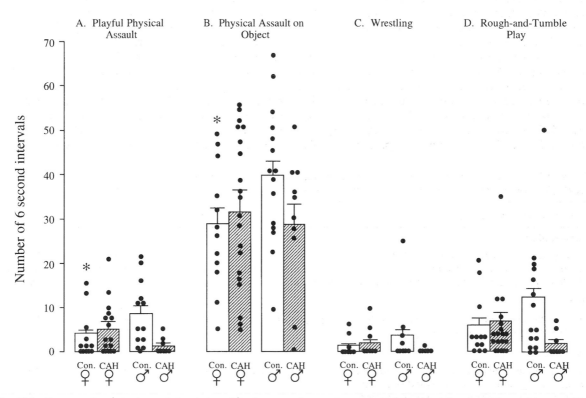

Fig. 10.8 Frequencies (number of 6-s intervals) of aggressive and rough-and-tumble play in girls and boys with congenital adrenal hyperplasia (CAH) as compared to non-CAH controls. Data are individual values, with means (+S.E.M.) indicated by histograms and bars. *P < 0.05 comparing control groups. (Redrawn from Hines and Kaufman 1994.)

male rhesus monkeys. Although studies of the sexual behaviour and sexual orientation of such women have produced variable findings, a number of studies has reported higher than expected frequencies of a homosexual or bisexual gender identity in CAH women (Ehrhardt *et al.* 1968a; Ehrhardt and Meyer-Bahlburg 1979; Money *et al.* 1984; Dittmann *et al.* 1992; Zucker *et al.* 1996). Some of these findings will be considered in greater detail below, in the section which explores homosexuality.

Androgen-insensitivity in males: the testicular feminization syndrome

In the *androgen insensitivity syndrome* (AIS) androgen receptors are virtually absent and hence the developing male foetus is unable to respond in the normal way to testosterone or other androgenic metabolites secreted by the testes. Under these conditions the Wolffian ducts and external genitalia fail to differentiate normally, so that the newborn has an essentially feminine genital appearance and is usually designated as a girl and raised as such. The vagina may be small and blind-ended, however, and these individuals lack a uterus and cervix. The testes remain endocrinologically functional and secrete a certain amount of oestrogen, as well as androgens during puberty. Oestrogen secretion may be sufficient to promote breast development but, in the absence of ovaries or normal development of the uterus, menstrual cycles fail to occur (Money and Ehrhardt 1972). As adults, males with AIS have a typically feminine appearance (Fig. 10.9) and feminine gender identity/role, with preferences for marriage and raising children. Two of the ten subjects reviewed by Money and Ehrhardt had successfully adopted children. Eighty per cent of subjects were exclusively heterosexual and none showed any incidence of lesbianism in adulthood. Thus, despite an XY genotype and differentiation of testes secreting androgens, AIS men develop psychosexually as women. They are sensitive to the feminizing effects of testicular oestrogens at puberty, with consequent development of a feminine body shape and breast size. However, if oestrogens are secreted by the testes during foetal life they have no apparent effect upon masculinization of the developing brain. This observation provides indirect evidence for the probable lack of importance of oestrogen in masculinizing the human brain during foetal life.

5-Alpha-reductase deficiency and male pseudohermaphroditism

The occurrence of 5-alpha-reductase deficiency dur-

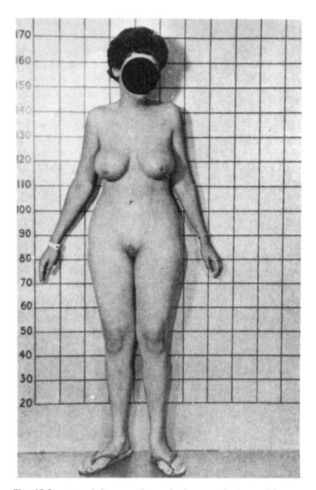

Fig. 10.9 An adult genetic male human being with androgen-insensitivity syndrome (AIS) showing the typically feminine appearance which results from this condition. (After Money and Ehrhardt 1972.)

ing foetal development in the human male results in incomplete differentiation of the external genitalia. A rudimentary, clitoris-like penis is formed and this may be hidden between the labial folds which fail to fuse and form a normal scrotum. The testes are concealed within these labia-like scrotal flaps (Fig. 10.10). The external genitalia of such a newborn boy therefore present a most unusual appearance, but they also differ anatomically from the newborn female genitalia. At puberty, the testes enlarge markedly and descend into the bilobed scrotum, whilst the rudimentary penis also grows, but remains very small by comparison with that of a normal man. Male pseudohermaphrodites also exhibit masculinization of other features at puberty (e.g., muscular development and deepening of the voice), as a result of testosterone secretion; there are considerable individual differences in the extent of these somatic changes. The behavioural consequences of 5-alpha-reductase deficiency in male pseudohermaphrodites have been studied in Dominica (Imperato-McGinley *et al.* 1974, 1979, 1982) and in New Guinea (Herdt and Davidson 1988) where a number of cases has occurred in isolated rural communities. Because Imperato-McGinley *et al.* originally reported this phenomenon (but see also Gajdusek (1964) in regard to the New Guinea population) it has become known as *Imperato-McGinley syndrome*. Imperato-McGinley *et al.* described 38 cases of the condition in four villages in the Dominican Republic. Thirty-three subjects were still alive at the time of the study and 19 of these were believed to have been raised unambiguously as girls. However, after masculinizing somatic and genital changes occurred at puberty, 18 of these subjects reportedly changed and adopted a male gender identity. These findings were used to bolster the proposition that testosterone plays a paramount role in formation of masculine gender-identity even when boys have been raised as girls prior to puberty (Imperato-McGinley *et al.* 1974, 1979). However, it is not the case that boys born with this condition go unrecognized in these village communities, or that such individuals are raised unambiguously as girls. They are even given a special name meaning 'balls at twelve' by families who are aware, for historical reasons, that pseudohermaphrodites are fundamentally male rather than female. Further, in an environment where sex roles are strictly defined and an infertile and unmarriageable woman is viewed as a social liability, it is not surprising that pseudohermaphrodites may adopt masculine clothes and habits after puberty once their genitalia and secondary sexual characters have developed to fit them for such a gender role.

Gilbert Herdt's studies of the Sambia, a tribe inhabiting the eastern highlands of Papua New Guinea, have shed much additional light on psychosexual differentiation in males with the 5-alpha-reductase deficiency syndrome (Herdt and Stoller 1985; Herdt and Davidson 1988). In addition to the usual two sexes (male (*aatmwul*) and female (*aambelu*)), the Sambia recognize a third category, the *kwolu-aatmwol* or *male-like thing*. Since the adoption of pigeon English by these people, another term has been coined to describe pseudohermaphrodites; the Sambia refer to them as *turnim-men* in recognition of the masculine changes which occur once they attain puberty. Sambian midwives are aware that kwolu-aatmwol or turnim-men babies present a different genital appearance at birth; the occurrence of a tiny penis hidden between oddly shaped labial folds allows them to be identified in most (but not all) cases. Such infants experience many difficulties as they grow and attempt to integrate themselves into Sambian society. Herdt and Davidson (1988) point out that 'shame and stigma afflict both pseudohermaphroditic children and their parents by association'. In some cases these children may not be as loved and securely attached to their parents as normal children. The parents and other village members refer to the child as being different (i.e. a turnim-man) and 'ambiguity in gender typing is present in the child's world from the earliest days of life'. Peers and playmates sometimes reinforce this feeling of difference by referring to the absence of a penis; turnim-men children may be the butt of cruel jibes. These problems are magnified when boys join the mens' cult and engage in ritualized homosexuality, as described earlier in this book (see pages 164–5). Boys at first act as fellators for older youths and unmarried young men in the mens' hut. Once boys reach puberty, however, they

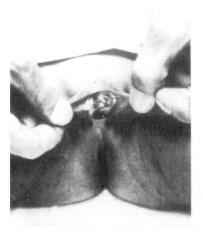 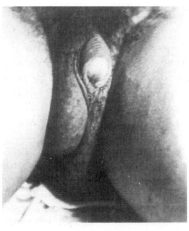

Fig. 10.10 Appearance of the external genitalia (left) in infancy and (right) at puberty in boys with 5-alpha-reductase deficiency (Imperato-McGinley syndrome). Enlargement of the testes, growth of pubic hair and incomplete growth of the penis has occurred in the pubertal example. (From Nelson 1995; photos by permission of Dr J. Imperato-McGinley.)

adopt the active role in such activities, being fellated by younger initiates. Turnim-men are not fellated, however. Their genitalia never grow to normal size (the penis remains small, being only 1.5 inches long when erect in one case recounted to Herdt). Such is the history of embarrassment experienced by these males and the fear of being 'shamed' in Sambian society, that they avoid exposing themselves to further ridicule. Since these homosexual activities, however bizarre they may seem from a westernized perspective, play an important role in readying young men for normal heterosexual life and future marriage, the Turnim-man again experiences an ambiguous situation as regards gender identity/role during puberty and early adulthood.

Herdt and Davidson (1988) identified 14 cases of pseudohermaphroditism among the Sambia (including eight deceased individuals) involving eight villages and a total population of 1700 individuals. They estimated that six pseudohermaphrodites were born per 1000 births in this population. Nine of the babies had been recognized as turnim-men and assigned as such at birth; they were reared as males. However, in five cases pseudohermaphrodites were wrongly assigned as female at birth and raised unambiguously as girls. Their histories provide a most interesting comparison with the Dominican studies. Four of these children raised as girls are known to have married but were rejected by their husbands once their genital ambiguity was discovered. The husbands reacted violently and the pseudohermaphrodites concerned were forced by public pressure to acknowledge their true (i.e. *turnim-man*) identity and to adopt a masculine gender role. In the case of 'Moragu', for instance, marriage took place at 18 years of age, before menarche (or its absence) was noted. Since Sambian women do not commence menstruation until an average of 19.2 years of age, the absence of menarche in this case would not have been viewed as unusual. However, when Moragu's husband, an 18 year old man from a neighbouring village, attempted intercourse with her, he was unable to penetrate. The next time he pulled back her grass skirt he 'saw a small penis in the middle of the labia, with the testes thick and undeveloped on both sides'. In a rage this man went to the elders of the village and threatened to kill Moragu. He subsequently left the village for some time to live down the shame of having wed a turnim-man. Moragu moved to a distant town, adopting a male gender role and westernized form of dress, but not engaging in sexual relationships beyond platonic friendships with women.

Herdt and Davidson (1988) also detail the histories of four pseudohermaphrodites raised as kwolu-aatmwol-males. The best studied example was a 31 year old named Sakulambei (Herdt and Stoller 1985). This individual persisted in homosexual activity in the mens' hut until he was post-pubertal. This unusual situation occurred, apparently, because he hoped that by ingesting semen he might become stronger and more masculine. Sambian boys are encouraged to believe that fellatio is essential for normal manly development (Herdt 1981). Sakulambei never changed to adopt the dominant (fellatad) role as he matured, however. He feared being shamed by his abnormal genital appearance. He reported the occurrence of penile growth and erection post-pubertally and had erotic fantasies involving both boys and women. This man ultimately married and gained some acceptance in the community as a holy man or shaman. He is reported to be 'moody and unpredictable' and has never fathered children; his only son is the result of an adulterous affair between his wife and a neighbour.

These two examples (Moragu and Sakulambei) indicate that pseudohermaphroditic children in Sambian society are profoundly influenced by social environment as regards the development of gender identity and gender role. Turnim men are not raised as girls, therefore, unless wrongly assigned to the female sex at birth. Where this occurs, the transition to a masculine gender role later in life (post-pubertally) is not accomplished smoothly; on the contrary it can be an extremely traumatic process because of society's rejection of the individual's attempts to live as a woman. Similar effects of social environment and difficulties in shifting to a male gender role have been described for pseudohermaphrodites raised as girls in Omani Arab communities (Al-Attia 1996).

Prenatal exposure to exogenous progestagens and oestrogens

Synthetic progestagens and the non-steroidal oestrogen diethylstilbestrol (DES) were formerly used for treating pregnant women, where there was a significant risk of abortion due to inadequate endogenous production of sex steroids. However, synthetic progestagens can exhibit androgenic properties, and some female children resulting from such pregnancies were found to have masculinized genitalia (Wilkins *et al.* 1958; Grumbach *et al.* 1959). Ehrhardt and Money (1967) reported that such girls exhibited behavioural masculinization, including tomboyishness and preferences for athletic pursuits, in a similar fashion to girls with congenital adrenal hyperplasia. Some of these findings have already been discussed (see Tables 10.1 and 10.2 and Fig. 10.8). In addition, Reinisch (1981) found that prenatal exposure to androgenizing progestins had significant effects upon aggressiveness in both male and female offspring (Fig. 10.11). Various synthetic progestagens had been administered to the mothers of these children, either singly or in combination (19-nor-17-ethynyltestosterone, ethynodiol, hydroxyprogesterone caproate, and medroxyprogesterone acetate). Each of the progestin-exposed subjects was compared with at least one sibling on a paper and pencil test designed to measure aggressive potential. This test (the Leifer-Roberts Response Hierarchy) requires subjects to choose between four options in response to a situation involving interpersonal conflict: 1. physical aggression; 2. verbal aggression; 3. withdrawal from the situation; and 4. non-aggressive coping with the other person. Boys or young men score significantly higher than girls or young women in this test. Reinisch also demonstrated a sex difference in aggressiveness of siblings who were not exposed to synthetic progestagens during gestation. Both boys and girls exposed to progestagens were significantly more aggressive than siblings of the same sex (subjects ranged from 6–18 years of age when tested: see Fig. 10.11).

The ability of oestrogen to organize the neural substrates which govern sexual behaviour in the rat was mentioned earlier in this chapter. In some species testosterone, secreted by the testes during foetal or neonatal life, acts as a prohormone and is aromatized to oestrogen by neurons in specific areas of the developing brain. In the rat, it is thought that the developing foetus is protected from maternal oestradiol, present in high concentrations during gestation, by the binding action of alpha-fetoprotein (Plaplinger and McEwen 1978). In the rhesus monkey, there is no evidence that oestrogens organize the developing nervous system in either sex. In any event, it has been noted that oestradiol is metabolized to oestrone (a less potent oestrogen) by the placenta in the rhesus monkey (Slikker *et al.* 1982). There is considerable doubt as to whether oestrogens, in the foetal circulation or produced centrally via aromatization of testosterone, play any physiological role in masculinizing the human brain. Some interest has centred around the question of whether the synthetic non-steroidal oestrogen diethylstilbestrol (DES), which is not neutralized metabolically and can bind to oestrogen receptors in the central nervous system, might influence psychosexual differentiation. Meyer-Bahlburg *et al.* (1984, 1995) have

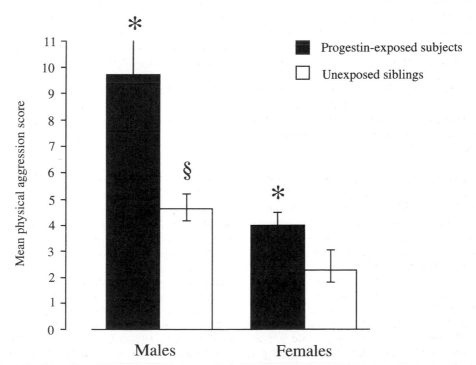

Fig. 10.11 Mean physical aggression scores for 8 male and 17 female human offspring, exposed *in utero* to synthetic progestins, compared to their unexposed sex-matched siblings. *indicates a significant difference between Progestin-exposed and unexposed subjects (P < 0.01). § indicates a significant difference between unexposed males and unexposed females (P < 0.02). (Redrawn from Reinisch 1981.)

reported that women whose mothers were treated with DES during pregnancy tend to show greater than expected tendencies towards a bisexual or homosexual (i.e. lesbian) gender identity. Data derived from four sets of DES-exposed women versus controls showed that masturbation fantasies, romantic or sexual daydreams, patterns of sexual responsiveness, and other measures are more likely to be homosexual or bisexual in nature (Kinsey Scale, K2-K6) than heterosexual (KO-K1) in the DES subjects. Meyer-Bahlburg *et al.* pointed out, however, that 'the extent to which bisexuality and homosexuality were increased in DES women was rather modest'. Only 14 out of 114 DES subjects scored between K2 and K6 on the Kinsey Scale on measures of lifelong sexual relations. Other measurements presented in the original reports of Meyer-Bahlburg *et al.* showed a relatively small DES effect; most of the subjects had a heterosexual gender identity/role. More detailed follow-up studies of 60 women whose mothers were treated with DES during pregnancy (Lish *et al.* 1992) failed to confirm the findings of Meyer-Bahlburg *et al.* (1984). Neither masculinization (nor defeminization) of childhood play nor gender identity/role in adulthood was affected by prenatal DES exposure. Indeed, in much earlier studies Hines (1981, 1982) had found no effects of DES upon childhood behaviour. Recent studies have failed to demonstrate any effects of prenatal exposure to DES upon cognitive development in women (Hines and Sandberg 1996). Thus, despite repeated attempts to implicate oestrogen in human psychosexual differentiation, the evidence is far from convincing that oestrogen (or aromatization of testosterone to oestrogen) plays a significant role in this process. We shall return to this topic in Chapter 13, in connection with the hormonal control of sexual behaviour in adult male primates.

The postnatal testosterone surge and psychosexual differentiation

In addition to the well-established prenatal effects of testicular hormones upon the development of sexually dimorphic behaviour in monkeys and human beings, it is also necessary to consider the possible effects of the *postnatal testosterone surge*. Males of several primate species have been shown to secrete higher levels of testosterone than females during early postnatal development (Rhesus and pigtail macaques: Robinson and Bridson 1978; Plant 1982; marmoset: Abbott and Hearn 1978; Dixson 1986b; chimpanzee: Winter *et al.* 1980; Man: Forest *et al.* 1976). In baby boys, circulating gonadotrophin

levels begin to increase during the second postnatal week, peak at about 3 months of age and then decline by 6 months of age (Forest 1990). Circulating testosterone levels follow a similar time course (Fig. 10.3). A diurnal rhythm of testosterone is well-established in infancy, so that nocturnal levels of the steroid greatly exceed those measurable in the morning hours, at least in the rhesus monkey (Plant 1982). How much of this impressive increase in circulating testosterone is biologically active and available to potential target tissues in the brain and genitalia is debatable. It has been reported that the amount of *free testosterone* (i.e. the small percentage not bound to sex-hormone-binding globulin) in baby boys does not necessarily parallel the postnatal surge of total testosterone measurable in blood. Huhtaniemi *et al.* (1986), for instance, measured salivary testosterone in boys during the first six postnatal months and reported a progressive decline in hormone levels from birth onwards. Since salivary testosterone levels are deemed to reflect the concentration of free steroid available in the circulation, it appears that less biologically active testosterone is present during the height of the surge. However, Huhtaniemi *et al.* did not measure circulating testosterone in their subjects; the relationship between the free and bound proportions of testosterone in infant male primates requires further study. As we shall see, there is now evidence that testosterone secretion during the postnatal period can influence pituitary function and genital development in male monkeys.

Pharmacological blockade of the pituitary–testicular axis, using luteinizing-hormone releasing hormone (LHRH) analogues, abolishes the postnatal testosterone surge. Such experiments, conducted using infant male rhesus monkeys and marmosets, have shown that when males reach puberty, levels of circulating LH and testosterone are lower than those measured in untreated controls. Hence, there is some delay in the onset of puberty (Mann *et al.* 1989; Lunn *et al.* 1994). The marmoset experiments were conducted on pairs of male twins, allowing the use of one twin as a control for effects of an LHRH antagonist (antide) administered to the co-twin. Antide treatment was given on post-partum days 1, 3 and 7, and then at 7-day intervals until 14 weeks of age. This procedure completely suppressed the normal increases in circulating testosterone which continue until approximately day 100 in this species. Male marmosets normally begin to show pubertal increases in testosterone at between 200 and 250 days of age (Dixson 1986b). However, in antide-treated subjects testosterone levels increased more gradually and were significantly lower than those measured in controls at between 43 and 70 weeks of

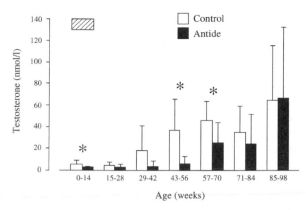

Fig. 10.12 Plasma concentrations of testosterone (T) in male marmosets treated in infancy with an LHRH antagonist (Antide: 10 mg kg^{-1} week^{-1} for 14 weeks as indicated by the hatched bar) as compared to untreated controls. Antide treatment (solid bars) suppresses the postnatal testosterone surge seen in controls (open bars) as shown by a significant decrease in plasma T at between 0–14 weeks of age. Antide-treated males also have lower T levels than controls during puberty. Data are means (+S.D.). *P < 0.05. (Redrawn from Lunn *et al.* 1994.)

age (Fig. 10.12; Lunn *et al.* 1994). Controls also exhibited higher LH levels during puberty. The mechanism by which LHRH analogues produce these effects is still uncertain; as well as occasioning changes in postnatal testosterone secretion such analogues might directly affect LHRH-containing neurons in the brain. Preliminary evidence, from experiments on the rhesus monkey, indicates that antide may interfere with development of NMA receptors on LHRH neurons during the postnatal period and hence reduce the LH response to NMA during puberty. Whether LHRH analogues might also impair LHRH neuronal systems in the brain in some more direct fashion remains unknown (Mann and Fraser 1996).

The behavioural sequalae of pituitary–testicular suppression in infancy have also been studied in marmosets and rhesus monkeys. Marmosets treated with antide in infancy show grossly normal patterns of copulatory behaviour in adulthood (Lunn *et al.* 1994). Whether more subtle effects of neonatal androgenization might occur is a more difficult question. It is interesting, for instance, that if male marmosets are castrated in infancy, then their sexual and aggressive behaviour in adulthood shows important differences from behaviour of males which were castrated either prepubertally or in adulthood (Dixson 1993c). Neonatally castrated males are intensely aggressive during pair-tests with females and never copulate, whereas prepubertally castrated males and adult castrates frequently mount and

thrust during such tests (Fig. 10.13). These effects are not immutable or the result of organizing influences of postnatal testicular hormones, since treating adult neonatally castrated marmosets with testosterone propionate activates mounting and reduces aggression during oppositely-sexed pair-tests (Table 10.3). These results indicate that neonatal castration has some effect upon the overall relationship between such males and females. Females respond to these castrates as if they are also females; mutual aggression occurs, with the result that many tests must be curtailed. In these circumstances it is hardly surprising that sexual behaviour is absent. Treatment of the neonatal castrate with testosterone in adulthood reverses these effects, but it is not clear whether this is due to peripheral changes (e.g., alterations in the male's cutaneous scent glands and urinary odour cues which effect his attractiveness to females), or central actions of testosterone on the male's sexual and social preferences, or both.

The behavioural effects of blocking the postnatal surge of testosterone in male rhesus monkeys are subtle and difficult to demonstrate. Sexually dimorphic patterns of behaviour such as mounting and rough-and-tumble play are not quantitatively affected by treating infant males with antide (Wallen *et al.* 1995). These authors found a slight increase in affiliative behaviour between yearling male rhesus monkeys and their mothers as a result of antide treatment. However, statistically significant effects emerged when antide-treated males were compared with antide plus testosterone-treated males, rather than with untreated male controls. The changes in affiliative behaviour reported by Wallen *et al.* are quite small and have not been confirmed in studies in which infant males were treated with the LHRH agonist meterelin to suppress the postnatal testosterone surge (Fig. 10.14). Eisler *et al.* (1993) have examined the behaviour of adult male rhesus monkeys which had been treated with an LHRH agonist during the first four postnatal months. Chronic agonist treatment had suppressed the postnatal secretion of testosterone and reduced LH and testosterone secretion during early puberty in these males. However, as adults, no effects on social rank were observed and the decrements in sexual behaviour reported by Eisler *et al.* (e.g., reduced masturbatory activity and some reduction in mounting frequencies with oestrogen-treated females) are not striking. Nor is there any suggestion that the sexual orientation and partner preferences of these males had been altered by reduction of early postnatal testosterone secretion.

The early postnatal increases in circulating gonadotrophins and testosterone which occur in male monkeys and in human males have been implicated

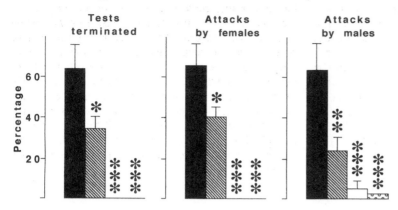

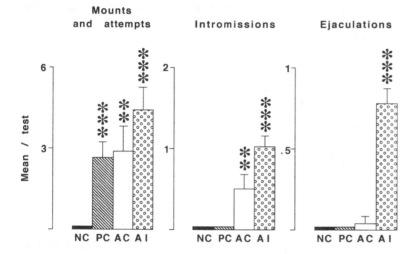

Fig. 10.13 (A) Sexual and (B) aggressive behaviour of adult male marmosets (*Callithrix jacchus*) castrated neonatally (NC: Closed bars), prepubertally (PC: hatched bars) or in adulthood (AC: open bars) during pair-tests with overiectomized, oestrogen-treated females. Behaviour of adult intact males (AI: stippled bars) is also shown. Data are means (+S.E.M.) per test. Statistical comparisons (Mann–Whitney U tests) are between NC males and males in the remaining groups. *P ≤ 0.05; **P ≤ 0.01; ***P ≤ 0.001. (Redrawn from Dixson 1993c.)

Table 10.3 Effects of testosterone propionate (TP) upon sexual behaviour of neonatally castrated (NC) male marmosets during paired encounters with ovariectomized, oestradiol-treated females

Male No.	Attempts +mounts	Mounts + penile erection	Mounts + pelvic thrusting	Intromission	Ejaculation
316	1.5	0.75	0.5	0	0
317	4.25	3.75	2.25	0	0
319	6.0	4.75	3.25	0.75	0.75
320	6.0	5.0	4.25	0	0
322	4.0	2.75	2.0	0	0
323	2.0	1.5	1.5	0	0
329	2.75	2.5	1.75	0	0
335	0.25	0	0	0	0
Mean ± SEM	3.3 ± 0.7**	2.6 ± 0.6**	1.9 ± 0.5**	0.1 ± 0.1	0.1 ± 0.1

From Dixson 1993d.
Data are mean frequencies per test for each NC male with four separate female partners. Before TP treatment no male mounted or attempted to mount a female.
**=P < 0.02 by comparison with pre TP treatment data (Wilcoxon test).

in the control of testicular descent (into the scrotum) and in production of Sertoli cells by the testes (Mann and Fraser 1996). In the rhesus monkey, the testosterone surge also stimulates growth of the infant's penis with resulting increases in its overall length and the separation of the prepuce from the glans (Fig. 10.15). In LHRH agonist-treated males, the glans and prepuce remain adherent for much longer and the penis presents a notably different appearance from that of a normal male (Fig. 10.16). This simple observation indicates that testosterone is biologically active and available to the genital tissues during early infancy. In human males the glans and prepuce do not separate at such an early phase of infant life. We do not know if the testosterone surge influences phallic growth in man, or whether differences in the magnitude and duration of the surge might affect the genitalia or development of behavioural (or cognitive) sex differences.

In monkeys there is the suspicion that testosterone has a priming action upon the developing penis, facilitating its responsiveness to testosterone during later life (i.e. at puberty). There is a parallel here between genital responsiveness and enhanced pituitary responsiveness later on in development. In neonatally castrated male marmosets, the penis fails to grow as large as in prepubertal castrates; treatment with testosterone in adulthood occasions only a limited increase in size. It is as if the sensitivity of the phallus to testosterone has been reduced by removal of the testes during early infancy (Dixson 1993d).

Hormones and homosexuality

In Chapter 6, it was emphasized that homosexuality

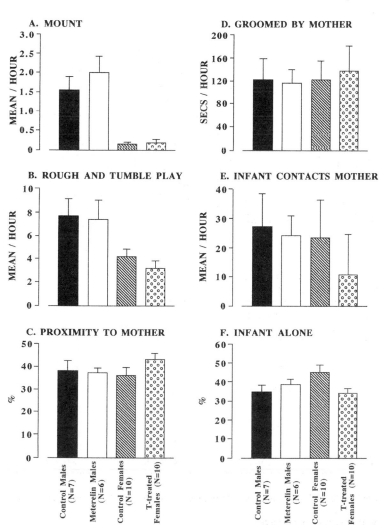

Fig. 10.14 Sociosexual and affiliative behaviour of infant rhesus monkeys, aged 9–12 months, living as members of captive, breeding groups. Data are means (+S.D.) for males treated with the LHRH agonist (meterelin) during their first six months, females treated with testosterone during their first six months and control groups of both sexes (untreated). Neither meterelin, nor testosterone treatment, had any significant effects on the following behavioural patterns A. Mounts; B. Bouts of rough-and-tumble play; C. Percentage of time spent <60 cm from mother; D. Duration of grooming received from mother; E. Infant maintains contact with mother (no. of times infant makes contact/no times infant breaks contact with mother); F. Percentage of scans infant observed alone. (Data from Nevison, Brown and Dixson 1997).

refers not to the use of particular motor patterns in sexual contexts (e.g. mounting or presenting the hindquarters), but to an erotosexual preference for a member of the same sex rather than of the opposite sex. The sociosexual activities of monkeys and apes, which frequently involve same-sex mounts, presentations, or genital manipulation, are not equivalent to human homosexuality. Nor does the fact that perinatal hormonal manipulations can enhance lordotic responsiveness in adulthood in male rodents indicate that such males are homosexuals. What is required is the demonstration that prenatal, or postnatal, hormonal mechanisms influence *partner preferences* in sexual contexts. Some recent studies of rats have demonstrated such effects. Neonatal treatment of males with an aromatase inhibitor (1,4,6-androstatriene-3,17-dione (ATD)) prevents conversion of testosterone to oestrogen during the critical period for the organization of ejaculatory

responsiveness and for defeminization of behaviour. As adults, such males exhibit marked deficits in ejaculatory ability during pair-tests with females, but enhanced lordotic responsiveness when mounted by male partners (Brand *et al.* 1991). These experiments confirm previous findings that oestradiol organizes neural substrates concerned with ejaculation during a critical period spanning the first 10 days of postnatal life, and that this effect occurs even though oestrogen treatment of the neonatally castrated male fails to stimulate a normal amount of phallic growth (Södersten and Hansen 1978). Likewise, it is oestrogen, the aromatization product of testosterone, that defeminizes the lordotic reflex in the male rat (Baum 1979). More recently it has been shown that oestradiol affects the development of partner preferences in the male rat. When intact males are given a choice of tethered partners in a three-compartment testing apparatus, they consistently exhibit a prefer-

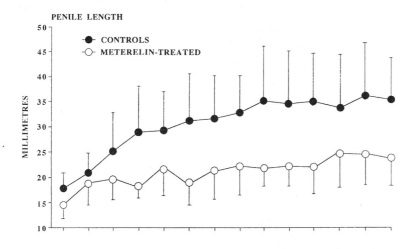

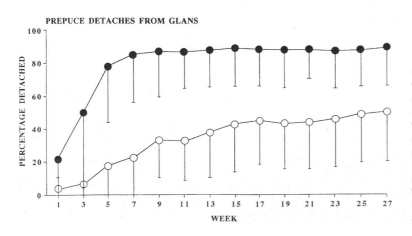

Fig. 10.15 Effect of the postnatal testosterone surge upon penile growth in infant male rhesus monkeys (*Macaca mulatta*). Blockade of testosterone secretion, using an LHRH agonist (meterelin), retards penile growth and detachment of the prepuce from the glans penis. Data are means (+S.D.) for 7 meterelin-treated and 7 control males. (Unpublished data: Brown, Nevision and Dixson.)

ence for an oestrous female over another intact male. Neonatal ATD treatment of male rats significantly reduces this preference, however (Fig. 10.17). Such males retain a bisexual potential in their behaviour; they lordose in tests with males but mount actively when tested with females (although they rarely or never ejaculate during these mounts) (Brand *et al.* 1991; Bakker 1996). Prenatal ATD treatment has very little effect on partner preference; frequencies of ejaculation are not significantly reduced and lordotic responsiveness is much less than in males treated neonatally with the aromatase inhibitor.

These experiments with rats are interesting because they indicate that a neonatal critical period exists for organization of the male's sexual preference for oestrous females and that oestrogen is the steroid responsible for the organization of the relevant neural substrates. The effect is not due to altered olfactory preferences; both ATD-treated males and controls prefer cage-bedding soiled by

oestrous females to bedding previously occupied by intact males. In the ferret also, prenatal and early postnatal treatment with testosterone or oestrogen affects sexual preference; female ferrets treated in this way are more likely to approach other females when given choice tests (Baum *et al.* 1990a). Unfortunately, no preference tests have been conducted using pseudohermaphroditic rhesus monkeys in order to determine, for example, whether prenatal androgenization might have affected their willingness to approach and interact sexually with other females rather than with males. Only free-access pair-tests have been performed and under these conditions androgenized female rhesus monkeys are not behaviourally defeminized (Phoenix *et al.* 1983). More might be learned if pseudohermaphroditic female monkeys could be observed as members of long-term social groups. In the only experiment conducted on this basis, Eaton *et al.* (1990) observed seven prenatally androgenized female Japanese macaques, living as members of a large social group,

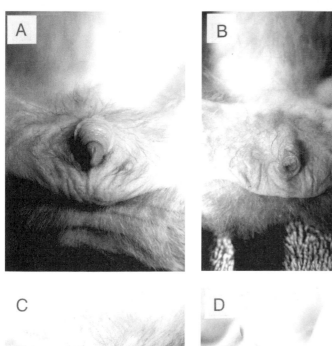

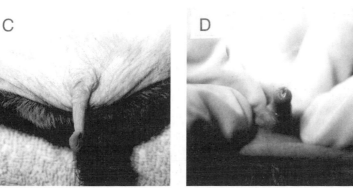

Fig. 10.16 External genitalia of (A, C) of normal infant male rhesus monkeys (*Macaca mulatta*), aged less than six months, compared to the genitalia of (B, D) LHRH-agonist treated males. Chronic agonist treatment during the first six months has blocked the postnatal surge of testosterone and retarded phallic growth. (Photographs by the author and Dr Gillian Brown.)

housed in an outdoor enclosure at the Oregon Primate Research Centre. Juvenile pseudoherm-aphrodites showed increased frequencies of mounting activity when compared with untreated females in the same group. The effect was most pronounced in four females which had received a higher dose of testosterone prenatally and in which masculinization of the external genitalia was more complete. Three other females which had been exposed to a lower testosterone dosage exhibited less masculinization of mounting behaviour and had retained patent vaginae, as well as possessing enlarged clitorides. There was no indication of changed sexual *orientation* in these juvenile macaques at two years of age. Unfortunately, no information is available concerning their sexual preferences and activities in adulthood.

Given that the available experimental studies of the non-human primates have little to tell us about possible developmental effects of hormones upon erotosexual preferences or homosexuality, what evidence is available in the human clinical literature? The effects of foetal exposure to synthetic progestagens, DES, and androgens (in CAH subjects) on the behaviour of girls and women have already been described. Some evidence links prenatal endocrine abnormalities to the greater likelihood of a homosexual/bisexual orientation. Data pertaining to the effects (or lack of effects) of prenatal exposure to DES have already been discussed. Congenital adrenal hyperplasia has also been shown, in several studies, to predispose women towards homosexual or bisexual fantasies or activities. It must be emphasized that most CAH patients are heterosexual; they

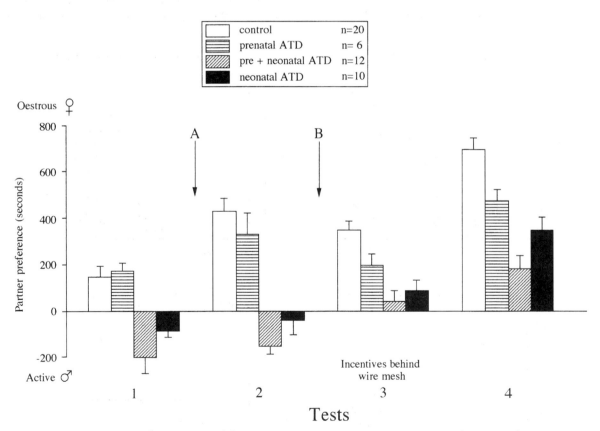

Fig. 10.17 Preference scores (means +S.E.M.) (for an oestrous female over a sexually active intact male) of intact adult male rats after perinatal ATD treatment (to block aromatization of testosterone to oestrogen) or control treatment. Males were tested four times (15 min/test) for partner preference behaviour in a three-compartment box. Preference scores were calculated by subtracting time spent with the sexually active male from time spent with the oestrous female. A positive score=preference for the female; a negative score=preference for the male. In tests 1, 2, and 4, sexual interactions with tethered incentives were possible; in test 3 sexual interactions were prevented by a wire mesh barrier. Between tests 1 and 2, males were pair-tested with an oestrous female (arrow A). Between tests 2 and 3, six pair-tests were carried out, three with an oestrous female and three with a sexually active male (arrow B). Redrawn and modified from Brand *et al.* 1991.)

may marry and raise children without complications. However, significant numbers of subjects have reported experiencing some homosexual/bisexual fantasies and/or participating in such activities (Ehrhardt *et al.* 1968b; Money *et al.* 1984; Dittmann *et al.* 1992). In a recent study of CAH girls and women, including saltwasting and simple virilizing subjects, Dittmann *et al.* found that their sexual orientation and experiences differed from those of their non-CAH sisters. More CAH patients reportedly wished for, or had experienced, a relationship with a female partner (at >16 years: 26% and at >21 years: 44%, as compared with 0% of the control group). Conversely, significantly more sisters than CAH subjects had ever 'been in love' with a male partner, were engaged in a long term relationship, or had experienced vaginal intercourse (Fig. 10.18).

The available evidence indicates that prenatal ex-

posure to androgenic hormones may facilitate erotosexual attraction to members of the same sex, rather than the opposite sex, in the human female. The term facilitate is appropriate here because *tendencies* towards homosexual/bisexual preferences are involved rather than an absolute determination of sex preferences by prenatal androgens (Money 1988; Dittmann *et al.* 1992; Meyer-Bahlburg 1993). It is also very difficult to apply the prenatal hormone theory when attempting to explain the genesis of homosexuality in men, or its equivalent (lesbianism) in women. Male and female homosexuals have normal genitalia and normally developed secondary sexual characteristics. There is no indication that the external genitalia of lesbians have been overexposed to androgens during foetal life, or that a generalized lack of androgenization has occurred in homosexual men. Levels of gonadal steroids and pituitary gonadotrophins in adult homosexuals fall

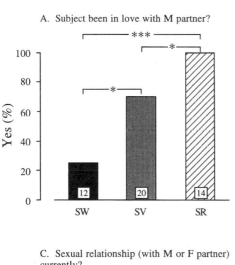

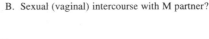

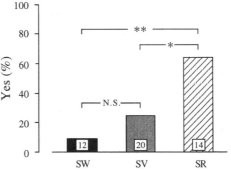

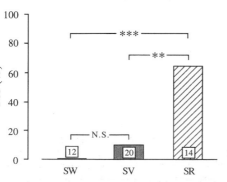

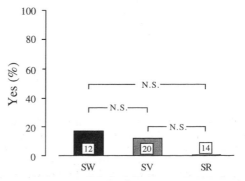

Fig. 10.18 Sexual preferences and activities in young women with 'salt-washing' (SW) or 'simple virilizing' (SV) forms of congenital adrenal hyperplasia, as compared with their sisters (SR). Numbers of subjects are given at the foot of each histogram. *P < 0.05; **P < 0.01; ***P < 0.001. (Redrawn and modified from Dittmann *et al.* 1992 and Dittmann 1998.)

within the normal range. The slight differences reported in some studies (e.g., lower levels of free plasma testosterone and higher levels of LH and FSH in 'effeminate homosexuals' than in heterosexuals) are of doubtful functional significance. The controversial view that male homosexuality is facilitated by an androgen deficit during foetal development has been espoused by Dörner and his co-workers (e.g., Dörner 1976, 1977, 1979, 1988; Dörner *et al.* 1980). Dörner also reported that homosexual men exhibit a positive feedback effect of oestrogen upon LH secretion; levels of LH first decrease (negative feedback) and then exhibit a marked surge after a single intravenous injection of 20 mg of the oestrogen 'Presomen'. Heterosexuals and bisexuals, by contrast, show only the negative feedback effect of oestrogen upon LH levels. Subsequently, Gladue *et al.* (1984) demonstrated a positive feedback effect of another oestrogen (Premarin) upon LH secretion in homosexual men. The response exceeds that found in heterosexual males, but is significantly lower than that seen in normal women (Fig. 10.19). As noted previously, the release of LH after an oestrogen challenge is not sexually dimorphic to the same extent in some primates as it is in rodents such as the rat. Positive feedback may be elicited in castrated adult male rhesus monkeys (Karsch *et al.* 1973) and in hypogonadal men (Kulin and Reiter 1976). Gladue *et al.* (1984) point out that their findings apply to men with a 'lifelong homosexual orientation' and 'might not apply to all male homosexuals'. Further, they acknowledge that what is demonstrated in Fig. 10.19 is 'an adult hormonal correlate of sexual orientation that is causally independent of sexual differentiation-a causal relation should not be inferred'.

One further variant of the prenatal hormone theory must be mentioned, as it relates to the genesis of male homosexuality. It has been suggested that marked stress during pregnancy might affect neuroendocrine development of the male foetus and hence predispose the individual towards a homosexual orientation. Dörner *et al.* (1980) advanced this hypothesis to explain a putatively higher frequency of homosexual births in Germany during, or just after, the second world war. Far-fetched as this theory might seem, it does have parallels with experiments on rats, showing that stressful events during perinatal life can influence sexual differentiation of the brain and behaviour (Matuszczyk *et al.* 1990; Ward 1972; Ward *et al.* 1994). In the rat, effects of prenatal stressors may be mediated via the opioid system (Kashon *et al.* 1992). Ellis *et al.* (1988) have reported that severe maternal stress can affect sexual orientation in human offspring. Bailey *et al.*

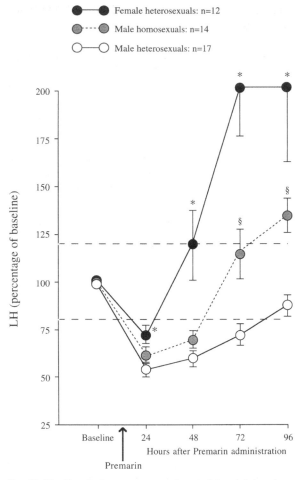

Fig. 10.19 Circulating concentrations of luteinizing hormone (LH) in homosexual men, heterosexual men and heterosexual women before and after a single injection of the synthetic oestrogen Premarin (arrow). Data are means (±S.E.M.). Dashed lines indicate the 95% confidence interval for baseline values for all groups. *P < 0.05 for female heterosexuals as compared with homosexual and heterosexual men at the same time-points. §P < 0.05 for comparisons between homosexual men and heterosexual men at the same time points. All groups show a decrease from baseline at 24 h. (Redrawn and modified from Gladue *et al.* 1984.)

(1991) found some correlation between 'stress-proneness' in mothers and births of effeminate male children. However, maternal stress did not show an overall correlation with either sexual orientation or childhood problems of gender conformity. Bailey *et al.* noted a strong familial influence upon the likelihood of having homosexual offspring; they suggest that familial and genetic factors might indeed be much more important than maternal stress as determinants of homosexuality.

The neuroanatomical basis of sexually dimorphic behaviour

One route by which testicular hormones might organize the neural substrates which govern sexually dimorphic patterns of behaviour is to alter the structure of such tissues. Structural sexual dimorphisms have now been identified in many areas of the vertebrate brain; examples are provided in Table 10.4 but this list is far from exhaustive. In some cases, causative links have been established between sex differences in brain structure, organizational effects of androgens (or oestrogens), and sex differences in behaviour. The best examples concern vocal control nuclei in the brains of male song birds (such as the canary) and spinal motor neurons which innervate the penile musculature (and hence influence sexual reflexes) in male rats. These examples will be discussed first, as they provide a sound basis for considering some of the more problematic sex differences in brain structure described in primates and other mammals.

Sex differences in vocal control systems in the brain

In vertebrate species where males, rather than females, emit loud courtship or territorial calls, sexual selection has favoured the evolution of specializations in the neurological control of the masculine vocal apparatus. Several cases have now been documented of marked sexual dimorphisms in brain structure which relate directly to vocal abilities in male fish, anurans, and birds. The male oyster toadfish (*Opsanus tau*) emits a 'boatwhistle' call by contracting the muscles of the gas-filled swimbladder. This call attracts females in breeding condition. The sonic motor nucleus which innervates the swimbladder muscles lies in the posterior medulla and anterior spinal cord. Neuronal cell bodies in the sonic motor nucleus of males are usually larger than those of females. Moreover, adult males may be divided into two types—those with larger neurons in the sonic motor nucleus which emit louder territorial calls and those with small neuronal cell bodies which are suspected to be satellite males (Fine *et al.* 1984). Development of the sonic motor nucleus occurs gradually and changes in neuronal size can proceed even in adulthood. It is not known how testicular hormones influence these changes. However, in the American toad (*Bufo americanus*), it is known that testosterone (rather than oestrogen) affects growth of neuronal cell bodies in the pretrigeminal nucleus which innervates the vocal apparatus. Only male toads vocalize and cell bodies in the male's pretrigeminal nucleus are larger than those which occur

in the female (Schmidt 1982).

More striking still, is the sexually dimorphic neural system which controls the vocalizations of male songbirds, such as the canary and zebra finch (Fig. 10.20). Several nuclei in the brain of the male are known to bind testosterone (the magnocellular nucleus of the anterior neostriatum: MAN; the caudal nucleus of the ventral hyperstriatum: HVc; the robust nucleus of the archistriatum: RA; the intercollicular nucleus: ICO; and the tracheosyringeal portion of the hypoglossal nucleus: nX11ts, which innervates the vocal organ). In the canary, and especially in the zebra finch, several of the song control nuclei (HVc, RA, and area X in Fig. 10.20) are much larger in adult males than in females (Nottebohm and Arnold 1976). Lesioning the RA or HVc in adult males impairs their ability to sing (Nottebohm *et al.* 1976). Females also possess the nuclei and fibre tracts found in the male brain, but the female structures are much smaller, particularly in the forebrain. The RA and HVc contain twice as many neurons in males and their RA neurons have larger cell bodies and dendritic fields. Organization of the song-control system is affected by both testosterone and oestrogen during a critical developmental period (Konishi and Gurney 1982). In the female zebra finch, the brain is sensitive to such organizational effects from the day of hatching until at least 40 days of age. Testosterone influences the larger numbers of neurons found in the masculinized RA and HVc, whilst oestradiol facilitates increase in the size of neuronal cell bodies. However, a most important observation concerns the plasticity of these hormone-sensitive systems in the adult brain, as has been shown in studies on the canary (Bottjer and Arnold 1984). Female canaries (unlike female zebra finches) can be induced to sing by treating them with testosterone in adulthood (Nottebohm 1980). This treatment also increases the volumes of the RA and HVc in the brains of these females. In male canaries seasonal changes in singing are pronounced; males sing during the spring mating season and in association with this, there are marked anatomical changes in the vocal control system (The RA and HVc increase in size: Nottebohm 1981). Seasonal changes in testicular hormone secretion are implicated in producing these effects upon the brain and behaviour. Changes in song patterns occur from year to year and Paton and Nottebohm (1984) have demonstrated that new neurons are incorporated into functional song circuits in the HVc in adulthood. These studies of the songbird brain therefore provide an example of *plasticity* and *reorganization* of sexually dimorphic neural circuitry in the adult brain. In this case, organization is not

Table 10.4 Examples of structural sexual dimorphisms in the mammalian nervous system

Structure	Species	Type of sexual dimorphism	Sources
Accessory olfactory bulb (AOB)	Rat	Volume of AOB and numbers of mitral cells greater in males than in females	Roos et al. 1988
Bed nucleus of accessory olfactory tract	Rat	Volume of nucleus and numbers of neurons greater in males than in females	Guillamón and Segovia 1993
Amygdala	Rat	Volume of medial amygdaloid nucleus 20% larger in males than in females	Mizukami et al. 1983b
	Squirrel monkey	Neuronal nuclei in medial amygdala larger in males than in females	Bubenik and Brown 1973
	Rat	Males have 25% more synapses than females which end on dendritic shafts in the medial amygdaloid nucleus	Nishizuka and Arai 1981
	Rat	Twice as many Vasopressin-containing neurons in the medial amygdala of males, than in females	DeVries 1995
Hippocampus	Mouse	In six inbred strains, females had lower densities of dentate granule cells than males of the same strain	Wimmer and Wimmer 1985
Preoptic area (POA) and anterior hypothalamus (AH)	Rat	More synapses of non-strial origin occur in the medial POA in females than in males	Raisman and Field 1971
	Rat	Sexually dimorphic nucleus (SDN–POA) is 2–4 times larger in males than in females	Gorski et al. 1978, 1980
	Rat	Parastrial nucleus of POA is larger in females than in males	Del Abril et al. 1990
	Gerbil	Sexually dimorphic area (SDA) and pars compacta (SDApc) are markedly larger in males than in females. Volume of left SDApc correlates with frequency of males ultrasonic courtship vocalizations	Commins and Yahr 1984 Holman and Rice 1996
	Guinea-pig	SDN–POA larger in males than in females	Byne and Bleier 1987
	Hamster	Dendritic density in part of the dorsomedial POA of males exceeds that found in females	Greenough et al. 1977
	Ferret	A small 'male nucleus' of the POA–AH is lacking in females	Tobet et al. 1986
	Long-tailed macaque	Golgi-staining shows a greater number of POA neurons having dendritic bifurcations and a greater number of spines in juvenile males than in females	
Interstitial nuclei of the anterior hypothalamus (INAH 1, 2, 3 and 4)	Man	INAH-1 (the SDN–POA) is larger and contains more neurons in males than in females; this dimorphism emerges between 4 years and puberty	Ayoub et al. 1983 Swaab and Fliers 1985;
		INAH-2 volume greater in men than in women	Swaab and Hofman 1988
		INAH-3 volume greater in men than in women	Allen et al. 1989
		INAH-3 In homosexual men, the volume of this nucleus is less than that of heterosexual men	Allen et al. 1989; Le Vay 1991 Le Vay 1991
Suprachiasmatic nucleus (SCN)	Rat	SCN volume greater in males than females	Robinson et al. 1986
	Rat	Greater numbers of synapses in males than in females	Guldner 1984
	Man	Vasopressin—containing area of SCN is rounded in shape in men, but elongated in women	Swaab et al. 1993

Table 10.4—*Continued*

Structure	Species	Type of sexual dimorphism	Sources
	Man	SCN is not sexually dimorphic but is larger in homosexual men than in heterosexual men	Swaab and Hofman 1990; Swaab et al. 1992
	Man	Younger men have more vipergic neurons in the SCN than young women, but this sex difference is reversed in middle age	Zhou et al. 1995b
Ventromedial hypothalamus (VMH)	Rat	Volume of VMH nucleus is greater in males than in females	Matsumoto and Arai 1983
	Rat	In the ventrolateral VMH males have 33% more axodendritic synapses than females	Matsumoto and Arai 1986a, b
Arcuate nucleus (ARC)	Rat	Females have approximately twice as many synapses on dendritic spines, whereas males have approximately twice as many synapses on perikarya	Matsumoto and Arai 1980
Vasopressin and oxytocin-containing nucleus (VON)	Pig	The VON, situated lateral to the third ventricle and dorsal to the SCN increases markedly in size at puberty and is largest in adult females, containing three times more neurons than in adult males	Eerdenberg 1991
Bed nucleus of stria terminalis (BNST)	Rat	Volume of medial posterior BNST is largest in males; volume of medial anterior BNST is largest in females	Del Abril et al. 1987
	Rat	Numbers of vasopressin immunoreactive neurons are 2 or 3 times greater in males than in females	DeVries 1995
	Man	Volume of medial posterior BNST is 2.5 times larger in men than in women	Allen and Gorski 1990
	Man	Vipergic innervation of central subdivision of the BNST has 44% greater volume in men than in women and is smaller in M $\rightarrow$ F transsexuals than in heterosexual men	Zhou et al. 1995a
Lateral septum (LS)	Rat	Vasopressinergic innervation is denser in males than in females	DeVries et al. 1984
Anterior commissure (AC)	Man	AC is 12% larger in females than in males	Allen and Gorski 1992
Corpus callosum (CC)	Man	Splenium (posterior fifth) of CC is larger and more bulbous in females than in males, both in adulthood and during foetal life	De Lacoste-Utamsing and Holloway 1982; De Lacoste et al. 1986
Spinal cord	Rat	Motor spinal nucleus of the bulbocavernosus (SNB) is significantly larger in males; the nucleus is much smaller or lacking in females	Breedlove and Arnold 1980, 1981
	Rat	Dorsolateral motor nucleus (DLN) is less dimorphic than the SNB, but is largest in males and innervates the ischiocavernosus muscles	Breedlove 1984
Hypogastric ganglion (HG)	Mouse	In new-born male mice, the HG, which innervates the internal genitalia, contains more neurons than in new-born females	Suzuki et al. 1982, 1983
Pudendal nerve	Rat	Cross sectional area of this nerve is greatest in males, being 3 times (sensory branch) and 5 times (motor branch) greater than in females. Males have less closely packed axons, and larger diameter axons, than females do	Moore and White 1996

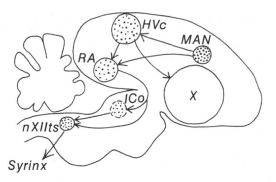

Fig. 10.20 The brain of a male songbird in sagittal view showing the major song-control nuclei and their interconnections. Stippled areas refer to nuclei which accumulate testosterone or its metabolites. HVc=Robust nucleus of the archistriatum; ICo=Intercollicular nucleus; MAN= Magnocellular nucleus of the anterior neostriatum; NXIIts =Tracheosyringeal portion of the hypoglossal nucleus; RA=Robust nucleus of the archistriatum; X=Area X. (Redrawn from Bottjer and Arnold 1984.)

immutable and limited strictly to an early period of brain development.

Spinal motor neurons innervating penile muscles

A second example of a sexually dimorphic neural system, organized by testicular hormones during perinatal life and having a well-defined function, is provided by the spinal motor neurons which innervate the striated penile muscles in the rat. Breedlove

and Arnold (1980) described two sexually dimorphic motorneuronal nuclei, situated in the fifth and sixth lumbar segments of the rat spinal cord—the spinal nucleus of the bulbocavernosus (SNB) and the dorsolateral nucleus (DLN). The SNB is markedly dimorphic; it contains more neurons and larger neurons in males, whereas adult females lack a coherent structure (Fig. 10.21). In the male, SNB neurons innervate the *levator ani* and *bulbocavernosus* muscles at the base of the penis; these muscles are present at birth in females, but subsequently they become atrophic. The DLN nucleus is less sexually dimorphic; some motor neurons in this nucleus supply the leg muscles in both sexes, whilst others innervate the *ischiocavernosus* penile muscles.

The functions of the bulbocavernosus and ischiocavernosus muscles in the rat have already been described in some detail (see pages 263–4). Contractions of the ischiocavernosus muscles cause flip-like dorsiflexions of the male's penis and are important in facilitating intromission. Contractions of the bulbocavernosus muscles produce flaring of the glans (penile cups); these assist the male in depositing his copulatory plug firmly against the female's cervix during the ejaculatory mount. Clearly, the motor neurons of the DLN and SNB spinal nuclei, which innervate the ischiocavernosus and bulbocavernosus muscles, play a most important role in facilitating penile reflexes in the rat.

The organization of SNB neurons during development in the male rat is controlled by testicular

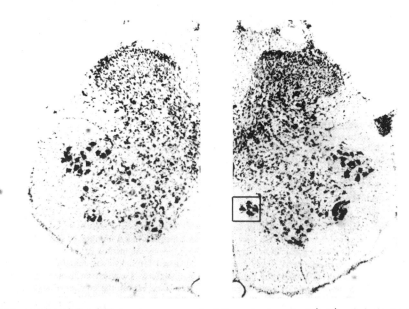

Fig. 10.21 Photomicrographs of the fifth lumbar segment of the spinal cord of (left) a female and (right) a male rat. The spinal nucleus of the bulbocavernosus (SNB) is recognizable as a discrete nucleus only in the male (as indicated by the box). (After Breedlove and Arnold 1981.)

androgens (Breedlove 1984; Arnold 1984). Testosterone propionate (TP) masculinizes the number of SNB neurons in the female rat if administered during the last week of gestation and during the first five or six postnatal days. The ability of TP to masculinize the *size* of neuronal cell bodies in the SNB also spans this perinatal period, but also continues beyond day 5, being effective in females treated up to day 11 after birth. Dihydrotestosterone propionate (DHTP) is also effective in masculinizing the SNB in the female rat during perinatal development. oestradiol benzoate is ineffective, however, indicating that the aromatization of testosterone to oestrogen is not involved in organizing this sexually dimorphic system in the spinal cord.

These experiments have shown that the processes which control neuronal cell number in the SNB of the male rat differ in some respects from processes which govern increase in cell size. Even in adulthood the SNB is sensitive to hormonal manipulations; castration results in a reduction in the size of SNB neuronal cell bodies. Treatment of castrated adult males or adult females with testosterone causes a 20% size increase in SNB neuronal cell bodies. The SNB takes up and concentrates testosterone (T) and dihydrotestosterone (DHT), but not oestradiol, in the adult male rat. Interestingly, whereas both T and DHT are effective in restoring flip and cup penile movements in castrated rats, oestradiol has no effect upon these sexual reflexes. The SNB provides a further example of hormonal effects upon sexually dimorphic neuronal systems in adulthood, so that organizational effects are not totally restricted to a critical period in the development of the central nervous system. The system exhibits some plasticity, and a capacity for synaptic reorganization in the adult, as is also the case for the sexually dimorphic neural system controlling song in the canary (Arnold 1990).

Sex differences in the primate brain, in comparative perspective

We shall now consider structural sex differences, which have been described in the brains of various primate species. The majority of these sexually dimorphic features have been described in the human brain (Table 10.4), whereas surprisingly few studies have involved the brains of non-human primates, despite the considerable potential for experimental investigations, particularly where monkeys are concerned. In contrast to the examples described so far, most of the human sex differences in Table 10.4 are poorly understood, both as regards hormonal influences upon their development and, most importantly, what their functional significance might be. Therefore, in the discussion that follows I shall refer wherever possible to parallel work on non-primate mammals (especially on rodents) which provides some useful comparisons with the limited information on sexual dimorphism in the human brain.

Sex differences in the corpus callosum

The corpus callosum (Fig. 10.22) comprises more than 200 million fibres connecting the two hemispheres in the human brain (Tomasch 1954). Attempts to relate callosal cross sectional area to

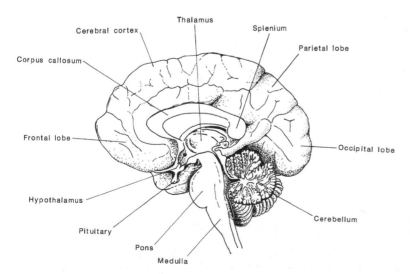

Fig. 10.22 The human brain in sagittal section, to show the position of the corpus callosum (including its posterior portion: the splenium) which connects the two hemispheres.

cognitive processes are not new. Spitzka, for example, measured the area of the corpus callosum in ten 'eminent men' and found it to be larger than in 'ordinary men'. Bean (1906) who cited Spitzka's observations, was unable to confirm them, but claimed that 'in dealing with the corpus callosum as a whole, it is found to be smaller in the Negro than in the Caucasian, just as the brain of the Negro is smaller than that of the Caucasian'. Such ideas now seem distasteful, as well as amusing in their naïvety. However, it is interesting to speculate how posterity will view more recent attempts to justify the notion that the corpus callosum differs between the sexes, rather than between races. This idea springs from anatomical studies conducted by De Lacoste-Utamsing and Holloway (1982) on the formalin-fixed brains of nine men and five women. These authors reported that the splenium (posterior fifth) of the corpus callosum has a greater sagittal sectional area in women than men. The difference, allowing appropriate correction for smaller brain size (due to smaller body size) in females, was not impressive and achieved statistical significance at the 0.08 level. Bulbosity of the splenium was quantified by measuring maximum splenial width and this measure was markedly larger in female than male brains ($P < 0.001$). Subsequent studies of 38 foetal and a further 16 adult brains by these authors indicated that splenial sexual dimorphism was measurable during prenatal life, as well as in adulthood (De Lacoste et al. 1986; Holloway and De Lacoste 1986). Yoshi et al. (1986) used magnetic resonance scanning procedures to measure the callosum in 19 male and 14 female subjects; they concluded that the splenium is more bulbous in females. Sexual dimorphism has also been described in a sample of 15 pongid brains, but not in other non-human primate species (De Lacoste and Woodward 1988).

Although these results appear to be consistent and convincing, a disturbing number of follow-up studies, using preserved material or imaging techniques on living brains, have failed to confirm the existence of a splenial sex difference. Bell and Variend (1985) examined 28 male and 16 female brains taken at post-mortem from children ranging from newborn to 14 years of age. They found no sexual dimorphism in the corpus callosum, despite the earlier report of such dimorphism in the foetal brain by De Lacoste et al. (1986). Witelson (1985), in a study of 42 subjects which showed that the corpus callosum is larger in left-handed people, noted in passing that she was unable to detect any sexual dimorphism. Studies by Weber and Weis (1986), Byne et al. (1988), Demeter et al. (1988), Going and Dixson (1990), and Emory et al. (1991) all failed to identify any differences between the sexes. These five studies have involved a total of 201 human brains. Emory et al. (1991) included 10 male-to-female transsexuals and 10 female-to-male transsexuals in their sample. No anatomical differences were measurable between transsexuals and controls ($N = 40$) in either splenial size or shape. This thorough study controlled for possible effects of handedness. Magnetic resonance techniques were employed, thus avoiding one criticism levelled at research on preserved brains, namely that fixation and shrinkage of neural tissue might affect the size of the corpus callosum.

Given these inconsistent results and the weight of negative findings, it should not be accepted uncritically that the splenium of the human corpus callosum exhibits sexual dimorphism in shape or in sagittal sectional area. Objections that negative findings are due solely to failure to use the same measurements of the callosum employed by De Lacoste-Utamsing and Holloway (1982), failure to correct for differences in male and female brain weights, or a preponderance of ageing subjects, are incorrect. Table 10.5 presents data on the human corpus callosum collected and analysed in accordance with studies which had produced a sex difference in splenial area. No such effects are seen in Table 10.5, however. The form factor analysis carried out in this study measures the circularity of an area and provides a more objective measure of bulbosity than maximum splenial width. The form factors for the whole callosum (CCFF) and for the splenium (SPFF) do not differ between the sexes.

Sex differences in the preoptic area and anterior hypothalamus

Three of the four interstitial nuclei of the anterior hypothalamus (the so-called INAH nuclei) in the human brain have been reported to exhibit sexual dimorphism, although there is some disagreement between the various reports. Swaab and Fliers (1985) were the first to identify the sexually dimorphic nucleus (SDN), equivalent to INAHI, which is larger in the brains of men than women (Fig. 10.23). The SDN is a tiny structure, however, containing between 30 000 and 40 000 cells in young adult men and about half that number in women of similar age. This sex difference develops only between four years of age and puberty, as can be seen in Fig. 10.24. Cell numbers in the SDN peak at between 2 and 4 years of age in both sexes, but neurons are then progressively lost from the female brain, resulting in the emergence of a larger SDN in males (Swaab and

Table 10.5 Measurements of the human corpus callosum (CC) indicating lack of sexual dimorphism in the size of the splenium (SP)

Measurements	Male brains (Mean ± SD)	Female brains (Mean ± SD)	t value
Number of Subjects	17.0	16.0	–
Age (years)	74.4 ± 11.3	82.0 ± 7.1	–
Hemispheric weight (g)	535.8 ± 46.1	476.0 ± 60.0	–
CC length (mm)	74.6 ± 5.6	75.8 ± 4.3	−0.69
CC circumference (mm)	204.0 ± 17.0	209.0 ± 17.0	−0.84
CC area (mm²)	631.0 ± 114.0	646.0 ± 100.0	−0.40
CC width (mm)	11.5 ± 1.9	11.4 ± 2.0	0.15
SP circumference (mm)	55.9 ± 5.1	55.1 ± 4.6	0.47
SP area (mm²)	185.0 ± 34.0	172.0 ± 35.0	1.08
CC width/CC length	0.149 ± 0.026	0.144 ± 0.028	0.53
SP area/CC area	0.298 ± 0.031	0.276 ± 0.022	2.35*
CC form factor	0.191 ± 0.028	0.189 ± 0.039	0.17
SP form factor	0.744 ± 0.103	0.724 ± 0.123	0.51

Data from Going and Dixson 1990.
Brain measurements have been corrected for sex differences in brain weight.
*P < 0.05.

Hofman 1988). It is possible, therefore, that post-natal and prenatal factors influence sexual differentiation of the human brain. The existence of a postnatal testosterone surge in baby male monkeys and in human males during the first 4–6 months of postnatal life has already been discussed. It remains

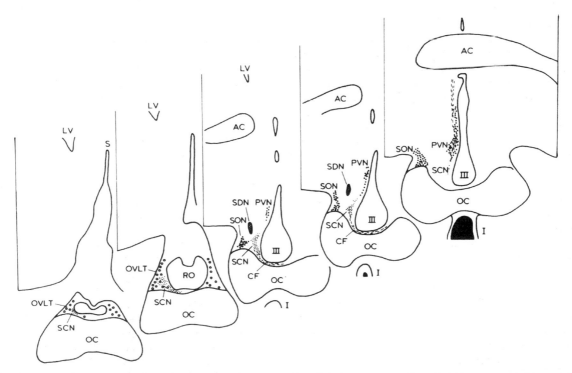

Fig. 10.23 Topography of the sexually dimorphic nucleus (SDN) in the human hypothalamus. III=Third ventricle; AC=Anterior commissure; 1=Infundibulum; LV=Lateral ventricle; OC=Optic chiasm; OVLT=Organum vasculosum of the lamina terminalis; RO=Recessus opticus; S=Septum; SCN=Suprachiasmatic nucleus; PVN=paraventricular nucleus, SON=supraoptic nucleus; CF=Commissural fibres of the suprachiasmatic nucleus. (After Swaab and Fliers 1985.)

unknown whether this postnatal event might play any role in protecting the male SDN from the process of pre-programmed cell death (*apoptosis*) which depletes the female homologue of cells later in infancy. In answer to the question 'why does it take so long for foetal or perinatal sex hormones to have an effect since sexual differentiation occurs only by the age of four?' Swaab *et al.* (1992) replied:

'If the sexual dimorphism of the human hypothalamus is actually due to these sexually dimorphic hormone peaks, which is something that still has to be established in the human brain, one might speculate that perhaps the receptors for sex steroids are not situated in the SDN but in other sexually dimorphic structures that are innervated by the SDN. The SDN fibres in females would only 'realize' that there is no target for them when they later innervate that structure and subsequently they would degenerate. Such a process might take a considerable amount of time.'

Both Allen *et al.* (1989) and Le Vay (1991) failed to confirm the occurrence of sexual dimorphism in INAHI in the human brain. Allen *et al.* reported that INAH 2 and 3 are both larger in male than in female brains. Le Vay also detected a sex difference in INAH3 but not in INAH2 in his studies of the human hypothalamus. Interestingly, Le Vay also reported that INAH3 was significantly smaller in the brains of homosexual men than in heterosexual men. As can be seen in Fig. 10.25, whilst these differences are statistically significant they are not particularly

striking. Some male homosexuals possess INAH3 nuclei of greater volume than those of heterosexual men; two of the six women depicted in Fig. 10.25 have larger INAH3 nuclei than a substantial number of the male subjects. These sex differences are not nearly as striking, therefore, as those previously described for brain nuclei governing song in canaries or spinal motor nuclei controlling penile reflexes in the rat. Nor can any particular functional significance for gender identity or sexual behaviour be attached to observations on the human INAH nuclei. Some of the subjects studied by Le Vay had died of AIDS-related diseases; this reservation applies to all of the brains of homosexual men included in Fig. 10.25. However, some of the heterosexual males in this study had also suffered from AIDS and it seems unlikely that the effect upon INAH3 was due solely to an AIDS-related neuropathy. No other studies have been reported and replication of Le Vay's findings would be most useful. Investigations by Swaab *et al.* (1992) of the SDN (INAHI) in homosexual and heterosexual men failed to reveal any differences in either volume or cell numbers. Nor was any effect of AIDS-related diseases upon SDN volume or cell number detected in these men.

Before structural sexual dimorphisms had been identified in the human hypothalamus, a sexually dimorphic nucleus of the preoptic area (SDN–POA) had been described in rat (Gorski *et al.* 1978) and subsequently in the guinea-pig (Bleier *et al.* 1982),

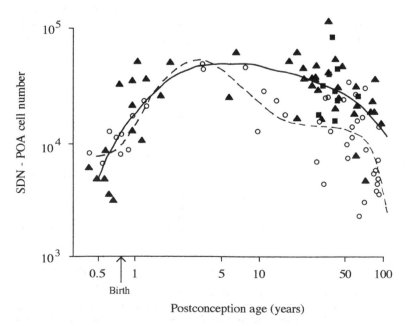

Fig. 10.24 Development and sexual differentiation of the sexually dimorphic nucleus (SDN) in the human hypothalamus. (▲) = males; (○) = females. Cell numbers peak at around 2–4 years postnatally, after which sexual differentiation occurs due to decreases in the cell numbers in the SDN of females. Cell numbers remain relatively stable in men until approximately 50 years of age. Cell numbers in the SDN of homosexual men (■) do not differ from those of heterosexuals. (Redrawn and modified from Swaab and Hofman 1988.)

gerbil (Commins and Yahr 1984), and ferret (Tobet *et al.* 1986). Sex differences in dendritic patterns in the POA had also been described in the rat (Raisman and Field 1971) and hamster (Greenough *et al.* 1977). Subsequently, Ayoub *et al.* (1983) demonstrated that higher frequencies of dendritic spines and dendritic bifurcations occur in the POA of juvenile male *Macaca fascicularis* than in juvenile females. A sexually dimorphic nucleus also exists in the POA of the quail (Panzica *et al.* 1987) and doubtless further examples remain to be described in other vertebrates (Table 10.4). The SDN–POA differs in size and structure in these various species. It is doubtful whether all these sexually dimorphic nuclei are homologous in the anatomical sense, and there is still less reason to assume that they all fulfil the same biological functions.

The POA/AHA continuum is undoubtedly important for the control of masculine copulatory behaviour in vertebrates (Larsson 1979; Hart and Leedy 1985; Sachs and Meisel 1988; Meisel and Sachs 1994). A full consideration of its functions is provided in Chapter 13, which deals with the hormonal control of sexual behaviour in adult male primates. At this stage it is important to note that lesions in the POA/AHA continuum, which destroy a substantial volume of neural tissue and not just sexually dimorphic nuclei, result in a profound decrease in copulatory behaviour in species such as the rat, dog, cat, goat, rhesus monkey, and marmoset. Sexual *interest* in the female also decreases, but is not necessarily totally abolished by such lesions. Thus far, with the exception of the gerbil (Yahr and Gregory 1993) it has proven exceedingly difficult to implicate the sexually dimorphic nuclei of the POA

in the control of masculine sexual behaviour. Yahr and Gregory have shown that cell-body lesions in either the medial or lateral portions of the male gerbil's SDN–POA cause marked decreases in mounting behaviour (Fig. 10.26). Such lesioned males continue to show some sexual interest in females during pair tests, however (Yahr, personal communication). There is also evidence that the medial and lateral portions of the SDN–POA affect masculine copulatory behaviour via different pathways (Yahr and Gregory 1993). Both regions, however, exert some action upon copulatory behaviour via their connections with the caudal part of the medial bed nucleus of the stria terminalis (Sayag *et al.* 1994). As we shall see in a moment, there is evidence of sexual dimorphism in the bed nucleus of the stria terminalis in the human brain.

Yahr and Gregory suggest that the medial portion of the sexually dimorphic nucleus of the POA in the gerbil may be homologous with the central part of the medial preoptic area (MPOA) nucleus in the rat. The MPOA nucleus in the rat is markedly sexually dimorphic, its volume being two or three times greater in adult males than in adult females (Gorski *et al.* 1978). Discrete lesions within the MPOA have been found to have no greater effects upon masculine copulatory behaviour than lesions elsewhere in the POA of the rat (Arendash and Gorski 1983). De Jong *et al.* (1989) noted some inhibition of sexual behaviour after making lesions in the sexually dimorphic area in the male rat, however. In the ferret, small lesions in the MPOA nucleus of the male caused some reduction in preintromission measures of sexual behaviour, whereas larger POA lesions produced the expected deficits in copulatory activity

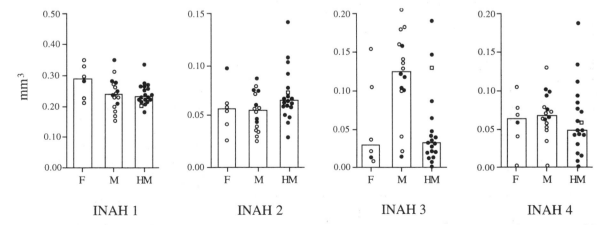

Fig. 10.25 Volumes of the four interstitial nuclei of the human hypothalamus (INAH 1, 2, 3 and 4) in females (F), presumed heterosexual males (M), and homosexual males (HM). ●=individuals who died of AIDS-related causes; □ =a bisexual male, who died of AIDS and is included with the homosexual sample. Data are individual volumes (in mm³) with overall median values for each group indicated by histograms. (Redrawn and modified from Le Vay 1991.)

(Cherry and Baum 1990). By contrast, lesions in the MPOA/AH of male ferrets increase the likelihood that they will exhibit feminine patterns of sexual preference in response to exogenous oestradiol. Castrated male ferrets treated in this way exhibit a greater likelihood to approach a stud male contained in the goal box of a 'T' shaped maze (Cherry and Baum 1990; Kindon *et al.* 1996). Similar increases in feminine responses, involving both proceptivity and receptivity, have been observed in response to oestradiol after POA lesions in castrated tom cats (Hart and Leedy 1983) and rats (Hennessey *et al.* (1986). In neither of these cases were the lesions restricted to sexually dimorphic POA structures, however.

Extensive studies of sexual differentiation in the MPOA of the rat have shown that this nucleus is organized by testicular hormones during late gestation and early postnatal life. It is known that aromatization of testosterone to oestradiol within the hypothalamus of the developing brain is important for the organization of the MPOA during this critical period (Dohler *et al.* 1984). Thus, the MPOA of the male rat might be important for inhibition of proceptivity and receptivity, representing part of the neural substrate by which testicular hormones defeminize the brain of the developing male. Other portions of this inhibitory circuitry include the septum (see pages 424–5 for a discussion of this topic), which also contains receptors for testosterone and oestrogen.

Interesting though these comparative studies of mammals other than primates certainly are, they tell us little about the functional significance of sexual dimorphism in the INAH nuclei of the human brain. If a striking dimorphism, such as occurs in the MPOA of the rat, cannot be linked to masculine copulatory patterns or sexual preferences, it is unlikely that the relatively tiny INAH nuclei could be of great significance in this regard. The best evidence implicating the SDN–POA of any mammal in the control of masculine copulatory behaviour pertains to the gerbil and the nucleus is especially large and complex in this rodent species. The MPOA nucleus is involved in inhibition of proceptivity in the male ferret and in inhibition of receptivity (lordosis), as well as proceptivity, in the male rat. How far these apparent defeminizing effects are relevant to primates is also debatable since, as discussed previously, androgens do not defeminize the developing brain in monkeys or human beings. If the human SDN–POA nucleus (INAHI) described by Swaab and Fliers (1985) is influenced by the postnatal testosterone surge in human infants, then much might be learned by studying effects of the surge upon hypothalamic development in male rhesus monkeys. The male rhesus monkey exhibits a postnatal testosterone surge and preliminary studies indicate that an SDN–POA is also present in this species (Swaab , personal communication). It should be recalled, however, that suppression of the pituitary–testicular axis in infant male rhesus monkeys

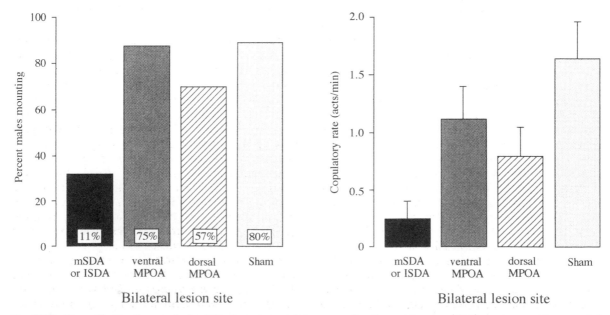

Fig. 10.26 The effects of neuronal cell-body lesions of the sexually dimorphic area (SDA) or surrounding medial preoptic area (MPOA) on mounting behaviour and copulatory rate in sexually experienced male gerbils. All males were given exogenous testosterone. (Redrawn from Yahr and Gregory 1993.)

by administration of LHRH analogues has produced little or no effect upon any aspect of sexually dimorphic behaviour.

Sex differences in other hypothalamic nuclei

The suprachiasmatic nucleus (SCN) in the human hypothalamus has been reported to be larger in male homosexuals than in heterosexual men (Swaab and Hofman 1990; Swaab *et al.* 1993). The SCN of homosexual men was found to contain approximately twice the number of vasopressin neurons (15 000 on average) and to occupy twice the volume of the SCN in heterosexuals. There is no implication that the SCN controls or contributes in any direct fashion to a homosexual orientation in men. This nucleus is important in the regulation of circadian rhythms and has been shown to exhibit both diurnal and seasonal changes in size in human beings. The volume of vasopressin cells in the human SCN is 2.4 times larger in the autumn months than during the summer (Hofman and Swaab 1993). It would be ridiculous to suggest that seasonal changes in sexual orientation occur in men and equally misguided to attach functional significance to the volume of the SCN as regards either masculine copulatory behaviour or erotosexual preferences. Lesions of the SCN have no effect upon sexual preferences in male rats (Kruijver *et al.* 1993). If some critical effect of androgens, or oestrogens, occurs in the SCN during development of the human brain it remains a mystery and the significance of the larger nucleus in homosexuals has yet to be explained.

Zhou *et al.* (1995a) have identified a marked sex difference in the volume of the central subdivision of the bed nucleus of the stria terminalis (BST_c) in the human hypothalamus. This is the more prominent subdivision of the BST, situated just dorsal to the anterior commissure (Fig. 10.27). A sex difference in the caudal BST had previously been described by Allen and Gorski (1990), who found this portion of the nucleus to be 2.5 times larger in men than women. However, Zhou *et al.* (1995a) demonstrated a much heavier innervation of the BST_c by VIPergic neurons in men than women and also showed that this effect was not apparent in men who were transsexuals. In the six male to female (M → F) transsexuals studied, the BST_c was similar in volume to the nucleus in heterosexual women.

The finding by Zhou *et al.* of a smaller BST_c in M → F transsexuals is as difficult to interpret in functional terms as previous reports of sexual dimorphism in the human SDN–POA, or the enlarged suprachiasmatic nucleus of homosexual men. Since all six transsexual subjects had been treated with oestrogen and an antiandrogen (cyproterone ac-

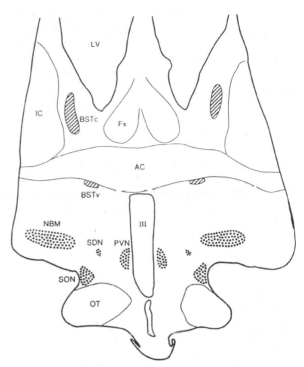

Fig. 10.27 Schematic frontal section through the human brain to show various sexually dimorphic structures discussed in the text. III=Third ventricle; AC=Anterior commissure; BSTc and BSTv=Central and ventral subdivisions of the bed nucleus of the stria terminalis; FX=Fornix; IC=Internal capsule; LV=Lateral ventricle; NBM=Nucleus basalis of Meynert; OT=Optic tract; PVN=Paraventricular nucleus; SDN = Sexually dimorphic nucleus; SON = supraoptic nucleus. (After Zhou *et al.* 1995a.)

etate) in adulthood, hormonal changes may have altered VIPergic expression in the BST_c of these subjects. The bed nucleus of the stria binds both testosterone and oestradiol in the brains of macaques and in other mammals. Further, lesions of this nucleus, which lies at a nodal point on the pathway connecting the amygdala and hypothalamus, are known to cause lengthening of mount and ejaculatory latencies (or failure of ejaculation) in male rats (Emery and Sachs 1976). Interesting though such effects on the male's *sexual performance* might be, there is no evidence that the BST plays a crucial role in *sexual preference* in animals, or in the gender identity of human beings. The erotosexual preferences of transsexuals are complex; many M → F transsexuals are attracted sexually to men but some find women sexually more attractive. All have the feeling of being *trapped in the body of the wrong sex*, however, and from an early age transsexuals wish to inhabit the body of the opposite sex. It is not known whether transsexuality itself influences postnatal differentiation of the BST_c or whether the neuroendocrine environment during some presumptive critical period is upset in individuals destined to become transsexuals.

Conclusions: sexual differentiation of the brain and behaviour

The evidence reviewed in this chapter points to an organizing action of testosterone upon brain development in primates, as well as in other mammals. Because primates are precocial (i.e. their infants are born well-developed and with their eyes open), the critical period for effects of testosterone is likely to occur during prenatal development. Experiments on rhesus monkeys support this hypothesis; prenatal androgenization of female rhesus monkeys masculinizes their playful and sociosexual behaviour as juveniles and enhances masculine copulatory patterns in adulthood. Yet defeminization of behaviour does not occur, so that pseudohermaphroditic female rhesus monkeys exhibit proceptivity and sexual receptivity in adulthood. Menstrual cycles are also normal; prenatal androgenization does not disrupt the hypothalamic regulation of gonodotrophin secretion, in contrast with some sub-primate mammals.

In the rhesus monkey, dihydrotestosterone (DHT) can substitute for testosterone in experiments on prenatal androgenization of the female. Since DHT is a non-aromatizable androgen, it seems less likely that testosterone exerts its organizing effects via conversion to oestrogen in the brain of this primate species. This again distinguishes the rhesus monkey from certain sub-primates (such as the rat) in which testosterone acts as a prohormone in the brain, being converted to oestrogen in order to masculinize (or defeminize) neural substrates which influence later behavioural development. Given the diversity of the extant primates, it is entirely possible that species exist in which aromatization of testosterone is important behaviourally; so few species have been studied that it would be rash to dismiss this possibility. However, in the case of man close parallels exist with rhesus monkeys as regards prenatal effects of testosterone upon masculinization of the brain. Girls exposed to high levels of androgen *in utero* as a result of congenital adrenal hyperplasia (CAH), tend to be more tomboyish, more interested in athletic pursuits, and less interested in feminine toys and pastimes than their non-CAH sisters. Although CAH subjects experience ovulatory menstrual cycles and may marry and have children in a perfectly normal manner, some report a greater than normal propensity towards lesbian/bisexual feelings and experiences. Perhaps, therefore, prenatal exposure of the brain to higher testosterone levels has some effect upon the development of sexual orientation in human beings. This is not a simplistic 'all or nothing' effect; genetic, neuroendocrine, and experiential factors during development all play their parts in the formation of an individual's gender identity/role.

The existence of a postnatal testosterone surge in baby boys and infant male monkeys remains an intriguing enigma as far as behavioural development is concerned. Pharmacological blockade of the pituitary–testicular axis of infant rhesus monkeys, using LHRH analogues, has provided one means of exploring the functions of this postnatal T surge. Markedly sexually dimorphic behavioural patterns, such as mounting or rough-and-tumble play, are not affected by blocking testosterone secretion in infant males. Affiliative behaviour may be slightly affected (Wallen *et al.* 1995) or unaffected by such treatments (Nevison *et al.*, 1997). Growth of the penis is retarded, and hormonal changes at puberty are somewhat delayed. Whether the postnatal T surge has similar significance for the development of the human male is not known and requires evaluation. Human babies are born in a less precocial state than are other anthropoids (despite the fact that their eyes are open at birth), so that there is some possibility that the critical period may extend into early postnatal life in man.

Attempts to account for the genesis of homosexuality in the human male, based upon reduced availability of testosterone during foetal development or reduced responsiveness of the brain to this androgen, remain far from convincing. Males born with androgen insensitivity syndrome (AIS) have feminized genitalia and, if raised as girls, usually adopt a female gender identity/role. Gooren and Cohen-Kettenis (1991) have documented the interesting case of partial AIS in a genetic male who was reassigned as female at 5 days of age. In adulthood (30 years of age) this individual applied for gender reassignment as a man. Despite the presence of an essentially feminine body type and external genitalia (the clitoris was slightly enlarged), psychological testing confirmed a strongly male gender identity and heterosexual orientation. Gooren and Cohen-Kettenis comment that 'with this information it is difficult to believe that in transsexuals or homosexuals, in whom there are no signs of defective virilization, their condition can be ascribed to a lower-than-normal androgenization of the brain'. The occurrence of positive feedback by exogenous oestrogen upon LH release has been identified as a neuroendocrine marker of male homosexuality in some studies (Dörner *et al.* 1980; Gladue *et al.* 1984) but not in others (Gooren 1986; Hendricks *et al.* 1989). If lack of androgenization during foetal life predisposes male homosexuals to exhibit a female-type release of LH in response to an oestrogen challenge, it is also puzzling that the partial AIS patient studied by Gooren and Cohen-Kettenis (1991) should have failed to manifest such a response until after he had been gonadectomized to remove the source of his feminizing oestrogens. The emergence of a positive feedback response subsequent to gonadectomy is not surprising; castrated men and castrated adult male rhesus monkeys produce an LH surge in response to an oestrogen challenge (Karsch *et al.* 1973; Kulin and Reiter 1976).

The hormonal organization of neural substrates which govern sexually dimorphic patterns of behaviour has been linked to sex differences in brain structure in a few cases. Most convincing are the studies of the vocal control nuclei in song birds, such as canaries and zebra finches, and studies of the spinal motor nuclei innervating the striated penile muscles of the rat. In these cases very marked sex differences in behaviour (only adult male canaries sing) or sexual reflexology (only the adult male rat possesses the muscles which control reflexive penile movements) are matched by a highly dimorphic organization of the central nervous system. However, the functional significance of many other reported sexual dimorphisms in CNS structures are much more problematic. The occurrence of a sexually dimorphic nucleus (SDN) in the preoptic area (POA), or anterior hypothalamic area (AHA), in various rodents, quail, ferrets, and human beings is a case in point. Because these sexually dimorphic nuclei are invariably larger in males than females, and because the POA/AHA continuum plays such a crucial role in masculine sexual behaviour, it is tempting to assume that an SDN–POA must be involved in sexual behaviour or sexual orientation. Neither of these assumptions is justifiable. The best evidence implicating the SDN–POA in copulatory behaviour (and scent-marking) concerns the male gerbil, in which the SDN is especially large and complex. Some of the tiny interstitial nuclei of the human anterior hypothalamus (INAH nuclei) may be larger in men than women but they have no known function. There is currently some disagreement in the published literature as to which of the four INAH nuclei are sexually dimorphic. The finding of a modest decrement in the volume of the INAH3 nucleus in homosexual men compared with heterosexual men (Le Vay 1991), requires confirmation. A statistical correlation between INAH3 vol-

ume and homosexuality does not indicate that the nucleus plays any role in human sexual orientation. Interestingly, the better characterized SDN–POA in man (INAH1) does not differ in either volume or cell numbers when heterosexual and homosexual men are compared (Swaab *et al.* 1992). Likewise, the occurrence of an enlarged suprachiasmatic nucleus in male homosexuals (Swaab and Hofman 1990) and a smaller than normal central division of the bed nucleus of the stria terminalis in male to female transsexuals (Zhou *et al.* 1995a) tells us little about how these conditions arise, beyond the possibility that a biological predisposition exists and that these anatomical differences might correlate in some way with broader biological determinants.

In Chapter 6, the distinction between the sociosexual behaviour of monkeys or apes and the nature of human homosexuality was discussed at some length. That chapter also emphasized the dual importance of genetic factors and experiential factors in shaping sexual (and sociosexual) development in primates. It is fruitless to debate whether sexual dimorphisms in behaviour are determined by genetics, sex differences in neuroendocrine development, or experiential factors from earliest infancy—all three play their part. Many homosexuals and transsexuals remember feeling 'different' from others as regards their gender identities from early childhood. The strength of such a biological predisposition towards a particular gender identity and sexual orientation in human beings has been indicated in many studies. Diamond and Sigmundson (1997), for instance, conducted a follow-up study of a classic, previously reported case of sex reassignment in infancy due to traumatic injury to the penis. A baby boy (aged 8 months) was accidentally injured during an operation to correct a tight foreskin (phimosis); the decision was then made to reassign the child surgically to the female sex. Despite earlier reports that transition to a female gender identity/role was successful in this case (Money and Ehrhardt 1972), at puberty this individual expressed the desire to revert to his original gender and successfully made the transition to living as a male. This is not to deny that sex reassignment in infancy can be successful, or that experiential factors are crucial for successful development of gender identity and sexual orientation in man. In the case of the 5-alpha-reductase deficiency syndrome discussed earlier in this chapter it was noted that some Sambian males with this condition were raised unambiguously as females and attempted to live out a female gender identity/role in their society (including marriage) until rejected by those around them. Others, recognized at birth to be 'turnim men' and raised as boys, adopted a male gender/identity role but were variously affected by the shame attaching to their condition of limited genital development (Herdt and Davidson 1988).

At this time we are far from understanding all the factors which govern the development of gender identity and sexual orientation in human beings. In the current state of knowledge it would, for instance, be premature to conclude that a single developmental pathway leads to homosexuality. In the next chapter we shall see that different factors such as diabetes, chronic alcoholism, or hyperprolactinaemia, may result in erectile dysfunction in men. It is likely that disorders of gender identity and erotosexual orientation can also arise for many different reasons.

Finally, we should consider the possibility that some plasticity might be retained in the postnatal human brain, as regards its responsiveness to androgen, so that prenatal organizational effects of testicular hormones need not be immutable. Such plasticity has been demonstrated in the canary brain, in association with the ability to learn and modify patterns of male song during successive annual mating seasons (Paton and Nottebohm 1984). In the pig

sexual differentiation of the hypothalamus 'continues during and after puberty' (Eerdenberg 1991) and in man differentiation of the SDN–POA emerges at between 4 years of age and the onset of puberty (Swaab and Hofman 1988). Environmental factors of many kinds might impinge upon the development of the human brain during childhood and adolescence. It might be naïve to believe that increased sexual feelings and activities during puberty depend solely upon activation by testosterone of preorganized neural substrates. Some maturation and reorganization of the brain might continue during the adolescent years. Puberty itself might represent another critical period for erotosexual development. It is interesting, for example, that male rhesus monkeys which have been castrated and adrenalectomized in infancy exhibit a progressive increase in circulating gonadotrophins when they reach pubertal age, even though no feedback by testicular hormones is possible (Plant 1994). Apparently, some androgen-independent maturation of the hypothalamic-pituitary axis occurs at this time; perhaps maturation of behaviourally relevant mechanisms is also in progress in the adolescent brain.

11   The ovarian cycle and sexual behaviour

From puberty onwards, female primates undergo cyclical changes in the secretion of hormones as a result of ovarian cycles, pregnancy, and lactation. These *recurrent endocrine events* (Herbert 1981) are associated with cyclical fluctuations in sexual activity. The female's sexual initiating behaviour (proceptivity) as well as her willingness to allow the male to copulate (receptivity) varies depending upon the stage of the ovarian cycle or whether she is pregnant or lactating. Hormonal changes might also influence non-behavioural components of female sexual attractiveness, such as sexual skin swellings or olfactory cues. Only at the end of reproductive life, during the post-menopausal period, are these recurrent endocrine effects upon sexual behaviour brought to a close. The task of this chapter is to consider how sexual behaviour is affected by hormonal changes during the ovarian cycle in prosimians, monkeys, apes, and human beings. Information on behaviour during pregnancy and the post-partum period will also be considered when it is useful for comparison with events during ovarian cycles. We are concerned here primarily with endogenous hormonal changes in gonadally intact females. The precise neuroendocrine mechanisms which govern proceptivity and receptivity, and the effects of ovariectomy and hormonal replacement, will be dealt with in the next chapter.

Ovarian cycles: proceptivity, receptivity, and attractiveness

Fluctuations in sexual activity during the ovarian cycle have been documented in many primate species; much useful information derives from field studies, as well as from observations on captive social groups, or animals paired for limited periods throughout the course of the female's cycle. To extract meaningful information on how female sexuality varies throughout the ovarian cycle it is necessary to make precise measurements of the behaviour

of both sexes, and to correlate these with the various endocrine events. Hormonal changes during the ovarian cycle are shown in Fig. 11.1, together with examples of possible changes in female proceptivity, receptivity and attractiveness. The schema is theoretical, therefore, and it is not intended to represent any particular species of monkey or ape. During the follicular phase of the cycle increasing amounts of oestrogens are secreted by the female's ovaries, culminating in a 'surge' of 17β-oestradiol at mid-cycle which triggers the pre-ovulatory surge of luteinizing hormone (LH) from the anterior pituitary gland. Once ovulation has occurred the ruptured follicle becomes converted into the corpus luteum, which secretes progesterone during the second half, or luteal phase, of the cycle. In many primate species, males mount and ejaculate most frequently during the follicular phase of the cycle, or show a peak of ejaculations during the peri-ovulatory period. In theory we may envisage primate species in which temporal patterns of female proceptivity, receptivity, and attractiveness vary in different ways throughout the ovarian cycle. Proceptivity and receptivity might increase at mid-cycle, for example, whilst the female's attractiveness remains relatively constant. By contrast, attractiveness might increase at mid-cycle, whereas female proceptivity and receptivity are unaltered. In both these circumstances, the frequency of the male's ejaculatory mounts might show a mid-cycle peak (Fig. 11.1). It follows that restriction, or virtual restriction, of copulation to the peri-ovulatory period in forms as widely separated in phylogeny as *Lemur catta* (Evans and Goy 1968), *Saimiri sciureus* (Wilson 1977a), and *Gorilla gorilla* (Nadler 1975) does not indicate that the underlying hormonal mechanisms are necessarily the same in each case.

The variations depicted in Fig. 11.1 are greatly simplified. It is possible, for instance, that proceptivity or attractiveness might increase earlier in the follicular phase, thus assisting the female to attract or locate males, whereas her receptivity might peak later in the cycle to coincide with ovulation. In the

(a) Hormonal changes

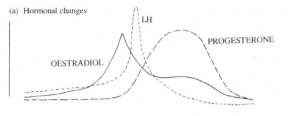

(b) Female shows pronounced cyclical changes in attractiveness (——) but not in proceptivity or receptivity (------).

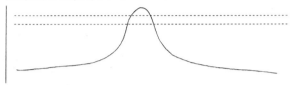

(c) Female shows pronounced cyclical changes in proceptivity or receptivity (------) but not in attractiveness (——).

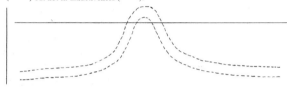

(d) Males mounting frequency with either female (b) or (c).

Fig. 11.1 Graphs to illustrate how changes in female sexuality (proceptivity, receptivity, and attractiveness) during the ovarian cycle may influence copulatory behaviour in the male. In these graphs 'attractiveness' refers to non-behavioural stimuli such as sexual skin or olfactory cues. (Redrawn and modified from Dixson 1983c.)

majority of non-primate mammals, such as rodents, ungulates, or canids, the restriction of mating activity to a peri-ovulatory oestrus is due primarily to effects of ovarian oestrogen and progesterone upon female sexual receptivity, although changes in proceptivity and attractiveness also contribute to the timing of sexual interactions (Beach 1976; Komisaruk 1978). However, among the anthropoid primates, such a rigid control of female receptivity by ovarian hormones does not occur. Indeed, as we shall see in the next chapter, even removal of the ovaries does not abolish sexual receptivity in some monkeys, any more than it does in the human female. The sections which follow, provide a comparative overview of how the ovarian cycle influences the expression of female sexuality in the prosimians and anthropoids.

The ovarian cycle and sexual behaviour in prosimians

The most detailed knowledge of hormonal changes during the ovarian cycle, and associated changes in sexual behaviour, in prosimians derives from studies of just three species; the greater galago (Eaton *et al.* 1973a), the ringtailed lemur (Evans and Goy 1968; Reynolds and Van Horn 1977), and the ruffed lemur (Foerg 1982b; Shideler *et al.* 1983). Eaton *et al.* (1973a) studied greater galagos which were then classified as *Galago crassicaudatus crassicaudatus*, but which are now assigned to the species *G. garnettii* as distinct from *G. crassicaudatus*. Major differences in reproductive biology occur between these two greater galago species (Dixson and Van Horn 1977; Dixson 1989), so that it is important to classify captive specimens correctly. Eaton *et al.* found that the ovarian cycle of *G. garnettii* lasted for 39–59 days (mean 44 days) and that the period of vaginal cornification (*vaginal oestrus*) varied from between 7–24 days (mean 12.4 days). A restricted phase of receptivity, lasting for 2–10 days (mean 5.8 days), occurred during the period of cornification. Behaviour was observed during pair-tests, the female being introduced daily into the male's home cage during periods of vaginal cornification and once per week at other times during the cycle. Copulations were initiated by males, which made low 'clucking' sounds and licked or sniffed the female's genitalia before attempting to mount. Receptive females responded to male investigations by crouching with the perineum raised and the tail deflected to the side. Non-receptive females avoided males, withdrawing rapidly or threatening and attacking them in response to persistent attempts at copulation. Often, females which were sufficiently attractive to elicit mount attempts by males were not necessarily sexually receptive. Although sexual receptivity lasted for 5.8 days on average, it increased gradually on days 1–3 and then declined, as indicated by changes in intromission latencies. The intromission latency averaged 14.7 min in tests conducted on Day 1 of the receptive phase, decreasing to 6.4 min (Day 2), 3.3 min (Day 3), and then lengthening again to an average of 8.8 min on the final day of receptivity (Day 6).

The male greater galago mounts in a dorso-ventral position, clasping the female around the abdomen and biting her gently on the back. Rapid pelvic thrusts are followed by slower, deeper thrusting movements once intromission is attained. Vaginal smears taken after the tests revealed that ejaculation had occurred at the end of such series of deep

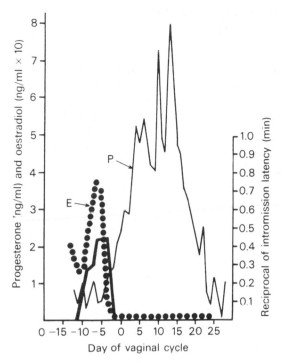

Fig. 11.2 Changes in circulating hormone levels and sexual receptivity during the ovarian cycle of Garnett's (greater) galago (*Galago garnettii*). E=oestradiol; P= progesterone. The reciprocal of intromission latency provides a useful index of sexual receptivity in this species (from Dixson 1983c); redrawn and modified from Eaton *et al*. 1973a.)

pelvic thrusts. To avoid pregnancies, therefore, partners were frequently separated and tests terminated once intromission was observed. Interestingly, in a subsequent study (Dixson and Van Horn 1977), in which pairs of *G. garnettii* were caged together permanently and copulations were undisturbed, the duration of vaginal cornification was shorter (range 5–12 days; mean 7.0 ± 2.04) than in the experiments of Eaton *et al*. (range 7–24 days; mean 12.4 ± 3.7). Therefore, it is possible that sexual stimuli influence the duration of vaginal oestrus in *G. garnettii*. Lengthy copulations are normal for these galagos and intromissions lasted for 13–260 min in the study by Eaton *et al*. Bouts of thrusting were repeated at approximately 20-min intervals and it is possible that more than one ejaculation occurred during a single intromission.

Figure 11.2 shows changes in plasma 17β-oestradiol and progesterone, as well as changes in vaginal cytology, during the ovarian cycle of the female greater galago, together with measurements of female sexual receptivity. During the period of vaginal cornification, circulating 17β-oestradiol

peaked at between 345–760 pg mL^{-1} (mean 519 ± 175 pg mL^{-1}). At this time, the vaginal epithelium consisted of large, angular anucleate cells; string mucus was absent from the vaginal smear and the external genitalia were swollen and reddish. The vaginal closure membrane, which is present during dioestrus, became perforate during vaginal oestrus. Females became sexually receptive, on average 4.7 days after the onset of vaginal cornification although there was considerable variation (range 2–16 days in 28 cases cited by Eaton *et al*. 1973). Sexual receptivity coincided with maximal levels of 17β-oestradiol and detectable levels of plasma progesterone. Timed matings indicated that most females conceived on days 2 and 3 of the receptive phase, when ovulation was presumed to have occurred. It is noteworthy, however, that female greater galagos remained sexually receptive for 1–2 days after 17β-oestradiol had passed its peak and had begun to decline (Fig. 11.2). After the last day of behavioural oestrus a metoestrus-type of vaginal smear was seen, consisting of scattered cornified cells with faint halos of granulated mucus, and plasma progesterone levels began to rise. Thereafter, leucocytes reappeared in the vaginal smear, a dioestrus condition was re-established, and plasma progesterone levels peaked at between 6–12.3 ng mL^{-1} during the 24-day luteal phase of the cycle.

The greater galago has been considered in some detail because excellent laboratory data exist for this nocturnal prosimian. Other nocturnal forms also copulate only during a limited period of female receptivity. In *Cheirogaleus medius*, for instance, copulation occurs during a 2–3-day period and the male sniffs or licks the female's external genitalia frequently before copulation (Foerg 1982a). A vaginal closure membrane also occurs in this species and mating begins one day after the vagina opens. In *Galago moholi*, copulations occur for only 1–3 days during the 32-day ovarian cycle (Doyle *et al*. 1967). The oestrous cycle of *Loris tardigradus* lasts for 29–40 days, during which copulations are confined to a 2-day period. The female scent-marks a great deal in the period preceding copulation, leaving a urine trail which the male follows and overmarks with his own urine (Izard and Rasmussen 1985). The potto (*Perodicticus potto*) has an oestrous cycle of 16 days on average (Ioannou 1966). However, field studies of pottos carried out in Gabon by Charles-Dominique (1977) have shown that male pottos frequently 'visit' and 'court' females in neighbouring home ranges, despite the fact that copulations are limited to a brief oestrus period. Charles Dominique placed liane 'bridges' bearing electrical recording devices within the home ranges of pottos

(Fig. 11.3) and was able to map their night-time movements with precision. He found that one male visited a female regularly throughout the year although she was not in oestrus. Regular visits continued at 2–4-day intervals and the animals might have communicated by olfactory signals, since they did not always interact directly. As the female entered oestrus, however, the male's visits increased in frequency until copulation occurred (Fig. 11.3).

Finally, mention must be made of the tarsier. Wright *et al.* (1986) studied sexual behaviour in six captive pairs of Bornean tarsiers (*Tarsius bancanus*) over a period of 8 months. Only seven copulations were observed and all were restricted to periods of vaginal cornification ranging in length from 1–3 days. Despite the nocturnal habits of these animals, females exhibited a proceptive sexual presentation display, standing with legs apart and looking towards the male from a distance of about 1 metre. Males frequently approached the female and sniffed her vulva, which became swollen and reddish during oestrus. 'Courtship' lasted for 1–2 h and males copulated once or twice each night. Copulations involved a single insertion with pelvic thrusting and lasted for 60–90 s. Although males continued to show sexual

Fig. 11.3 Upper: By the ingenious use of liane 'bridges' and electrical recording devices, Charles-Dominique monitored the movements of free-ranging pottos (*Perodicticus potto*) in the Gabonese rainforest. Lower: These monitoring studies showed that a male potto repeatedly courted and visited a female prior to mating with her during the brief period of sexual receptivity. (Redrawn and modified from Charles-Dominique 1977.)

interest in females during the 1–2 days of metoestrus after vaginal cornification, females were sexually unreceptive and refused mount attempts. During subsequent periods of dioestrus or pregnancy no sexual interactions were observed.

Turning now to the diurnal prosimians, detailed information on ovarian cycles and sexual behaviour exist for *Lemur catta* and *Varecia variegata*, as well as less detailed, but valuable, field observations on *Propithecus verreauxi*. In these species there is also a greater opportunity to study hormonal effects upon visual displays used during sexual encounters and to examine the possible influences of gregarious forms of social organization upon sexual activities. Fig. 11.4 shows patterns of plasma 17β-oestradiol and progesterone measured by Van Horn and Resko (1977) during the 39 day ovarian cycle of *Lemur catta*. These workers found that a pre-ovulatory peak of oestradiol (mean 120 pg mL^{-1}) preceded the onset of female sexual receptivity by 26.3 ± 6.8 h, and that copulations occurred only during a period averaging 21.66 h. Female receptivity varied in duration from 0 h (the female was receptive during a single pair-test) to 43.67 h maximum. These data on captive *L. catta* accord with observations made in Madagascar by Jolly (1966) on free-ranging troops of ringtails, but subsequent fieldwork by Koyama (1988) has shown that the receptive period lasts for as little as 4 h in some females. Both free-ranging and captive *L. catta* frequently scent-mark in association with periods of sexual activity. Females rub the anogenital region against substrates and males may engage in 'stink fights'—rubbing the tail against glands situated on the shoulder and wrist. The tail is then flicked to disperse the scent (Fig. 3.9, page 41). The cutaneous glands are most conspicuous in males and a horny spur on the wrist gland is used to gouge branches, leaving a visible mark, as well as facilitating scent deposition. Evans and Goy (1968), who studied *L. catta* in captivity, proposed that such displays might facilitate pheromonal communication and assist reproductive synchrony in ringtail groups. Evans and Goy also observed that females were only sexually receptive during the period of vaginal cornification, when the external genitalia were swollen and the vulval lips assumed a pinkish hue. Receptive females adopted a lordotic posture, equivalent to the response seen in rodents and some other mammals (Fig. 5.12, page 103). However, sexual receptivity lasted for 10 h or less; this is much shorter than the times recorded in the later studies of Van Horn and Resko (1977). One major difference between these two laboratory studies is that Van Horn and Resko separated their animals before ejaculations had occurred whereas Evans and Goy allowed matings to run their full course. Perhaps, as previously discussed for studies of greater galagos, terminating copulations prematurely in these prosimians leads to some extension of the receptive period. In the case of the ringtail, Van Horn and Resko (1977) also question whether the failure of males to ejaculate and to deposit a copulatory plug, might have extended the female's oestrus. The copulatory plug in *L. catta*, and in some other Lemur species, is large and firm in consistency (see Fig. 8.15, page 237). Perhaps the presence of the plug, distending the upper portion of the vagina and pressing against the cervix, might trigger neural mechanisms which reduce sexual receptivity? Alternatively, some chemical factor might be involved. In the guinea-pig receptivity is reduced by coital stimulation of the vagina and cervix or by artificial stimulation using a glass rod (Goldfoot and Goy 1970). In the ringtailed lemur, it is possible that intromission influences the duration of receptivity whilst the deposition of a copulatory plug has a more pronounced effect. Clearly, much remains to be learned about the control of sexual receptivity in the ringtailed lemur and in other prosimian primates. However, it is apparent that the brief period of sexual receptivity in *L. catta* begins after levels of circulating 17β-oestradiol have peaked and begun to decline (Fig. 11.4). The timing

of receptivity probably coincides, therefore, with the LH surge which triggers ovulation. It is noteworthy also that progesterone levels begin to rise gradually about 3–4 days after the oestradiol peak, so that facilitation of oestrogenic stimulation of receptivity by progesterone in *L. catta* seems to be unlikely (Van Horn and Resko 1977).

The sexual behaviour of the ruffed lemur, (*Varecia variegata*) has been studied in captivity by Foerg (1982b) and Shideler *et al.* (1983). Throughout most of the year the vaginal opening is closed in female ruffed lemurs and the testes of adult males are small and inactive. Both sexes show marked changes in their reproductive physiology and behaviour during the restricted annual mating season. Swelling of the vulval region and vaginal opening is followed, in 3–5 days, by vaginal cornification, which lasts for 1–3 days. Although captive males frequently sniff or lick the genitalia and urine of females during the early stages of vulval tumescence, females are aggressive and refuse any mount attempts at this time. Behavioural oestrus (i.e. female sexual receptivity) is restricted to the later stages of vaginal oestrus and copulations occur only during a period of 4–8 h (maximum 12 h: Foerg 1982b). Shideler *et al.* (1983) obtained further information on seven female and five male *V. variegata* and were able to measure

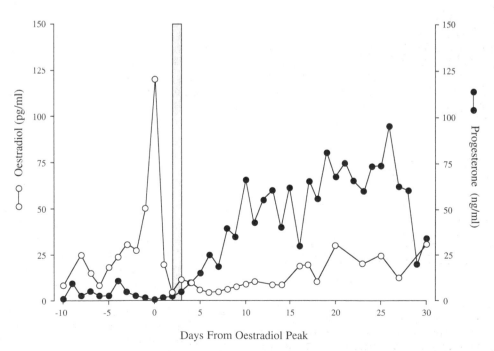

Fig. 11.4 Changes in circulating hormone levels and sexual receptivity during the ovarian cycle of the ringtailed lemur (*Lemur catta*). Open circles show mean levels of oestradiol; closed circles show mean levels of progesterone; the vertical bar indicates the period during which females are sexually receptive to males during pair-tests. (Redrawn and modified form Van Horn and Resko 1977.)

patterns of urinary oestrogen excretion, as well as sexual and associated patterns of behaviour. Receptive females adopted a 'crouch presentation' posture in response to male contact and this receptive (lordotic) posture occurred the day after the peak of urinary oestrogen excretion (Fig. 11.5). It is important to note that in their original publication Shideler *et al.* (1983) aligned data on peak urinary oestrogen levels and female crouch presentations on the same day for statistical purposes. In reality, oestrogen peaked one day before the day of female receptivity and females mated at the expected time of ovulation.

Shideler *et al.* (1983) comment on the relative absence of proceptive displays in ruffed lemurs. Females which self-groomed their genitalia and made eye-contact with males might have been exhibiting some type of invitational behaviour, but this is uncertain. Males were attracted by the swollen female genitalia, however, and visual as well as olfactory cues might have been important in stimulating their mounting behaviour.

Information on ovarian cycles and sexual behaviour in other diurnal prosimians is less detailed. Kappeler (1987), who studied four captive pairs of *Lemur coronatus*, found that the ovarian cycle lasted for 34 days on average, but all copulations occurred on a single day. Among the Indriidae, field studies of the sifaka (*Propithecus verreauxi*) have shown that this species lives in small groups, each containing several adult males and the same number of females (Jolly 1966, 1972; Richard 1976, 1978). The mating

season occurs between January and March in Madagascar. Swelling and pinkness of the vulval region is seen in females during February, when males begin to show increased frequencies of scent-marking, 'roaming' behaviour (extending their ranges into areas occupied by neighbouring troops), intermale aggression, and inspections of females (Richard 1976). All copulations observed by Richard occurred during just 4 days in March, however, when several males attempted to mate with each female during her receptive period. Females rejected the majority of male attempts, but when accepting the male they adopted a squatting position-clinging to the vertical trunk of a tree and with the tail deviated slightly to one side. No obvious invitational displays were used by these female *P. verreauxi*; attractiveness primarily involved chemical cues and females presumably exerted some degree of 'choice' by selective refusals or acceptances of males' mounting attempts. More recent fieldwork by Brockman and Whitten (1996) has shown that sifakas mate during periods when oestradiol levels are highest, as indicated by measurements of faecal oestradiol in five subjects. Oestradiol was elevated for 10–15-day periods, but ejaculatory mounts were only observed during periods of from 1–3 days. Females mated with a number of partners during oestrus. However, they also mated when not in oestrus, after a new male had joined the social group. Brockman and Whitten suggest that in the sifaka 'hormone-sexual behaviour disjunctions may occur....associated with group male membership and female preferences for certain males'.

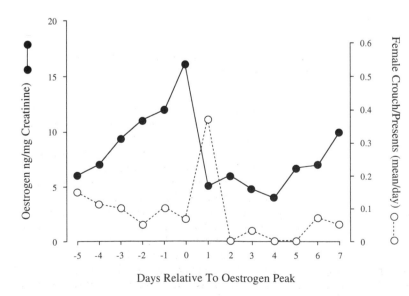

Fig. 11.5 Mean levels of urinary oestrogen and mean daily scores of sexually receptive crouch presentations in female ruffed lemurs (*Varecia variegata*). In this species receptive females adopt a distinct crouch presentation posture in response to being contacted by a male. The timing of receptivity in relation to oestrogen excretion is discussed in the text. (Redrawn and modified from Shideler *et al.* 1983.)

Conclusions: the ovarian cycle and sexual behaviour in prosimians

In all prosimian species for which data on ovarian cyclicity and sexual behaviour exist, copulation is usually restricted to the peri-ovulatory phase of the cycle during the period of vaginal cornification. Interestingly, some extension of 'behavioural oestrus' occurs in certain species (e.g., *G. garnettii* and *P. verreauxi*), so that matings are possible outside the strict confines of the ovulatory period. In other species, copulation is tightly harnessed to ovulation and receptivity lasts for less than 24 h (e.g. *Lemur catta*: Evans and Goy 1968). Studies of hormones and behaviour in several species indicate that receptivity occurs after the pre-ovulatory oestradiol peak, when oestrogen levels are beginning to fall (*Lemur catta*: Van Horn and Resko 1977; *Varecia variegata*: Shideler *et al.* 1983) and coincides with the probable timing of the luteinizing hormone (LH) surge. It is unfortunate that few measurements of LH during ovarian cycles have been reported for any prosimian species. In various non-primate mammals it is known that oestrogen *primes* the neural substrates which control female receptivity and that progesterone exerts an initial *facilitatory* action upon receptivity followed, in some cases, by an *inhibitory* action, so that receptivity is curtailed (Young 1961; Lisk 1978; Morali and Beyer 1979; Blaustein *et al.* 1990). It is not known whether comparable phenomena occur in any prosimian species; indeed in *L. catta*, increases in circulating progesterone occur after the receptive period, and it is possible that oestrogen alone facilitates the behavioural changes (Van Horn and Resko 1977). There is also the intriguing possibility that copulation contributes in some way to the termination of receptivity; natural matings are associated with shorter periods of behavioural oestrus in *G. garnettii* and *L. catta*, for instance.

In nocturnal and diurnal prosimian species, female attractiveness to males increases before the onset of sexual receptivity, and it is thought that hormonal effects upon vaginal, urinary, or cutaneous olfactory cues might mediate such effects. Given the non-gregarious nature of the social organization in many nocturnal prosimians (Bearder 1987), the need for efficient communication of reproductive condition by indirect methods has presumably favoured such mechanisms. As noted in Chapter 4 (see page 90), olfactory cues might also contain important information regarding the genetic constitution of the marking animal and thus influence mate choice. In diurnal prosimians, such as various *Lemur* species or in *Propithecus*, it appears

that olfaction still plays a key role in sexual contexts. Given that many of these species have very restricted mating seasons, chemical cues might also facilitate reproductive synchrony, both within and between the sexes. We shall return to this topic in the final chapter which deals with socioendocrinology and sexual behaviour.

Finally, we should note that few studies of prosimians have produced evidence that the ovarian cycle affects proceptive behaviour. A notable exception is provided by the spectral tarsier (*Tarsius spectrum*), in which females in oestrus frequently invite copulations by presenting to males (Wright *et al.* 1986). Despite the apparent absence of proceptivity in other species it is possible, for instance, that females of nocturnal species alter their ranging behaviour and patterns of scent-marking in association with the follicular phase of the ovarian cycle. This possibility remains to be examined by field research. Studies of captive mouse lemurs (Buesching *et al.* 1998) have shown that females scent-mark and 'trill-call' more frequently during the peri-ovulatory phase of the cycle (see Fig. 5.10, page 102). Clearly, sexual receptivity is under a well-structured control by cyclical changes in ovarian hormones in prosimians, a fact which has been appreciated for over 60 years (Zuckerman 1932). To this extent, therefore, female prosimians may be said to exhibit 'oestrus', although a behavioural dissection of attractiveness, receptivity, and proceptivity is more informative in revealing the mechanisms that control sexual interactions.

The ovarian cycle and sexual behaviour in New World monkeys

New World monkeys differ from other primates in a number of facets of their ovarian anatomy and endocrine physiology. Some understanding of these differences is required in order to appreciate how ovarian hormones influence sexual behaviour in New World primates. All primate ovaries contain similar component tissues—follicles, corpora lutea, interstitial glandular tissue, and accessory luteal tissue (Koering 1974, 1987). Although a number of primordial follicles begin to enlarge during the follicular phase of the ovarian cycle, selection of just one *dominant* follicle occurs, whilst the remainder undergo atresia. It is the dominant follicle that is responsible, both for the development of the viable ovum, and for secretion of the oestradiol required to trigger the pre-ovulatory LH surge (Knobil 1974). In some primates, atretic follicles are converted into accessory corpora lutea (due to luteinization of granulosa cells: Corner 1940; Koering 1979) and into interstitial glandular tissue (which arises mainly from the theca interna of the follicle: Mossman and Duke 1973). Such interstitial glandular tissue occurs in large quantities in the ovaries of New World primates (e.g. *Alouatta*, *Ateles*, *Saimiri*: Koering 1974, 1986). Extensive luteinization of ovaries, owing to formation of accessory corpora lutea, has also been described in some New World forms (*Cebus*: Nagle and Denari 1983; *Aotus*: Hertig *et al*. 1976). Perhaps it is for these reasons that the ovaries of New World monkeys are larger, in relation to their body weights, than those of other primates (Wislocki 1932; Hill 1972).

It is also possible that interstitial glandular tissue and accessory luteal tissue might contribute to the very high levels of circulating ovarian steroids which occur in New World primates. Circulating levels of progesterone are very high in *Saimiri*, *Cebus*, *Aotus*, *Callithrix*, and *Saguinus* (Robinson and Goy 1986) and greatly exceed those measured in Old World monkeys or apes. It is unlikely that such high levels of ovarian steroids result from slower metabolic clearance rates, or a greater binding capacity of plasma globulins. Large quantities of progestagenic and oestrogenic metabolites are excreted in the urine of New World monkeys, such as *Aotus lemurinus* (Bonney *et al*. 1979; Bonney and Setchell 1980), *Saguinus oedipus* (Brand 1981), *Cebuella pygmaea* (Ziegler *et al*. 1990), and other species (Hodges *et al*. 1979). In the common marmoset (*Callithrix jacchus*) sex hormone binding globulin (SHBG) occurs in the plasma and oestradiol is transported principally in the bound state. Progesterone, however, is not transported by SHBG in the marmoset and the very high concentrations of plasma progesterone which occur during the luteal phase of the ovarian cycle ($>100\,\mathrm{mg}\;\mathrm{mL}^{-1}$: Kendrick and Dixson 1983) are transported, either in the unbound state, or loosely bound to plasma albumen (Hodges *et al*. 1983). The functional significance of high progesterone levels relates to the lower sensitivity of target organs to the steroid, due, at least in part, to the lower concentrations of receptors for this hormone (Chrousos *et al*. 1982). A similar hypothesis has been advanced to account for the very high output of cortisol by the adrenal glands of squirrel monkeys, marmosets, and some other New World primates (Cassorla *et al*. 1982; Pugeat *et al*. 1984). As we shall see in the next chapter, the adrenal cortex is also responsible for secreting sex steroids, including progesterone, androstenedione, and oestrone, and these occur in higher concentrations in New World primates than

in Old World monkeys. These peculiarities of New World primate endocrinology must be taken into account when considering the possible central and peripheral actions of steroids upon their sexual behaviour.

A further contrast between New world monkeys and the Old World anthropoids, concerns the occurrence of the menstrual cycle. Very few New World primates menstruate, in contrast with the catarrhines for which menstrual cycles are the norm. Cyclical shedding of the uterine endometrium occurs in *Cebus apella* (Nagle *et al*. 1979; Nagle and Denari 1983), but blood loss is much less than in Old World monkeys, apes and women. In other cebids, menstruation either does not occur (e.g. in *Saimiri* and *Aotus*), or it is very difficult to detect externally (as in *Ateles*: Goodman and Wislocki 1935; Eisenberg 1976). None of the callitrichid primates menstruate. Morphological changes in the female external genitalia are absent in most New World monkeys and there is no sexual skin. Where swelling of the external genitalia occurs, as in some female squirrel monkeys (Wilson 1977b), or in *Alouatta palliata* (Glander 1980; Jones 1985), it may provide an approximate indication of the likely occurrence of ovulation. Cyclical changes in vaginal cornification have also been identified in a few genera, such as *Saimiri* (Denniston 1964), *Cebus* (Nagle and Denari 1983), and *Alouatta* (*A. caraya*: Colillas and Coppo 1978) but, in many cases, there is no cornification cycle (e.g., *Aotus*; *Callithrix*; *Saguinus*: Dixson 1983a; Epple and Katz 1983). Hence, in the majority of New World primates, measurements of circulating or urinary steroids are required to monitor ovarian cycles accurately. Considerable intraspecific as well as interspecific variability in cycle lengths has been reported in various studies. Intraspecific variations may reflect, in part, the effects of those methods used to obtain vaginal lavages or blood samples upon the hypothalamic–pituitary–ovarian axis. Such effects might be most pronounced when animals are not fully habituated to the sampling procedures. In the common marmoset (*Callithrix jacchus*), for instance, estimates of ovarian cycle length were originally shorter (mean 16.5 days: Hearn and Lunn 1975) than in subsequent studies using a larger number of well-habituated subjects (mean 28 days: Harding *et al*. 1982; Hearn 1983; Kendrick and Dixson 1983). The initial studies probably included cycles with inadequate luteal phases and consequent failure to sustain normal secretion of progesterone. Preslock *et al*. (1973) reported that in *Saguinus oedipus* ovarian cycles average 15.5 days, whereas Brand (1981) found that longer cycles, averaging 22.7 days, were more common. In squirrel monkeys (*Saimiri*) a vast range of cycle lengths (7–25 days) has been reported by different authors (Denniston 1964; Castellanos and McCombs 1968; Hutchinson 1970; Wolf *et al.* 1977). In this case, shorter cycles appear to be the norm for squirrel monkeys (e.g. 9 days: Wolf *et al.* 1977; 6–12 days: Diamond *et al.* 1984) and longer cycles probably result from protracted follicular phases with low oestrogen levels and consequent delay (or failure) of ovulation. In Nagle and Denari's (1983) studies of *Cebus apella*, cycle lengths were also variable, ranging from 13–28 days, but 69.4% of the 104 cycles measured by these workers fell within the range 18–23 days, with a mean duration of 20.8 days.

Experiments on ovarian cycles and sexual behaviour in New World primates must, therefore, address the possible effects of handling or sedating animals and collecting blood samples, as well as social conditions (housing and testing procedures), upon ovarian function. Quantitative measurements of sexual activity during isolated pair-tests might also differ from observations made on permanent social groups, where intrasexual competition and mate choice add considerable complexity to the situation.

The first New World monkey studied in any detail to provide information on the ovarian cycle and sexual behaviour was the squirrel monkey. Squirrel monkeys were more readily available at that time because of their frequent use in a variety of biomedical research programmes (Rosenblum and Cooper 1968). Initially, it was not appreciated that the genus *Saimiri* contains at least four species. Maclean (1964) divided squirrel monkeys into two types according to whether the pelage forms a rounded (Roman arch) or pointed (Gothic arch) arrangement over each eye. Subsequently, Hershkovitz (1984) identified *Saimiri boliviensis* as belonging to the Roman arch type, whilst Gothic arch forms could be divided into three species: *S. sciureus*, *S. ustus*, and *S. oerstedi*. In the wild, squirrel monkeys live in multimale–multifemale troops which vary considerably in size (e.g. 10–35 individuals per troop in small forests in Central America but up to 120–300 per troop in undisturbed Colombian forests: Baldwin and Baldwin 1981). Females occupy central positions within the troop, whereas adult and sub-adult males remain on the periphery, except during the annual mating season (Baldwin and Baldwin 1981). Squirrel monkeys mate between June–August (the dry season) in southern parts of their range, and six months earlier in more northern areas such as in Panama. Males may become 'fatted' during the mating season and increase in weight by 85–222 g around the upper

torso and arms (Mendoza *et al.* 1978; Baldwin and Baldwin 1981). Marked seasonal rhythms of spermatogenesis occur in male squirrel monkeys (Dumond and Hutchinson 1967), and there are seasonal changes in circulating oestrogens and progesterone in adult females (Diamond *et al.* 1984). In captive groups, there is some evidence that, as females enter mating condition, they prefer to approach and associate with fully fatted adult males (Mendoza *et al.* 1978). However, it appears that the majority of copulations are male-initiated. Males often display by pacing, looking at the female, opening the mouth as if to vocalize, and flattening or wiggling their ears (Baldwin and Baldwin 1981). Early work on captive squirrel monkey groups had shown that spermatozoa were detectable in vaginal lavages at approximately 8-day intervals during periods of maximal cornification (Rosenblum *et al.* 1967). Subsequently, Wilson (1977a) carried out more detailed behavioural studies on captive pairs and groups of squirrel monkeys. Using behavioural criteria, she found that the 'oestrous cycle' had a modal length of 8 days and that copulation occurred on one day (rarely on 2 days) during the cycle. Pair-tests conducted across 32 cycles revealed that male partners mounted 5.4 times on average during 30-minute encounters with females on the day of 'oestrus' and 2.4 times on the next day. Ejaculations occurred only on the day of 'oestrus', however, except in six cases where an ejaculation was also recorded on the next day. Figure 11.6 shows that although males occasionally attempted to mount at earlier stages of

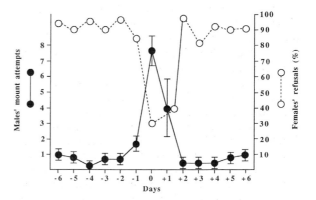

Fig. 11.6 Sexual behaviour during the ovarian cycle of the squirrel monkey (*Saimiri sciureus*). Mean frequencies of males' mounting attempts are graphed in relation to the percentages of attempts refused by females during pair-tests. Refusals decrease markedly on the day when most copulations occur (day 0) and on day +1; for the rest of the cycle males rarely attempt to mount and females refuse almost all of their attempts. (Redrawn from Wilson 1977a.)

the cycle, females refused 90–100% of such attempts. The increase in mounting activity during 'oestrus' correlates with a sharp decrease in frequencies of females' refusals, indicating an increase in female sexual receptivity at this time. Since 89% of the 216 mounts observed by Wilson were male-initiated there was little evidence of increased female proceptivity during oestrus, although females occasionally presented to males. Wilson's (1977a) studies also showed that sexual activity was limited to an annual mating season of 6.8–19.7 weeks, as was the case in the wild (8–12 weeks: Dumond and Hutchinson 1967; Baldwin and Baldwin 1981).

Only one detailed study of hormonal changes during the ovarian cycle of the squirrel monkey has been reported (Wolf *et al.* 1977) and no concurrent studies of sexual behaviour were undertaken. The ovarian cycle was measured at 8–9 days, as judged by interpeak intervals of circulating 17β-oestradiol and progestagens. Peak levels of oestradiol averaged 503 pg mL^{-1} and progestagen levels peaked at 399 ng mL^{-1} during the 6-day luteal phase of the cycle. Wolf *et al.* verified that 85% of the progestagen measured was, in fact, progesterone. In a second study Mendoza *et al.* (1978) recorded higher levels of oestradiol than Wolf et al. (up to 1200 pg mL^{-1}). Moreover, there is evidence that sexual activity itself can influence patterns of oestradiol secretion in the squirrel monkey. When females copulate on the day of the LH surge, or on the following day, plasma oestradiol levels are higher than in isosexually-housed females (Yeoman *et al.* 1987).

The case of the squirrel monkey, and of changes in sexual behaviour during its exceptionally short ovarian cycle, has been dealt with at some length because for many years it was the only New World monkey to have received detailed study. However, subsequent laboratory work on other New World primates showed that copulation is not necessarily restricted to the peri-ovulatory period. Indeed, during laboratory pair-tests, copulations occur throughout the ovarian cycle in *Aotus lemurinus* (Dixson 1983a), *Callithrix jacchus* (Kendrick and Dixson 1983), *Saguinus oedipus* (Brand and Martin 1983) and in captive groups of *Leontopithecus rosalia* (Stribley *et al.* 1987). Mating has also been recorded during pregnancy in all these species (*Aotus*: Dixson 1983a; *S. oedipus*: Brand and Martin 1983; *L. rosalia*: Kleiman and Mack 1977; *C. jacchus*: Dixson and Lunn 1987). This is not to say that ovarian hormones have no effect upon sexual interactions in these cases. Some examples, drawn from the field literature, will illustrate that remarkable changes in female attractiveness and behaviour occur in association with ovarian cycles. Then we shall consider

laboratory studies which have sought to determine how ovarian cycles influence proceptivity, receptivity and attractiveness.

Pioneering field studies conducted by Carpenter on the mantled howler monkey (*Alouatta palliata*: Carpenter 1934, 1965) indicated that females form consortships with a succession of males in the troop during an 'oestrus' of 3–4 days in duration. The onset of mating activity is signalled by the female approaching the male, which she does 'persistently and repeatedly', but both sexes may invite copulation by making rhythmic tongue-protrusion displays. Females may also present the rump as a proceptive behaviour and males frequently sniff and lick the urine or external genitalia of females before mounting. Further fieldwork on *A. palliata* showed that the mating peak coincides with some swelling and pinkness of the female's vulva. These genital changes occur at intervals of 11–24 days (mean 16.3 days: Glander 1980). Jones (1985) also recorded an average cycle length of 15–16 days, with peaks of sexual activity lasting from 1–3 days in free-ranging groups. Mantled howlers do not have a restricted mating season in the wild.

Unfortunately, no endocrine data are available on ovarian cyclicity in howler monkeys. Colillas and Coppo (1978) found that cyclic changes in vaginal cornification occurred every 20 days in captive black howlers (*A. caraya*). It seems likely that the marked increase in sexual activity seen in *A. palliata* during vulval swelling involves hormonal effects upon female attractiveness (olfactory cues) as well as behaviour. Proceptivity certainly increases at this time, but concrete evidence for changes in female receptivity during this putative oestrus period is lacking. There is, for instance, no indication that males find females sufficiently sexually attractive at other times of the cycle, or that females refuse to copulate when not in 'oestrus'.

In spider monkeys (*Ateles*) ovarian cycles last for 24–27 days with a menstrual period of 3–4 days (Goodman and Wislocki 1935; Harms 1956). In the wild, spider monkeys live in communities of up to 30 individuals but for most of the time these are split into flexible sub-groups containing 3–8 monkeys (Van Roosmalen and Klein 1988). Evidence pertaining to seasonal breeding is very limited; in Surinam, a birth peak occurs from November to February in the black spider monkey (*A. paniscus*). Van Roosmalen and Klein's field studies of *A. paniscus* show that females initiate the majority of matings and that each female copulates with a number of males during 8–10 day periods of heightened sexual activity. Females range widely at this time and may respond to the 'long calls' of males in distant subgroups by approaching and associating with them. Females also mark branches and foliage with urine. Klein (1971), who conducted field studies on long-haired spider monkeys (*A. belzebuth*) and black-handed spider monkeys (*A. geoffroyi*), observed that males lick, sniff or drink female urine and might spend up to five minutes examining an area where a female has micturated. Klein also observed that both sexes may head-shake, wrestle, or make panting sounds before mating and that the female often presents her hindquarters to the male or sits in his lap.

Closely related to spider monkeys is the woolly spider monkey (*Brachyteles arachnoides*). This, the largest New World monkey, once inhabited the magnificent Atlantic coastal forests of Brazil from Bahia southwards to Sao Paulo and Parana. Destruction of these forests has reduced the range of *Brachyteles* to a small scattering of sites where 250–400 monkeys live in multimale–multifemale groups numbering 25–30 individuals (Aguirre 1971; Nishimura *et al.* 1988). In her field studies of *Brachyteles* at Barriero Rico in Sao Paulo State, Milton (1984, 1985a) found that a group may be divided into sub-groups with a fusion–fission type of social organization. The most consistent sub-groupings contained 3–5 adult females with offspring and these occupied stable home ranges, whereas adult males ranged more widely, either alone or in groups of 2–8 individuals. Copulations occur at any time of year but there is a mating peak during the early rainy season in October and November. Data on the ovarian cycle are limited (Strier and Ziegler 1994), but it is known that copulations take place during a restricted period (36–48 h) and that a single female can mate with four or five different males sequentially (Aguirre 1971; Milton 1985a). Before the commencement of mating, however, the female increases her day range and begins to give twittering vocalizations, and to deposit urine more frequently by means of 'urine-washing' displays. Milton (1985a) observed that 7–9 males arrived in the foraging area of a female which behaved in this way. The males formed a mating aggregation around the female and 31 copulations occurred, larger males taking precedence during the earlier phase of the mating period. Towards the end of the two-day mating period the larger males sniffed the female's genitalia, but they did not mount her and gradually moved away. This occurred despite the fact that the female followed males and solicited copulations. Apparently, the mating aggregation dissolved due to a waning of female sexual attractiveness rather than because the female was no longer proceptive or receptive.

Periods of heightened sexual activity have also been recorded in females of some other New World species, such as the woolly monkey (*Lagothrix lagotricha*: Nishimura 1988) and the black-capped capuchin (*Cebus apella*: Janson 1984). It is notable that both proceptivity and female sexual attractiveness increase at such times. Black-capped capuchins live in multimale–multifemale troops and females exhibit a restricted period of sexual activity lasting for 5–6 days (Janson 1984). During this time, the female solicits copulations with the highest-ranking adult male in the troop, following him, vocalizing (a 'soft whistle' rising in pitch to a 'hoarse whine'), and making a grimacing facial expression. Janson (1984) found that these proceptive displays were most commonly observed over a 4-day period, during which the highest-ranking male mated every day. However, on day 5, the female began to solicit subordinate males, whilst the highest-ranking male followed and guarded her. By day 6 this guarding behaviour had ceased and up to six subordinate males then mated with the female. As in the previous description of the sexual behaviour of the woolly spider monkey, it appears that less dominant or smaller males have access to the female once her period of maximal attractiveness has passed, and that she remains sexually receptive at this time. Janson's field observations on *C. apella* are supported by observations of Welker *et al.* (1990) on a captive group at the University of Kassel in Germany. Welker *et al.* found that females directed 81% of observed courtship episodes (N = 738) at the dominant group male, 14% at four other adult males and 5% at two sub-adult males. The dominant male accounted for 52% of observed copulations. Linn *et al.* (1995) also report increases in proceptivity at mid-cycle in captive groups of *C. apella*; high-ranking females were most successful in obtaining copulations with dominant males in these social groups.

Laboratory work has shown that the ovarian cycle of *C. apella* lasts for 20 days, on average, with a follicular phase of 8 days and a luteal phase of 11 or 12 days. Nagle and Denari (1983) established that plasma levels of 17β-oestradiol during the follicular phase increase from $50–150$ pg mL^{-1} to a pre-ovulatory peak of $420–600$ pg mL^{-1}. A surge of LH was measurable 24 h after the oestradiol peak. Progesterone, which remained at low levels (5 ng mL^{-1}) during the follicular phase, rose to 50 ng mL^{-1} during the period of the LH surge and then to $60–100$ ng mL^{-1} during the mid-luteal phase of the cycle. Unfortunately, no study has attempted to correlate endocrine events during the cycle with the most interesting changes in patterns of sexual behaviour which are assuredly underpinned, in some fashion, by hormonal mechanisms.

The most detailed experimental studies of the ovarian cycle and sexual behaviour in a New World primate species have involved the common marmoset (*Callithrix jacchus*: Kendrick and Dixson 1983; Dixson and Lunn 1987; Dixson 1992). These studies will be described in some detail, because they are relevant to those sections of the next chapter which explore the effects of ovarian steroids upon the brain and behaviour in female primates. Figure 11.7 shows profiles of plasma progesterone and testosterone recorded for seven female common marmosets. These females were pair-tested with the same male partners, three or four times per week, for periods between 44 and 68 days (Kendrick and Dixson 1983). Ovarian cycles ranged from 24–30 days in duration (mean 27.7 days) and it was estimated that 13 ovulatory cycles (1–3 cycles per female) had occurred during the study period. Ovulation was predicted to have occurred 24 h before the rise in progesterone above 10 mg mL^{-1}. The peri-ovulatory phase of the cycle was taken to include two tests before ovulation and two tests after it (i.e. 3–4 days either side of ovulation). The luteal phase of each cycle was calculated as lasting from the time plasma progesterone rose above 10 mg mL^{-1} until hormone levels had fallen again to remain consistently below this level. Luteal phases ranged from 14–24 days in length, with a mean duration of 19 days. The remainder of the cycle, during which plasma progesterone was at low levels, was taken to be the follicular phase (range 4–12; mean 8.7 days). Also shown in Fig 11.7 are the average frequencies of proceptive tongue-flicking by these female marmosets towards their male partners. Female common marmosets invite copulation by remaining immobile, staring at the male, and making rhythmic tongue protrusion movements. Tongue-flicks can be counted relatively easily and it is clear from Fig 11.7A that proceptivity increases markedly during the peri-ovulatory period and decreases or disappears completely in most females once the luteal phase of the cycle has begun.

Rhythmic changes in receptivity also occurred during ovarian cycles in these females, but were much less pronounced than changes in proceptivity. This is made clear in Fig. 11.8 which shows frequencies of proceptivity and receptivity during the follicular, peri-ovulatory, and luteal phases of the cycle for females in each of the seven pairs studied. Females tongue-flicked at males significantly more frequently during the peri-ovulatory phase than at other times during the cycle. They remained sexually receptive throughout the cycle, however, although they refused slightly more mount-attempts by males during tests conducted in the luteal phase of the cycle. Males mounted and ejaculated at all phases of the ovarian cycle and, despite increased frequencies of these patterns during the peri-ovulatory phase (and

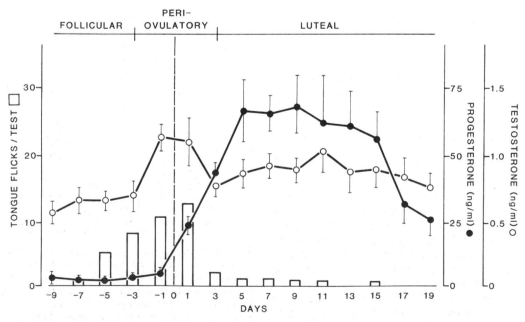

Fig. 11.7 Overall means (±S.E.M.) for plasma levels of progesterone (•) and testosterone (○) during the ovarian cycle of the common marmoset (*Callithrix jacchus*), together with measurements of proceptive tongue-flicking (open bars) shown during pair-tests. Data refer to seven females. (Redrawn and modified from Kendrick and Dixson 1983.)

a reduction in the males' post-ejaculatory intervals), there was no indication of a limited period of female receptivity or 'oestrus' such as occurs in prosimian primates.

The contribution made by the female's proceptive displays to the male's increased sexual performance during the peri-ovulatory phase is shown in Table 11.1. By dividing each female's peri-ovulatory phase into tests when proceptive tongue-flicking was present and those when it was absent, it was found that five measurements of the males' sexual activity all improved significantly during tests when females behaved proceptively. Thus, males mounted and ejaculated more frequently, their post-ejaculatory intervals were shorter, and the durations of their penile erections after ejaculation were greater in these circumstances (Table 11.1). Males also tongue-flicked more frequently during tests with proceptive females; both sexes use tongue-flicking as a precopulatory display in this callitrichid species. Therefore, it appears that the female's proceptivity during her peri-ovulatory phase, rather than her increased receptivity, or some non-behavioural cue, has the most significant effect upon the male's sexual behaviour during pair-tests. Indeed, when the data for peri-ovulatory phase tests which lacked female proceptive tongue flicking were compared with data collected during the follicular and luteal phases there was no longer any significant variation in the sexual activity of the male partners (Kendrick and Dixson 1983).

Pair-testing monkeys to observe sexual interactions under laboratory conditions has numerous advantages. All behavioural patterns may be observed and recorded accurately from behind a one-way mirror during restricted periods of time (30 min per test in the experiments described above). Social complexity is reduced to a minimum, and blood samples can be obtained from the female at other times during the test-day, in order to monitor the course of her ovarian cycle. However, pair-tests may also distort hormone-behaviour interrelationships. Although marmosets mount and ejaculate at all stages of the female's ovarian cycle during pair-tests, it seems unlikely that this occurs in permanent social groups of these monkeys in the wild. The male's sexual arousal may be much greater during pair-tests, for instance, because he is isolated during the rest of the day and has only a limited period available in which to copulate. In the wild, *Callithrix jacchus* normally lives in small family groups containing a single breeding female. In N.E. Brazil, groups of *C. jacchus jacchus* (the subspecies studied by the author in the laboratory) range in size from 3–13 individuals (mean 8.6: Hubrecht 1984). Hubrecht

found that up to three adult males and two adult females occurred in these groups, and similar findings have been reported for other marmoset species (Stevenson and Rylands 1988). The breeding female gives birth to twins once or twice each year and a peak of sexual activity occurs two or three weeks after parturition (Dixson and Lunn 1987). Female marmosets are adapted to maintain a very high reproductive rate and there is no lactational suppression of ovarian cyclicity such as occurs in many catarrhine species (McNeilly *et al.* 1981). Measure-

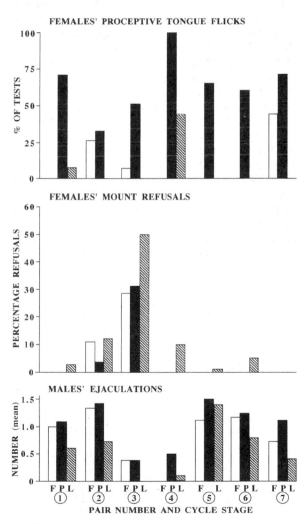

Fig. 11.8 Mean percentages of pair-tests during which female common marmosets (*Callithrix jacchus*) tongue-flick proceptively at males, percentages of males' mount attempts refused by females, and frequencies of ejaculatory mounts during the follicular phase (F: open bars), peri-ovulatory phase (P: closed bars) and luteal phase (L: hatched bars) of the ovarian cycle. Data are for seven females and their male partners. (Redrawn from Kendrick and Dixson 1983.)

ments of circulating or urinary luteinizing hormone and ovarian steroids show that ovulations and conceptions occur between two and three weeks after parturition in some callitrichids (*Saguinus oedipus*: Ziegler *et al.* 1987; *Cebuella pygmaea*: Ziegler *et al.* 1990; *Callimico goeldii*: Ziegler *et al.* 1989). Naturally enough, the increase in sexual activity which occurs at this time has often been referred to as a 'post-partum oestrus'. Soini (1987) has provided an account of postpartum oestrus in a group of pigmy marmosets (*Cebuella pygmaea*) which he studied in the Amazonian forests of eastern Peru. Two males were present in the family group, but one mate guarded the female and prevented copulations by the second male. Nine days after the female had given birth, both males showed a marked increase in olfactory inspections of the female's genitalia. However, the female responded to the males' approaches by vocalizing aggressively and making a distinctive genital presentation posture with the tail raised. Such postures occur in dominance-related contexts in a variety of callitrichid species and are not equivalent to the sexual presentations of other monkeys. Soini observed no copulations until the 15th and 16th day after parturition. The dominant male made tongue-protrusion and tongue vibrating displays before mounting the female. Interestingly, studies of captive *Cebuella* have shown that the pre-ovulatory LH surge occurs, on average, 15.6 days after parturition, at the precise time when Soini (1987) pinpointed the occurrence of 'behavioural oestrus'.

How do ovarian hormones influence sexual activity during the post-partum period in callitrichid primates? To investigate this question, we used a video-monitoring system to record all incidences of sexual and associated behavioural patterns during 24–30 days after parturition in eight groups of common marmosets. Blood samples were obtained at regular intervals from females, in order to measure changes in plasma LH and progesterone throughout the post-partum period and to correlate these with the behavioural data (Dixson and Lunn 1987). In six of the eight groups studied, females exhibited a pre-ovulatory LH surge between days 10 and 18 (mean $\pm$ S.E.M.: 13.8 ± 1.3 days). Frequencies of mounts and ejaculations by their male partners increased during the post-partum period and were highest during the peri-ovulatory phase (defined as ± 3 days from the pre-ovulatory LH peak (day 0)). However, as can be seen in Fig. 11.9, mating was not totally restricted to the time around ovulation and occurred also during the pre-ovulatory phase and the post-ovulatory phase of the post-partum period.

Males in the six groups first attempted to mount females between days 1–10 after parturition (mean 4.8 ± 1.5 days). In all groups, males initiated the first mounts (Table 11.2) and used tongue-flicks as precopulatory invitations before females exhibited any proceptivity. However, before the peri-ovulatory phase began, ejaculatory mounts were less frequent, particularly during the first 5–7 days after parturition. During this pre-ovulatory phase, females were more aggressive to males and refused or terminated more of their mounts It should be kept in mind that multiple factors, such as trauma to the female's reproductive tract and discomfort after giving birth, her subsequent maternal behaviour, and endocrine changes after parturition, might all influence behavioural responsiveness and receptivity to the male. The effects measured were quite variable and in some groups ejaculations occurred as early as day 1 or 2 after parturition (Table 11.2).

If we next consider the peri-ovulatory phase, this was associated in all groups with increased frequencies of mounts and ejaculations. Decreases in female aggressiveness towards males, coupled with

Table 11.1 Sexual behaviour of male marmosets during the peri-ovulatory phase of the female's cycle and the effect of her proceptive tongue-flicking display

	Tests with proceptive tongue-flicks				Tests without proceptive tongue-flicks			
Male	No M	No E	PEI (s)	ED (s)	No M	No E	PEI (s)	ED (s)
1	2.00	1.20	1455.0	1565.0	1.00	1.00	1800.0	340.0
2	4.25	1.50	1026.5	201.0	3.13	1.38	1030.0	148.5
3	25.00	0.50	–	–	19.00	0.00	–	–
5	6.20	1.60	520.0	110.0	2.83	1.33	827.5	82.5
6	4.30	1.67	532.0	205.0	1.50	1.00	1774.5	190.3
7	4.00	1.18	323.5	170.2	3.00	1.00	504.0	133.5
Mean	7.63	1.28	771.4	450.2	5.08*	0.95*	1187.2*	152.2*

*P < 0.05.
Means are given for all parameters excepting the PEI where medians are used.
M=mounts; E=ejaculations; PEI=post-ejaculatory interval; ED=post-ejaculatory erection duration.
Data are from Kendrick and Dixson 1983.

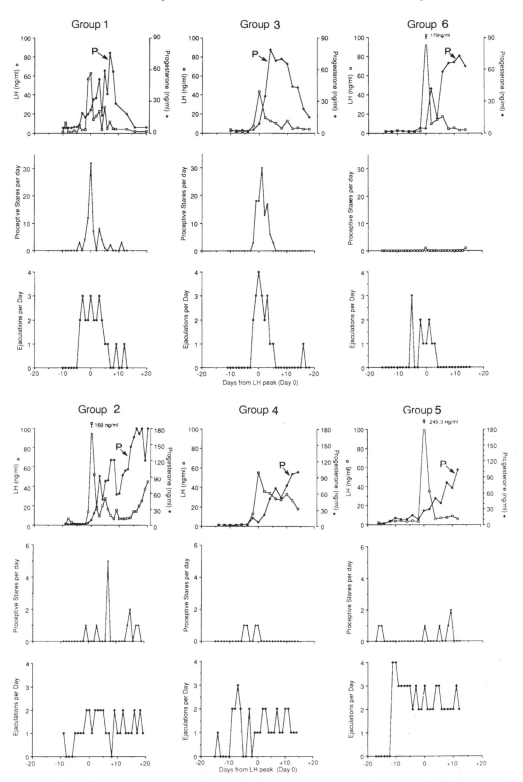

Fig. 11.9 Hormonal and behavioural changes during the post-partum period in six captive family groups of common marmosets (*Callithrix jacchus*). In groups 1, 3 and 6 females did not conceive; in groups 2, 4, and 5 conceptions occurred. For each female, circulating levels of luteinizing hormone (LH) and progesterone (P) are shown, together with daily frequencies of proceptive staring displays by females and ejaculatory mounts by their male partners. Daily measurements of behaviour and hormone levels are plotted with respect to the occurrence of the pre-ovulatory LH peak (Day 0). (Redrawn from Dixson and Lunn 1987.)

heightened proceptivity and subtle changes in receptivity might have combined to enhance the male's sexual activity at this time. Because there were important differences between pairs, data on hormonal changes and sexual interactions in the individual groups are presented in Fig. 11.9. Three females failed to conceive during the study (Groups 1, 3, and 6 in the upper section of Fig. 11.9). Marked peaks in frequencies of proceptive staring displays in two females (Groups 1 and 3) coincided with the day of the LH surge, and in all three groups, males ejaculated most frequently during the peri-ovulatory period and showed a marked decrease in ejaculations during the luteal phase of the female's cycle. By contrast, in the three groups where females became pregnant (Groups. 2, 4, and 5 in the lower section of Fig. 11.9), males began to ejaculate earlier in the follicular phase and continued to do so throughout the peri-ovulatory phase and into early pregnancy. Examination of data from individual family groups failed to reveal any consistent differences in proceptivity, receptivity, or aggression between these two sets of females during the post-ovulatory phase. It seems likely, therefore, that non-behavioural (sexually attractive) cues continue to operate in female marmosets after they have conceived, resulting in maintenance of sexual activity by males. In an adaptive context, such a mechanism may assist the newly pregnant female to maintain a positive relationship with the male (or males) in her group which subsequently assist in transporting her twin offspring.

Little has been said so far about the mechanisms which control female sexual attractiveness in marmosets during the post-partum period. However, changes in female attractiveness certainly occur at this time and have much greater effects upon the male's copulatory behaviour than those observed during isolated pair-tests. Since callitrichids make extensive use of cutaneous secretions and urine for olfactory communication, it is possible that chemical cues might influence female sexual attractiveness. The videotape studies of captive marmoset groups revealed that frequencies of genital scent-marking behaviour by both sexes showed no consistent changes during the first post-partum month. Although males' olfactory inspections of the females' genitalia and scent-marks increased during the peri-ovulatory phase, these effects were not statistically significant. However, frequency measurements of this kind on captive groups are of limited value, because the observation cage is presumably permeated by the odour of the animals and repeated close contact with chemical cues might be rendered unnecessary. It is more relevant to determine whether depriving males of olfactory inputs has any measurable effects upon their sexual activity. Figure 11.10 shows the results of such an experiment. In four family groups of common marmosets, the olfactory bulbs of the breeding males were transected while their female partners were in the final stages of gestation. Males were returned to their family groups on the day of the operation; they recovered rapidly and continued to feed normally. Olfactory discrimination tests of the ability of these males to identify the presence of caproic acid demonstrated the effectiveness of the operations, which had disconnected inputs from the main olfactory system and from the vomeronasal organ. Videotape recordings of sexual activity and hormonal measurements were made throughout the post-partum period in these four groups (Dixson 1992). Figure 11.10 shows average daily frequencies of ejaculatory mounts in the four anosmic male marmosets compared with data for the six intact males studied in the previous experiment. Interestingly, the anosmic males ejaculated more frequently than intact males, and they began to mount and ejaculate much earlier during the

Table 11.2 Day of first occurrence of selected features of sexual behaviour during the post-partum period in family groups of common marmosets

| Behaviour pattern | Group number | | | | | | Mean ± SEM |
	1	2	3	4	5	6	
Male: pre-copulatory display	1	6	8	5	2	10	5.33 ± 1.40
Female: proceptivity	6	9	10	9	2	10	7.66 ± 1.28
Mount or mount attempt	6	1	8	2	2	10	4.83 ± 1.51
Male initiates mount	6	1	8	2	2	10	4.83 ± 1.51
Female initiates mount	7	9	10	10	18	11	10.83 ± 1.53
Ejaculation	7	1	10	2	7	11	6.33 ± 1.66
Day of LH peak	11	10	12	16	18	16	13.83 ± 1.32

After Dixson and Lunn 1987.
Day 1 is the day of parturition

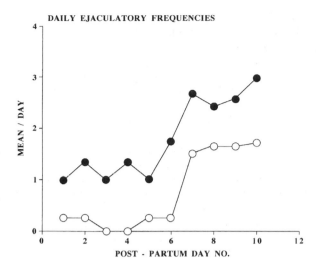

Fig. 11.10 Effects of anosmia (olfactory bulb transection) upon ejaculations during the post-partum period in male common marmosets (*Callithrix jacchus*). Data are daily means for anosmic males (filled circles: N=4) and intact males (open circles: N=6) for the first 10 days after their female partners gave birth. (Redrawn from Dixson 1992.)

post-partum period. Thus, olfactory inputs are not essential for activation of masculine sexual arousal and copulatory behaviour under these conditions, but they are important in enabling the male to co-ordinate his sexual activity with the female's endocrine status. Although some females are more aggressive during the early post-partum phase, and reject the male's mount attempts more readily at this time, anosmic males are sufficiently persistent to succeed in mounting and ejaculating. Intact males, by contrast, find their female partners less attractive during the early post-partum phase and are less likely to initiate mounts, or to persist if their attempts are refused. Figure 11.10 also shows that both the anosmic and intact groups show a gradual increase in ejaculatory activity after the earliest post-partum period (e.g., note data for post-partum days 6–10). Presumably, behavioural cues from the females influenced these changes in male sexual activity, since they also occurred in anosmic males which were unable to detect changes in female odour. Changes in the appearance of the female genitalia are most unlikely to be relevant in this context, given that genital swellings do not occur in marmosets.

Conclusions: the ovarian cycle and sexual behaviour in New World monkeys

Despite the occurrence of well-defined periods of sexual activity in some New World monkeys, often referred to as 'oestrus' by field workers (e.g., in *Alouatta*: Carpenter 1934; Jones 1985; *Cebus*: Janson 1984), there are major differences in the hormonal control of female sexual behaviour between New World primates and the prosimians described in the previous section. As we have seen, female lemurs and many other prosimians are sexually receptive only during the peri-ovulatory phase of the ovarian cycle, when the vaginal epithelium becomes cornified due to oestrogenic stimulation. Female prosimians exhibit 'oestrus' in the sense originally defined by Heape (1900). Although Heape acknowledged that oestrus might occur at other times than around ovulation in mammals, he made a number of mutually contradictory statements which preclude a clear definition of his concept. It is precisely for this reason that it is misleading to say that New World monkey females exhibit 'oestrus', whether in the wild or in captivity. In the wild, for instance, females may continue to invite copulations after their sexual attractiveness to males has begun to wane during the post-ovulatory period (*Brachyteles arachnoides*: Milton 1985a; *Cebus apella*: Janson 1984). In captivity, copulations may occur throughout the ovarian cycle (e.g., in *Callithrix jacchus*: Kendrick and Dixson 1983; *Saguinus oedipus*: Brand and Martin 1983; *Aotus trivirgatus*: Dixson 1983a) or, in some species, during pregnancy. Although a post-partum mating peak occurs in many callitrichid primates, laboratory studies have shown that changes in female receptivity are relatively subtle at this time. Changes in female proceptivity and attractiveness play a greater role in co-ordinating the male's sexual activity and thus in creating a peri-ovulatory peak in sexual activity (Dixson and Lunn 1987; Dixson 1992; Jurke *et al.* 1995).

If, indeed, sexual receptivity is controlled less rigidly by ovarian hormones in New World monkeys than in prosimian primates, why should this be so? Why are ovarian events more strongly linked to changes in proceptivity or sexual attractiveness in some of these monkeys? Firstly, there are some similarities between the two primate groups in the importance of olfactory cues in the control of female sexual attractiveness. New World monkeys, like prosimians, make extensive use of cutaneous glandular secretions and urine for scent-marking purposes (e.g., Epple and Lorenz 1967; Epple *et al.* 1993). Many New World monkeys, like prosimians, also possess a functional vomeronasal organ and an accessory olfactory system (Hunter *et al.* 1984), so that chemical stimuli are expected to play a key role in neuroendocrine contexts, as is the case in various non-primate mammals. Although the New World forms all live in social groups and, with the exception of *Aotus*, are diurnal in behaviour, they have retained mechanisms of olfactory communication which originated in nocturnal, non-gregarious ancestral primates. However, the New World monkeys do exhibit more complex types of social communication than most prosimians and their visual displays are much more highly developed, in association with their diurnal habits. It is possible that during the evolution of such social complexity, in New World monkeys as among anthropoids in general, selection has favoured mechanisms which allow females a greater degree of choice in sexual contexts. The ability to attract a number of males in the period leading up to ovulation might be elaborated in monkeys living in complex social groups. Invitational displays can be used more selectively by the female and choice may operate by the expression of proceptivity, dissociated from a rigid hormonal control of receptivity. Females might, for instance, invite more dominant males to copulate during the fertile period, but solicit or allow copulations by lower-ranking individuals at other times (e.g., in *Brachyteles*: Milton 1985a; *Cebus*: Janson 1984). Given that several adult males are present in groups of some callitrichids, the ability of females to remain attractive and receptive during gestation might be important in strengthening affiliative relationships and encouraging paternal care of the resulting offspring. Primarily monogamous New World monkeys, such as callitrichids or *Aotus*, might also benefit during the phase of pair formation from the ability to dissociate sexual receptivity from hormonal control. In captivity, newly formed pairs of marmosets or owl monkeys certainly copulate much more frequently than established pairs. In the wild, it is known that copulations occur between individuals in different social groups of marmosets; such behaviour might represent part of the process of forming new breeding units.

The menstrual cycle and sexual behaviour in catarrhine primates

All catarrhine species studied to date have a menstrual cycle, approximately one month in duration, although cycle lengths vary considerably, both at the individual level and between species. The amount of menstrual flow is also highly variable. Average cycle lengths of 24–27 days occur in *Presbytis entellus* (David and Ramaswami 1969; David and Rao 1969; Chowdhury *et al*. 1980), 30–31 days in *Cercopithecus aethiops* (Rowell 1970b; Hess *et al*. 1979), 30–34 days in *Cercocebus* spp. (*C. albigena*: Rowell and Chalmers 1970; *C. atys*: Aidura *et al*. 1981), 30–35

days in *Papio* spp. (e.g., *P. cynocephalus* and *P. anubis*: Wildt *et al*. 1977), and 31–36.7 days in *Pan troglodytes* (McArthur *et al*. 1981; Reyes *et al*. 1975; Coe *et al*. 1979). In some cases, menstruation is difficult to identify externally, but erythrocytes are detectable in vaginal lavages (e.g., in *Cercocebus albigena* and *Cercopithecus aethiops*).

The neuroendocrine control of the menstrual cycle and comparative information on hormonal changes during cycles in monkeys and apes have been extensively reviewed (e.g. Robinson and Goy 1986; Hotchkiss and Knobil 1994). It should be kept in mind that many studies have concentrated upon the most typical patterns of hormonal cyclicity exhibited by females, yet menstrual cycles may become irregular under certain conditions and the follicular phase is particularly susceptible to modulation by environmental factors. Longer follicular phases occur, for example, if female baboons are isolated from their social groups or if they are subject to aggression by other group members (Rowell 1970a). Irregular and anovulatory cycles, involving reduced secretion of gonadotrophins and oestradiol, also occur in some seasonally breeding species once the mating season is over (e.g., in the rhesus monkey: Koering 1986).

The rhesus monkey

The physiology of the menstrual cycle has been particularly well-studied in the rhesus monkey and this species has also been used for numerous experiments on the hormonal control of sexual behaviour. We shall, therefore, consider how the menstrual cycle influences patterns of sexual behaviour in the rhesus monkey, and then examine comparative data on other Old World monkeys. The menstrual cycle of the rhesus monkeys lasts, on average, for 25.5–29.5 days (Robinson and Goy 1986). Circulating concentrations of 17β-oestradiol increase from 50–100 pg mL^{-1} in the early follicular phase, to 150–200 pg mL^{-1} as the dominant follicle matures, and peak at 350 pg mL^{-1} before ovulation. The oestradiol peak occurs 9–15 h before the peak of luteinizing hormone which triggers ovulation approximately one day later (Weick *et al*. 1973). As can be seen in Fig. 11.11, levels of follicle-stimulating hormone (FSH) also peak at the same time as LH. Oestradiol levels remain low during the luteal phase of the menstrual cycle in the rhesus macaque, but in a related species (*Macaca arctoides*) there is a secondary, but smaller, rise in oestradiol during the luteal phase (Wilks 1977), as is also the case for the great apes and the human female. Progesterone concentrations are low during the follicular phase of the rhesus monkey's menstrual cycle (<0.5 ng mL^{-1}), but they begin to rise 1–2 days before ovulation and then attain maximal levels (4–6 ng mL^{-1}.) during the luteal phase, between days 15–22. Progesterone levels then decline before menstruation (Fig. 11.11). Whilst the characteristics of the menstrual cycle can be adequately defined with respect to changing concentrations of plasma oestradiol, progesterone, and gonadotrophins, these form part of a more complex picture. Hormonal activity will also depend upon transport mechanisms (such as availability of sex-hormone-binding globulin) upon the availability of receptors in target tissues and intracellular mechanisms of hormone metabolism. All these factors might alter during the course of the cycle. Other hormones besides those shown in Fig. 11.11 also exhibit rhythmic fluctuations during the menstrual cycle. During the luteal phase of the cycle substantial concentrations of 17α-hydroxypro-gesterone and progesterone occur in the circulation of the rhesus monkey and both steroids follow a similar pattern (Bosu *et al.* 1972). A mid-cycle elevation of testosterone is also present in the rhesus monkey, as in some other anthropoids including the human female.

Field studies have shown that female rhesus monkeys undergo menstrual cycles during an annual mating season. Periods of heightened sexual and associated behaviour occur during menstrual cycles and these so-called 'oestrous' periods last on average for 5.2 days (Loy 1971), 9.2 days, (Carpenter 1942), or 11 days (Kaufmann 1965). Lindburg (1983) also found that these periods averaged 9.4 days but a tremendous range of durations was possible (from 1–25 days in five groups he studied). He pointed out that 'the period of oestrus is relatively long in this species, and certainly much longer than the interval during which fertilization may take place'. The criteria used by fieldworkers with respect to 'oestrus' in female rhesus monkeys have included occurrences of grooming between the sexes, following and proximity, continued close association between a pair (i.e. consortship), as well as proceptivity and mounting. However, this broad-spectrum approach is of limited value in determining the contributions made by ovarian hormones to female receptivity, proceptivity or sexual attractiveness. Detailed exploration of these questions has involved experiments on captive rhesus monkeys.

Classical pair-test studies conducted in the 1960s and 1970s showed that in female rhesus monkeys sexual receptivity is not restricted to the peri-ovulatory phase of the menstrual cycle. However, many pairs copulate more frequently during the follicular phase or show a mid-cycle peak in ejaculations (Michael *et al.* 1967; Everitt and Herbert 1972; Goy and Resko 1972). Considerable variability occurs between pairs of rhesus monkeys, as is shown in Fig. 11.12. Some males exhibit a peak in mounts and ejaculations during the female's premenstrual period, as well as during the peri-ovulatory phase. In other cases, copulations occur at similar frequencies throughout the cycle, with little or no evidence of rhythmic fluctuations in the male's behaviour. Moreover, some pairs continue to copulate during pregnancy and a secondary peak of ejaculations occurs between the 6th–10th week of gestation (Bielert *et al.* 1976). Continued copulation during early pregnancy has also been noted in field studies of the rhesus monkey (Loy 1971; Lindburg 1983), but the proximate mechanisms underlying this phenomenon are not understood. Czaja and Bielert (1976) point out that a positive ratio of oestradiol 17β to progesterone continues during the first 8 weeks of gestation, and this might contribute to the

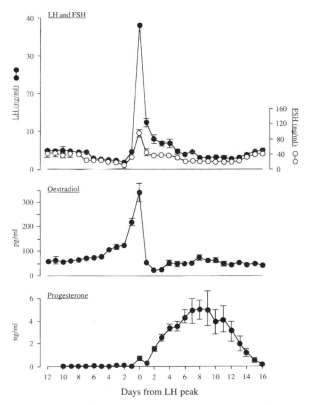

Fig. 11.11 Plasma concentrations (Mean ± S.E.M.) of luteinizing hormone (LH), follicle-stimulating hormone (FSH), oestradiol, and progesterone throughout the menstrual cycle of the rhesus monkey (*Macaca mulatta*). Data have been plotted with respect to the timing of the pre-ovulatory LH peak (Day 0). Each point is based upon seven measurements for FSH 19 for LH, 11 for oestradiol, and 7 for progesterone. (Redrawn from Knobil 1974, after Krey *et al.*)

peak in ejaculations. Yet sexual interactions cease during the second half of gestation despite a high oestradiol/progesterone ratio.

Attempts have been made to define how changes in female receptivity, proceptivity, or attractiveness during the menstrual cycle might influence the male's mounting or ejaculatory frequencies. One measure of receptivity often employed is the *male success ratio* (the proportion of the male's mounting attempts which are accepted by the female). Decreases in this measure have been noted during the luteal phase; this might contribute to the reduction in mounts and ejaculations at this time (Michael 1968). However, the changes are relatively subtle and are unlikely, by themselves, to account for alterations in the male's behaviour. Johnson and Phoenix (1978), for instance, found that females accepted 94–100% of all mounting attempts and showed no cyclical changes in receptivity. Proceptive patterns, including sexual presentations, head-bobs, head-ducks, and hand-reaches (which are variously summed together in different studies) might occur at any time during the menstrual cycle and vary

tremendously between pairs (Michael 1968; Johnson and Phoenix 1978). Higher frequencies of presentations during the peri-ovulatory period and of other invitations (head-bobs, head-ducks and hand-reaches) during the follicular and peri-ovulatory phases of the cycle have been reported by Zumpe and Michael (1983). Females may also seek proximity with the male (prox behaviour) most often during the peri-ovulatory phase of the menstrual cycle (Goy and Resko 1972; Czaja and Bielert 1975) and fear-grimace at him less (Goy and Resko 1972). The percentage of female solicitations which elicit a mounting response from the male (the *female success ratio*) increases at mid-cycle and decreases during the luteal phase in some pairings (Michael 1968). Female proceptivity might thus contribute to increases in the male's sexual activity, but it also appears that females are more attractive to males during the follicular and peri-ovulatory cycle phases. Not only do males respond more readily to female invitations at these times, but they initiate more mounts and ejaculate more frequently. The time required to attain ejaculation is reduced and rates of

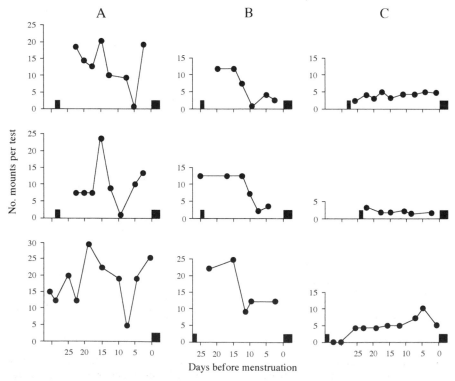

Fig. 11.12 Variability in copulatory behaviour during the menstrual cycle of the rhesus monkey (*Macaca mulatta*). Each graph refers to pair-test data for a single female and her male partner. Horizontal bars indicate the occurrence of menstruation. A. In some pairings mounts peak at mid-cycle and during the pre-menstrual period. B. Mounts occur most frequently during the follicular phase, and decrease during the luteal phase of the cycle. C. Sexual activity is at low levels and exhibits little cyclical change. (Redrawn from Everitt and Herbert 1972.)

intromitted pelvic thrusting (sexual performance index: Zumpe and Michael 1983) are greater than occur during the luteal phase.

These findings indicate that there are changes in female sexual attractiveness and proceptivity during the menstrual cycle in rhesus monkeys, but that their importance in determining copulatory frequencies during pair-tests varies considerably between individuals. Many attempts have been made to identify hormone-dependent stimuli which affect sexual attractiveness in female rhesus monkeys. The obvious candidates are olfactory stimuli (genital odours) and visual cues (cyclical changes in sexual skin coloration). A thorough discussion of the mechanisms by which oestrogen and progesterone affect female

attractiveness is included in the next chapter (see pages 354–7). However, it is relevant to note here that depriving male rhesus monkeys of their sense of smell does not alter the rhythmic changes in their ejaculatory frequencies (and latencies) during their female partner's menstrual cycles (Goldfoot *et al.* 1978a). Olfactory cues alone do not explain the increased sexual attractiveness of female rhesus monkeys during the follicular phase or at mid-cycle (Fig. 11.13).

As long ago as 1935 Ball and Hartman had reported that female 'sexual excitability' in rhesus monkeys increased before ovulation and decreased thereafter, although copulations occurred throughout the menstrual cycle. They summed female presentations, attempts to attract the male's attention by threatening an outsider ('threatening away'), and approaches to the male, to give an index of female excitability. However, they provided no quantitative data for any of these measurements. Since this early

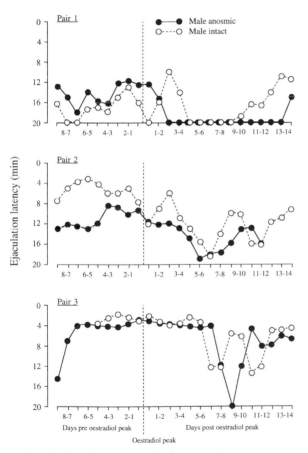

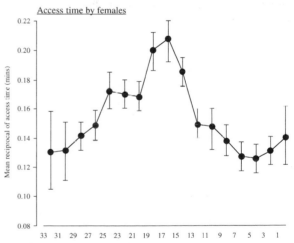

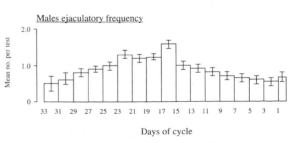

Fig. 11.13 Latency to first ejaculation during successive menstrual cycles in three rhesus monkeys (*Macaca mulatta*) pairs before (○) and after (●) males had been rendered anosmic by destruction of the nasal epithelium. (Data are plotted with respect to the occurrence of the pre-ovulatory oestradiol peak.) (Redrawn from Goldfoot *et al.* 1978a.)

Fig. 11.14 The operant behaviour of five female rhesus monkeys (access times) and the sexual behaviour of their male partners (ejaculatory frequencies) throughout the course of 32 menstrual cycles (mean cycle length=29.7 ± 0.8 days) plotted as days from menstruation. Access time =latency to first response plus time spent pressing to release the male. (Redrawn from Keverne 1977.)

attempt to relate changes in female sexual motivation to the phases of the menstrual cycle, much more sophisticated experiments have been reported. Operant techniques have shown that female rhesus monkeys will bar-press to gain access to a male partner most rapidly around the expected time of ovulation (Fig. 11.14). When the female is separated from her male partner by a wire mesh door she seeks proximity to him (by remaining less than 1 foot away from the partition) for longer during the peri-ovulatory period (Czaja and Bielert 1975). If rhesus monkeys are pair-tested, but the male is tethered to restrict his sphere of movement, females are more proceptive during the follicular phase than during the luteal phase of the cycle (Pomerantz and Goy 1983). These experiments indicate that hormonal changes might be exerting central effects upon the female's proceptive behaviour, although the possibility that females might be responding in some way to subtle changes in the behaviour of male partners still cannot be dismissed. Cyclical changes in female proceptivity are even more dramatic, however, if rhesus monkeys are tested in social groups and provided with adequate space in which to exhibit their social and sexual behaviour. Wallen *et al.* (1984) conducted a study in which each of four vasectomized adult male rhesus monkeys was introduced daily into an established group of nine adult females. Tests lasted for 30 min and the order of introduction of males was rotated throughout the study period (during the annual mating season). Blood samples were collected from females for measurement of circulating 17β-oestradiol, testosterone, and progesterone. This testing procedure produced much more focusing of sexual activity around the peri-ovulatory period of the cycle, as can be seen from Fig. 11.15. Mounts, intromissions, and ejaculations increased in frequency on the day of the oestradiol peak, remaining elevated for a further 2 days (i.e. during the putative fertile period). Wallen *et al.* found that ejaculations only occurred in the period extending from 4 days before until 5 days after the oestradiol peak. This is reminiscent of what occurs in free-ranging rhesus monkeys (as discussed earlier) but is in marked contrast with the results of free-access pair-test studies. Females played a crucial role in initiating copulations during this period, and there were differences in the timing of certain proceptive patterns in relation to the day of the oestradiol peak. Frequencies of several patterns began to increase several days before ovulation (e.g. approach, follow, establish proximity to the male). Such behaviour might be important in facilitating a consortship between the pair. By contrast, some female invitational behaviour (e.g., hand-slap and threatening away) showed their first significant

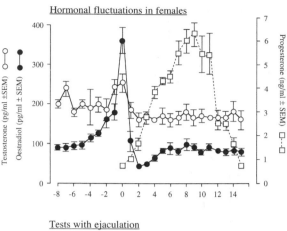

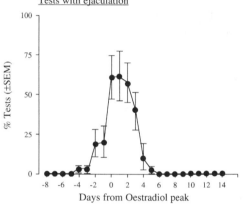

Fig. 11.15 Mean serum levels of oestradiol (●), progesterone (□) and testosterone (○) during menstrual cycles in nine female rhesus monkeys (*Macaca mulatta*), together with the mean percentage of tests with ejaculation. Data are plotted with respect to the day of the pre-ovulatory oestradiol peak (Day 0). Behavioural tests involved interactions between a single male and the *group* of females, rather than the use of free-access pair-testing procedures. (Redrawn from Wallen *et al.* 1984.)

elevation in frequency on day 0 (the day of the oestradiol peak), during the period of most intense copulatory activity. These points are made clear with reference to Fig. 11.16, which compares daily frequencies of approaches and hand-slaps by females in relation to the preovulatory oestradiol peak. None of these behavioural frequencies correlated significantly with plasma testosterone levels across the cycle, whereas female proceptivity correlated strongly and positively with oestradiol levels. It appears that female proceptivity can provide the male with precise information about her reproductive status; much more so than during free-access pair-tests. Changes in non-behavioural components of female attractiveness were much more difficult to discern in the studies by Wallen *et al.* Males sniffed the fe-

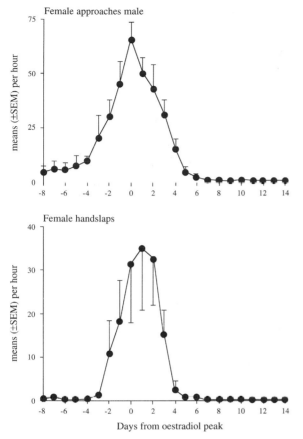

Fig. 11.16 Proceptivity in female rhesus monkeys observed as a group, in a large compound and with a single male present during the test period. Peaks of proceptivity are more pronounced under these conditions than is usually the case during free-access pair-tests. Data are mean hourly frequencies of female initiated approaches to the male and hand-slap displays in relation to the day of the pre-ovulatory oestradiol peak. (Redrawn from Wallen *et al*. 1984.)

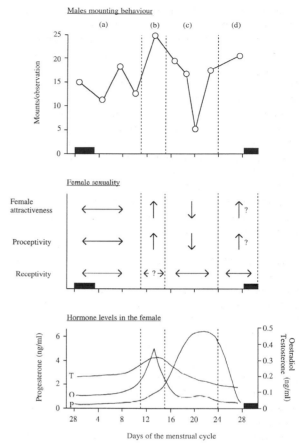

Fig. 11.17 Summary of information concerning circulating hormone levels and sexual behaviour during the menstrual cycle of the rhesus monkey (*Macaca mulatta*). O= oestradiol; P=progesterone; T=testosterone. Phases of the cycle are (a) follicular; (b) peri-ovulatory; (c) luteal, and (d) pre-menstrual. Dark bars=menstruations. (Redrawn and modified from Dixson *et al*. 1973.)

males' genitalia infrequently; the greatest frequency $(0.8 \pm 0.3 \, h^{-1})$ occurred on the day of the oestradiol peak, but the effect was not statistically significant. It is unknown whether males received olfactory information via a more indirect route, or whether subtle changes in female sexual skin coloration affected attractiveness. On the face of it, however, the female's proceptive behaviour appeared to be the primary mechanism which co-ordinated sexual activity during the peri-ovulatory phase of the menstrual cycle.

Figure 11.17 summarizes the results of studies on sexual behaviour during the menstrual cycle of the rhesus monkey. The summary shows that, whilst males may mount and ejaculate at any stage, they do so most frequently during the peri-ovulatory period of the female's cycle, when levels of oestradiol (and

testosterone) are highest. Mounts and ejaculations occur very much less frequently in the luteal phase, when circulating progesterone increases, but sometimes show a secondary pre-menstrual rise. Females are sexually receptive throughout the menstrual cycle, but proceptivity and attractiveness increase during the late follicular phase and peak around the time of ovulation. Proceptivity and attractiveness fall markedly in the luteal phase of the cycle and high circulating progesterone levels might be inhibitory in this context. Studies by Wallen *et al*. implicate the increases in circulating oestradiol (rather than testosterone) in triggering changes in female proceptivity during the follicular phase of the cycle. Equally, oestradiol might affect female attractiveness, both at this time and during the pre-menstrual period when progesterone levels are waning (Fig. 11.17). The precise neuroendocrine mechanisms underlying these behavioural events will be explored in more detail in the next chapter.

The menstrual cycle and sexual behaviour in other Old World monkeys

Rhythmic changes in sexual activity during the menstrual cycle occur in many other Old World monkeys besides the rhesus macaque, although to date none has been studied so intensively to tease apart the behavioural and hormonal mechanisms. Mounts and ejaculations tend to be most frequent in the follicular phase or during the peri-ovulatory period, rather than during the luteal phase. This applies whether behaviour is observed in the wild (W), or in captivity during free-access pair-tests (P), operant tests (O), or in social groups (G); *Colobus badius* (W): Struhsaker 1975; *Presbytis entellus* (W): Jay 1965. *Miopithecus talapoin* (G): Scruton and Herbert 1970; (W): Rowell and Dixson 1975; *Erythrocebus patas* (G): Rowell and Hartwell 1978; *Macaca fascicularis* (P): Zumpe and Michael 1983; (W): Van Noordwijk 1985; *Macaca nemestrina* (G): Tokuda *et al.* 1968; (O): Eaton 1973b; (P): Eaton and Resko 1974a; *Macaca sylvanus* (W): Taub 1980; *Macaca silenus* (P): Lindburg *et al.* 1985; (W): Kumar and Kurup 1985; *Macaca fuscata*: (G): Enomoto *et al.* 1979; *Macaca nigra* (G): Dixson 1977; *Macaca maurus* (W): Matsumura 1993; *Papio ursinus* (W): Saayman 1970; (P): Bielert 1986; *Papio cynocephalus* (W): Hausfater 1975; *Papio hamadryas* (W): Kummer 1968; *Theropithecus gelada* (W): Dunbar 1978; *Mandrillus sphinx* (G): Dixson *et al.* 1993; *Cercocebus albigena* (G): Chalmers and Rowell 1971; (W): Wallis 1983.

Do rhythmic changes in sexual behaviour occur during menstrual cycles in all Old World monkey species? The studies cited above refer to just 17 Cercopithecoid species, yet I believe that they are representative of the total picture. Further studies of the African *Cercopithecus* monkeys and of Asiatic colobine species would be valuable. However, it is probable that such studies would confirm that cyclical changes in proceptivity and attractiveness provide the major driving force resulting in peaks of male copulatory activity. Ovarian hormones play an important role in these respects, but have much less impact upon sexual receptivity. One or two apparent exceptions deserve special scrutiny. In *Macaca arctoides*, rhythmic changes in sexual activity during menstrual cycles are negligible when males are given free-access tests with single females, or tests with trios of females (Slob *et al.* 1978). Of 28 behavioural patterns analysed in these studies the only statistically significant change was a mid-cycle increase in

female presentations in response to contact by the male. A different picture emerged, however, when a permanent social group of 34 stumptail macaques was studied (Murray *et al.* 1985). Although males copulated throughout the female's cycle, ejaculations were significantly more frequent during the peri-ovulatory phase. Dominant males, in particular, were more sexually active with females during the peri-ovulatory phase of the menstrual cycle than were subordinate males in the social group. The reasons for the greater mating success of high-ranking males in this study were not clear. However, other studies have shown that the presence of dominant males in a captive stumptail group can suppress sexual activity in low-ranking males. Removal of higher-ranking males stimulates copulatory behaviour of subordinates and females also become more proceptive towards lower-ranking males under these conditions (Estep *et al.* 1988). As in the experiments by Wallen *et al.* (1984) on rhesus macaques, therefore, relationships between the female's cycle and sexual behaviour might actually become clearer when studies involve the social group, rather than isolated pairs of animals observed during encounter tests. We have no field data on the sexual behaviour of stumptail macaques during the female's menstrual cycle. The prediction is, however, that peri-ovulatory increases in copulation will occur under natural conditions, primarily involving high-ranking males, and stimulated by either increased female attractiveness or proceptivity (or both).

Female stumptail macaques exhibit very little change in sexual skin during the menstrual cycle. Visual cues regarding the time of ovulation are probably absent, therefore, although olfactory cues might exist, because males sniff the female's genitalia before copulation (Murray *et al.* 1985). In most *Cercopithecus* species and South East Asian colobines, however, changes in visual or chemical cues around the time of ovulation are difficult for human observers to identify. In the case of the vervet monkey (*Cercopithecus aethiops*), it has been reported that ovulations are 'concealed'; no changes in the female's genital morphology or behaviour indicate her peri-ovulatory status (Andelman 1987). Despite this, examination of data from Andelman's field studies shows that males attempt to mount and copulate most frequently when females are likely to conceive (Fig. 11.18). It seems, therefore, that some signal (or signals) from females must prompt higher copulatory frequencies at this time. Chemical cues might be involved or some subtle shift in female proceptivity. Laboratory studies of vervet monkeys have shown that males intromit and ejaculate with ovariectomized females more frequently when the latter have been treated with 17β-oestradiol. These effects are apparently a result of increased female attractiveness because little or no changes in female proceptivity or receptivity were measurable (Girolami 1989).

Since many Old World monkeys do possess prominent sexual skin swellings as indicators of the female's reproductive status, it will be appropriate to conclude this section by considering an example for which there are detailed laboratory and field data on sexual behaviour: the chacma baboon (*Papio ursinus*). Bielert (1986) has obtained valuable information on sexual behaviour in captive *P. ursinus*, by using free-access pair-tests to examine behavioural changes throughout the menstrual cycle. Thirty two menstrual cycles were monitored in eight females and these ranged from 29–39 days in length (median 35 days). Sexual skin swelling during the follicular phase occurred over a period of 16–29 days from the first day of menstruation until the day of 'breakdown' (initial loss of turgor of the sexual skin). The follicular phase varied in length, therefore, but had a median duration of 21 days. The luteal phase, during which the sexual skin underwent progressive detumescence, varied from 9–18 days (median 14 days) until the onset of menses. Because no hormonal data were collected, it was not possible to pinpoint the occurrence of ovulation. Other studies have shown that most (but not all) ovulations occur on the last day of maximal turgescence (38.5%) as well as on the day of breakdown (26.9%). Females may ovulate at other times, however, including the 2–5 day period which precedes sexual skin breakdown (17.8% of cycles: Wildt *et al.* 1977).

Bielert tested the same pairs on alternate days throughout four cycles in each female. Figure 11.19 shows the clear rhythmic changes in male's mounting activity, intromissions and ejaculations during these cycles. Frequencies of ejaculation peaked 4 days before sexual skin breakdown. Although copulations occurred throughout the cycle, they were much more frequent during the late follicular phase than in the luteal phase, whereas males failed to ejaculate during the majority of luteal phase tests. This marked rhythm in the male's sexual performance was not due to changes in female receptivity, however, as females continued to accept all the male's mounting attempts throughout the cycle (Fig. 11.20). By contrast, the male's acceptance of female presentations varied markedly across the cycle and peaked between days −7 to −2 before sexual skin breakdown. This measurement, which gives an index of the female's attractiveness, indicates that she is most attractive to her partner during the final week of maximum sexual skin swelling, but that attractiveness wanes just before sexual skin breakdown. Female proceptivity might have contributed to this increase in the male acceptance ratio, since frequencies of presentations were higher during the follicular phase of the cycle than during the luteal phase. Presentations during which the female turned to look at the male partner ('eye-contact proceptivity') showed more marked cyclical changes than presentations without eye contact (Fig. 11.20). Eye-contact constitutes a particularly sensitive measurement of female sexual initiating behaviour. Both types of proceptivity showed a dip in frequency on break-

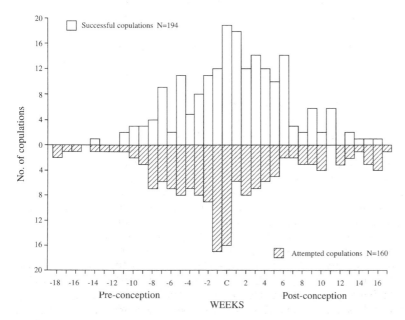

Fig. 11.18 Distribution of successful copulations and rejected copulation attempts in three groups of free-ranging vervet monkeys (*Cercopithecus aethiops*), in relation to the probable timing of conceptions (c). (Redrawn from Andelman 1987.)

down days −7 to −3 which is when the males' acceptance ratio was at its highest. One interpretation of these data is that females are so attractive to their male partners at this time that invitational displays are less necessary. This leaves us with the possibility that the sight or smell of the sexual skin and genitalia might control the peak in female attractiveness. Males investigate and sniff the female's genitalia more at mid-cycle than at other times; this is particularly so when males' responses to their partners' first presentation during the test are examined (Fig. 11.21). Hence the female's proceptivity might amplify non-behavioural cues but, when the female is maximally attractive, repeated presentations might be rendered unnecessary. It is certainly the case that male chacma baboons find the sight of

the female's fully swollen sexual skin arousing sexually. This has been demonstrated by attaching an artificial swelling to an ovariectomized female and observing increased levels of masturbation in males caged nearby but not in physical contact with her (Girolami and Bielert 1987).

The picture that emerges from Bielert's laboratory studies of *Papio ursinus* is one of rhythmic changes in female attractiveness and proceptivity during the menstrual cycle which enhance the males' sexual activity during the peri-ovulatory period. The female remains sexually receptive throughout the menstrual cycle and hormonal effects on her receptivity are probably negligible. The broader functions of attractiveness and proceptivity once again become clearer when social groups of baboons are studied, rather than isolated pairs (Hall and Devore 1965; Saayman 1970; Hausfater 1975). Saayman's

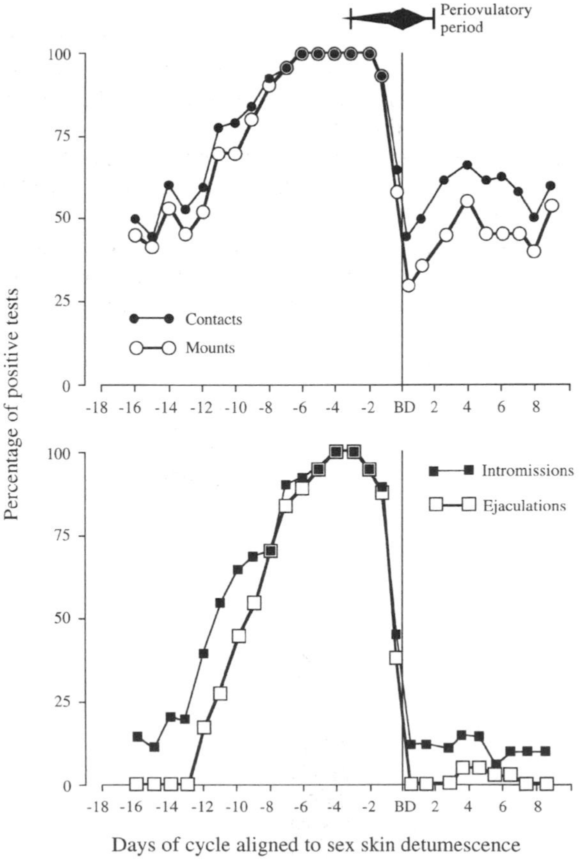

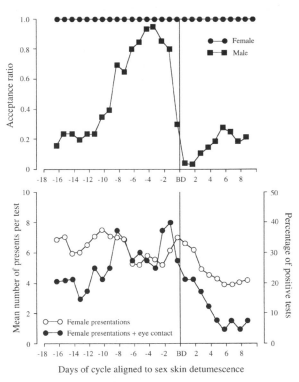

Fig. 11.19 Sexual behaviour during the menstrual cycle of the chacma baboon (*Papio ursinus*), as measured during laboratory pair-tests conducted on alternate days throughout four cycles in eight females. Data are two-day running means for (upper) males' contacts and mounts and (lower) intromissions and ejaculations. Data are aligned to the day of sexual skin breakdown (BD) in females and the putative peri-ovulatory period is indicated by a horizontal bar. (Redrawn and modified from Bielert 1986.)

Fig. 11.20 Sexual behaviour during the menstrual cycle of the chacma baboon (*Papio ursinus*), as measured during laboratory pair-tests conducted on alternate days throughout four cycles in eight females. Data are two-day running means for (upper) acceptance ratios in males and females and (lower) percentages of tests in which females exhibit proceptive hindquarter presentations or presentations and eye-contact with their male partners. Data are aligned to the day of sexual skin breakdown (BD) which included the probable time of ovulation (as indicated on Fig. 11.19). (Redrawn and modified from Bielert 1986.)

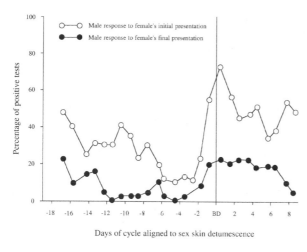

Fig. 11.21 Sexual behaviour during the menstrual cycle of the chacma baboon (*Papio ursinus*) as measured during laboratory pair-tests conducted on alternate days throughout four cycles in eight females. Data are two-day running means for males' olfactory investigations in response to females' initial, and final, sexual presentations during tests. Data are aligned to the day of sexual skin breakdown (BD) which includes the probable time of ovulation (as indicated in Fig. 11.19). (Redrawn and modified from Bielert 1986.)

(1970) study of a troop of 77 chacma baboons in the Honnet Nature Reserve in the Northern Transvaal is especially useful, as he specifically addressed the question of how the menstrual cycle affects sexual interactions. There were 31 adult females in the troop, three adult males, fifteen sub-adult males and eight juvenile males. Males of all three age-classes mounted females at all stages of the menstrual cycle, but the three adult males accounted for the major proportion of mounts with fully swollen fe-

males. Whereas, adult males only ejaculated with females at the height of the follicular phase, sub-adults also ejaculated at earlier stages of the follicular phase and during the luteal phase of the cycle (Fig. 11.22). These findings are reminiscent of those of Hall and Devore (1965) and Hausfater (1975), who showed that higher-ranking males accounted for the majority of ejaculations around the expected time of ovulation, whilst lower-ranking individuals were more active at earlier stages of the swelling

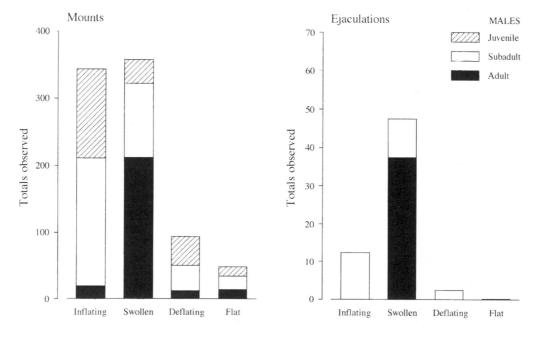

Fig. 11.22 Total numbers of mounts and ejaculations by adult, sub-adult, and juvenile males in a free-ranging troop of Chacma baboons (*Papio ursinus*) in relation to changes in females' sexual skin swellings during menstrual cycles. (Redrawn and modified from Saayman 1970.)

cycle. This might result both from competition between males for access to attractive females at full swelling and from increased proceptivity of females towards higher-ranking partners. In Saayman's study, females presented most frequently to the three adult males and 'adult males initiated mounts almost exclusively with females at the height of the follicular phase'.

The proceptive displays of female chacma baboons are certainly more complex under natural conditions than during laboratory pair-tests. Saayman noted that females became more mobile during the early follicular phase of the menstrual cycle, as the sexual skin gradually increased in size. Such females were often observed moving through the troop and actively approaching males. These approaches were distinctive and involved a 'direct and rapid gait' accompanied by attempts to make eye-contact with the male concerned. The female used a distinctive 'eye-face', with eyebrows raised and ears flattened, and her approach behaviour usually culminated in a sexual presentation posture. Females soliciting in this fashion during the early follicular phase were more likely to elicit a response from sub-adult males or juveniles—adult males rarely mounted in response to the females' invitations. The behaviour of both sexes changed, however, during the height of the follicular phase of the menstrual

cycle when the females were at full swelling. Females with fully swollen sexual skins were less mobile, according to Saayman (1970), and tended to form close associations (i.e. consortships) with the adult males in the troop. Although adult males in general responded to relatively few presentations by mounting females, they were most responsive when females had full swellings (Fig. 11.23). The processes that govern consortship formation and copulation in baboons during the peri-ovulatory phase of the menstrual cycle are likely to be complex. Male rank is certainly a factor and Fig. 4.3 (page 56) shows data from Hausfater's (1975) study of *Papio cynocephalus* which clearly demonstrate the positive correlation between male rank and mating success at the expected time of ovulation. However, females might also exert choice, and the freedom from a rigid control of sexual receptivity by ovarian hormones might facilitate this aspect of female sexuality. Receptivity is by no means obligatory or passive, however. Saayman (1970) noted that a great deal of subtle visual and postural communication preceded many mount attempts. Outright refusals do occur, and Smuts (1985) describes an example where a female olive baboon refused 42 consecutive copulation attempts by a male which attempted to replace an older male as her consort during the fertile phase of her menstrual cycle. In Smut's study, conducted

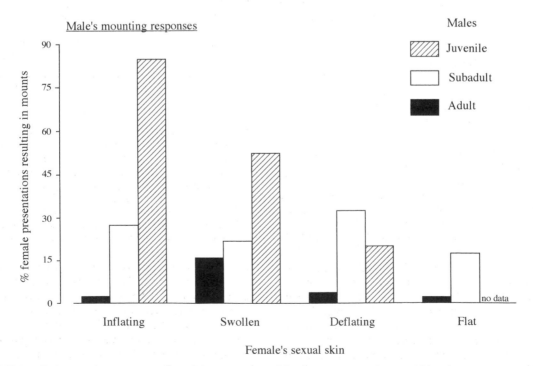

Fig. 11.23 Males' mounting responses (i.e. the percentage of females' presentations resulting in mounts by males) in relation to the stage of the menstrual cycle and extent of female sexual skin swelling in a free-ranging troop of chacma baboons (*Papio ursinus*). (Redrawn and modified from Saayman 1970.)

on a troop of baboons near the town of Gilgil in Kenya, females formed long-lasting relationships ('friendships') with specific males. Such relationships might have affected the likelihood that consortship would occur between the pair when the female's sexual skin became swollen. Female choice, and not simply male rank, determined the patterning of these sexual relationships.

The menstrual cycle and sexual behaviour in the great apes

During the 1930s and 1940s, the sexual behaviour and menstrual cycle of the chimpanzee (*Pan troglodytes*) was studied extensively by Robert Yerkes, William Young, and their colleagues (Yerkes and Elder 1936; Yerkes 1939; Young and Yerkes 1943; Young and Orbison 1944). These pioneering studies still provide valuable information and insights for present-day students of sexual behaviour in the great apes. Yerkes and Young had to rely upon changes in the morphology of the sexual skin to gauge the time-course of the follicular and luteal phases of the cycle. It became clear that the menstrual cycle lasted for approximately 35 days and that the female's sexual skin underwent a rapid phase of tumescence during the early follicular phase, remained fully swollen for an extended period (12 days approximately), and then underwent rapid detumescence during the luteal phase and pre-menstrual period. Yerkes and Elder (1936) suggested that ovulation might occur at mid-cycle or slightly afterwards. Modern studies have shown that ovulation usually occurs somewhat after mid-cycle and that the pre-ovulatory LH peak precedes sexual skin breakdown by 3 days (Fig. 11.24). Female chimpanzees reach maximal sexual skin swelling 9 days or so before they ovulate and the prediction would be that they should be sexually attractive to males and engage in copulation well before they are capable of ovulating. This is indeed the case, and it is instructive to consider Young and Orbison's (1944) comparison of sexual activity during the follicular and luteal phases of the menstrual cycle. These experiments involved seven female chimpanzees and three males. Each female was paired repeatedly with the same male for 30-min observation periods throughout her menstrual cycle. The male and female were placed in adjacent cages and a dividing door was opened to start each test session. Copulations occurred throughout the cycle, but were more than five times more frequent during the follicular phase than during the luteal phase of the cycle (Fig. 11.25). Certain measurements of female sexual or

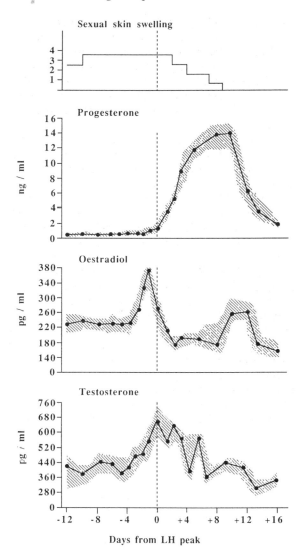

Fig. 11.24 Serum levels of oestradiol, progesterone, and testosterone, and changes in female sexual skin swelling during the menstrual cycle of the chimpanzee (*Pan troglodytes*). Data (means ± S.E.M., shaded areas) are aligned to the day of the pre-ovulatory luteinizing hormone (LH) peak, which was determined for these subjects but is not shown in the figure. In the chimpanzee, the LH peak occurs several days before sexual skin breakdown. (Redrawn and modified from Nadler *et al.* 1985.)

social interest in the male were also greater during the follicular phase. Females were more likely to wait next to the dividing door before the commencement of a test, they presented to males more frequently during the test, and moved to occupy the same cage as their male partners most often during the follicular phase. Young and Orbison also found that males groomed females significantly more frequently during the follicular phase of the cycle, whereas females did not exhibit any change in

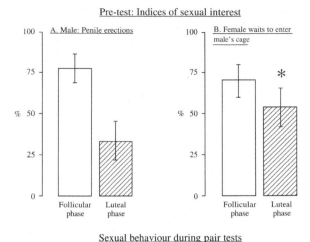

Pre-test: Indices of sexual interest

Sexual behaviour during pair tests

Fig. 11.25 Behaviour before and during pair-tests between adult male and female chimpanzees, to show differences between tests conducted in the follicular (open bars) versus luteal phase (cross-hatched bars) of the menstrual cycle. Males show more penile erection displays pre-test during the follicular phase and females are more likely to wait by the entrance to the male partner's cage. During pair-tests, female presentations and frequencies of copulation are greater during the follicular phase than during the luteal phase of the cycle. Date are means (+ S.E.M.) for seven females and three males. *P < 0.05. (Data are from Young and Orbison 1944.)

grooming activity. Yerkes and Elder had previously studied the sexual behaviour of 18 chimpanzees (13 females and 5 males), including the same subjects observed by Young and Orbison. Yerkes and Elder attempted to identify a period of 'oestrus' or heightened sexual arousal in female chimpanzees, and they believed this to occur during several days at maximal swelling. However, they noted considerable individual differences in the patterning of sexual interactions between pairings. A dominant assertive male could initiate copulation at any stage of the cycle, whereas, when a female was paired with a less

forceful individual, she was more likely to initiate sexual activity during her peri-ovulatory phase. Yerkes and Elder concluded that 'the female chimpanzee is cautious, discreet, and accommodating at all times and as a rule she seeks to avoid arousing the male to violence. Whether or not in the oestral phase she may offer herself to the male, at his command or when pursued by him, for sexual intercourse. Males differ markedly in sexual selectiveness, preferences, dominance, and aggressiveness'. One of the males studied, appropriately named *Pan*, was 'highly selective' and usually refused to mate unless the female was at maximal swelling, whereas a younger and less sexually experienced individual (Bokar) often mated at other cycle stages. One female (Lia) 'exhibited exceptionally subservience to the male and a willingness to mate at any time in the sexual cycle', whilst another, (Mimi) 'gave evidence of low sexual desire' or, in the case of Nira, 'was extremely timid and unwilling to mate'.

Despite these complications, Yerkes and Elder were able to demonstrate that females exhibited greater initiative during tests conducted at maximum sexual skin swelling. Females were the first to transfer and enter the partner's cage during 85% of tests conducted during maximum swelling, compared with 65% of tests conducted at other stages of the menstrual cycle. Clearly, in some females, 'initiative' was high at all times (perhaps motivated by a desire for social as well as sexual contact?), but was greatest during maximal swelling. These authors noted that at full swelling the 'maximally receptive' female behaved in a distinctive fashion when the dividing door was opened to commence the test. Given access to the male such a female 'rushes to him eagerly, prostrates herself before him, and awaits sexual union'. It was this type of behaviour, rather than the act of copulation itself, which prompted Yerkes and Elder to identify a period of heightened sexual arousal or 'oestrus' in their female subjects. They stressed that 'ordinarily a male can effect sexual union if he so desires whatever the cycle phase of the female and irrespective of her receptivity. This may obscure oestrus, but it does not disprove its occurrence'.

The problem with this interpretation of female sexual motivation or 'oestrus' is that it is extremely difficult to disentangle the effects of female attractiveness from female sexual behaviour during free-access pair-tests. The male chimpanzee, in full view of the female in an adjoining cage, often communicates with her by postures, gestures, and vocalizations before the test commences. The male's sexual invitations are directed specifically towards females at maximum swelling. The male may seat himself 'in readiness for copulation and awaiting the approach of the consort'. Seated he may 'beat the ground with his palms or extend an arm with palm uppermost toward the female'. Penile displays also form a prominent part of the male's sexual invitations to swollen females, so that 'resting on his legs, with his arms extended crutch-like, he may thrust his body forward toward the other animal with the erect penis conspicuous' (Yerkes and Elder 1936). I have observed exactly the same display patterns in all-male groups of chimpanzees housed next to females, but separated from them by a corridor approximately 6 ft. wide between their cages. The males were induced to enter the corridor for a short period each day in order that their cage might be cleaned. At such times, copulations were possible between the cage bars. All copulations observed between a group of five males and five females over a six-month period occurred exclusively when females were at maximum swelling. In almost every case the copulations were initiated by the males; females approached the bars and backed towards the males in response to their penile displays. These qualitative observations confirm Yerkes and Elder's contention that where the female chimpanzee has a greater measure of control over the encounter, mating will occur during the peri-ovulatory period. Given the potent attractiveness of the male's pre-copulatory displays, it should not surprise us if some swollen female chimpanzees stay close to the adjoining door of the male's cage before commencement of a mating test (Young and Orbison 1944, see Fig. 11.25), or enter his cage more rapidly and present avidly in these circumstances (Yerkes and Elder 1936). Since males rarely display in this fashion towards females at other stages of the menstrual cycle, it is not possible to judge how sexually responsive non-swollen females might be to male invitations. An experimental approach to this problem is feasible. By attaching an artificial swelling to the rump of an ovariectomized chimpanzee (in the manner described for chacma baboons by Girolami and Bielert 1987) one might measure the interplay between female attractiveness and female sexual responsiveness to the male in the absence of oestrogenic stimulation. No such experiment has been reported.

Field studies, conducted chiefly in Tanzania, confirm that increases in female sexual attractiveness during the period of sexual skin swelling play a major role in regulating sexual interactions among chimpanzees. Field studies have also greatly improved our understanding of the complex mating tactics employed by both sexes, and of the importance of age and dominance rank as determinants of mating success (McGinnis 1979; Tutin 1979, 1980; Goodall 1968, 1986; Hasegawa and Hiraiwa-Hasegawa 1990; Takahata 1990). The question of

mating tactics was reviewed in Chapter 4; as a reminder that female chimpanzees can influence the course of sexual encounters, it is worthwhile to reconsider Fig. 4.5 (see page 60) which deals with consortship formation. Tutin's detailed analysis of how consortships begin, or fail to materialize, makes it clear that female chimpanzees exert a major influence on the likelihood that matings will occur. Thus, in the wild as well as in captivity, there are subtle effects of the menstrual cycle upon the female chimpanzee's behaviour, as well as major effects upon her sexual attractiveness.

Until recently, much less was known about the sexual behaviour of the pigmy chimpanzee (the bonobo: *Pan paniscus*) than about its larger and more widely distributed sister species. However, fieldwork by Kano (1992), Furuichi (1987, 1992), and others, as well as studies of captive bonobos, have provided some picture of sexual activity during the menstrual cycle. Female bonobos differ from female chimpanzees in displaying a longer swelling phase and a less abrupt detumescence of the swelling during the second half of the menstrual cycle. Dahl (1986) estimates that female bonobos exhibit swelling for 73.5% of the cycle, as compared to 51.5% in the chimpanzee. Dahl's observations on captive animals agree with studies of free-ranging bonobos, which indicate that swellings are prolonged and semi-permanent. Bonobos are also very active sexually, and originally it was thought that menstrual cycles have little effect upon copulatory frequency in these animals. However, detailed field observations by Furuichi (1987) revealed that the firmness (degree of turgidity) of the swelling, as well as its overall size, shows cyclical changes which correlate with copulatory frequencies. Loss of firmness is associated with decreased attractiveness of the female, with a corresponding decrease in the male's ejaculatory mounts (Fig. 11.26). Sexual attractiveness therefore alters in the expected fashion during the menstrual cycle. The evidence for cyclical changes in receptivity and proceptivity is very limited by comparison. Females are receptive throughout the menstrual cycle; increased proceptivity at the height of swelling, just before detumescence, has been recorded in high-ranking captive females (Dahl 1987).

Unlike chimpanzees and bonobos, neither the gorilla nor the orang-utan exhibits a sexual skin swelling at mid-menstrual cycle. In the gorilla, only slight tumescence of the labia occurs (Fig. 11.27), whilst in the orang-utan there are no changes in the morphology of the female's external genitalia except during pregnancy, when a swelling is apparent (Lippert 1974). Female gorillas are highly proceptive at mid-cycle, when the labia are tumescent, and this

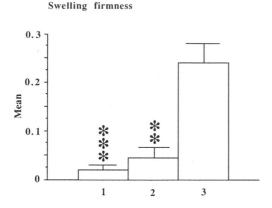

Swelling firmness

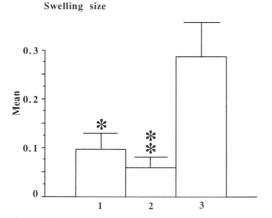

Swelling size

Fig. 11.26 Frequencies of ejaculatory mounts by males in relation to the degree of firmness and size of females' sexual skin swellings in a free-ranging community of pigmy chimpanzees (*Pan paniscus*). Increasing firmness and swelling size are rated on a scale of 1–3 and data are shown as means (+S.E.M.) per hour. *P=0.05; **P=0.01; ***P=0.001. (Data from Furuichi 1987.)

increase in proceptivity has been observed both in captive animals, during pair-tests (Nadler 1975, 1976), and in free-ranging groups (Harcourt *et al.* 1980). Females engage the male in eye-contact and also present to him, sometimes backing up forcefully to the male in their attempts to initiate copulation (Fig 11.28). In most cases, matings are confined to a period of just one to four days at mid cycle (Fig. 11.29). Although females are proceptive at other stages of the cycle, the female 'success ratio' (i.e. the proportion of presentations accepted by the male) peaks sharply during labial tumescence, as is also shown in Fig 11.29. This finding is consistent with the idea that female gorillas are also more sexually attractive when their labia are swollen.

The restriction of copulation to a brief phase at mid-cycle in the gorilla might depend upon some

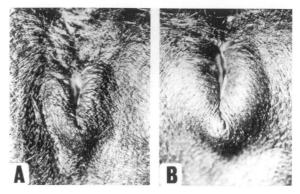

Fig. 11.27 The external genitalia of an adult female western lowland gorilla (*Gorilla g. gorilla*) showing A. minimal swelling of the labia; B. maximal swelling of the labia at mid-cycle. (After Dixson 1981.)

combination of greater female attractiveness and heightened proceptivity. The marked increase in proceptivity argues for some central effect of ovarian hormones upon the female's behaviour. Both oestradiol and testosterone are present at higher circulating concentrations during the period of labial tumescence, but correlative studies alone cannot resolve the question of whether these hormones act upon the brain to alter proceptivity in gorillas. We have seen that Yerkes and Elder formed the opinion that when the female chimpanzee is able to control sexual interactions there is greater restriction of

Fig. 11.28 A female western lowland gorilla (*Gorilla g. gorilla*) initiates copulation by backing forcefully towards a male during a pair-test. (Author's drawing, after Nadler 1976.)

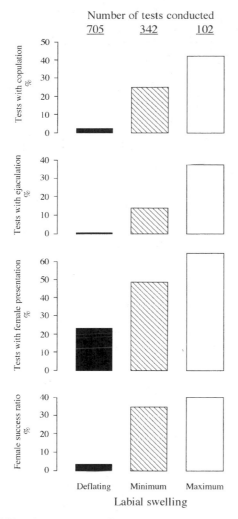

Number of tests conducted
705   342   102

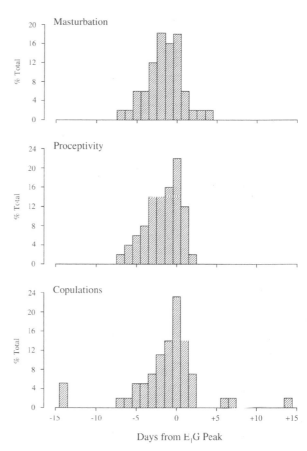

Fig. 11.29 Changes in sexual interactions during the menstrual cycle of the western lowland gorilla (*Gorilla g. gorilla*). Data were collected during pair-tests conducted when the females' labia were deflating or at minimum and maximum swelling. (Redrawn from Dixson 1981; after Nadler 1975.)

Fig. 11.30 Frequencies of masturbation and of proceptive and copulatory behaviour of female orang-utans (*Pongo pygmaeus*) during restricted-access pair-tests (RATs) with males. Data are normalized with respect to the day of peak levels of urinary oestrone glucuronide in the females. When female orang-utans have control of sexual initiation, as is the case during RATs, then mid-cycle peaks of sexual activity are much more pronounced than during free-access pair tests. (Redrawn and modified from Nadler 1988.)

copulations to the fertile phase of the cycle. Sexually assertive and dominant male chimpanzees often copulate at other times, however, and females modify their sexual responses dependent upon their partner's behaviour. In the gorilla, where sexual initiative rests largely with the female, the expected peak in copulations during the peri-ovulatory period occurs in most captive pairings. However, Nadler *et al.* (1983) have observed some pairings where males were more assertive sexually so that, in one case, a pair copulated during 11 consecutive days of the cycle.

In the orang-utan, adult males are often highly assertive sexually and, in captivity, copulations occur throughout the female's menstrual cycle (Nadler 1977), which lasts for approximately 4 weeks (Collins *et al.* 1975). These findings refer to behaviour during free-access pair-tests, where the female is introduced into a male's home cage for limited periods throughout her cycle. However, the use of *restricted-access* tests, during which a barrier separates the partners and only the female is small enough to pass underneath, has produced markedly different results. When female orang-utans control sexual initiative, both female proceptivity and copulations peak during the mid-menstrual cycle (Fig. 11.30). It is also interesting that captive female orang-utans show cyclical changes in auto erotic (masturbatory) behaviour during restricted-access pair-tests. Such findings have not been reported in any other ape species; this might indicate that female sexual responsiveness is affected by hormonal changes in the late follicular and peri-ovulatory period, when levels of plasma oestradiol and testosterone are highest (Fig. 11.30).

Nadler's (1988) studies of captive orang-utans, and of behaviour during restricted-access pair-tests, parallel field observations of orang-utans in the wild by Schurmann (1982). Adult females tend to approach and associate with mature territorial males during the fertile phase of the ovarian cycle and to initiate sexual interactions at this time. Adolescent females also approach dominant males and can be remarkably persistent, even to the extent of 'slapping disinterested adult males or pinching their genitals' (Galdikas 1995).

Sexual behaviour during the human menstrual cycle

Information obtained from studies of the menstrual cycle and sexual behaviour in Old World monkeys and apes is useful when considering the extensive, and often contradictory, literature on the human menstrual cycle and sexuality. Firstly, many monkeys and apes copulate at times when conception is unlikely; female receptivity is not rigidly controlled by hormonal changes during the peri-ovulatory phase of the ovarian cycle. As discussed in Chapter 5, human beings have inherited from our anthropoid ancestors this capacity to dissociate sexual activity from obligatory hormonal control. There has been no 'loss of oestrus' during human evolution. Secondly, although rhythmic changes in sexual activity do occur during menstrual cycles in monkeys and apes, they are subject to much individual variation. The reader might wish to refer again to Fig. 11.12, which shows how some rhesus monkeys, tested under laboratory conditions, show follicular and/or mid-cycle peaks of copulatory behaviour, whereas other pairings lack such obvious rhythms. In some cases copulations increase in frequency before menstruation. Women and their male partners (no less than rhesus monkeys or chimpanzees) might also be expected to show individual differences in the frequency and temporal patterning of sexual behaviour during the menstrual cycle. Experiential factors and the quality of relationships with their male partners will vary between women. The amounts of oestradiol, progesterone, and luteinizing hormone secreted during menstrual cycles also exhibit important inter-individual differences (Whalen 1975) and there is always the possibility that some women are more sensitive than others to hormonal fluctuations. In many studies of sexual behaviour during the human menstrual cycle circulating hormone levels were not measured, although this is necessary to determine the exact stage of the cycle.

A third important lesson learned from studies of the non-human primates concerns the need for precision in defining and measuring female sexuality. Distinguishing between hormonal effects upon the female's sexual behaviour and effects of hormones upon her sexual attractiveness is of paramount importance (Herbert 1970). There is also an important distinction to be made between active, sexual initiation by the female (proceptivity) and her willingness to receive the male and to allow intravaginal ejaculation (i.e. receptivity: Beach 1976). In some anthropoid species, such as baboons and chimpanzees, changes in female attractiveness largely determine the follicular-phase and peri-ovulatory increases in male-initiated mounts and ejaculations. Whether ovarian hormones act upon the female's brain to alter her sexual receptivity in these species is much more problematic. However, we have seen that central effects of ovarian hormones upon female proceptivity are likely in New World monkeys (such as the marmoset) and among Old World catarrhines (such as the rhesus macaque, gorilla, and orang-utan). Increases in proceptivity during the follicular phase, or at mid-cycle, might be influenced by heightened oestradiol levels, or by the mid-cycle rise in testosterone. Reduced proceptivity during the luteal phase of the cycle might reflect falling amounts of oestradiol and/or testosterone coupled with an inhibitory action of progesterone. The premenstrual increases in sexual behaviour seen, for example, in some rhesus monkey pairings could result from the fall in progesterone which occurs before menstruation, and release from its inhibitory effects upon attractiveness or proceptivity. Neuroendocrine mechanisms which govern cyclical changes in proceptivity, receptivity, and attractiveness will be considered in detail in the next chapter. As regards the human menstrual cycle and sexual behaviour, studies of non-human primates indicate that hormonal effects upon proceptivity might be much more important than effects upon female sexual receptivity. In the case of the human female, we may hypothesize that measurements of sexual thoughts and feelings, as well as auto-erotic behaviour (masturbation), might correlate with measurements of proceptive patterns. Some authors doubt whether the terminology used to describe sexuality in female monkeys and apes can be applied to human beings.

Small (1993), for instance, states that 'Beach's three categories of sexual behaviour describing female sexuality of other animals—attractive, proceptive, and receptive—don't really fit human females anyway, and consequently shouldn't be used to describe the evolution of human female sexuality'. On the contrary, it seems to me that the best course would be to establish a common ground between measurements of female sexuality in women, females of other primate species, and female animals in general. Otherwise human sexuality becomes an isolated and quasi-unique phenomenon, cut off from its evolutionary heritage with the non-human primates. The use of ill defined terminology, such as 'libido', 'sexual drive', or 'sexual gratification', coupled with a failure to distinguish between behaviour initiated by the female and that initiated by the male, has characterized many studies of the human menstrual cycle and sexual behaviour.

Scores of studies have addressed the question of how the menstrual cycle might affect sexuality in human beings, but no consistent picture has emerged. Although it is commonplace to see particular studies cited because they have found a mid-cycle peak in sexual activity (e.g., Udry and Morris 1968; Adams et al. 1978) there is, in fact, no consensus on this point. Some results from the study by Adams *et al.* are shown in Fig. 11.31. It is of considerable interest that peri-ovulatory peaks of female-initiated (proceptive) and auto-erotic activity were reported. These are the measurements we might

expect to reveal cyclical changes and mid-cycle peaks. However, re-analysis of these data failed to confirm the original findings (Kolodny and Bauman 1979). Likewise, although Udry and Morris (1968) reported that copulation occurred most often during the peri-ovulatory period, their data have also been interpreted to indicate that the peak occurs post-menstrually (James 1971). In her thesis on female sexual arousability in relation to the menstrual cycle, Schreiner-Engel (1980) reviewed 32 studies then available and found that peaks of sexual activity had been recorded at mid-cycle (in 8 cases), pre-menstrually (17 cases), during menses (4 cases), and post-menstrually (18 cases).

Given that changes in sexual activity during the human menstrual cycle are extremely subtle, precision of behavioural and hormonal measurement is essential, yet very difficult to achieve in practice. A most detailed and careful analysis has been conducted by Sanders, Bancroft, and their colleagues (Sanders and Bancroft 1982; Sanders *et al.* 1983; Bäckström *et al.* 1983; Bancroft *et al.* 1983) on 55 women. It is notable, however, that 19 of these subjects were receiving treatment at a clinic for pre-menstrual syndrome (PMS) whereas 18 others suffered from less severe pre-menstrual symptoms. Blood samples were collected two or three times each week from these women in order to measure levels of plasma oestradiol, progesterone, testosterone, and androstenedione. Luteinizing hormone levels were also measured, but only in the women

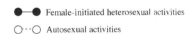

●—● Female-initiated heterosexual activities

○--○ Autosexual activities

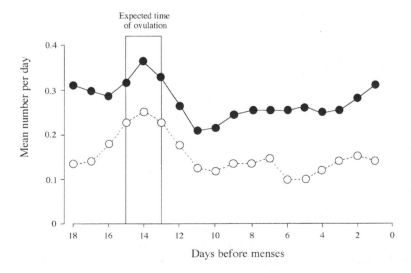

Fig. 11.31 Peri-ovulatory increases in female-initiated sexual activities and in autosexual behaviour in 29 women. Few studies of the human menstrual cycle have shown such clear relationships between sexual behaviour and the expected time of ovulation, as is discussed in the text. (Redrawn and modified from Adams, Gold and Burt, 1978.)

undergoing treatment for PMS in the clinic. On the basis of these endocrine measurements, the menstrual cycle was divided into six parts (early, mid-, and late follicular and luteal phases). The late follicular phase included the pre-ovulatory increases in oestradiol and LH; hence, the peri-ovulatory period fell within this portion of the cycle (Fig. 11.32). The women completed daily diaries, recording on visual analogue scales their self ratings of sexual feelings and thoughts, as well as mood and physical symptoms (e.g. breast tenderness and swelling). Principal component scores, derived from these analogue-scale data, showed that in these women feelings of well-being, physical distress, and sexuality all varied in a cyclical but non-identical fashion. Feelings of well-being were higher in the mid- and late follicular phases of the cycle than during the luteal phase; these changes were most obvious in women suffering from PMS, for whom well-being was lowest in the late luteal phase before menstruation. All women, including the non-PMS group, exhibited cyclic changes in physical distress, so that symptoms such as breast tenderness and swelling peaked in the pre-menstrual phase of the cycle. Once the influences of mood changes and physical symptoms had been accounted for, sexual thoughts and feelings were found to be most positive during the mid-follicular phase and (in women with PMS) during the late luteal phase of the cycle (Fig. 11.32). Scores of 'sexuality' did not achieve significance when data for

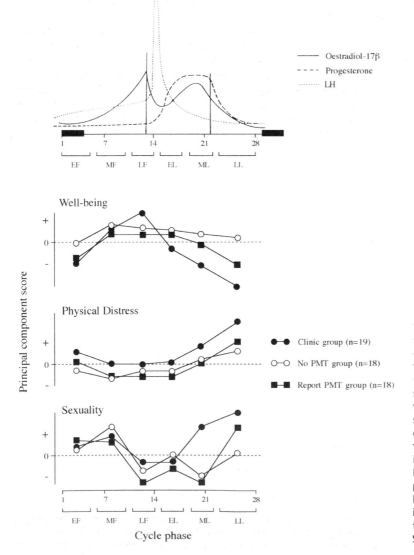

Fig. 11.32 Changes in feelings of well-being, physical distress, and sexuality during the menstrual cycle of women, some of whom are being treated for pre-menstrual tension (PMT: clinic group) or report PMT symptoms (Report PMT group), whereas others have no such symptoms (no PMT group). The upper graph shows hormonal changes during the menstrual cycle and its division into six hormonally distinct phases: EF, MF, LF=early, mid and late follicular phases; EL, ML, LL=early, mid and late luteal phases. The results are discussed in the text. (Redrawn and modified from Sanders and Bancroft 1982 and Sanders *et al.* 1983.)

the three groups of women were analysed separately, although combining the three data sets resulted in a significant cyclical pattern. Likewise, frequencies of sexual activity with a partner and of female-initiated (or mutually initiated) sexual activity also peaked during the mid-follicular phase, although the numbers of subjects for whom data were available were more limited (Bancroft *et al.* 1983). Frequencies of masturbation and orgasm exhibited similar cyclical changes, although these were not statistically significant (see Fig. 11.33). A peak in masturbation also occurred in the late luteal phase of the cycle, before menstruation.

Pre and post-menstrual peaks in sexual activity might be conditioned by cultural factors, since some couples might be reluctant to engage in intercourse during menstruation and may practise abstinence at this time. A pre-menstrual increase in sexual activity could occur in anticipation of abstinence during menstruation and likewise intercourse might increase after the period of menstrual abstinence. Sanders and Bancroft (1982) acknowledge this possibility, but they point out that 'it is not unusual to find women whose sexual interest or preparedness to engage in sexual intercourse is restricted to a brief period of the cycle....in such women the peak

is nearly always perimenstrual. It is difficult to account for that pattern on the basis of abstinence in view of the long periods of unenforced abstinence for the rest of the cycle'. We have also seen that in some of the non-human primates, such as the rhesus monkey, certain individuals exhibit pre-menstrual increases in copulatory activity (see Fig. 11.12). Whether these are due to changes in the female's behaviour or her attractiveness requires clarification. However, where the human species is concerned, it is possible that physiological changes at the approach of menstruation have effects upon sexual thoughts and feelings in at least some women. Cultural factors, such as the desire to engage in sexual activity because conception is unlikely at the approach of menstruation, or because of anticipated menstrual abstinence are unlikely to provide the whole explanation.

Increases in breast sensitivity occur during the luteal phase of the cycle (Robinson and Short 1977) and this, coupled with increased 'pelvic congestion' during the pre-menstrual period, might exert feedback upon sexuality (Sanders and Bancroft 1982). However, women with PMS who experience unpleasant physical symptoms before menstruation may nonetheless show peaks of sexual thoughts and feel-

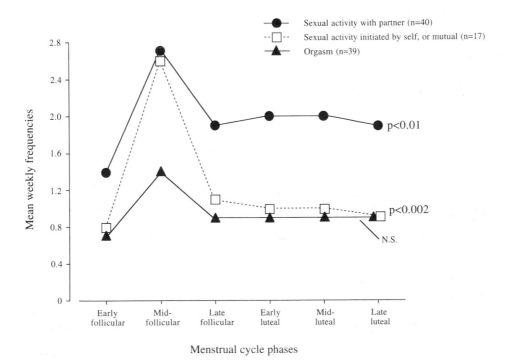

Fig. 11.33 Mean weekly frequencies of women's sexual activity and orgasm during the menstrual cycle. The cycle is divided into six phases in the same way as shown in Fig . 11.32. Sexual activity with a partner and sexual initiation (by self or by partner) show statistically significant peaks during the mid-follicular phase. Frequencies of masturbation and orgasm also increase at this time, but the changes are not statistically significant. (Data are from Bancroft *et al.* 1983.)

ings at this time (Fig 11.32). Some other sensory changes, such as greater olfactory acuity, occur at mid-cycle rather than pre-menstrually in women (Henkin 1974). It is not known whether subtle sensory and perceptual changes during the follicular phase of the human menstrual cycle have any effects upon sexual activity and no studies have been conducted on other primate species.

Some workers have attempted to assess changes in womens' sexual arousal by recording alterations in vaginal blood flow, or volume, and labial temperature in response to erotic stimuli. Schreiner-Engel *et al.* (1981) found that vaginal responsiveness was greater during the follicular and luteal phase than around the time of ovulation. Hoon *et al.* (1982), by contrast, found no changes during the menstrual cycle. Slob *et al.* (1996) report an initial ('first test') correlation between labial temperature and menstrual cycles in women; women in the follicular phase of the cycle showed greater temperature increases in response to an erotic video than did women in the luteal phase.

The possible effects, or lack of effects, of oestradiol, progesterone, and testosterone upon sexuality in women will be considered in greater detail in the next chapter. It has been suggested, however, that a low, threshold level of oestradiol, rather than fluctuations in levels of the hormone during the menstrual cycle, might play some role in maintaining sexual interest (Persky et al. 1978a; Bancroft 1984). Negative effects of progesterone upon sexual interest have been mooted by some authors, largely on the basis of decreased sexual enjoyment and interest in some women using progestagen-based oral contraceptives (Grant and Pryse-Davies 1968; Bancroft 1984). Women who suffer from PMS might be more sensitive to the negative side-effects of such preparations, including effects upon mood; this might, in turn, adversely affect sexual interest (Cullberg 1972). As regards testosterone, there is a little evidence that the mid-cycle increases in testosterone levels affect sexuality in women, but once again the findings vary between studies. Persky *et al.* (1978b) reported a correlation between frequencies of intercourse and womens' mid-cycle testosterone levels. Morris *et al.* (1987) obtained similar results and found that sexual activity correlated with levels of total testosterone and free (i.e. unbound) testosterone at mid-cycle. Van Goozen *et al.* (1997) found that free testosterone (whether at mid-cycle or across the whole cycle) was positively correlated with frequencies of intercourse in women. Levels of several androgens (testosterone, androstenedione, and dehydroepiandrosterone) correlated positively with masturbation in these subjects. Bancroft *et al.* (1983),

however, found no mid-cycle increase in sexual intercourse and no relationship between testosterone levels and behaviour at any specific stage of the cycle. They did note that women with the highest mean testosterone levels masturbated more frequently. There is also an interesting relationship between plasma testosterone levels and sexual interest in women during lactation. Alder *et al.* (1986) found that plasma testosterone levels were consistently lower during post-partum weeks 4–22 in breast-feeding women who complained of reduced sexual interest, compared with those who reported no reductions in sexual interest. Circulating oestradiol was not correlated with levels of sexual interest in these women. Therefore, some positive relationship between testosterone levels and womens' sexual activity or interest has been found in a number of studies. However, the results of testosterone treatment of women who complain of less sexual enjoyment or interest have been far from straightforward, as we shall see in the next chapter.

Some remarks on the question of 'concealed ovulation'

The fact that women can remain receptive throughout their menstrual cycles and lack obvious signals of impending ovulation, such as sexual skin swellings, has lead to the suggestion that *concealed ovulation* represents an evolutionary strategy in human beings. Various authors have debated the reasons why ovulation might be concealed, either from males or from the females themselves (Alexander and Noonan 1979; Burley 1979). For example, if males are unable to determine the likely fertile period, they might be more likely to engage in prolonged mate-guarding and monogamy, since this offers greater 'certainty' of paternity (Alexander and Noonan 1979). Extension of such arguments to the non-human primates is supported, in part, by the observation that most monogamous monkeys and apes lack signals of ovulation such as sexual skin (Sillen-Tullberg and Møller 1993). The white-handed gibbon is a notable exception to this rule, however (Dahl and Nadler 1992). Moreover, a great many monkeys which lack monogamous mating systems, such as most of the guenons (*Cercopithecus* spp.) and langurs (*Presbytis* spp.), do not have swellings. In such cases, it has been argued that concealed ovulation and multiple partner matings enable females to impose 'uncertainty of paternity' upon males. This might, in turn, reduce the likelihood that males will engage in tactics which are disadvantageous to females, such as infanticide (Hrdy 1979, 1981).

In evaluating the concept of concealed ovulation, particularly as it is applied to non-human primates, it is necessary to point out that neither extended periods of female receptivity nor the absence of sexual skin swellings, necessarily indicate that males are denied information concerning female reproductive status. Throughout this chapter, numerous examples have been described of increases in female initiating behaviour (proceptivity) and attractiveness (including possible olfactory cues as well as visual signals) which occur during the follicular and/or peri-ovulatory phases of the ovarian cycle. Indeed, the rhythmic changes in sexual activity which occur during ovarian cycles in the vast majority of monkeys and apes result mainly from shifts in female proceptivity and attractiveness, irrespective of whether sexual skin is present. It is tempting, for example, to assume that ovulation is concealed in monkeys such as marmosets and tamarins (Stribley *et al.* 1987) or in vervets (Andelman 1987) because no changes in the females' genitalia are discernible during the peri-ovulatory period. Yet, detailed behavioural and endocrine studies have shown that peri-ovulatory increases in female proceptivity and/or attractiveness certainly occur in callitrichids

(*Callithrix jacchus*: Dixson and Lunn 1987; Dixson 1992, 1993a; *Saguinus oedipus*: Ziegler *et al.* 1993; *Cebuella pygmaea*: Converse *et al.* 1995; *Callimico goeldii*: Jurke *et al.* 1995). In the vervet monkey, peaks in mating attempts and copulations have been recorded at times when conception is most likely (see Fig. 11.18, which is based on Andelman's field data). Presumably, males are receiving non-behavioural or proceptive cues from females, although the nature of these remains to be determined. In the monogamous owl monkey matings occur throughout the ovarian cycle, as is the case in Andelman's studies of vervets. Female owl monkeys are receptive at all stages of the cycle during free-access pair-tests (Dixson 1983a). Yet, despite the fact that copulations are exceedingly infrequent in captive family groups of owl monkeys, they consistently result in conceptions. Subtle and as yet undetermined cues must exist which co-ordinate such sexual activity.

For the above reasons, I believe that the bulk of the information reviewed in this chapter supports the notion that it is more fruitful to ask why conspicuous advertisements of ovulation have developed in some primates, rather than arguing that their absence must reflect an evolutionary strategy of concealed ovulation (Burt 1992).

12 The neuroendocrine regulation of sexual behaviour in the adult female

The last chapter explored relationships between the ovarian cycle and patterns of sexual activity amongst the prosimians and anthropoids. A clear distinction emerged between the two sub-orders; prosimian females exhibit oestrus, being dependent upon peri-ovulatory changes in ovarian hormones for the control of their sexual receptivity, whereas among the anthropoids a relaxation of such rigid effects of hormones upon receptivity has occurred during the course of evolution. Nonetheless, there are rhythmic changes in sexual activity during ovarian cycles in monkeys and apes; subtle effects have also been measured by some students of the human menstrual cycle. Higher copulatory frequencies during the follicular or peri-ovulatory phase of the ovarian cycle in anthropoids result mainly from increased female attractiveness and sexual invitations (proceptivity) to males. Social factors, as well as hormones, influence female sexuality and there are marked interspecific differences, for instance between forms which possess sexual skin swellings and those which lack striking visual signals of female reproductive status.

This chapter examines how hormones affect either the central nervous system or genitalia to influence behaviour and sexual attractiveness in female primates. The adrenal cortices, as well as the ovaries, secrete oestrogens, androgens, and progestagens; one section of this chapter deals with the role played by the adrenal glands in female sexuality. We shall also explore what is known about the neural substrates that govern sexual behaviour in females, comparing and contrasting the limited evidence on primates with more extensive information on the brain and sexual behaviour in rodents, and other non-primate mammals.

Peripheral effects of hormones upon female sexuality
Possible effects upon visual and tactile cues
The origins and evolution of sexual skin swellings in some of the catarrhine primates, including various macaques, mangabeys, baboons and the chimpanzee, were reviewed in Chapter 7. There it was noted that sexual skin swelling is stimulated by oestrogen and antagonized by progesterone (e.g. in the baboon: Gillman 1940; Gillman and Stein 1941). Circumstantial evidence indicates that the swollen sexual skin is highly attractive to males (e.g. in the talapoin: Dixson and Herbert 1977; black ape of Celebes: Dixson 1977; Chimpanzee: Goodall 1986). The increased sexual activity which occurs during the follicular and peri-ovulatory phases of the menstrual cycle in species which possess swellings could be due to oestrogenic enhancement of female attractiveness, whilst detumescence of the swelling during the luteal phase might cause a reduction in attractiveness to males. However, oestrogen and progesterone might act upon visual qualities of the sexual skin, olfactory (vaginal) cues, or tactile responsiveness of the swelling to copulatory stimulation by the male. These possible peripheral effects of ovarian hormones make it difficult to interpret central effects of the steroids upon female proceptivity or receptivity. The only clear-cut experiments to have demonstrated that female swellings are visually sexually arousing to males have been carried out by Girolami and Bielert (1987) using the chacma baboon (*Papio ursinus*). Singly-housed male chacma baboons show little sexual interest when an ovariectomized female is placed in visual, olfactory, and auditory contact with them. However, when the female is provided with an artificial sexual skin swelling, closely resembling that of an intact female at mid-menstrual cycle, the males exhibit an increased frequency of seminal emissions (see Fig. 7.29, page 205). These changes are due solely to altered female appearance; ovariectomized females fitted with artificial swellings do not exhibit changes in behaviour which might, in turn, influence the male's arousal (Girolami and Bielert 1987). These results support Bielert's (1986) contention that much of the mid-cycle peak in sexual activity seen during pair-tests in chacma baboons results from enhanced

female attractiveness to the male, whereas receptivity is little affected by hormones. It is unfortunate that none of the studies using artificial sexual skin swellings in chacma baboons have allowed males direct access to the stimulus female. It would be interesting to know whether artificial swellings elicit mounting or whether, upon closer inspection of the female, other cues (olfactory, tactile, or behavioural) might be required by males. The large size of the swelling seems to be crucial, whilst the evidence that its pink/reddish colour is important is less robust (Bielert *et al.* 1989).

Effects of ovarian hormones upon genital sensitivity have been elegantly demonstrated in the rat (as reviewed by Pfaff 1980). The cutaneous receptive field of the pudendal nerve enlarges when ovariectomized rats are treated with 17β-oestradiol. Tactile feedback from the perineal area is important since, when the male mounts and makes pelvic thrusting movements upon the receptive female, she adjusts her position and exhibits lordosis, thus facilitating intromission. Oestradiol facilitates receptivity via changes in peripheral responsiveness, as well as by influencing the brain in the female rat. Similar mechanisms probably exist in other female mammals which exhibit lordosis, including some prosimians (e.g. *Lemur catta*, *Propithecus verreauxi*, and *Varecia variegata*: see page 101 for a discussion of lordosis in prosimian primates). Although female anthropoids do not display a lordotic reflex, Zuckerman (1932) suggested that tactile sensitivity of the oestrogen-dependent sexual skin swelling might increase at mid-cycle and that this, in turn, might influence the female's sexual behaviour. No experimental studies have been carried out to explore this interesting question. In the New World squirrel monkey oestradiol enhances the responsiveness of mesencephalic and pontine neurons to artificial vaginal stimulation (Rose and Michael 1978), but these studies did not examine tactile sensitivity of the skin surrounding the genital area.

Possible effects upon olfactory cues

The decreases in sexual activity which occur after ovariectomy in certain anthropoid primates are doubtless due to reductions in female attractiveness, as well as changes in behaviour (Herbert 1970; Dixson 1983c). This finding applies to the rhesus monkey, in which the adult female has a reddish sexual skin but no swelling, as well as to species in which females have oedematous sexual skin. Ovariectomized rhesus monkeys may continue to solicit copulations from male partners during laboratory pair-tests and to accept mounting attempts

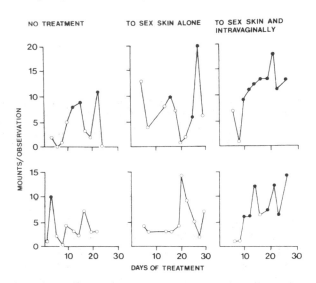

Fig. 12.1 Effects of applying oestrogen directly to the genitalia of ovariectomized rhesus monkeys (*Macaca mulatta*) upon the sexual behaviour of their male partners. Black circles = ejaculations. Data refer to two females and are redrawn from Dixson *et al.* (1973), after Herbert (1970).

(Herbert 1970; Chambers and Phoenix 1987). In the female stumptail, which lacks all but minor changes in perineal coloration during the menstrual cycle, females remain proceptive and receptive after ovariectomy. Frequencies of mounts, intromissions and ejaculations by male partners all decline significantly, however, indicating a decrease in female sexual attractiveness (Slob *et al.* 1978).

The rhesus monkey has been extensively studied in an attempt to determine how ovarian hormones affect female attractiveness. Much attention has focused upon the role of vaginal odour in this context and a review of the evidence is necessary because of conflicting results obtained by different workers. Application of an oestrogen-containing cream directly to the vagina of ovariectomized rhesus monkeys leads to a marked increase in ejaculations by their male partners (Fig. 12.1). The effects of oestradiol upon vaginal cues and sexual attractiveness are reversible by intravaginal administration of minute amounts of progesterone (Fig. 12.2). At least part of the reason for rhythmic changes in sexual activity during the menstrual cycle of the rhesus monkey relates to peripheral effects of oestrogen and progesterone upon sexually attractive vaginal cues. Whether these hormones exert their effects by changing genital odour is debatable, however. Michael and Keverne (1968) used operant conditioning experiments to show that male rhesus monkeys, rendered anosmic by plugging the nasal passages,

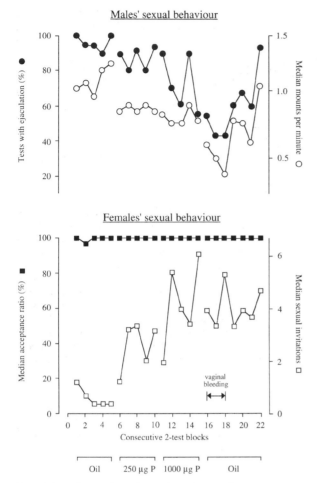

Males' sexual behaviour

Females' sexual behaviour

vaginal bleeding

Consecutive 2-test blocks

Oil   250 µg P   1000 µg P   Oil

Fig. 12.2 Effects of applying low doses of progesterone intravaginally to ovariectomized, oestrogen-treated rhesus monkeys upon sexual interactions during laboratory pair-tests. Behavioural measurements are (upper) for the male: (○) median mounts per minute; (●) percentage of tests with ejaculation and (lower) for the female: (□) median sexual invitations; (■) median acceptance ratio. Control tests involved placement of arachis oil (vehicle) intravaginally rather than arachis oil and progesterone (250 µg; 1000 µg) as indicated at the foot of the figure. (Redrawn and modified from Baum *et al.* 1976.)

showed increased bar-pressing to gain access to oestrogen-treated females once the plugs were removed. Subsequently, Michael and Keverne (1970) found that vaginal lavages obtained from ovariectomized, oestradiol-treated females enhanced sexual attractiveness when applied to the rumps of untreated females. The active portion of the lavage consisted of a mixture of aliphatic acids (Curtis *et al.* 1971) which result from bacterial breakdown of cervical secretions, although other substances in the lavages may also have contributed to their be-

havioural potency (Keverne 1976). Synthetic mixtures of the necessary aliphatic acids were shown to influence behaviour and a *vaginal pheromone* or *copulin* was postulated to exist in rhesus monkeys and some other primates, including man (Michael *et al.* 1975).

Persuasive as these experimental findings may be, it would be incorrect to assume that male rhesus monkeys are dependent upon olfactory stimuli from females in order to assess their endocrine condition. Goldfoot *et al.* (1978a) showed that anosmic male rhesus monkeys continue to show rhythmic changes in ejaculatory frequencies during their female partners' menstrual cycles (See Fig. 11.13, page 335). In this study, three males were rendered anosmic by placing cotton swabs soaked in 10% formalin directly in contact with the olfactory epithelium; inability of the subjects to perform an olfactory discrimination task confirmed that this method was effective. Hence, cyclical changes in ejaculation frequency during pair-tests are not dependent solely upon changes in sexually attractive olfactory cues from females.

The validity of the pheromonal interpretation of sexual attractiveness in female rhesus monkeys can also be questioned on other grounds. Beauchamp *et al.* (1976) have pointed out that releaser pheromones, such as those studied in insects, have pronounced and relatively species-specific effects upon behaviour. The vaginal odour cues described in rhesus monkeys do not meet these criteria (Goldfoot 1981). Much of the original experimentation involved only six males, and it is possible that learning had enhanced their responsiveness to olfactory stimuli. Subsequent studies using 12 naïve subjects, each tested with two female partners, showed that synthetic copulins were effective in three males, partially effective in six, and had no behavioural effects in three (Michael *et al.* 1977b). Goldfoot *et al.* (1976), using 19 male rhesus monkeys and 27 ovariectomized females, were unable to demonstrate any consistent effect of vaginal lavages obtained from oestrogen-treated donors. The presence of ejaculate in the vaginae of females was also shown to cause an increase in aliphatic acid levels; this variable had not been controlled for in some previous experiments. Further, in unmated females, aliphatic acid levels actually increased during the *luteal* phase of the cycle, several days after ovulation, as confirmed in other studies by Michael and Bonsall (1977). Paradoxically, aliphatic acid levels peaked at a time when frequencies of sexual interaction usually decrease in rhesus monkeys.

In an excellent review on olfactory communication and sexual attractiveness in the rhesus monkey

Goldfoot (1981) reported stimulatory effects of artificial odours upon male sexual activity. Some effects of copulins might be due to their novelty and not to physiological mechanisms. This is reminiscent of an unexpected positive effect of model sexual skin swellings which were coloured black, rather than a normal pink or reddish hue, upon sexual arousal in male chacma baboons (Bielert *et al.* 1989). In the rhesus monkey it is doubtful whether vaginal odours influence attractiveness under normal conditions. They may play an ancillary role as part of a pattern of stimuli which control attractiveness. A simple, but rarely considered, explanation concerns the fact that oestrogen enhances vaginal lubrication, whilst progesterone antagonizes this effect. It has long been acknowledged that oestrogen promotes formation of vaginal transudate lubrication in women (Money 1961), but no experiments on other primate species have addressed this question. It is interesting, however, that in their studies of ovariectomized rhesus monkeys Chambers and Phoenix (1987) pre-treated females with a vaginal lubricant jelly. This was done 'because vaginal conditions in long-term ovariectomized females might make intromission aversive to the female and difficult for the male'. Some of these females remained attractive, as well as proceptive and receptive, to their male partners.

Experiments on rhesus monkeys have been invaluable in arousing interest in the subject of olfactory communication among primates. However, scent-marking displays using urine and cutaneous glandular secretions are much more prevalent among the prosimians and New World primates than among catarrhines such as the rhesus monkey (see Table 7.4, on page 184). Epple's studies of the marmosets and tamarins have shown that these New World monkeys can discriminate the individual identity, sex, and social status of various group members solely on the basis of olfactory cues (Epple 1974, 1976; Epple *et al.* 1993). Chemical cues from females are not essential to stimulate males to copulate, however. Sexually experienced male marmosets continue to mate after transection of the olfactory bulbs (Dixson 1992, 1993a) but they are less discriminating than normal males (see Fig. 11.10, page 331). Evidence for a sexually attractive urinary cue in the cottontop tamarin has been obtained by Ziegler *et al.* (1993). Urine collected from females during the peri-ovulatory phase of the ovarian cycle stimulated mounts by males when applied to the rumps of non-ovulatory females. The limited evidence available indicates a significant role for olfactory cues in the control of female attractiveness in New World monkeys. Circumstantial evidence also points strongly to such effects in prosimian primates. Fur-

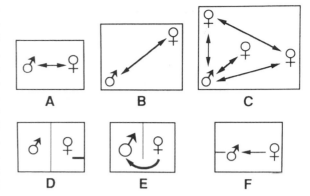

Fig. 12.3 Procedures for testing sexual behaviour in captive primates. (A). The standard free-access pair-test. (B) Enlargement of the pair-test area, providing greater freedom of movement and choice for both partners. (C) Studies of captive groups in large cages or outdoor enclosures to recreate some social complexity and competition. (D) Operant procedures: the female 'bar-presses' to gain access to the male. (E) Restricted-access tests: in sexually dimorphic species such as the orang-utan, only the female is small enough to pass beneath a barrier separating the sexes. (F) Tethered male tests: the male is harnessed and can only move within a restricted area. (After Dixson 1990a.)

ther investigations of olfactory communication and sexual attractiveness (rather than sexual recognition) in prosimians and New World monkeys would be most worthwhile.

Central effects of hormones upon female sexuality
Some comments on methods of behavioural observation and measurement

We saw in the last chapter that female sexual receptivity and proceptivity vary depending upon the conditions under which behaviour is observed and measured. Sexual behaviour may be studied by pair-testing animals for restricted periods in the laboratory, by observing captive groups, or by conducting field studies under more natural conditions. All these methods can yield valuable information and each has particular strengths and limitations (Fig. 12.3). Within the confines of a laboratory cage, large males of sexually dimorphic species (e.g. rhesus and stumptail macaques, baboons, and orangutans) can dominate sexual encounters during pair-tests and suppress female proceptivity. Various methods have been adopted to deal with these problems, including operant procedures, tethered male tests, and enlarging the testing area, as detailed in Fig. 12.3. Rhythmic changes in proceptivity during ovarian cycles, with highest frequencies during the follicular phase or peri-ovulatory period, are most readily seen under these conditions (e.g., in the rhesus monkey: Wallen 1982; Pomerantz and Goy 1983 or the orang-utan: Nadler 1982). The same is true if social groups are studied, so that an element of competition between females is encouraged (talapoins: Scruton and Herbert 1970; pigtail macaques: Goldfoot 1971; rhesus monkeys: Wallen and Winston 1984). Observations of free-ranging primates also show that proceptivity often increases during the period when females are likely to conceive (e.g., Barbary macaque: Taub 1980; gorilla: Harcourt *et al.* 1980; orang-utan: Schurmann 1982; brown capuchin: Janson 1984; woolly spider monkey: Milton 1985a). Whilst social factors may influence proceptivity under certain conditions (e.g., preference for high-ranking males in talapoins: Dixson and Herbert 1977, or in brown capuchins: Janson 1984; individual preferences in pigtails: Eaton 1973, or baboons: Smuts 1985), there is increasing evidence that hormones have direct effects upon the brain and sexual behaviour both in anthropoid primates and in prosimians. Where the anthropoids are concerned, however, the principal effects of steroids involve proceptivity rather than receptivity. The following sections present evidence linking particular ovarian and adrenal steroid hormones to proceptive and receptive behaviour in female primates.

Central effects of oestrogens upon sexual behaviour

In the few prosimian primates which have been studied, ovariectomy abolishes sexual receptivity and this may be at least partially restored by treatment with 17β-oestradiol. When ovariectomized greater galagos are treated with a large dose of oestradiol, copulations only occur during a maximum of 50% of tests (Fig. 12.4). Females accept the males' mounting attempts more readily, but fail to exhibit tail deviation and other postural adjustments which facilitate intromission. Males therefore often fail to intromit, particularly when females are paired with unfamiliar males rather than with partners with which copulation has occurred during previous tests (Fig. 12.4). Long-term, ovariectomized subjects were used in these experiments; it is possible that they were less sensitive to hormonal stimulation than younger females. Lipschitz (1988, 1994) was more successful in inducing receptivity in ovariectomized lesser galagos (*Galago moholi*) by injecting 17β-oestradiol. Only Lipschitz has observed activation of proceptivity in response to oestradiol in the galago, possibly be-

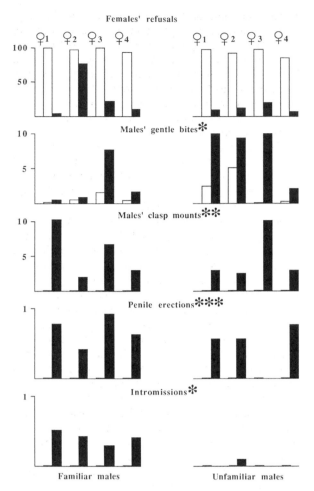

Fig. 12.4 Effects of administering oestradiol benzoate to ovariectomized Garnett's galagos (*Galago garnettii*) upon sexual interactions with familiar and unfamiliar males, during pair-tests. Data are means per test for 8 tests pre-hormone treatment (open bars) and 8 tests during which females received 40 mg day^{-1} oestradiol benzoate by subcutaneous injection. *P < 0.05; **P < 0.01; ***P < 0.001, for various behaviours of the eight male partners. (Redrawn and modified from Dixson 1978).

cause he used a large cage for pair-tests which allowed females sufficient freedom to exhibit invitational behaviour.

Reynolds and Van Horn (1977) administered 17β-oestradiol (subcutaneously in silastic capsules) to eight photo-inhibited female ringtailed lemurs (*Lemur catta*). Five females copulated and vaginal cornification occurred in four of these animals. Periods of cornification were erratic, however, calling into question the effectiveness of the oestrogen dosage employed. Males showed no significant increases in tail-marking behaviour or genital inspections of females during pair-tests. It is possible that photo-inhibition of the ovarian cycle in the ring-tailed lemur is not physiologically equivalent to ovariectomy, since photo-inhibition affects the sensitivity of neural mechanisms which control sexual behaviour in some other mammals (e.g., the male hamster). Unfortunately, no other studies have been conducted using ringtailed lemurs or other prosimian species to examine effects of oestrogens upon receptivity or proceptivity.

Turning to the anthropoid primates, a number of studies has shown that ovariectomy reduces sexual interactions, and that oestradiol treatment reverses such effects (e.g., chimpanzee: Young and Orbison 1944; rhesus monkey: Herbert 1970; *Chacma baboon*: Saayman 1973; *talapoin*: Dixson and Herbert 1977, *cynomolgus macaque*: Glick *et al.* 1982). There is an obvious requirement to control for peripheral effects of oestradiol upon female attractiveness, if central effects upon proceptivity or receptivity are to be demonstrated. In practice this can be very difficult to achieve. Some workers have attempted to distinguish between peripheral and central effects of oestradiol upon the behaviour of ovariectomized rhesus monkeys or baboons. Michael and Saayman (1968), for instance, concluded that a 5 μg dose of oestradiol administered subcutaneously was effective in activating receptivity, via a central action, whereas the same dosage applied intravaginally enhanced attractiveness but not receptivity. Although intravaginally treated females were highly attractive to males they terminated more mounts (*after* ejaculation had occurred) and were more aggressive to their partners during pair-tests. However, a reappraisal of the data shows that all the females (N = 3) receiving intravaginal oestrogen (5 μg) initiated more copulations than those receiving lower dosages, and that female refusals of males' mounting attempts did not increase. Hence there is little indication of reduced proceptivity or receptivity in ovariectomized females which were treated with oestradiol via the vaginal route, whereas results of this method of treatment upon sexual attractiveness are unequivocal and agree with those obtained in other studies (Herbert 1966, 1970).

Saayman (1973) also attempted to distinguish between peripheral and central effects of oestradiol in chacma baboons. He treated ovariectomized females with graded dosages of oestradiol (subcutaneously implanted silastic capsules producing between 10–200 μg of hormone per day), released them into a free-ranging troop and observed sexual interactions. Surprisingly, he reported that neither the dosage of oestradiol, nor the degree of sexual skin swelling, correlated positively with the males' mounting responses or females' sexual presenta-

tions. Social variables confound any simplistic interpretations of these results, however. These females apparently did not interact sexually with dominant adult males in the troop, and instead copulated with sub-adults or juveniles. Females with partial swellings showed the highest frequencies of proceptivity (presentations to males), but since males initiated more mounts with fully swollen females, these results are also consistent with effects of oestrogen upon sexual attractiveness, as well as upon sexual initiating behaviour. Indeed, seven of the nine recorded ejaculations involved two females with fully swollen sexual skins which had received the highest doses of oestradiol (80 μg and 200 μg per day). Subsequent research has demonstrated the importance of the size of the female swelling in signalling her attractiveness (Bielert and Anderson 1985; Girolami and Bielert 1987), whereas increases in female eye-contact proceptivity during the late follicular phase of the menstrual cycle imply that central effects of ovarian hormones upon sexual initiating behaviour also occur at this time (Bielert 1986).

Chambers and Phoenix (1987) have shown that ovariectomized rhesus monkeys exhibit large individual differences in their sexual behaviour during pair-tests. 'High performance' females were found to be consistently more proceptive and attractive than 'low performance' females, even after the latter had been treated with oestradiol (Fig. 12.5). However, oestradiol stimulated proceptivity in both groups, whereas sexual receptivity was not affected. These experiments support the view that proceptivity, rather than receptivity, is enhanced by the increased amounts of oestradiol secreted during the follicular and peri-ovulatory phases of the menstrual cycle (Czaja and Bielert 1975; Wallen 1982; Pomerantz and Goy 1983).

Experiments using the common marmoset have provided unequivocal evidence for effects of ovarian hormones upon proceptivity. In this species, copulations occur throughout the ovarian cycle when animals are pair-tested in the laboratory, and ovariectomy has remarkable little effect upon female attractiveness or receptivity under these conditions (Kendrick and Dixson 1983, 1984b). Despite the importance of olfactory cues for sexual behaviour in family groups of marmosets (Dixson 1992), isolated males are sufficiently sexually aroused to copulate during pair-tests, irrespective of whether the female is intact or ovariectomized (Fig. 12.6). The sexes are of equal size and males do not dominate sexual encounters in the same fashion as occurs in sexually dimorphic species, such as the rhesus monkey. These facts have made the marmoset an ideal subject in which to examine central effects of ovarian hormones upon the female's sexual behaviour. Proceptive females remain immobile or crouch, stare at the male, and tongue-flick or lip-smack. Sexual presentations do not occur and these facial displays and crouching postures do not serve to draw the male's attention directly to the female genitalia. Proceptivity is markedly increased when ovariectomized marmosets are treated with 17β-oestradiol (subcutaneously implanted silastic capsules) at dosages which produce peri-ovulatory levels of circulating hormone (Fig. 12.7). Receptivity is slightly enhanced by oestradiol, so that ovariectomized females refuse or terminate fewer of their male partners' mounts in these circumstances. These effects are very small, however, when compared with changes in proceptivity (Dixson 1986c; Kendrick and Dixson 1985a). There is no doubt that the marked effects of 17β-oestradiol upon proceptivity shown in Fig. 12.7 reflect a central action, rather than a peripheral effect upon attractiveness. Thus, the stimulatory effect of oestradiol upon proceptivity may be blocked by hypothalamic lesions in the marmoset (Kendrick and Dixson 1986; Dixson and Hastings 1992). As in the rhesus monkey, ovariectomized marmosets differ markedly in their proceptivity and in their behavioural responsiveness to oestradiol (Fig. 12.8). In

all cases oestradiol stimulates proceptive behaviour, however, whilst effects on receptivity (refusals or pre-ejaculatory terminations of males' attempts or mounts) are negligible by comparison.

Intrahypothalamic implantation of oestrogen has rarely been attempted in primates, although it has proven valuable in mapping the hormone-sensitive substrates which control receptivity in other mammals (e.g. in rodents: Pfaff 1980; Pfaff and Schwartz-Giblin 1988; in sheep: Domanski *et al.* 1972a). Michael (1969, 1971) reported stimulatory effects of intrahypothalamic implants of diethylstilboestrol upon behaviour in ovariectomized rhesus monkeys. However, leakage of hormone into the general circulation produced peripheral effects (vaginal cornification and reddening of the sexual skin) which hampered the interpretation of the results. Preliminary data on intracranial effects of 17β-oestradiol are available for one species only, the common marmoset. Intrahypothalamic implants of crystalline oestradiol activated proceptive behaviour (crouching or tongue-flicking displays) in ovariectomized marmosets (Fig. 12.9).

Finally, it is important to record that experiments involving oestrogen replacement in female primates have typically utilized subcutaneous (sc) implantation of silastic capsules, or sc and intramuscular injections, resulting in chronic elevations of circulating hormone. In ovariectomized rats and guinea pigs *pulsatile* administration of 17β-oestradiol is highly effective in priming the neural substrates which are sensitive to progesterone-facilitation of lordosis (Sodersten *et al.* 1981; Wilcox *et al.* 1984). Pulses of 17β-oestradiol also induce progesterone receptors in the preoptic area and mediobasal hypothalamus, although at lower concentrations than after a single injection of oestradiol benzoate (in the guinea-pig: Blaustein *et al.* 1990). Oestrogenic induction of proceptivity in primates, such as in the female marmoset does not involve a secondary facilitation by progesterone. Nonetheless, it would be of interest to determine whether discrete pulses of 17β-oestradiol, which mimic the short term changes in hormone levels seen in intact females, might be more effective in stimulating proceptivity.

Whilst there is increasing evidence that oestrogen stimulates proceptivity in monkeys, such as the marmoset and rhesus macaque, there is currently no agreement about whether oestrogen exerts central effects upon sexuality in women. Although the vaginal dryness and discomfort which occurs after ovariectomy in women is reversible by oestrogen treatment, various studies have failed to demonstrate effects of oestrogen replacement upon sexual interest in women (e.g. Utian 1975; Sherwin *et al.*

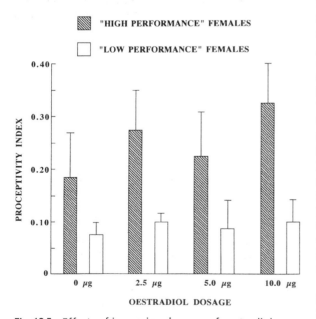

Fig. 12.5 Effects of increasing dosages of oestradiol upon proceptivity in ovariectomized rhesus monkeys. Before treatment 'high performance' females were consistently more proceptive and attractive to male partners than were 'low performance' females. This difference between the two groups of females persisted during oestrogen replacement. (Redrawn from Chambers and Phoenix 1987.)

1985). However, Dennerstein and Burrows (1982) reported that oestrogen improved both sexual interest and vaginal lubrication. Although the traditional view has been that ovariectomy has no significant effect upon the neural mechanisms which control sexual behaviour in the human female (Kinsey *et al.* 1953), it is possible that oestrogens are centrally active in at least some subjects. One recalls that in monkeys the effects of ovariectomy and oestradiol treatments vary tremendously between individuals (see Fig. 12.8, for example). Individual differences in hormonal sensitivity might also occur in human beings.

Central effects of progestagens upon sexual behaviour

In a number of non-primate mammals it is known that progesterone exerts central effects upon receptivity and also upon proceptivity in certain species. Progesterone facilitates sexual receptivity when administered to *oestrogen-primed* ovariectomized females (e.g., in the rat, hamster, guinea-pig, or beagle). Inhibitory effects of progesterone upon receptivity also occur subsequent to facilitation; the hormone has a *biphasic action* upon the female's sexual behaviour (Lisk 1978; Morali and Beyer 1979; Etgen *et al.* 1990). No evidence exists for facilitation

of proceptivity or receptivity by progesterone in primates. In some circumstances, progesterone can facilitate the luteinizing hormone surge in oestrogen-primed rhesus monkeys or women (Hotchkiss and Knobil 1994), but no behavioural effects have been recorded. However, there is evidence that progestagens might directly inhibit sexual behaviour, as well as attractiveness, in certain primate species. Michael (1968) recorded a decrease in the percentage of males' mounting attempts accepted by female rhesus monkeys during the luteal phase of the menstrual cycle. This suggests that receptivity declines slightly when plasma progesterone levels are maximal. A similar small, but statistically significant, decrease in sexual receptivity was recorded during the luteal phase of the ovarian cycle in the common marmoset (Kendrick and Dixson 1983). However, Johnson and Phoenix (1978) could not confirm the findings on rhesus monkeys, because in their study females accepted virtually all (94–100%) of males' mounting attempts during pair-tests, irrespective of the cycle stage. Administration of very large dosages of progesterone to ovariectomized rhesus monkeys can cause a *refusal reaction* (Michael *et al.* 1968) but not if physiological amounts of the hormone are employed (Baum *et al.* 1976). Subcutaneous implantation of silastic capsules containing progesterone into ovariectomized marmosets, sufficient to raise plasma

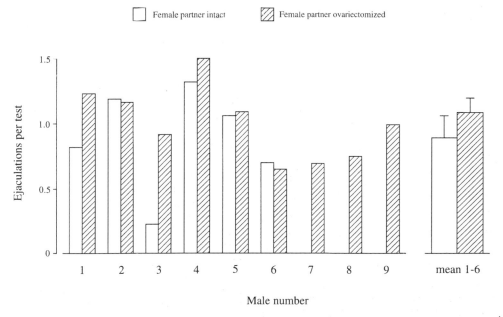

Fig. 12.6 Effects of ovariectomizing female partners upon ejaculatory frequencies by male marmosets (*Callithrix jacchus*) during pair-tests. Data are means per test for females as intacts (open bars) and post-ovariectomy (hatched bars). For pairs 7, 8, and 9 only post-ovariectomy data were collected. On the right hand side of the figure means (+S.E.M.) for pairs 1–6 are shown, since both pre and post-ovariectomy data are available for these pairs. (Redrawn from Kendrick and Dixson 1984b.)

progesterone levels to those seen in the luteal phase of the ovarian cycle, causes a marked decrease in proceptivity and a significant increase in refusals (Fig. 12.10). Females treated in this way refuse, on average, 25% of their male partner's attempts to mount. Males are still successful in 75% of attempts, however, so that copulations occur during most tests. These effects of progesterone upon proceptivity and receptivity seem to involve the brain, because intravaginal administration of the steroid does not alter attractiveness or change behaviour (Kendrick and Dixson 1985a). There is, of course, the possibility that intravaginal administration of progesterone fails to affect chemical cues in the urine or secretions of the circumgenital cutaneous glands, and that these might influence female sexual attractiveness.

Progestagens might also have some inhibitory effects upon sexual behaviour in the human female. Some women taking oral contraceptives containing progestagens experience sexual problems (Grant and Pryse Davies 1968; Royal College of General Practitioners 1974; Bancroft 1989). It is possible that individual differences in neural sensitivity to progestagens play some part in these behavioural effects. Just as monkeys vary in their behavioural sensitivity to ovarian steroids (e.g. with respect to effects of 17β-oestradiol upon proceptivity—see Fig. 12.8), the same might be true in women.

Central effects of androgens upon sexual behaviour

Plasma testosterone levels are highest during the peri-ovulatory phase of the ovarian cycle in a number of primates, such as the owl monkey (Bonney *et al.* 1980), common marmoset (Kendrick and Dixson 1983), rhesus monkey (Hess and Resko 1973), gorilla (Nadler *et al.* 1979), chimpanzee (Nadler *et al.* 1985)

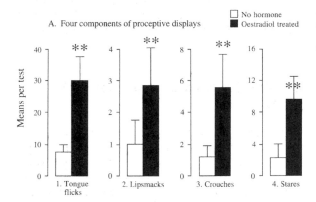

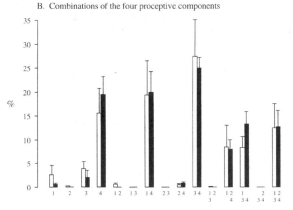

Fig. 12.7 Effects of oestradiol 17β upon proceptivity in ovariectomized marmosets. A. Oestradiol treatment results in significant increases in four components of proceptivity (1. tongue-flicks; 2. lip-smacks, 3. crouches and 4. stares), **P < 0.01, Wilcoxon tests. However, the ways that these four proceptive elements are combined during proceptive displays do not change after oestradiol treatment (B.—lower part of figure). Data are means (+S.E.M.) for nine females and their male partners. (Redrawn from Dixson 1986c.)

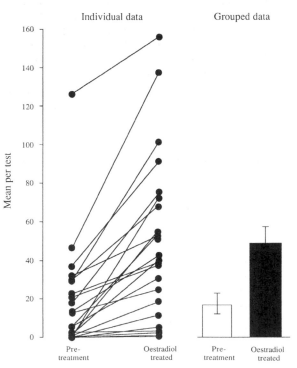

Fig. 12.8 Individual differences in proceptive tongue-flicking behaviour and in the effects of oestradiol treatment upon proceptivity in ovariectomized common marmosets (*Callithrix jacchus*). Data are means/test during paired encounters with males. Individual scores for 23 females are shown on the left hand side of the figure; the overall group mean (+S.E.M.) is shown on the right. (Author's unpublished observations.)

and man (Judd and Yen 1973). Such increases in testosterone might enhance female sexual responses or contribute to mid-cycle increases in proceptivity in certain species. In the common marmoset, for instance, females with the highest mean levels of plasma testosterone during their cycles show the greatest frequencies of proceptive tongue-flicking (Kendrick and Dixson 1985a). A similar positive correlation between high mean levels of testosterone and increases in sexuality (masturbation) has been reported in some studies of women (e.g., Bancroft et al. 1983). However, psychosocial influences can readily obscure correlations between androgens and sexual behaviour (e.g. Bancroft et al. 1991). Also, women with polycystic ovarian syndrome (characterized by excessive androgen secretion) have been reported to be sexually more assertive (Gorynski and Katz 1977).

Treating ovariectomized rhesus monkeys with testosterone propionate stimulates their proceptivity (Herbert and Trimble 1967). However, exogenous testosterone has no effect on sexual behaviour in ovariectomized marmosets (Fig. 12.10). In the studies of rhesus monkeys, the dosages required to activate sexual behaviour produced plasma hormone levels which were far higher than normal (Michael and Zumpe 1977). Michael et al. (1978) conducted elegant experiments in which they treated ovariectomized rhesus monkeys with sequential, graded doses of oestrogen, testosterone, and progesterone, in order to mimic normal menstrual cycles. They studied the sexual behaviour of females during these artificial menstrual cycles and verified that it was essentially normal. Omission of testosterone from the injection regime had no effect upon the females' proceptivity (as measured by operant techniques), indicating that mid-cycle increases in circulating testosterone, of ovarian origin, are unlikely to have a significant influence upon behaviour (Fig. 12.11).

There is an older clinical literature which points to positive effects of androgen upon sexual 'desire' or 'libido' in women (e.g., Salmon and Geist 1943; Sopchak and Sutherland 1960). Carney et al. (1978) found that sexually unresponsive women showed a greater improvement when sexual counselling was combined with testosterone, rather than diazepam (tranquillizer) treatment. Interpretation of such results in favour of an effect of testosterone has been criticized (Sanders and Bancroft 1982), but a number of more recent studies support the view that androgens do play some physiological role in womens' sexuality (e.g., Sherwin et al. 1985; Sherwin and Gelfand 1987; Sherwin 1988; Van Goozen et al. 1997).

The adrenal glands and sexual behaviour in female primates

Since the adrenal cortex secretes androgens and other 'sex steroids', it is possible that these might influence female proceptivity or receptivity. Figure 12.12 shows metabolic pathways for the production of androgens, oestrogens, and progestagens in the human adrenal cortex. Androstenedione secreted by the adrenal cortex is converted to testosterone in the bloodstream (man: Baird et al. 1968; rhesus monkey: Resko 1971). Considerable debate has cen-

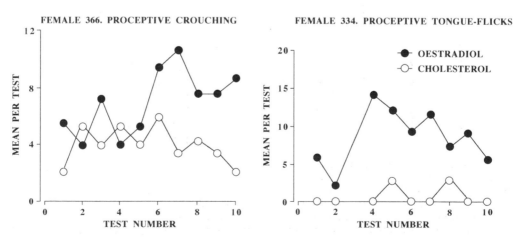

Fig. 12.9 Intrahypothalamic implants of 17β-oestradiol activate proceptivity in ovariectomized common marmosets (*Callithrix jacchus*). Data refer to two females and show effects of bilateral implants of cholesterol (control: ○) or oestradiol (●) in the anterior hypothalamus upon frequencies of crouching or tongue-flicking during pair-tests with males. (Author's unpublished observations.)

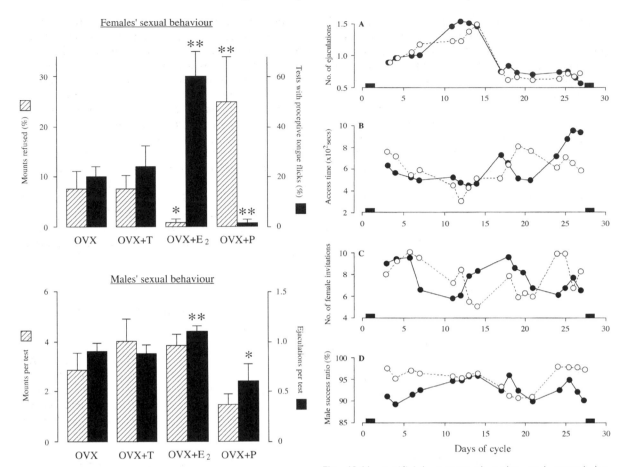

Fig. 12.10 Effects of testosterone, (T), 17β-oestradiol (E₂) and progesterone (P) administered to ovariectomized common marmosets (N=8) upon sexual interactions during laboratory pair-tests. Upper: shows data on female receptivity (hatched bars: per cent males' mounts refused by females) and proceptivity (closed bars: per cent tests with proceptive tongue-flicks). Lower: shows behaviour of male partners; hatched bars: mounts; closed bars: ejaculations. Data are means +S.E.M. *P < 0.05; **P < 0.01, Wilcoxon test compared to the no hormone (OVX) condition. (Redrawn and modified from Kendrick and Dixson 1985a.)

Fig. 12.11 Artificial menstrual cycles and sexual behaviour in rhesus monkeys (*Macaca mulatta*). Ovariectomized females were treated daily with various dosages of steroid hormones to recreate the changes which normally occur during the menstrual cycle. ● = artificial cycles with oestradiol, progesterone and testosterone; ○ = artificial cycles with oestradiol and progesterone alone. Omission of testosterone from the replacement regime had no consistent effects on A. numbers of ejaculations, B. access time by females during operant tests, C. female proceptivity to male partners, and D. the male success ratio. Data are two-day running means. Black bars indicate menstruations. (Redrawn from Michael *et al*. 1978.)

tred upon whether these androgens might stimulate sexual behaviour in monkeys or in women.

Suppression of the adrenal cortex, using the synthetic glucocorticoid dexamethasone, leads to a decrease in proceptivity in ovariectomized rhesus monkeys and might also influence receptivity (as indicated by increased refusals of the male's attempts to mount during pair-tests). Adrenalectomy has similar effects upon the female's sexual behaviour. In these studies (Everitt and Herbert 1971; Everitt *et al*. 1972), the behavioural effects of adrenal suppression, or ablation, were reversible by treating

females with testosterone propionate, or with larger doses of androstenedione (Fig. 12.13). Throughout these experiments, ovariectomized rhesus monkeys were treated with oestradiol benzoate (25 μg day⁻¹ by subcutaneous injection) to maintain their sexual attractiveness to their male partners. This complicates interpretation of the results. However, it seems unlikely that exogenous androstenedione or testosterone augmented sexual behaviour via aromatization to oestrogen, because females were already receiving a substantial dose of oestradiol. Subsequent experiments by Martensz and Everitt (1982) showed that passive immunization against testosterone also reduced proceptivity and receptivity in ovariectomized rhesus monkeys. Finally, small unilateral implants of testosterone propionate into the anterior hypothalamus reversed the effects of ovariectomy and adrenalectomy upon sexual behaviour in female rhesus monkeys (Fig. 12.14).

These experiments indicate a role for adrenal androgens in the control of sexual behaviour in female rhesus monkeys. However, some negative findings must also be considered. Nor should findings on rhesus monkeys be interpreted as being applicable to primates in general. Johnson and Phoenix (1976) found that dexamethasone treatment caused decreases in proceptivity in ovariectomized rhesus monkeys, but the effect was not statistically significant. The methods used to observe behaviour, which involved pairing each female for 10 min with eight males in succession, might have obscured more subtle effects upon female sexual initiating behaviour. Lovejoy and Wallen (1990) administered dexamethasone to intact female rhesus monkeys to examine the effect of removing adrenal steroids upon behaviour during menstrual cycles. This most interesting approach revealed that the sexual behaviour of both sexes was unaffected by suppression of adrenocortical activity. In a second macaque species, the stumptail (*Macaca arctoides*), neither dexamethasone suppression nor adrenalectomy has any affect upon proceptivity or receptivity in ovariectomized females (Baum *et al.* 1978; Goldfoot *et al.* 1978b). Nor is the remarkable retention of sexual behaviour seen in female marmosets after removal of the ovaries due to continued secretion of adrenal hormones. Adrenalectomy has no effect upon proceptivity, receptivity, or attractiveness in these monkeys (Dixson 1987c), whilst oestradiol treatment stimulates proceptivity in the absence of adrenal sex steroids (Fig. 12.15). In women, adrenalectomy might be associated with reductions in 'libido' (Waxenburg *et al.* 1959; Waxenburg *et al.* 1960; Drellich and Waxenburg 1966). However, a precise role for adrenal androgens in womens' sexuality has yet to be confirmed (Bancroft 1981, 1989).

The hypothalamic basis of sexual receptivity and proceptivity

Experiments using lesions, hormone implants and electrical stimulation or recording techniques have established that the hypothalamus plays a crucial role in regulating sexual behaviour, as well as gonadotrophin secretion, in non-primate mammals. In

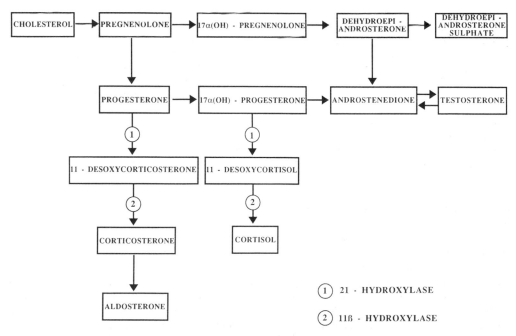

Fig. 12.12 Metabolic pathways for steroid hormone synthesis in the human adrenal cortex. Deficiencies in 21-hydroxylase or 11β-hydroxylase during foetal development cause over-production of androgens and are responsible for the occurrence of congenital adrenal hyperplasia (CAH.)

primates, attention has focused on the hypothalamic–pituitary–ovarian axis, particularly in the rhesus monkey because its menstrual cycle closely resembles that of the human female. Extensive hypothalamic lesions do not disrupt pulsatile secretion of luteinizing hormone or follicle-stimulating hormone by the anterior pituitary gland of the rhesus, provided that the mediobasal region, and especially the arcuate nucleus, remains functional (Fig. 12.16; Knobil 1974; Hotchkiss and Knobil 1994). The hypothalamic control of gonadotrophin secretion in rhesus monkeys therefore differs markedly from that reported for some mammals, such as the rat (Freeman 1988), or sheep (Goodman 1988), in which lesions in the rostral hypothalamus disrupt LH se-

cretion. We may expect that hypothalamic mechanisms which influence sexual behaviour in female anthropoids will also differ in important ways from those which control oestrus in other mammals. In sheep, lesions in the anterior hypothalamus, just caudal to the optic chiasm, block sexual receptivity whereas oestradiol implants in this same region activate receptivity in ovariectomized ewes pre-treated with progesterone (Domanski *et al.* 1972a, b). Anterior hypothalamic lesions impair receptive (lordotic) responses to oestradiol and progesterone in the ovariectomized rat (Law and Meagher 1958), whereas ventromedial hypothalamic lesions disrupt both proceptivity and receptivity (Edwards and Matthews 1977; Pfaff 1980). Yet, lesions in some

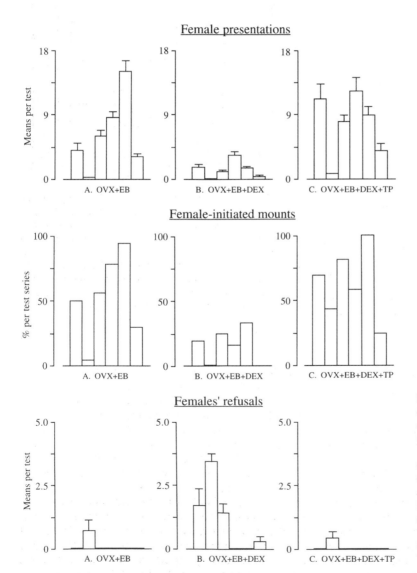

Fig. 12.13 The effect of: A. 25 μg day^{-1} oestradiol; B. oestradiol +0.5 mg kg^{-1} day^{-1} dexamethasone; C. oestradiol + dexamethasone + 100 μg–200 mg day^{-1} testosterone, on the sexual behaviour of six ovariectomized rhesus monkeys. Adrenal suppression, using dexamethasone, suppresses proceptivity and also leads to an increase in females' refusals of males' mounting attempts. (Redrawn and modified from Everitt and Herbert 1969.)

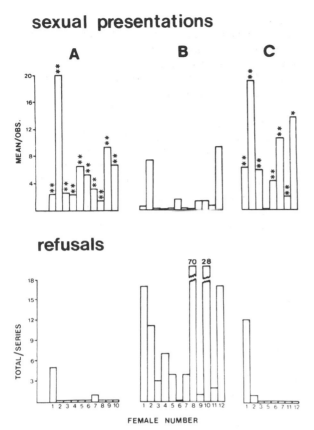

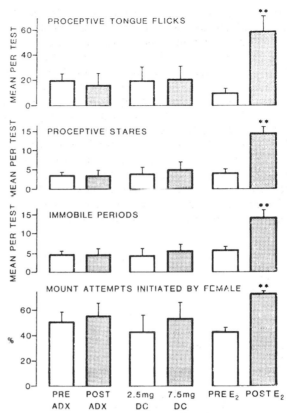

Fig. 12.14 Effects of unilateral anterior hypothalamic implants of testosterone propionate (TP) upon proceptivity (sexual presentations) and receptivity (refusals of males' mounting attempts) in female rhesus monkeys (*Macaca mulatta*). Females were treated with oestradiol throughout the experiment. A. Females ovariectomized; B. Ovariectomized and adrenalectomized; C. Ovariectomized, adrenalectomized, and implanted with TP in the hypothalamus. *P < 0.05; **P < 0.01, Wilcoxon test. (From Dixson 1983c; after Everitt and Herbert 1975.)

Fig. 12.15 Effects of adrenalectomy (ADX), deoxycorticosterone (DC) and oestradiol 17β (E2) upon proceptivity in ovariectomized common marmosets (*Callithrix jacchus*). Data are means (+S.E.M.) per test for six females (pre and post ADX; pre and post E2) and seven females (pre and post the DC dosage increase from 2.5–7.5 mg week⁻¹). ADX females were maintained on twice daily injections of 6 mg hydrocortisone hemisuccinate. **P = 0.025, Wilcoxon test. Pre-condition (open bars) and post-condition (stippled bars) for each comparison. (Redrawn from Dixson 1987c.)

areas enhance the display of lordosis (e.g., medial preoptic area: Powers and Valenstein 1972). Given that ovarian hormones influence proceptivity in anthropoids much more strongly than they affect receptivity, it is interesting that lesions in the anterior and medial hypothalamus of the female marmoset may abolish sexual initiating behaviour whilst sparing the female's ability to accept the male and allow intravaginal ejaculation (Kendrick and Dixson 1986). Before considering the limited information available on neural mechanisms which govern sexual behaviour in female anthropoids, it will be useful to review studies on sub-primate mammals and especially rodents such as the rat, hamster, and guinea pig. Far more is known about the neural basis of sexual behaviour, and especially sexual receptivity,

in rodents than in any primate species.

Pfaff (1980) and his colleagues at Rockefeller University have mapped the neural pathways that control the sexually receptive lordotic posture of the female rat (Fig. 12.17). Tactile stimulation by the male plays a key role in eliciting lordosis in the female of this species. As the male mounts, he palpates the female's flanks with his paws and makes pelvic thrusting movements against her perineum; the cutaneous stimulation thus provided reaches the brainstem (medullary reticular formation and lateral vestibular nucleus), as well as the midbrain central grey via the anterolateral columns of the spinal cord. These sensory inputs do not elicit a lordotic reflex, however, unless the female's brain has been exposed to the correct endocrine milieu; oestrogen is required to prime hypothalamic neurons and progesterone to facilitate this oestrogenic action upon lordotic responses. One crucial target area for these hormones is the ventromedial hypothalamus; neu-

rons in the ventromedial nucleus and just lateral to this nucleus accumulate oestradiol. In the rat, two descending pathways from the ventromedial hypothalamus terminate in the midbrain central grey and facilitate lordotic responsiveness during oestrus.

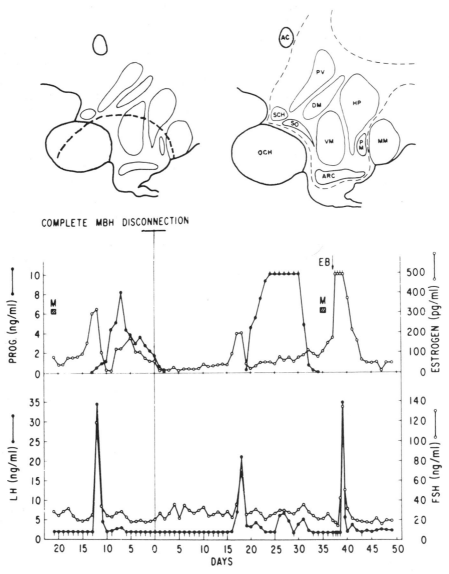

Fig. 12.16 Disconnection of the mediobasal hypothalamus does not abolish gonadotrophin secretion and ovarian cycles in female rhesus monkeys (*Macaca mulatta*). The upper part of the figure shows a schematic drawing of a parasagittal section through the hypothalamus of the rhesus monkey, with the various structures labelled, together with a section through the hypothalamus of a female after placement of a lesion (using the Halasz knife) to disconnect the mediobasal hypothalamus from surrounding structures. ARC=arcuate nucleus; DM=dorsomedial nucleus; HP= posterior hypothalamic nucleus: PV=paraventricular nucleus; SCH=suprachiasmatic nucleus; SO=supraoptic nucleus; VM=ventromedial nucleus; MM=medial mammillary body; PM=premammillary area; AC=anterior commissure; OCH= optic chiasm. The lower portion of the figure shows the time courses of circulating gonadotrophins (●=LH; ○=FSH) and ovarian steroids (●=progesterone; ○=oestrogen) before and after lesion placement in this particular female. Menstrual periods are indicated by M. The arrow on day 38 indicates the subcutaneous injection of oestradiol benzoate. (After Knobil 1974.)

One pathway descends directly through the medial hypothalamus and the other, and more important tract, projects laterally to reach the midbrain by a circuitous route, as can be seen in Fig. 12.17. Lesions in the midbrain central grey which damage the terminal areas of these hypothalamic projections cause a reduction in lordosis (Sakuma and Pfaff 1980). The excitability of cells in the central grey is increased by oestrogenic stimulation of ventromedial hypothalamic neurons. The midbrain central grey neurons project in turn to the reticular formation in the lower brainstem. Signals from the midbrain are transferred to the spinal cord via contacts with reticulospinal neurons in the ventral *nucleus gigantocellularis*. Descending pathways in the spinal cord innervate motor neurons in the ventral roots, from thoracic 12 through to sacral 1; these motor neurons control two sets of deep-back muscles (the *lateral longissimus* and *transversospinalis*) which elevate the female's rump during lordosis.

Lordosis is affected by both progesterone and oestrogen; oestrogen-priming results in an increased concentration of hypothalamic cytosol progestin receptors in the rat (Moguilewsky and Raynaud 1979) and the guinea pig (Blaustein and Feder 1979). Experiments on guinea pigs have shown that proges-

terone treatment of oestrogen-primed females causes progestin receptor concentrations in cell nuclei to increase, but that numbers then decrease by the end of oestrus, as do concentrations of cytosol receptors. These events may correspond to the biphasic actions of progesterone, firstly facilitating and then inhibiting lordosis (Blaustein *et al.* 1990). Experiments involving intracranial implantation of progesterone into oestrogen-primed ovariectomized rats have shown that the ventromedial hypothalamus (VMH) is also a target site for progesterone in the facilitation of sexual behaviour. Both lordosis and proceptivity (hopping, darting, and ear-wiggling) are stimulated by progesterone and the effect is not due to activation of adrenal hormone secretion (Barfield *et al.* 1983). These authors also concluded that the *inhibitory* effects of progesterone upon sexual behaviour are mediated by its effects upon VMH neurons, and that only those oestrogen-primed females which had received progesterone implants in the VMH (as compared with the preoptic area or midbrain) displayed refractoriness to the behavioural effects of progesterone when it was injected subcutaneously 24 h later. Interspecific differences occur among rodents, so that in the hamster progesterone exerts its stimulatory effect on sexual

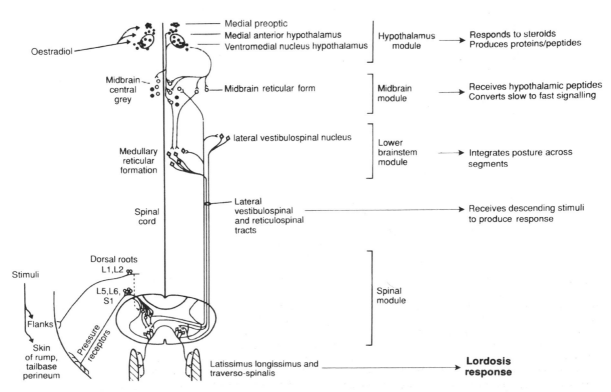

Fig. 12.17 Summary diagram of the neuronal circuitry which controls lordosis in the female rat. (after Wilson 1993; adapted from Pfaff 1980.)

behaviour by influencing neurons in the VMH and the ventral tegmental area (VTA). Hamsters are more dependent upon the sequential actions of oestradiol and progesterone to induce lordosis than are rats and, interestingly, the effect of progesterone upon the VTA involves membrane effects of the steroid, rather than a genomic action as in the VMH (De Bold and Frye 1994).

Throughout this brief overview of the hormonal control of sexual behaviour in female rodents, attention has focused upon hypothalamic mechanisms. However, the hypothalamus is not the only region of the brain to bind ovarian steroids and nor is it a 'sex centre', because proceptive and receptive patterns require the integration of a number of neural mechanisms. Even at the simplistic level of the lordotic reflex this has proven to be the case (Fig. 12.17); clearly the proceptive displays of monkeys and apes must involve more complex circuitry and include the processing of visual signals from potential mates, rather than reflexive posturing in response to tactile inputs. In addition to the VMH, uptake of oestradiol also occurs in the preoptic area and midbrain, as well as in the lateral septum, corticomedial division of the amygdala, and bed nucleus of the stria terminalis (in the rat: Pfaff and Keiner 1973). Essentially the same distribution of oestrogen-concentrating neurons is found in the rhesus monkey brain (Pfaff *et al.* 1976; see pages 414–16 for a more detailed discussion of this topic). The distribution of oestrogen-binding neurons in the vertebrate brain has remained relatively conservative during vertebrate evolution, but the behavioural significance of these phenomena must be rather different in anthropoid primates from those in the rodents considered so far.

In addition to facilitatory mechanisms, mechanisms that inhibit the expression of lordosis also exist in the rodent brain. As an example, electrical stimulation of the medial preoptic area (MPOA) interrupts lordosis in the female rat, whilst excitotoxic lesions in this region enhance lordosis. In each case, the effects of these manipulations upon the female's proceptivity are the reverse of those which involve receptivity (Sakuma 1994). Forebrain lesions or 'knife-cuts' which sever pathways between the septum and hypothalamus serve to disinhibit the lordosis reflex in the rat and also render females more responsive to hormonal stimulation of their proceptivity (Yamanouchi and Arai 1990). Clearly, the overall control of lordosis and proceptivity must be more complex in the rat than the circuitry depicted in Fig. 12.17 would suggest. Nor is tactile stimulation obligatory to elicit the response in all species; in the hamster, for example, the proximity of a male is sufficient to trigger a pronounced lordotic response in the oestrous female.

Having considered the endocrine mechanisms and neural pathways that govern sexual behaviour in female rodents, we may now turn to the more limited information available for primates. It is likely, but unproven, that similar neural mechanisms to those which control lordosis in the rat probably operate in female prosimian primates. Females of some prosimian species exhibit a lordotic response when contacted by the male and in all species receptivity is limited in duration and equivalent to oestrus in sub-primate mammals. In the few prosimians that have been studied, oestradiol alone is sufficient to induce receptivity; facilitatory and inhibitory actions of progesterone have not been demonstrated. In the rat, oestrogen alone is also sufficient to induce receptivity in ovariectomized females, if a sufficiently large dose is employed (Davidson *et al.* 1968). It is possible that during prosimian evolution, selection has favoured an oestrogen-only control mechanism, but more experiments will be required to address this question before we conclude that progesterone has no role to play in prosimian sexual behaviour. Among the anthropoids, females do not exhibit lordosis and their receptivity is not rigidly yoked to changes in oestrogen and progesterone secretion during the peri-ovulatory phase of the cycle. This calls into question whether the hypothalamic mechanisms governing receptivity in rodents have any counterpart in the anthropoid brain. Despite Pfaff's (1980) not unreasonable view that neuranatomy has remained relatively conservative during vertebrate evolution and that 'we generally expect similar principles of neural control in humans as in experimental animals', there are reasons to believe that the hypothalamus of monkeys plays only a minor role in controlling sexual receptivity and is much more important where proceptive patterns are concerned. A series of experiments on the common marmoset, involving either ovariectomized or ovariectomized-adrenalectomized females treated with 17β-oestradiol, showed that thermal lesions in the anterior or medial hypothalamus reduced or abolished proceptive responses (Fig. 12.18 and 12.19), whereas receptivity was unaffected in most cases (Kendrick and Dixson 1986; Dixson and Lloyd 1988; Dixson 1990b). In three females which sustained substantial damage to the ventromedial region of the hypothalamus, a 25–42% increase in terminations of males' mounts occurred, but this effect was reversed by oestradiol (Dixson 1990b). Subsequent experiments showed that excitotoxic lesions in the anteromedial hypothalamus (made using ibotenic acid) also depress or abolish proceptive behavioural patterns in female marmosets, whilst sparing sexual receptivity (Figs 12.20 and 12.21). These results provide evidence that selective damage to neuronal cell bodies within the marmoset hypothalamus, rather than to fibres of passage or terminals of extrinsic neurons, is is sufficient to disrupt proceptivity (Dixson and Hastings 1992).

The data obtained from marmosets indicate that a distinction may be made at the hypothalamic level between the control of proceptivity and of receptivity in these monkeys. Anteromedial lesions severely

compromise the female's invitational behaviour, whilst having minimal effects upon her willingness to accept copulation. Experiments with female macaques have shown that electrical stimulation of the VMH or preoptic area/anterior hypothalamus evokes sexual presentations towards male partners (Koyama et al. 1988). Some results from these studies are shown in Fig. 12.22. Single-unit recordings have also been made from hypothalamic neurons of female macaques as they engage in proceptive sexual presentations and during copulation (Aou et al.

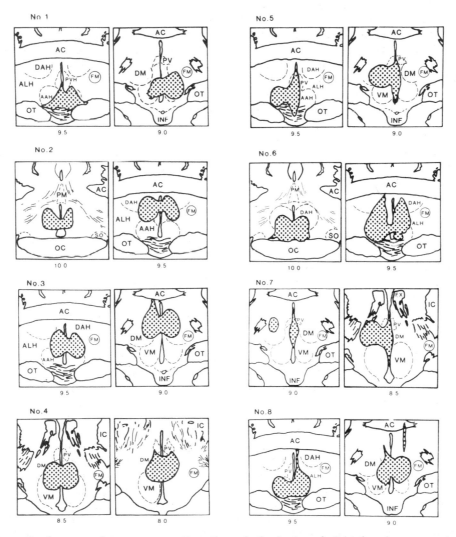

Fig. 12.18 Schematic diagrams of transverse sections through the brains of eight female marmosets to show the positions of thermal lesions (stippled areas) and nuclei (broken lines) in the hypothalamus. The level of each section, anterior to interaural zero, is shown in millimetres. Effects of these lesions upon sexual behaviour are shown in Fig. 12.19. AAH=anterior hypothalamus; AC=anterior commissure; ALH=lateral hypothalamus; DAH=dorsal anterior hypothalamus; DM=dorsomedial nucleus; FM=medial forebrain bundle; FX=fornix; IC=internal capsule; INF= infundibulum; OC=optic chiasm; OT=optic tract; PM=median preoptic nucleus; PV=paraventricular nucleus; SO= supraoptic nucleus; VM=ventromedial nucleus. (Adapted from Kendrick and Dixson 1986.)

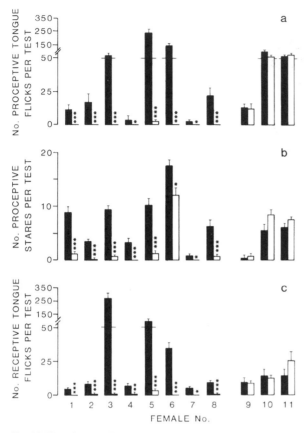

Fig. 12.19 Effects of antero-medial hypothalamic lesions upon the sexual behaviour of female common marmosets (*Callithrix jacchus*). Thermal lesions (as shown in Fig. 12.18) were made in ovariectomized, oestradiol-treated marmosets (females 1–8) and sham lesions in controls (females 9–11). Effects of these procedures upon (a) proceptive tongue-flicking, (b) proceptive stares, and (c) tongue-flicking during copulation (so-called 'receptive' tongue-flicking) are shown for each female before (closed bars) and after (open bars) a lesion or sham operation. *P < 0.05; **P < 0.01; ***P < 0.001; Mann-Whitney U test. (Redrawn from Kendrick and Dixson 1986.)

1988). Some VMH neurons exhibit increased firing rates under both conditions, whilst the firing rates of POA neurons typically decrease during female invitations and during copulation. Thus, at the neurophysiological level, a difference again emerges between proceptivity and receptivity. It is not, however, justifiable to interpret these results in terms of the VMH playing a role in sexual receptivity akin to that described for the rat. Interpretation of these studies is also complicated by the fact that interspecific pairings were employed (*M. mulatta* vs. *M. fuscata*), involving animals at unknown stages of the menstrual cycle, which were partially restrained in primate chairs.

In an attempt to clarify differences in the neuroendocrine control of sexual behaviour which exist between anthropoid primates, such as the rhesus monkey and sub primate females, such as the rat Wallen (1990) proposed that a distinction should be made between the *desire* of the female to engage in copulation and her *ability* to do so. This argument states that the ovariectomized female rat lacks the ability to mate because oestrogen and progesterone are essential to facilitate her lordotic posture—which, in turn, is necessary to facilitate intromission. Intromission in a female monkey does not require her to adopt a reflexive posture, however, although it is necessary for her to remain immobile if the male is to mount and thrust successfully. Interesting as this observation is, it should not obscure the fact that the activation of proceptivity and receptivity in the rat is dependent upon ovarian hormones, whilst neither female invitations nor acceptances are strictly dependent upon the ovaries in monkeys, apes, or women. There is a major difference between anthropoids and most sub-primate mammals in this respect, and this difference is not simply related to hormonal facilitation of mating postures. In mammals such as sheep, which lack marked rump elevation to enable male penetration, females are nonetheless unreceptive after ovariectomy. They refuse to stand still for the male as an oestrous female normally would. It is also misleading to equate proceptivity in monkeys and apes with 'oestrus', such as occurs in sub-primate mammals. Although proceptivity is facilitated by oestrogen, it is not necessarily limited to the peri-ovulatory period of the ovarian cycle and in some species occurs in the total absence of ovarian stimulation. This situation is quite different from that seen in rodents.

In his classic paper on 'sexual attractivity, proceptivity, and receptivity in female mammals' Beach (1976) pointed out the male presents some 'optimal combination of multisensory stimulation for the elicitation of such (proceptive) behaviour'. Among the anthropoid primates visual cues are probably paramount in this respect. The importance of eye-contact between the sexes as part of precopulatory behaviour, and the existence of eye-contact proceptivity in various monkeys and apes, has been previously described (see page 97). Such eye-contact displays occur, for instance, when a female marmoset crouches, stares at, and tongue-flicks towards her male partner before copulation. Careful observation of these patterns has revealed that females may crouch or stare in response to the sight of the male, but in the apparent absence of establishing eye-contact with him. However, the female's tongue-flicking display is elicited specifically by eye-contact between

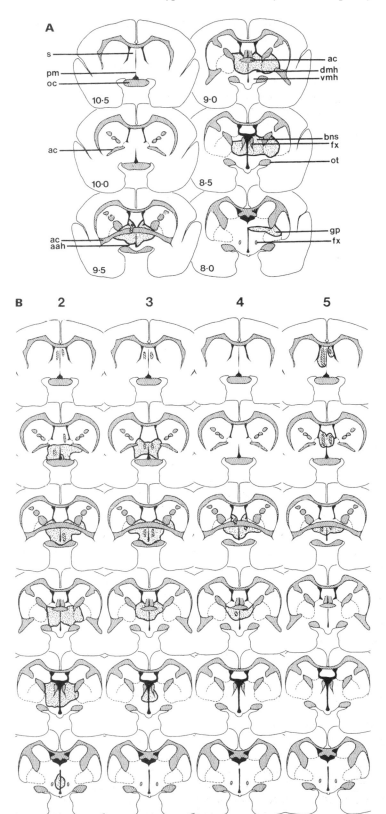

Fig. 12.20 A. Schematic diagrams of coronal sections through the brain of a female common marmoset (number 1) to show the extent of neuronal degeneration (stippled areas) produced by the excitotoxin, ibotenic acid. The level of each section anterior to interaural zero (8.0–10.5) is shown on the left-hand side. Major structures are labelled: aah = anterior hypothalamus; ac = anterior commissure; bns = bed nucleus of stria terminalis; dmh = dorsomedial hypothalamus; fx = fornix; gp = globus pallidus; oc = optic chiasm; ot = optic tract; pm = median preoptic nucleus; s = septum; vmh = ventromedial hypothalamus. Cross-hatching indicates position of cannula tracks. B. Composite diagrams to show the extent of ibotenic acid-induced hypothalamic lesions in female numbers 2 to 5. Behavioural effects of these lesions are shown in Fig. 12.21. (After Dixson and Hastings 1992.)

the partners and oestradiol is only effective in stimulating proceptive tongue-flicking under such conditions (Fig. 12.23). It is not essential for the male to be sexually active or to direct facial displays towards the female, however. Oestrogen-treated females continue to invite the male and to tongue-flick in response to his gaze, even after placement of lesions in the male's hypothalamus which reduce or abolish his sexual behaviour (Dixson and Lloyd 1989). It is possible that the reductions of female proceptivity which occur after thermal or excitotoxic hypothalamic lesions (Figs 12.19 and 12.21) damage neural substrates which respond to complex visual inputs processed initially within the cerebral cortex. The

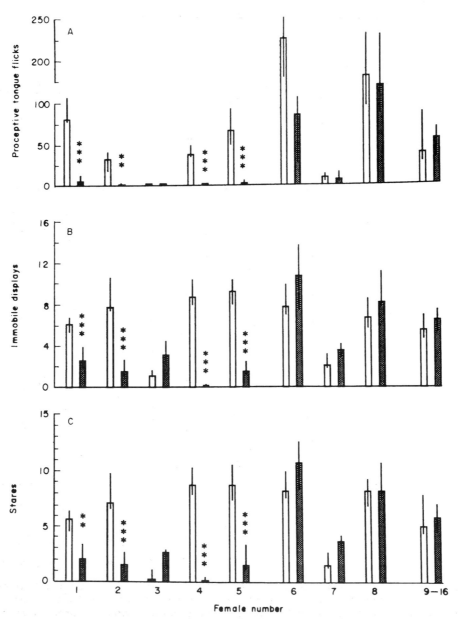

Fig. 12.21 Effects of ibotenic acid lesions in the hypothalamus (females 1–5), non-lesions (procedures failed to produce lesions: females 6–8) and sham lesions (saline infused controls: females 9–16) upon A. proceptive tongue-flicking, B. proceptive immobile displays, and C. proceptive staring by ovariectomized, oestrogen-treated female marmosets (*Callithrix jacchus*). Histograms show median (±interquartile ranges) frequencies for each individual, pre-operatively (open histograms: 10 mating tests) and post-operatively (closed histograms: 10 mating tests). **P < 0.02; ***P < 0.002; Mann-Whitney U test. (After Dixson and Hastings 1992.)

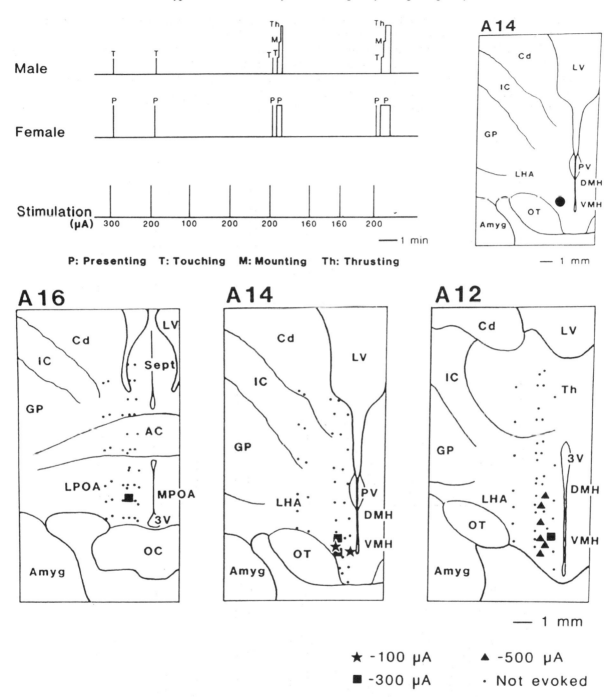

Fig. 12.22 Upper: Example of presenting evoked by electrical stimulation of the ventromedial hypothalamus in a female macaque. In each case presenting preceded the male's contacts with the female. Lower: Overview of stimulation sites and effects upon proceptivity (sexual presentations) in a representative female macaque. The symbols indicate the threshold required to evoke a response. Small dots show the sites where presenting was not elicited. Locations of the sites stimulated are shown on the closest stereotaxic plane. Numbers on the left side of each diagram indicate the distance (mm) from the interaural line. AC, anterior commissure; Amyg, amygdaloid complex; Cd, caudate nucleus; DMH, dorsomedial hypothalamic nucleus; GP, globus pallidus; IC, internal capsule; LHA, lateral hypothalamic area, LPOA lateral preoptic area; LV, lateral ventricle, MPOA, medial preoptic area; OC, optic chiasm, OT, optic tract; PV, periventricular nucleus; Sept, septal nucleus; Th. thalamus; VMH, ventromedial hypothalamic nucleus; 3V, third ventricle. (After Koyama *et al.* 1988.)

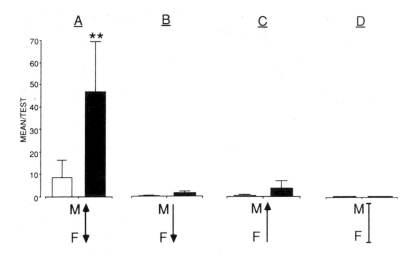

Fig. 12.23 Eye-contact proceptivity in the female marmoset (*Callithrix jacchus*). Mean (± S.E.M.) frequencies per test of proceptive tongue-flicking by eight ovariectomized marmosets before (open bars) and after (closed bars) treatment with 17β-oestradiol.

Tongue-flick frequencies were measured under four conditions: A. during or immediately after periods of eye-contact between male (M) and female (F). B. Male looks towards female. C. female looks towards male. D. no eye-contact between partners. **P = 0.01. Wilcoxin tests by comparison with conditions B, C, and D. (After Dixson and Lloyd 1989.)

following section therefore explores how the relevant visual stimuli are analysed at the cortical level, and further processed within the limbic system, to call forth an appropriate proceptive response from the female. This constitutes an attempt to provide a *neural model* of proceptivity in primates.

A neural model of proceptivity in female primates

Although studies of the primate brain which deal specifically with female proceptivity are few in number, other sources of information are available which are highly relevant to understanding the neural basis of sexual behaviour. For instance, much is known about how the brain processes sensory stimuli and about the role of the limbic system in emotionality and motivation. We might speculate that some of these brain processes must also participate in the control of sexual activity. Firstly, it will be useful to examine how the brain deals with complex visual stimuli, since eye-contact between the sexes and the exchange of precopulatory visual displays is so important in many anthropoids. In monkeys it is known that further processing of inputs received by the primary (occipital) visual cortex occurs in the temporal lobe. Gross *et al.* (1972) were the first to show that cells exist in the inferotemporal cortex of macaques which respond specifically to the sight of faces. Subsequent work by a number of groups has mapped the distribution of such *face cells*; Perrett *et al.* (1992) have reviewed this information and Fig.

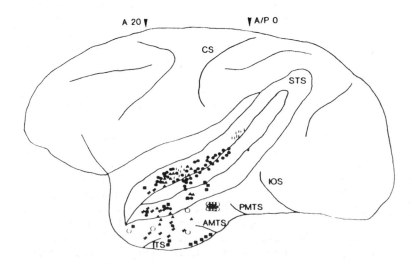

Fig. 12.24 Location of cells in the macaque temporal cortex which are selective for faces. The drawing, of the left side of the rhesus macaque brain, is based upon studies by Perrett *et al.* (●); Harries (I); Hasselmo (#); Rolls (▲), Tanaka (○) and Yamane (■). Abbreviations: STS, superior temporal sulcus; IOS, inferior occipital sulcus; CS, central sulcus; ITS, inferior temporal sulcus; AMTS, anterior medial temporal sulcus; PMTS, posterior medial temporal sulcus. (After Perrett *et al.* 1992.)

12.24 is reproduced from their work. Electrophysiological recording studies have shown that populations of cells fire in response to the sight of particular faces or facial features, such as the eyes or mouth. Some cells respond to changes in facial expression or facial orientation (e.g. profile vs. full-face view). Some neurons also encode for body postures and movements. Figure 12.25 shows an interesting example, from Perrett *et al.* (1992), of neurons which fire when the gaze is directed downwards or when changes in body posture are combined with a downward shift in head orientation. Face cells are not confined to the primates, however; Kendrick has shown that they also occur in the temporal cortex of sheep (Kendrick and Baldwin 1987; Kendrick 1994). One group of cells codes for the presence and size of horns in the ram, 'possibly allowing for a rapid estimation of the perceived

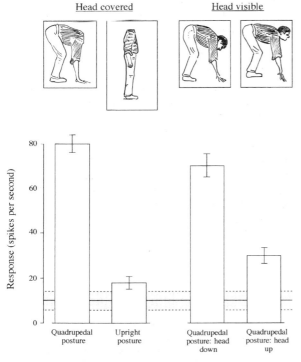

Fig. 12.25 Example of the response of a single cell in the macaque inferotemporal cortex which fires in response to changes in complex visual stimuli. This cell is sensitive to head and body postures indicative of attention directed downwards. The visual stimuli used are represented in the upper part of the figure. Lower: mean (±S.E.M.) responses of the cell, showing that it responds preferentially to the quadrupedal posture, rather than the vertical body posture (when the head is covered), and to head down, rather than head up positions. The stippled area denotes levels of spontaneous activity shown by the neuron. (Redrawn and modified from Perrett *et al.* 1992.)

animal's sex or position in the dominance hierarchy' (Kendrick and Baldwin 1987). Given that many male primates display striking secondary sexual facial characters (e.g. the cheek flanges of the orang-utan or brilliantly coloured mask of the mandrill: see Fig. 7.22, page 196), it would be interesting to know whether specific cortical neurons exist which code for these features. No data are available, as only macaques have been studied in the laboratory. However, it seems likely from the sheep experiments that cortical mechanisms exist to encode for features specifically relevant to sexual recognition and (perhaps) to sexual attraction. We may hypothesize that eye-contact proceptivity, such as occurs in the chacma baboon or marmoset (Bielert 1986; Dixson and Lloyd 1989), involves visual processing in the temporal cortex. Perrett's group has shown that face cells in the superior temporal sulcus (STS) occur in 3–5 mm 'patches'; grouping of face cells might also occur in other parts of the temporal cortex, but this remains to be determined. However, the STS also has outputs to the parietal cortex arranged in repeating 3–5 mm patches. Since the parietal cortex is concerned with spatial awareness (i.e. awareness of one's position in relation to surrounding objects) such pathways may provide links between socially relevant cues and cortical mechanisms serving for spatial awareness and attention (Perrett *et al.* 1992).

Although the temporal cortex analyses facial and postural visual stimuli in a highly complex fashion, it is not believed to assign emotional significance to such information. The inferotemporal cortex has many reciprocal connections with the underlying amygdala, which in turn has reciprocal connections with the hypothalamus. Some of these pathways are shown diagrammatically in Fig. 12.26. This attempts to map neural substrates which control proceptive behaviour in female monkeys. This diagram borrows heavily from research on the neural basis of emotion (e.g., Rolls 1986, 1990) and from general models of motivated behaviour (Mogenson *et al.* 1980). The rationale behind the model is discussed in the following sections, commencing with the role played by the amygdaloid complex in primate social and sexual behaviour.

Damage to the amygdala and overlying temporal cortex has long been known to cause profound changes in the social, sexual, and aggressive behaviour of monkeys (Klüver and Bucy 1939). Macaques exhibit hyper-oral behaviour, tameness, and inappropriate sexual responses (such as mounting inanimate objects) as a result of such lesions. Amygdalectomy severely compromises social behaviour, such that free-ranging animals are unable to sustain a normal group-living existence (vervet monkey: Kling *et al.* 1970). It is important to note that lesions of the temporal polar cortex produce Klüver–Bucy type effects (e.g., in the cat: Green *et al.* 1957), and that in many primate studies both the amygdala and temporal cortex have been damaged to varying degrees. Weiskrantz (1956) was the first to suggest that bilateral amygdalectomy (accompanied by damage to the medial temporal polar cortex) made it 'difficult for reinforcing stimuli whether positive or negative to become established or recognized'. Monkeys suffering from the *Klüver–Bucy syndrome* are probably not so much hyperphagic or hypersexual as *indiscriminate* in their responses to important environmental stimuli, such as food or potential sexual partners. This lack of discrimination probably results from two factors. Firstly, any damage to the temporal cortex which occurs during amygdalectomy compromises the ability of a monkey to recognize conspecifics and socially relevant visual cues. Secondly, amygdalectomy removes many of the projection sites for these visually sensitive neurons in the temporal cortex. Information processed by the inferotemporal cortex is transferred mainly to the lateral nucleus of the amygdala, with some projections also to the basal nucleus. Within the amygdala itself, the lateral nucleus has many projections to the basal nucleus which, in turn, has back projections to various levels of the visual cortex (Fig. 12.27; Amaral *et al.* 1992). Amygdaloid lesions therefore damage mechanisms concerned with the further processing of emotionally relevant visual information (Rolls 1992).

In some unknown fashion the amygdala creates a linkage between complex sensory inputs and their emotional or motivational significance. Electrophysiological recording studies of the macaque amygdala have revealed single neurons which respond to visual stimuli (including the sight of faces), whilst others are responsive to auditory, gustatory, or somatosensory stimulation. Such neurons are important in stimulus reinforcement learning, as Weiskrantz originally suggested. Approximately 20% of visually responsive neurons fire primarily in response to stimuli associated with positive reinforcement (food reward: Rolls 1990). In man, calcification of the amygdala, resulting from Urbach–Wiethe disease has been associated (in a single female patient) with inability to discriminate the emotion of fear in facial representations, or to tell the difference between subtle variations in facial cues (Adolphs *et al.* 1994). It is not known whether processing of facial information of an affiliative or specifically sexual nature is compromised by Urbach–Wiethe disease, and in any case it is extremely difficult to assess the precise extent of amygdaloid damage which occurs in such patients.

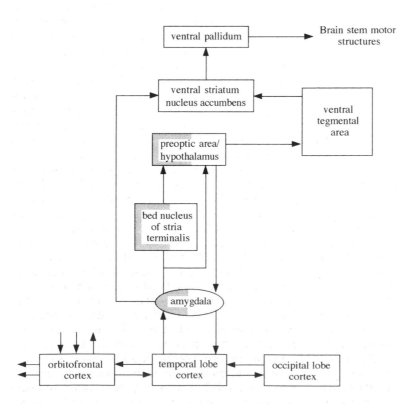

Fig. 12.26 Some of the possible neural pathways which control proceptive behaviour, and process visual inputs relevant to proceptive responses, in female primates. In this simplified scheme only major interconnecting fibre tracts are indicated (e.g. many interconnections of the orbitofrontal cortex are omitted). Areas which contain neurons binding oestradiol are stippled. Further explanation is provided in the text. (Redrawn from Dixson 1990a.)

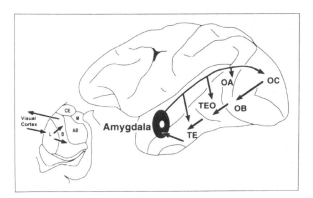

Fig. 12.27 Schematic illustration showing the relationship of the amygdala with the visually related cortices of the temporal and occipital lobes. Visual information is processed in a hierarchical fashion in cortical regions at progressively greater distances. (OB OA TEO TE) from the primary visual cortex (OC). The amygdaloid complex receives a substantial input from the highest level of the hierarchy (area TE). This information enters the amygdala via the lateral nucleus (see the drawing of the amygdaloid complex at lower left). A projection back to the visual cortex originates in the basal nucleus (B) and this projection extends to all levels of the visually related cortex in the temporal and occipital lobes. Within the amygdala, prominent projections between the lateral (L) and basal nuclei potentially 'close the loop' on amygdaloid interconnections with the visual cortex. (after Amaral *et al*. 1992.)

Controlled experiments involving effects of selective lesions to various amygdaloid nuclei in monkeys are required to examine these problems but none have been reported. Only the effects of gross lesions involving the whole amygdala have been examined. Spies *et al.* (1976), for instance, showed that bilateral destruction of the amygdala in female rhesus monkeys led to decreases in their proceptive (prox) behaviour but not to other measures of sexual activity during pair-tests. Unfortunately, the questions of eye-contact and facial displays before copulation were not examined in this study.

Much remains to be learned about how the primate amygdala analyses sensory signals and contributes to decision-making processes in relation to patterns of social and sexual behaviour. In posing this question, it is necessary to ask how sensory inputs are processed within the amygdala itself and via which outputs this information reaches other parts of the brain. The amygdala has reciprocal connections with the hypothalamus, via the ventral amygdalofugal pathway and stria terminalis; these pathways are known to be important in the regulation of sexual behaviour and neuroendocrine processes (Larsson 1979; Sachs and Meisel 1988). Autoradiographic studies of the brains of rodents and

macaques have shown that neurons in the medial amygdala, bed nucleus of the stria terminalis, ventromedial hypothalamus, and preoptic area preferentially take up and bind oestradiol (Bonsall *et al.* 1986; Michael and Bonsall 1990). It is possible, therefore, that oestradiol facilitates proceptivity by acting at a number of sites within the cortical-amygdaloid-hypothalamic circuitry. No studies, for instance using intracranial implants of oestradiol, have adequately addressed these questions in primates.

Neural mechanisms which link motivation to action probably share common properties for a variety of behavioural responses (Mogenson *et al.* 1980). For example, descending pathways from the hypothalamus to the ventral tegmentum play a crucial role in the activation of maternal responses, as well as masculine copulatory behaviour (in the rat: Numan 1988; Sachs and Meisel 1988). Information is relayed from the ventral tegmental area, via mesolimbic dopaminergic neurons, to the ventral striatum. Interestingly, the amygdala projects to the striatum, where cells which are responsive to the sight of faces have also been identified (Rolls and Williams 1987). Mogenson (1984) suggested that communication between the amygdala and ventral striatum might play a significant role in limbic-motor integration and these portions of his model have been included in Fig. 12.26. Within the ventral striatum, the nucleus accumbens in particular may interface between motivational systems and the motor systems required to execute voluntary movements and thus to express behaviour.

The orbito-prefrontal cortex (OFC) has also been included in the circuitry depicted in Fig. 12.26 although, for the sake of clarity, its interconnections with many brain regions have been omitted from the diagram. The OFC has reciprocal connections with the inferotemporal cortex and receives inputs from the medio-dorsal nucleus of the thalamus (which in turn receives afferents from the amygdala and temporal lobe). The OFC projects, via the striatum, to motor centres and there are also strong connections to limbic areas (cingulate cortex, amygdala, and hypothalamus), as well as with the autonomic nervous system (Rolls 1986; Goldman-Rakic 1987). Lesions of the OFC are known to cause marked changes in social competence in human patients and have been reported to cause antisocial and hypersexual behaviour in some individuals. In monkeys, orbito prefrontal lesions impair the ability to 'unlearn' responses that were previously rewarded by food. Such monkeys perform poorly on go/no go tasks or object-reversal trials (Rolls 1986). Bilateral surgery involving the OFC in man (e.g., to remove tumours) can result in lack of self motivation and forward-planning ability, or the tendency to react slowly and obsessively. Nauta (1971) suggested that prefrontal cortical damage might produce a state of *interoceptive agnosia*, an inability to link emotional and autonomic responses in an appropriate fashion when engaging in, or planning, social behaviour. Indeed, Damassio (1991) has proposed that the prefrontal cortex evolved to facilitate selection of advantageous responses in complex social situations. He describes patients with OFC damage as exhibiting a form of *acquired sociopathy*.

In memory associated with visual stimulus reward in macaques, the amygdala is entirely dependent upon its interaction with the frontal lobes, either by direct projections or indirectly via projections through the medio-dorsal thalamus (Gaffan *et al*. 1993). Electrophysiological recordings throughout the caudal OFC have detected bimodal neurons which respond to both vision with olfaction and taste with olfaction (Rolls and Baylis 1994). Rolls (1986, 1990) has proposed that OFC neurons influence emotional behaviour by allowing the animal to adapt flexibly to environmental changes and to alter responses, rather than persisting with unrewarding patterns of behaviour. This aspect of cortical regulation has been included in Fig. 12.26, because it enables some conception of how a female might alter her proceptivity dependent upon social and other environmental constraints. Hence, a female monkey might suppress her proceptivity in a particular testing situation, or in the presence of a more dominant female in her social group. She might also shift the direction of her proceptive displays if a previously preferred or more attractive male becomes available. The prediction is that OFC lesions should have major effects upon the ability of monkeys to regulate their proceptivity in this way. Unfortunately, no experiments have been conducted to examine this specific question in non-human primates.

It will be appreciated that the model described here, and depicted in Fig. 12.26, is simplistic. However, no better scheme has been advanced to account for the neural control of proceptivity in anthropoid primates. It remains my hope that this framework will generate criticism and encourage badly needed experiments in this area of behavioural neuroscience. Even though knowledge of how the brain controls sexual behaviour in primates is limited, it is nonetheless clear that such mechanisms must differ in many ways from those which govern sexual receptivity and lordosis in rodents (Pfaff 1980; Pfaff and Schwartz-Giblin 1988). Successful mapping of the lordotic circuitry has involved multidisciplinary studies, bringing together expertise in neurophysiology, neuroanatomy, endocrinology,

molecular biology, and behaviour. The same global approach has yet to be applied to any aspect of reproductive behaviour in primates. However, some additional insights may be gained by considering the effects of various neurotransmitters, or drugs which affect neurotransmission in the brain, upon patterns of sexual behaviour in females.

Neurotransmitters and sexual behaviour
Monoaminergic neurotransmitters

The catecholamines dopamine (DA) and noradrenaline (NA) and the indoleamine (5-HT or serotonin) have been implicated in numerous physiological processes such as pituitary regulation, sleep, feeding, locomotor activity, and aggression. Therefore it would be incorrect to explain the actions of these neurotransmitters solely in relation to mechanisms of sexual behaviour. The neuroanatomical distribution of monoamine neurons is quite similar in the brains of rodents and primates (Figs 12.28 and 12.29). With some important exceptions (such as the tuberoinfundibular dopaminergic neurons) groups of cell bodies are located in the midbrain, pons, and medulla. The axons of these neurons project rostrally, often forming distinct *bundles* (e.g., the ventral and dorsal noradrenergic bundles; mesolimbic and nigrostriatal dopaminergic tracts) with terminal fields in numerous areas of the forebrain, including regions relevant to sexual functions (e.g., the limbic system and orbito-prefrontal cortex).

It is interesting that some areas of the brain which receive a rich monoaminergic innervation also take up and concentrate gonadal hormones (Fig. 12.29; see also Sar and Stumpf 1981). Selective uptake of oestradiol by catecholaminergic neurons has been described including, for example, nuclear uptake of oestradiol by dopaminergic neurons in the tuberoinfundibular system. In the female rat noradrenaline plays an important role in the regulation of the pre-ovulatory LH surge by ovarian hormones (Kalra and Kalra 1983; Crowley 1986). Since the female's sexual receptivity is limited to the periovulatory period in this species, it is possible that similar noradrenergic mechanisms are concerned in the timing of both ovulation and sexual behaviour. According to the schema advanced by Etgen and her co-workers (1990), treatment of the ovariectomized rat with oestrogen and progesterone elevates release of NA in or near the ventromedial hypothalamus, in response to the stimulation which occurs during mating. As the male mounts and thrusts, the tactile stimuli which elicit the lordotic reflex (Pfaff 1980) also result in release of NA within the female's hypothalamus. Noradrenaline levels in dialysates ob-

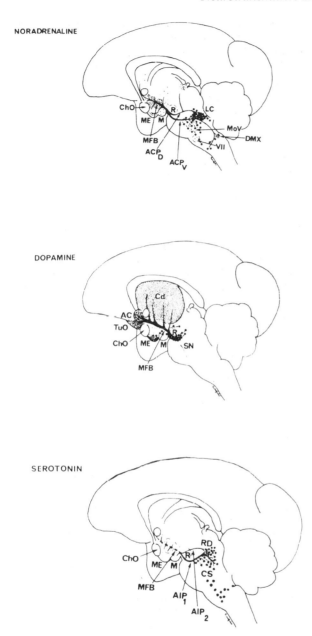

NORADRENALINE

DOPAMINE

SEROTONIN

Fig. 12.28 Sagittal view of the monkey brain showing the organization of cell bodies, major ascending fibre pathways and terminal fields of neurons containing noradrenaline, dopamine, and 5-HT (serotonin). AC=nucleus accumbens; ACPD=ascending dorsal catecholamine pathway; ACPV=ascending ventral catecholamine pathway; AIP$_1$; AIP$_2$=ascending indoleamine pathways; Cd=caudate nucleus; Cho=optic chiasm; CS=nucleus centralis superior; DMX=nucleus dorsalis motorius nervi vagi; LC=locus coeruleus; M=corpus mammillaris; ME=median eminence; MFB=medial forebrain bundle; MOV=nucleus motorius nervi trigemini; R=nucleus ruber; RD=nucleus raphe dorsalis; SN=substantia nigra; TuO=Tuberculum olfactorium; VII=nucleus nervi facialis. (After Schofield and Dixson 1982.)

tained from the ventromedial hypothalamus of the female rat show a marked peak in response to a stimulus male, provided the female has been pretreated with oestrogen and progesterone (Fig. 12.30). Hormonal treatment does not cause NA release in the absence of copulatory stimulation, however. One mechanism by which oestrogen and progesterone might enhance NA responsiveness is by increasing the availability of tyrosine hydroxylase (TH), the enzyme required for synthesis of the neurotransmitter. However, Etgen *et al.* (1990) have shown that ovarian hormones do not augment NA biosynthesis. As an alternative, it is possible that oestrogen and progesterone might act by removing inhibitory influences upon noradrenergic neurons (e.g., involving 5-HT or endogenous opioid peptides), or by increasing the density of oxytocin receptors—with consequent stimulation of NA release by oxytocin in response to tactile stimuli during copulation (Etgen and Karkanias 1994). Some of these possibilities will be discussed in more detail below.

Lesioning studies have shown that destruction of the ventral noradrenergic bundles, which innervate the ventromedial hypothalamus (see Fig. 12.29 of NA distribution in the rat), also reduces lordotic responsiveness in the female rat, even when appropriate doses of oestrogen and progesterone have been administered (Hansen *et al.* 1980). Proceptivity persists, however; indicating once again that the neural substrates which underlie proceptivity differ in important ways from those which control reflexive receptive postures. As far as primates are concerned, the possible effects of noradrenergic mechanisms upon sexual behaviour in females are quite unknown. Since monkeys, apes, and human females do not exhibit lordosis in response to oestrogen and progesterone, there is little value in speculating whether any parallel exists between effects of NA upon receptivity in rodents and anthropoids. Nor is there much information about NA and proceptivity in female monkeys or apes. Although clomipramine causes rapid declines in proceptivity and receptivity in female monkeys, it enhances both NA and 5-HT transmission so that its effects might be mediated via a serotonergic mechanism (Everitt 1977).

The effects of dopamine upon proceptivity and receptivity have also been studied much more extensively in rodents than in primates. Dopamine (DA) has been shown to have an inhibitory effect upon lordosis in the female rat. The DA antagonist pimozide enhances lordosis in ovariectomized, oestrogen-primed females (Everitt *et al.* 1974). Subsequent work has implicated the D$_2$ receptor sub-type in the mechanism by which DA inhibits lordosis (Grierson *et al.* 1988). The picture is not straightforward, however, because some dopamine agonists have been reported to firstly stimulate, and then to inhibit the female's lordotic response (e.g. Foreman and Hall 1987; Grierson *et al.* 1988; Pfaus and Everitt 1995).

It has been suggested that such effects occur because certain drugs stimulate inhibitory DA autoceptors at lower doses and then act upon postsynap-

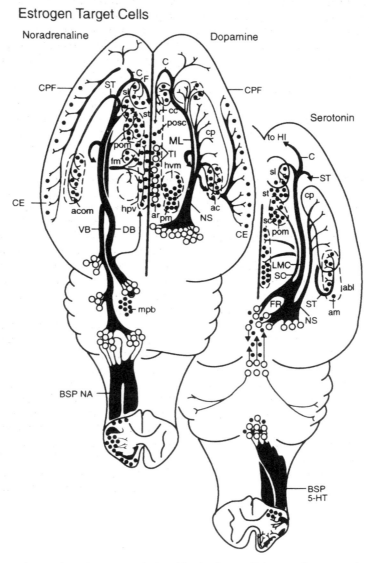

Fig. 12.29 Monoamine-oestrogen target neuron relationships in the rat CNS. Neural systems identified by histochemical fluorescence methods containing noradrenaline (NA), dopamine (DA), or serotonin (5-HT) are shown from left to right, respectively. Oestrogen target neurons were determined by autoradiography. Dopaminergic neurons displaying nuclear uptake of oestradiol 17β have been found among the tuberoinfundibular dopamine neurons (TI). abl= basolateral amygdala; ac=central amygdala; acom=cortico-medial amygdala; am=medial amygdala; ar=arcuate nucleus; BSP 5-HT—bulbospinal serotonin systems; BSP NA=bulbospinal NA system; C=cingulum; CC=crus cerebri; CE=entorhinal cortex; cp=caudate putamen; CPF=piriform cortex; DB=dorsal ascending NA bundle; F=fornix; fm= paraventricular nucleus (magnocellular region); FR = fasciculus retroflexus; hpv = periventricular nucleus; Hl = hippocampus; hvm=ventromedial hypothalamus; LMC=lateral mesencephalo-cortical serotonin tract; ML=mesolimbic dopamine system; Mpb=medio-lateral parabranchial nucleus; NS=nigroneostriatal dopamine system; pm=ventral premammillary nucleus; pom=medial preoptic nucleus; posc-=suprachiasmatic nucleus; sl=lateral septum; so= supraoptic nucleus; ST=stria terminalis; st=nucleus interstitialis stria terminalis; TI=tuberoinfundibular dopamine system; VB=ventral ascending NA bundle; ○=monoaminergic neurons. ●=oestrogen target neurons. (From Brown 1994; after Grant and Stumpf 1975.)

tic receptors at higher doses. The site (or sites) at which these presynaptic and postsynaptic effects occur in the brain is not known (Ahlenius 1993). The ventromedial hypothalamus is a possible candidate, since microinfusion of DA into the VMH stimulates lordosis whilst treatment with a DA agonist (pergolide) maintains lordosis in VMH lesioned females (Foreman and Moss 1979; Mathews *et al.* 1983). It is also interesting that DA enhances the more active proceptive (hopping, darting, and ear-wiggling) responses of the female rat (Caggiula *et al.* 1978; Wilson 1993).

Although stimulation of sexual behaviour by dopamine and dopamine agonists has been recorded in male monkeys and in human males (see pages 441–2), there have been no experiments to investigate dopaminergic mechanisms and sexual activity in female primates. This is unfortunate, and it remains unknown whether dopamine influences receptivity in female monkeys or women. Since the proceptive displays of some primates involve immobile postures (e.g., sexual presentation in many monkeys or staring and crouching in the marmoset), it is theoretically possible that dopamine agonists might decrease proceptivity by encouraging motor activity. Conversely, those proceptive displays which involve movement (e.g., running from the male before presenting, as in the talapoin, or tongue-flicking, as in the marmoset), might be stimulated by drugs which enhance dopaminergic activity. Relatively simple psychopharmacological experiments could easily resolve such questions and the results might have greater relevance to the human female than experiments on rats.

Turning to serotonin (5-HT), this monoamine has traditionally been held to have inhibitory effects upon sexual behaviour in females (e.g. Meyerson *et al.* 1985). However, many earlier studies employed

drugs that were not specific enough in their effects upon 5-HT (Wilson 1993); nor was it appreciated that this transmitter acts upon a variety of receptors (16 receptor types and sub-types for 5-HT have been described so far: Peroutka 1995). The production of drugs which act selectively upon particular 5-HT receptors has enabled more detailed analysis of how 5-HT affects lordosis in oestrogen-treated or oestrogen/progesterone-treated ovariectomized rats. The inhibition of lordosis might involve a 5-HT$_1$-receptor mediated mechanism, since 8-OH DPAT, which is an agonist acting at the 5HT$_{1A}$ receptor, inhibits receptivity in females treated either with oestrogen alone, or in combination with progesterone (Ahlenius *et al.* 1986). Conversely, activity of 5-HT via 5-HT$_2$ receptors has a facilitatory effect upon lordosis (Ahlenius 1993). Turnover of 5-HT and densities of 5-HT receptors in the brain are influenced by ovarian steroids (Biegon *et al.* 1982; James *et al.* 1989). Wilson (1993) suggests that the inhibitory effect of 5-HT upon lordosis might involve 5-HT receptors in the VMH, whilst stimulation of lordosis, via 5-HT$_2$ receptors, might occur outside this region of the hypothalamus.

Psychopharmacological studies involving female rhesus monkeys also point to an inhibitory role for 5-HT in the control of proceptivity and receptivity, but few experiments have been conducted and none has involved drugs specific for 5-HT$_1$ or 5-HT$_2$ receptor sub-types. Ovariectomized, oestrogen-treated rhesus monkeys become less proceptive and receptive when treated with 5-hydroxytryptophan (5-HTP), the precursor of 5-HT (Gradwell *et al.* 1975). These authors also showed that the deficits in sexual behaviour which result from adrenalectomy in rhesus monkeys are reversible, either by treating females with androgens (as previously discussed), or with a drug that depletes 5-HT (parachlorophenylalanine:

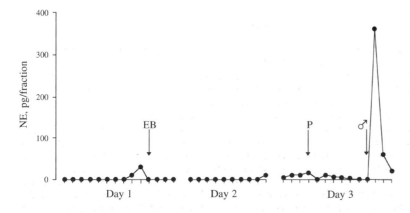

Fig. 12.30 Profile of nordrenaline content of dialysates collected from the ventromedial hypothalamus of a female rat which showed high levels of lordosis when tested with a stimulus male. The female had been pretreated with oestradiol benzoate (EB) and progesterone (P) to induce sexual receptivity. Testing with the sexually active male began 4 h after P treatment and continued for 75 min. (Redrawn from Etgen *et al.* 1990.)

PCPA). Further studies are required to examine the possible involvement of various 5-HT receptor subtypes in proceptivity as well as receptivity in female primates. Such studies are potentially important because drugs that act at 5-HT$_2$ receptors (trazodone, for example) are in demand as treatments for depression. The possible side-effects of these serotonergic drugs upon sexuality are poorly understood at present (Kellett 1993).

Neuroactive peptides

Neuroactive peptides occur in many areas of the central and peripheral nervous system; more than 100 probably exist and the list grows with each succeeding year (Hökfelt *et al.* 1980a; Burgen *et al.* 1980; Hökfelt 1991; Brown 1994). Certain peptides were originally described in the gut and only later identified as putative neurotransmitters in the brain (e.g. cholecystokinin (CCK), vasoactive intestinal polypeptide (VIP), and substance P: Dockray and Gregory 1980). A variety of neuroactive peptides has been identified in the innervation of the genitalia and some might be involved in the control of penile erection (see pages 431–4 for a discussion of this problem). In the brain, some peptidergic neurons also contain an aminergic or other transmitter (e.g. CCK in the mesolimbic dopaminergic pathway: Hökfelt *et al.* 1980b). Close associations between aminergic and peptidergic systems also occur in the arcuate nucleus and median eminence, where DA and NA form part of the control system for LHRH release. Interactions between brain peptides and steroid hormones are important for the control of sexual behaviour, as well as for regulation of pituitary function. Just as oestrogen affects the sensitivity of the anterior pituitary gland to LHRH (e.g. in the rhesus monkey: Nakai *et al.* 1978; Chappell *et al.* 1981), so steroids might modulate the effects of peptides upon behaviour at the hypothalamic level and elsewhere (Herbert 1977, 1993). In the rat, bewildering numbers of neuroactive peptides have been reported to facilitate lordosis when infused into the hypothalamus or midbrain central grey of ovariectomized females (e.g. LHRH, substance P, CCK, and oxytocin: see review by Pfaff *et al.* 1994). It is hard to imagine that all these peptides are implicated in sexual behaviour under physiological conditions. Where the primates are concerned, there is a serious dearth of experimental evidence on peptidergic mechanisms in relation to proceptivity or receptivity. However, it will be worthwhile to consider several peptides that play important roles in the context of reproductive physiology, because

they might also be implicated in the control of sexual behaviour.

Luteinizing-hormone-releasing hormone (LHRH)

The neuroanatomical distribution of LHRH in the hypothalamus of monkeys and rodents is broadly similar; diffuse populations of neurons are found in the preoptic area/anterior hypothalamus and in the ventromedial region (e.g. squirrel monkey: Barry 1979; baboon: Marshall and Goldsmith 1980; rat: Setalo *et al.* 1976; Hoffman *et al.* 1978). Barry's (1979) immunocytochemical studies of female squirrel monkeys showed that morphological changes occur in both these major groups of LHRH neurons during the ovarian cycle. In the rhesus monkey, destruction of the rostral hypothalamus does not disrupt LH secretion (Plant *et al.* 1979), whereas the menstrual cycle is abolished after lesions in the basal hypothalamus, but may be restored by infusing LHRH at hourly (*circhoral*) intervals into the arcuate region (Nakai *et al.* 1978). It will be recalled that female rhesus monkeys may show pronounced peaks of proceptivity during the peri-ovulatory phase of the cycle, assuming sexual behaviour is observed under conditions which allow females to exert a degree of initiative (e.g. Wallen *et al.* 1984). The implication that LHRH, as well as oestradiol, might be involved in these behavioural effects is strengthened by the observation that exogenous LHRH potentiates proceptivity in ovariectomized, oestrogen-primed marmosets (Kendrick and Dixson 1985b). In these experiments small silastic capsules (6–16 mm long) containing oestradiol were implanted subcutaneously into eight ovariectomized marmosets to partially stimulate their proceptivity. When LHRH (25 μg) was administered by intramuscular injection to these females, a further marked activation of proceptive staring and tongue-flicking displays was observed 2 h later (Fig. 12.31). Control injections of saline had no effect on behaviour, and plasma concentrations of cortisol and progesterone did not change significantly after LHRH treatment. Therefore, it is likely that LHRH stimulated proceptivity in these female monkeys by acting upon the brain rather than via changes in adrenocortical steroid secretion.

The results shown in Fig. 12.31 provided the first demonstration that LHRH may influence proceptivity in a primate species. However, the ability of LHRH to potentiate receptivity (lordosis) had previously been documented in the rat. The lordotic reflex is enhanced in oestrogen-primed, ovariectomized rats by administering LHRH peripherally

(Moss and McCann 1973), or by infusing the decapeptide into the hypothalamus (Moss and Foreman 1976), the midbrain central grey (Sakuma and Pfaff 1980), or the vertebral canal (Sirinathsinghji 1983). Electrophysiological recording studies have shown that the number of neurons in the ventromedial nucleus of the hypothalamus that respond to mechanical stimulation of the vagina and cervix increases during oestrogen/progesterone treatment. These neurons are also responsive to iontophoretic application of LHRH (Chan *et al.* 1984). In the marmoset, the surge of oestradiol

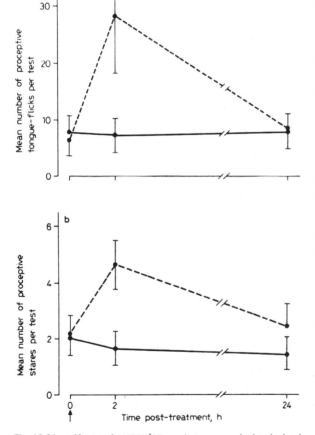

Fig. 12.31 Effects of LHRH (25 μg intramuscularly: dashed line) or saline (solid line) upon (a) proceptive tongue-flicking and (b) proceptive stares by ovariectomized, oestrogen-primed marmosets (*Callithrix jacchus*). **P < 0.01, Wilcoxon test for comparisons between tests at time 0 (before LHRH treatment, indicated by an arrow) and 2 h after LHRH injection. Data are means ($\pm$S.E.M.) for eight females. (After Kendrick and Dixson 1985b.)

which occurs before ovulation might, therefore, increase the number (or sensitivity) of LHRH-responsive neurons that influence proceptivity. To date, no studies have examined the behavioural effects of infusing LHRH directly into the hypothalamus of female monkeys.

Oxytocin

As well as its classic functions as a posterior pituitary hormone which is released during suckling and is responsible for milk ejection in female mammals, oxytocin has also been implicated in the control of both sexual and maternal behaviour in the rat, sheep, and other mammals (Keverne 1995; Pedersen 1997). Neuronal cell bodies which synthesize oxytocin are found in the supraoptic (SO) and paraventricular (PV) hypothalamic nuclei. As well as transporting oxytocin to the posterior pituitary gland these neurons in the SO and PV project to the hippocampus, amygdala, nucleus accumbens, substantia nigra, and other brain areas as shown in Fig. 12.32. In the ovariectomized rat, oxytocin enhances lordotic responses in oestrogen/progesterone-treated females when it is infused directly in the VMH (lordosis duration is increased) or the POA (frequency and duration of lordosis are increased) (Caldwell *et al.* 1986; Schultze and Gorzalka 1991). It appears that both oestrogen and progesterone pre-treatments are required for oxytocin to fully exert its behavioural effects. Administration of the oxytocin antagonist ornithine vasotocin inhibits both receptivity and proceptivity in oestrogen/progesterone-treated rats (Witt and Insel 1991). Work by Gorzalka *et al.* (1991), implicates the C-terminal tripeptide region of the oxytocin molecule in the stimulation of lordosis. Oestrogen stimulates production of oxytocin receptors in the VMH whilst progesterone in some way promotes redistribution of receptors to the lateral VMH (Schumacher *et al.* 1990). The lateral VMH receives a rich noradrenergic innervation and oxytocin increases noradrenaline release, thus facilitating the female's lordotic response (Etgen and Karkanias 1994). Figure 12.33 shows the effect of oxytocin upon NA release in the VMH of female rats; only those females which were pre-treated with oestrogen and progesterone show significant elevations of NA in dialysates collected from the ventromedial hypothalamus (Vincent and Etgen 1993). Etgen and Karkanias (1994) have constructed a model to account for interactions between ovarian steroids, oxytocin, and noradrenaline in the control of lordosis in the female rat and this is reproduced here as Fig. 12.34. Included in this schema is the positive effect of NA upon oxytocin release, which might include an α1-adrenoceptor-mediated mechanism (Parker and Crowley 1993). Also included is the ability of oestrogen to up-regulate α1-adrenoceptors in the hypothalamus of the female rat (Etgen and Karkanias 1990). Etgen and Karkanias (1994) therefore speculate that 'oxytocin and noradrenaline augment each others' secretion, forming a local positive feedback loop within the VMH'.

There is no evidence linking oxytocin to the control of proceptivity or receptivity in female primates, but this might simply reflect the lack of any experiments to investigate the problem. Coitus and orgasm can lead to oxytocin release in women (Carmichael *et al.* 1987). If this peptide plays any role in primate sexual behaviour it probably does not operate via a progesterone-dependent mechanism as in the female rat. We have seen that progesterone does not synergize with oestrogen to stimulate receptivity or proceptivity in female monkeys or apes. Not only is lordosis lacking in anthropoids, but the hormonal mechanisms which affect sexual behaviour are also different from those which operate in rodents. It remains to be seen whether the oxytocin-induced changes in NA release shown in Fig. 12.33 find any parallel in the control of sexual behaviour in female primates.

Opioid peptides

Chronic drug abuse involving opiates such as heroin adversely affects sexual desire and performance. The evidence on this question is better documented for men than for women (Cushman 1972; Parr 1976; Huws and Sampson 1993). Where the brain's natural opioid systems are concerned, β-endorphin plays a special role in reproductive contexts and has been implicated in the control of gonadotrophin and oxy-

tocin secretion, as well as in the regulation of sexual and maternal behaviour (Dyer and Bicknell 1989; Keverne 1995). In the rat, infusion of β-endorphin into the cerebral ventricles of oestrogen/progesterone-primed, ovariectomized females suppresses lordosis; this effect is reversible by treatment with the opiate antagonist naloxone (Weisner and Moss 1989). Intraventricular infusions of β-endorphin (at a dosage of 1 μg) also suppress

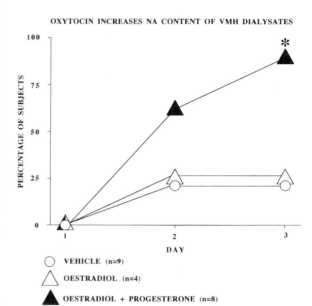

Fig. 12.33 Effects of oxytocin (daily intraperitoneal injection of one I.U.) upon noradrenaline content of dialysates collected from the ventromedial hypothalamus (VMH) of female rats treated with oestrogen alone ($\triangle$), oestrogen and progesterone ($\blacktriangle$) or vehicle control ($\circ$). *P < 0.05, Fisher's exact probability test. (Data are from Vincent and Etgen 1993.)

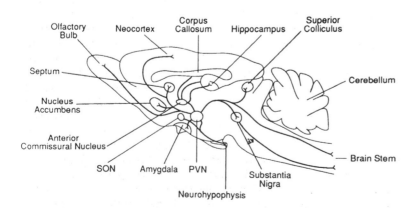

Fig. 12.32 Schematic representation of the projections of the oxytocin neurons in the rat brain. (From Brown 1994; after Argiolas and Gessa 1991.)

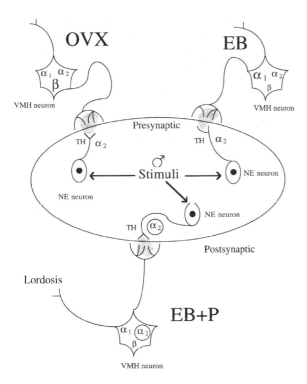

Fig. 12.34 Model by Etgen *et al.* of presynaptic and postsynaptic changes in hypothalamic noradrenergic (NE) function induced in ovariectomized (OVX) rats by a priming dose of oestradiol benzoate (EB) alone, or by combined EB and progesterone (EB+P), in doses sufficient to induce lordosis. Stimulus-induced release of noradrenaline is elevated in EB+P rats relative to either OVX or EB females. Tyrosine hydroxylase (TH) activity is unaffected by ovarian steroids in either noradrenergic neuronal cell bodies or in hypothalamic terminal fields. The symbol size for noradrenergic receptor sub-types denotes their functional coupling to cyclic AMP synthesis rather than receptor number. An increase or decrease in symbol size relative to OVX represents an increase or decrease, respectively, in a receptor's contribution to the total cyclic AMP response to noradrenaline. The (α₂) indicates a change in receptor function, from inhibition to facilitation of cyclic AMP accumulation. Since it is unknown whether hypothalamic α_2 receptors are presynaptic, postsynaptic or both, they have been drawn in both locations. (Redrawn from Etgen *et al.* 1990.)

proceptivity in female rats.

Weisner and Moss (1986, 1989) demonstrated that bilateral infusions of β-endorphin into the ventromedial hypothalamus and medial reticular formation led to a reduction in lordosis and virtually abolished the proceptivity of female rats. Infusions at other sites in the brain (POA, midbrain central grey, amygdala, and caudate nucleus) had no significant effects upon sexual behaviour. By using selective opioid receptor antagonists to block the suppressive effects of β-endorphin upon receptivity and proceptivity Weisner and Moss (1986) also showed that β-endorphin acts at μ_1 receptors (and partly at δ receptors) to suppress sexual behaviour.

There is currently no agreement about the mechanism by which various opioid drugs affect receptivity and proceptivity in the female rat; some of these drug effects are stimulatory and others inhibitory, depending upon the receptor type and area of the brain affected (Pfaus and Everitt 1995). Since β-endorphin inhibits LHRH and hence reduces LH secretion (Meites *et al.* 1979), it is reasonable to suggest that its behavioural effects might also involve the LHRH system. Indeed, infusion of LHRH into the midbrain central grey has been reported to reverse the suppression of lordosis produced by prior injection of β-endorphin (Sirinathsinghji *et al.* 1983). Against the LHRH hypothesis is the observation that concurrent infusion of LHRH does not prevent the ability of β-endorphin to suppress either lordosis or proceptivity in the female rat (Fig. 12.35). In these experiments peptides were infused into the cerebral ventricles and could potentially reach a number of brain areas. Other experiments, using male rats, have demonstrated that the site at which β-endorphin is injected into the brain, and the methods of testing employed, crucially affect the behavioural outcome (see pages 438–9 for a discussion of this problem). Infusing β-endorphin into the POA of the male rat suppresses mounting behaviour, but this effect does not occur if the infusion is made after the first mount in a series has been accomplished (Herbert 1993). I do not know of any comparable studies in female rodents to determine, for instance, if β-endorphin fails to suppress proceptivity or lordosis if it is infused after the female has commenced a sequence of responses with a male partner. Nor have any experiments been reported to examine effects of β-endorphin upon proceptivity or receptivity in female monkeys or apes. Some experiments on rodents indicate that effects upon sexual behaviour are relatively specific and not due to some generalized action of β-endorphin involving locomotor activity (Gessa *et al.* 1979; Meyerson 1981). It is important to bear in mind, however, that opioid peptides are also implicated in affiliative behaviour and reward, so that treating talapoin monkeys or rhesus macaques with opiate antagonists (naloxone or naltrexone) leads to increased grooming behaviour in members of social groups (Meller

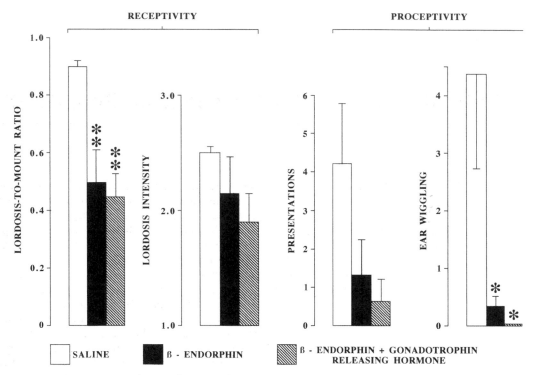

Fig. 12.35 Effects on receptivity (lordosis) and proceptivity (ear-wiggling and presentations) in female rats of intra cerebro-ventricular infusions of saline (1.0 μl), β-endorphin (1.0 μg), or a combination of gonadotrophin-releasing hormone (100 ng) plus β-endorphin (1.0 μg). *P < 0.01; **P < 0.005, versus saline scores. (Redrawn and modified from Weisner and Moss 1989.)

et al. 1980; Martel *et al.* 1993). Finally, it should be noted that some authors express scepticism concerning the physiological validity of reported effects of β-endorphin on sexual behaviour in the rat. Södersten *et al.* (1989), for instance, are concerned about the high dosage levels of β-endorphin used in some studies. These authors were unable to demonstrate suppression of sexual behaviour in female rats after intracerebral injections of β-endorphin into either the ventricles or in the midbrain central grey. They concluded from this and from other evidence that 'β-endorphin may not play a physiologically important role in inhibiting sexual behaviour in female rats'.

13 Hormones and sexual behaviour in the adult male

Seasonal changes in hormones and sexual behaviour

Although many monkeys and apes mate, conceive, and give birth throughout the year, examples of seasonal cycles of sexual activity have been well documented both for anthropoids and for various prosimian primates (Lancaster and Lee 1965; Van Horn and Eaton 1979; Lindburg 1986). Table 13.1 provides examples of seasonal cycles of testicular activity (spermatogenesis and/or androgen secretion) and associated changes in sexual behaviour in prosimians and in New World and Old World monkeys. As far as we know, such seasonal changes do not occur either in the lesser apes or in the four great ape species. Subtle seasonal effects cannot be ruled out in some cases but, on current field evidence, it is clear that the major changes which occur in species such as the ringtailed lemur, squirrel monkey, or rhesus macaque during annual mating seasons do not find any parallel among the apes (Siamang: Chivers 1974; Orang-utan: Rijksen 1978; Chimpanzee: Goodall 1986; bonobo: Kano 1992; gorilla: Schaller 1963; Harcourt et al. 1980). Despite occasional reports of circannual changes in plasma testosterone in the human male (e.g. Reinberg et al. 1975) and of higher frequencies of sexual violence during the summer months in N. America (Michael and Zumpe 1983), there is little evidence of any annual physiological rhythm underlying human sexual behaviour.

As an example of how seasonal changes in testicular hormone secretion correlate with sexual activity we can consider the Japanese macaque (*Macaca fuscata*) which mates during the autumn months in its native habitat and also in Oregon, where Rostal et al. (1986) studied a troop of 316 monkeys maintained in a large outdoor enclosure. During the non-mating season (spring and summer months) plasma testosterone (T) was low (<2.0 ng mL^{-1}) in 12 adult males studied. However, plasma T rose dramatically, reaching 5–6 ng mL^{-1}, during August and September, before the onset of copulatory behaviour (Fig. 13.1). More modest increases in plasma

dihydrotestosterone (DHT) were also measurable at this time. Rostal et al. noted that plasma T rose 1 or 2 months before sexual activity increased in the troop. Similar delays between increases in plasma testosterone levels and activation of sexual behaviour have also been noted in the rhesus macaque (Gordon et al. 1976; Lindburg 1983). Reddening of the male's scrotal and perineal sexual skin, which is stimulated by testicular hormones (Vandenbergh 1965), occurs before the onset of mounting in both the rhesus and the Japanese macaque. Rostal et al. (1986) also noted that males began to carry their tails in a distinctive manner (raised and flexed forwards over their backs) prior to the onset of mating activity. Similar behaviour also occurs in high-ranking rhesus monkeys. In Japanese macaques, circulating concentrations of T and DHT began to increase as day length declined. Rostal et al. also compared circannual rhythms of plasma androgens to seasonal shifts in temperature and rainfall, but they were of the opinion that day length provided the most likely proximate cue triggering increased activity of the hypothalamic–pituitary–testicular axis. For seasonally breeding primates which live sufficiently far from equatorial latitudes, changes in photoperiod are probably crucial in timing the onset of the mating season (Lindburg 1986). In the rhesus monkey, a six month shift in the timing of the mating and birth seasons occurs as a result of transfer from the northern to the southern hemisphere (Bielert and Vandenbergh 1981). Such transfers, with their associated changes in day length, have also been reported to alter the timing of breeding seasons in various Malagasy prosimians, as well as in the squirrel monkey (Lindburg 1986). Van Horn's experimental studies on captive ringtailed lemurs demonstrated that reducing illumination by as little as 30 min per day (from 12.5 h of artificial light: 11.5 h darkness to 12.0 h light: 12.0 h dark) was sufficient to bring females into oestrus. Reduction of photoperiod also stimulated testicular activity in male ringtails (Van Horn and Eaton 1979; Van Horn 1980). For many monkeys which live on or near the equator, seasonal shifts in day length are minimal.

Table 13.1 Circannual changes in copulatory behaviour and testicular function in some seasonally breeding and non-seasonally breeding primate species

Species	Mating season?	Sources	Testicular function	Sources
Seasonal breeders				
Microcebus murinus	During September in SE Madagascar	Martin 1973	Longer day length stimulates increases in testicular volume, spermatogenesis and plasma testosterone in captive males	Petter-Rousseaux 1980; Perret 1992
Propithecus verreauxi	During March in S. Madagascar.	Richard 1974, 1976	No data	
Lemur catta	During April in S. Madagascar	Jolly 1966	Reduction of day length stimulates testicular growth and secretion of testosterone during annual mating season in captive males	Van Horn and Eaton 1979; Van Horn 1980
Saimiri sciureus	From June–August during the dry season in Colombia, Brazil and Peru	Baldwin and Baldwin 1981	A marked seasonal rhythm of spermatogenesis occurs; plasma testosterone levels can exceed 30 ng mL^{-1} during the mating season	Dumond and Hutchinson 1967; Wiebe *et al.* 1984
Cercopithecus aethiops	During the dry season, from May–October in Kenya	Struhsaker 1967a	Seasonal cycle of spermatogenesis described in the St. Kitts population	Conoway and Sade 1969
Erythrocebus patas	From June–August in Kenya	Chism *et al.* 1983, 1984	Plasma testosterone increases from 3.0 to 12.0 ng mL^{-1} and testes volume from 6.4–15.8 cc in mating vs birth season (captive colony)	Bercovitch 1996
Macaca fuscata	From September to April in Japan	Kawai *et al.* 1967	Marked seasonal cycles of testes size, spermatogenesis and plasma testosterone	Nigi *et al.* 1980; Rostal *et al.* 1986
M. mulatta	From October to February, with a peak in November/December in India (Dehra Dun)	Lindburg 1971	Marked increases in testicular volume, spermatogenesis and in circulating testosterone during annual mating season	Sade 1964; Zamboni *et al.* 1974; Gordon *et al.* 1976
M. radiata	Peak in November/December in India (Bangalore)	Rahaman and Parthasarathy 1969b	Seasonal cycle of testosterone occurs	Glick 1979
Miopithecus talapoin	During the dry season, from December to February in Cameroon	Rowell and Dixson 1975	No circannual changes in plasma testosterone or testicular volume have been reported in captivity. No field data are available	Keverne 1979b
Mandrillus sphinx	During the dry season, from July–October in Gabon	Dixson *et al.* 1993	Qualitative observations indicate that testicular size and sexual skin coloration are maintained throughout the year	Wickings and Dixson 1992a, b

Table 13.1—*Continued*

Species	Mating season?	Sources	Testicular function	Sources
Non-seasonal breeders				
Cebus albifrons	Believed to mate throughout the year; there is no birth season in Colombia, Peru or Panama	Robinson and Janson 1987	Plasma testosterone shows individual variations, ranging from 5–110 ng mL^{-1} but there is no circannual rhythm (*C. apella*: captive colony)	Nagle and Denari 1983
Alouatta spp.	Believed to be sexually active throughout the year; there is no restricted birth season	Carpenter 1934 Crockett and Eisenberg 1987	No data	
Cercocebus albigena	Mates throughout the year in Uganda with no obvious peak in frequency	Chalmers 1968b; Wallis 1983	No data	
Papio ursinus, *P. anubis*	Mating, conceptions and births occur throughout the year	Hall and Devore 1965; Saayman 1970; Smuts 1985	No seasonal changes in testicular function have been reported	
Colobus badius	In the Kibale forest, Uganda, copulations and births occur all year round	Struhsaker 1975	No data	
Pongo Pygmaeus	No evidence for seasonal changes in mating behaviour	Rijksen 1978; Galdikas 1981	No seasonal changes in testicular function have been reported	
Pan paniscus	Mating occurs throughout the year	Kano 1992	No Data	
Pan troglodytes	Mating occurs throughout the year. No significant mating peak has been reported	Goodall 1986; Nishida *et al.* 1990	No seasonal changes in testicular function have been reported	
Gorilla gorilla	Copulations, conceptions and births occur at all times of the year	Harcourt *et al.* 1980	No seasonal changes in testicular function have been reported	

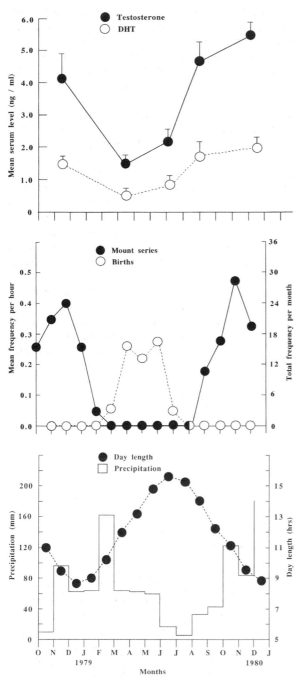

dry season and births during the wet season when food is more abundant (e.g. in various *Cercopithecus* species: Butynski 1988; talapoin monkey: Gautier-Hion 1968; Rowell and Dixson 1975; Squirrel monkey: Dumond and Hutchinson 1967). The exact nature of the proximate cues which time the onset of sexual activity in these monkeys is not known. Changes in rainfall might exert their effects via a secondary cue, such as changes in vegetation (and hence in nutrition), or very small changes in photoperiod might still be significant (Lindburg 1986). Moreover, the co-ordination of endocrine and sexual activity within monkey troops can involve 'fine-tuning' by social cues and is not solely a response to climatic changes. If female rhesus monkeys are treated with oestradiol during the non-mating season, males exhibit hormone-dependent reddening of the sexual skin, resumption of spermatogenesis and activation of mounting behaviour (Vandenbergh 1969). In free-ranging talapoins, troops in the same area of forest differ in the timing of onset of the mating season, although reproductive synchrony *within* each troop is well co-ordinated (Rowell and Dixson 1975). Hence, it is possible that the onset of testicular activity, ovarian cyclicity, and sexual behaviour during the annual mating seasons of macaques, guenons, squirrel monkeys, and other species is triggered by social, as well as by climatic factors (Bielert and Vandenbergh 1981; Herndon 1983; Lindburg 1986).

Effects of castration and testosterone replacement

Causative relationships between androgens and the activation or maintenance of sexual behaviour in male primates have been examined in only six species (rhesus monkey: Wilson and Vessey 1968; Phoenix 1973; Phoenix *et al.* 1973; Michael and Wilson 1973; stumptail: Schenck and Slob 1986; talapoin: Dixson and Herbert 1977; cynomolgus macaque, *M. fascicularis*: Michael *et al.* 1986; Michael and Bonsall 1990; common marmoset: Dixson 1990c, 1993d; and Man: Bancroft and Skakkebaek 1979; Davidson *et al.* 1979; Bancroft 1989). This section examines the effects of castration, whether surgically or by means of chemical suppression of testicular activity, and exogenous testosterone treatments upon patterns of sexual behaviour.

Castration of sexually experienced, adult male rhesus monkeys produces a gradual decline in sexual behaviour which is reversible by administration of testosterone propionate (Fig. 13.2). Frequencies of ejaculation and intromission are the first patterns affected after castration (by 2–4 weeks: Michael and

Fig. 13.1 The seasonal cycle of reproduction in semi-free-ranging Japanese macaques (*Macaca fuscata*). Upper: Mean (±S.E.M.) concentrations of circulating testosterone and dihydrotestosterone (DHT) in adult males. Middle: Mean monthly frequencies of mounts and births. Lower: Mean monthly levels of rainfall and measurements of day length. (Redrawn and modified from Rostal *et al.* 1986.)

In such cases breeding cycles are often related to changes in precipitation—mating occurs during the

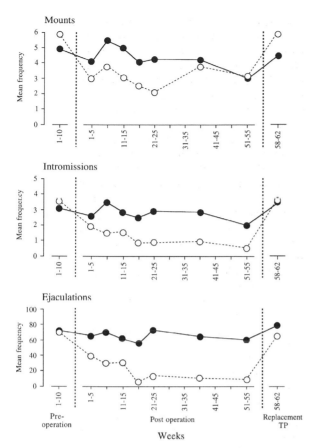

Fig. 13.2 Effects of castration and hormonal replacement (using testosterone propionate: TP) upon the sexual behaviour of adult male rhesus monkeys (*Macaca mulatta*) during pair-tests with females. ○=castrated males (N=10); ●=sham operated controls (N=9). (Redrawn and modified from Phoenix *et al*. 1973.)

havioural patterns were less than those observed for eight intact males, no statistically significant differences were measured between the two groups. However, only one castrate exhibited an ejaculatory response and, whilst all seven castrates achieved intromission, intromission frequencies were significantly less than for intact males (Dixson 1990c). These effects were not caused simply by a failure to achieve full penile erection. In their studies of castrated rhesus monkeys, Michael and Wilson (1973) filmed copulations and noted that these 'showed the penis bending on contact with the female's perineum'. By contrast, castrated male marmosets maintained full erections and made vigorous attempts to intromit. Deficits in genital sensory feedback, or in the central processing and responsiveness to such stimuli, provide more likely explanations of the effects on intromission and ejaculation shown in Fig. 13.4.

Studies of small captive groups of talapoin monkeys have shown that sexually experienced males castrated in adulthood rarely mount females and fail to intromit when observed 6–12 months after surgery (Dixson and Herbert 1977). Treatment of castrates with testosterone resulted in restoration of sexual

Wilson 1973), whereas mounting behaviour declines more gradually. Individual variability is marked, and five out of ten males studied by Phoenix (1973) were still capable of intromission, and three showed ejaculatory responses, one year after the operation. Loy (1971) observed a castrated male rhesus monkey on Cayo Santiago which was still capable of mounts, intromissions, and the ejaculatory pause characteristic of intact males seven years after the operation (Fig. 13.3).

In the common marmoset, castration of sexually inexperienced adult males results in lower frequencies of intromissions and ejaculations, compared with intact inexperienced males, when pair-tests are conducted 6–10 months after the operation (Fig. 13.4). It is interesting that the seven castrates observed in this study frequently mounted females, exhibited full penile erections, and made pelvic thrusting movements. Although the frequencies of these be-

Fig. 13.3 A castrated adult rhesus male monkey (*Macaca mulatta*) which was still capable of copulating and exhibiting the ejaculatory pause typical of intact males, seven years after the operation. The female 'reaches back' in response to the male's ejaculatory pause. (Author's drawing, after Loy 1971.)

behaviour, but the effects of hormone treatment were dependent upon the dominance ranks of the males. Top-ranking males mounted females and ejaculated after receiving a 25-mg subcutaneous implant of testosterone, whereas subordinate males failed to copulate. Subordinates did exhibit more frequent penile erections, however, and masturbated to ejaculation, whereas none of the castrates had ejaculated before testosterone treatment (Fig. 13.5). It seems that castration does not totally abolish sexual interest, because all castrated males frequently look at the sexual skin swellings of oestradiol-treated females (Dixson and Herbert 1977). However, only intact dominant males, or dominant castrates treated with testosterone, are able to seek proximity to females and then to mate with them.

Despite earlier assertions to the contrary, it is now established that testicular hormones also play an important role in the maintenance of sexual thoughts, sexual interest, and copulatory behaviour in the adult human male (Bancroft 1989). Kinsey *et al.* (1953), in reviewing the largely anecdotal evidence then available on the effects of castration, concluded that 'the frequency and intensity of [sexual] response may or may not be reduced by castration. The psychological effects of such an operation may make it difficult for some of the males to make socio-sexual adjustments'. Kinsey *et al.* also noted large individual variations in the effects of castration upon sexual behaviour in human males, 'including the case of one male who was normally active thirty years after castration'. Modern controlled studies have shown that treatment of hypogonodal men with testosterone undecanoate increases sexual arousal and copulatory behaviour and that such results are not due to placebo effects or to non-specific actions of androgen upon general 'vigour' or well-being

(Fig. 13.6). Men, no less than monkeys, exhibit individual differences in behavioural responses after castration, and although the psychological impact of castration is doubtless greater in human males, we have seen that social factors also influence copulatory activity and responsiveness to testosterone in castrated monkeys. A study by Skakkebaek *et al.* (1981) demonstrates that withdrawal of androgen therapy from the hypogonodal male leads to declines in sexual thoughts and sexual activity; this occurred after 3–4 weeks in the case depicted in Fig. 13.6. These declines are not affected by placebo treatment but are reversed after 1–2 weeks of treatment with testosterone undecanoate. Androgens also affect penile erection in the human male, but the hormonal effect depends very much upon the circumstances in which erections occur. Bancroft and Wu (1983) found that hypogonodal men exhibit a similar erectile response to eugonodal men when viewing erotic films. Testosterone replacement has no significant effect upon penile erection in these circumstances. However, hypogonodal men are less able to produce an erectile response during erotic fantasy and this deficit is somewhat improved by testosterone replacement (Fig. 13.7). Moreover, the occurrence of penile erection during sleep (nocturnal penile tumescence or NPT) is markedly affected by testosterone in the human male (O'Carroll *et al.* 1985). The significance of these observations for understanding of the neural mechanisms which govern penile erection will be considered in more detail later in this chapter.

Antiandrogens and sexual behaviour

Several studies have examined the effects of syn-

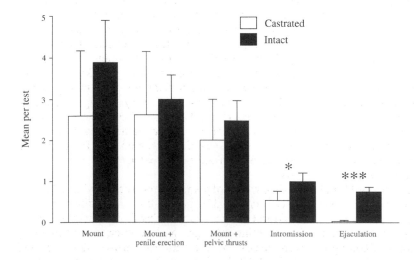

Fig. 13.4 Measurements of copulatory behaviour in intact and castrated sexually inexperienced adult male marmosets (*Callithrix jacchus*). Data are means (+S.E.M.) per test for seven castrates (open bars) and eight intacts (closed bars) during pair-tests with ovariectomized, oestradiol-treated females. *P < 0.05; ***P < 0.001. Mann–Whitney U test. Tests were conducted 6–10 months after castration. (Redrawn from Dixson 1990c.)

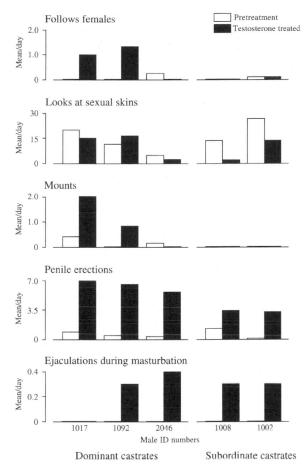

Fig. 13.5 Effects of testosterone upon the sexual behaviour of dominant and subordinate castrated male talapoin monkeys (*Miopithecus talapoin*) living as members of captive social groups. Open bars show mean frequencies of behaviour per day (240 min observations) in males 6–12 months after castration. Closed bars show mean daily frequencies in the same subjects after subcutaneous implantation with testosterone (25 mg). (Redrawn and modified from Dixson and Herbert 1977.)

thetic antiandrogens upon sexual and associated behavioural patterns in male primates. Cyproterone acetate, which has antiandrogenic and antigonadotrophic properties, has been used clinically for treatment of hypersexuality and sex-offenders (e.g. Bancroft *et al.* 1974; Davies 1974; Laschet and Laschet 1975). Michael *et al.* (1973) observed decreases in sexual behaviour in two male rhesus monkeys treated with cyproterone acetate, whereas a third male was unaffected. Slob and Schenck (1981) conducted more detailed studies involving seven adult male stumptail macaques. These males were treated with cyproterone acetate by intramuscular injection (100 mg per week increasing to 210 mg per week) over the course of 12 weeks. Males were

pair-tested with females, twice weekly, in order to record sexual interactions. Although decreases in plasma testosterone and testicular volume occurred (similar to those reported in human males) sexual behaviour did not decline. However, castration of male stumptails also has little effect on their sexual behaviour during the first 12 weeks after the operation (Schenck and Slob 1986), which might account for the lack of change after cyproterone. Slob and Schenck (1981) also suggest that effects of cyproterone acetate in hypersexual human males might be due to its anxiolitic-tranquillizing properties rather than to antiandrogenic effects. An alternative hypothesis concerns the hyperprolactinaemic action of cyproterone acetate (e.g. in female rhesus monkeys: Herbert *et al.* 1977) because hyperprolactinaemia is associated with impotence in man (Thorner 1977). Gynecomastia has been noted as a side-effect of cyproterone treatment in the human male (Gräf *et al.* 1974).

Medroxyprogesterone acetate (MPA: marketed under the name 'Depo-provera') is a long-acting synthetic progestin with antiandrogenic properties which has been used for the treatment of male sex offenders (Money 1970; Kiersch 1990). Zumpe and Michael (1988) administered MPA to adult male cynomolgus monkeys (*Macaca fascicularis*) and recorded sexual interactions during pair-tests with familiar females. Pronounced declines in plasma testosterone occurred as a result of MPA treatment, accompanied by a decrease in ejaculatory frequencies which achieved statistical significance by the second week (Fig. 13.8). Decreases in mounting were not as marked as the effects upon ejaculation. However, subsequent studies by Zumpe *et al.* (1992) showed that decreases in mount attempts and lengthening of the latency to the first mount attempt are more marked in MPA-treated cynomolgus monkeys than in males which have been surgically castrated. This difference is thought to be due to the active suppression by MPA of central mechanisms regulating sexual motivation. Yawning, which is androgen-dependent in male rhesus monkeys (Phoenix 1973), showed an immediate fall in frequency during week 1 of MPA treatment in cynomolgus monkeys. Deputte *et al.* (1994) have also demonstrated an inhibitory effect of the antiandrogen hydroxyflutamide upon yawning frequencies in castrated, testosterone propionate-treated male rhesus monkeys.

Male cynomolgus monkeys treated with MPA display considerable individual variability in behavioural effects. Some males show very little change in mounting attempts and potency, as indicated by frequencies of intromitted thrusts (Zumpe and Michael 1988). The female partners used in these studies also significantly influenced the males' sexual responses, and it was notable that female proceptivity showed consistent declines throughout the course of the study. Withdrawal of MPA did not reverse the effects on the male's sexual behaviour or yawning frequencies, despite the recovery in plasma testosterone levels. In a subsequent study Michael and Zumpe (1993) demonstrated a direct, central action of MPA upon sexual behaviour in male cynomolgus monkeys. Eight males were castrated and treated with testosterone (implanted subcutaneously in silastic capsules), in order to maintain consistent circulating levels of the hormone. Six males were subsequently treated with MPA (40 mg. per week by I.M. injection), whilst two subjects served as controls. Treated males showed significant decreases in ejaculations and mount attempts by 5–6 weeks after commencement of MPA. The authors interpret these results in terms of the ability of the synthetic progestin to accumulate in neuronal nuclei of the hypothalamus and preoptic area (Rees *et al.* 1986), these regions of the brain play a crucial role in the control of sexual behaviour (Larsson 1979; Sachs and Meisel 1988).

Suppression of the pituitary-testicular axis, using an LHRH antagonist, also has effects upon sexual behaviour in male monkeys which differ, in a quantitative fashion, from those observed after surgical castration. Wallen *et al.* (1991) studied seven adult male rhesus monkeys living as members of a large captive social group (N = 74, including 35 intact adult females). All males were treated simultaneously with a single dose of the powerful LHRH antagonist antide (30 mg kg^{-1} by subcutaneous injection) during the height of the annual mating season. Treatment brought about a prompt suppression of luteinizing hormone and testosterone secretion, which lasted for at least three weeks, and an equally prompt decline in ejaculatory frequencies during the first post-treatment week (Fig. 13.9). By week 4, only one male continued to ejaculate. Mounting frequencies similarly declined from week 1 onwards. These effects were more rapid in onset and more extreme than those recorded during pair-test experiments to evaluate effects of surgical castration in the rhesus monkey (Phoenix *et al.* 1973; Michael and Wilson 1973). Wallen *et al.* (1991) suggest that this may be due to the greater complex-

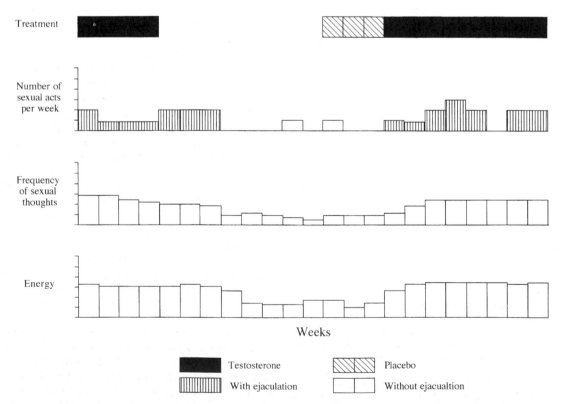

Fig. 13.6 The effects of testosterone undecanoate treatment in a hypogonadal man aged 40, castrated one year earlier for testicular neoplasm. Sexual thoughts and activities decline about 3 weeks after stopping testosterone treatment. There is no response to placebo, but a rapid response within one or two weeks of restoring testosterone. (Redrawn from Bancroft 1989; after Stakkebaek *et al.*, 1981.)

ity of the social environment employed during their studies, citing the greater effects of ovarian hormones upon female proceptive behaviour in the group situation, as compared with the isolated pair-test (Wallen *et al.* 1984). However, an alternative explanation, not favoured but not ruled out by Wallen *et al.*, is that antide directly affects central mechanisms which govern masculine sexual behaviour. This might indeed be the case, given the effects of LHRH upon copulatory behaviour in male rats (Dorsa *et al.* 1981) and its possible influence on sexual arousal in the human male (e.g. Evans and Distiller 1979). Moreover, castrated male rhesus monkeys, living as members of semi-free-ranging social groups, sometimes show remarkable retention of copulatory behaviour years after surgery (Wilson and Vessey 1968; Loy 1971: see Fig. 13.3). Social competition does not necessarily hasten the behavioural effects of androgen deprivation, therefore. Antide is a small molecule and a long-acting peptide capable of crossing the blood–brain barrier and of reaching neural substrates which influence copulatory behaviour. Further experiments may demonstrate whether it affects sexual arousal and copulatory behaviour by such an action, which is more potent than that of antiandrogenic drugs such as

cyproterone acetate. Administration of another LHRH antagonist, 'Nal-Flu', to healthy young men results in suppression of testosterone secretion after one week and, by 4–6 weeks, there are 'statistically significant decreases in the frequency of sexual desire, sexual fantasies, and intercourse (Bagatell *et al.* 1994).

Behavioural effects of metabolites of testosterone

Testosterone is metabolized in the brain, as well as in peripheral target organs, to yield a variety of steroid molecules (Fig. 13.10). Of principal interest to the student of sexual behaviour is the conversion of testosterone to dihydrotestosterone (DHT) via 5-alpha-reductase activity and the aromatization of testosterone to oestrogens, and particularly oestradiol. The 5-alpha-reductase system is especially important in peripheral tissues, such as the accessory sexual organs and the penis. Aromatization of androgens takes place in the brain; it is a phylogenetically ancient process and occurs for instance in the turtle brain (androstenedione conversation to oestrone: Callard *et al.* 1977) and in the rhesus monkey (Flores *et al.* 1973) and man (Naftolin *et al.* 1975). It is relevant to ask, therefore, whether testosterone exerts its effects upon sexual behaviour directly or via metabolic conversion products, such as DHT, oestradiol, and other steroids shown in Fig. 13.10. In some birds and mammals, aromatization of androgen to oestrogen plays an important role in the activation and/or maintenance of sexual behaviour in the adult male. This is the case in the dove (Steimer and Hutchison 1981), the Japanese quail (Balthazart *et al.* 1990), rat (Larsson 1979), ram (Parrott 1978), red deer stag (Fletcher 1978), and ferret (Baum *et al.* 1990b). The non-aromatizable androgen DHT activates penile reflexes and maintains peripheral structures in castrated rats, but fails to restore their sexual behaviour (Feder 1971). However, the aromatization hypothesis is by no means universally applicable to mammals; DHT will restore copulatory activity after castration in Swiss–Webster mice (Luttge and Hall 1973), guinea-pigs (Aslum and Goy 1974), and King–Holtzman rats (Olsen and Whalen 1984). Moreover, in the castrated rat, oestradiol benzoate (EB) given in combination with DHT is more effective in restoring copulatory behaviour than EB alone, indicating some synergism between oestrogenic and androgenic effects (Larsson *et al.* 1973). Among the primates, there is a little evidence that aromatization of testosterone by the brain has a significant effect upon masculine sexual behaviour. However,

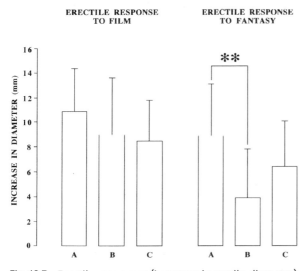

ERECTILE RESPONSE TO FILM   **ERECTILE RESPONSE TO FANTASY**

Fig. 13.7 Erectile responses (increases in penile diameter) in A. Normal men (N=8); B. Hypogonadal men (N=8) and C. Hypogonadal men treated with testosterone (N=8). Erectile responses to erotic film were similar in the three groups whereas hypogonadal men were less responsive than controls during erotic fantasy. Testosterone improved erectile responsiveness during fantasy in hypogonadal men. Although this effect was not significant statistically the latencies to erectile response were significantly reduced by testosterone. Data are means ($\pm$S.D.) for each group. **P < 0.01, t-test. (Data are from Bancroft and Wu 1983.)

very few experiments have been conducted and only rhesus, cynomolgus, and stumptail macaques have been studied thus far. In the rhesus monkey, treatment of the castrated adult male with 17β-oestradiol does not maintain sexual behaviour and, in contrast to the rat, oestradiol has some inhibitory effect upon the ability to gain intromission. This may be due to the genital oedema caused by oestradiol treatment in male rhesus monkeys (Michael *et al.* 1990). Phoenix (1976) found that 19-hydroxytestosterone, an aromatization product of testosterone, had no effect upon sexual behaviour in castrated rhesus monkeys. More recently Zumpe *et al.* (1993) examined the effects of the non-steroidal aromatase inhibitor fadrozole on the sexual behaviour of castrated, testosterone-treated cynomolgus monkeys. Fadrozole at a daily dosage of $0.25\,\mathrm{mg\,kg^{-1}}$ inhibited more than 98% of the conversion of testosterone to oestradiol and its accumulation by hypothalamic nuclei. Interestingly, males treated in this manner did show a decrease in ejaculatory frequencies and mount attempts, as well as an increase in mount latency and ejaculatory latencies during pair-tests with ovariectomized, oestradiol-treated females (Fig. 13.11). However, oestradiol treatment improved ejaculatory frequencies in only three of the six fadrozole-treated males, whereas mounting frequencies remained unchanged. Zumpe

et al. conclude that aromatization of testosterone to oestradiol (E_2) does, indeed, play a role in mediating the effects of testosterone upon copulatory behaviour in the male cynomolgus monkey, but that exogenous oestradiol 'does not fully duplicate the behavioural affects of E_2 produced locally in the brain by aromatization of T.' If this is the case, it may partly explain the failure of exogenous oestradiol to maintain or restore sexual activity in castrated male rhesus monkeys (Michael *et al.* 1990). Baum *et al.* (1990b) also suggest that aromatization of testosterone to oestradiol has more powerful effects upon copulatory behaviour in the male ferret than can be achieved by administering oestradiol exogenously.

Several studies have examined the effects of DHT upon sexual behaviour in castrated male macaques. A stimulatory action of DHT has been reported in castrated rhesus monkeys by Phoenix (1974b) and Cochran and Perachio (1977). Phoenix administered dihydrotestosterone propionate (DHTP: a synthetic and longer-acting version of DHT) to 10 castrated adult male rhesus monkeys at a dosage of $1\,\mathrm{mg\,kg^{-1}}$. Even before treatment, six males achieved intromissions and three ejaculated during some pair-tests. However, DHTP significantly improved the performance of castrates, such that nine out of ten males ejaculated during an average of 53% of behavioural

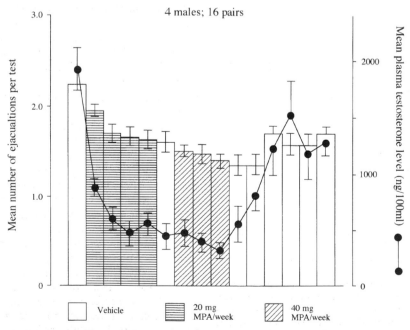

Fig. 13.8 Changes in plasma testosterone (•) and ejaculatory activity (histograms) of intact adult male cynomolgus monkeys (*Macaca fascicularis*) when treated with medroxyprogesterone acetate (MPA). Data are means ($\pm$S.E.M.) for 16 plasma samples in the case of testosterone levels, and 64×1 hour pair-tests in the case of ejaculatory activity. (Redrawn from Zumpe and Michael 1988.)

tests. Males increased in weight during DHTP treatment and, although hormonal effects upon the genitalia were not recorded it seems very likely that they would have occurred (e.g. growth of seminal vesicles and prostate, because these males began to exhibit seminal plug deposition, and growth of the penile spines, because these are restored by DHT in the rat (Feder 1971)). However, Phoenix postulated that the behavioural effects produced by DHTP were likely to be due to its central actions, rather than to purely genital effects.

Phoenix (1974b) and Cochran and Perachio (1977) did not measure circulating DHT in male rhesus monkeys, and it is not known whether hormone replacement produced physiological or pharmacological levels of the steroid. In more recent studies using castrated cynomolgus monkeys, Michael *et al.* (1986) compared the behavioural effects of testosterone propionate (TP) and DHTP upon sexual behaviour at dosages which produced plasma concentrations within the physiological range for intact males. Some of the results of their experiments are shown in Fig. 13.12. Whereas TP treatment caused an increase in ejaculatory frequencies and a decrease in the latency to mount and to attain ejaculation in the four males studied, DHTP was without significant effect, even at dosages which produced supranormal levels of circulating DHT. Michael *et al.* (1986) consider that the dosages of DHTP used by Phoenix (1974b) probably produced levels of plasma DHT up to tenfold higher than those which occur in intact adult male rhesus monkeys. Michael *et al.* also treated castrated rhesus monkeys with increasing dosages of either TP or DHTP and found that only TP was effective in restoring ejaculations. Nor does treatment of castrated stumptail macaques with DHTP, either alone or in combination with EB, restore their sexual behaviour (Schenck and Slob 1986).

Our incomplete knowledge of the behavioural potency of testosterone's metabolites in male non-human primates is matched by incomplete information on the human male. Gooren (1985) has reported some evidence that dihydrotestosterone undecanoate can substitute for testosterone undecanoate in the treatment of sexual dysfunction (in hypogonadal men). However, Gooren found no effect of the aromatase inhibitor testolactone or of the antioestrogen tamoxifen upon sexual interest in normal men. Testolactone was also without effect upon sexual thoughts or behaviour in men concurrently treated with an LHRH antagonist and testosterone enanthate (Bagatell *et al.* 1994). Exogenous oestrogen tends to depress rather than to enhance sexual interest in the human male, and it is unlikely that oestradiol is centrally active in stimulating sexuality (Bancroft *et al.* 1974; Bancroft 1989). The only compelling evidence that aromatization of androgen to oestrogen plays a significant role in sexual behaviour in primate males derives from the study by Zumpe *et al.* (1993) of effects of fadrazole in male rhesus monkeys. It is important to recall, however, that very few primate species have been studied so far and that, among sub-primates, the importance of aromatization can vary between species within a single mammalian order (e.g. among rodents). No studies have been conducted with prosimians or New World monkeys to examine effects of DHT, oestradiol, or other androgenic metabolites, upon sexual activity in adult males. It is possible that comparative studies will bring to light examples of oestrogenic effects upon masculine sexual behaviour in non-human primates. Aromatization occurs only

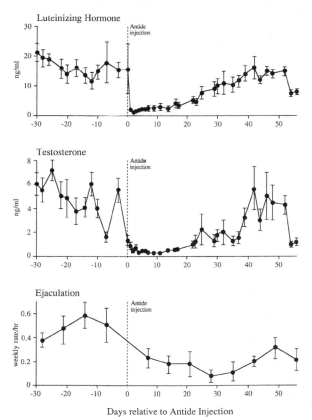

Fig. 13.9 Effects of treating intact male rhesus monkeys (*Macaca mulatta*) with an LHRH antagonist (antide) upon levels of circulating LH and testosterone, and upon their sexual activity (ejaculations) with females in a captive social group. Data are means (±S.E.M.) for seven males before and after a single subcutaneous injection (30 mg kg⁻¹) of antide. (Redrawn and modified from Wallen *et al.* 1991.)

in those brain areas where the aromatase enzyme is concentrated; it thus provides another means of targeting the effects of testosterone to specific regions (such as the preoptic area). However, a fundamental question—why central aromatization of testosterone to oestradiol has been incorporated into the control of masculine sexual 'appetite' and ejaculatory responsiveness in some cases (e.g. in the rat, red deer stag, or ferret) and not others—has not been answered.

Taken in broad perspective, the effects of castration and testosterone replacement upon sexual behaviour in primates are comparable with those reported in other mammals, including rodents, dogs, cats, rabbits, goats, rams, and stags (reviews: Larsson 1979; Meisel and Sachs 1994). Ejaculation and intromission are the most androgen-sensitive components of copulatory behaviour. Mounting and various forms of sexual interest (such as following the female and looking at or sniffing her genitalia) can persist for longer periods. Marked interspecific

and individual differences occur, as the following examples show. Some buck rabbits cease to mount by 1–2 weeks after castration whereas others continue to do so for several months (Beyer *et al.* 1980). Although intromission frequencies decline rapidly after castration in cats, Rosenblatt and Aronson (1958) found that one individual was capable of intromitting 3.5 years after the operation. Seven out of eight male red Sokoto goats studied by Hart and Jones (1975) still mounted and exhibited ejaculatory responses one year after castration (the eighth animal became lame and was dropped from the experiment). Sexually experienced dogs (beagles) continue to intromit 3.5 years after castration, although the duration of the copulatory lock does decrease during the first six post-operative months (Beach 1970). It appears that the retention of sexual interest and copulatory ability in castrated monkeys, such as rhesus or stumptail macaques, or in castrated and hypogonadal men, is not exceptional when compared with that in many other mammals. Testosterone has

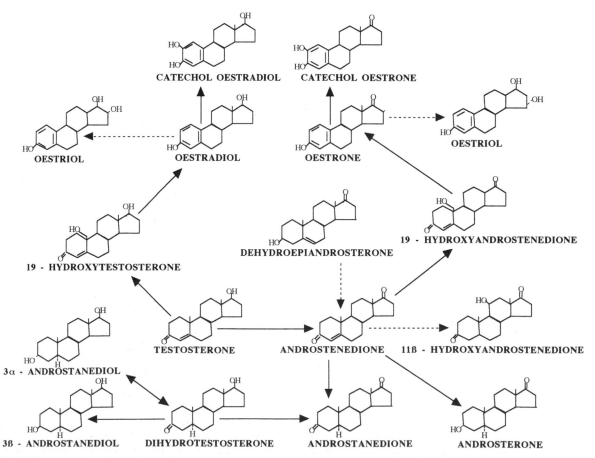

Fig. 13.10 Structural formulae and metabolic pathways for testosterone and related steroid hormones. Metabolic pathways known to occur in the brain are indicated by solid arrows. (Redrawn and modified from Luttge 1979.)

important effects upon sexual interest and copulatory competence in male primates just as in males of non-primate mammals. The relative enlargement of the neocortex and frontal lobes which has occurred during primate evolution has not invariably been associated with a lessening importance of testicular hormones in the control of sexual activity.

Sources of individual variability in sexual behaviour

Individual differences in circulating androgen levels

Attempts have been made to correlate individual differences in physical constitution or plasma testosterone levels with frequencies of sexual activity. In a study of professional musicians (Nieschlag 1979) showed that bass singers have a more athletic body build, higher testosterone levels, and greater weekly frequency of ejaculations than tenors. However, individual variations in circulating testosterone do not correlate with differences in sexual activity in male guinea pigs (Grunt and Young 1952), rats (Damassa *et al.* 1977), rhesus monkeys (Robinson *et al.* 1975; Phoenix *et al.* 1977; Chambers *et al.* 1982), and stumptails (Slob *et al.* 1979) or with orgasmic frequency in human males (Kraemer *et al.* 1976). No correlation has been demonstrated between individual differences in the capacity of the brain to bind testosterone and individual differences in sexual activity (guinea pig: Harding and Feder 1976). In the ram, however, greater occupation of oestradiol receptors in the preoptic area of the brain has been noted in highly sexually active males compared with 'low-performance' males (Alexander *et al.* 1993).

Circulating testosterone levels vary depending upon the time of day, being highest at night in

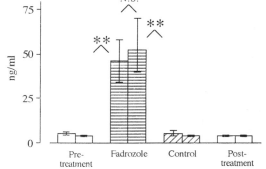

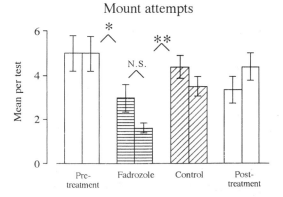

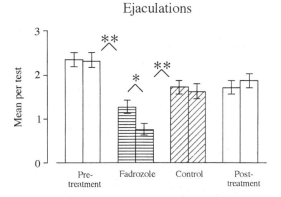

Fig. 13.11 Effects of the aromatase inhibitor fadrazole upon plasma testosterone and sexual behaviour in castrated, testosterone-treated, cynomolgus monkeys (*Macaca fascicularis*). Data are means (+S.E.M.) for 6 males before and after (open bars) treatment with fadrazole (horizontally-hatched bars) or vehicle (diagonally-hatched bars) via subcutaneously implanted osmotic minipumps. Each histogram gives data for 2 weeks. *P < 0.05; **P < 0.01; N.S.=not statistically significant. (Redrawn and modified from Zumpe *et al.*, 1993.)

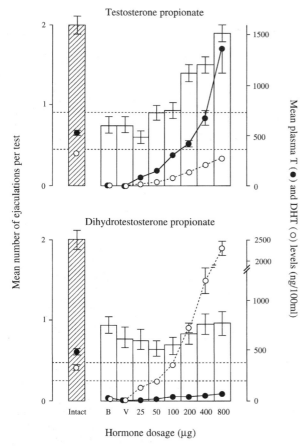

Fig. 13.12 Effects of treating castrated adult male cynomolgus monkeys (*Macaca fascicularis*) with either (upper) testosterone propionate (TP) or (lower) dihydrotestosterone propionate (DHTP) upon circulating levels of androgens and frequencies of ejaculations during pair-tests with females. Data are means (±S.E.M.) for four males each tested with four different females. Cross-hatched bars provide data for ejaculations when males were intact, open bars provide post-castration data. B=pre-treatment baseline; V=vehicle baseline. Dosages of TP and DHTP are given on the horizontal axis. (Redrawn from Michael *et al.* 1986.)

various diurnal primate species (e.g. rhesus monkey: Goodman *et al.* 1974; ringtailed lemur: Van Horn *et al.* 1976; Man: Faiman and Winter 1971). In non-human primates, plasma testosterone levels are highest during those times of the diurnal cycle when males are sexually inactive or asleep. This observation also applies to the only nocturnal anthropoid, the owl monkey, in which plasma testosterone peaks during the day when these animals are inactive (Dixson and Gardner 1981). In man, daily frequencies of sexual activity vary depending upon cultural considerations, and it is by no means the case that sexual activity is greatest at night in all human societies (Ford and

Beach 1952). Chambers *et al.* (1982) examined sexual behaviour and circulating hormone levels (testosterone, luteinizing hormone, oestradiol, and cortisol) in male rhesus monkeys at 0900 h and 2100 h. The highest levels of testosterone and luteinizing hormone were measured in blood samples collected at 2100 h (2 h after the lights in the animal rooms were turned off). By contrast, plasma cortisol and oestradiol levels were higher during the lighted phase, at 0900 h. The frequencies with which males contacted females, mounted, and gained intromission were all higher at 0900 h than at 2100 h, despite the lower levels of circulating testosterone and LH at that time (Fig. 13.13). Although there is no simplistic relationship between diurnal changes in circulating testosterone and frequencies of sexual behaviour in male rhesus monkeys, there has been some suggestion that the magnitude of the range between highest and lowest testosterone levels during the day relates to sexual activity. Michael *et al.* (1984a) kept male rhesus monkeys under a controlled photoperiod (14 h light: 10 h dark) for two years. Seasonal changes in sexual behaviour continued under these conditions and Michael *et al.* demonstrated that the daily range in plasma testosterone (measured at 0800 h, 1600 h and 2200 h) was smallest during the mating season, when ejaculatory frequencies reached their peak. Testosterone levels did not alter as a result of sexual activity, however. Thus, some males exhibited declines in ejaculatory behaviour during the non-mating season at a time when plasma testosterone levels peaked in nocturnal blood samples. Michael *et al.* (1984a) discuss the unlikely possibility that high nocturnal testosterone levels might have some inhibitory effect on potency in male rhesus monkeys. Alternatively, there might be some change in neural substrates underlying ejaculatory responsiveness during the non-mating season which does not depend upon testosterone.

Circulating levels of testosterone in male primates and other mammals are probably sufficient to maintain full sexual interest and copulatory behaviour, independent of short-term fluctuations (due to episodic secretion), diurnal changes, or acute effects of sexual stimuli. It is the case, however, that *pulsatile* administration of testosterone to castrated male rats is more effective in maintaining their sexual behaviour, so that the *episodic secretion* of testosterone which occurs in intact males might be significant in this regard (Taylor *et al.* 1990). If testosterone is maintained above a certain *threshold* level males remain sexually potent. Castrated rats continue to copulate normally if plasma testosterone is maintained at less than one third of circulating concentrations found in the intact male (Damassa *et*

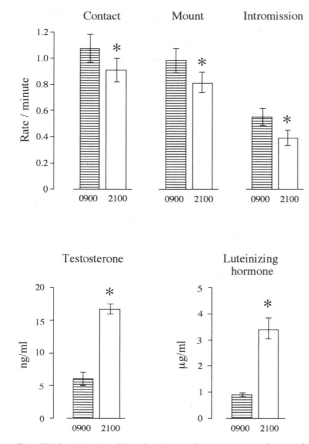

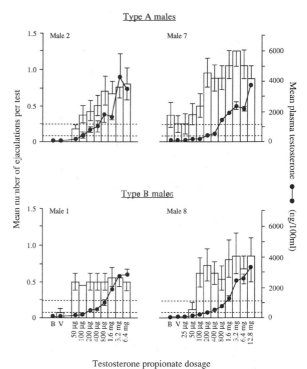

Fig. 13.13 Relationships between hormones and sexual behaviour in adult male rhesus monkeys (*Macaca mulatta*) when pair-tested with females at two different times of day (0900 h and 2100 h). Data are means (±S.E.M.) for rates min⁻¹ of contacts, mounts, and intromissions and for levels of testosterone and LH (ng mL⁻¹). *P < 0.05. (Redrawn and modified from Chambers *et al.* 1982.)

Fig. 13.14 Behavioural sensitivity of castrated adult male rhesus monkeys (*Macaca mulatta*) to increasing replacement doses of testosterone propionate (TP). Type A males (nos. 2 and 7) show a graded increase in ejaculations (open bars) in response to gradual increases in circulating testosterone (●). Type B males (nos. 1 and 8) exhibit an all-or-none type response, with ejaculatory frequencies showing a marked increase once a certain threshold level of circulating testosterone has been reached. Data are means (±S.E.M.). B=pre-treatment baseline; V=vehicle baseline. Horizontal dashed lines give the upper and lower physiological plasma testosterone range. (Redrawn and modified from Michael *et al.* 1984b.)

al. 1977). In the rhesus monkey, castrated adult males may show behavioural sensitivity to very low dosages of TP. At daily dosages of 50–100 μg, some males exhibit increased frequencies of intromitted thrusts and ejaculations, despite the fact that plasma testosterone is below the normal range for intact subjects (Michael *et al.* 1984b). Michael *et al.* were also able to identify two types of male as regards their responsiveness to TP. Type A males showed a graded increase in ejaculatory frequencies as the dosage of TP was increased and plasma T levels rose accordingly. Type B males showed an all-or-none type of response; ejaculatory frequencies increased at a certain threshold dosage of TP but then did not rise any further. Representative examples of these two types of male responsiveness are shown in Fig. 13.14. This experiment is valuable, because it helps

to explain some of the variability shown by castrated rhesus monkeys in response to exogenous testosterone (Phoenix 1973; Phoenix *et al.* 1973; Michael and Wilson 1973). As Michael *et al.* put it, 'much depends on the type of male that happens to be selected for an experiment'. Males might be more sensitive to some peripheral action of testosterone (e.g. on penile rigidity and hence on ability to intromit) or to its central effects upon sexual interest. Both types of effect might be facilitated by hormone treatment. No experiments on primates have determined whether individual differences in sexual behaviour might correlate with differences in testosterone binding in various parts of the brain. As mentioned previously, no such correlation was found in a study of male guinea pigs (Harding and Feder 1976).

Adrenal androgens

Because the adrenal cortex secretes androgens, it is possible that these might assist maintenance of sexual behaviour after castration. This seems unlikely because plasma testosterone decreases rapidly after castration (by six hours in the guinea pig: Resko 1970b) and occurs at very low concentrations in castrated monkeys. The role of adrenal androgens (androstenedione and dehydroepiandrosterone) in male rhesus monkeys does not correspond to that described previously for females of this species (see pages 363–5). A greater neural sensitivity to circulating testosterone must exist in female rhesus monkeys. However, Michael *et al.* (1984b) point out that as *some* castrated male rhesus monkeys are very sensitive to fluctuations in plasma testosterone when levels are low (Fig. 13.14) 'it is not impossible that fluctuations in adrenal androgen secretion might be sufficient to influence potency. Whether adrenal secretion could actually sustain the potency of certain males after castration must await the results of adrenalectomy.' No such experiments have been conducted on primates. However, the combined effects of castration and adrenalectomy have been studied in several mammals, and the results indicate that adrenal androgens do not contribute significantly to the maintenance of sexual behaviour, in hamsters (Warren and Aronson 1957), rats (Bloch and Davidson 1968), cats (Cooper and Aronson 1958), or dogs (Schwartz and Beach 1954; Beach 1970).

The role of previous sexual experience

Previous sexual experience can ameliorate the behavioural effects of castration in some mammals (cat: Rosenblatt and Aronson 1958; guinea pig: Valenstein and Young 1955). Such considerations are less important in species such as the rat. Although male rats reared in social isolation show less sexual activity during initial sex tests after castration than experienced animals, these differences diminish rapidly as the sexual performance of inexperienced males improves (Larsson 1979). Individual variability in retention of sexual behaviour in castrated rhesus monkeys, cynomolgus monkeys, and stumptails is probably not a function of previous sexual experience, since experienced animals were used in the various studies reviewed above. Sexually inexperienced male marmosets are able to copulate with female partners during initial pair-tests. Castrated, inexperienced males mount and make pelvic thrusts as frequently as intact males, although their ability to intromit and ejaculate is impaired (Fig. 13.4). Inexperienced males, whether intact or castrated, occasionally mount the female from the side or at the anterior end. It is unknown whether sexually experienced males might retain the ability to intromit and ejaculate for longer after castration than the naïve subjects shown in Fig. 13.4.

In Chapters 6 and 10, the important effects of early social experiences upon the development of copulatory behaviour were described. If male rhesus monkeys are reared in social isolation for the first 6–12 months, then they rarely show adequate sexual behaviour in adulthood. Yet, some isolate-reared males will mount, thrust, and ejaculate if given access to a dummy female (see Fig. 6.15, page 159). The neuroendocrine mechanisms which control copulatory patterns are functional in such males; the problem lies in their reduced ability to relate socially and sexually to female conspecifics.

The role of the female partner

The importance of individual sexual preferences and favouritism as determinants of mate choice in primates was discussed in Chapter 4 (see pages 85–8). There is no doubt, for example, that male rhesus monkeys exhibit sexual preferences for certain female partners when a choice is available (Herbert 1968; Everitt and Herbert 1969). Effects of female partners upon ejaculatory frequencies, and other measurements of male sexual behaviour, subsequent to castration and hormone treatments have been noted by various workers (e.g. in the stumptail macaque: Slob and Schenck 1981; Schenck and Slob 1986; rhesus macaque: Michael *et al.* 1984b; cynomolgus macaque: Zumpe and Michael 1985). An especially striking example of effects of female partners upon ejaculatory frequencies after castration in male rhesus monkeys has been described by Goy (1992), and his results are reproduced here in Fig. 13.15. Ten males were pair-tested before castration and for up to 55 weeks after castration, with a number of female partners. Some females were consistently more attractive and elicited a greater frequency of ejaculatory mounts than others. In Fig. 13.15, the ejaculatory performances of the 10 males with the top four *optimal* females is compared with their behaviour during tests with the four least attractive *suboptimal* partners. As Goy pointed out, 'the effect of these optimal females on the postcastrational ejaculatory performance of the males is striking, and it acts to retard the rate of decline'. These results show that 'the attractiveness of the female is generalizable from one sample of males to another. It is something the female carries with her and brings to the pairing situation'. Here is an

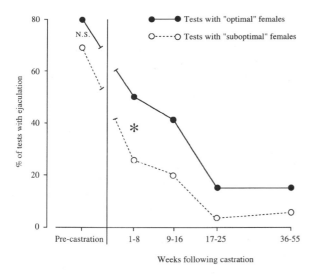

Fig. 13.15 Effect of female sexual attractiveness upon the post-castrational sexual performances of adult male rhesus monkeys (*Macaca mulatta*). Data are means for the percentages of tests in which males (N = 10) ejaculated during pair-tests with the most attractive females (N = 4; optimal partners: ●-●) as compared to the least attractive females (N = 4; suboptimal partners: ○-○). *P < 0.01; N.S. = not statistically significant. (Redrawn from Goy 1992.)

important variable affecting the castration-induced decline in ejaculatory performance in male rhesus monkeys. The qualities which promote individual differences in female attractiveness under these conditions are not understood and cannot be explained solely by female hormonal status.

Long-term familiarity between subjects used in experiments to study sexual behaviour can also have pronounced effects upon sexual performance. Male rhesus monkeys exhibit progressive declines in potency when tested with the same female partners regularly for two years. The effect is reversible, however, by provision of fresh partners, unfamiliar to the males. Physical contact is not essential to produce a deficit in male potency. When male rhesus monkeys are caged individually, but they have visual, auditory, and olfactory contact with females for one year they ejaculate significantly less during pair-tests with these partners than males which have had no visual contact with the females (Zumpe and Michael 1984). These authors suggest that the *familiar partner phenomenon* might have some relevance to behaviour under natural conditions, since it might encourage male emigration and incest avoidance (the role of familiarity in incest avoidance is discussed on pages 88–90). It is also relevant to note that, when conducting experimental studies of primate sexual behaviour, allowance must be made for

the possible effects of visual contact between the sexes upon their sexual behaviour during subsequent pair-tests.

Genetic variability between males

In the guinea pig and mouse, genetic variability has been shown to influence individual differences in sexual activity and its decline after castration—high or low levels of potency occur in intact males and these individual differences persist when males are castrated and treated with equivalent doses of testosterone propionate (Grunt and Young 1953; Valenstein *et al.* 1955). Mice produced by artificial selection (B6D2F1 males: McGill and Manning 1976) exhibit exceptional retention of copulatory behaviour after castration which is not due to compensatory adrenal secretion or experiential factors (Thompson *et al.* 1976; Manning and Thompson 1976). These results represent 'the most consistent retention of sexual behaviour by a group of animals following castration recorded in the literature for any species,' including primates with their more highly developed neocortex and frontal lobes. After castration, B6D2F1 male mice exhibit a *difficult period* of sexual adjustment. During the early post-castration weeks, frequencies of ejaculation decline, and ejaculatory latencies increase; these effects then gradually reverse themselves and potency returns to normal, or near normal, levels (Fig. 13.16). McGill (1978) pointed out that the *difficult period* might be due to two possibilities 'which are perhaps not mutually exclusive'. Firstly, castrated males might be learning in the early weeks after the operation to readjust their behaviour and to maintain the level of sensory input required for ejaculation. Secondly, some reorganization might occur of neural substrates required for ejaculation, which is not dependent upon testicular hormones or experiential factors. McGill showed that the second possibility is the important variable, by denying castrated males any opportunity to copulate for 16 weeks after the operation, and then comparing their ejaculatory frequencies and latencies to those of intact males or castrates which had received continuous testing. The results clearly show (Fig. 13.17) that, in the absence of post-operative sexual experience, castrated males had passed through their difficult period and regained similar levels of ejaculatory responsiveness to those measured in castrates, or intacts, subjected to continuous testing. Interestingly, a fourth group of intact males subjected to 16 weeks of sexual inactivity showed greater deficits in ejaculatory latencies and frequencies than the castrated males (Fig. 13.17). The nature of the changes which occur in neural

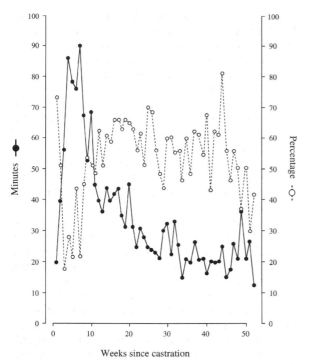

Fig. 13.16 The 'difficult period' for recovery of copulatory performance which occurs after castration in male B6D2F1 mice. Data are means per week for post-castration ejaculatory latencies (●) and percentages of tests with ejaculatory responses (○). (Redrawn and modified from McGill and Manning 1976.)

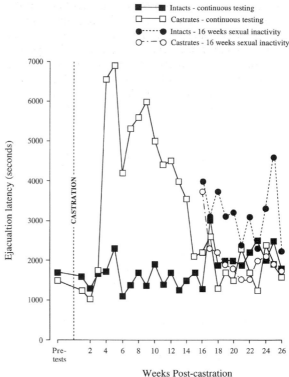

Fig. 13.17 Ejaculation latencies of castrated or intact groups of B6D2F1 male mice which were continuously tested (once per week) or given 16 weeks of sexual inactivity before testing resumed. The significance of these data is discussed in the text. (Redrawn from McGill 1978.)

substrates during the difficult period in castrated B6D2F1 male mice remains unknown. This problem provides a fascinating opportunity to study the central mechanisms that govern ejaculation, and the role played by testosterone in the maintenance of such mechanisms, in male mammals.

Effects of age

Quantitative studies of the effects of the ageing process upon sexual function in male primates are limited to just two species—man and the rhesus monkey. Declines in sexual vigour occur throughout adult life in men (Kinsey *et al.* 1948) and old male rhesus monkeys exhibit less copulatory behaviour than younger adults (Phoenix and Chambers 1982). Kinsey *et al.* found that total frequencies of sexual intercourse (marital and extra-marital) declined from a mean of 4.27 per week at 16–20 years of age to 1.8 at 41–45 years and 1.09 per week in men aged 56–60 years (Fig. 13.18). They also noted that in the younger age-groups, between 16–30 years, some

married men reported very high frequencies of intercourse (up to 25 times per week), whereas by 50 years of age the maximum was 14, and by 60 it was 3 episodes per week. By 60 years of age, approximately 6% of males reported that they no longer engaged in intercourse. Erectile responses also declined throughout adulthood. Younger men, in their late teens or early twenties, are capable of maintaining erection for almost one hour prior to ejaculation, whereas by 41–45 years of age this declines to 31 min and by 66–70 years it is just 7 min (median values: Fig. 13.18). Kinsey *et al.* identified a gradual decrease in various parameters of 'erotic responsiveness' in men with age (Fig. 13.18) and an increased occurrence of erectile impotence. Whereas impotence was reported by <1% of men under 35 years of age, by 55 years 6.7% of men were impotent, and this had increased to 55% by the time men were 75 years old (Fig. 13.19). Erectile dysfunction is a major concern for those men who seek medical advice for sexual problems, particularly in the group between

40–60 years of age (Foreman and Doherty 1993). Foreman and Doherty, in reviewing this subject, pointed out that 'it is incorrect to assume that patients seeking medical help for erectile dysfunction are primarily elderly. Although the *frequency* of erectile dysfunction is highest in the oldest population, the *impact* of impotence on the lives of men can be greater in younger age groups when sexual drive is higher'.

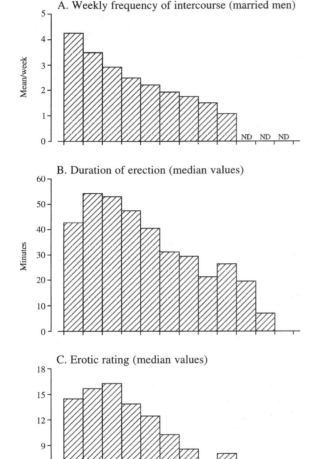

Fig. 13.18 Effects of age upon sexual activity and erotic responsiveness in men. A. Weekly frequencies of sexual intercourse; B. Duration of penile erection; C. Erotic responsiveness. (Data from Kinsey *et al.* 1948.)

The decline in sexual vigour throughout adulthood is a gradual process which effects almost every aspect of sexuality. Ejaculations become less powerful and the volume of ejaculate diminishes. The post-ejaculatory interval increases in duration. Erections during sleep and upon waking become less frequent, indicating that the underlying mechanisms which govern erection are changing, as well as erotic responsiveness to sexual stimulation. Nor should these changes be ascribed simply to failing health in older men; various studies have shown that age itself plays the major role in mediating changes in male sexuality (Martin, C. 1981; Brecher 1984; reviewed by Bancroft 1989).

Ageing rhesus monkeys also show declines in sexual behaviour. Chambers *et al.* (1982) compared sexual behaviour of eight males aged 22 years with eight subjects aged approximately 12 years. Each subject was pair-tested with eight ovariectomized, oestradiol-treated partners. Percentages of tests with penile erections, mounts, intromissions, and ejaculations were lower for the older males (Fig. 13.20) although the two groups did not differ as regards latencies to first mount, intromission, and ejaculation.

Phoenix and Chambers (1982) attempted to improve sexual performance in ageing male rhesus monkeys by providing them with novel partners and with younger females (4.5 years old) during laboratory pair-tests. These manipulations had little effect upon frequencies of copulatory behaviour. However, old males did contact young females more frequently and the percentage of tests in which penile erections occurred was also greater under these conditions. Although the data are far from complete, one difference between older rhesus monkeys and older men concerns the occurrence of erectile dysfunction. Impotence does not appear to be a problem for ageing rhesus monkeys. However, it must be kept in mind that very few older monkeys have been studied and that rhesus males have often been tested with a variety of partners rather than with a single female.

How far are the declines in sexual activity in ageing men or rhesus monkeys connected to changes in testosterone secretion or to mechanisms of steroid transport, uptake, and metabolism in target tissues? Young and old adult male rhesus monkeys do not differ as regards levels of circulating testosterone, oestradiol, and cortisol (Chambers *et al.* 1982). Although the index of free testosterone did not differ between young and old males a significantly higher percentage of circulating testosterone in older males was transported by testosterone-binding globulin

(TeBG: Chambers *et al.* 1981). Moreover, the percentage of testosterone bound to TeBG in the nine older males studied correlated negatively with various measures of their sexual behaviour, including frequencies of intromission and ejaculation (Fig. 13.21).

Despite the negative correlation between bound testosterone and sexual activity in older male rhesus monkeys reported by Chambers *et al.* (1981), exogenous testosterone has no effect upon the sexual behaviour of such males (Phoenix and Chambers 1982, 1988). The same is true in ageing male guinea pigs (Jacubczak 1964). The declines in sexual behaviour and erectile function which occur throughout adulthood in human males (see Figs. 13.18 and 13.19) are gradual, and are not associated with progressive declines in circulating testosterone. Various studies of older men have reported a lowering of circulating testosterone (from approximately 50 years of age onwards: Harman 1983). This includes men in good health (Deslypere and Vermeulen 1984; Tsitouras and Hagen 1984) although not all studies agree on this point (e.g. Harman and Tsitouras 1980). Whilst it is conceivable that some older men might benefit from exogenous testosterone treatments, it seems far more likely that a constellation of factors contributes to sexual decline in the ageing male and that most men continue to produce sufficient amounts of testosterone. Psychosocial factors are probably very important; some older men may feel that society 'expects' that they will lose interest in sexual activities with advancing years (Bancroft 1989). This view is not shared by all human societies (Winn and Newton 1982) and nor can it explain the age-related declines in copulatory behaviour in rhesus monkeys (Phoenix and Chambers 1982). These authors discussed some evidence that older male rhesus monkeys show an accentuation of their preferences for particular females and that 'certain subtle aspects of the behaviour of these preferred females may be critical in causing old males to respond'. Subsequent work (Chambers and Phoenix 1984) showed that pairing old males with their preferred partners did improve sexual performances markedly-to levels found in much younger subjects. This approach was much more successful than purely physiological manipulations, such as administration of apomorphine or yohimbine, both of which would be expected to enhance sexual arousal via central effects upon dopaminergic or noradrenergic mechanisms (Chambers and Phoenix 1989). This is not to dispute the probable importance of physiological changes with age and their involvement in sexual decline. The ability of peripheral tissues to bind testosterone declines with age, for example (e.g. in the skin of the scrotum). Central deficits might also occur, and age-related changes in a whole variety of neurological systems which influence sexual arousal, erection, and copulatory behaviour are probably implicated in decreased sexual responsiveness. Even at the level of the penis itself, age takes its toll. Thus, intracavernosal injections of the smooth muscle re-

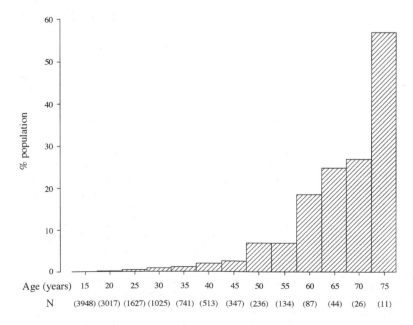

Fig. 13.19 Occurrence of impotence with increasing age in men. Ages (in years) and sample sizes are given on the horizontal axis. Percentages of subjects exhibiting impotence in each age group are indicated by histograms. (Data from Kinsey *et al.* 1948.)

laxant papaverine, which causes erection and is used as a treatment for impotence, are effective at lower dosages in younger men. Older patients often need larger doses of papaverine (Crowe and Qureshi 1991).

Peripheral versus central effects of androgens upon male sexuality

Declines in sexual activity after castration might be influenced by peripheral changes in androgen-dependent features, such as penile morphology, accessory sexual organs, secondary sexual adornments, and changes in muscular vigour, as well as by effects involving the brain and spinal cord. There is no doubt that the central effects of testosterone and its various metabolites play a crucial role in maintaining sexual interest and copulatory behaviour. In the castrated male rat, for example, oestradiol treatment restores most aspects of copulatory behaviour in the absence of peripheral effects upon the genitalia (Larsson 1979; Sachs and Meisel 1988). Conversely, the non-aromatizable androgen DHT will maintain peripheral structures and sexual reflexes in castrated rats, but fails to activate sexual behaviour (Feder 1971). The peripheral effects of androgens do not, by themselves, maintain sexual activity. However, androgenic effects upon the male's genitalia and secondary sexual characteristics do have important consequences for behaviour as the following examples show.

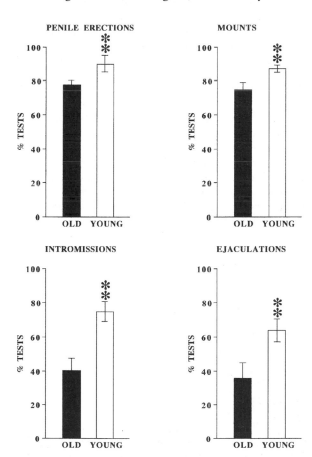

Fig. 13.20 Comparisons of sexual activity in old male rhesus monkeys (aged 22 years: closed bars) and young males (aged 12 years: open bars). Data are mean (+S.E.M.) percentages of tests with penile erections, mounts, intromissions and ejaculations. Males were each pair-tested with eight different ovariectomized, oestrogen-treated partners. **P < 0.01. (Redrawn and modified from Chambers *et al.* 1982.)

Androgens and penile morphology

The glans penis of many primates is clothed in small, androgen-dependent epidermal spines which overlie tactile receptors in the dermis (e.g. in rhesus and talapoin monkeys: Dixson and Herbert 1974; Herbert 1974). The evolution of penile spines was reviewed in Chapter 9; here we are concerned with their significance in copulatory behaviour. Beach and Levinson (1950) postulated that atrophy of the penile spines in the castrated male rat impairs penile sensitivity and hence contributes to declines in copulatory behaviour. It is now appreciated that the central effects of testosterone, and its aromatization products, play a far more important role in copulatory behaviour in the male rat. However, it is entirely possible that the small Type 1 penile spines which occur in the rat, as in many primate species, might contribute in a subtle fashion to the sensory processes which facilitate intromission and ejaculation. The importance of tactile feedback from the penis for successful copulation has been established in experiments on several mammalian species. Local anaesthesia of the glans penis, or transection of the dorsal penile nerves, causes a marked impairment of erection and intromission in the rat (Carlsson and Larsson 1964; Adler and Bermant 1966; Larsson and Södersten 1973). Desensitization of the glans penis does not prevent erection in the cat or rabbit, but cats fail to orient correctly when mounting, or to intromit, whilst rabbits exhibit delays in intromission and ejaculation (Aronson and Cooper 1968; Agmö 1976). Local anaesthesia of the glans penis in the human male has been reported to retard ejaculation

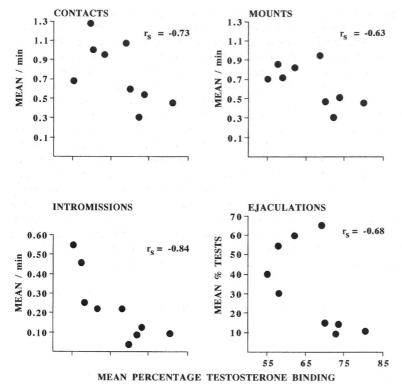

Fig. 13.21 Relationships between percentages of testosterone binding by testosterone binding globulin (TeBG) in the circulation, and measurements of sexual behaviour in old male rhesus monkeys (*Macaca mulatta*). Data refer to nine subjects; rank correlation coefficients are all statistically significant (P < 0.05). (Redrawn and modified from Chambers *et al.* 1981.)

(Johnson 1968). In the male rhesus monkey, transection of the dorsal penile nerves does not prevent erection or intromission. However, such males make ataxic pelvic thrusting movements and rarely ejaculate (Herbert 1973).

A series of experiments on marmosets has provided some insights into the precise role played by penile spines in sexual behaviour (Dixson 1986a, 1988, 1991a). Application of the local anaesthetic

lidocaine to the penis of the male marmoset has no effect upon erectile ability or mount latency. Males mount females promptly during pair-tests and make vigorous attempts to intromit. However, they usually fail to execute the deep pelvic thrust required to attain intromission (Table 13.2). Should intromission occur, then males usually ejaculate normally and there is no difference between the ejaculatory latencies of treated and control males. It is possible,

Table 13.2 Effects of penile anaesthesia upon sexual behaviour in male marmosets

Behavioural measurement	Control tests Mean ± SEM	Lidocaine treatment Mean ± SEM	Wilcoxon matched-pairs test
Mount latency (s)	5.03 ± 1.8	5.7 ± 2.9	N.S.
*Intromission latency (s)	15.00 ± 4.0	185.7 ± 72.8	P = 0.02
Duration of first mount (s) (excluding intromission)	7.4 ± 2.8	116.4 ± 54.8	P < 0.01
**Ejaculatory latency (s)	5.3 ± 0.3	6.5 ± 1.2	N.S.
Ejaculations/test	1.1 ± 0.06	0.4 ± 0.14	P = 0.01
Mounts and mount attempts/test	1.5 ± 0.2	3.7 ± 1.2	N.S.
Intromissions/test	1.1 ± 0.06	0.5 ± 0.12	P = 0.01

Data are from 10 males except *(N=7) **(N=5) where only males which intromitted or ejaculated under both experimental conditions provided data for statistical analysis.
Times are given in seconds (s).
N.S. = not statistically significant.
Data are from Dixson 1986a.

therefore, that lidocaine impairs cutaneous tactile feedback required for intromission, whilst sparing pressure receptors deeper in the penis which facilitate ejaculation once intromission has occurred. Consistent with this interpretation is the observation that dorsal penile nerve transection (DPNT) greatly reduces, or abolishes, both intromission and ejaculation in the male marmoset (Fig. 13.22). All three males shown in Fig. 13.22 continued to mount and make well-orientated pelvic thrusts after DPNT. However, the deep pelvic thrust required to achieve intromission was rarely observed and, on the few occasions when males 1 and 3 intromitted, they continued to thrust rapidly, rather than making a few, slower pelvic thrusts such as normally occur. Only two ejaculations were observed in the DPNT males, whereas control subjects continued to mount, intromit, and ejaculate at normal frequencies. In the marmoset, therefore, intromission depends upon adequate cutaneous sensory feedback from the erect penis during bouts of rapid pelvic thrusts which

serve to locate the vaginal orifice. These observations do not support the view expressed by Agmö (1976) that 'in more highly evolved mammals the process of intromission is less dependent upon sensory feedback'. It seems likely that at least some of the relevant cutaneous receptors are situated in proximity to the penile spines of the marmoset. Selective removal of the spines in gonadally intact males (by application of thioglycollate cream) resulted in an increased duration of pre-intromission pelvic thrusting and a smaller (but statistically significant) increase in the duration of intromitted thrusts (Fig. 13.23). The seven males used in this study continued to intromit and ejaculate during postoperative tests, although three exhibited partial intromissions (failure to achieve complete insertion) during some tests. This had not occurred before spine removal and was not observed in sham-operated subjects.

Penile spines are concave at the base and each contains a papilla of dermal tissue. No information

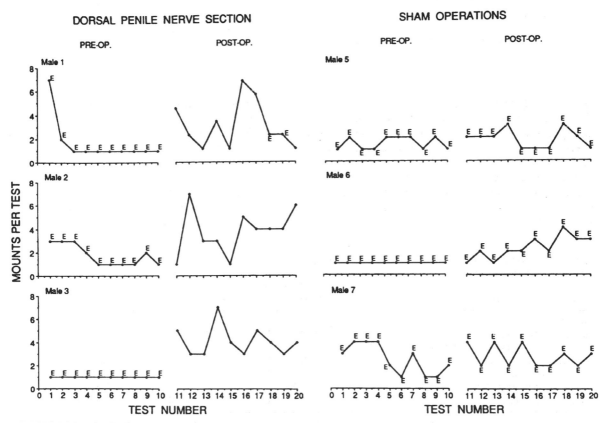

Fig. 13.22 Effects of bilateral dorsal penile nerve transection, or sham operations, on the temporal patterning of mounts and occurrence of ejaculation in male common marmosets (*Callithrix jacchus*). Ten pair-tests were conducted before (pre-op) and after (post-op) nerve transection or a sham operation. Data=mounts per test; E=occurrence of an ejaculation during the test. (After Dixson 1988.)

is available concerning the distribution of dermal tactile receptors in the glans of the marmoset. In man, glabrous skin contains two types of rapidly-adapting receptors (Meissner's and Pacinian corpuscles) and two slow-adapting receptor types (Merkel-cell-neurite complexes and Ruffini endings: Iggo 1982; Brown 1986; Halata and Munger 1986). The functions of these mechanoreceptors in facilitating intromission and ejaculation are not fully understood (Sachs and Meisel 1988; Meisel and Sachs 1994). It is of interest that rapidly-adapting receptors are concentrated in the proximal region of the glans penis of the cat, which is also the region bearing penile spines in this species (Johnson *et al.* 1986). In the rat, rapidly-adapting mechanoreceptors in the glans penis are extremely sensitive to the to and fro deflection of penile spines (Johnson and Halata 1991). Removal of the penile spines in the marmoset, using a depilatory cream, might also have damaged receptors in the underlying dermal papillae (Dixson 1991a). However, it seems reasonable to conclude that the spines, and associated dermal receptors, play a subtle (but not indispensable) role in sensory feedback during copulation. Androgen-dependent spines on the glans are less important behaviourally than originally envisaged (Beach and Levinson 1950), but they are significant, nonetheless. Only sexually experienced marmosets were used in these experiments, and they were tested with familiar female partners. The effects of spine removal might be greater in sexually naïve males, or

males tested with unfamiliar females. Penile deafferentation in the rat has more severe effects upon copulatory behaviour in sexually inexperienced males (Dahlof and Larsson 1978).

Very large Type 2 and Type 3 penile spines occur on the glans penis of certain prosimians, such as galagos, *Avahi*, and *Phaner* (see page 250). These robust penile spines are also androgen-dependent (Fig. 13.24), but it is unlikely that they serve to enhance tactile sensory feedback to the male during copulation. They might assist the male to maintain insertion during prolonged copulations, or impart tactile stimulation of sexual significance to the female partner (see pages 260–3). Removal of the much smaller Type 1 spines found in the marmoset has no effect upon the female's behavioural responses during intromission (struggling movements, turning the head, and termination of intromission by the female: Dixson 1991a).

Castration can lead to loss of muscle-mass and to a general decline in strength and vigour. Although the force of pelvic thrusting may decrease after castration (in the rabbit: Beyer *et al.* 1980), there is no evidence that this contributes to post-castrational declines in sexual behaviour in primates. More important is the observation that the striated muscles at the base of the penis atrophy after castration (Wainman and Shipounoff 1941; Sachs 1983). In the rat, the *ischiocavernosus* and *bulbocavernosus* muscles control penile reflexes which are important for intromission and during ejaculation (Sachs 1982,

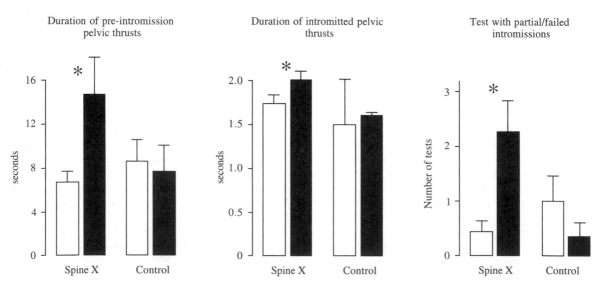

Fig. 13.23 Effects of penile spine removal or a sham operation (controls) upon copulatory behaviour in male common marmosets (*Callithrix jacchus*). Data are means (±S.E.M.) for seven males pre- and post-spine removal and eight males pre- and post-sham operations. Pre-operative data: open histograms; post-operative data: closed histograms. *P ≤ 0.05, Wilcoxon test. (Data are from Dixson 1991a.)

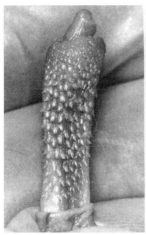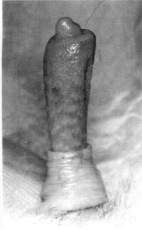

Fig. 13.24 Androgen-dependent robust penile spines on the glans penis of Garnett's galago (*Galago garnettii*). Left: intact male; Right: castrated male.

1983; Sachs and Meisel 1988). Both the striated muscles and the vascular mechanisms regulating penile erection are androgen-sensitive in the rat (Leipheimer and Sachs 1993). There have been no comparable studies of the effects of androgens on penile reflexes or striated muscle mass in any primate species. Castration can also cause a general decrease in size of the glans penis (e.g. in the greater galago: Fig. 13.24). However, in the few anthropoids examined it seems that testosterone stimulates growth of the penis during puberty, but is not required to maintain its adult dimensions (e.g. in the talapoin: Dixson and Herbert 1974). In the beagle, by contrast, the glans penis decreases in size after castration and this affects the male's ability to maintain the genital lock which occurs during copulation (Beach 1970).

The accessory sexual organs

On an historical note, the James-Lange theory of the emotions attributed great importance to the notion that sensory feedback from the organs of the body might be involved in the genesis of particular emotional states. On this basis, the growth and expansion of the seminal vesicles, prostate, and other accessory sexual organs which occurs in the pubescent male, or in the adult male monkey during the mating season, might contribute to the genesis of sexual arousal at the central level. No experimental evidence supports this view. Likewise, atrophy of the accessory sexual organs after castration probably has very little direct effect upon behaviour. Removal of the seminal vesicles, dorsolateral prostate, Cow-

pers glands, *vas deferens*, and *levator ani* muscles from male rats does not impair copulatory behaviour (Beach and Wilson 1963; Tissell and Larsson 1979). Removal of the prostate in man, due to carcinoma or benign hypertrophy, might be associated with impotence in some cases, but this is due to secondary, neurological damage rather than to absence of the prostate itself.

Secondary sexual adornments

The question of how androgen-dependent secondary sexual traits in male primates might influence male sexual attractiveness, and hence affect mating success, has never been addressed experimentally. Various secondary sexual traits were described in Chapter 7 and discussed in relation to sexual selection by intermale competition and female choice. In a few cases, we know that these secondary sexual characters atrophy after castration of the adult male. Such androgen-dependent structures include the cutaneous scent glands of the male ringtailed lemur (Andrimiandra and Rumpler 1968) and greater bushbaby (Dixson 1976) and the red sexual skin of the male rhesus monkey (Vandenbergh 1965) and hamadryas baboon, in which the cape of hair on the shoulders and back is also lost after castration (Zuckerman and Parkes 1939). In the squirrel monkey males become fatted during the annual mating season (Dumond and Hutchinson 1967) and this weight increase is partly stimulated by rising levels of plasma testosterone (Nadler and Rosenblum 1972). In captive groups of squirrel monkeys, females approach fully fatted males more frequently (Mendoza *et al.* 1978), whilst in the wild there is evidence that the largest male is preferred by females and experiences greatest mating success (Boinski 1987). This is a rare example where evidence is available concerning the question of masculine secondary sexual traits and female choice in primates. Experimental manipulations of masculine adornments in avian species (such as widow birds and swallows: reviewed by Andersson 1994) have largely vindicated Darwin's (1871) views concerning the evolution of extravagant secondary sexual characters in male animals. The question of how non-behavioural cues might influence male sexual attractiveness in primates deserves much more attention.

Central effects of androgens upon male sexuality

Although androgens exert important peripheral effects in the adult male, it is the central actions of testosterone and its metabolites which are principally responsible for the activation and maintenance of sexual interest and copulatory behaviour. The relatively slow declines in copulatory behaviour which occur after castration, and the gradual improvement of sexual activity caused by testosterone replacement, indicate that these central effects of testosterone are probably *genomic* in nature. Steroids might exert effects upon target neurons via several mechanisms involving binding to cell surface receptors or to intracellular receptors. Traditionally, the route involving intracellular receptors has been described as the *genomic* effect of a steroid, such as testosterone. Testosterone forms a receptor–

steroid complex which acts upon the genome to cause formation of messenger RNA and protein synthesis. Conversely, more rapid effects of steroid hormones in neurons have been ascribed to *non-genomic* mechanisms. Recently this distinction has begun to crumble, with the realization that steroids binding to cell-surface receptors can in some cases be coupled to second messenger systems which provide an alternative route to the cell nucleus via DNA binding proteins (McEwen 1991, 1994). Nonetheless, given that the genome, contained in the cell nucleus, is a key target for androgens, it is important to determine where in the brain the various neurons are situated that take up and concentrate testosterone and its metabolites within their

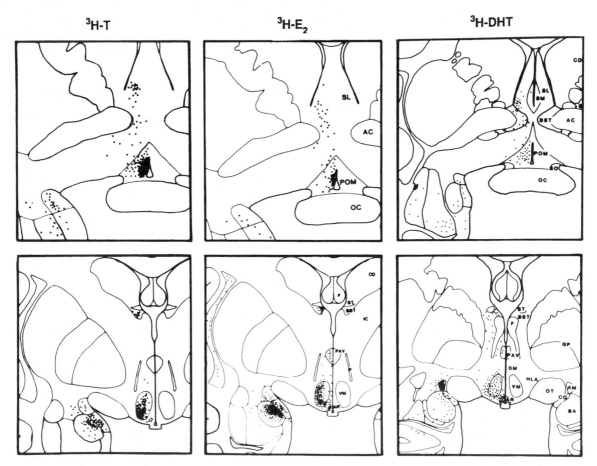

Fig. 13.25 Maps of labelled neurons in the brains of castrated male rhesus monkeys 60 min after administration of 5 mCi ^{3}H-T, 5 mCi ^{3}H-E$_2$ or 2 mCi ^{3}H-DHT. Upper maps are at the level of the preoptic region and lower maps are at the level of the infundibulum (levels of ^{3}H-DHT are slightly different). Each dot represents six labelled neurons. Abbreviations: AC, anterior commissure; AM, medial amygdaloid nucleus (n.); AR, arcuate n.; BA, basal accessory amygdaloid n.; BST, bed nucleus of stria terminalis; CD, caudate; CO, cortical amygdaloid n.; DM, dorsomedial hypothalamic n.; F, fornix; GP, globus pallidus; HLA, lateral hypothalamic area; IC, internal capsule; OC, optic chiasm; OT, optic tract; PAV, paraventricular n.; POM, medial preoptic n.; SL, lateral septal n.; SM, medial septal n.; so, supraoptic n.; ST, stria terminalis; VM, ventromedial nucleus. (After Michael and Bonsall 1990.)

nuclei. This problem has been extensively studied in the male rhesus monkey by means of autoradiography and high-performance liquid chromatography (Rees *et al.* 1988; Bonsall *et al.* 1989; Michael and Bonsall 1990).

Tritiated testosterone (^{3}H-T) accumulates in higher concentrations within specific brain regions in the castrated male rhesus monkey. Labelling indices (percentage of neurons labelled with tritiated steroid) range from 20–84% in the medial preoptic nucleus, the hypothalamus (anterior hypothalamic region, ventromedial nucleus, premammillary nucleus, and arcuate nucleus) the septum (lateral nucleus), amygdala (cortical, medial, and accessory basal nuclei), the bed nucleus of the stria terminalis, and the intercalated mammillary nucleus (Fig. 13.25). However, some of this labelling consists of the metabolic products of testosterone rather than testosterone itself. After entering the neuron ^{3}H-T is either reduced to ^{3}H-DHT or aromatized to ^{3}H-

E_2. High-pressure liquid chromatographic (HPLC) measurements of nuclear radioactivity in various regions has revealed three types of responsive tissue (Fig. 13.26). In Type 1 tissues (the preoptic area, hypothalamus, and bed nucleus of the stria) 48–70% of nuclear radioactivity is in the form ^{3}H-E$_2$. In the pituitary gland, the septum and some other brain regions 35–85% of radioactivity is in the form ^{3}H-T (Type II tissues). Finally, in the genital tract (seminal vesicles, prostate, and penis) 5-alpha-reductase activity results in 86–99% of nuclear radioactivity being expressed in the form of ^{3}H-DHT (Bonsall *et al.* 1989). Further experiments (Rees *et al.* 1988; Bonsall and Michael 1989; Michael, Rees and Bonsall 1989) have examined the effects of pretreatment of castrated male rhesus monkeys with either oestradiol (E$_2$) or DHT upon the ability of various brain regions to accumulate tritiated testosterone. Treatment of castrated males with oestradiol benzoate 3 h before administration of ^{3}H-T, results

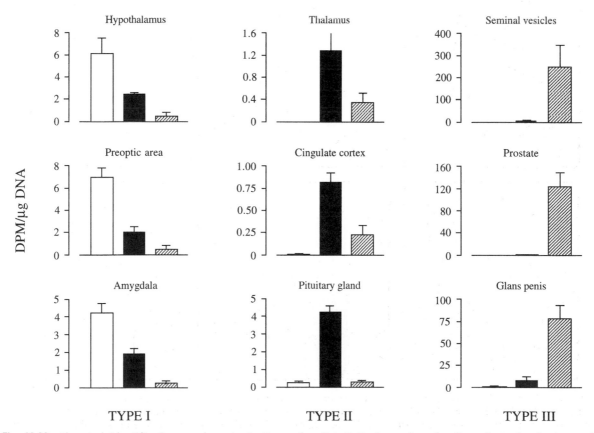

Fig. 13.26 Chemical identifications and concentrations of radioactivity in nuclear fractions from the brains and peripheral tissues of six adult, castrated male rhesus monkeys 60 min after the injection of 5 mCi ^{3}H-T. In type I tissues, most of the radioactivity was in the form of ^{3}H-E$_2$ (open columns), in type II tissues, most of the radioactivity was in the form of ^{3}H-T (closed columns), and in type III tissues, most of the radioactivity was in the form of ^{3}H-DHT (hatched columns). Vertical bars give standard errors of the means. (Redrawn from Bonsall *et al.* 1989.)

in a different pattern of nuclear testosterone binding than that seen in control subjects. Autoradiography shows that the numbers of nuclear-labelled neurons are reduced by 35% in the anterior hypothalamus and ventromedial hypothalamic nucleus and by 65% in the accessory basal nucleus and cortical nucleus of the amygdala. Reductions in some other areas (e.g. the medial preoptic nucleus and bed nucleus of the stria) are also measurable, but not statistically significant. Hence, it appears that in some areas of the male rhesus monkey's brain, such as the hypothalamus and amygdala, testosterone exerts a major proportion of its effects via aromatization and subsequent binding to E_2 receptors. Conversely, in certain other regions, such as the arcuate nucleus, lateral septal nucleus and pituitary gland, testosterone acts primarily in its unchanged state or after 5-alpha reduction to DHT. Pre-treatment of castrated males with DHTP blocks or reduces nuclear binding of ^{3}H-T in the arcuate, lateral septal, premammillary, and intercalated mammillary nuclei. At these sites T acts principally via androgen receptors. Much less blocking by DHTP pre-treatment occurs in the medial preoptic, ventromedial, and accessory basal amygdaloid nuclei, suggesting again that T acts mainly via aromatization and E_2 binding in these regions of the brain (Fig. 13.27).

The experiments reviewed above have provided much greater insights into mechanisms of regional testosterone metabolism and binding in the brain of the rhesus monkey. However, from the behavioural prospective it must be remembered that only T itself adequately restores mounting, intromission and ejaculatory activity in the castrated male, whilst DHT is ineffective in physiological dosages (in the cynomolgus macaque at least) as is exogenous E_2. The experiments of Zumpe *et al.* (1993) with the aromatase inhibitor fadrazole indicate a possible role for E_2 in sexual behaviour in the male rhesus monkey. However, this conclusion is only valid if one accepts that neuronal aromatization of T to E_2 is in some way behaviourally potent, whilst exogenous administration of E_2 is not. Given the widespread uptake of E_2, T, and DHT in the central nervous system of the rhesus monkey, it is clear that these steroids probably have much wider physiological significance than in the control of sexual behaviour alone. Regulation of pituitary gonadotrophin secretion is important in this context so that, for example, testosterone slows the frequency of pulsatile secretion of luteinizing hormone in castrated rhesus monkeys (Plant and Dubey 1984). Nor is androgen binding restricted to the brain, because DHT uptake is also widespread in the spinal cord of the rhesus monkey. Uptake occurs in the nociceptive system, as well as in the visceral and somatic

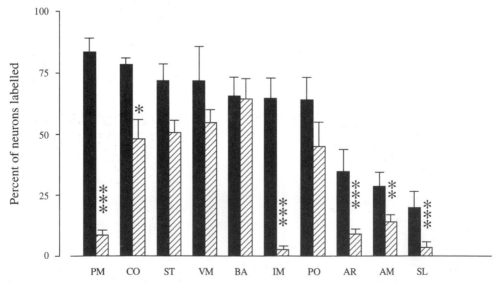

Fig. 13.27 Mean percentages of neurons labelled in a given region according to a 1% Poisson labelling criterion following the administration of ^{3}H-T to control (N=4) (closed columns) and to DHTP-pretreated (N=4) (hatched columns) castrated male rhesus monkeys. Significant reductions in DHTP-pretreated males compared with controls are given by asterisks: *P < 0.05; **P < 0.01; ***P < 0.001. Abbreviations: PM, premammillary nucleus (n.); CO, cortical amygdaloid n.; ST, bed n. of stria terminalis; VM, ventromedial n.; BA, basal accessory amygdaloid n.; IM, intercalated mammillary n.; PO, medial preoptic n.; AR, arcuate n.; AM, medial amygdaloid n.; SL, lateral septal n. Vertical bars give standard errors of the means. (Redrawn from Michael *et al.* 1989.)

motor systems (Sheridan and Weaker 1981). This has given rise to speculation concerning the possible spinal effects of DHT upon penile reflexes (as in the male rat: Gray *et al.* 1980) as well as its role in pain perception.

From what has been said so far, it is clear that the declines in sexual activity which occur after castration in male mammals reflect widespread central actions of testicular hormones, as well as associated peripheral androgenic effects. Recently, C-fos mapping studies have shown that various areas of the brain, which take up and concentrate testosterone and its metabolites, are indeed activated during copulatory behaviour. Immunocytochemical staining of the nuclear protein product (FOS) produced by an immediate-early gene (C-fos) enabled Robertson *et al.* (1991) and Baum and Everitt (1992) to examine effects of sexual stimuli on neuronal activity in specific areas of the male rat's brain. Copulation resulted in expression of FOS in neuronal cell nuclei in the preoptic area, medial amygalada, bed nucleus

of the stria terminalis, and some other brain regions (Fig. 13.28). Access to the female (without mounting her) stimulated FOS expression in these regions, but the effect was greatest in males which had achieved a series of intromissions and which had attained ejaculation. Unfortunately, this most informative technique has not been applied to studies of sexual behaviour in primates.

Another more traditional approach to studying the role of steroid-sensitive regions of the brain in sexual behaviour is to implant androgen directly into these regions after castration. Such experiments have shown that testosterone implants in the preoptic area/anterior hypothalamus activate copulatory behaviour or courtship displays in castrated rats and doves (Davidson 1966; Smith, E. *et al.* 1977; Hutchison 1978). Whilst this approach has yet to be applied to studies of primate sexual behaviour, investigators have employed lesioning and electrical stimulation or recording techniques to study central mechanisms in several monkey species. Some of

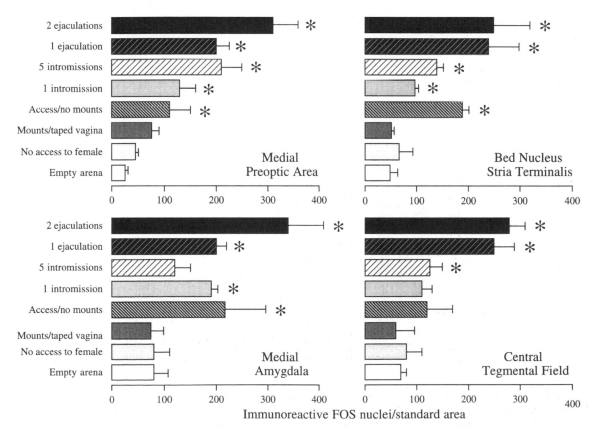

Fig. 13.28 Effects of different amounts or types of interaction with an oestrous female on the numbers of FOS-immunoreactive nuclei counted in different brain regions of male rats killed 1 hour after each treatment. Data are means (+S.E.M.) for between 3–6 males in the various treatment groups. *P < 0.005, Duncan's multiple range test for comparisons with the empty arena group. (Redrawn and modified from Baum and Everitt, 1992.)

these studies will be considered below, and compared with the much more extensive literature on brain mechanisms and sexual behaviour in non-primates.

The preoptic area and hypothalamus

It is appropriate to begin this discussion with the preoptic area and hypothalamus, since lesions in the medial preoptic area/anterior hypothalamic continuum (MPOA/AHA) cause pronounced deficits in sexual behaviour among vertebrates in general (Larsson 1979; Sachs and Meisel 1988). Larsson (1979) views the POA as 'a final common pathway that is activated by a combination of arousing stimuli from the olfactory and vomeronasal systems; the gustatory, auditive, visual, and tactile sensory systems and circulatory hormones'. Testosterone might affect the male's sexual behaviour by influencing neural inputs to the POA, as well as its responsiveness to such inputs. In the rhesus monkey, testosterone causes a slowing of electroencephalographic activity in the POA (and some hypothalamic regions) in response to penile stimulation (Mangat *et al*. 1978a, b). Rapid increases in POA responsiveness to olfactory stimuli (including urine from oestrous females) also occur in castrated rats after testosterone treatment (Pfaff and Pfaffman 1969). Androgen enhances the male's preference for the urine of oestrous females in this species (Le Magnen 1952). Testicular hormones might influence the male's perception of sexually attractive stimuli, as well as his responsiveness to such cues. The relative importance of potentially arousing stimuli must vary among male mammals. We might expect primary olfactory and vomeronasal stimuli to be important in prosimians, for instance, whereas visual stimuli might be more important in the monkeys, apes, and man.

Electrolytic lesions in the MPOA/AHA abolish intromission and ejaculation in male rhesus monkeys (Slimp *et al*. 1978) as is also the case in other mammals (Rat: Heimer and Larsson 1966; cat: Hart *et al*. 1973; dog: Hart 1974; and goat: Hart 1986). The effects of such lesions occur promptly, rather than gradually, as after castration, and are not due to impairments of pituitary–testicular function. Damage to neurons intrinsic to the MPOA/AHA is sufficient to produce these effects upon masculine copulatory behaviour. Injection of ibotenic acid, an excitotoxin which destroys neuronal cell bodies but spares fibres of passage, into the MPOA of the rat abolishes copulatory behaviour (Hansen *et al*. 1982), and similar deficits occur in male marmosets after

microinfusion of ibotenic acid into the POA/AHA junction (Lloyd and Dixson 1989).

It would be incorrect to identify the MPOA or MPOA/AHA as a 'sex centre'; not all forms of sexual activity are abolished by lesions in this area. Male rhesus monkeys continue to masturbate and ejaculate in their home cages after MPOA/AHA lesions and will bar press to gain access to a female partner even though unable to copulate. Likewise, various authors have recorded that male rats bearing MPOA/AHA lesions continue to follow and sniff females and to make abortive mounting attempts. Lesioned males fail to make the deep pelvic thrust required to attain intromission, however (Hansen 1982a). Everitt and Stacey (1987) demonstrated that male rats trained under a second-order schedule of reinforcement continued to work for access to an oestrous female after receiving excitotoxic lesions in the MPOA. Castration, by contrast, markedly reduced instrumental responses in male rats, an effect which was reversed by implanting males subcutaneously with silastic capsules containing testosterone. Hart (1974) noted that dogs with MPOA lesions sometimes mounted females and made shallow pelvic thrusting movements, but then appeared to lose interest. Lesioned cats rarely mounted and then failed to exhibit the treading and pelvic thrusting patterns which precede intromission (Hart *et al*. 1973). Male goats continued to exhibit flehmen and to mount females after MPOA lesions, although mounts were often poorly orientated (Hart 1986).

In the case of the rat, lesion studies have been interpreted as showing that this region of the brain is concerned primarily with 'the execution of genito-pelvic reflex patterns associated with copulation and less with its initiation/procurement phases' (Hansen 1982a). Lesions in the MPOA/AHA continuum thus damage what Beach (1956) referred to as the *intromission and ejaculatory mechanism* rather than the *arousal mechanism* in the male. Whilst I understand this argument I do not entirely agree with it. Arousal and orientation to the female are also affected by these lesions, hence the failure of male dogs, cats, goats, or rats to mount females in a correct position or to persist in their attempts to copulate. Penile anaesthesia or dorsal penile nerve transection also reduces the ability of the male to perform the reflexive deep pelvic thrust required for intromission (marmoset: Dixson 1986a, 1988), yet such males are highly aroused sexually, mount frequently, and persist in their (often fruitless) attempts to gain intromission. It might be more accurate to view lesions in the MPOA/AHA as severing linkages between arousal and copulatory mechanisms, rendering the male less able to respond appropriately to sexual stimuli provided by the female, as well as unable to execute correctly patterns of pelvic thrusting and intromission (Dixson 1983c). Some experiments involving hypothalamic lesions in male marmosets might serve to amplify this argument. Figure 13.29 shows thermal lesions situated beneath the anterior commissure, damaging principally the anterodorsal and paraventricular hypothalamic nuclei in four adult male marmosets. These males (Nos. 4–7) formed part of an experimental group of 10 animals which were lesioned in the POA, AHA, and medial hypothalamus in order to study effects upon precopulatory and copulatory behaviour (Lloyd and Dixson 1988). Lesions beneath the anterior commissure in males 4–7 were most effective in producing deficits in sexual behaviour (Fig. 13.30). Males with more rostral or caudal lesions were not so severely affected. These lesions are not equivalent to MPOA lesions in male rats, but they are situated in the continuum between the POA and AHA and produce similar behavioural effects to those recorded by Slimp *et al.* (1978) in male rhesus monkeys. Examination of Fig. 13.30,

reveals that males 4–7 showed decreases in precopulatory, as well as copulatory, aspects of sexual behaviour. Males 5, 6 and 7 showed significantly less anticipatory penile erections before introduction of a female partner into the test cage. Whilst all males continued to investigate females after placement of lesions, frequencies of anogenital investigation were reduced—significantly so for males 6 and 7. Mounting was greatly reduced or abolished and none of the four males succeeded in intromitting during post-operative tests. Only male 5 was observed to make pelvic thrusting movements. Hence, in these cases there is a lack of appropriate interest and arousal, as well as failure to thrust correctly and to attain intromission. Nor are effects of hypothalamic lesions limited to deficits in sexual behaviour in male marmosets. Aggression towards strange male intruders also decreases as a result of these lesions (Lloyd and Dixson 1989). Most interestingly, whilst male marmosets normally carry their twin offspring (Ingram 1977), lesioned males show profound deficits in carrying behaviour, as well as copulatory activity, during the female's post-partum lactational period (Fig. 13.31). Lesioned males are not debilitated in

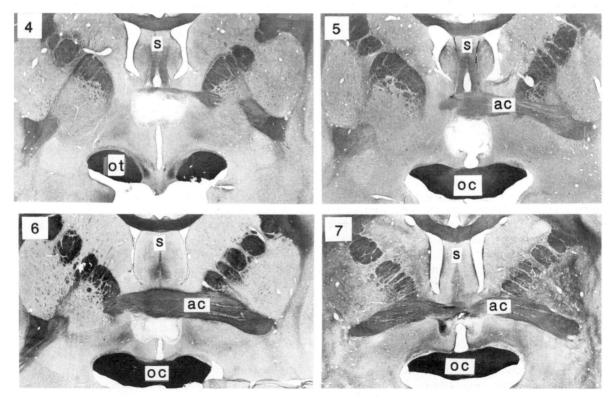

Fig. 13.29 Transverse sections of the brains of four adult male common marmosets (nos. 4, 5, 6, and 7) to show thermal lesions of the anterior hypothalamus. Effects of these lesions upon sexual behaviour are discussed in the text, and shown in Fig. 13.30. AC=anterior commissure; OC=optic chiasm; OT=optic tract; S-septum. (After Lloyd and Dixson 1988.)

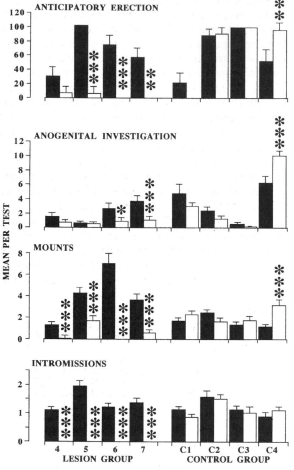

Fig. 13.30 Effects of thermal lesions of the anterior hypothalamus (males 4–7) or sham lesions (males C1–C4) upon anticipatory erections, anogenital investigation of the female, mounts, and intromissions by male common marmosets. Histograms show mean (+S.E.M.) frequencies per test for each male pre- (solid bars) and post-lesion or sham (open bars). *P < 0.05; **P < 0.01; ***P < 0.001, Mann-Whitney U test. (Redrawn and modified from Lloyd and Dixson 1988.)

any way and twin offspring frequently transfer from their mothers and attempt to contact their fathers. However, males bearing hypothalamic lesions do not respond to the distress calls of infants, and dislodge them from their bodies should they succeed in transferring (Lloyd and Dixson, unpublished observations). These observations indicate, that although male marmosets continue to show normal social behaviour in some contexts (e.g. in allogrooming of female partners: Lloyd and Dixson 1988), they are less able to respond appropriately to conspecifics in sexual, aggressive, or parental contexts as a result of receiving hypothalamic lesions and especially lesions

in the rostral hypothalamus.

The importance of the POA during the *pre-copulatory* phase of sexual activity is also indicated by electrophysiological measurements of neuronal activity in male macaques before and during copulation. Yoshimatsu (1983) and Oomura *et al.* (1984) made single-unit recordings from hypothalamic neurons in macaque males. An operant paradigm was used to measure sexual arousal. The male was restrained within sight of a female and could gain access to her by pressing a bar. During this operant task POA neurons showed a high firing rate *which decreased after acquisition of the female and during copulation.* Some POA neurons showed a transient increase in firing in relation to penile erection before copulation occurred. By contrast, *neuronal activity in the dorsomedial and posterior hypothalamus increased during mounting, intromission, and pelvic thrusting* (Fig. 13.32). Yoshimatsu (1983) therefore proposed that a *dual control mechanism* might exist in the hypothalamus; the POA being concerned primarily with sexual arousal and precopulatory behaviour whilst the dorsomedial and posterior hypothalamus constitutes part of the copulatory mechanism. It is, of course, possible that male monkeys may show some important differences from rats in the preoptic/hypothalamic regulation of sexual behaviour. In the male rat, electrical stimulation of a small region (0.43 mm^3) encompassing the POA causes exaggerations of copulatory behaviour. Males ejaculate more frequently and latencies to ejaculation and interejaculatory intervals are shortened (Merari and Ginton 1975). Stimulus-bound effects upon sexual behaviour have also been reported in male rhesus monkeys (Perachio *et al.* 1979). However, the effective stimulation sites differed from those described for the rat. When male rhesus monkeys were tested with oestradiol-treated females, stimulation of the POA caused males to mount and thrust for longer than during normal copulation. No ejaculations occurred, despite earlier reports of erection and ejaculation during electrical stimulation of the POA in restrained males (Robinson and Mishkin 1966). Perachio *et al.* (1979) found that stimulation of the dorsomedial hypothalamus of the male rhesus monkey evoked complete copulatory patterns, involving increased mount durations, greater numbers of pelvic thrusts per mount, and higher thrust rates and ejaculatory frequencies (Fig. 13.33). In isolated males, stimulation of the anterior, ventromedial, and posterior hypothalamus evoked penile erections yet such stimulation had no effect upon sexual behaviour when males were paired with females. Similarly, earlier studies on the cerebral representation of penile erection in isolated, restrained squirrel monkeys by Maclean and Ploog (1962) implicated the POA, paraventricular nucleus, supraoptic nucleus, lateral hypothalamus, and various extra-hypothalamic sites in erection. It is doubtful whether all these sites are relevant to integrated patterns of copulatory behaviour, however, particularly because squirrel monkeys make such extensive use of penile erection for social communication in non-sexual contexts (see page 152 and Fig. 6.8).

The evidence reviewed above, implicates the dorsomedial nucleus of the hypothalamus and the posterior hypothalamus in the integration of copulatory thrusting patterns and ejaculation in male monkeys. The POA, by contrast, plays some greater role in precopulatory orientation towards the female and in the transition from precopulatory behaviour to mounting of the female. Whether these same mechanisms are activated during sociosexual mounts by males or females is unknown. A major output *from the POA* is provided by the medial forebrain bundle

(MFB), a complex pathway containing ascending as well as descending components. Descending connections from the POA, which course thorough the lateral hypothalamus, are most important and Halasz knife cuts in this region abolish, or greatly reduce, copulation in the male rat (Paxinos and Bindra 1971). No comparable experiments have been reported for male primates. Ascending systems, including aminergic inputs to the hypothalamus and other regions, are damaged by MFB knife cuts; this complicates interpretation of the behavioural effects. It is interesting, therefore, that Kendrick (1983) has shown that the refractory periods of neurons projecting *from* the POA into the MFB increase after castration in the male rat. Testosterone propionate reverses this effect after 5 days of treatment; this time-course of testosterone's effect on refractory periods correlates with the time required for TP to restore mounting and intromission in castrated rats (Kendrick 1983). Various authors have noted

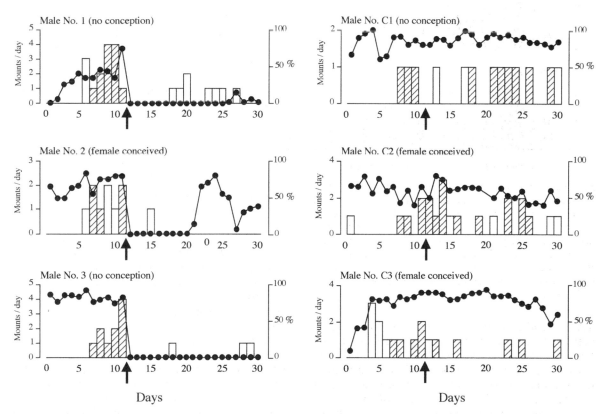

Fig. 13.31 Effects of thermal lesions of the anterior hypothalamus upon sexual and paternal behaviour in male common marmosets. Left: Data on three family groups of marmosets in which breeding males were lesioned on day 11 (arrowed) after their female partners had given birth to twins. Data are the daily frequencies of paternal care (% time male spends carrying the twins=●) and sexual behaviour (bars=mounts; cross-hatched bars=days on which ejaculatory mounts occurred) for each of the three males. Right: Data on three family groups in which breeding males received a sham-operation (arrowed) on day 11 during the females' post-partum period. (Unpublished observations by Lloyd and Dixson.)

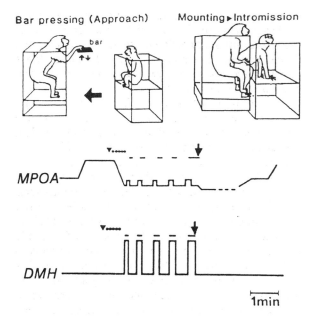

Bar pressing (Approach)   Mounting ▸ Intromission

MPOA

DMH

1min

Fig. 13.32 Upper: Experimental arrangement by which a male macaque bar-presses to activate a motor drive which brings the female's cage towards him. Each bar press causes the female to move 5 cm; ultimately the pair are close enough for mating to occur. Lower: Schematic summary of firing rates of medial preoptic area (MPOA) and dorsomedial hypothalamic (DMH) neurons in the male macaque during bar pressing and copulation. ▼ = shutter opened so that male and female are in visual contact; ●●● = male bar-presses to move female's cage nearer; short dashes = mounts with intromission and pelvic thrusting; vertical arrow = ejaculation. Note that MPOA neurons fire mainly during the pre-copulatory phase, when the male is bar-pressing and immediately before each mount. DMH neuronal activity, by contrast, is lowest during the pre-copulatory phase and increases markedly during intromission/pelvic thrusting sequences. (After Oomura *et al.* 1984.)

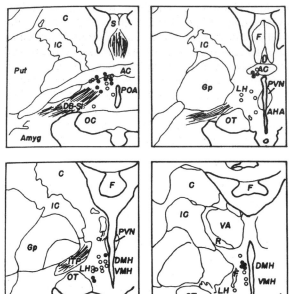

Fig. 13.33 Electrical stimulation of the hypothalamus and sexual behaviour in the male rhesus monkey (*Macaca mulatta*). Closed circles represent locations of electrode placements from which mounting behaviour was elicited. Open circles represent stimulation sites from which mounting was not elicited. The locations of the stimulation sites are shown in an estimated position relative to the closest stereotaxic plane of reference. Abbreviations: AC, anterior commissure; Amyg, amygdaloid complex; C, caudate nucleus; DB-SI, diagonal band-substantia innominata; DMH, dorsomedial nucleus; F, fornix; Gp, globus pallidus; IC, internal capsule; ITP, inferior thalamic peduncle; LH, lateral hypothalamus; OC, optic chiasm; OT, optic tract; POA, preoptic area; Put, putamen; PVN, paraventricular nucleus; R, reticular nucleus; S, septum; VA, ventral anterior nucleus of the thalamus; VMH, ventromedial nucleus. (After Perachio *et al.* 1979.)

that some days must elapse before sexual behaviour is stimulated by implantation of TP, or testosterone, into the POA of castrated male vertebrates (leopard frog: 3–6 days, Wada and Gorbman 1977a; capon: 4–7 days; Barfield 1969; rat: 5–6 days; Davidson 1966). Hutchison (1978) suggested that different populations of neurons which influence sexual responses may differ in their sensitivity to testosterone. Kendrick's (1983) observations on POA efferents to the MFB may be cited in support of this view. Kendrick *et al.* (1981) also studied testosterone-sensitive neurons in the stria terminalis, projecting from the corticomedial amygdala to the POA. Castration of male rats resulted in lengthening of the refractory periods of these neurons and TP treatment for 7 days reversed this effect. This time-

course correlated with restoration of the ejaculatory response by TP in the castrated rat. Functional relationships between the amygdala and POA are considered in greater detail below.

Finally, it is important to note that the POA has extensive neuronal connections and influences the autonomic nervous system, as well as playing a role in pituitary regulation. Pfaff (1980) has therefore cautioned that POA lesions can result in pulmonary oedema, due to systemic vasoconstriction, and that disruption of parasympathetic mechanisms and failure of penile erection could account for deficits in sexual behaviour. However, the effects of POA lesions cannot be explained solely on the basis of ill health or failure to achieve erection (and hence intromission). Male rhesus monkeys with

POA/AHA lesions continue to exhibit erections (Slimp *et al.* 1978). Moreover, the POA plays a vital role in the integration of sexual arousal and emission of sperm in species which lack a penis and which employ external fertilization (e.g. male leopard frogs: Wada and Gorbman 1977b; sunfish: Demski *et al.* 1975; salmon: Satou *et al.* 1984). The involvement of the POA in masculine sexual behaviour is a phylogenetically ancient phenomenon, pre-dating the evolution of the mammalian penis.

Influences that impinge upon hypothalamic mechanisms

Amygdala and stria terminalis

Bilateral lesions of the temporal lobes, which damage the amygdala and overlying regions of the cortex, induce a condition known as the Klüver–Bucy syndrome. When lesioned in this way rhesus monkeys become docile, mouth objects, and exhibit aberrant mounting patterns (Klüver and Bucy 1939). Equivalent effects of gross temporal lobe damage have been described in the cat (Green *et al.* 1957) and in man (Terzian and Dalle Ore 1955). In the cat, damage to the temporal cortex alone, rather than to the amygdala, produced changes in sexual behaviour (Green *et al.* 1957). Male cats attempted to mount inanimate objects, or rats and guinea pigs, indicating a lack of discrimination in the choice of sex partners rather than hypersexuality in these subjects (Larsson 1979). In Chapter 12 (see pages 376–9), the role of the temporal cortex and amygdala in the control of proceptivity in female primates was discussed in some detail. Many of the arguments advanced to account for cortical processing of sensory inputs and the interfacing of such inputs with limbic mechanisms are also applicable to the male. In monkeys, it is known that inputs received by the primary (occipital) cortex are further processed in the inferotemporal cortex, where specialized neurons encode for facial features and body postures. The inferotemporal cortex has reciprocal connections with the underlying amygdala and the amygdala has reciprocal connections with the POA and hypothalamus, via the stria terminalis. In some fashion this two-way traffic between the cortex, amygdala and hypothalamus results in the generation of appropriate emotional responses to relevant sensory inputs. We might imagine a male monkey responding to complex visual cues from a proceptive female, such as a sexual presentation coupled with a facial display and eye-contact. This might be accompanied by a non-behavioural cue, such as a swollen sexual skin. Nor is the amygdala limited to receipt of visual inputs; auditory, tactile, and olfactory stimuli all have access to the amygdala and hence (via its numerous interconnections) to other parts of the limbic system. Destruction of the temporal lobes damages the male's ability to analyse and respond appropriately to the total gestalt of these sensory inputs. Viewed in this way, it is scarcely surprising that gross temporal lobe damage or amygdalectomy severely comprises social behaviour, and not just sexual behaviour, in macaques (Horel 1988).

Neurons in the corticomedial division of the amygdaloid nuclei project via the stria terminalis to the POA and hypothalamus. The retrocommissural branch of the stria includes the bed nucleus of the stria terminalis and projects to the POA/AHA. The supracommissural branch of the stria additionally projects to the ventromedial hypothalamus and ventral premammillary nucleus (Nauta and Haymaker 1969). Since the corticomedial amygdala and bed nucleus of the stria terminalis are regions which take up and concentrate testosterone, experiments have been conducted to examine the role of these structures in masculine copulatory behaviour. Virtually all such studies have involved rodents, rather than primates. Bilateral electrolytic lesions of the corticomedial amygdala, or bed nucleus of the stria, cause deficits in copulatory behaviour (lengthening of the mount and ejaculatory latencies or failure to ejaculate in male rats: Harris and Sachs 1975; Emery and Sachs 1976). Similar effects occur if knife cuts are made dorsal to the POA, thus severing connections with the amygdala and also the hippocampus (Szechtman *et al.* 1978). Baum *et al.* (1982) report that implantation of dihydrotestosterone propionate into the corticomedial amygdala of castrated, oestradiol-primed rats stimulates ejaculatory responses. Previous work by Kendrick and Drewett (1979) had shown that certain neurons projecting from the corticomedial amygdala into the stria terminalis are testosterone-sensitive. Perhaps some of these neurons are responsive to the 5-alpha-reduction product of testosterone, so that DHT, as well as oestradiol, exerts some effect upon sexual activity in the male rat.

In the rat, olfactory stimuli are probably of much greater importance in activating neurons in the corticomedial division of the amygdala, and hence in influencing the POA/AHA, than they are in higher primates. The corticomedial amygdala receives projections from the accessory olfactory bulb and from the main olfactory system. Olfactory bulb lesions have variable effects upon sexual behaviour in male rats, but have been reported to result in inconsistent mounting activity and lengthened ejaculatory latencies (Heimer and Larsson 1967). In the marmoset, severing the olfactory bulbs does not reduce sexual activity in the male but it does reduce his ability to co-ordinate ejaculatory frequency with respect to the female's peri-ovulatory period (Dixson 1992, 1993a). Most interestingly, Baum and Everitt (1992) have shown that olfactory inputs, possibly of vomeronasal origin, influence C-fos expression in the corticomedial amygdala of the male rat during sexual activity. Thermal lesions were made unilaterally in the olfactory peduncle and lesioned males were allowed to mount oestrous females. C-fos expression was increased in the corticomedial amygdala ipsi-lateral to the olfactory peduncular lesion, provided that males mounted but did not intromit and ejaculate during sex tests (lidocaine anaesthesia of the glans penis was employed to reduce the likelihood that intromission could occur). Whilst such effects might also occur in certain primates, and in particular various prosimians and New World monkeys in which the vomeronasal system is functional, it is probably more important to consider effects of visual inputs upon amygdaloid activity in higher primates. A report by Adolphs *et al.* (1994) indicates that the human amygdala plays an essential role in recognition of the emotional content of facial expressions. A female patient suffering from nearly complete bilateral destruction of the amygdala (due to Urbach–Wiethe disease) provided Adolphs *et al.* with a rare opportunity to examine responses to facial expressions indicating happiness, surprise, fear, anger, disgust, and sadness. The patient showed reduced ability to recognize multiple emotions in a single facial expression and a marked inability to recognize fearfulness. Ability to recognize individual identity was not impaired however, indicating that this aspect of facial communication is separable at the neural level from the analysis of the emotional content of facial expression.

The septum

Despite the fact that testosterone is taken up and concentrated in the lateral septum of male primates, very little is known about the functions of the septum in relation to primate sexual behaviour. Electrical stimulation of the septum induces penile erection in the squirrel monkey (Maclean and Ploog 1962), but in view of the peculiar penile display behaviour of this species (Ploog 1967), it would be incorrect to interpret septal mechanisms solely in terms of sexual response. In the rat, septal lesions do not affect masculine patterns of copulatory behaviour (Goodman *et al.* 1969). However, septal lesions do result in a remarkable disinhibition of feminine, lordotic, responses in the male rat. Lordo-

sis, in response to oestrogen stimulation, is potentiated in gonadectomized male rats by septal lesions (Nance *et al.* 1975a). These authors found that a similar potentiation of lordosis by septal lesions also occurred in the female rat (Nance *et al.* 1975b). However, the inhibitory mechanism appears to be stronger in the male and is not disinhibited by castration and oestrogen-treatment unless the septum is damaged. Yamanouchi and Arai (1985) further explored this inhibitory mechanism in the castrated male rat by means of Halasz knife cuts made dorsal to the POA, severing its connections with the septum. Males subjected to such anterior roof deafferentation (ARD) displayed proceptive behaviour (ear-wiggling and hopping) and lordosis when treated with oestradiol and progesterone. Yamanouchi and Arai postulate that a forebrain system exists in the male rat that inhibits the expression of proceptive and receptive behavioural patterns. This inhibitory system also exists in the female brain, but is much more readily disinhibited in the female by the action of oestrogen and progesterone during oestrus. These inhibitory effects impinge upon the POA and hypothalamus. In this regard it is of interest that lesions of the POA potentiate lordosis in the rat (Powers and Valenstein 1972). Such lesions presumably remove an inhibitory influence which suppresses the ventromedial hypothalamic mechanisms concerned in the control of the lordotic reflex. Yamanouchi and Arai (1985) suggest that one of the organizing actions of testicular hormones during neural development in the male rat may be to strengthen forebrain mechanisms inhibitory to the expression of feminine-type sexual behaviour. By this means behavioural defeminization of the brain might occur.

Do such inhibitory mechanisms occur in the brains of primates? In Chapter 10, which dealt with sexual differentiation of the brain, it was argued that androgen masculinizes, but does not necessarily defeminize the developing brain of anthropoids. Female rhesus monkeys exposed to testosterone propionate *in utero*, exhibit proceptivity and receptivity in adulthood (see pages 283–4). The anthropoid primates do not employ reflexive lordotic postures during copulation and the neural mechanisms underlying sexual receptivity are, therefore, probably substantially different from those which occur in rodents. It is entirely possible that in prosimians the ARD operation might disinhibit receptive mechanisms in males in the same fashion as described for the male rat. Female monkeys and apes *do display proceptive behaviours*, however, and we do not know if a forebrain inhibitory mechanism regulates their expression in either sex. As in so many areas of primate sexuality, we are faced with an absence of the necessary experiments to explore fundamental questions relating to behaviour.

The midbrain

In the last chapter it was argued that neural mechanisms which link motivation to action probably share common properties for a variety of behaviours (Mogenson *et al.* 1980). Pathways from the POA and hypothalamus, situated in the medial forebrain bundle, terminate in (or traverse) the ventral and dorsolateral tegmentum in the midbrain (Conrad and Pfaff 1975). These pathways play a crucial role in the control of maternal behaviour and masculine patterns of copulatory behaviour in the rat (Numan 1994; Meisel and Sachs 1994) and might be implicated in the control of proceptivity in female primates (see pages 376–80 and Fig. 12.26). Bilateral lesions in the dorsolateral tegmentum have effects upon copulatory behaviour which closely resemble those seen after MPOA lesions in male rats (Hansen and Gummersson 1982). Such males follow and sniff females, for example, but fail to copulate. Males with a unilateral tegmental lesion combined with a contralateral MPOA lesion display similar deficits in copulatory behaviour (Brackett and Edwards 1984). It is important to recall, however, that the medial forebrain bundle contains fibres both to and from the MPOA, so that lesions might affect efferents from the tegmental area as well as MPOA efferents to the midbrain. Excitotoxic lesions in the central tegmental field of the male rat, made using quinolinic acid which selectively damages neuronal cell bodies, block C-fos activation in the MPOA during mounting and intromission (Baum and Everitt 1992). These observations indicate the importance of the central tegmental field in processing somatosensory information from the genitalia, and in conveying such inputs to the MPOA via the medial forebrain bundle. Likewise, electrical stimulation of the central tegmental field does not significantly affect the precopulatory behaviour of the male rat, but does influence the copulatory mechanism. Males require fewer mounts and intromissions to achieve ejaculation, perhaps because sensory inputs from the penis are facilitated by electrical stimulation of the central tegmental field (Shimura and Shimokochi 1991).

The neural control of pelvic thrusting

In the last chapter, I discussed the neural mechanisms which are believed to link motivation to ac-

tion for various behavioural patterns, with special reference to the control of proceptive sexual behaviour in female primates. Information relayed from the ventral tegmental area via mesolimbic dopaminergic neurons reaches the ventral striatum, which interfaces with the motor control systems required for voluntary movement. The same schema, described on pages 376–80 and represented in Fig. 12.26, applies to the male during that phase when he approaches the female, engages in precopulatory communication with her and prepares to mount. The motor cortex of the brain plays an executive role in controlling the speed, force, and precision of such voluntary movements. Human patients who have suffered stroke damage to the precentral gyrus (the seat of the motor cortex: Fig. 13.34) suffer from impairments involving precision movements (e.g. of the hands), but are still able to construct 'motor programs' for their execution. The somatotropic organization of the primary and supplementary motor cortex in a macaque is shown in Fig. 13.35. Large areas are devoted to the control of the limbs, fingers, and toes, so essential for locomotion, feeding, social, and sexual activities. From the precentral gyrus the fibres of the pyramidal tract descent to the spinal cord, also innervating the striatum, thalamus, pontine system, and midbrain (Fig. 13.36). In addition, the various regions of the motor cortex communicate with many other cortical areas, often via reciprocal connections (Hepp-Reymond 1988). The neural mechanisms governing voluntary movements, which enable a male monkey to mount and grasp the female in an appropriate copulatory position, are immensely complicated, since they involve postural adjustments in response to tactile and proprioceptive feedback, as well as the smooth operation of motor programmes. A simplified model of pathways involved in the planning, programming, and execution of voluntary movements is reproduced here in a modified form (Fig. 13.37).

In the context of masculine copulatory behaviour, it is important to ask whether the pelvic thrusting movements required to gain intromission, and to ensure that ejaculation occurs intravaginally, are controlled by voluntary or by reflexive neural mechanisms? Pelvic thrusting movements are often exceedingly rapid. In the rat, for example, each pre-intromitted pelvic thrust takes just 50 ms to perform, whilst in the rabbit each thrust lasts approximately 80 ms (Beyer and Gonzalez-Mariscal 1994). It is unlikely that these rapid and regular pelvic oscillations result from voluntary motor control. Beyer and Gonzalez-Mariscal propose a reflexive model to account for pelvic thrusting, based upon neural models for the generation of rhythmic motor patterns in general (Delcomyn 1980). In this schema, command neurons trigger the motor pattern and communicate via interneurons which generate rhythmic activity (the central pattern generator or CPG) in the motorneurons controlling the necessary muscle groups. In the rat, command neurons are believed to be situated in the preoptic region and the same might be true of other mammals. We have seen, however, that in male macaques neurons in the dorsomedial and posterior hypothalamus are active during copulatory thrusting movements (Yoshimatsu 1983). The CPG regulating rapid thrusting is believed to reside in the spinal cord, as is the case for the CPG which controls rhythmic limb movements during locomotion (Carew 1985). Pelvic thrusting responses can sometimes be elicited in spinal subjects (dog: Hart, B. 1968; Man: Comarr and Gunderson 1975). Testicular hormones apparently have little or no effect upon these reflexive patterns, except for the rabbit, where castration results in a decrease in pelvic thrusting frequency and amplitude. Testosterone propionate restores mounting behaviour but a longer period of hormone treatment is required to reinstate the normal thrusting pattern (Contreras and Beyer 1979; Beyer and Gonzalez-Mariscal 1994).

It is noteworthy that pelvic thrusting movements in the rabbit and rat are suppressed once the male attains an intromission. Sensory feedback from the penis during the pre-intromission phase of the mount enables the male to locate the vaginal orifice, and to institute the deep pelvic thrust required for insertion. In the rabbit, a single intromission results in ejaculation, whilst in the rat a series of brief intromissions occurs. However, among the primates, all species yet studied continue to make pelvic thrusts during intromission. The intromitted thrusts are slower and deeper than those which occur before insertion (see Chapter 5, pages 116–18). In the marmoset, for instance, males thrust consistently at a rate of 4.87 times s^{-1} before intromission, but this decreases to 1.54 thrusts s^{-1} when intromission is attained (Kendrick and Dixson 1984a). Males of some primate species may alter the thrusting rate during intromission, or remain immobile for periods in between bouts of thrusting movements (e.g. in the greater galago: Eaton *et al.* 1973a; Orang-utan: Nadler 1977; Woolly spider monkey: Milton 1985a). Sensory feedback from the penis is crucial for the co-ordination of patterns of intromitted pelvic thrusts. If the dorsal penile nerves are transected,

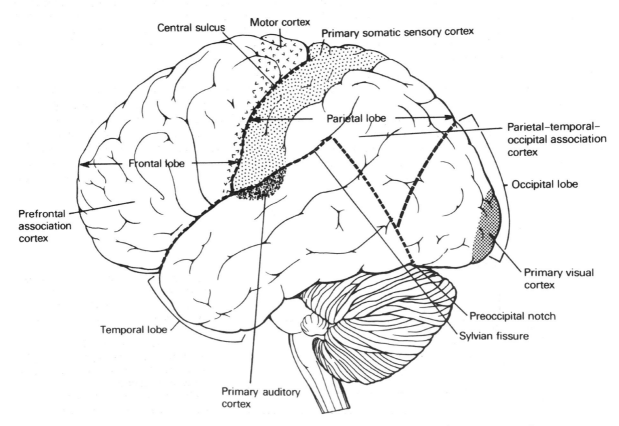

Fig. 13.34 Major divisions of the human cerebral cortex, in lateral view. (After Kelly 1985.)

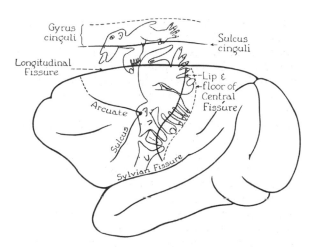

Fig. 13.35 Somatotropic organization of primary and supplementary motor areas in the macaque brain. The central and longitudinal fissures are 'opened out', with a dotted line representing the floor of the fissure. (From Hepp-Reymond 1988; adapted from Woolsey *et al.*)

thrusting becomes ataxic during intromission and ejaculation rarely occurs (rhesus monkey: Herbert 1973). By contrast, the rapid and shallow pelvic thrusts which precede intromission do not require tactile feedback from the penis for their co-ordination (Marmoset: Dixson 1986a, 1988). The transition to a slower, deeper pattern of pelvic thrusting during intromission probably involves a strong measure of voluntary motor control, rather than a purely reflexive control by spinal CPG interneurons.

Sensations arising in the penis during copulation are conveyed to the brain via two ascending somatic sensory systems—the dorsal column-medial lemniscal system and the anterolateral system (Fig. 13.38). The dorsal column (DC) system transmits tactile

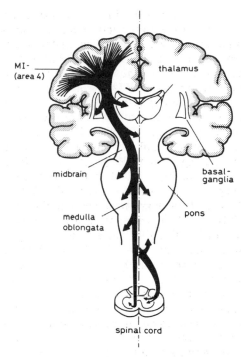

Fig. 13.36 Diagram showing fields of origin, terminal regions, and components of the pyramidal tract (PT). The cortical connections to subcortical structures are made partly by collaterals of PT axons and partly by parapyramidal axons. (After Hepp-Reymond 1988; adapted from Phillips and Porter.)

and proprioceptive inputs, including impulses from the penile mechanoreceptors, as well as proprioceptive feedback from the male's muscles and joints as he performs pelvic thrusts. The anterolateral (AL) system is concerned primarily with nociceptive (painful) and thermal sensations, although it deals also with crude tactile inputs. Both systems are

Programming   Execution

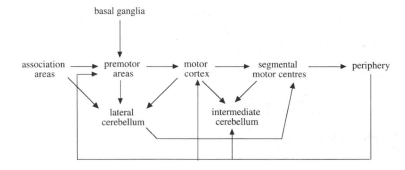

Fig. 13.37 Schematic diagram of information flow between cortical areas and subcortical structures during the planning, programming, initiation and execution of voluntary movements. On the left: cortical connections and loops participating in programming. On the right: structures involved in execution of movements. (Redrawn from Hepp-Reymond 1988; based on Allen and Tsukahara.)

relevant in sexual contexts, since the penis receives thermal and mechanical stimulation during intromission. Although the anatomical arrangements of the DC and AL systems exhibit some important differences, both convey somatosensory inputs to the ventral posterior lateral nucleus and posterior nuclear group of the thalamus and thence to the somatic sensory cortex (areas SI, SII and posterior parietal cortex) for further processing. The AL system also projects to the midbrain tectum and to the brainstem reticular formation. As regards processing of thermal stimuli, a distinct nucleus has now been identified in the posterior thalamus of the cynomolgus macaque and of man which is specific for nociceptive and temperature sensation (Craig *et al*. 1994). Within the somatic sensory cortex, sensations arising from the various areas of the body are represented in an orderly fashion; a *homunculus* is produced in

which receptor-rich regions are particularly well represented (e.g. the face, tongue, and index finger in man: Fig. 13.39). Sensations from the penis are processed within an area of the somatosensory cortex situated in the sagittal fissure which separates the two hemispheres, as is shown also in Fig. 13.39. Electrophysiological recording studies conducted by Mountcastle and by Kaas and his co-workers, have shown that the somatosensory cortex is organized into vertical columns and that each neuron in a column of cells is activated by the same type of sensation, arising from virtually the same receptor sensory fields in the periphery (Fig. 13.40). The cortex also consists of horizontally arranged histological layers; classically six layers have been recognized but more might occur in some regions of the cortex. Each cortical layer projects to (or receives inputs from) different regions of the brain. Layer

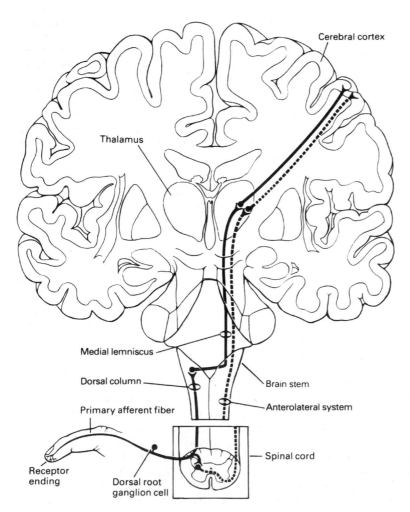

Fig. 13.38 The organization of sensory pathways in the human central nervous system. The dorsal column-medial lemniscal system (solid line) is concerned primarily with tactile sensations. The anterolateral system (broken line) mediates painful sensations and (to a much lesser extent) tactile stimuli (After Martin 1985.)

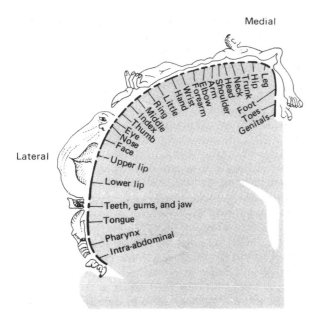

Fig. 13.39 Projection of the body surface onto a transverse section of the postcentral gyrus of the human parietal cortex. Note that tactile stimuli from the genitalia are represented in somatosensory cortex situated in the mid-sagittal fissure which separates the two hemispheres. (From Kandel 1985; adapted from Penfield and Rasmussen.)

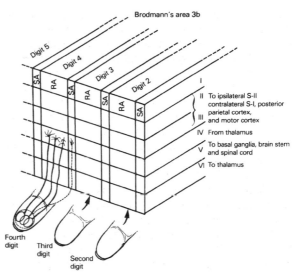

Fig. 13.40 Vertical and horizontal organization of somatosensory cortex. Vertically, the cortex is organized into columns, running from the surface to the white matter. Horizontally, the cortex is organized into layers, receiving inputs from certain brain regions and projecting to other regions. This example deals with processing of rapidly adapting (RA) and slowly adapting (SA) tactile receptor inputs by Brodmann's area 3b. The importance of tactile feedback from the penis for the timing of pelvic thrusting movements, and the control of erection, intromission, and ejaculation is discussed in the text. (After Kandel 1985; adapted from Kaas *et al.*)

IV, for instance, receives inputs from the thalamus whilst layers II and III project to a number of cortical areas (Fig. 13.40). Projections to the posterior parietal cortex are especially important, since in this area tactile sensations, proprioceptive feedback, and visual inputs are integrated to provide the animal with an overall perception of its bodily position in relation to the surrounding world. Likewise, transcortical communication with the temporal lobe is important for the integration of sensory inputs which are then processed by the amygdala and hypothalamus in order to determine their emotional significance.

This brief description of sensory processing by the brain is relevant to discussions of copulatory behaviour because it is likely that the intravaginal pelvic thrusting rhythm is strongly influenced by cortical sensory-motor mechanisms, and not solely by a spinal central pattern generator. We may hypothesize that the rapid, shallow, pelvic oscillations preceding an intromission are mediated by spinal mechanisms. However, with the attainment of intromission a barrage of sensory inputs is received from the penis, involving movement, pressure changes, and thermal stimulation of the glans and shaft. The

male shifts to a slower, deeper pattern of pelvic movements, sometimes pausing, or altering position slightly. These manoeuvres depend upon processing of tactile and proprioceptive inputs by the brain and fall under the executive control of the cortex.

In an earlier section of this chapter it was noted that male rhesus monkeys with lesions in the POA/AHA continue to masturbate and ejaculate, despite their failure to mount females during pair-tests (Slimp *et al.* 1978). Some authors have cited these findings as evidence that lesioned males continue to exhibit sexual interest and arousal. However, this interpretation may be incorrect. The POA/AHA lesioned males studied by Slimp *et al.* masturbated whilst alone in their home cages, rather than in response to female proximity during pair-tests. From the preceding discussion it is evident that the spinal–thalamic—cortical pathways required for perception of penile stimulation are still operative in such lesioned males. Assuming that masturbation is perceived as a pleasant sensation by isolated male monkeys, there is no reason to assume that POA/AHA lesions should affect this be-

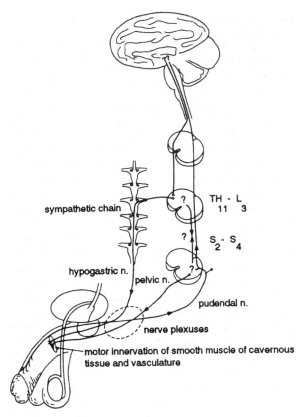

sympathetic chain

TH - L
11   3

S₂ - S₄

hypogastric n.

pelvic n.

pudendal n.

nerve plexuses

motor innervation of smooth muscle of cavernous
tissue and vasculature

Fig. 13.41 Schematic representation of the sympathetic, parasympathetic and somatic innervation of the human penis. The roles of these pathways in the control of penile erection are described in the text. (From Wagner and Kaplan 1993; after Wagner and Green 1981.)

haviour, or that it is strictly equivalent to human masturbation which is often preceded and accompanied by erotic fantasy.

The neural control of erection and ejaculation
The innervation of the penis

Erection and detumescence of the penis result from a complex interaction between parasympathetic, sympathetic, and somatic neural mechanisms. The anatomical arrangements and functions of the various neural pathways differ between species (Creed *et al.* 1991; Andersson and Wagner 1995). In Man (Fig. 13.41), the parasympathetic supply arises in the 2nd–4th sacral segments of the spinal cord (Eckhard 1869; Brindley *et al.* 1986), travels in the pelvic nerves to the pelvic plexus, and thence in the cavernous nerves to the penis itself. This parasympathetic supply plays a crucial role in causing penile erection, owing to dilatation of the penile arteries

and increased blood flow into the corpora cavernosa (dog: Dorr and Brody 1967; man: Lue and Tanagho 1988; Newman and Northup 1981). The somatic neural supply of the penis involves the pudendal nerves which supply motor fibres to the striated penile musculative, as well as receiving sensory fibres from the dorsal penile nerves. We have already considered the functions of the striated penile muscles in Chapter 9 (see pages 263–5), but a brief resumé is necessary here. In the rat the ischiocavernosus muscles are responsible for the *flip* reflexive movements of the penis, which assist the male in gaining intromission, whereas the bulbocavernosus muscles control the *cupping* or flaring of the glans penis, which occurs at ejaculation and which helps to pack and seal the copulatory plug against the cervix of the female (Sachs 1982; Hart and Melese-D'Hospital 1983). These penile reflexes are controlled by motor neurons whose cell bodies are situated in the lumbar region of the spinal cord and which are androgen-sensitive (Breedlove and Arnold 1980, 1981). In man, motor neurons innervating the perineal musculature originate in a nucleus situated in the sacral spinal cord which was first described by Onufrowicz (1901), and hence is known as *Onuf's nucleus*. In neither the rat nor man do the striated penile muscles play an *essential* role in penile erection. After their removal in the rat, erection is maintained by vascular mechanisms (Sachs 1982). However, in some mammals the striated penile muscles play an important role in maintaining high pressures within the corpora cavernosa of the penis during erection (dog: Hart 1972a; goat: Beckett *et al.* 1972; stallion: Beckett *et al.* 1973). There have been no experimental studies of the functions of the striated penile muscles in any primate species. It seems likely that the ischiocavernosus muscles might assist elevation of the penis before intromission, and that contractions of the bulbocavernosus muscles during ejaculation facilitate expulsion of semen from the penis. In man, a *bulbocavernosus reflex* can be elicited by squeezing the glans penis; sensory feedback via the dorsal penile nerves results in reflexive contractions of the striated penile muscles. It is possible that this reflex is activated during intromission and that it facilitates additional stiffening of the penis due to increased intracavernosal pressure (Bancroft 1989). Thus, although the pudendal nerves and striated penile muscles are not essential for erection in man, they probably play some ancillary role in the process, as well as being important for ejaculatory responses.

The third component of the penile nerve supply concerns the sympathetic outflow of the thoracic spinal cord, which is situated between segments T10 to T12 in man. Branches of this sympathetic innervation enter the hypogastric, pelvic, cavernous, and pudendal nerves and might influence detumescence of the penis, as well as erection (see Fig. 13.42, which summarizes the neural pathways which control penile tumescence and detumescence). Preganglionic fibres from the ventral roots of thoracic segments 10–12 enter the paravertebral sympathetic chain or the superior hypogastric plexus. Postganglionic axons from the paravertebral sympathetic chain project to the penis in the pelvic, cavernous and pudendal nerves. Sympathetic fibres from the superior hypogastric plexus travel in the hypogastric nerves and (after making synaptic connections within the pelvic plexus) in the cavernous nerves to the penis (Fig. 13.42). The classic electrophysiological studies of Semans and Langworthy (1938), on the cat, showed that stimulation of sympathetic pathways in the lumbar chain and the hypogastric and

pudendal nerves causes detumescence of the penis. In the non-erect penis, sympathetic tone plays a role in the maintenance of flaccidity; thus administration of ganglionic blocking agents or transection of the sympathetic supply results in protrusion of the non-erect penis (in the rabbit: Sjöstrand and Klinge 1979). Erection involves the thoracic—lumbar sympathetic innervation, therefore, as well as the parasympathetic sacral innervation of the penis. Interestingly, these sympathetic mechanisms play some role in mediating the effects of psychogenic stimulation upon penile erection. In both the cat and the dog, males continue to exhibit penile erection in response to the proximity of an oestrous female, even after removal of the sacral spinal segments which control (parasympathetic) reflexive erections (Root and Bard 1947; Muller 1902). More recently, Courtois et al. (1993) have demonstrated the differential roles played by the sympathetic and parasympathetic systems in the control of penile erection in the rat. Male rats were given spinal lesions which destroyed the sacral connections required for reflexive penile responses (flips, cups, and erections). When these same subjects received electrical stimulation of the POA, erections occurred, being mediated via sympathetic outflow from the thoracic-lumbar segments, above the site of the spinal lesions. These results, summarized in Fig. 13.43, demonstrate the importance of the central control of the sympathetic innervation upon erection. Signals from the brain might serve both to disinhibit the sympathetic mechanisms which maintain the penis in its flaccid state and perhaps to activate excitatory pathways. The distinction between these two possibilities has still to be resolved by experiment. Further, it remains unclear how the parasympathetic sacral outflow which controls 'reflexive' erectile responses might be integrated with sympathetic activity during sexual behaviour in the spinally intact male. Wagner and Kaplan (1993) indicate uncertainty concerning this possible relationship by inserting a question mark over a possible pathway linking sympathetic and parasympathetic centres in the human spinal cord (see Fig. 13.41).

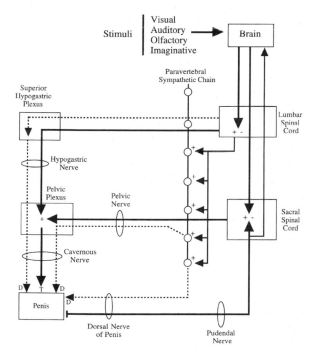

D = Penile detumescence
T = Penile tumescence
+ = synaptic excitatory mechanisms
- = synaptic inhibitory mechanisms

Fig. 13.42 Diagram summarizing the peripheral and central pathways in the mammalian nervous system which control penile tumescence (T) and detumescence (D). Synaptic excitatory mechanisms are indicated by + and inhibitory mechanisms by - symbols at the appropriate points. (Redrawn from Andersson and Wagner 1995; after De Groat and Steers 1988.)

Neurotransmitters, erection, and detumescence

Part of the mechanism by which the sympathetic nervous system maintains the penis in a flaccid condition involves the release of noradrenaline, which causes smooth muscle cells in the corpora cavernosa, and in the walls of the penile arterial supply, to contract (Benson et al. 1980; Giuliano et al. 1993; Andersson and Wagner 1995). Both alpha-adrenoceptors and beta-adrenoceptors occur in the penis but alpha-adrenoceptors occur at higher concentrations, exceeding concentrations of beta-adrenoceptors by a factor of 10 in the human corpus cavernosum (Levin and Wein 1980). As regards sub-types of the alpha-receptor, *in vitro* studies, using strips of human cavernosal tissue, have demonstrated that blockade of alpha$_1$-adrenoceptors (using prazosin) causes relaxation of the smooth musculature and is more effective in this regard than alpha$_2$ receptor blockers (such as yohimbine: Hedlund and Andersson 1985; Christ *et al*. 1990). Conversely, *in vivo* experiments have shown that intra-corporal injections of the alpha-adrenoceptor agonist metariminol, or of noradrenaline itself, cause the penis to detumesce. These procedures have been used to treat priapism in the human male (Brindley 1984; De Meyer and De Sy 1986). Despite the proven importance of the sympathetic neural supply to the penis in the control of detumescence and erection, it has been shown that alpha or beta receptor blockade does not affect penile erection in normal male volunteers (Wagner and Brindley 1980). Some non-adrenergic factor is therefore also involved in mediating affects of the sympathetic nerves upon the penis.

Acetylcholine occurs in the excitatory, parasympa-

thetic innervation of the penis, as in many other parasympathetic nerves throughout the body. Histochemical studies have shown that acetylcholinesterase is present in penile nerves (Dail 1993) and the presence of muscarinic receptors has been demonstrated in the erectile tissues (Godec and Bates 1984; Lepor and Kuhar 1984). It has been estimated that approximately 45 000 muscarinic binding sites occur on each smooth muscle cell in the human corpora cavernosa. There are 15 times as many alpha-adrenoceptors on these same cells, however (Andersson and Wagner 1995).

The involvement of the sacral parasympathetic nerves in producing penile erection cannot be explained solely by the release of acetylcholine. Blockade of muscarinic receptors by atropine does not prevent erection in response to electrical stimulation of the sacral roots in the baboon (Brindley and Craggs 1975) or in response to pelvic nerve stimulation in the dog (Andersson *et al*. 1984), although in the latter case the erectile response was reduced. Nor does intravenous infusion of atropine prevent erection in the human male, either in response to tactile stimulation of the penis, or when viewing erotic material (Wagner and Brindley 1980). Conflicting evidence concerning the role played by

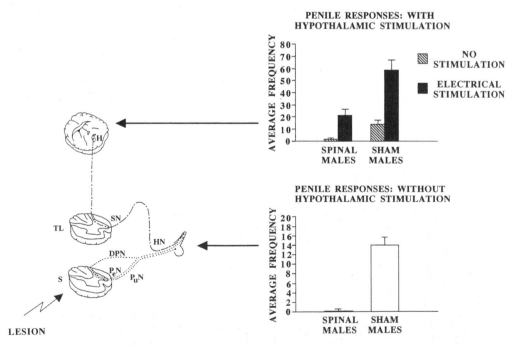

Fig. 13.43 Male rats with lesions in the sacral region (S) of the spinal cord, which block reflexive penile responses (lower histograms), still exhibit penile erection in response to electrical stimulation of the preoptic area/anterior hypothalamus (H) (upper histograms). This effect is mediated via sympathetic outflow at the thoracic / lumbar (TL) level of the cord which reaches the penis via the splanchnic (SN) and hypogastric nerves (HN). DPN=dorsal penile nerve; PeN=pelvic nerve; PuN=pudendal nerve. (Redrawn and modified from Courtois *et al*. 1993.)

acetylcholine in penile erection has been obtained from experiments on animals. For example, infusion of acetylcholine into the penile arterial supply may cause erection in rabbits and dogs (Sjöstrand and Klinge 1979; Carati *et al.* 1988). In monkeys also, acetycholine may stimulate erection when injected into the corpora cavernosa. This effect is not antagonized by atropine, but it is blocked by combined administration of nicotinic and muscarinic antagonists (Stief *et al.* 1989).

Erection involves increased arterial inflow to the erectile bodies of the penis as well as relaxation of the smooth muscles in the corpora cavernosa. It is now known that release of nitric oxide (NO) from the vascular and cavernosal endothelium plays a key role in causing relaxation of the smooth muscles (Ignarro *et al.* 1987; Palmer *et al.* 1987). It appears that at least part of the search for a non-cholinergic/non-adrenergic factor involved in causing penile erection has been solved by the realization that NO acts as a neurotransmitter in the penis as well as in many other parts of the body (Burnett *et al.* 1992).

Several neuroactive peptides occur in the penis, including vasoactive intestinal polypeptide (VIP), neuropeptide Y (NPY), calcitonin gene-related peptide (CGRP), substance P, and the endothelins. The possible involvement of these peptides in the control of erection and/or detumescence of the penis has been much debated. Vasoactive intestinal polypeptide has potent vasodilatory properties (Said and Mutt 1970), so the discovery of VIP in nerves supplying the pudendal arteries and erectile tissues of the penis provoked considerable interest (Lundberg

et al. 1980; Willis *et al.* 1983; Polak *et al.* 1981; see Fig. 13.44). Radioimmunoassay of plasma VIP showed that it occurs at higher concentrations in blood drawn from the corpora cavernosa than in the peripheral circulation of a variety of mammals, even when the penis is flaccid (e.g. in the puma, cheetah, Barbary sheep, wallaby, and chimpanzee: Dixson *et al.* 1984). Indeed, manual stimulation of the non-erect penis of the wallaby causes a large increase in plasma VIP, whereas electrical stimulation of erection (in the absence of tactile input) has much less effect on levels of the peptide (Fig. 13.45). Despite reports that infusion of VIP into the penile arteries enhances erection (in the dog: Andersson *et al.* 1984) or that cavernosal VIP levels increase during erection in man (Virag *et al.* 1982), it seems that release of VIP alone does not cause penile erection. VIP occurs in the same parasympathetic penile nerves as acetylcholine (e.g. in the rat: Dail *et al.* 1983). In some other contexts, such as the control of salivary secretion and pancreatic secretion, VIP and acetylcholine act synergistically to bring about their physiological effects. The same may be true in the penis.

Neuropeptide Y (NPY) co-localizes with noradrenaline in sympathetic nerve fibres and participates in vasoconstrictor mechanisms in a number of regions of the body (Grundemar and Hkkansson 1993). Although NPY occurs in the human corpora cavernosa (Adrian *et al.* 1984) its possible role in penile detumescence, or in the maintenance of flaccidity, has yet to be demonstrated (Andersson and Wagner 1995). Likewise, calcitonin gene-related peptide (CGRP) occurs in the corpora cavernosa of

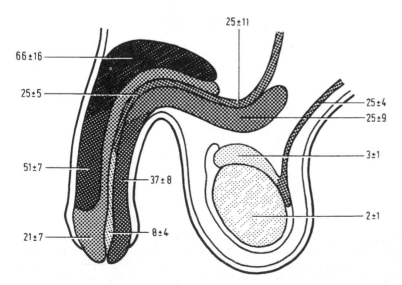

Fig. 13.44 Schematic representation of the quantitative distribution of vasoactive intestinal polypeptide (VIP: expressed as pmol g^{-1} tissue), in the external genitalia of the human male. (After Polak *et al.* 1981.)

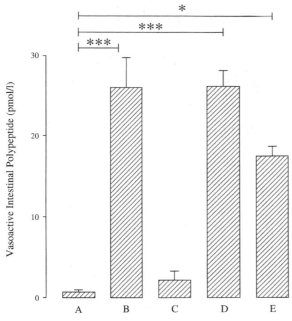

Fig. 13.45 Plasma concentrations of vasoactive intestinal polypeptide (VIP) in the corpora cavernosa penis of the wallaby (*Macropus rufogriseus*). Histograms show mean (±S.E.M.) concentrations of VIP in A (N=6), the flaccid penis (no tactile stimulation); B. (N=7), after manual extension of the flaccid penis; C. (N=6) during electrical stimulation of erection (no tactile stimulation); D. (N=7), during manually stimulated erection; E. (N=4), electrical stimulation combined with tactile stimulation of erection. One-way analysis of variance revealed a significant difference between treatments (P < 0.001). * P < 0.05; *** P < 0.001; Scheffé test. (Redrawn from Dixson *et al.* 1984.)

the human penis. Intracorporal injection of CGRP causes erection in man (Stief *et al.* 1991) and relaxation of smooth muscles accompanied by increased arterial inflow in monkeys (Stief *et al.* 1993). Substance P is only sparsely represented in the vasculature and erectile tissues of the human penis, but is found in greater concentrations in the skin of the glans penis, both in man (Gu *et al.* 1983) and in the rat (Lamano Carvalho *et al.* 1986). Current evidence does not support a role for substance P in penile erection, but it may be involved in neural mechanisms which serve for sensory feedback from the glans.

Finally, it is important to mention the endothelins (Et^{-1}, Et^{-2} and Et^{-3}), because they might be involved in modulating the effects of noradrenaline on penile detumescence and flaccidity. Endothelial cells lining the human corpora cavernosa contain Et^{-1} (Saenz De Tejada *et al.* 1991) and Et^{-1} causes long-lasting contractions of corpora cavernosa and penile blood vessels *in vitro* (man: Lau *et al.* 1991;

man and rabbit: Holmquist *et al.* 1992). These effects are reversible by exposing the tissues to VIP plus carbachol (a muscarinic cholinergic agonist), as demonstrated by Holmquist *et al.* (1990). Andersson and Wagner (1995) stress the importance of these observations, given that acetycholine and VIP colocalize in the same penile nerves and that nitric oxide synthase has also been identified in the same neurons as VIP (in human corpora cavernosa: Jünemann *et al.* 1993). It is possible, therefore, to envisage a link between cholinergic (and perhaps VIPergic) mechanisms and relaxation of penile smooth muscles, due to release of nitric oxide, to cause penile erection. Given that VIP is released by tactile stimulation of the penis, it might play a role in the maintenance of erection during intromission (Dixson *et al.* 1984). Once copulation is over, detumescence occurs in response to sympathetic control involving noradrenaline (and perhaps NPY), as well as the localized action of endothelin within the erectile tissues to maintain the penis in a quiescent state.

Erectile dysfunction

Previous sections have provided an overview of the complex neuro-vascular mechanisms which control penile erection. Erectile dysfunction is not uncommon in human males; indeed it has been reported to be on the increase (Wagner and Green 1981; Buvat *et al.* 1990; Riley *et al.* 1993; Wagner and Kaplan 1993). If one adopts the definition of impotence as an inability to produce a penile erection, or to sustain erection, sufficient for sexual intercourse to occur, it is exceedingly difficult to quantify the frequency of this condition in the general population. Understandably, many men are reluctant to admit to the problem or to discuss it with a doctor. The causes of impotence are legion and it is fruitless to assume that any particular form of treatment will always be effective. Certain hormonal deficiencies, such as hypogonadism, result in reduced sexual interest and arousal, so that impotence is a secondary consequence of the condition. In other cases, desire and arousal may be normal but the ability to develop or sustain erection is impaired for organic reasons, such as disease (e.g. diabetes, arteriosclerosis, multiple sclerosis), accidental injuries, (e.g. to the spinal cord or pelvic neural supply), or as a side-effect of drug treatment (e.g. to relieve hypertension). Impotence may also result from *psychogenic* causes, such as depression, and from interpersonal problems involving fear of failure in sexual

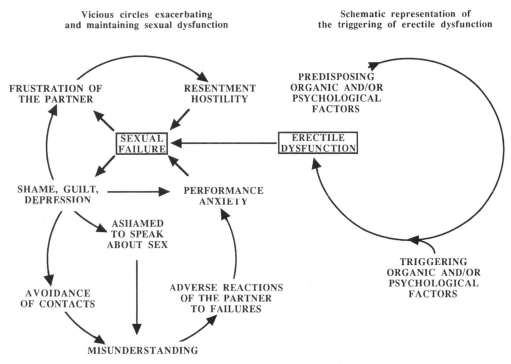

Vicious circles exacerbating
and maintaining sexual dysfunction

Schematic representation of
the triggering of erectile dysfunction

Fig. 13.46 Factors which lead to erectile dysfunction in men and which reinforce and maintain problems of sexual dysfunction. (Redrawn and modified from Buvat *et al.* 1990.)

contexts. Indeed, erectile dysfunction may be reinforced by psychological factors even if it has some organic basis, so that in practice the distinction between *organic* and *psychogenic impotence* is an artificial one (Fig. 13.46). In the past, it was not appreciated that hyperprolactinaemic men become impotent as a result of their endocrine pathology and that this could be reversed by bromocryptine treatment. Likewise, some men fail to maintain erections due to leakage of blood from the corpora cavernosa into the venous outflow from the penis (Wagner and Green 1981). Such patients might be misdiagnosed as suffering from psychogenic impo-

tence when the causes of their erectile problems are organic. It is likely that further organic causes of impotence remain to be discovered. Table 13.3 summarizes the known causes of erectile dysfunction in the human male. For excellent reviews of this subject the reader should consult Wagner and Green (1981), Batra and Lu (1990), Buvat *et al.* (1990), Wagner and Kaplan (1993) and Bancroft and Gutierrez (1996).

For some patients, the desire for a rapid remedy to long-lasting impotence can be a paramount concern. In cases of purely organic impotence, which are unresponsive to hormone or drug treatments

Table 13.3 Some known causes of erectile dysfunction in the human male

General	Alcoholism, Cirrhosis, Diabetes, Renal failure
Iatrogenic	Side-effects of taking drugs, such as anti hypertensive and antipsychotic drugs. In rare cases, smoking is associated with impotence
Endocrine	Hypogonadism, Hypo and Hyperthyroidism, Acromegaly, Hyperprolactinaemia, Cushing syndrome, Adrenal insufficiency
Vascular	Arteriosclerosis, Obstruction of the aortic bifurcation, Selective occlusions in the hypogastric and internal pudendal arteries, Venous incompetence
Neurological	Temporal tumour, Multiple sclerosis, Epilepsy, Spinal cord disease or injuries, Peripheral neuropathies, Selective damage to the somatic or autonomic pelvic nerves, in the spinal cord, or brainstem
Psychological	Depression, Anxiety and fear of failure in sexual contexts

Information from Buvat *et al.* 1990; Wagner and Green 1981; Barnes and Harvey 1993; and Huws 1993.

(e.g. diabetic neuropathy), penile implants have sometimes proven useful. In a small number of cases where a fault exists in the arterial supply to the penis, or in the cavernosal/venous drainage, corrective vascular surgery has been used to restore potency (Wagner and Green 1981). Paraplegic patients have been implanted with electrical stimulators which improve bladder control, as well as restoring erectile function (Brindley 1988). Pharmacological approaches can improve erectile responsiveness and sexual desire in some men (e.g. yohimbine: Morales *et al.* 1987; Susset *et al.* 1989; yohimbine in combination with naxolone: Charney and Heninger 1986), and there is the hope that increasing knowledge of neuropharmacology will furnish clinicians with more effective treatments (e.g. using 5-HT_{1A}, or 5HT_2 receptor agonists: Foreman *et al.* 1989; Foreman and Doherty 1993). For the moment, however, the most effective pharmacological method of inducing erection in impotent men involves injection of smooth muscle relaxants directly into the corpora cavernosa of the penis. Virag (1982) pioneered this approach by demonstrating that cavernosal injections of the smooth muscle relaxant papaverine produce penile erection in impotent men. Fourteen patients (including seven diabetics) received two or more such injections and 'nine of these described a significant improvement in penile rigidity and four reported a return to normal sexual life'. A variety of substances has subsequently been used for intracavernosal treatments, including phenoxybenzamine, phentolamine, prostaglandin E^{-1}, and VIP (reviewed by Wagner and Kaplan 1993). The use of cavernosal injections for the treatment of impotence is not without drawbacks. Anxiety concerning this technique can cause increased sympathetic tone and constriction of the penile vasculature; cavernosal injections might then prove ineffective (Buvat *et al.* 1986). Prolonged erections (priapism) sometimes occur after injection of vasoactive drugs into the corpora cavernosa (Gilbert and Gingell 1991). Priapism must be treated before irreversible damage occurs to the penile vasculature. With long-term treatments, involving multiple injections into the corpora cavernosa, there is increased risk of fibrosis of the erectile tissues. Nonetheless, the injection treatment for impotence has proven beneficial and is widely used (Wagner and Kaplan 1993). It is especially encouraging that this technique can also assist men with psychogenic impotence. In some cases 'the reassurance that erection is still possible, which follows pharmacologically induced erection in the consulting room, is sufficient to resolve their erectile dysfunction, especially if they are able to use the induced erection for inter-

course' (Riley and Riley 1993).

Most recently, a new drug has been developed ('Viagra') which facilitates erection, but which may be taken orally rather than by cavernosal injection. This non-invasive treatment clearly offers some advantages.

Emission and ejaculation

Emission involves the pooling of semen (spermatozoa plus secretions of the sexual accessory glands) in the posterior urethra. Closure of the of the bladder neck occurs, in order to prevent the semen from entering the bladder (so-called *retrograde ejaculation*), rather than being expelled from the penis (Koraitim *et al.* 1977). Ejaculation itself involves rapid and forceful contractions of the striated penile muscles, which propel semen along the urethra. The sympathetic nervous system, which innervates the penis and influences erection and detumescence, also supplies the internal genitalia, controls the process of emission and assists closure of the bladder neck (Benson, 1994). Removal of the lumbar sympathetic chain interferes with these processes and has been reported to cause ejaculatory failure (due to absence of emission) in several mammalian species (guinea pig: Bacq 1931; dog: Beach 1969). Contractions of the internal genitalia which result in emission of semen, are controlled both by descending influences from the brain upon the sympathetic nervous system and by afferent stimuli from the penis during intromission. Beyer *et al.* (1982) measured seminal vesicle pressures in intact male rats before and during mounts and intromissions. As can be seen in Fig. 13.47, pressure within the seminal vesicles begins to rise during a mount without intromission. However, insertion of the penis results in a further, marked reflexive increase in pressure. Finally, after a series of such intromissions and pressure changes, a sudden contraction of the seminal vesicles causes emission of fluid into the urethra. Beyer and Gonzalez-Mariscal (1994) therefore suggest that a gradual recruitment of preganglionic neurons occurs with sustained penile activation. Testicular hormones play some role in this process since, in the rat at least, the reflexive responses disappear rapidly after castration, although the male is still capable of mounting the female.

Alpha adrenergic receptors are involved in mediating the effects of the sympathetic nervous system upon emission, and hence upon ejaculation of semen. Drugs which act as alpha-adrenoceptor antagonists interfere with the ejaculatory response in the human male (e.g. indoramin: Pentland *et al.* 1981;

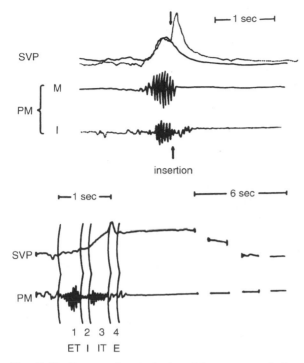

Fig. 13.47 Changes in seminal vesicle pressure during copulatory behaviour in the male rat. Upper: Tracings to show recordings of the pelvic movements (PM) of a male during a mount (M) and intromission (I). Changes in recordings of seminal vesicle pressure (SVP) during the mount and intromission are also shown, with the point of insertion being indicated by an arrow. Lower: PM and accompanying changes in SVP across a copulatory series which includes mounting, extra-vaginal pelvic thrusting (ET), intromission (I), intra-vaginal pelvic thrusting (IT) and seminal emission (E). (After Beyer *et al.* 1982.)

phenoxybenzamine: Homonnai *et al.* 1984). Erection and orgasm are not necessarily affected by these drugs (e.g. Pentland *et al.* 1981) and intracavernosal injection of alpha blockers can indeed cause erections to occur. 'Orgasm', the pleasurable sensation which normally accompanies ejaculation in the human male, does not depend upon either emission or ejaculation for its occurrence (Money 1961; Bancroft 1989). The central events which underlie the experience of orgasm in men and women remain unknown. Likewise, the neurological basis of the refractory period which occurs after ejaculation in many male mammals has yet to be fully explained. In the rat, an absolute refractory period occurs after ejaculation, during which time the male is unresponsive to the female. This is succeeded by a period of relative refractoriness, when arousing stimuli can cause the male to resume mounting. Among the primates refractory periods are often more labile

than described for the rat (see pages 104–7 for a discussion of this), and events such as the sight of another male copulating, or the approach of an attractive female, can reinstate copulatory behaviour within a remarkably short time. Nonetheless, refractoriness does occur and increases with age, at least in the human male (Bancroft 1989).

Neurotransmitters and sexual behaviour

The neuroanatomical distributions and functions of aminergic and peptidergic transmitters were discussed in Chapter 12, in relation to their roles in the control of sexual behaviour in female primates. This section provides a review of how various neuroactive peptides, catecholamines (dopamine and noradrenaline), and serotonin (5-HT) are thought to influence sexual behaviour in adult male monkeys and apes, as well as in the human male. Since the picture is far from complete where primates are concerned, I shall once again rely heavily upon the literature dealing with rodents and other mammals, in order to make comparisons with primates where these are warranted.

Neuroactive peptides
β-Endorphin

β-Endorphin plays a most important role in regulating both pituitary gonadotrophin secretion and various aspects of reproductive behaviour. In general its actions are inhibitory, serving to reduce gonadotrophin secretion and to stimulate prolactin release in response to social stress, or during nonmating periods in seasonally breeding mammals (Lincoln and Ssewannyana 1989; Herbert 1993). β-Endorphin also suppresses mounting activity in male rats when infused into the POA (Hughes *et al.* 1987a, 1990). It is likely that increased activity of this opioid peptide is important in the reduction of sexual behaviour which occurs in low-ranking males of various primate species. Evidence supporting this view derives from studies of captive groups of talapoin monkeys. Levels of β-endorphin in the cerebrospinal fluid of low-ranking males are higher than those found in dominant individuals (Martensz *et al.* 1986). The opioid antagonist naloxone stimulates luteinizing hormone secretion when administered to dominant male talapoins, whereas subordinates fail to respond (Fig. 13.48). The lower levels of plasma testosterone and suppression of sexual behaviour which occur in subordinate male talapoins may be

A. C.S.F. ß ENDORPHIN LEVELS

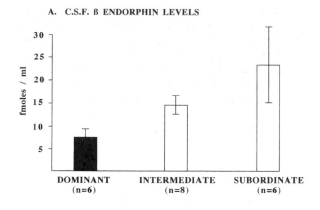

B. SERUM LH RESPONSE TO NALOXONE CHALLENGE

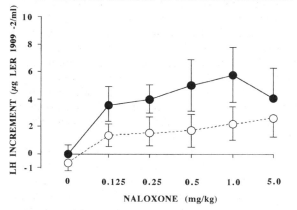

Fig. 13.48 Correlations between dominance rank, levels of β-endorphin in the cerebrospinal fluid (CSF) and the serum luteinizing hormone (LH) response to naloxone challenge in male talapoin monkeys (*Miopithecus talapoin*). Upper: Levels of β-endorphin in the CSF of subordinate male talapoins are three times higher than in dominant males. Lower: The increment (mean ± S.E.M.) in serum LH concentrations 20 min after various doses of naloxone were administered to dominant males (●, N=4) or males of intermediate and low rank (○, N=11) in captive social groups. (Redrawn and modified from Keverne 1992.)

due to altered activity of the opiate system dependent upon social environment (Keverne 1992).

How does β-endorphin produce its effects upon sexual behaviour? The results of infusing this opioid into the POA of the male rat resemble, in some respects, the effects of POA lesions upon sexual behaviour. Males given tiny doses of β-endorphin (10–40 picomol) exhibit normal precopulatory sexual arousal—they follow and investigate the female, but fail to mount (Hughes *et al*. 1987a). These males will also continue to perform an operant task to gain access to the female, despite their inability to make the behavioural transition from sexual arousal to

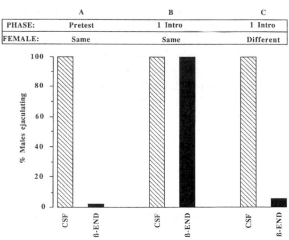

Fig. 13.49 Selective effects of β-endorphin on the male rat's sexual behaviour after bilateral infusions into the preoptic area (POA) at various phases of sexual interaction. The inhibitory effect, shown in A, is prevented (B) by allowing one intromission before infusion, but not (C) if the female is changed for another partner. (Redrawn from Stavy and Herbert 1989.)

copulation (Hughes *et al*. 1990). Stavy and Herbert (1989) made the most interesting observation that β-endorphin's effects within the POA depend upon the *stage of the sexual interaction* at which the peptide is administered. If β-endorphin is infused into this area *after* the male has achieved his first intromission during a pair-test, then mounts, intromissions, and ejaculation continue normally, provided the *same* female partner is present. However, substitution of a *new* partner after the first intromission, with subsequent infusion of β-endorphin into the male's POA, suppresses mounting behaviour once more. These results, summarized in Fig. 13.49, point to a crucial role for β-endorphin in enabling the male to make the correct transitions in a series of sexually motivated behavioural patterns, rather than in the control of copulatory movements *per se*. They also encourage a re-interpretation of the results of POA lesion studies, as stressed by Herbert (1993). Perhaps, once a mount or mount-series has begun, the POA is not required for the *continuation of copulation*, at least so long as the male interacts with the same female partner, and further neural elements constituting the copulatory mechanism must lie elsewhere in the hypothalamus? This notion recalls Yoshimatsu's (1983) observation that POA neurons in male macaques fire primarily during precopulatory orientation towards the female, whereas neurons in the dorsomedial and posterior hypothalamus are activated during mounting and pelvic thrusting sequences (see Fig. 13.32).

Given that neurons of the β-endorphin system, centred in the arcuate region, project widely in the brain to the limbic system and other regions, it is important to ask whether this opioid influences sexual activity at sites other than the POA. In previous sections, the importance of the temporal cortex and amygdala in processing sexually relevant sensory stimuli was stressed. The corticomedial division of the amygdala receives and processes olfactory stimuli, whilst the basolateral amygdala receives temporal inputs and might thus be more concerned with the processing of non-chemical cues, such as visual stimuli (Price *et al.* 1987). Whilst no experiments have examined effects of infusing β-endorphin into the amygdala of primates, the relevance of β-endorphin to the processing of olfactory stimuli by the amygdala has been examined in the rat. Infusions of β-endorphin into the amygdala of the male rat result in decreased investigatory behaviour before copulation. Such males continue to follow and sniff females, but these investigatory behaviours are reduced in frequency. However, although copulation is delayed in such circumstances, once the male begins to mount he performs a series of mounts with intromission, culminating in ejaculation (McGregor and Herbert 1992). It would be most interesting to know if β-endorphin plays a similar role in the processing of precopulatory sexual stimuli in male primates, where visual cues from the female, rather than chemical cues, are of considerable importance in eliciting sexual interest and copulation.

It has been suggested that neuroactive peptides in the limbic system function as the vocabulary of a 'chemical language' (Peaker 1992; Herbert 1993). Thus, β-endorphin, depending upon its site of release and action (e.g. in the amygdala or POA), enables the male to attend preferentially to sexually stimulating cues from a female and then to shift from investigation to copulation with a specific partner. The application of this experimental approach to other peptides involved in sexual behaviour constitutes a promising area for future research.

The evidence pertaining to the role of opioid peptides in the sexual behaviour of male primates is limited and contains some negative findings. Naloxone and naltrexone, which are opioid antagonists with preferential (but not specific) affinity for the Mu receptor, do not potentiate sexual behaviour in either rhesus monkeys (Naloxone: Abbott *et al.* 1984) or talapoins (Naltrexone: Meller *et al.* 1980). However, when given in combination with yohimbine, naloxone caused penile erections lasting for 60 min or longer in healthy human volunteers (Charney and Heninger 1986). Whether naloxone enhanced sexual interest under these conditions is an unanswered

question. Systemic administration of naloxone affects opioid receptors throughout the body and is a relatively crude method of assessing sexual effects. Unfortunately, naloxone may cause nausea in these circumstances (e.g. in rhesus monkeys), so that lack of sexual responsiveness is understandable. In the rat, however, some studies have found that naloxone stimulates copulatory behaviour both in experienced and in sexually sluggish males (e.g. Myers and Baum 1979).

Luteinizing-hormone-releasing hormone

There is some evidence that LHRH stimulates sexual behaviour in male rats and in primate males, but it is far from conclusive and the site(s) within the central nervous system which respond to LHRH are not known. In the female rat LHRH has been reported to potentiate lordosis in the oestrogen-primed female when infused into the hypothalamus, midbrain central grey, or vertebral canal (see pages 384–5). It seems that the decapeptide can affect sexual arousal (e.g. potentiation of proceptivity in the female marmoset, Kendrick and Dixson 1985b), as well as reflexive elements of female copulatory behaviour. In male rats, some studies show that LHRH causes a reduction in the mount and ejaculatory latencies of castrated males, provided that some testosterone is present (Moss 1978). Dorsa *et al.* (1981) report potentiation of sexual behaviour by a potent LHRH agonist in the male rat. The findings in primates are far from convincing, however. Preliminary work involving intact adult male rhesus monkeys showed no effect of systematically administered LHRH (Everitt *et al.* 1981). Intact and fully active males might not be suitable subjects for evaluation of the possible effects of LHRH upon sexual behaviour. The effects of a single injection of antide (an LHRH antagonist) upon the sexual behaviour of male rhesus monkeys have already been described (Wallen *et al.* 1984). Ejaculatory and mount frequencies began to decrease just one week after antide treatment, and more rapidly than would be expected if the males had been castrated. A central inhibitory effect of this LHRH antagonist upon sexual behaviour might occur in the male rhesus monkey, but this possibility requires further study. In man, treatment of LHRH-deficient patients with the decapeptide has been reported to increase sexual interest. This change in sexual interest occurred *before* plasma testosterone began to rise, and so it might have been due to a direct central action of LHRH (Mortimer *et al.* 1974). Attempts to use LHRH for the treatment of impotence have not produced conclusive results (Benkert 1975) and in normal men LHRH failed to enhance erectile responses to erotic stimuli (Evans and Distiller 1979). There is a need for controlled studies to determine the significance of LHRH in the control of sexual behaviour in male primates. Experiments using monkeys, to examine effects of infusing LHRH or its antagonists into the brain, would be valuable in this regard.

Oxytocin

In the rat, concentrations of oxytocin in the cerebrospinal fluid increase after ejaculation (Hughes *et al.* 1987b), whilst peripheral and central (intraventricular) administration of this octapeptide stimulates copulatory behaviour (Arletti *et al.* 1992). There is no evidence pertaining to effects of oxytocin upon sexual behaviour in male primates. The importance of oxytocin for maternal and sexual behaviour in female mammals is discussed on pages 385–6.

Prolactin

Hyperprolactinaemia is associated with impotence in the human male (Perryman and Thorner 1981; Franks *et al.* 1978) and with reductions in sexual interest as well as in erectile function. The condition is reversible by treatment with bromocryptine, a dopamine agonist which enhances the dopaminergic inhibition of prolactin secretion. Although erectile function improves if hyperprolactinaemic men receive sex therapy, bromocryptine treatment leads to further improvements in sexual interest (Schwartz *et al.* 1982; Bancroft *et al.* 1984). Bancroft (1989) suggests that the primary effect of hyperprolactinaemia in the human male 'is to reduce sexual interest in a manner comparable to androgen deficiency'. This view is supported by animal experiments. Treatment of male rats with the dopamine antagonist domperidone (which acts selectively at the pituitary level to block dopaminergic inhibition of prolactin secretion) brings about a gradual decrease in copulatory behaviour. The same effect occurs if males are given pituitary grafts (beneath the kidney capsules) which release supra-physiological amounts of prolactin (Bailey and Herbert 1982). High circulating levels of prolactin can influence the brain, and levels of prolactin in the cerebrospinal fluid increase markedly under these conditions (in the rhesus monkey: Herbert *et al.* 1982). Some authors have attempted to explain the effects of hyperprolactinaemia upon masculine sexual behaviour in terms of adverse effects upon LHRH or dopaminergic mechanisms (Doherty *et al.* 1981; Doherty *et al.* 1986), although more recent experiments render the dopamine hy-

pothesis unlikely (Doherty *et al.* 1989). Nor can decreases in circulating testosterone in hyperprolactinaemic males be held responsible for the behavioural effects since, in studies on male rats, sexual behaviour is still affected when testosterone is administered exogenously. A more likely explanation concerns the effects of hyperprolactinaemia upon testosterone-sensitive neuronal systems in the brain which are involved in sexual arousal and copulatory behaviour. Electrophysiological studies have shown that the refractory periods of testosterone-sensitive neurons in the stria terminalis, projecting from the amygdala to the POA, increase in hyperprolactinaemic male rats and resemble those found in castrated males (Fig. 13.50). This result was obtained despite implanting males with silastic capsules containing testosterone, in order to maintain circulating levels of the hormone within the normal range (Kendrick and Dixson 1984c). The gradual decreases in sexual interest and copulatory behaviour which occur in hyperprolactinaemic men and male rats probably result from decreased central sensitivity to testosterone. How high levels of prolactin produce such effects, whether by a direct or indirect action, remains to be determined.

Other pro-opiomelanocortin-derived peptides

β-Endorphin belongs to a family of peptides derived from pro-opiomelanocortin (POMC). Other products of this precursor molecule, such as alpha-melanocyte-stimulating hormone (alpha-MSH) and adrenocorticotrophin (ACTH), can influence sexual behaviour or sexual reflexes in male mammals. Intraventricular infusions of ACTH4–10 or of alpha-MSH in male rabbits cause spontaneous erections and ejaculations (Bertolini *et al.* 1975). In the male rat, infusion of alpha-MSH into the POA enhances sexual arousal and copulatory activity, resulting in a reduction of mount and ejaculatory latencies (Hughes *et al.* 1988, 1990) Therefore, some products of POMC can have effects upon sexual behaviour opposite to those described previously for β-endorphin. Co-release of POMC peptides might bring about complex effects, depending upon the balance between the various molecular types. Hughes *et al.* (1988) examined this problem by infusing various ratios of β-endorphin/alpha-MSH into the POA of male rats. A shift from inhibition of mounting behaviour at higher ratios of β-endorphin to alpha-MSH gave way to essentially normal mounting activity when the ratio of alpha-MSH to β-endorphin was increased. These results provide an interesting view of how POMC peptides might produce a spectrum of behavioural effects, depending upon the ratios released from nerve terminals. Unfortunately, none of these POMC products has been studied to determine their possible involvement in primate sexual behaviour.

Monoaminergic neurotransmitters
Dopamine

Interest in dopamine (DA) as a neurotransmitter having effects upon sexuality arose form the observation that patients treated with l-DOPA, to relieve

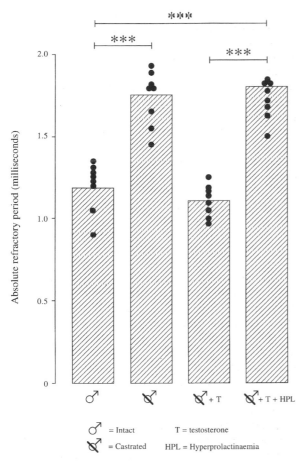

Fig. 13.50 Either castration or hyperprolactinaemia lengthen the refractory periods of testosterone-sensitive neurons in the male rat's brain. Absolute refractory periods are shown of corticomedial amygdala neurons projecting to the medial preoptic/anterior hypothalamic area in gonadally intact and castrated rats, and in castrates treated with testosterone (T) or testosterone plus four pituitary grafts under the kidney capsules in order to produce hyperprolactinaemia (HPL). Data are from eight rats in each group. Bars indicate the overall mean and the dots individual means. ***P < 0.001, Scheffé test. (Redrawn and modified from Kendrick and Dixson 1984c.)

the symptoms of Parkinson's disease, sometimes reported an increase in sexual interest (Bowers *et al.* 1971; Brown *et al.* 1978). In a small number of cases (less than 1%: Goodwin 1971) men receiving l-DOPA became hypersexual. These clinical observations have provided the basis for a wide range of animal experiments on the effects of dopamine upon sexual arousal, copulatory behaviour, and penile reflexes. Most of this work has involved the male rat, but recently there have been some studies involving male monkeys and attempts to evaluate effects of dopamine agonists upon sexual responses in the human male.

Apomorphine, which acts at both D_1 and D_2 receptors, has been reported to stimulate penile erection in both impotent and normally functional men (Lal *et al.* 1989). Although apomorphine failed to enhance copulatory behaviour in male rhesus monkeys (Everitt *et al.* 1981; Chambers and Phoenix 1989), there is good evidence that this dopamine agonist can potentiate sexual arousal and erectile responses in the rhesus monkey (Pomerantz 1990). Pomerantz employed a most effective experimental paradigm which allowed male rhesus monkeys to see, hear, and smell a potential female partner in the absence of physical contact. Under these conditions, males exhibited varying degrees of penile erection, yawned, masturbated, and made purse-lip invitational displays towards the stimulus female. At lower doses (25–100 μg kg^{-1}) apomorphine stimulated yawning whereas at dosages of 50–200 μg kg^{-1} males showed increased frequencies of penile erection and masturbation in the presence of a female (Fig. 13.51). At high dosages, exceeding 200 μg kg^{-1}, the males showed increased stereotypic behaviour, especially oral hyperkinesia (mouthing and licking of a chain attached to the testing cage). In subsequent experiments Pomerantz (1991) showed that quinelorane (LY163502), which is a selective D_2 agonist, also stimulated erections, masturbation, and yawning responses. These effects were more pronounced at lower dosages (2.5 and 5.0 μg kg^{-1}) and tended to decrease when larger amounts of the drug were employed (25 μg kg^{-1}). Administration of the DA antagonist domperidone (which is thought not to cross the blood—brain barrier) failed to block these actions of quinelorane. Pre-treatment of males with haloperidol, which enters the brain and antagonizes the action of quinelorane centrally, prevented stimulation of yawning, erection, and masturbatory behaviour. It appears that central (rather than peripheral) stimulation of D_2 receptors is involved in mediating effects of quinelorane upon behaviour and erection. It is of interest that neither apomorphine nor quinelorane treatments significantly affected the

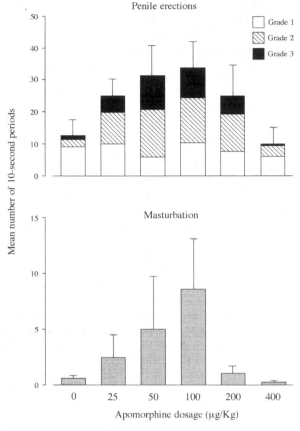

Fig. 13.51 Effects of apomorphine on penile erection and masturbation in male rhesus monkeys (*Macaca mulatta*), caged within sight of a stimulus female.
Upper: Dose-dependent effects of apomorphine on penile tumescence (grade 1: protrusion of glans; grade 2: partial erection; grade 3: fully erect). Data are means (+S.E.M.) for 8 males, for the number of 10 second intervals during which responses occurred.
Lower: Dose-dependent effects upon masturbatory behaviour (means+S.E.M.) for 8 males. (Redrawn and modified from Pomerantz 1990.)

precopulatory facial displays of these male rhesus monkeys. Nor is it known whether the drugs might have affected copulatory activity, since the sexes were not allowed to interact physically during these experiments.

Experiments on male rats have established that DA plays a stimulatory role in both the arousal and copulatory components of sexual behaviour (Bitran and Hull 1987; Agmö and Fernandez 1989; Everitt 1990; Hull *et al.* 1990; Wilson 1993). Dopaminergic neurons in the midbrain (cell-body group A9 in the *zona compacta* of the *substantia nigra* and cell-body group A10 in the *ventral tegmentum*) project rostrally in the medial forebrain bundles to innervate the *striatum*, *nucleus accumbens*, amygdala, and limbic cortex (see Fig. 12.28 and Fig. 12.29, pages 381–2 for the neuroanatomical distributions of the various DA cell-body groups and fibre tracts).

Lesions which damage either the nigro-striatal system (originating in A9: Brackett *et al.* 1986), or mesolimbic system (originating in A10 and projecting to nucleus accumbens: Hull *et al.* 1990), impair copulatory behaviour in male rats. Testosterone interacts with these dopaminergic systems, so that DA turnover is reduced by castration and is restored by testosterone in both the POA (Gunnett *et al.* 1986) and the nucleus accumbens (Mitchell and Stewart 1989). Microdialysis studies have demonstrated that DA is released in the POA of the male rat during precopulatory exposure to an oestrous female (Hull *et al.* 1995). It has also been shown that DA activity increases in the *nucleus accumbens* as the male rat mounts and intromits, but then falls during the post-ejaculatory refractory period (Pfaus *et al.* 1990; Pleim *et al.* 1990). Both D_1 and D_2 receptors might be involved in mediating the effects of DA upon arousal, whereas D_2 receptors are probably more important for expression of dopaminergic effects upon copulatory activities (such as mount and intromission frequencies: see Wilson 1993 for review). The central effects of DA agonists upon yawning and penile erection might involve the paraventricular nucleus, since infusions of apomorphine into this region induce these responses in male rats (Melis *et al.* 1987). However, spinal projections of the dopaminergic system might also be involved in effects upon erection, given that such connections exist between the diencephalon and lumbar and spinal cord (Skakerberg *et al.* 1982) and infusions of the agonist lisuride into the vertebral canal affect sexual responses (Hansen 1982b).

Despite these important effects of various dopaminergic pathways upon sexual arousal, erection, and copulatory behaviour, dopamine agonists have proven of limited value for the treatment of sexual dysfunction in the human male. One reason is that agonists have widespread effects in the central nervous system and their side effects (such as dizziness, nausea, and hypotension) are often counterproductive (Lal *et al.* 1989; Segraves *et al.* 1991).

Noradrenaline

The effects of the sympathetic nervous system and of noradrenaline (NA) upon penile tumescence have been described in an earlier section of this chapter. At a central level, however, there is ample evidence that NA might mediate sexual arousal. Treatment of male rats with the alpha$_2$-adrenoceptor antagonist yohimbine, enhances sexual activity by reducing the ejaculatory latency and shortening the post-ejaculatory interval (Clark *et al.* 1984, 1985a). These authors also found that stimulation of central alpha$_1$-adrenoceptors by clonidine increased sexual arousal, whilst alpha$_1$-adrenoceptor blockade (by prazosin) had the reverse effect (see also Clarke 1991). Yohimbine was effective in castrated males, as well as in intact males in which the penis had been anaesthetized (Clark *et al.* 1984, 1985b). Testosterone is therefore not essential for yohimbine to exert its central effects upon masculine sexual arousal, and nor do these effects depend upon sensory feedback from the penis during sexual activity. Such findings (reviewed by Wilson 1993) indicate that noradrenergic mechanisms influence sexual arousal by blockade of alpha$_2$ and stimulation of alpha$_1$-adrenoceptors in the brain.

Yohimbine has been used for the treatment of impotence in human males, but with varying amounts of success (Riley 1994; Rowland *et al.* 1997). Interestingly, in man, the occurrence of nocturnal penile tumescence (NPT) which accompanies rapid eye movement (REM) sleep, is associated with a reduction in sympathetic activity, especially in the *locus coeruleus*, a major cell-body group of the noradrenergic system, situated in the *pons* but projecting widely to forebrain areas including the neocortex of primates. Perhaps the reduction of sympathetic activity which occurs during REM sleep disinhibits the sympathetic tone, which affects the penis and thus results in NPT (Bancroft 1995). The cell bodies of the *locus coeruleus* are normally under tonic alpha$_2$-adrenergic inhibition (Quintin *et al.* 1986).

Serotonin (5-HT)

Cell bodies of neurons containing 5-HT are situated in the dorsal and medial raphe. Fibre tracts from these cell-body groups project rostrally to terminate in the amygdala, hypothalamus, septum, and limbic cortex. Traditionally, 5-HT has been viewed as exerting inhibitory effects upon various components of masculine sexual behaviour (Meyerson *et al.* 1985; Bitran and Hull 1987; Foreman *et al.* 1989). Destruction of 5-HT neurons (using the neurotoxin 5–7 dihydroxytryptamine) increases sexual responsiveness. Larsson *et al.* (1978) lesioned the medial ascending 5-HT fibre tract in this way, and found that the operation improved the sexual responsiveness of castrated male rats treated with testosterone. Increasing the activity of 5-HT neurons (e.g. by administration of substances which inhibit re-uptake of the transmitter from the synaptic cleft, or drugs that increase 5-HT release) leads to reductions of sexual behaviour in male rats. Fenfluramine (a 5-HT releaser) has been reported to reduce sexual interest in patients treated for obesity (Pinder *et al.* 1975), whilst similar effects can occur in patients taking 5-HT re-uptake inhibitors to relieve depression (e.g. setraline: Doogen and Caillard 1988).

At least 16 receptors and sub-types of 5-HT receptors have been identified so far (Peroutka 1995). As these receptors become better known, and pharmacological agents are synthesised which have specific effects upon the various types, so effects of 5-HT upon sexual behaviour might be better understood. 5-HT$_{1A}$ autoreceptor agonists, such as 8-OHDPAT, have a stimulatory effect upon sexual behaviour in the male rat (Ahlenius and Larsson 1987). However, Pomerantz *et al.* (1993) reported that 8-OHDPAT inhibits penile erection in the male rhesus monkey (in response to the proximity of a female but in the absence of physical contact), but enhances ejaculatory responses during pair-tests. These authors found that administration of a 5-HT$_{1C/1D}$ agonist (M-CPP) stimulated erection in male rhesus monkeys. This is of interest because Trazodone, which is used to treat depression and has 5-HT$_{1C/1D}$ agonist activity, sometimes causes priapism in male patients (Warner *et al.* 1987) and clitoral priapism in women (Riley: personal communication). Experiments on rats also indicate that the 5-HT$_{1C}$ receptor sub-type is important in mediating effects of 5-HT upon penile erection (Berendsen *et al.* 1990).

14 Socioendocrinology and sexual behaviour

Two broad strands of enquiry have been explored throughout this book. Firstly, I have tried to construct some picture of how sexual behaviour has been shaped by selective forces during primate evolution. This has involved comparisons of mating systems and mating tactics, as well as an exploration of copulatory patterns, genital morphology, and sperm competition in the various primate lineages. Secondly, I have examined the proximate mechanisms by which hormones and neural mechanisms govern the expression of sexual behaviour in males and females, both during development and in adulthood. In this final chapter, the two strands of enquiry become entwined as we consider socioendocrinology in relation to primate sexual behaviour. As stated by Bercovitch and Ziegler (1990) 'A primary goal of socioendocrinology is to understand the links between social environment, hormones, and behaviour because they modulate the reproductive success of individuals. This perspective provides a framework for connecting evolutionary biology with reproductive endocrinology'. Such a framework, as it relates specifically to sexual behaviour, is depicted in Fig. 14.1. An individual's social relationships affect not only its sexual behaviour, but also neuroendocrine processes which, in turn, impinge upon sexual activities. Sexual behaviour and neu-

roendocrine status affect reproductive success. In adulthood, for example, lower-ranking individuals in the social framework differ physiologically and behaviourally from more dominant individuals. In those species which are seasonal breeders, reproductive synchrony might be achieved in response to social cues as well as to non-social environmental cues. During development an individual's social environment, including its interactions with its mother, close kin, and other group members, might have repercussions for functioning of the hypothalamic-pituitary-gonadal axis during puberty and in adulthood. These issues, and related questions, are considered in the following sections.

Social rank and neuroendocrine function in male primates

In the last chapter, it was noted that circulating concentrations of plasma testosterone do not predict levels of sexual arousal or copulatory activity in male primates, provided that a sufficient threshold level of hormone is present to maintain sexual behaviour. Nor, in most cases, do measurements of circulating testosterone indicate which adult males will become aggressive or attain high-rank within a social group. Under conditions where rank orders are *well-established*, so that agonistic behaviour is stabilized, basal levels of testosterone do not correlate with dominance in captive Japanese macaques (Eaton and Resko 1974b), rhesus monkeys (Gordon *et al*. 1976), baboons (Sapolsky 1982), or vervets (Raleigh and McGuire 1990). If plasma testosterone is measured before group formation, the levels do not correlate with the ranks attained by the various males once they are introduced into the group situation (rhesus monkeys: Rose *et al*. 1975; squirrel monkeys: Mendoza *et al*. 1979; talapoins: Eberhart and Keverne 1979). This is not surprising, because there is no simple cause-and-effect relationship between testosterone and aggression in male primates; effects exist but they are less robust than effects of

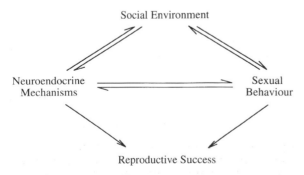

Fig. 14.1 Interactions between social environment, neuroendocrine mechanisms and sexual behaviour which influence reproductive success. (Modified from Bercovitch and Ziegler 1990.)

androgens upon copulatory behaviour, so that minor fluctuations in circulating testosterone are irrelevant as predictors of future aggressiveness or submissiveness in monkeys (Dixson 1980).

The situation is somewhat different when one considers *unstable* hierarchies and relationships between aggression and circulating testosterone in male primates. Various studies have shown that under conditions of social instability dominant, aggressive males tend to exhibit elevated testosterone levels whereas testosterone levels might decrease in subordinates (Sapolsky 1993). Most studies of the rhesus monkey, for example, indicate that testosterone levels increase in the winners of aggressive encounters in unstable social groups, whilst testosterone levels decline in losers (Rose *et al.* 1972, 1975; Bernstein *et al.* 1974). There is one report, however, of increased testosterone following aggression in both dominant and subordinate rhesus monkeys (Perachio 1978) some studies have reported increases of salivary testosterone in man after winning sports contests, including non-athletic contests such as chess tournaments (Dabbs 1993).

Sapolsky (1982, 1985, 1987, 1993) has conducted valuable fieldwork on the effects of social and other environmental stimuli upon pituitary–testicular and pituitary–adrenal function in adult male baboons (*Papio anubis*). His research provides some insights into the mechanisms which underpin the neuroendocrine strategies of dominant versus subordinate males; the superior coping mechanisms of dominants in stressful circumstances might contribute to their survival, and ultimately to their greater reproductive success.

When male baboons are immobilized by use of anaesthetic darts, plasma levels of luteinizing hormone (LH) and testosterone decline over the next 8 hours. Dominant males exhibit a transient increase in testosterone levels during the first hour of immobilization, despite declines in circulating LH—plasma testosterone in dominant males tends to be more resistant to suppression than that in subordinates (Fig. 14.2). The fall in LH levels is mediated in part by opioids—if males are treated with the opioid antagonist naloxone immediately after darting, stress-induced declines in LH are prevented (Sapolsky 1987). The transient increase in testosterone in dominant males is due to two factors. Firstly, increased activity in the sympathetic nervous system enhances testosterone secretion, although whether this occurs as a result of increased blood flow through the testes or by some direct stimulatory action upon the Leydig cells is unknown (Sapolsky 1987). Pre-treatment of dominant males with chlorisondamine, to block release of noradrenaline and

adrenaline by sympathetic neurons, reduces the early rise in plasma testosterone. Secondly, dominant males are more resistant to the negative effects of stress-induced glucocorticoid secretion upon testicular function. High levels of cortisol reduce testicular sensitivity to LH and hence testosterone secretion declines; this effect is more pronounced in subordinate males (Sapolsky 1985). An overview of these various effects of anaesthetic darting and immobilization upon LH and testosterone in high and low-ranking baboons is shown in Fig. 14.3.

Increased secretion of cortisol during threatening and stressful situations is adaptive since glucocorticoids mobilize glucose reserves, suppress inflammatory responses to tissue damage, and improve cardiovascular tone in situations demanding a *fight* or *flight* response (Sapolsky 1993). However, prolonged (i.e. chronic) elevations of circulating cortisol can be damaging to the brain and the immune system. It is significant, therefore, that studies have shown that low-ranking individuals in stable primate groups often exhibit higher levels of circulating cortisol than

A. Testosterone

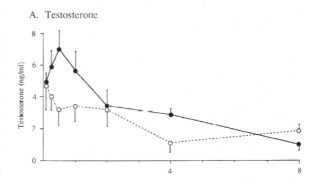

B. Luteinizing Hormone

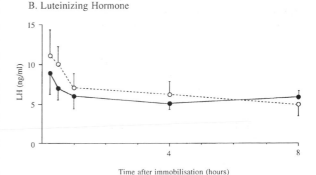

Time after immobilisation (hours)

Fig. 14.2 Effects of darting and immobilization of free-ranging male baboons upon circulating concentrations of testosterone and luteinizing hormone in high-ranking (•) and low-ranking (○) individuals. Data are means (±S.E.M.). (Redrawn and modified from Sapolsky 1987.)

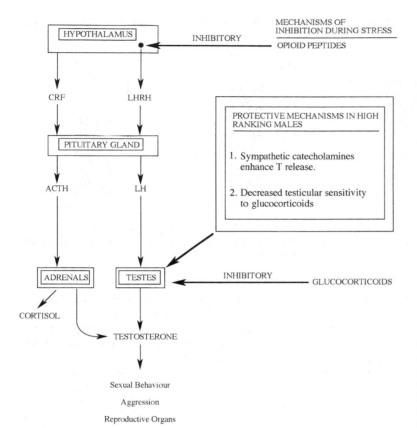

Fig. 14.3 Schematic diagram of some possible mechanisms by which stressful stimuli affect the hypothalamic–pituitary–testicular and adrenal axes in high-ranking, and low-ranking, adult male baboons. (Modified from Sapolsky 1987.)

dominant males (e.g. in the talapoin: Keverne *et al.* 1982; squirrel monkey: Manogue *et al.* 1975; and baboon: Sapolsky 1989). Dominant males are better able to respond to stressful situations by additional secretion of cortisol and to return to baseline levels after such increases have occurred. In the cynomolgus monkey (*Macaca fascicularis*), dominant males tend to have smaller adrenal glands than do subordinates (Shively and Kaplan 1984), whereas in the baboon there is preliminary evidence that dominant males have the highest circulating lymphocyte counts. The implication is that lower basal levels of glucocorticoids are commensurate with a healthier immune system in such males (Sapolsky 1993).

Social rank and secondary sexual traits in male primates

In Chapter 7 the phenomenon of delayed secondary sexual development in male primates was described, with reference to the effects of inter-male competition in multimale–multifemale, polygynous and dis-

persed mating systems. Examples included failure of full sexual skin development in mature male mandrills, suppression of cheek flange growth in orangutans, delayed nasal growth in proboscis monkeys, and slowing of secondary sexual changes in pelage coloration in various monkeys (see Table 7.7, page 197, for a list of examples). In this section, I shall describe what little is known about the socioendocrinology of secondary sexual suppression with reference to the male mandrill and the male orangutan.

In a semi-free ranging group of mandrills living under natural conditions in Gabon two morphological variants of adult male were identifiable. Group-associated or fatted males had a stocky appearance and possessed large fat deposits around the rump and flanks. These males had brilliant red and blue sexual skin on the face and rump. Solitary or peripheral males, by contrast, were non-fatted in appearance, with greater head—body length, and a more subdued sexual skin coloration. Non-fatted individuals also had smaller testes and lower circulating levels of testosterone than age-matched, group-associated adults, as can be seen in Fig. 14.4. It

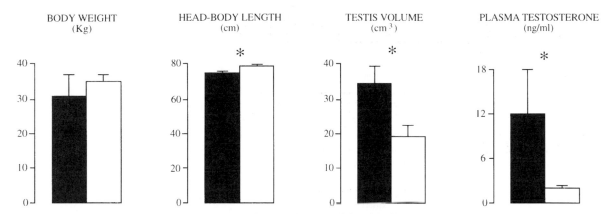

Fig. 14.4 Comparisons of body weight, head–body length, volume of the left testis and plasma testosterone levels in (closed bars) three fatted adult male mandrills (*Mandrillus sphinx*) and (open bars) three non-fatted individuals. Data are means (+S.E.M.), *P < 0.05, Mann-Whitney U test. (From Dixson 1997a.)

remains unknown whether non-fatted males are fertile, but despite their smaller testes it should not be assumed that they suffer from a global suppression of reproductive function. Such males are sexually active (they mate opportunistically) and ejaculate and produce seminal fluid; sufficient testosterone is present to maintain copulatory behaviour and functional accessory sexual glands. Unfortunately, there are no data available on spermatogenesis in these males. Nor should this type of suppression be viewed as an irreversible phenomenon. In the one case where dominant, fatted males were removed from the social group, this resulted in increased fattedness, sexual skin coloration, and sexual activity in a previously subordinate male. The intense intermale competition which occurs in the mating system of this species (see pages 53–4) might result in some individuals opting for a slower rate of secondary sexual development. The longer and leaner appearance of non-fatted males might be accentuated by an extended period of skeletal growth and delayed epiphysial fusion. This may have disadvantages, in that non-fatted males are less competitive or less sexually attractive to females. However, a longer term benefit may accrue to a suppressed individual in lessening the risk of damaging aggressive conflicts with dominant males, delaying investment in costly secondary sexual adornments, and yet allowing some occasional opportunities for matings. It has also been suggested that high testosterone levels carry certain 'costs' in immunological terms, and that immunosuppressive effects of the hormone can, for example, increase the susceptibility of males to various parasitic infections (Folstad and Karter 1992; Saino and Møller 1994; Wedekind and Folstad 1994). We still know too little about the endocrinology of the male mandrill to say how this most interesting

secondary sexual suppression is controlled. An examination of hypothalamic–pituitary—adrenal function would be relevant, for instance, to determine whether heightened glucocorticoid secretion occurs and whether this interferes with testicular physiology in non-fatted males.

The adult male orang-utan also exhibits an impressive array of secondary sexual traits, including massive size, long hair, fatty cheek flanges, and a large laryngeal sac, which is important for the production of the territorial great call. Captive male orang-utans reach puberty and begin to exhibit spermatogenesis at approximately 6–7 years of age, but the secondary sexual traits described above do not develop immediately (Dixson *et al.* 1982). If young males are reared in captivity without a fully developed male being present, then the cheek flanges, throat sac, and other structures grow promptly from puberty onwards and are fully expressed by 9–10 years of age. The presence of a dominant (i.e. fully developed) adult male suppresses the growth of secondary sexual characteristics in younger individuals for between 3 and 7 years (Kingsley 1982, 1988). Kingsley measured urinary oestrogen and testosterone excretion in 20 male orang-utans of varying ages and included both non-flanged and flanged adult males in her sample. Urinary excretion of both oestrogen and testosterone was greatest in flanged adult males; the testosterone data are shown in Fig. 14.5. Some of the males in this study were breeders, whereas others had not sired offspring; there is some suggestion that the breeding males engaged in dominance displays towards subordinates. However, the precise nature of the cues which govern intermale suppression is not known. Visual cues and vocalizations seem the most likely stimuli in this context. In the wild, fully developed territorial males vocalize (using the long call) and engage in visual displays (such as tree-swaying) as well as reacting aggressively towards sub-adult or non-flanged adult males. The latter appear to be less attractive to females, but they attempt coercive matings on an opportunistic basis (see pages 49–50 for a discussion of mating tactics in orang-utans). It seems likely that Kingsley's observations on secondary sexual suppression in captive male orang-utans provide an accurate reflection of socioendocrinological events under natural conditions. However, we still know very little about how the presence of a fully developed male affects the neuroendocrine system of subordinates. Graham and Nadler (1990) suggest that fertility might be inhibited in subordinates and that such inhibition might continue in certain cases even after secondary sexual characters have developed. I think this is unlikely, however. Subordinate male orang-utans which I have examined, lacked cheek flanges, yet they had functional testes (sperm counts were made using semen samples obtained by electroejaculation (Dixson *et al.* 1982). Further, one of these males mated and impregnated a female despite his non-flanged status. Graham and Nadler

(1990) also propose several possible mechanisms by which suppression of secondary sexual development might occur.

1. Inhibition of adrenarche and adrenal androgen secretion which accompanies puberty in man and chimpanzees. Adrenal androgen secretion has been implicated in growth of the pubic hair in human beings; whether it might play some role in secondary sexual development in the orang-utan is unknown.

2. A stress response 'affecting endorphin/ ACTH/cortisol secretion' and/or

3. Hyperprolactinaemia with associated inhibition of the pituitary-gonadal axis.

None of these possibilities has been explored by experiment. Nor do we know precisely what mechanisms underlie the suppression of secondary sexual development in a number of other primate species. Because such males may be fertile and sexually active it is possible that androgen-sensitive tissues such as the cheek flanges of the orang-utan or the

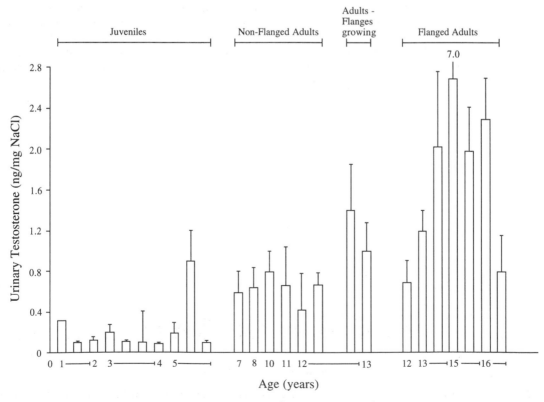

Fig. 14.5 Urinary testosterone levels in male orang-utans (*Pongo pygmaeus*): correlations with age and with the presence, or absence, of secondary sexual characteristics (cheek flanges) in adults. (Redrawn and modified from Kingsley 1982.)

sexual skin of the mandrill are desensitized in some way. Receptors might be compromised, or perhaps the required conversion of testosterone to dihydrotestosterone (DHT) is inhibited by 5-alpha-reductase deficiency. Even the simple experiment of treating a suppressed male with a large dose of DHT, or testosterone, has yet to be attempted.

Female social rank, the ovarian cycle, and fertility

The menstrual cycle is disrupted by a variety of stressful factors, including illness (chimpanzees: Yerkes 1943; Goodall 1986), being captured or transported after capture (baboon: Gillman and Gilbert 1946), high levels of exercise and restricted food intake (e.g. in female athletes, ballet dancers, and those on stringent diets: Pirke *et al.* 1989). However, it appears that in some instances psychological factors affect the cycle directly, rather than via some secondary agency such as increased competition for nutritional resources. There is good evidence that low social rank, receipt of aggression, or the stress of social isolation, can all affect the duration of the menstrual cycle in female baboons. The first conclusive demonstration of such effects was by Rowell (1970a), who studied the menstrual cycles of *Papio anubis* living in a captive social group. She observed that when females were moved into (or out of) the group at the time of menstruation, then the

ensuing follicular phase of the cycle, as indicated by sexual skin swelling, was longer than normal (Fig. 14.6). The follicular phase of the cycle, rather than the luteal phase, was most sensitive to social stressors, such as receipt of aggression from other females, and in extreme cases the sexual skin remained flat for up to three months.

Long follicular phases also occur in low-ranking female mandrills, and a correlation exists between female rank and duration of the swelling phase of the menstrual cycles in these animals (Fig. 14.7). Lower-ranking females encounter aggression more frequently, but an additional factor is the nutritional status of these animals because lower-ranking females also spend less time feeding than others. It appears that low rank might be associated with a delay in maturation of the dominant (ovulatory) follicle within the ovary and with lower circulating levels of oestradiol, as reflected by the smaller size of the sexual skin swelling. The mechanisms which might produce such effects will be discussed in a moment. One consequence of disrupting follicular development might be that lower-ranking females will require more menstrual cycles in order to conceive. Some evidence to support this contention has been obtained by Wasser and Starling (1988) who studied three free-ranging troops of yellow baboons (*Papio cynocephalus*) in the Mikumi National Park, Tanzania. Females sometimes formed coalitions in order to attack others. Females were most likely to

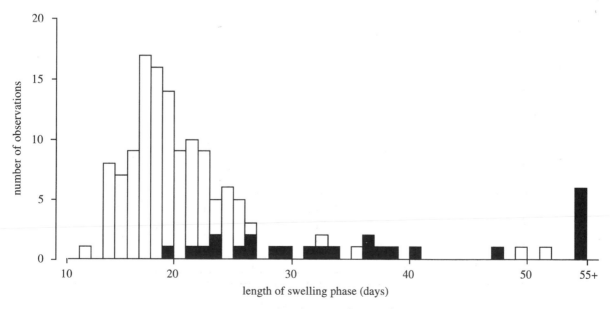

Fig. 14.6 Effects of social stress upon the length of the follicular (swelling) phase of the menstrual cycle in captive baboons. Lengths of the swelling phases for the sexual skins of seven females were recorded during all cycles. Black bars indicate cycles in which the female was moved into or out of the group, or had been attacked by other group members. (Redrawn and modified from Rowell 1970a.)

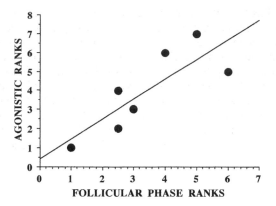

Fig. 14.7 Correlation between the length of the follicular phase of the menstrual cycle, as reflected by sexual skin swelling, and agonistic rank in semi-free-ranging mandrills (*Mandrillus sphinx*) in Gabon. $r_s = 0.829$, $P < 0.05$. (Author's unpublished observations.)

be targeted by attack coalitions in late pregnancy or soon after giving birth. They often exhibited 'an increased number of cycles to conception and longer interbirth intervals, as well as spontaneous abortion, premature birth, and prolonged gestation.' Packer *et al.* (1995) report that high-ranking female baboons at the Gombe Stream Reserve in Tanzania exhibit reduced interbirth intervals. However, they did not find that the number of menstrual cycles before conception was greater in lower-ranking than in dominant females—on the contrary, these authors point to a higher frequency of miscarriages and delayed age of first conception in dominant females as being indicative of the *physiological costs* associated with high rank. They suggest that higher levels of androgens in dominant females might be responsible for fertility problems but there is no evidence that such androgenization had actually occurred. These findings have also been disputed by other field workers (Wasser 1995).

Irregular ovarian cycles or total suppression of ovarian function might result from a variety of neuroendocrine mechanisms. Receipt of aggression or the stress associated with low rank or nutritional constraints might be expected to affect the hypothalamic–pituitary–ovarian axis directly, or via the adrenal glands and glucocorticoid secretion. Female yellow baboons at Mikumi in Tanzania rarely conceive during the dry season, for example, and Wasser (1996) has obtained evidence that lower progesterone levels and luteal phase deficiency (LPD) is the cause of reduced fertility at this time. Harsh environmental conditions, and especially lack of food in the dry season, affect the ovarian cycle and pro-

duce a condition resembling the LPD seen in some infertile women. Hyperprolactinaemia is also associated with infertility in the human female (Thorner and Besser 1977). In the talapoin monkey, subordinate females in captive social groups are hyperprolactinaemic and fail to exhibit a preovulatory surge of LH in response to an exogenous oestrogen challenge (Fig. 14.8). Treatment of such females with bromocryptine reduces prolactin levels and restores their ability to display the positive feedback response to oestradiol (Bowman *et al.* 1978). It is not known how far these results, obtained by studying ovariectomized females in small captive groups, might apply to free-ranging talapoins. However, it is known that high levels of aggression occur during the annual mating season of free-ranging talapoins and also that some females are still carrying offspring of the previous year at this time. Infants represent more than 40% of maternal body weight, so that to transport and nurture the offspring must be an energetically costly exercise. It is by no means certain that all females breed during each year (Rowell and Dixson 1975).

High prolactin levels also occur during lactation, and lactational amenorrhea occurs in monkeys and apes, as well as in the human female (Short 1976). Death of an infant is known to shorten this period of ovarian quiescence and thus to reduce the duration of the interbirth interval in a number of species. The relationship between mother and infant has powerful effects upon her neuroendocrine status and her return to sexual activity once weaning is underway. Theories of parent-infant conflict (Trivers 1972) predict that it might be adaptive, from the infant's perspective, to prolong the mother's lactation period. Indeed there is evidence that subordinate female rhesus monkeys may suckle their infants for longer and that this, in turn, results in a prolongation of the period of lactational amenorrhea (Gomendio 1990). Subordinate lactating female baboons take longer to resume menstrual cyclicity than do dominant females in their troops (Packer *et al.* 1995).

Coplan *et al.* (1996) have obtained evidence for long-term effects of poor maternal care upon neuroendocrine function in the offspring of bonnet monkeys. Female bonnet macaques with small infants were subjected to a variable foraging regime, involving periods when they had to perform operant tasks or dig for hidden food, alternating with periods when food was more readily available. These mothers exhibited 'inconsistent, erratic, and sometimes dismissive rearing behaviour'. Interestingly, the infants of these female monkeys exhibited higher than normal levels of corticotrophin releasing factor (CRF) in the cerebrospinal fluid (CSF) when sampled at 2 years of age. Coplan *et al.* (1996) acknowledge that caution is required when interpreting this result, as the animals studied were substantially younger when CSF was collected than those in the control groups. With this reservation in mind, there is an indication that poor maternal care might have a long-term effect upon the hypothalamic–pituitary–adrenal axis of the developing offspring.

Positive effects of maternal care upon the neuroendocrine development of offspring have been described in male rats. Male rat pups emit a chemical cue which stimulates higher frequencies of perineal licking by their mothers (Moore and Morelli 1979). This perineal stimulation is associated with greater development of the sexually dimorphic nucleus of the bulbocavernosus (SNB) in male pups (larger numbers of neurons occur in such males: Moore *et al.* 1992). Because the SNB controls reflex-

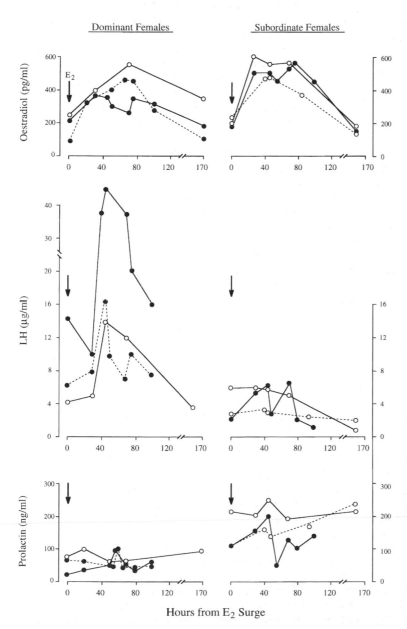

Fig. 14.8 Effects of positive oestrogen feedback (upper section, oestrogen-treated females receive a further two oestrogen implants — arrow indicates time point) on luteinizing hormone (LH) release in dominant (N = 3) or subordinate (N = 3) female talapoin monkeys (*Miopithecus talapoin*). Subordinate females, in receipt of high aggression, fail to show the characteristic preovulatory surge of LH in response to an oestrogen challenge and have higher levels of prolactin than dominant females. (Redrawn from Keverne 1992; after Bowman *et al.* 1978.)

ive penile movements which occur during copulation in male rats (see pages 263–5 and Fig. 9.14 for a discussion of this), maternal behaviour might have long-term effects on sexual behaviour and reproductive success in the male rat. Very little is known about how maternal behaviour might influence neuroendocrine development, including the characteristics of the postnatal testosterone surge, in male monkeys or in human infants.

Little is known about the involvement of the hypothalamic–pituitary–adrenal axis in stress-induced suppression of fertility in female monkeys or apes. Chronic elevation of plasma cortisol is associated with suppression of LH pulsatility in the rhesus monkey (in castrated males: Dubey and Plant 1985) and with amenorrhea in women (Suh *et al.* 1988). Interestingly, suppression of adrenocortical secretion in the intact female rhesus monkey (by dexamethasone treatment) does not disrupt the menstrual cycle. Levels of oestradiol, LH and progesterone tend to be higher in these adrenal-suppressed females, although only in the case of progesterone is there a statistically significant effect (Lovejoy and Wallen 1990). Perhaps the hypothalamic–pituitary–adrenal axis exerts some type of brake upon ovarian activity, so that increased adrenocortical activity in a subordinate female might lead to irregular cycles or to amenorrhea.

A striking example of reproductive suppression of subordinate females by the presence of a dominant female is provided by the common marmoset (*Callithrix jacchus*). In free-ranging or captive family groups of this primate species it is usual for just one dominant female to reproduce, whilst subordinate females (often her own daughters) do not breed and are subject to varying degrees of ovarian suppression. Suppression of LH secretion, and hence of ovulation, is maintained via non-behavioural (olfactory) cues and by as yet incompletely-defined behavioural cues from the dominant, reproductive female. Barrett *et al.* (1990) showed that when reproductively suppressed female marmosets are re-

moved from their family groups, activation of LH secretion is delayed if subordinates remain exposed to olfactory cues from dominants (Table 14.1). Callithrichids possess a functional vomeronasal organ and accessory olfactory system, as do many other New World monkeys and prosimians. It is possible that primer pheromones activate this pathway to clamp the production of LH whilst maturing daughters remain in their natal groups. Chemical cues alone are insufficient, however, so that other signals from the dominant female, including perhaps visual displays and agonistic behaviour, must play some role in maintaining reproductive suppression. Two neuroendocrine mechanisms have been implicated in suppression of LH secretion in female marmosets. Firstly, subordinate females are exquisitely sensitive to the effects of tiny amounts of oestradiol secreted by their ovaries. Oestradiol exerts a negative feedback upon LH secretion in subordinate females (Abbott 1988; Abbott *et al.* 1990). When ovariectomized subordinate marmosets are exposed to a low dose of oestradiol (sufficient to produce circulating hormone levels of $350\,\mathrm{pg\,mL^{-1}}$) marked suppression of LH occurs; dominant, ovariectomized females fail to exhibit such negative feedback (Fig. 14.9). The second mechanism resulting in suppression of LH secretion and ovulation in subordinate marmosets concerns the endogenous opioid peptides. Ovariectomy is not, by itself, usually sufficient to disinhibit LH secretion in a subordinate marmoset if she remains in the family group; removal of negative feedback by ovarian oestradiol is only successful if the female is also removed from the presence of the dominant female. However, treatment of subordinate ovariectomized females with the opioid antagonist naloxone causes an increase in plasma LH concentrations (Abbott *et al.* 1990; Abbott 1993). Likewise, if an ovariectomized, oestradiol-treated female marmoset receives aggression from a dominant female (in a pair-test), this causes a prompt decrease in plasma LH, resulting from a reduction in the amplitude of the pulses of pituitary LH,

Table 14.1 Effects of social isolation, with or without presence of group odour, upon ovulation in subordinate female marmosets

Experimental group	Number of females studied	Experimental conditions	Time to onset of ovulation (days)
Control group 1	8	Isolated, without olfactory cues from home group	10.8 ± 1.4
Control group 2	5	Isolated, without olfactory cues from home group	10.4 ± 0.8
Scent-transfer group	8	Isolated, but olfactory cues from home group maintained	$31.0 \pm 6.4*$

*$P < 0.05$. Data are from Barrett *et al.* 1990.

rather than a lessening in pulse frequency (O'Byrne *et al.* 1988). Pre-treatment of such females with naloxone prevents the stress-induced decline in LH secretion (Fig. 14.10).

Subordinate female marmosets exhibit remarkably little sexual behaviour (Abbott 1993). Given that proceptivity in the female marmoset is influenced by increased oestradiol secretion and LHRH activity during the peri-ovulatory period (Kendrick and Dixson 1985b; Dixson and Lunn 1987), suppression of these mechanisms in the subordinate female might affect both behaviour and fertility. However, it may be incorrect to assume that subordinate female marmosets are completely suppressed at the hypothalamic level, and that the LHRH pulse generator is inactive, due to opioid-induced suppression, or to negative feedback by oestradiol (or both). Abbott's latest research on this problem, conducted at the Wisconsin Primate Centre, shows that subordinate females continue to produce pulses of LHRH, measurable in fluid collected from the extracellular

spaces surrounding the pituitary stalk (i.e. not directly from the portal vessels themselves). The significance of these observations is at present unclear (Abbott *et al.* 1997).

Suppression of ovulation has been documented in several other callitrichid species (e.g. subordinate cotton-top tamarins: French *et al.* 1984; Abbott *et al.* 1993; and saddleback tamarins: Epple and Katz 1984) but not in subordinate golden lion tamarins (French *et al.* 1989). In the golden lion tamarin, behavioural sexual suppression, rather than physiological suppression of fertility, is sufficient to prevent daughters becoming pregnant whilst they remain in their natal groups.

Social environment and reproductive synchrony

Some co-ordination of reproductive function between the sexes is essential if fertile matings are to occur; this is especially so in those primates where mating is constrained by seasonal factors (Lindburg 1986). In addition to cues from the physical environment, such as changes in photoperiod or rainfall, evidence has gradually accumulated that social factors can profoundly affect the co-ordination of endocrine status and behaviour in seasonally breeding primates. In the talapoin monkey, for example, neighbouring troops in the wild may exhibit significant differences in the fine-tuning of reproductive status; patterns of sexual swelling and copulation are tightly co-ordinated within troops, but differ between troops in adjacent home ranges (Fig. 14.11). Communication between groups in the wild is minimal, hence social co-ordination of the mating season operates primarily within the troop; in Fig. 14.11 the two troops are approximately two weeks out of phase in the timing of the mating season. In captivity, treatment of ovariectomized female talapoins with oestradiol results in increased levels of circulating testosterone in adult males living in the same group (Keverne 1987). One type of co-ordination which might occur under natural conditions is for females which enter reproductive condition to stimulate changes in hypothalamic–pituitary–testicular activity in male conspecifics. Several examples of this type may be cited. Firstly, among the prosimians Perret's (1985, 1992, 1995) studies of the lesser mouse lemur (*Microcebus murinus*) provide an excellent example of how cues from the physical environment (changes in photoperiod) and social environment (olfactory stimuli from pro-oestrous females) combine to stimulate testosterone secretion in males during the mating season. Long day-length (>12 h)

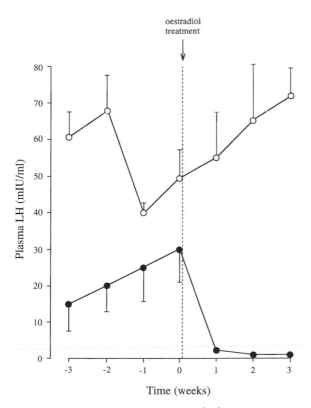

Fig. 14.9 Plasma luteinizing hormone (LH) concentrations in dominant and subordinate female common marmosets (*Callithrix jacchus*) before and after treatment with oestradiol 17β (subcutaneous silastic implant). Data are means (±S.E.M.) for three dominant (○) and five subordinate (●) females. (Redrawn from Abbott *et al.* 1990.)

stimulates gonadal activity in mouse lemurs. Males also show increased levels of plasma testosterone when housed with oestrous females in captivity (Perret 1985). Merely exposing a male to female urine over a four-week period activates testosterone secretion, provided that the donor female is in prooestrous (Fig. 14.12). In the natural environment, where the sexes occupy separate home ranges, and direct contact is limited, scent-marking by females has an important stimulatory action upon central (dominant) males in the population. These effects begin during the pro-oestrous period some days before a female becomes sexually receptive. Her scent-marks serve both to attract males and to stimulate testicular activity in potential mates.

A second well-documented case of increased testicular activity in response to female stimuli concerns the rhesus monkey (Vandenbergh 1969; Vandenbergh and Drickamer 1974). Three semi-free-ranging groups of rhesus monkeys on Le Cueva

Island (Puerto Rico) were utilized for Vandenbergh and Drickamer's (1974) study. Two mid-ranking females in Group 1 were ovariectomized and returned to their group. Subsequently, these females were treated with oestradiol benzoate during the non-mating season, at a time of year when copulations do not occur and the male's sexual skin, covering the rump and genitalia, is pale owing to reduced testicular activity (Vandenbergh 1965). Hormonal treatment of the two females brought about a marked activation of copulatory behaviour among the nine adult males in Group 1, as well as androgen-dependent reddening of the sexual skin (Fig. 14.13). Most interestingly, intact females in this group also entered breeding condition earlier than normal and the subsequent birth season was one month earlier than expected. These effects did not occur in the two control groups. In Vandenbergh and Drickamer's study it was not possible to capture males in order to measure changes in circulating

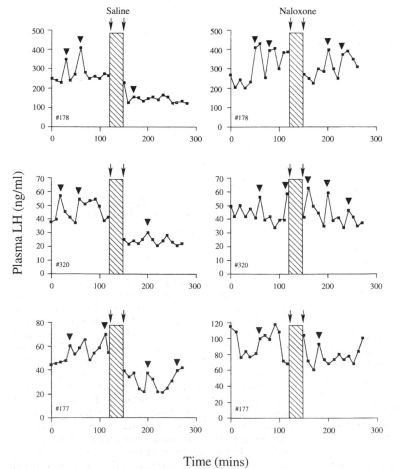

Time (mins)

Fig. 14.10 Plasma luteinizing hormone (LH) profiles for three ovariectomized, oestradiol-primed common marmosets (*Callithrix jacchus*) during physical restraint to collect rapid serial blood samples before, and after, a 30 minute aggressive encounter test with a female conspecific (cross-hatched bar). The responses following receipt of aggression alone (left column) and when females were treated with naloxone (1 mg / kg) immediately before, and after, the encounter test (right column) are illustrated for each subject. LH pulses are indicated by pointers. (Redrawn from O'Byrne, Lunn and Dixson, 1989).

A. Percentages of Females with Medium/Large Swellings

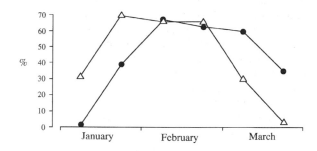

B. Distribution of Copulations

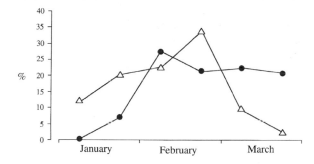

Fig. 14.11 A. Percentages of females showing medium / large sexual skin swellings and B. distribution of copulations during the mating season in two free-ranging troops of talapoin monkeys (*Miopithecus talapoin*) occupying adjacent home ranges in the Cameroonian rainforest. (Redrawn from Rowell and Dixson 1975.)

A. Exposure to anoestrous female urine

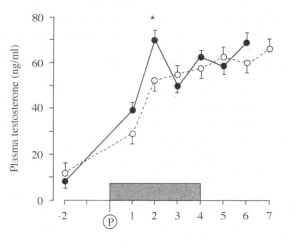

B. Exposure to pro-oestrous female urine

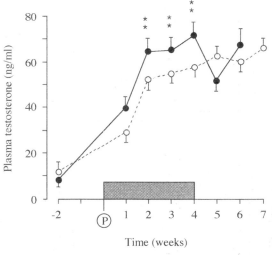

Fig. 14.12 Effects of long photoperiod and female urinary cues upon plasma testosterone (T) levels in adult male mouse lemurs (*Microcebus murinus*). A. Control males (○) showed increased T in response to extended artificial day length (beginning at P) and sustained for 4 weeks (as indicated by the horizontal bar.) They were exposed to non-odourized air, whereas experimental males (●) received air which had been odourized with urine from an anoestrous female. There was only one week during which experimental males had higher T levels (*P < 0.05.) B. The same experimental arrangement was used except that experimental males were exposed to air permeated with urinary cues from a pro-oestrous female. This produced significant increases in T during weeks 2, 3, and 4 as compared with the control group (**P < 0.01.) (Redrawn and modified from Perret 1992.)

testosterone. However, subsequent work has shown that male rhesus monkeys, like male talapoins, exhibit increased levels of plasma testosterone when exposed to sexually attractive females in a social group (Rose *et al.* 1972). In a captive all-male group of rhesus monkeys, activation of sexual behaviour did not occur during the mating season unless the group was housed next to a heterosexual group in which sexual activity increased in the normal way (Gordon and Bernstein 1973). Some increases in plasma testosterone and sexual behaviour were measured in a subsequent study involving an isolated all-male group (Gordon *et al.* 1978). These studies have been interpreted as indicating the ability of male rhesus monkeys to respond directly to cues in the physical environment, which initiate hypothalamic—pituitary—gonadal activity. However, additional exposure to females is required to fully activate gonadal recrudescence and copulatory behaviour during the mating season (Lindburg 1986).

As well as these more detailed studies of female-stimulated increases in male testicular physiology, there are several other reports which have been less well documented. In the ruffed lemur (*Varecia variegata*) testicular volume increases during the annual mating season and reaches a maximum to coincide with the female's brief oestrus (Izard 1990). In the mandrill, reddening of an isolated male's sexual skin (mid nasal strip and perineal field) occurred when he was allowed visual (and auditory) contact with a heterosexual group during the mating season (author's unpublished observations). The phenomenon is probably a general one, and might be

demonstrable in other seasonally breeding primates. Although pheromonal cues have been shown to be important in the mouse lemur, it is likely that behavioural stimuli, or non-behavioural visual signals, (e.g. female sexual skin swellings) are important in seasonally breeding monkeys.

A second mechanism that might facilitate reproductive co-ordination between the sexes concerns effects of male stimulation upon activation or synchronicity of ovarian cycles in female conspecifics. Numerous examples of such effects can be cited for birds and mammals. The courtship displays of the male ringdove are important for co-ordination of his

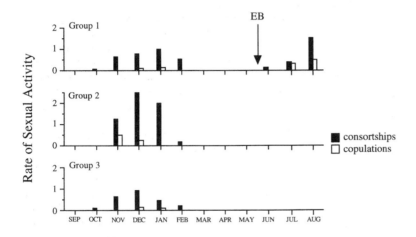

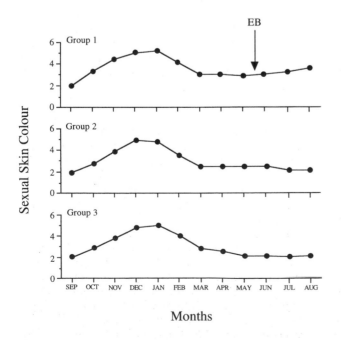

Fig. 14.13 Monthly changes in frequencies of consortships (closed bars), copulations (open bars) and sexual skin coloration (filled circles) in adult males in three semi-free-ranging groups of rhesus monkeys (*Macaca mulatta*). The ovariectomized females in group 1 were implanted with oestrodiol benzoate (EB) in the non-mating season, as indicated. (Redrawn and modified from Vandenbergh and Drickamer 1974.)

female partner's endocrine status and incubation behaviour (Lehrman 1965). In canaries the presence of a male accelerates nest building by the female, an effect which is probably mediated via increased secretion of ovarian oestrogen (Hinde 1965). Effects of male pheromones upon oestrus synchrony have been well-documented in mice (Bronson 1989). Pheromonal stimuli also influence the onset of oestrous in pigs and in sheep (the 'ram effect': Signoret *et al.* 1984). The roaring vocalizations given by a harem-holding stag, during the annual rut, advance the date of oestrus in the hinds in his group (McComb 1987).

Few comparable examples of male effects upon female ovarian cyclicity have been identified among the primates, although this probably reflects the relative lack of experimental work in this area. Male ringtailed lemurs employ complex scent-marking displays which might play some role in social facilitation of oestrus synchrony among females of this species (Jolly 1966; Evans and Goy 1968). Synchronicity of reproduction in a captive colony of greater galagos (*Galago crassicaudatus*) maintained under constant photoperiod (12 h light: 12 h dark) and a constant diet, presumably resulted from social factors, but the contribution made by stimuli from males was not assessed (Izard 1990). Dominant males in free-ranging bushbabies and pottos visit and court females in adjacent home ranges. Again the possibility exists that such behaviour might stimulate ovarian activity (see Fig. 11.3 page 318, showing frequencies of visiting and courtship behaviour by a male potto). Further experimental investigation of the possible effects of male primer pheromones upon female reproductive status in prosimians would be most valuable.

There is very little evidence for effects of male stimulation on reproductive function in monkeys or apes. To examine this problem, Vandenbergh and Post (1976) sterilized six adult females (by oviductal ligation) in several semi-free-ranging groups of rhesus monkeys. The females accordingly failed to conceive during the annual mating season, and were removed from their groups during the non-mating season when their menstrual cycles had ceased. Three females were paired with testosterone-treated adult males and three with sham-operated males. These males had been removed from different groups before the experiment, and were unfamiliar to the females. Although testosterone treatment had measurable effects on the males' mounting and aggressive behaviour, their female partners exhibited minimal changes in sexual behaviour and no activation of ovarian cyclicity. Thus, no male influence upon female reproductive status was measurable.

However, the conditions of the experiment were unusual, in that pairs of animals were studied, rather than social groups.

When squirrel monkeys are caged in heterosexual pairs before the onset of the annual mating season, males respond by entering *breeding readiness* and their plasma testosterone levels increase. Females, however, fail to show activation of ovarian function under these conditions (Mendoza and Mason 1989). Yet, in a later study, Mendoza and Mason (1991) demonstrated that if unfamiliar females are housed together in triads for four weeks and then a strange male is introduced to the group, breeding readiness is induced in females by exposure to the male. Females exhibit decreased cortisol secretion when housed together, levels of plasma oestrogen rise within 24 h of introduction of a male, and luteal phase increases in circulating progesterone are measurable six days later (Fig. 14.14). Mendoza and Mason therefore suggest that female–female priming of the neuroendocrine system is important as a prelude to male-induced activation of ovarian cyclicity.

These findings using laboratory groups of squirrel monkeys might be relevant to understanding reproductive co-ordination in rhesus monkey groups. When female rhesus monkeys are housed in isosexual groups containing intact and ovariectomized individuals, treatment of the ovariectomized females with oestradiol during the non-breeding season fails to advance the onset of menstrual cyclicity in intact animals (Ruiz de Elvira *et al.* 1983). This finding is the converse of that described above by Vandenbergh and Drickamer (1974), but in that study adult males were present in the social group when females received oestradiol. Ruiz de Elvira *et al.* (1983) therefore suggest that the presence of males might be a prerequisite for activation of ovarian cyclicity in females. In Vandenbergh and Post's (1976) study, failure of testosterone-treated males to activate ovarian function in females during the non-breeding season might have been because the monkeys were housed in pairs; female–female priming might also be required, as reported for the squirrel monkey.

Synchrony of ovarian cyclicity in group-housed females has been described for a number of mammals, and this phenomenon has been especially well-documented for mice, hamsters, and rats. In the rat, airborne pheromonal cues are sufficient to produce such effects (McClintock 1978); preovulatory odours shorten or phase-advance the cycles of recipients whereas ovulatory odours exert an opposite effect (McClintock 1984). Oestrous synchrony has been mooted for a number of prosimian primates (Izard 1990) and especially for the ringtailed lemur (Jolly 1966; Pereira 1991), but the existence of interfemale pheromonal effects has yet to be demonstrated. Some menstrual synchrony, or at least reduction of irregular cycle lengths, has been reported to occur among females of several monkey species (e.g. *Macaca fascicularis*: Wallis *et al.* 1986; *Papio hamadryas*: Zinner *et al.* 1994) and in captive chimpanzees (Wallis 1985). Tight co-ordination of conception cycles is also inferred for certain species with a restricted birth season (e.g. *Cercopithecus aethiops*: Schapiro 1985; *Saimiri sciureus*: Milton and Johnston 1984).

The advantages of tight co-ordination of matings, conceptions and subsequent births in some seasonally breeding primates might relate to predation pressure and to seasonal changes in food availability. If all females give birth during the same period of the year, then all are carrying young offspring at this time and the whole troop will move more slowly as a consequence. A single female with a young infant might be more vulnerable to predation. Milton and Johnston (1984) use this reasoning to explain why 80% of squirrel monkey births occur during a 10-day time-span in the wild. Births are also clustered during the wet season in squirrel monkeys, at a time of year when food availability and nutritional support for lactating females is improving. With some exceptions (e.g. *Erythrocebus patas*) birth seasons, or birth peaks, in primates which exhibit reproductive seasonality, tend to occur during wetter periods (Lindburg 1986). Hence, the

various neuroendocrine mechanisms which promote synchrony of sexual activity and conceptions may be advantageous in terms of increased infant survival. For primates with very long interbirth intervals, such as the 3–4 year intervals typical of the great apes and human hunter–gatherers, timing of mating and birth seasons with respect to annual cycles of food availability becomes less feasible. In this regard, it might be significant that all of the hominoids are non-seasonal breeders.

Despite lack of seasonality among the apes and man, there have been many reports that menstrual synchrony can occur in human females. McClintock (1971) was the first to describe synchrony among women living in a college dormitory. A number of later studies confirmed her findings, although the effect might be greatest when women are both friends and room-mates (Weller and Weller 1993; Weller *et al.* 1995). One study has reported synchronization of cycles in response to olfactory cues alone (axillary secretions collected from donor females at various stages of the menstrual cycle: Preti *et al.* 1986). An axillary organ consisting of large apocrine glands occurs in the African apes, as well as in man; it would be interesting to know whether its secretions have any effects upon their reproductive physiology. Given the rarity with which a number of females cycle simultaneously in free-ranging groups of gorillas or chimpanzees, it is difficult to envisage the selective advantage of mechanisms which result in menstrual synchrony. There might, however, be

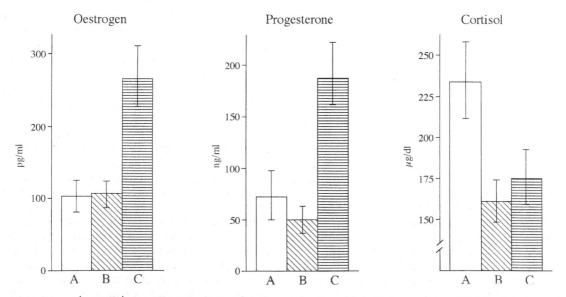

Fig. 14.14 Mean (±S.E.M.) concentrations of circulating oestrogen, progesterone and cortisol in female squirrel monkeys (*Saimiri sciureus*) when A. housed individually for 2 weeks, B. housed in female triads for 4 weeks and C. 4 weeks after the introduction of a male to the female triad. The significance of prior grouping of females for induction of 'breeding readiness' by exposure to a male is discussed in the text. (Redrawn and modified from Mendoza and Mason, 1991.)

some advantage to such mechanisms during adolescence, when cycles are irregular and anovulatory. This possibility is discussed below, in the section dealing with social factors and puberty.

In Chapter 4, the concept of the operational sex ratio was discussed in relation to sexual selection and mating tactics in primates. The operational sex ratio refers to the numbers of females which are in breeding condition versus numbers of sexually active males at any given time in a particular mating system. When a high degree of menstrual synchrony exists among females, so that they ovulate within a relatively short time-span, the likelihood that an individual male could monopolize sexual access to all females is accordingly reduced. Multiple partner copulations and sperm competition are probably favoured under conditions of female synchrony. It was also noted in Chapter 4 that breeding seasonality and concentration of sexual activity within discrete mating seasons is much less common among polygynous primates (i.e. one-male units such as gorillas, geladas and hamadryas baboons), than among those which form multimale–multifemale mating systems (e.g. rhesus macaques or talapoins). The possible *disadvantages* of menstrual synchrony for females in a polygynous mating system have been highlighted by Zinner *et al.* (1994), in a study of captive hamadryas baboons. No seasonality of reproduction occurred in this captive colony during a 12-year study period, but examples did occur when more than one female exhibited sexual skin swelling and was sexually active. The degree of female swelling synchrony within one-male units influenced the probability that conception would occur. A female was more likely to conceive if her swelling phase did not coincide with that of other females in the one-male unit (Fig. 14.15). Zinner *et al.* argue that 'sperm might be a limited resource in the one-male reproductive units of hamadryas baboons'. Other factors such as male preferences for particular females and competition between females might be equally or more important, however. Whatever the factors involved, the findings shown in Fig. 14.15 raise the interesting possibility that females might engage in a certain degree of *ovarian asynchrony*, depending upon the dynamics of the mating system.

This question was examined by Pereira (1991), in relation to the mating season of the ringtailed lemur (*Lemur catta*). Pereira pointed out that despite the brevity of the mating season in this species (1–3 weeks in the wild) and the fact that individual females are receptive for less than 24 h, two animals very rarely come into oestrus on the same day (Jolly 1966; Koyama 1988; Sauther 1991), as is shown in Fig. 14.16. One possible explanation is that within the overall pattern of breeding synchrony in ring-

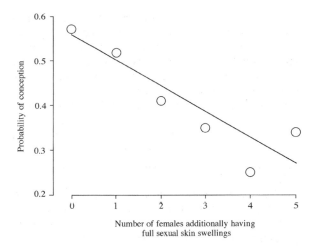

Fig. 14.15 Probability of conception in female hamadryas baboons (*Papio hamadryas*) which were members of a captive one-male unit. As the number of females which simultaneously had swollen sexual skins increased, so the probability of conception was accordingly reduced. (Kendall's r=0.87; P < 0.015). (Redrawn and modified from Zinner *et al.* 1994.)

tailed lemur groups, pheromonal mechanisms exist that enable individuals to advance, or retard, the timing of oestrus with respect to the ovarian status of conspecifics. McClintock's findings on phase-shifting of oestrous cycles in rats, with respect to airborne pheromonal signals from pre-ovulatory or ovulatory females, have already been referred to. The scent-marking displays of female ringtails, involving anogenital cutaneous glands vaginal secretions and urine, might fulfil a similar pheromonal role. The adaptive significance of these small, but consistent, asynchronies in the timing of oestrus might be that female–female competition is reduced and that individuals are able to exert a greater degree of mate choice during their limited period of sexual receptivity.

A less extreme example of ovarian asynchrony during an annual mating season is provided by the mandrill. Fourteen females in a captive social group exhibited menstrual cycles during the 5-month season, and 12 are known to have conceived. The timing of maximum sexual skin tumescence and sexual skin breakdown for each female is plotted in Fig. 14.17. These cycles are not tightly synchronized and the spacing between ovulations enables dominant males in the social group to mate-guard individual partners, and then to transfer attention to another female once sexual skin breakdown has occurred. However, some synchrony is also apparent, because the majority of females ovulate during July or August and there is also a tendency for mothers and daughters to exhibit closer coupling of their ovarian cycles (see for example female 2 and her daughters 2C and 2D; Female 17 and her daughter 17B in Fig. 14.17). Just as menstrual synchrony can be greatest among women who are close friends, and between mothers and daughters (Weller and Weller 1993; Weller *et al.* 1995), so perhaps synchrony is increased among female monkeys belonging to the same matriline.

Effects of social stimuli upon pubertal development

Experiments on mice have shown that primer pheromones contained in the urine of adults, can exert profound effects upon pubertal development and timing of the first ovulation in females (Vandenbergh 1983; Bronson 1989). Exposure to male odour accelerates puberty in female mice, but sometimes tactile and pheromonal cues act together to produce such effects (Bronson and Maruniak 1975). Rapid release of LH followed by heightened levels of circulating oestradiol are measurable in young females in these circumstances (Bronson and Desjardins 1974). It is also well-established that urinary pheromones act via the vomeronasal organ and accessory olfactory system to advance, or retard, pubertal development in mice (Reynolds and Keverne 1979; Kaneko *et al.* 1980).

It is likely, but still unproven, that primer pheromonal effects upon pubertal development occur in prosimians. Olfactory communication is highly developed among the prosimians, which include many nocturnal forms. Likewise, the prosimians have a well-developed vomeronasal organ and accessory olfactory bulbs in the brain. Izard (1990) has obtained evidence that exposure of young female galagos to the presence of an adult male advances the onset of first oestrus. Females were paired with adult proven breeding males, or young peer males, and oestrus onset was determined by occurrence of vaginal cornification (Fig. 14.18). Whether olfactory cues alone, or in combination with tactile and other cues from males, are responsible for the advancement of puberty in female galagos is unknown. In the wild, females occupy home ranges which overlap extensively with the range of at least one dominant 'A' male (Bearder 1987). They are exposed to urinary cues from such males, distributed by urine-washing and other forms of scent-marking behaviour. Males regularly visit neighbouring females and courtship begins long before a female comes into oestrus. The potential for male primer pheromonal effects to occur is strong, therefore. Likewise, the propensity of young females to be philopatric and to remain in areas scent-marked by the mother and by other breeding females, might be associated with inhibitory effects upon pubertal development. Although there is evidence that disruption of the oestrous cycle occurs if female mouse lemurs are exposed to overcrowding (Perret 1982), the specific role of pheromonal cues upon age of first oestrus in prosimians has not been investigated.

Among monkeys and apes, as well as in the hu-

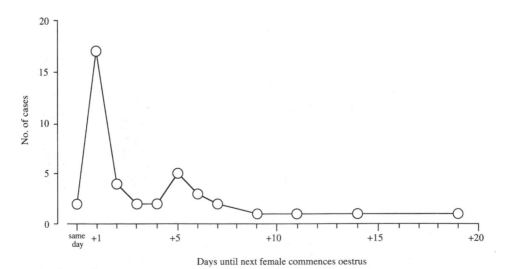

Fig. 14.16 Asynchrony in the timing of oestrus in female ringtailed lemurs (*Lemur catta*) during their brief annual mating season. Despite the overall pattern of breeding synchrony within ringtail groups, females rarely become sexually receptive on the same day; usually oestrous periods are separated by a one-day interval, and longer intervals occur in many cases. (Data from Pereira 1991.)

man female, a period of *adolescent sterility* commonly occurs (Hartman 1931), so that irregular, anovulatory cycles precede the fertile cycles typical of maturity. Figure 14.19 provides data on intermenstrual intervals and occurrences of ovulation in 59 laboratory-housed rhesus monkeys which were monitored between 18 months and 5 years of age by Bercovitch and Goy (1990). The average age of menarche (i.e. first menstruation) in these females was 29.8 months (range 20.1–42.4 months). Intervals between successive menstruations were initially highly irregular, but then stabilized rapidly. The average interval between menarche and first ovula-

tion was 378 days (range for age at first ovulation was 31.7–53.3 months). Nor was the first ovulation necessarily followed by regular and ovulatory cycles in these animals. Bercovitch and Goy (1990) stress that these data were obtained from a laboratory colony, in which monkeys were housed singly, or with one or more companions of a similar age. In captive rhesus monkey groups, housed outdoors, females first ovulate at either 2.5 or 3.5 years of age; the gap of 1 year occurs because of the annual reproductive cycle of this species (Schwartz *et al.* 1988; Wilson and Gordon 1989). There is evidence that the dominance rank of females is positively correlated with age at first ovulation in the rhesus monkey (Schwartz *et al.* 1985). This is not so for the stumptail macaque, however (Nieuwenhuijsen *et al.* 1988). There is very little information about how social factors might influence the duration of adolescent sterility in monkeys; the problem is clearly important, since earlier maturation might affect lifetime reproductive success. The daughters of dominant females have been found to produce their first offspring at an earlier age in several macaque species (Harcourt 1987) but the possible impact of social factors, as well as nutritional status, upon pubertal development in dominant daughters is not understood.

Adolescent sterility also occurs in the great apes,

Fig. 14.17 Timing of maximal sexual skin swelling during the annual mating season in 14 adult female mandrills (*Mandrillus sphinx*) living as members of a semi-free-ranging group in Gabon. Horizontal bars show the duration of full swellings; vertical lines indicate occurrences of sexual skin breakdown. Broken bars, in the case of female 17 and her daughter 17B, indicate temporary detumescence and resumption of swelling. Cycles are not tightly synchronized; however, the majority of females ovulate during July/August and there is a tendency also for females in the same matriline to exhibit synchrony (e.g. female 2 and her daughters 2C and 2D). (Author's unpublished observations.)

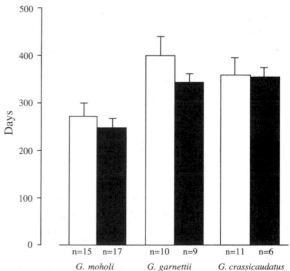

Fig. 14.18 Age at puberty (mean ± S.E.M. in days) in female galagos (*G. moholi; G. garnettii; G. crassicaudatus*) which were paired with young peer males (open bars) or with proven breeding males of adult age (closed bars). Puberty was determined by dating the first occurrence of a fully cornified vaginal smear. (Redrawn and modified from Izard 1990.)

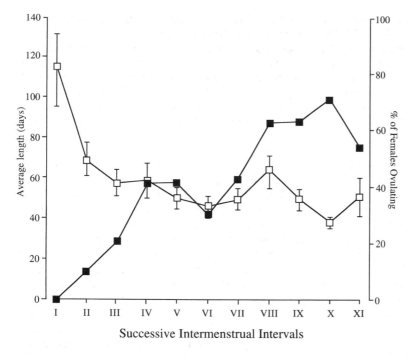

Fig. 14.19 Average length (mean ±S.E.M. in days) of intermenstrual intervals (□) in 59 laboratory-housed female rhesus monkeys (*Macaca mulatta*) aged between 1.5 and 5 years, together with the percentages of females ovulating during successive cycles (■). (Redrawn and modified from Bercovitch and Goy 1990.)

as well as in the human female. Table 14.2 provides information on the age of menarche and duration of adolescent sterility in the orang-utan, gorilla, chimpanzee and man. Tutin (1980) has pointed out that captive chimpanzees tend to reach menarche earlier and to have a shorter duration of adolescent sterility than their counterparts in the wild. A superior diet in captivity might be an important contributory factor. If premenarchial rhesus monkeys are fed a high-fat diet, then they exhibit earlier sexual maturation (Schwartz *et al.* 1988). It has been mooted that there is a positive relationship between age at menarche, fat reserves, and body weight in the human female (Frisch and McArthur 1974). The progressive decline in age at menarche in industrial societies during the present century is also believed to be because of superior nutrition; girls now reach menarche at 12.5–13.0 years of age rather than at

Table 14.2 Age at menarche and the duration of adolescent sterility in the great apes and man

Species		Age at menarche (years)	Duration of adolescent sterility (months)	Sources
Orang-utan	(C)	5.08–7.08	–	Markham 1995
	(C)	Approx. 7.0	–	Van der Werff ten Bosch 1982
	(W)	?	12.0+*	Galdikas 1995
Chimpanzee	(C)	7.3–10.2 (mean 8.9)	4–17 (mean 11.25)	Young and Yerkes 1943
	(C)	7.0–10.75 (mean 8.75)	–	Smith *et al.* 1975
	(W)	10.0+	13–34 (median 26)	Tutin 1980
Gorilla	(C)	6.0–7.0	mean 26.0	Dixson 1981
	(W)	6.5–8.5	10.4–18.4	Harcourt *et al.* 1980
Man	!kung Hunter gatherers	mean 16.9	mean 17.0	Howell 1979 Johnson and Everitt 1988
	W. Europe	10.5–15.5	24.0+	Marshall and Tanner 1969

C=captive animals; W=wild populations.
*Galdikas states that wild orang-utans may not give birth until 14–15 years of age. An extended period of adolescent sterilty might occur in the wild, but its duration is unknown.

16.9 years, on average, as in the hunter–gatherer society of the !Kung bushmen (Howell 1979). Do social factors influence age at menarche, or the lengthy period of adolescent sterility, in the human female? Surbey (1990) has reviewed this field and has provided evidence from her own studies for effects of family composition and stressful social factors upon the timing of human menarche. Girls whose fathers were absent from the family, or who lacked both parents, tended to reach menarche earlier and the effect did not depend upon socioeconomic factors. Surbey concluded that physical exposure to a *related* male delays menarche, whereas physical exposure to an *unrelated* man (such as a stepfather) has the reverse effect. The nature of the cues which mediate these effects are unknown. However, some role for olfactory stimuli should not be discounted, given that male axillary secretions have been reported to influence the duration of the menstrual cycle in women (Cutler *et al.* 1986). In this regard it is pertinent to mention the possible significance of the 'axillary organ' which occurs in the African apes and man. The axillary organ consists of enlarged apocrine glands (see Fig. 7.18, page 188) and is more prominent in adult males of *Gorilla*, *Pan*, and *Homo* than in adult females. Male axillary odour is also different, being pungent in silverback gorillas, for example, and distinguishable from the odour of the female even under field conditions (Schaller 1963). No function for the axillary organ has been identified. However, it is possible that it might affect ovarian physiology in fe-

males, assisting the onset of regular ovulatory cycles in pubescent females. This is pure speculation, of course, but it might be worthwhile to collect data on effects of male proximity upon menstrual cycles in the gorilla. Silverback gorillas establish their breeding units by taking or attracting females from established one-male groups. It is common for young mature females to transfer in this way (Harcourt *et al.* 1980). Perhaps the stressful experience of transfer is counteracted by male stimuli which stabilize the female's cycle? Likewise female transfers between chimpanzee communities might involve socioendocrinological effects upon their menstrual cycles. Perhaps the influences of strange vs. familiar males upon menarche in human females are also inheritances from our evolutionary past?

Although adolescent sterility does not occur in pubescent male primates, it is usual for an extended period of adolescence to be interposed between the juvenile and adult stages of life (Tanner 1962). Changes that occur during adolescence in males include a 'growth spurt', increased genital development and growth of the secondary sexual characteristics. Pubertal increases in testosterone stimulate certain of these changes, as well as activating sexual and associated patterns of behaviour.

An important question concerns the possible effects of social environment upon the timing and extent of adolescent changes in male anthropoids. In the male rhesus monkey, for example, individual differences in adolescent development are quite marked, and some males begin to produce sperma-

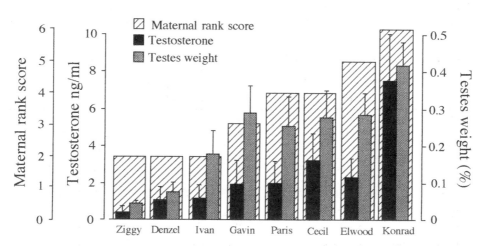

Fig. 14.20 Correlations between maternal dominance in rhesus monkeys (*Macaca mulatta*) and levels of serum testosterone (T) and testes weights in adolescent male offspring during the annual mating season. Data are means (+S.E.M.) for eight adolescent males (named on the horizontal axis) living as members of four captive breeding groups. Total maternal dominance scores (minimum score=2; maximum=6) were positively correlated with T levels in adolescent sons (r_s=0.935, P < 0.005) and with their testes/body weight percentages (r_s=0.833, P < 0.01). (From Dixson and Nevison 1997.)

tozoa and to copulate at between 3.0 and 3.5 years of age (Rose *et al.* 1978; Bercovitch and Goy 1990). Agonistic rank among adolescent male rhesus monkeys is positively correlated with testicular growth and plasma testosterone levels at the onset of the annual mating season (Bercovitch 1993). These effects are non-linear, however, being most pronounced when comparisons are made between individuals at opposite poles of the hierarchy. It is possible that affiliative interactions between adolescent males and females might accelerate masculine development. Yet, in a recent study of this problem, allogrooming and sexual interactions between adolescent rhesus monkeys and adult females in their social groups bore no relationship to the extent of reproductive maturation in males (Dixson and Nevison 1997). However, this study did demonstrate positive correlations between the dominance ranks of the *mothers* of adolescent males and levels of circulating testosterone, testes weights, growth of the baculum, and maintenance of body weights in sons during the annual mating season (Fig. 14.20).

The findings summarized in Fig. 14.20 point to possible effects of maternal rank, as well as intermale agonistic rank, in determining reproductive maturation during adolescence in the male rhesus monkey. Similar effects probably occur in some other primates and might be significant in terms of the reproductive tactics and reproductive success of maturing males. In the stumptail macaque, for example, DNA typing studies have shown that males born to high-ranking mothers begin to reproduce earlier, and sire significantly more offspring during the first four years of sexual activity, than do the sons of low-ranking mothers (Paul, Kuester and Arnemann 1992). It is also possible that some of the remarkable effects of social suppression upon the development of masculine secondary sexual characteristics discussed earlier in this chapter and in Chapter 7 (see Table 7.7, page 197), might be more prevalent among males born to low-ranking mothers. No studies have yet been reported which adequately address this question, however.

Short-term effects of copulatory stimuli upon gonadal function
Male primates

Exposure of male mouse lemurs, rhesus monkeys, squirrel monkeys, or talapoins to the presence of attractive females stimulates gradual increases in testosterone secretion. In the mouse lemur, pheromonal cues alone are sufficient to produce this effect, as was described earlier (Fig. 14.12). However, does copulation itself trigger gonadotrophin or testosterone secretion in male primates? In various non-primate mammals copulation, or the sight or odour of the female, results in rapid increases in plasma testosterone in males (e.g. hamster: Macrides *et al.* 1974; rat: Purvis and Haynes 1974; mouse: Macrides *et al.* 1975; rabbit: Saginor and Horton 1968; bull: Katongole *et al.* 1971). Increased LH secretion has also been noted in some experiments (e.g. bull: Katongole *et al.* 1971; rat: Kamel *et al.* 1975) but not in others (e.g. bull: Gombe *et al.* 1973; rat: Spies and Niswender 1971). There is the theoretical possibility that rapid elevations in plasma testosterone might exert a feedback effect upon sexual arousal or future copulatory performance in some species. As an example, intraperitoneal administration of testosterone one hour before mating in male rats results in shortening of the ejaculatory latency (Mälmnas 1977). The spinal mechanisms that control penile reflexes are highly sensitive to testosterone (T) and dihydrotestosterone (DHT) and are activated by 24 h after T or DHT administration to castrated males (Gray *et al.* 1980). Testosterone also has rapid effects upon hypothalamic responses to olfactory stimulation in the castrated rat (Pfaff and Pfaffmann 1969). Precopulatory increases in circulating testosterone might also influence sexual arousal in some strains of mice (Batty 1978). However, it is doubtful whether such effects have any general relevance among mammals and, as we shall see below, the evidence pertaining to primates is far from convincing.

Copulation has been found to have no short-term effects upon plasma LH or testosterone levels in either male rhesus monkeys (Fig. 14.21) and human males (Stearns *et al.* 1973). However, post-coital increases in plasma testosterone in male rhesus monkeys were detected by Herndon *et al.* (1981). Multiple ejaculations have no effect upon serum testosterone in the stumptail macaque (Goldfoot *et al.* 1975). Nor is testosterone affected by copulation in the male common marmoset (Dixson and Schofield, unpublished observations). Likewise, masturbation does not affect plasma testosterone in the human male (Ismail *et al.* 1972), although at least one study has found that testosterone levels rose after masturbation (Purvis *et al.* 1976). Attempts to demonstrate some relationship between sexually arousing stimuli (erotic film) and pituitary–testicular function in man have proven equivocal. Lincoln (1974) found no effects on plasma testosterone and LH levels, whereas Pirke *et al.* (1974) reported an increase in testosterone levels. More recently, Stoleru *et al.* (1993) have reported significant increases in the pulsatile secretion of LH, and in

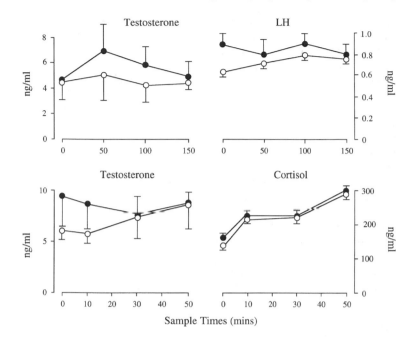

Fig. 14.21 Effects of copulation upon plasma levels of testosterone (T), luteinizing hormone (LH) and cortisol in adult male rhesus monkeys (*Macaca mulatta*) Data are means (±S.E.M.) for 10 males in each series. Samples were collected before (time 0) and after a 10-min behaviour test during which copulation and ejaculation occurred (●). In control tests, no female was present (○). Significant increases in cortisol levels occurred at 10, 30, and 50 min in both the control and ejaculatory series (P < 0.01, Newman–Keuls). T and LH levels were not significantly affected by these procedures. (Data from Phoenix *et al.* 1977.)

circulating testosterone in young men, as a result of viewing an erotic film. Some of these results are shown in Fig. 14.22. In the chacma baboon, adult males exhibit increased frequencies of seminal emissions and heightened serum testosterone levels in response to the proximity of a swollen female. It is not known whether this results from masturbatory activity or from visual cues alone (Bielert and Van Der Walt 1982).

Overall, the evidence concerning short-term effects of sexual stimulation upon pituitary–testicular function in male primates is limited and much of it is conflicting. Increases which occur during annual mating seasons in monkeys such as the rhesus macaque are probably due to longer-term female influences, as well as to direct responses to cues from the physical environment. As in so many areas of primate sexuality, however, we must acknowledge that only a handful of species has been examined. Whether copulatory stimuli might elicit rapid changes in plasma LH and testosterone in male prosimians has never been investigated, for instance.

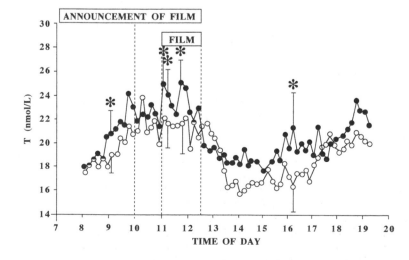

Fig. 14.22 Mean serum testosterone (T) levels in nine men sampled at 10-min intervals during experimental sessions (●) when a sexually arousing film was shown and control sessions (○). Points at which T levels differed significantly across sessions are indicated by *(P < 0.05) and for these points bars indicate S.E.M. (Redrawn and modified from Stoleru *et al.* 1993.)

Female primates

As far as is known all primates are spontaneous ovulators. Copulation-induced triggering of the pre-ovulatory LH surge, such as occurs in the rabbit (Harris 1965) and in certain carnivores (e.g. the ferret, racoon, and domestic cat: Asa 1987), has not been described for any primate species. However, the distinction between spontaneous and induced ovulation is not necessarily an absolute one. It is possible for copulatory stimuli to influence neuroendocrine responses in cases where females normally ovulate without the occurrence of mating. The female rat is a spontaneous ovulator, for example, but under conditions of constant light, release of LH and ovulation can be triggered by vagino–cervical stimulation. In normal circumstances coital stimuli provided by the male affect prolactin secretion in the female rat; enhanced prolactin secretion is necessary for support of the corpus luteum during pregnancy.

Scattered throughout the literature on primate reproduction are occasional references to possible effects of mating upon pituitary–ovarian function. In the squirrel monkey, for example, ovulation occurs in the absence of a male. However, if a female squirrel monkey mates on the day of the LH surge, or on the following day, oestradiol levels are greatly increased (to a mean of 2595 pg mL-1). In the absence of a male, circulating oestradiol levels average $330\,\mathrm{pg\,mL}^{-1}$ on the day of the LH surge and decline to $128\,\mathrm{pg\,mL}^{-1}$ one day later (Yeoman *et al.* 1985; Alexander *et al.* 1991). The significance of these effects is unclear, but it is known that the squirrel monkey has the shortest ovarian cycle of any primate species (averaging 9 days in length) and that the cycle is peculiarly sensitive to a range of factors in the social and physical environment.

There have been no detailed studies of effects of copulation upon menstrual cyclicity in Old World monkeys or apes. In women, there are reports that regular sexual activity might be associated with fertile patterns of menstrual cyclicity whereas sporadic (less than weekly) coitus is associated in some cases with shortening of the luteal phase (Cutler *et al.* 1979, 1980). Cutler *et al.* suggest that, for some women at least, coital stimuli might assist support of the corpus luteum. Copulation might not be required to produce such effects, however, because in women with aberrant cycle lengths (defined as <26

days or >32 days), exposure to male axillary secretions alone is sufficient to normalize unusually short, or long, cycles in a significant number of cases (Cutler *et al.* 1986). These observations are intriguing, and one is left wondering whether subtle effects of tactile, olfactory, or other cues might have some effects upon the menstrual cycle in macaques and baboons, or among the great apes?

One group of primates in which coital stimuli almost certainly have neuroendocrine effects in females is the prosimians. Although no formal experiments have been conducted to examine this problem, comparison of different reports shows that the duration of oestrus is shortened if female ringtailed lemurs and greater galagos mate and receive ejaculations in the normal way. In laboratory studies where intromissions occurred, but matings were terminated prior to ejaculation, females displayed consistently longer periods of receptivity (e.g. *L. catta*: Evans and Goy 1968; *G. crassicaudatus*: Eaton *et al.* 1973a). The evidence pertaining to these observations was reviewed in Chapter 11 (see pages 317–19) and will not be repeated here. For some prosimians there might be parallels with the phenomenon of coitally-induced oestrus termination, such as occurs in the guinea pig (Goldfoot and Goy 1970). In the guinea pig, the male's copulatory plug exerts some chemical effect which shortens the period of female receptivity. It has been suggested that prostaglandins present in the males' accessory sexual secretions might shorten the life of the corpus luteum in the female and bring about inhibitory effects upon her behaviour. In prosimians, we have no evidence that progesterone can produce the type of biphasic effect upon female receptivity which occurs in the guinea pig. The limited data available indicate that oestrogen alone governs the onset of receptivity in the female ringtailed lemur and galago. It is possible that coitus might have some direct (i.e. tactile) effect upon the duration of behavioural oestrus. In the butterfly *Pieris rapae*, for example, stretch receptors in the female's *Bursa copulatrix* respond to the male's ejaculate and this leads to a reduction in female sexual receptivity (Sugawara 1979). In the ringtailed lemur, it is conceivable that the tactile stimulation received by the female during mating, or the presence of the male's large copulatory plug, might shorten her receptive period. Speculation is no substitute for experiment, however.

References

Abbott, D. H. (1988). Natural suppression of fertility. *Symp. Zool. Soc. Lond.*, **60**, 7–28.

Abbott, D. H. (1993). Social conflict and reproductive suppression in marmoset and tamarin monkeys. In *Primate social conflict*, (eds W. A. Mason and S. P. Mendoza), pp. 331–372. State University of New York Press, New York.

Abbott, D. H. and Hearn, J. P. (1978). Physical, hormonal and behavioural aspects of sexual development in the marmoset monkey, *Callithrix jacchus*. *J. Reprod. Fertil.*, **53**, 155–166.

Abbott, D. H., Holman, S. D., Berman, M., Neff, D. A., and Goy, R. W. (1984). Effects of opiate antagonists on hormones and behavior of male and female rhesus monkeys. *Arch. Sex. Behav.*, **13**, 1–25.

Abbott, D. H., George, L. M., Barrett, J., Hodges, J. K., O'Byrne, K. T., Sheffield, J.W., *et al.* (1990). Social control of ovulation in marmoset monkeys: a neuroendocrine basis for the study of infertility. In *Socioendocrinology of primate reproduction*, (eds T. E. Ziegler and F. B. Bercovitch), pp. 135–158. Wiley–Liss, New York.

Abbott, D. H., Barrett, J., and George L. (1993). Comparative aspects of the social suppression of reproduction in female marmosets and tamarins. In *Marmosets and tamarins: systematics, behaviour and ecology*, (ed. A. B. Rylands), pp. 151–163. Oxford University Press.

Abbott, D. H., Saltzman, W., Schultz-Darken, N. J., and Smith, T. E. (1997). Specific neuroendocrine mechanisms not involving generalized stress mediate social regulation of female reproduction in cooperatively breeding marmoset monkeys. *Ann. N. Y. Acad. Sci.*, **807**, 219–238.

Abegglen, J. J. (1984). *On socialization in hamadryas baboons: a field study*. Bucknell University Press, Lewisburg.

Adams, D. B., Gold, A. B., and Burt, A. D. (1978). Rise in female sexual activity at ovulation blocked by oral contraceptives. *New Eng. J. Med.*, **299**, 1145–1150.

Adler, N. T. (1969). Effects of the male's copulatory behavior on successful pregnancy of the female rat. *J. Comp. Physiol. Psychol.*, **69**, 613–622.

Adler, N. and Bermant, G. (1966). Sexual behavior of male rats: effects of reduced sensory feedback. *J. Comp. Physiol. Psychol.*, **61**, 240–243.

Adolphs, R., Tranel, D., Damasio, H., and Damasio, A. (1994). Impaired recognition of emotion in facial expressions following bilateral damage to the human amygdala. *Nature Lond.*, **372**, 669–672.

Adrian, T. E., Gu, J., Allen, J. M., Tatemoto, K., Polak, J. M., and Bloom, S. R. (1984). Neuropeptide Y in the human male genital tract. *Life Sci.*, **35**, 2643–2648.

Agmö, A. (1976). Mating in male rabbits after anaesthesia of the glans penis. *Physiol. Behav.*, **17**, 435–437.

Agmö, A. and Fernandez, H. (1989). Dopamine and sexual behavior in the male rat: a re-evaluation. *J. Neural Trans.*, **77**, 21–37.

Agoramoorthy, G. and Mohnot, S. M. (1988). Infanticide and juvenilicide in langurs in Hanuman langurs (*Presbytis entellus*) around Jodhpur, India. *Hum. Evol.*, **3**, 279–296.

Aguirre, A. C. (1971). O mono *Brachyteles arachnoides* (E. Geoffroy). Situaçåo atual da espécie no Brasil. Academia Brasileira de Ciências, Rio De Janeiro.

Ahlenius, S. (1993). The effect of dopaminergic and serotonergic agents on the lordosis response in female rats. In *Sexual pharmacology*. (eds. A. J. Riley, M. Pect, and C. A. Wilson), pp. 59–72. Clarendon Press, Oxford.

Ahlenius, S. and Larsson, K. (1987). Evidence for a unique pharmacological profile of 8-OH-DPAT by evaluation of its effects on male rat sexual behaviour. In *Brain 5-HT$_{1A}$ receptors*. (eds. C. T. Dowish, S. Ahlenius, and P. H. Hutson), pp. 185–198. Ellis Horwood, Chichester.

Ahleniuis, S., Fernandez-Guasti, A., Hjorth, S., and Larsson, K. (1986). Suppression of lordosis behaviour by the putative 5-HT receptor agonist 8-OH-DPAT in the rat. *Eur. J. Pharmacol.*, **124**, 361–363.

Aidura, D., Badaivi, M., Tahiri-Zagret, C., and Robyn, C. (1981). Changes in concentrations of serum prolactin, FSH, oestradiol and progesterone and of the sex skin during the menstrual cycle in the mangabey monkey (*Cercocebus atys lunulatus*). *J. Reprod. Fertil.*, **62**, 475–481.

Al-Attia, H. M. (1996). Gender identity and role in a pedigree of arabs with intersex due to 5 alph-reductase 2 deficiency. *Psychoneuroendocr.*, **21**, 651–657.

Alcock, J. (1984). *Animal behavior*. 3rd ed. Sinauer, Sunderland, Massachusetts.

Alder, E., Cook, A., Davidson, D., West, C., and Bancroft, J. (1986). Hormones, mood and sexuality in lactating women. *Br. J. Psychiat.*, **148**, 74–79.

Aldrich-Blake, F. P. G. (1968). A fertile hybrid between two *Cercopithecus* species in the Budongo Forest, Uganda. Folia Primatol., **9**, 15–21.

Alecozay, A. A., Casslén, B. G., Riehl, R. M., *et al.* (1989). Platelet-activating factor (PAF) in the human luteal phase endometrium. *Biol. Reprod.*, **41**, 578–586.

Alexander, B. M. Perkins, A., Van Kirk, E. A., Moss, G. E., and Fitzgerald, J. A. (1993). Hypothalamic and hypophyseal receptors for estradiol in high and low sexually performing rams. *Horm. Behav.*, **27**, 296–307.

Alexander, R. D. (1964). The evolution of mating behaviour in arthropods. *Symp. Roy. Ent. Soc. Lond.*, **2**, 78–94.

Alexander, R. D. (1974). The evolution of social behavior. *Ann. Rev. Ecol. Syst.*, **5**, 325–383.

Alexander, R. D. and Noonan, K. M. (1979). Concealment of ovulation, parental care and human evolution. In *Evolutionary biology and human behavior: an anthropological perspective*. (eds. N. A. Chagnon and W. Irons), pp. 436–453. Duxbury Press, North Scituate, Massachusetts.

Alexander, S. E., Yoeman, R. R., Williams, L. E., Akscl, S., and Abee, C. R. (1991). Confirmation of ovulation and characterization of luteinizing hormone and progesterone secretory patterns in cycling, isosexually housed Bolivian squirrel monkeys (*Saimiri boliviensis boliviensis*). *Am. J. Primatol.*, **23**, 55–60.

Allen, C. (1961). Perversions, sexual. In *The Encyclopaedia of sexual behaviour*. (eds. A. Ellis and A. Abarbanel), pp. 802–811. Corsano, London.

Allen, L. S. and Gorski, R. A. (1990). Sex difference in the bed nucleus of the stria terminalis of the human brain. *J. Comp. Neurol.*, **302**, 697–706.

Allen, L. S. and Gorski, R. A. (1992). Sexual orientation and the size of the anterior commissure in the human brain. *Proc. Natl. Acad. Sci. U.S.A.*, **89**, 7199–7202.

Allen, L. S., Hines, M., Shryne, J. E., and Gorski, R. A. (1989). Two sexually dimorphic cell groups in the human brain. *J. Neurosci.*, **9**, 497–506.

Allen, M. L. (1977). Sexual response and orgasm in the female chimpanzee (*Pan troglodytes*). Unpublished Master's thesis, University of Oklahoma.

Allen, M. L. (1981). Individual copulatory preference and the 'strange female effect' in a captive group-living male chimpanzee. *Primates*, **22**, 221–236.

Allen, M. L. and Lemmon, W. B. (1981). Orgasm in female primates. *Am. J. Primatol.*, **1**, 15–34.

Alterman, L., Doyle, G. A., and Izard, M. K. (1995). (eds.). *Creatures of the dark: the nocturnal prosimians*. Plenum, New York.

Altmann, J. (1990). Primate males go where females are. *Anim. Behav.*, **39**, 193–195.

Altmann, J., Altmann, S. A., Hausfater, G., and McCuskey, S. A. (1977). Life history of yellow baboons: physical development, reproductive parameters and infant mortality. *Primates*, **18**, 315–330.

Altmann, J., Hausfater, G., and Altmann, S. A. (1988). Determinants of reproductive success in savannah baboons, *Papio cyncocepthalus*. In *Reproductive success*. (ed. T. H. Clutton-Brock), pp, 403–418. University of Chicago Press.

Altmann, J., Alberts, S. C., Haines, S. A., Dubach, J., Muruthi, P., Coote, T., *et al.* (1996). Behavior predicts genetic-structure in a wild primate group. *Proc. Nat. Acad. Sci. U.S.A.*, **93**, 5797–5801.

Altmann, S. A. (1962). A field study of the sociobiology of the rhesus monkey, *Macaca mulatta. Ann. N.Y. Acad. Sci.*, **102**, 338–435.

Altmann, S. A. (ed) (1967). *Social communication among primates*. University of Chicago Press.

Altmann, S. A. and Altmann, J. (1970). *Baboon ecology: African field research*. University of Chicago Press.

Alvarez, F. (1973). Periodic changes in the bare skin areas of *Theropithecus gelada. Primates*, **14**, 195–199.

Amaral, D. A., Price, J. L., Pitkanen, A., and Carmichael, S. T. (1992). Anatomical organization of the primate amygdaloid complex. In *The amygdala: neurobiological aspects of emotion, memory and mental dysfunction*. (ed. J. Aggleton), pp. 1–66. Wiley-Liss, New York.

Amos, W., Twiss, S., Pomeroy, P. P., and Anderson, S. S. (1993). Male mating success and paternity in the grey seal, *Halichoerus grypus*: a study using DNA fingerprinting. Proc. Roy. Soc. Lond. B., **252**, 199–207.

Andelman, S. J. (1987). Evolution of concealed ovulation in vervet monkeys (*Cercopithecus aethiops*). *Am. Nat.*, **129**, 785–799.

Anderson, C. M. and Bielert, C. (1994). Adolescent exaggeration in female catarrhine primates. *Primates*, **35**, 283–300.

Andersson, K. E. and Wagner, G. (1995). Physiology of penile erection. *Physiol. Revs.*, **75**, 191–236.

Andersson, M. (1982). Female choice selects for extreme tail length in a widow bird. *Nature Lond.*, **299**, 818–820.

Andersson, M. (1994). *Sexual selection*. Princeton University Press, Princeton, New Jersey.

Andersson, P. O., Björnberg, J., Bloom, S. R., and Mellander, S. (1984). Hemodynamics of pelvic nerve induced penile erection in the dog: possible mediation by vasoactive intestinal polypeptide. *J. Physiol. Lond.*, **350**, 209–224.

Andriamasimanana, M. (1994). Ecoethological study of free-ranging aye-ayes (*Daubentonia madagascariensis*) in Madagascar. *Folia Primatol.*, **62**, 37–45.

Andrimiandra, A. and Rumpler, Y. (1968). Rôle de la testosterone sur le déterminisme des glandes brachiales et anté brachiales chez le *Lemur catta. Behavior*, **65**, 309–319.

Aou, S., Oomura, Y., and Yoshimatsu, H. (1988). Neuron activity of the ventromedial hypothalamus and the medial preoptic area of the female monkey during sexual behaviour. *Brain Res.*, **455**, 65–71.

Arendash, G. W. and Gorski, R. A. (1983). Effects of discrete lesions of the sexually dimorphic nucleus of the preoptic area or other preoptic regions on the sexual behavior of male rats. *Brain Res. Bull.*, **10**, 147–154.

Argiolas, A. and Gessa, G. L. (1991). Central functions of oxytocin. *Neurosci. Biobehav. Revs.*, **15**, 217–231.

Arletti, R., Benelli, A., and Bertolini, A. (1992). Oxytocin involvement in male and female sexual behavior. *Ann. N. Y. Acad. Sci.*, **652**, 180–193.

Arnold, A. P. (1984). Androgen regulation of motor neuron size and number. *Trends in Neurosci*. July 1984, pp. 249–242.

Arnold, A. P. (1990). Hormonally induced synaptic reorganization in the adult brain. In *Hormones, brain and behaviour in vertebrates*, Vol. 1. (ed. J. Balthazarle), pp. 82–91. Karger, Basel.

Aronson, L. R. and Cooper, M. L. (1968). Desensitization of the penis and sexual behavior in cats. In *Perspectives in reproduction and sexual behavior*. (ed. M. Diamond), pp. 51–82. Indiana University Press, Bloomington.

Asa, C. S. (1987). Reproduction in carnivores and ungulates. In *Psychobiology of reproductive behavior: an evolutionary perspective*. (ed. D. Crews), pp. 258–290. Prentice-Hall, Englewood Cliffs, New Jersey.

Ashwell, M., Cole, T. J., and Dixon, A. K. (1985). Obesity: new insight into anthropometric classification of fat distribution shown by computed tomography. *Brit. Med. J.*, **290**, 1691–1694.

Aslum, P. and Goy, R. W. (1974). Actions of esters of testosterone, dihydrotestosterone, or estradiol on sexual behavior in castrated male guinea pigs. *Horm. Behav.*, **5**, 207–217.

Aureli, F., Schino, G., Cordischi, C., Cozzolino, R., Scucchi, S., and Van Schaik, C. (1990). Social factors affect secondary sex ratio in captive Japanese macaques. *Folia Primatol.*, **55**, 176–180.

Austad, S. N. (1984). Evolution of sperm priority patterns in spiders. In *Sperm competition and the evolution of animal mating systems*. (ed. R. L. Smith), pp. 223–249. Academic Press, New York.

Austin, C. R. (1951). Observations on the penetration of sperm into the mammalian egg. *Aust. J. Sci. Res. B.*, **4**, 581–596.

Aykroyd, O. E. and Zuckerman, S. (1938). Factors in sexual-skin oedema. *J. Physiol. Lond.*, **94**, 13–25.

Ayoub, D. M., Greenough, W. T., and Juraska, J. M. (1983). Sex differences in dendritic structure in the preoptic area of the juvenile macaque monkey brain. *Science*, **219**, 197–198.

Bachmann, C. and Kummer, H. (1980). Male assessment of female choice in hamadryas baboons. *Behav. Ecol. Sociobiol.*, **6**, 315–321.

Bäckström, T., Sanders, T., Leask, R., Davidson, D., Warner, P., and Bancroft, J. (1983). Mood, sexuality, hormones and the menstrual cycle. II. Hormone levels and their relationship to the premenstrual syndrome. *Psychosom. Med.*, **45**, 503–507.

Bacq, Z. M. (1931). Impotence of the male rodent after sympathetic denervation of the genital organs. *Am. J. Physiol.*, **96**, 321–330.

Bagatell, G. J., Heiman, J. R., Rivier, J. E., and Bremner, W. J. (1994). Effects of endogenous testosterone and estradiol on sexual behavior in normal young men. *J. Clin. Endocr. Metab.*, **78**, 711–716.

Bailey, D. J. and Herbert, J. (1982). Impaired copulatory behavior of male rats with hyperprolactinaemia induced by domperidone or pituitary grafts. *Neuroendocrinology* **35**, 186–193.

Bailey, J. M. and Pillard R. C. (1991). A genetic study of male sexual orientation. *Arch. Gen. Psychiat.*, **48**, 1089–1096.

Bailey, J. M., Willerman, L., and Parks, C. (1991). A test of the maternal stress theory of human male homosexuality. *Arch. Sex Behav.*, **20**, 277–293.

Bailey, J. M., Pillard, R. C., Neale, M. C., and Agyei, Y. (1993). Heritable factors influence sexual orientation in women. *Arch. Gen. Psychiat.*, **50**, 217–223.

Baird, D., Horton, R., Longcope, C., and Tate, J. F. (1968). Steroid prehormones. *Perspect. Biol. Med.*, **11**, 384–421.

Baker, R. R. and Bellis, M. A. (1988). 'Kamikaze' sperm in mammals? *Anim. Behav.*, **36**, 936–939.

Baker, R. R. and Bellis, M. A. (1989a). Elaboration of the kamikaze sperm hypothesis: a reply to Harcourt. *Anim. Behav.*, **37**, 865–867.

Baker, R. R. and Bellis, M. A. (1989b). Number of sperm in human ejaculates varies in accordance with sperm competition theory. *Anim. Behav.*, **37**, 867–869.

Baker, R. R. and Bellis, M. A. (1993a). Human sperm competition: ejaculate adjustment by males and the function of masturbation. *Anim. Behav.*, **46**, 861–885.

Baker, R. R. and Bellis, M. A. (1993b). Human sperm competition: ejaculate manipulation by females and a function for the female orgasm. *Anim. Behav.*, **46**, 887–909.

Bakker, J. (1996). Sexual differentiation of the brain and partner preference in the male rat. Ph.D. thesis, University of Rotterdam.

Bakwin, H. (1973). Erotic feelings in infants and young children. *Am. J. Dis. Childhood.*, **126**, 52–54.

Baldwin, J. D. (1968). The social behavior of adult male squirrel monkeys (*Saimiri sciureus*) in a seminatural environment. *Folia Primatol.*, **9**, 281–314.

Baldwin, J. D. and Baldwin, J. I. (1973). The role of play in social organization: comparative observations on squirrel monkeys (*Saimiri*). *Primates*, **14**, 369–381.

Baldwin, J. D. and Baldwin, J. I. (1981). The squirrel monkeys, genus *Saimiri*. In *Ecology and behavior of neotropical primates*, Vol. 1. (eds. A. F. Coimbra-Filho and R. A. Mittermeier), pp. 277–330. Academia Brasileira de Ciências, Rio de Janeiro.

Baldwin, J. D. and Baldwin, J. I. (1989). The socialization of homosexuality and heterosexuality in a non-western society. *Arch. Sex. Behav.*, **18**, 13–29.

Ball, J. and Hartman, C. G. (1935). Sexual excitability as related to the menstrual cycle in the monkey. *Amer. J. Obst. and Gynec.*, **29**, 117–119.

Balmford, A., Jones, I. L., and Thomas, A. L. R. (1993). On avian asymmetry: evidence of natural selection for symmetrical tails and wings in birds. *Proc. Roy. Soc. Lond. B.*, **252**, 245–251.

Balthazart, J., Foidart, A., Surlemont, C., and Harada, N. (1990). Preoptic aromatase in the quail: behavioral, biochemical and immunocytochemical studies. In *Hormones, brain and behaviour in vertebrates*, Vol. 2. (ed. J. Balthazart), pp. 45–162. Karger, Basel.

Bancroft, J. (1974). Deviant sexual behaviour: modification and assessment. Clarendon Press, Oxford.

Bancroft, J. (1981). Hormones and human sexual behaviour. *Br. Med. Bull.*, **37**, 153–158.

Bancroft, J. (1984). Hormones and human sexual behavior. *J. Sex. and Marital Therapy.*, **10**, 3–21.

Bancroft, J. (1989). *Human sexuality and its problems*. Churchill Livingstone, Edinburgh.

Bancroft, J. (1995). Are the effects of androgens on male sexuality noradrenergically mediated—some consideration of the human. *Neurosci. Biobehav. Revs.*, **19**, 325–330.

Bancroft, J. and Gutierrez, P. (1996). Erectile dysfunction in men with and without diabetes mellitus: a comparative study. *Diabetic Medicine*, **13**, 84–89.

Bancroft, J. and Skakkeback, N. E. (1979). Androgens and human sexual behaviour. In *Sex, hormones and behaviour. Ciba Found Symp.* **62**, 209–220. Excerpta Medica, Amsterdam.

Bancroft, J. and Wu, F. C. W. (1983). Changes in erectile responsiveness during androgen therapy. *Arch. Sex. Behav.*, **12**, 59–66.

Bancroft, J., Tennent, G., Loucas, K., and Cass, J. (1974). The control of deviant sexual behaviour by drugs: behavioural changes with oestrogens and antiandrogens. *Br. J. Psychiat.*, **125**, 310–315.

Bancroft, J., Sanders, D., Davidson, D. W., and Warner, P. (1983). Mood, sexuality, hormones and the menstrual cycle. III. Sexuality and the role of androgens. *Psychosom. Med.*, **45**, 509–516.

Bancroft, J., O'Carroll, R., McNeilly, A., and Shaw, R. W. (1984). The effects of bromocriptine on the sexual behaviour of hyperprolactinaemic man: a controlled case study. *Clin. Endocr.*, **21**, 131–137.

Bancroft, J., Sherwin, B. B., Alexander, G. M., Davidson, D. W., and Walker, A. (1991). Oral contraceptives, androgens, and the sexuality of young women: II. the role of androgens. *Arch. Sex. Behav.*, **20**, 121–135.

Barash, D. P (1977). *Sociobiology and behavior*. Elsevier North Holland Inc., New York.

Barfield, R. J. (1969). Activation of copulatory behavior by androgen implanted into the preoptic area of the male fowl. *Horm. Behav.*, **1**, 37–51.

Barfield, R. J. and Krieger, M. S. (1977). Ejaculatory and postejaculatory behavior of male and female rats: effects of sex hormones and electric shock. *Physiol. Behav.*, **19**, 203–208.

Barfield, R. J., Rubin, B. S., Glaser, J. H., and Davis, P. G. (1983). Sites of action of ovarian hormones in the regulation of oestrous responsiveness in rats. In *Hormones and behavior in higher vertebrates*. (eds. J. Balthazart, E. Prove and R. Gilles), pp. 1–17. Springer, Berlin.

Barnes, T. R. E. and Harvey, C. A. (1993). Psychiatric drugs and sexuality. In *Sexual pharmacology* (eds. A. J. Riley, M. Peet and C. Wilson), pp. 176–196. Oxford Scientific Publications, Clarendon Press, Oxford.

Barraclough, C. A. (1966). Modifications of the C.N.S. regulation of reproduction after exposure of prepubertal rats to steroid hormones. *Rec. Prog. Horm. Res.*, **22**, 503–539.

Barratt, C. L. R., Bolton, A. E., and Cooke, I. D. (1990). Functional significance of white blood cells in the male and female reproductive tract. *Hum. Reprod.*, **5**, 639–648.

Barrett, E. B. M. (1985). The ecology of some nocturnal arboreal mammals in the rainforest of peninsula Malaysia. Unpublished Ph.D. Thesis, University of Cambridge.

Barrett, J., Abbott, D. H., and George, L. M. (1990). Extension of reproductive suppression by pheromonal cues in subordinate female marmoset monkeys, *Callithrix jacchus*. *J. Reprod. Fertil.*, **90**, 411–418.

Barrington, E. J. W. (1963). An introduction to general and comparative endocrinology Clarendon Press, Oxford.

Barry, J. (1979). Immunofluorescence study of the preoptico-terminal LRH tract in the female squirrel monkey during the estrous cycle. *Cell. Tiss. Res.*, **198**, 1–13.

Bartlett, T. Q., Sussman, R. W., and Cheverud, J. M. (1993). Infant killing in primates: a review of observed cases with specific reference to the sexual selection hypothesis. *Am. Anthropol.*, **95**- 958–990.

Bateman, A. J. (1948). Intra-sexual selection in Drosophila. *Heredity* **2**, 349–368.

Batra, A. K. and Lu, T. F. (1990). Physiology and pathology of penile erection. *Ann. Revs. Sex Res.*, **1**, 251–263.

Batra, S. K. (1974). Sperm transport through vas deferens:

review of hypotheses and suggestions for a quantitative model. *Fertil. Steril.*, **25**, 186–202.

Battalia, D. E. and Yanagimachi, R. (1980). The changes in oestrogen and progesterone levels trigger adovarian propulsive movement of the hamster oviduct. *J. Reprod. Fertil.*, **59**, 243–247.

Batty, J. (1978). Acute changes in plasma testosterone levels and their relation to measures of sexual behaviour in the male house mouse. (*Mus musculus*). *Anim. Behav.*, **26**, 349–357.

Bauers, K. A. and Hearn, J. P. (1994). Patterns of paternity in relation to male social rank in the stumptailed macaque. *Behaviour*, **129**, 149–171.

Baum, M. J. (1979). Differentiation of coital behavior in mammals: a comparative analysis. *Neurosci. Behav. Rev.*, **3**, 265–284.

Baum, M. J. and Everitt, B. J. (1992). Increased expression of C-fos in the medical preoptic area after mating in male rats: role of afferent inputs from the medial amygdala and midbrain central tegmental field. *Neurosci.*, **50**, 627-646.

Baum, M. J., Keverne, E. B., Everitt, B. J., and Herbert, J. (1976). Reduction of sexual interaction in rhesus monkeys by a vaginal action of progesterone. *Nature Lond.*, **263**, 606–608.

Baum, M. J., Everitt, B. J., Herbert, J., and Keverne, E. B. (1977). Hormonal basis of proceptivity and receptivity in female primates. *Arch. Sex. Behav.*, **6**, 173–192.

Baum, M. J., Slob, A. K., De Jong, F. H., and Westbrock, D. L. (1978). Persistence of sexual behavior in ovariectomized stumptail macaques following dexamethasone treatment or adrenalectomy. *Horm. Behav.*, **11**, 323–347.

Baum, M. J., Tobet, S. A., Starr, M. S., and Bradshaw, W. G. (1982). Implantation of dihydrotestosterone propionate into the lateral septum or medial amygdala facilitates copulation in castrated male rats given estradiol systemically. *Horm. Behav.*, **16**, 208–223.

Baum, M. J., Erskine, M. S., Kornberg, E., and Weaver, C. E. (1990a). Prenatal and neonatal testosterone exposure interact to affect differentiation of sexual behavior and partner preference in female ferrets. *Behav. Neurosci.*, **104**, 183–198.

Baum, M. J., Carroll, R. S., Cherry, J. A., and Tobet, S. A. (1990b). Steroidal control of behavioral, neuroendocrine and brain sexual differentiation: studies in a carnivore, the ferret. *J. Neuroendocr.*, **2**, 401–418.

Baumgardner, D. J., Hartung, T. G., Sawrey, D. K., Webster, D. G., and Dewsbury, D. A. (1982). Muroid copulatory plugs and female reproductive tracts: a comparative investigation. *J. Mammal.*, **63**, 110–117.

Bayliss, W. M. and Starling, E. H. (1902). The mechanism of pancreatic secretion. *J. Physiol.*, **28**, 325.

Beach, F. A. (1956). Characteristics of masculine 'sex drive'. In *The Nebraska symposium on motivation*. Vol. 4 (ed. N. E. Lincoln), pp. 1–32. University of Nebraska Press, Nebraska.

Beach, F. A. (1967). Cerebral and hormonal control of reflexive mechanisms involved in copulatory behaviour. *Physiol. Revs.*, **47**, 289–316.

Beach, F. A. (1968a). Factors involved in the control of mounting behavior by female mammals. In *Perspectives in reproduction and sexual behavior: a memorial to William C.*

Young. (ed. M. Diamond). Indian University Press, Bloomington.

Beach, F. A. (1968b). Coital behavior in dogs. 3. effects of early isolation on mating in males. *Behaviour*, **30**, 218–238.

Beach, F. A. (1969). Coital behavior in dogs. VII. Effects of sympathectomy in males. *Brain Res.*, **15**, 243–245.

Beach, F. A. (1970). Coital behavior in dogs. VI. Long term effects of castration upon mating in the male. *J. Comp. Physiol. Psychol. Monogr.*, **70**, 1–32.

Beach, F. A. (1974). Human sexuality and evolution. In *Reproductive behavior* (eds. W. Montagna and W. Sadler), pp. 333–365. Plenum, New York.

Beach, F. A. (1975a). Variables affecting 'spontaneous' seminal emission in rats. *Physiol. Behav.*, **15**, 91–95.

Beach, F. A. (1975b). Hormonal modifications of sexually dimorphic behavior. *Psychoneuroendocr.*, **1**, 3–23.

Beach, F. A. (1976). Sexual attractivity, proceptivity and receptivity in female mammals. *Horm. Behav.*, **7**, 105–138.

Beach, F. A. (1981). Historical origins of modern research on hormones and behavior. *Horm. Behav.*, **15**, 325–376.

Beach, F. A. and Holz, A. M. (1946). Mating behavior of male rats castrated at various ages and injected with androgen. *J. Exp. Zool.*, **101**, 91–142.

Beach, F. A. and Levinson, G. (1950). Effects of androgen on the glans penis and mating behavior of castrated rats. *J. Exp. Zool.*, **114**, 159–171.

Beach, F. A. and Wilson, J. R. (1963). Mating behavior in male rats after removal of the seminal vesicles. *Proc. Nat. Acad. Sci. USA* **49**, 624–626.

Beach, F. A., Kuehn, R. E., Sprague, R. H., and Anisko, J. J. (1972). Coital behavior in dogs. XI. Effects of androgenic stimulation during development on masculine mating responses in females. *Horm. Behav.*, **3**, 143–168.

Beams, H. W. and King. R. L. (1933). Sperm storage of seminal vesicles. *J. Urol.*, **29**, 95

Bean, R. B. (1906). Some racial peculiarities of the negro brain. *Am. J. Anat.*, **5**, 353–432.

Bearder, S. K. (1987). Lorises, bushbabies and tarsiers: diverse societies in solitary foragers. In *Primate Societies* (eds D. Cheney, R. Seyfarth, B. Smuts, R. Wrangham, and T. Struhsaker), pp. 11–24. University of Chicago Press.

Bearder, S. K. and Doyle, G. A. (1974). Ecology of bushbabies, *Galago senegalensis* and *Galago crassicaudatus*, with some notes on their behaviour in the field. In *Prosimian biology*. (eds. R. D. Martin, G. A. Doyle and A. C. Walker), pp. 109–130. Duckworth, London.

Bearder, S. K. and Martin, R. D. (1979). The social organization of a nocturnal primate revealed by radio tracking. In *A handbook of biotelemetry and radio tracking* (eds. C. J. Amlaner, Jr and D. W. Macdonald), Pergamon, Oxford.

Bearder, S. K., Honess, P. E., and Ambrose, L. (1995). Species diversity among galagos with special reference to mate recognition. In *Creatures of the dark: the nocturnal prosimians*. (eds L. Alterman, G. A. Doyle, and M. K. Izard), pp. 331–352. Plenum Press, New York.

Beatty, R. A. (1960). Fertility of mixed semen from different rabbits. *J. Reprod. Fertil.*, **19**, 52–60.

Beatty, R. A., Bennett, G. H., Hall, J. G., Hancock, J. L., and Stewart, D. L. (1969). An experiment with heterospermic insemination in cattle. *J. Reprod. Fertil.*, **19**, 491–501.

Beauchamp, G. (1989). Canine tooth size variability in primates. *Folia Primatol.*, **52**, 148–155.

Beauchamp, G., Doty, R. L., Moulton, D. G., and Mugford, R. A. (1976). The pheromone concept in mammalian chemical communication: a critique. In *Mammalian olfaction, reproductive processes and behavior*. (ed. R. L. Doty), pp. 143–160. Academic Press, New York.

Beckett, S. D., Hudson, R. S., Walker, D. F., Vachon, R. I., and Reynolds, T. M. (1972). Corpus cavernosum penis pressure and external penile muscle activity during erection in the goat. *Biol. Reprod.*, **7**, 359–364.

Beckett, S. D., Hudson, R. S., Walker, D. F., Reynolds, T. M., and Vachon, R. I. (1973). Blood pressure and penile muscle activity in the stallion during coitus. *Am. J. Physiol.*, **225**, 1071–1075.

Bedford, J. M. (1974). The biology of primate spermatozoa. In *Reproductive biology of the primates*. (ed. W. P. Luckett), pp. 97–119. Karger, Basel.

Bell, A. D. and Variend, S. (1985). Failure to demonstrate sexual dimorphism of the corpus callosum in childhood. *J. Anat. Lond.*, **143**, 143–147.

Bell, A. P. and Weinberg, M. S. (1978). *Homosexualties: a study of diversity among men and women*. Mitchell Beazley, London.

Benkert, O. (1975). Clinical studies on the effects of neurohormones on sexual behavior. In *Sexual behavior: pharmacology and biochemistry*. (eds. M. Sandler and G. L. Gessa), pp. 297–305. Raven Press, New York.

Bennett, E. L. and Sebastian, A. C. (1988). Social organisation and ecology of proboscis monkeys (*Nasalis larvatus*) in mixed coastal forest in Sarawak. *Int. J. Primatol.*, **9**, 233–255.

Benson, G.S. (1994). Male sexual function: erection, emission and ejaculation. In *The physiology of reproduction*, Vol. 1. (eds. E. Knobil and J. D. Neill), pp. 1489–1506. Raven Press, New York.

Benson, G. S., McConnell, J., Lipschultz, L. J., and Corriere, J. N. (1980). Neuromorphology and neuropharmacology of the human penis. *J. Clin. Invest.*, **65**, 506–513.

Berard, J. D., Nürnberg, P., Eplen, J. T., and Schmidtke, J. (1993). Male rank, reproductive behavior and reproductive success in free-ranging rhesus macaques. *Primates*, **34**, 481–489.

Bercovitch, F. B. (1986). Male rank and reproductive activity in savanna baboons. *Int. J. Primatol.*, **7**, 533–550.

Bercovitch, F. B. (1988). Coalitions, cooperation and reproductive tactics among adult male baboons. *Anim. Behav.*, **36**, 1198–1209.

Bercovitch, F. B. (1989). Body size, sperm competition, and determinants of reproductive success in savanna baboons. *Evolution*, **43**, 1507–1521.

Bercovitch, F. B. (1991). Mate selection, consortship formation, and reproductive tactics in adult female savanna baboons. *Primates*, **32**, 437–452.

Bercovitch, F. B. (1992). Re-examining the relationship between rank and reproduction in male primates. *Anim. Behav.*, **44**, 1168–1170.

Bercovitch, F. B. (1993). Dominance rank and reproductive maturation in male rhesus monkeys (*Macaca mulatta*). *J. Reprod. Fertil.*, **99**, 113–120.

Bercovitch, F. B. (1996). Testicular function and scrotal coloration in patas monkeys. *J. Zool. Lond.*, 93–100.

Bercovitch, F. B. and Berard, J. D. (1993). Life history costs and consequences of rapid reproductive maturation in female rhesus monkeys. *Behav. Ecol. Sociobiol.*, **32**, 103–109.

Bercovitch, F. B. and Goy, R. W. (1990). The socioendocrinology of reproductive development and reproductive success in macaques. In *Socioendocrinology of primate reproduction.* (eds. T. E. Zigler and F. B. Bercovitch), pp. 59–93. Wiley-Liss, New York,

Bercovitch, F. B. and Nürnberg, P. (1996). Socioendocrine and morphological correlates of paternity in rhesus macaques. *J. Reprod. Fertil.*, **107**, 59–68.

Bercovitch, F. B. and Ziegler, T. E. (1990). Introduction to socioendocrinology. In *Socioendocrinology of primate reproduction.* (eds. T. E. Ziegler and F. B. Bercovitch), pp. 1–9. Wiley-Liss, New York.

Berendson, H. G. G., Jenck, F., and Broekkamp, C. L. E. (1990). Involvement of 5-HT$_{1C}$ receptors in drug-induced penile erections in rats. *Psychopharmacology*, **101**, 57.

Berlese, A. (1925). *Gli insetti*, Vol. 2. Societa Editrice Libraria, Milan.

Bernstein, I. S. (1968). The lutong of Kuala Selangor. *Behaviour* **32**, 1–16.

Bernstein, I. S. (1970). Primate status hierarchies. In *Primate behavior*, Vol. 1. (ed. L. A. Rosenblum), pp. 71–109. Academic Press, New York.

Bernstein, I. S. (1981). Dominance: the baby and the bathwater. *Behav. Brain Sci.*, **3**, 419–458.

Bernstein, I. S. and Gordon, T. P. (1980). Mixed taxa introduction, hybrids and macaque systematics. In *The macaques: studies in ecology, behavior and evolution.* (ed. D. G. Lindburg), pp. 125–147. Van Nostrand Reinhold, New York.

Bernstein, I. S. and Sharpe, L. (1966). Social roles in a rhesus monkey group. *Behaviour.*, **26**, 91–103.

Bernstein, I. S., Rose, R. M., and Gordon, T. P. (1974). Behavioural and environmental events influencing primate testosterone levels. *J. Hum. Evol.*, **3**, 517–525.

Berthold, A. A. (1849). Transplantation der Hoden. *Arch. Anat. Physiol.*, **16**, 42–46.

Bertolini, A., Gessa, G. L., and Ferrari, W. (1975). Penile erection and ejaculation: a central effect of ACTH-like peptides in mammals. In *Sexual behavior: pharmacology and biochemistry.* (eds. M. Sandler and G. L. Gessa), pp. 247–257. Raven Press, New York.

Bertram, B. C. R. (1975). The social systems of lions. *Sci. Am.*, **232**, 54–65.

Bertrand, M. (1969). The behavioral repertoire of the stumptail macaque. *Bibliotheca Primatol.*, **11**, 1–273. Karger, Basel.

Beyer, C. and Gonzalez-Mariscal, G. (1994). Effects of sex steroids on sensory and motor spinal mechanisms. *Psychoneuroendocrinology*, **19**, 517–527.

Beyer, C., Velazquez, J., Larsson, K., and Contreras, J. L. (1980). Androgen regulation of the motor copulatory pattern in the male New Zealand white rabbit. *Horm. Behav.*, **14**, 170–190.

Beyer, C., Contreras, J. L., Larsson, K., Olmedo, M., and Morali, G. (1982). Pattern of motor and seminal vesicle activities during copulation in the male rat. *Physiol. Behav.*, **29**, 495–500.

Bieber, I., Dain, H. J., Dince, P. R., *et al.* (1962). *Homosexuality: a psychoanalytic study.* Basic Books, New York.

Biegon, A., Fishette, C. T., Rainbow, T. C., and McEwen, B. S. (1982). Serotonin receptor modulation by estrogen in discrete brain nuclei. *Neuroendocrinology*, **35**, 287–291.

Bielert, C. (1986). Sexual interactions between captive adult male and female chacma baboons (*Papio ursinus*) as related to the female's menstrual cycle. *J. Zool. Lond.*, **209**, 521–536.

Bielert, C. and Anderson, C. M. (1985). Baboon sexual swellings and male response: a possible operational mammalian supernormal stimulus and response interaction. *Int. J. Primatol.*, **6**, 377–393.

Bielert, C. and Girolami, C. (1986). Experimental assessments of behavioral and anatomical components of female chacma baboon (*Papio ursinus*) sexual attractiveness. *Psychoneuroendocrinology*, **11**, 75–90.

Bielert, C. F. and Goy, R. W. (1973). Sexual behavior of male rhesus: effects of repeated ejaculation and partner's cycle stage. *Horm. Behav.*, **4**, 109–122.

Bielert, C. and Vandenbergh, J. G. (1981). Seasonal influences in births and male sex skin coloration in rhesus monkeys (*Macaca mulatta*) in the southern hemisphere. *J. Reprod. Fertil.*, **62**, 229–233.

Bielert, C. and Van der Walt, L. A. (1982). Male chacma baboon (*Papio ursinus*) sexual arousal: mediation by visual cues from female conspecifics *Psychoneuroendocrinology*, **7**, 31–48.

Bielert, C., Czaja, J. A., Eisele, S., Scheffler, G., Robinson, J. A., and Goy, R. W. (1976). Mating in the rhesus monkey (*Macaca mulatta*) after conception and its relationship to oestradiol and progesterone levels throughout pregnancy. *J. Reprod. Fertil.*, **46**, 179–187.

Bielert, C., Howard-Tripp, M. E., and Van der Walt, M. A. (1980). Environmental and social factors influencing seminal emission in chacma baboons (*Papio ursinus*). *Psychoneuroendocrinology*, **5**, 287–303.

Bielert, C., Girolami, L., and Jowell, S. (1989). An experimental examination of the colour component in visually mediated sexual arousal of the male chacma baboon (*Papio ursinus*). *J. Zool. Lond.*, **219**, 569–579.

Bingham, H. C. (1928). Sex development in apes. *Comp. Psychol. Monogr.*, **5**, 1–165.

Bingham, H. C. (1932). Gorillas in a native habitat. *Carnegie Inst. Wash. Publ.*, **426**, 1–66.

Birkhead, T. R. (1988). Behavioural aspects of sperm competition in birds. *Adv. Study Behav.*, **18**, 35–72.

Birkhead, T. R. and Hunter, F. M. (1990). Mechanisms of sperm competition. *Trends Ecol. Evol.*, **5**, 48–52.

Birkhead, T. R. and Møller, A. P. (1992). *Sperm competition in birds*. Academic Press, London.

Birkhead, T. R., Hunter, F. M., and Pellat, J. E. (1989). Sperm competition in the zebra finch, *Taenopygia guttata*. *Anim. Behav.*, **38**, 935–950.

Birmingham, A. T. (1970). Sympathetic denervation of the smooth muscle of the vas deferens. *J. Physiol.*, **206**, 645–661.

Bitran, D. and Hull, E. M. (1987). Pharmacological analysis of male rat sexual behavior. *Neurosci. Biobehav. Rev.*, **11**, 365–389.

Blanchard, R. and Bogaert, A. F. (1996a). Homosexuality in men and number of older brothers. *Am. J. Psychiat.*, **153**, 27–31.

Blanchard, R. and Bogaert, A. F. (1996b). Biodemographic comparisons of homosexual and heterosexual men in the Kinsey interview data. *Arch. Sex. Behav.*, **25**, 551–579.

Blanchard, R. and Klassen, P. (1997). H-Y antigen and homosexuality in men. *J. Ther. Biol.*, **185**, 373–378.

Blanchard, R., Zucker, K. J., Bradley, S. J., and Hume, C. S. (1995). Birth order and sibling sex ratio in homosexual male adolescents and probably prehomosexual feminine boys. *Devel. Psychol.*, **31**, 22–30.

Blandau, R. J. (1945). On factors involved in sperm transport through the cervix uteri of the albino rat. *Am. J. Anat.*, **73**, 253–272.

Blaustein, J. D. and Feder, H. H. (1979). Cytoplasmic progestin-receptor in guinea pig brain: characteristics and relationship to the induction of sexual behavior. *Brain Res.*, **169**, 481–497.

Blaustein, J. D., Olster, D. H., Delville, Y., Nielsen, K. H., Tetel, M. J., and Turcotte, J. C. (1990). Hypothalamic sex steroid hormone receptors and female sexual behavior: new insights from immunocytochemical studies. In *Hormones, brain and behavior in vertebrates*, Vol. 2. (ed. J. Balthazart), pp. 75–90. Karger, Basel.

Bleier, R., Byne, W., and Siggelkow, I. (1982). Cytoarchitectonic sexual dimorphisms of the medial preoptic and anterior hypothalamic areas in the guinea pig, rat, hamster and mouse. *J. Comp. Neuro.*, **212**, 118–130.

Bloch, G. J. and Davidson, J. M. (1968). Effects of adrenalectomy and experience on post castration sexual behavior in the male rat. *Physiol. Behav.*, **3**, 461–465.

Boelkins, R. C. And Wilson, A. P. (1972). Intergroup social dynamics of the Cayo Santiago rhesus (*Macaca mulatta*) with special reference to changes in group membership by males. *Primates*, **13**, 125–140.

Bogaert, A. F., Bezeau, S., Kuban, M., and Blanchard, R. (1997). Pedophilia, sexual orientation and birth order. *J. Abnormal Psychol.*, **106**, 331–335.

Bogart, M. H., Cooper, R. W., and Benirschke, K. (1977). Reproductive studies of black and ruffed lemurs, *Lemur macaco macaco* and *L. variegatus*. *Int. Zoo Yb.*, **17**, 177–183.

Boggess, J. (1984). Infant killing and male reproductive strategies in langurs (*Presbytis entellus*). In *Infanticide: comparative and evolutionary perspectives*. (eds. G. Hausfater and S. B. Hrdy), pp. 283–310. Aldine, New York.

Bohlen, J. G., Held, J. P., Sanderson, M. D., and Ahlgren, A. (1982). The female orgasm: pelvic contractions. *Arch. Sex. Behav.*, **11**, 367–386.

Boinski, S. (1987). Mating patterns in squirrel monkeys (*Saimiri oerstedii*). *Behav. Ecol. Sociobiol.*, **21**, 13–21.

Bolwig, N. (1959). A study of the behaviour of the chacma baboon, *Papio ursinus*. *Behaviour*, **14**, 136–163.

Bonney, R. C. and Setchell, K. D. R. (1980). The excretion of gonadal steroids during the reproductive cycle of the owl monkey (*Aotus trivirgatus*). *J. Steroid. Biochem.*, **12**, 417–421.

Bonney, R. C., Dixson, A. F., and Fleming, D. (1979). Cyclic changes in the circulating and urinary levels of ovarian steroids in the adult female owl monkey. *J. Reprod. Fertil.*, **56**, 271–280.

Bonney, R. C., Dixson, A. F., and Fleming, D. (1980). Plasma concentrations of oestradiol – – 17β, oestrone, progesterone and testosterone during the ovarian cycle of the owl monkey (*Aotus trivirgatus*). *J. Reprod. Fertil.*, **60**, 101–107.

Bonnier, G. and Trulsson, S. (1939). Selective fertilization in poultry. *Hereditas*, **25**, 65–76.

Bonsall, R. W. and Michael, R. P. (1989). Pretreatments with 5α-dihydrotestosterone and the uptake of testosterone by cell nuclei in the brains of male rhesus monkeys. *J. Steroid Biochem.*, **33**, 405–411.

Bonsall, R. W., Rees, H. D., and Michael, R. P. (1986). ^{3}H-estradiol and its metabolites in the brain, pituitary gland and reproductive tract of the male rhesus monkey. *Neuroendocrinology*, **43**, 98–109.

Bonsall, R. W., Rees, H. D., and Michael, R. P. (1989). Identification of radioactivity in cell nuclei from brain, pituitary gland and genital tract of male rhesus monkeys after the administration of [^{3}H] testosterone. *J. Steroid Biochem.*, **32**, 599–608.

Booth, A. H. (1957). Observations on the natural history of the olive colobus monkey *Procolobus verus* (Van Beneden). *Proc. Zool. Soc. Lond.*, **129**, 421–430.

Booth, J. (1979). Sexual differentiation of the brain. *Oxford Revs. Repro. Biol.*, **1**, 58–158.

Borries, C., Sommer, V., and Srivastava, A. (1991). Dominance, age and reproductive success in free-ranging female Hanuman langurs (*Presbytis entellus*). *Int. J. Primatol.*, **12**, 231–258.

Boskoff, K. J. (1978). Indicators of reproductive cyclicity in the prosimian primate, *Lemur fulvus*: an analysis of vaginal and genital changes. Unpublished M. A. thesis, Duke University, U.S.A.

Bosu, W. T. K., Holmdahl, T. H., Johansson, E. D. B., and Gemzell, C. (1972). Peripheral plasma levels of oestrogens, progesterone and 17α- hydroxyprogesterone during the menstrual cycle of the rhesus monkey. *Acta. Endocr.*, **71**, 755–764.

Boswell, J. (1980). *Christianity, social tolerance and homosexuality*. University of Chicago Press.

Bottjer, S. W. and Arnold, A. P. (1984). Hormones and structural plasticity in the adult brain. *Trends in Neurosci.* May 1984, pp. 168–171.

Bowers, M. B., Van Woert, M., and Davis, L. (1971). Sexual behavior during L-dopa treatment for Parkinsonism. *Am. J. Psychiat.*, **127**, 1691.

Bowlby, J. (1969, 1973, 1980). *Attachment and loss*, 3 vols. Basic Books, New York and Hogarth Press, London.

Bowlby, J. (1990). *Charles Darwin: a biography*. Hutchinson, London.

Bowman, L. A., Dilley, S. R., and Keverne, E. B. (1978). Suppression of oestrogen-induced LH surge by social subordination in talapoin monkeys. *Nature Lond.*, **275**, 56–58.

Brackett, N. L. and Edwards, D. A. (1984). Medial preoptic connections with the midbrian tegmentum are essential for male sexual behavior. *Physiol. Behav.*, **32**, 79–84.

Brackett, N. L., Iuvone, P. M., and Edwards, D. A. (1986). Midbrain lesions, dopamine and male sexual behavior. *Behav. Brain Res.*, **20**, 231–240.

Bradbury, J. W. (1977). Lek mating behavior in the hammer-headed bat. *Z. Tierpsychol.*, **45**, 225–255.

Bradbury, J. W. and Gibson, R. (1983). Leks and mate choice. In *Mate choice*. (ed. P. P. G. Bateson), pp. 109–138. Cambridge University Press.

Brand, H. M. (1981). Urinary oestrogen excretion in the female cotton-topped tamarin (*Saguinus oedipus oedipus*). *J. Reprod. Fertil.*, **62**, 467–473.

Brand, H. M. and Martin, R. D. (1983). The relationship between urinary estrogen excretion and mating behavior in cotton-topped tamarins, *Saguinus oedipus oedipus*. *Int. J. Primatol.*, **4**, 275–290.

Brand, T., Kroonen, J., Mos, J., and Slob, A. K. (1991). Adult partner preference and sexual behavior of male rats affected by perinatal endocrine manipulations. *Horm. Behav.*, **25**, 323–341.

Brecher, E. M. (1984). *Love, sex and aging. A consumer's union report*. Little, Brown, Boston.

Breed, W. G. (1981). Unusual anatomy of the male reproductive tract in *Notomys alexis (Muridae)*. *J. Mammal.*, **62**, 373–375.

Breed, W. G. (1990). Copulatory behavior and coagulum formation in the female reproductive tract of the Australian hopping mouse, *Notomys alexis*. *J. Reprod. Fertil.*, **88**, 17–24.

Breedlove, S. M. (1984). Steroid influences on the development and function of a neuromuscular system. *Progr. Brain Res.*, **61**, 147–169.

Breedlove, S. M. and Arnold, A. P. (1980). Hormone accumulation in a sexually dimorphic motor nucleus in the rat spinal cord. *Science*, **210**, 564–566.

Breedlove, S. M. and Arnold, A. P. (1981). Sexually dimorphic motor nucleus in the rat lumbar spinal cord: response to adult hormone manipulation, absence in androgen-insensitive rats. *Brain Res.*, **225**, 297–307.

Brenner, R. M. and Slayden, O. D. (1994). Cyclic changes in the primate oviduct and endometrium. In *The physiology of reproduction*, Vol. 1. (2nd edition) (eds. E. Knobil and J. D. Neill), pp. 541–569. Raven Press, New York.

Brereton, A. R. (1994). Return-benefit spite hypothesis: an explanation for sexual interference in stumptail macaques (*Macaca arctoides*). *Primates*, **35**, 123–136.

Brindley, G. S. (1984). A new treatment for priapism. *Lancet* ii, 220–221.

Brindley, G. S. (1988). The actions of the parasympathetic and sympathetic nerves in human micturition, erection and seminal emission, and their restoration in paraplegic patients by implanted electrical stimulators. *Proc. Roy. Soc. Lond. B.*, **235**, 111–120.

Brindley, G. S., and Craggs, M. D. (1975). *J. Physiol.* **256**, 55p.

Brindley, G. S., Polkey, C. E., Rushton, N., and Cardozo, L. (1986). Sacral anterior root stimulation for bladder control in paraplegia. *J. Neurol. Neurosurg. Psychiat.*, **49**, 1104–1114.

Brockman, D. K. and Whitten, P. L. (1996). Reproduction in free-ranging *Propithecus verreauxi*: estrus and the relationship between multiple partner matings and fertilization. *Am. J. Phys. Anthropol.*, **100**, 57–69.

Bronson, F. H. (1968). Pheromonal influences on mammalian reproduction. In *Reproduction and sexual behavior*. (ed. M. Diamond), pp. 344–365. Indiana University Press, Bloomington.

Bronson, F. H. (1989). *Mammalian reproductive biology*. University of Chicago Press.

Bronson, F. H. and Desjardins, C. (1974). Circulating concentrations of FSH LH, estradiol and progesterone associated with acute, male-induced puberty in female mice. *Endocrinology*, **94**, 1658–1668.

Bronson, F. H. and Maruniak, J. A. (1975). Mali-induced puberty in female mice: evidence for a synergistic action of social cues. *Biol. Reprod.*, **13**, 94–98.

Brooks, D. E. (1990). Biochemistry of the male accessory glands. In *Marshall's physiology of reproduction*, Vol. 2 (fourth edition) (cd. G. E. Lamming), pp. 569–690. Churchill Livingstone, Edinburgh.

Brown, A. G. (1986). Sensory function. In *Scientific basis of dermatology: a physiological approach*. (eds. A. J. Thody and P. S. Friedmann), pp. 74–88. Churchill Livingstone, Edinburgh.

Brown, E., Brown, G. M., Kofman, O., and Quarrington, B. (1978). Sexual function and affect in Parkinsonian men treated with L-Dopa. *Am. J. Psychiat.*, **135**, 1552–1555.

Brown, K. S. (1981). The biology of *Heliconius* and related genera. *Ann. Rev. Entomol.*, **26**, 427–456.

Brown, R. E. (1994). *An introduction to neuroendocrinology*. Cambridge University Press.

Bruce, K. E. and Estep, D. Q. (1992). Interruption and harassment during copulation by stumptail macaques, *Macaca arctoides*. *Anim. Behav.*, **44**, 1029–1044.

Brun B., Cranz., C., Clavert, A., and Rumpler, Y. (1987). A safe technique for collecting semen from *Lemur fulvus mayottensis*. *Folia Primatol.*, **49**, 48–51.

Bubenick, G. A. and Brown, G. M. (1973). Morphologic sex differences in primate brain areas involved in regulation of reproductive activity. *Experientia* **15**, 619–621.

Buchanon, D. B., Mittermeier, R. A., and Van Roosmalen, M. G. M. (1981). The saki monkeys, genus *Pithecia*. In *Ecology and behavior of neotropical primates*, Vol. 1. (eds. A. F. Coimbra-Filho and R. A. Mittermeier), pp. 391–417. Academia Brasileira de Ciências, Rio de Janeiro.

Budnitz, N. and Dainis, K. (1975). *Lemur catta*; ecology and behavior. In *Lemur biology* (eds. I. Tattersall and R. W. Sussman), pp. 219–235. Plenum, New York.

Buechner, H. K. and Schloeth, R. (1965). Ceremonial mating behavior in Uganda kob (*Adenota kob thomasi* Neumann) *Z. Tierpsychol.*, **22**, 209–225.

Buesching, C. D., Heistermann, M., Hodges, J. K., and Zimmermann, E. (1998). Multimodal oestrus advertisement in a small nocturnal prosimian, *Microcebus murinus*. *Folia Primatol*, **69**, 295–308.

Buhrich, N., Bailey, J. M., and Martin, N. G. (1991). Sexual orientation, sexual identity and sex-dimorphic behaviours in male twins. *Behav. Genet.*, **21**, 75–96.

Burgen, A., Kosterlitz, H. W., and Iversen, L. L. (1980) (eds.). *Neuroactive peptides: a Royal Society discussion meeting*. The Royal Society, London.

Burley, N. (1979). The evolution of concealed ovulation. *Am. Nat.*, **114**, 835–858.

Burnett, A. L., Lowenstein, C. J., Bredt, D. S., Chang, T. S. K., and Snyder, S. H. (1992). Nitric oxide: a physiologic mediator of penile erection. *Science* **257**, 401–403.

Burt, A. (1992). 'Concealed ovulation' and sexual signals in primates. *Folia Primatol.*, **58**, 1–6.

Burton, F. D. (1971). Sexual climax in the female *Macaca mulatta*. In *Proc. IIIrd Int. Congr. Primatol.* pp. 180–191. Karger, Basel.

Busse, C. D. and Estep, D. Q. (1984). Sexual arousal in male pigtailed monkeys (*Macaca nemestrina*): effects of serial matings by two males. *J. Comp. Psychol.*, **98**, 227–231.

Butynski, T. M. (1982). Harem male replacement and infanticide in the blue monkey (*Cercopithecus mitus stuhlmanni*) in the Kibale Forest, Uganda. *Am. J. Primatol.*, **3**, 1–22.

Butynski, T. M. (1988). Guenon birth seasons and correlates with rainfall and food. In *A primate radiation: evolutionary biology of the African guenons.* (eds. A Gautier-Hion, F. Bouliere, J. P. Gautier and J. Kingdon), pp. 284–322. Cambridge University Press.

Buvat, J., Buvat-Herbaut, M., Dehaene, J. L., and Lemaire, A. (1986). Is intracavernous injection of papaverine a reliable screening test for vascular impotence? *J. Urol.*, **135**, 476–478.

Buvat, J., Buvat-Herbaut, M., Lemaire, A., Marcolin, G., and Quittelier, E. (1990). Recent developments in the clinical assessment and diagnosis of erectile dysfunction. *Ann. Revs. Sex Res.*, **1**, 265–308.

Bygott, J. D. (1972). Cannibalism among wild chimpanzees. *Nature Lond.*, **238**, 410–411.

Byne, W. and Bleier, R. (1987). Medial preoptic sexual dimorphisms in the guinea pig. I. *An investigation of their hormonal dependence. J. Neurosci.*, **7**, 2688–2696.

Byne, W., Bleier, R., and Houston, L. (1988). Variations in human corpus callosum do not predict gender: a study using magnetic resonance imaging. *Behav. Neurosci.*, **102**, 222–227.

Caggiula, A. R., Herndon, J. G., Scanlon, R., Greenstone, D., Bradshaw, W., and Sharp, D. (1978). Dissociation of active from immobility components of sexual behavior and sensorimotor responsiveness. *Brain Res.*, **172**, 505–520.

Caldecott, J. O. (1981). Findings on the behavioural ecology of the pigtailed macaque. *Malay. Appl. Biol.*, **10**, 213–220.

Caldwell, J. D., Prange, A. J., and Pedersen, C. A. (1986). Oxytocin facilitates the sexual receptivity of estrogen-treated female rats. *Neuropeptides* **7**, 175–189.

Callard, G. V., Petro, Z. and Ryan, K. J. (1977). Identification of aromatase in the reptilian brain. *Endocrinology*, **100**, 1214–1218.

Cameron, J. L. (1988). The role of increased cortisol secretion in the suppression of GnRH pulse generator activity in monkeys placed on a regime of reduced food intake. *Society for Neuroscience*. Abstract 425.

Cameron, J. L. (1989). Influence of nutrition on the hypothalamic-pituitary-gonadal axis in primates. In *The menstrual cycle and its disorders: influences of nutrition, exercise and neurotransmitters.* (eds. K. M. Pirke, W. Wuttke, and U. Schweiger), pp. 66–78. Springer, Heidelberg.

Cameron, J. L., Helmreich, D. L., Parfitt, D. B., and Nosbisch, C. (1989). Slowing of pulsatile LH and testosterone secretion after one day fast in adult male rhesus monkeys. In *Serono Symposium on neuroendocrine regulation of reproduction*. Abstract. Papa, California.

Camperio-Ciani, A. (1984). A case of infanticide in a free-ranging group of rhesus monkeys (*Macaca mulatta*) in the Jackoo Forest, Simla, India. *Primates*, **25**, 372–377.

Cant, J. G. H. (1978). Population survey of the spider monkey, *Ateles geoffroyi*, at Tikal, Guatemala. *Primates*, **19**, 525–535.

Cant, J. G. H. (1981). Hypothesis of the evolution of human breasts and buttocks. *Am. Nat.*, **117**, 199–204.

Carati, C. J., Creed, K. E., and Keogh, E. J. (1988). Vascular changes during erection in the dog. *J. Physiol. Lond.*, **400**, 75–88.

Carew, T. J. (1985). Posture and locomotion. In *Principles of neural science* (2nd edition) (eds. E. R. Kandel and J. H. Schwartz), pp. 478–486. Elsevier, New York.

Carlsen, E., Giwercman, A., Keiding, N., and Skakkebaek, N. E. (1992). Evidence for decreased quality of semen during past 50 years. *Brit. Med. J.*, **305**, 609–613.

Carlsson, S. G. and Larsson, K. (1964). Mating in male rats after local anesthetization of the glans penis. *Z. Tierpsychol.*, **21**, 854–856.

Carmichael, M. S., Humbert, R., Dixen, J., Palmisano, G., Greenleaf, W., and Davidson, J. M. (1987). Plasma oxytocin increases in the human sexual response. *J. Clin. Endocr. Metab.*, **64**, 27–31.

Carney, P. A., Bancroft, J., and Matthews, A. (1978). The combination of hormone and counselling in the treatment of female sexual unresponsiveness. *Br. J Psychiat.*, **133**, 339–346.

Caro, T. M. (1976). Observations on the ranging behaviour and daily activity of lone silverback mountain gorillas. *Anim. Behav.*, **24**, 889–897.

Caro, T. M. and Bateson, P. (1986). Organization and ontogeny of alternative tactics. *Anim. Behav.*, **34**, 1483–1499.

Carpenter, C. R. (1934). A field study of the behavior and social relations of howling monkeys. *Comp. Psychol. Monogr.*, **10**, 1–168.

Carpenter, C. R. (1935). Behavior of red spider monkeys in Panama. *J. Mammal.*, **16**, 171–180.

Carpenter, C. R. (1940). A field study in Siam of the behavior and social relations of the gibbon (*Hylobates lar*). *Comp. Psychol. Monogr.*, **16**, 1–212.

Carpenter, C. R. (1941). The menstrual cycle and body temperature in two gibbons (*Hylobates lar*). *Anat. Rec.*, **79**, 291–296.

Carpenter, C. R. (1942). Sexual behavior of free ranging rhesus monkeys (*Macaca mulatta*). I. specimens procedures and behavioral characteristics of estrus. II. periodicity of estrus, homosexual, autoerotic and nonconformist behavior. *J. Comp. Psychol.*, **33**, 113–162.

Carpenter, C. R. (1965). The howlers of Barro Colorado island. In *Primate behavior: field studies of monkeys and apes.* (ed. I. DeVore), pp. 250–291. Holt, Rinehart, and Winston, New York.

Cartmill, M. (1974). Rethinking primate origins. *Science*, **184**, 436–443.

Casperd, J. M. (1996). *Primate Eye* No. 58 supplement: current primate field studies.

Cassorla, F. G., Albertson, B. D., Chrousos, G. P., Booth, J. D., and Renquist, D. (1982). The mechanism of hypercortisolemia in the squirrel monkey. *Endocrinology*, **111**, 448–451.

Castellanos, H. and McCombs, H. L. (1968). The reproductive cycle of the New World monkey: gynecologic problems in a breeding colony. *Fertil. Steril.*, **19**, 213–227.

Catchpole, H. R. and Fulton, J. F. (1943). The oestrus cycle in *Tarsius:* observations on a captive pair. *J. Mammal.,* **24**, 90–93.

Chalmers, N. R. (1968a). Group composition, ecology and daily activities of free-living mangabeys in Uganda. *Folia Primatol.,* **8**, 247–262.

Chalmers, N. R. (1968b) The social behaviour of free living mangabeys in Uganda. *Folia Primatol.,* **8**, 263–281.

Chalmers, N. R. (1968c). The visual and vocal communication of free living mangabeys in Uganda. *Folia Primatol.,* **9**, 258–280.

Chalmers, N. and Rowell, T. E. (1971). Behaviour and reproductive cycles in a captive group of mangabeys. *Folia Primatol.,* **14**, 1–14.

Chambers, K. C. and Phoenix, C. H. (1984). Restoration of sexual performance in old rhesus macaques paired with a preferred female partner. *Int. J. Primatol.,* **5**, 287–297.

Chambers, K. C. and Phoenix, C. H. (1987). Differences among ovariectomized female rhesus macaques in the display of sexual behavior without and with estradiol treatment. *Behav. Neurosci.,* **101**, 303–308.

Chambers, K. C. and Phoenix, C. H. (1989). Apomorphine, (—)-deprenyl, and yohimbine fail to increase sexual behavior in rhesus males. *Behav. Neurosci.,* **103**, 816–823.

Chambers, K. C., Hess, D. L., and Phoenix, C. H. (1981). Relationship of free and bound testosterone to sexual behavior in old rhesus males. *Physiol. Behav.,* **27**, 615–620.

Chambers, K. C., Resko, J. A., and Phoenix, C. H. (1982). Correlation of diurnal changes in hormones with sexual behavior and age in male rhesus macaques. *Neurobiol. Aging* **3**, 37–42.

Chan, A., Dudley, C. A., and Moss, R. L. (1984). Hormonal and chemical modulation of ventromectial hypothalamic neurons responsive to vaginocervical stimulation. *Neuroendocrinology,* **38**, 328–336.

Chance, M. R. A., Emory, G. R., and Payne, R. G. (1977). Status referents in long-tailed macaques (*Macaca fascicularis*): precursors and effects of a female rebellion. *Primates,* **18**, 611–632.

Chang, M. C. (1951). Fertilizing capacity of spermatozoa deposited into the fallopian tubes. *Nature Lond.,* **168**, 697–698.

Chapais, B. (1983). Matriline membership and male rhesus reaching high ranks in their natal troops. In *Primate social relationships: an integrated approach.* (ed. R. A. Hinde), pp. Blackwell, Oxford.

Chapman, T., Liddle, L. F., Kalb, J. M., Wolfner, M. F., and Partridge, L. (1995). Cost of mating in Drosophila melanogaster females is mediated by male accessory gland products. *Nature Lond.,* **373**, 241–244.

Chappell, S. C., Resko, J. A., Norman, R. L., and Spies, H. (1981). Studies in rhesus monkeys on the site where estrogen inhibits gonadotropins: delivery of 17β-estradiol to the hypothalamus and pituitary gland. *Endocrinology,* **52**, 1–8.

Charles-Dominique, P. (1977). *Ecology and behavior of nocturnal primates.* Duckworth, London.

Charles-Dominique, P. and Bearder, S. K. (1979). Field studies of lorisid behavior: methodological aspects. In *The study of prosimian behavior.* (eds. G A. Doyle and R. D. Martin), pp. 567–629. Academic Press, New York.

Charney, D. S. and Heninger, G. R. (1986). Alpha2 -adrenergic and opiate receptor blockade, synergistic effects on anxiety in healthy subjects. *Arch. Gen. Psychiat.,* **43**, 1037–1041.

Cheney, D. L. and Seyfarth, R. M. (1990). *How monkeys see the world.* University of Chicago Press.

Cheney, D. L., Seyfarth, R. M., Andelman, S. J., and Lee, P. C. (1988). Reproductive success in vervet monkeys. In *Reproductive success* (ed. T. H. Clutton-Brock). University of Chicago Press.

Cheng, K. M., Burns, J. T., and McKinney, F. (1982). Forced copulation in captive mallards. III. sperm competition. *Auk,* **100**, 302–310.

Cherry, J. A. and Baum, M. J. (1990). Effects of lesions of a sexually dimorphic nucleus in the preoptic area/anterior hypothalamus area on the expression of androgen and oestrogen-dependent sexual behavior in male ferrets. *Brain Res.,* **522**, 191–203.

Chevalier-Skolnikoff, S. (1974). Male-female, female-female, and male-male sexual behavior in the stumptail monkey, with special attention to the female orgasm. *Arch. Sex. Behav.,* **3**, 95–116.

Chism, J. and Rowell, T. E. (1986). Mating and residence patterns of male patas monkeys. *Ethology,* **72**, 31–39.

Chism, J., Olson, D. K., and Rowell, T. E. (1983). Diurnal births and perinatal behavior among wild patas monkeys: evidence for an adaptive pattern. *Int. J. Primatol.,* **4**, 167–184.

Chism, J., Rowell, T. E., and Olson, D. K. (1984). Life history patterns of female patas monkeys. In *Female primates: studies by women primatologists.* (ed. M. F. Small), pp. 175–190. Alan R. Liss, New York.

Chivers, D. J. (1974). The siamang in Malaya. *Contrib. Primatol.,* **4**, 1–335. Karger, Basel.

Chowdhury, S. R., Saran, R. K., and Harish, C. (1980). Plasma levels of estradiol and progesterone during normal menstrual cycle in langur monkey (*Presbytis entellus entellus*). *Indian J. Physiol. Pharmacol.,* **24**, 364–366.

Christ, G. J., Maayani, S., Valcic, M., and Melman, A. (1990). Pharmacological studies of human erectile tissue: characteristics of spontaneous contractions and alternations and ?-adrenoceptor responsiveness with age and disease in isolated tissue. *Br. J. Pharmacol.,* **101**, 375–381.

Chrousos, G. P., Renquist, D., Brandon, D., Barnard, D., Fowler, D., Loriaux, D. L., and Lipsett, M. B. (1982). The squirrel monkey: receptor-mediated end-organ resistance to progesterone? *J. Clin. Endocr. Metab.,* **55**, 364–368.

Ciochon, R. L. and Chiarelli, A. B. (1980). (eds) *Evolutionary biology of the New World monkeys and continental drift.* Plenum Press, New York.

Clark, A. B. (1982). Scent marks as social signals in *Galago crassicaudatus.* 11. discrimination between individuals by scent. *J. Chem. Ecol.,* **8**, 1153–1165.

Clark, A. B. (1985). Sociality in a nocturnal 'solitary' prosimian: *Galago crassicaudatus. Int. J. Primatol.,* **6**, 581–600.

Clark, J. T., Smith, E. R., and Davidson, J. M. (1984). Enchancement of sexual motivation in male rats by yohimbine. *Science,* **225**, 847–848.

Clark, J. T., Smith, E. R., and Davidson, J. M. (1985a).

Evidence for the modulation of sexual behavior by ?-adrenoceptors in male rats. *Neuroendocrinology*, **41**, 36–43.

Clark, J. T., Smith, E. R., and Davidson, J. M. (1985b). Testosterone is not required for the enhancement of sexual motivation by yohimbine. *Physiol. Behav.*, **35**, 517–521.

Clark, W. E. Le Gros (1959). *The antecedents of man*. Edinburgh University Press.

Clarke, I. J., Scaramuzzi, R. J., and Short, R. V. (1976). Sexual differentiation of the brain: endocrine and behavioural responses of androgenized ewes to oestrogen. *J. Endocr.*, **71**, 171–176.

Clarke, J. T. (1991). Suppression of copulatory behavior in male rats following central administration of clonidine. *Neuropharmacology*, **30**, 373–382.

Clarke, M. R. (1981). Aspects of male behavior in mantled howlers (*Alouatta palliata* Gray) in Costa Rica. *Am. J. Primatol.*, **54**, 209.

Clarke, M. R. (1983). Infant killing and infant disappearance following male takeovers in a groups of free-ranging howling monkeys (*Alouatta palliata*) in Costa Rica. *Am. J. Primatol.*, **5**, 241–247.

Clayton, R. B., Cogura, J., and Kraemer, H. C. (1970). Sexual differentiation of the brain: RNA metabolism in female rats. *Nature Lond.*, **226**, 810–812.

Clewe, T. H. and DuVall, W. M. (1966). Observations on frequency of ejaculation of squirrel monkeys, *Saimiri sciureus*. *Am. Zool.*, **6**, 411 (abstract).

Clutton-Brock, T. H. (1974). Primate ecology and social organization. *Nature Lond.*, **250**, 539–542.

Clutton-Brock, T. H. (1989). Mammalian mating systems. *Proc. Roy. Soc. Lond. B.*, **236**, 339–372.

Clutton-Brock, T. II. (1991). The evolution of sex differences and consequences of polygyny in mammals. In *The development and integration of behaviour: essays in honour of Robert Hinde*. (ed. P. Bateson), pp. 229–253. Cambridge University Press.

Clutton-Brock, T. H. and Albon, S. D. (1979). The roaring of red deer and the evolution of honest advertisement. *Behaviour*, **69**, 145–170.

Clutton-Brock, T. H. and Harvey, P. H. (1976). Evolutionary rules and primate societies. In *Growing points in ethology*. (eds. P. P. G. Bateson and R. A. Hinde), pp. 195–237. Cambridge University Press.

Clutton-Brock, T. H. and Harvey, P. H. (1977). Primate ecology and social organization. *J. Zool. Lond.*, **183**, 1–39.

Clutton-Brock, T. H. and Parker, G. A. (1995). Sexual coercion in animal societies. *Anim. Behav.*, **49**, 1345–1365.

Clutton-Brock, T. H., Harvey, P. H., and Rudder, B. (1977). Sexual dimorphism, socionomic sex ratio and body weight in primates. *Nature Lond.*, **269**, 797–800.

Clutton-Brock, T. H., Albon, S. D., and Guiness, F. E. (1988). Reproductive success in red deer. In *Reproductive success*. (ed. T. H. Clutton-Brock), pp. 325–343. University of Chicago Press.

Cochran, C. A. and Perachio, A. A. (1977). Dihydrotestosterone effects on dominance and sexual behaviors in gonadectomized male and female rhesus monkeys. *Horm. Behav.*, **8**, 175–187.

Coe, C. L., Connolly, A. C., Kraemer, H. C., and Levine, S.

(1979). Reproductive development and behavior of captive female chimpanzees. *Primates*, **20**, 571–582.

Coelho, A. M. Jr, Coelho, L. S., Bramblett, C A., Bramblett, S. S., and Quick, L. B. (1976). Ecology, population characteristics, and sympatric association in primates: a socio-bioenergetic analysis of howler and spider monkeys in Tikal, Guatemala. *Yb. Phys. Anthropol.*, **20**, 96–135.

Coimbra-Filho, A. F. and Mittermeier, R. A. (1981). (eds). *Ecology and behavior of neotropical primates*, Vol. 1. Academia Brasileira de Ciências, Rio de Janeiro.

Colillas, O. and Coppo, J. (1978). Breeding *Alouatta caraya* in Centro Argentino de Primates. In *Recent advances in primatology*, Vol. 2. *conservation*. ed. D. J. Chivers and W. Lane Petter), pp. 201–214. Academic Press, London.

Collings, M. R. (1926). A study of the cutaneous reddening and swelling about the genitalia of the monkey *Macacus rhesus*. *Anat. Rec.*, **33**, 271–287.

Collins, D. A. (1981). *Social behaviour and patterns of mating among adult yellow baboons (Papio c. cynocephalus)*. Unpublished Ph.D. Thesis, University of Edinburgh.

Collins, D. A., Busse, C. D., and Goodall, J. (1984). Infanticide in two populations of savanna baboons. In *Infanticide: comparative and evolutionary perspectives*. (eds. G. Hausfater and S. B. Hrdy), pp. 193–215. Aldine, New York.

Collins, D. C., Graham, C. E., and Preedy, J. R. K. (1975). Identification and measurement of urinary estrone, estradiol-17β, estriol, pregnanediol and androsterone during the menstrual cycle of the orang-utan. *Endocrinology*, **96**, 93–101.

Colmenares, F. (1990). Greeting behaviour in male baboons, 1: communication, reciprocity and symmetry. *Behaviour*, **113**, 81–116.

Colmenares, F. (1991). Greeting behaviour between male baboons: oestrous females, rivalry and negotiation. *Anim. Behav.*, **41**, 49–60.

Comarr, A. E. and Gunderson, B. B. (1975). Sexual function in traumatic paraplegia and quadriplegia. *Am. J. Nurs.*, **75**, 250–255.

Commins, D. and Yahr, P. (1984). Adult testosterone levels influence the morphology of a sexually dimorphic area in the Mongolian gerbil brain. *J. Comp. Neuro.*, **224**, 131–140.

Conoway, C. H. and Koford, C. B. (1965). Estrous cycles and mating behavior in a free-ranging band of rhesus monkeys. *J. Mammal.*, **45**, 577–588.

Conoway, C. H. and Sade, D. S. (1965). The seasonal spermatogenic cycle in free-ranging rhesus monkeys. *Folia Primatol.*, **3**, 1–12.

Conoway, C. H. and Sade, D. S. (1969). Annual testis cycle of the green monkey (*Cercopithecus aethiops*) on St. Kitts, West Indies. *J. Mammal.*, **50**, 833–835.

Conrad, L. A. and Pfaff, D. W. (1975). Axonal projections of medial preoptic and anterior hypothalamic neurons. *Science*, **190**, 1111–1114.

Contreras, J. L. and Beyer, C. (1979). A polygraphic analysis of mounting and ejaculation in the New Zealand white rabbit. *Physiol. Behav.*, **23**. 939–943.

Converse, L. J., Carlson, A. A., Ziegler, T. E., and Snowdon, C. T. (1995). Communication of ovulatory state to mates by female pygmy marmosets, *Cebuella pygmaea*. *Anim. Behav.*, **49**, 615–621.

Cooper, M. C. and Aronson, L. R. (1958). The effect of adrenalectomy on the sexual behavior of castrated male cats. *Anat. Rec.*, **131**, 544.

Cooper, T. G. and Hamilton, D. W. (1977). Phagocytosis of spermatozoa in the distal vans deferens in rats. *Am. J. Anat.*, **150**, 247–268.

Coplan, J. D., Andrews, M. W., Rosenblum, L. A., Owens, M. J., Friedman, S., Gorman, J. M., *et al.* (1996). Persistent elevations of cerebrospinal fluid concentrations of corticotropin-releasing factor in adult nonhuman primates exposed to early-life stressors: implications for the pathophysiology of mood and anxiety disorders. *Proc. Nat. Acad. Sci. U.S.A.*, **93**, 1619–1623.

Cords, M. (1987). Forest guenons and patas monkeys: male-male competition in one-male groups. In *Primate societies*. (eds. B. Smuts, D. Cheney, R. Seyfarth, R. Wrangham and T. Struhsaker), pp. 98–111. University of Chicago Press.

Cords, M. (1988). Mating systems of forest guenons: a preliminary review. In *A primate radiation: evolutionary biology of the African guenons* (eds A. Gautier-Hion, F. Boulière, J. P. Gautier and J. Kingdon), pp. 323–339. Cambridge University Press.

Cords, M., Mitchell, B. J., Tsingalia, H. M., and Rowell, T. E. (1986). Promiscuous mating among blue monkeys in the Kakamega forest, Kenya. *Ethology*, **72**, 214–226.

Corner, G. W. (1940). Accessory corpora lutea in the ovary of the monkey *Macaca rhesus*. *Anales de la Facultad de Medicina de Montevideo*, **25**, 553–560.

Courtois, F. J., Macdougall, J. C., and Sachs, B. L. (1993). Erectile mechanism in paraplegia. *Physiol. Behav.*, **53**, 721–726.

Cowlishaw, G. and Dunbar, R. I. M. (1991). Dominance ranking and mating success in male primates. *Anim. Behav.*, **41**, 1045–1056.

Craig, A. D., Bushnell, M. C., Zhang, E. T., and Blomquist, A. (1994). A thalamic nucleus specific for pain and temperature sensation. *Nature Lond.*, **372**, 770–773.

Cramer, D. L. (1977). Craniofacial morphology of *Pan paniscus*: a morphometric and evolutionary appraisal. *Contrib. Primatol.*, **10**, 1–64. Karger, Basel.

Creed, K. E., Carati, C. J., and Keogh, E. J. (1991). The physiology of penile erection. *Oxford Revs. Repro. Biol.*, **13**, 73–95.

Creighton, S. J. (1985). An epidemiological study of abused children and their families in the United Kingdom between 1977–1982. *Child Abuse and Neglect.*, **9**, 441–448.

Crockett, C. M. and Eisenberg, J. F. (1987). Howlers: variations in group size and demography. In *Primate societies*. (eds. B. Smuts, D Cheney, R. Seyfarth, R. Wrangham and T. Struhsaker), pp. 54–68. University of Chicago Press.

Crook, J. H. (1966). Gelada baboon herd structure and movement: a comparative report. *Symp. Zool. Soc. Lond.*, **18**, 237–258.

Crook, J. H. (1972). Sexual selection dimorphism, and social organization in the primates. In *Sexual selection and the descent of man*. (ed. B. G. Campbell), pp. 231–281. Aldine, Chicago.

Crook, J. H. and Aldrich-Blake, P. (1968). Ecological and behavioural contrasts between sympatric ground dwelling primates in Ethiopia. *Folia Primatol.*, **8**, 192–227.

Crook, J. H. and Crook, S. J. (1988). Tibetan polyandry: problems of adaptation and fitness. In *Human reproductive behaviour: a Darwinian perspective*. (eds. L. Betzig, M. Borgerhoff Mulder, and P. Turke), pp. 97–114. Cambridge University Press.

Crowe, M. J. and Qureshi, M. J. H. (1991). Pharmacologically induced penile erection (PIPE) as a maintenance treatment for erectile impotence: a report of 41 cases. *Sex. and Mar. Therapy.*, **6**, 273–285.

Crowley, W. R. (1986). Reproductive neuroendocrine regulation in the female rat by central catecholamine-neuropeptide interactions: a local control hypothesis. *Ann. N. Y. Acad. Sci.*, **474**, 423–436.

Cullberg, J. (1972). Mood changes and menstrual symptoms with different gestagen estrogen combinations. *Acta. Psychiat. Scand. Suppl.*, 236.

Cummins, J. M. and Woodall, P. F. (1985). On mammalian sperm dimensions *J. Reprod. Fertil.*, **75**, 153–175.

Cuny, T. H. (1956). *Ivan Pavlov: the man and his theories*. Paul S. Erikson, New York.

Curtin, R. and Dolhinow, P. (1978). Primate social behavior in a changing world. *Am. Sci.*, **66**, 468–475.

Curtis, R. F., Ballantine, J. A., Keverne, E. B., Bonsall, R. W., and Michael, R. P. (1971). Identification of primate sexual pheromones and properties of synthetic attractants. *Nature Lond.*, **232**, 396–398.

Cushman, P. (1972). Sexual behavior in heroin addiction and methadone maintenance. *N.Y. State J. Med.*, **72**, 1261–1265.

Cutler, W. B., Garcia, C. R., and Krieger, A. M. (1979). Luteal phase defects: a possible relationship between short hyperthermic phase and sporadic sexual behavior in women. *Horm. Behav.*, **13**, 214–218.

Cutler, W. B., Garcia, C. R., and Krieger, A. M. (1980). Sporadic sexual behavior and menstrual cycle length in women. *Horm. Behav.*, **14**, 163–172.

Cutler, W., Preti, G., Krieger, A., Huggins, G. R., Garcia, C. R. and Lawley, H. J. (1986). Human axillary secretions influence womens' menstrual cycles: the role of donor extract from men. *Horm. Behav.*, **20**, 463–473.

Czaja, J. A. and Bielert, C. (1975). Female rhesus sexual behavior and distance to male partner: relation to stage of the menstrual cycle. *Arch. Sex. Behav.*, **4**, 583–597.

D'Alessio, V. (1996). Born to be gay? *New Scientist* Sept. 28th, 1996, pp. 32–35.

Dabbs, J. M. (1993). Salivary testosterone measurements in behavioural studies. *Ann. N. Y. Acad. Sci.*, **694**, 177–183.

Dahl, J. F. (1985). The external genitalia of female pigmy chimpanzees. *Anat. Rec.*, **211**, 24–28.

Dahl, J. F. (1986). Cyclic perineal swelling during the intermenstrual intervals of captive female pygmy chimpanzees (*Pan paniscus*). *J. Hum. Evol.*, **15**, 369–385.

Dahl, J. F. (1987). Sexual initiation in a captive group of pygmy chimpanzees (*Pan paniscus*). *Primate Report.*, **16**, 43–53.

Dahl, J. F. (1988). External genitalia. In *Orang-utan biology* (ed. J. H. Schwartz), pp. 133–144. Oxford University Press, New York.

Dahl, J. F. and Nadler, R. D. (1992). The external genitalia of female gibbons, *Hylobates (H.) lar. Anat. Rec.*, **232**, 572–578.

Dahl, J. F., Gould, K. G., and Nadler, R. D. (1993). Testicle size of orang-utans in relation to body size. *Am. J. Phys. Anthrop.*, **90**, 229–236.

Dahlof, L. G. and Larsson, K. (1978). Copulatory performances of penile desensitized male rats as a function of prior social and sexual experience. *Behav. Biol.*, **24**, 492–497.

Dail, W. G. (1993). Autonomic innervation of the male reproduction genitalia. In *The autonomic nervous system. Nervous control of the urogenital system*, Vol. 6. (ed. C. A. Maggi), pp. 69–101. Harwood, London.

Dail, W. G., Moll, A., and Weber, K. (1983). Localization of vasoactive intestinal polypeptide in penile erectile tissue and in the major pelvic ganglion of the rat. *Neuroscience*, **10**, 1379–1386.

Daly, M. and Wilson, M. (1983). *Sex evolution and behavior*, (2nd edn). Wadsworth, Belmont, California.

Daly, M. and Wilson, M. I. (1985). Child abuse and other risks of not living with both parents. *Ethol. Sociobiol.*, **6**, 197–210.

Daly, M. and Wilson, M. I. (1996). Violence against stepchildren. *Current Directions in Psychol. Sci.*, **5**, 77–81.

Damasio, A. (1991). Concluding comments. In *Frontal lobe function and dysfunction.* (eds. H. A. Levin, H. M. Eisenberg, and A. L. Bention), pp. 401–407. Oxford University Press, New York.

Damassa, D. A., Smith, E. R., Tennent, B., and Davidson, T. (1977). The relationship between circulating testosterone levels and male sexual behavior in rats. *Horm. Behav.*, **8**, 275–286.

Dantchakoff, V. (1938). Role des hormones dans la manifestation des instincts sexuelles. *C. R. Hebd. Séanc. Acad. Sci. Paris.*, **206**, 79–86.

Darwin, C. (1851). A monograph of the sub-class Cirripedia, Vol. 1. the Lepadidae., London.

Darwin, C. (1854). A monograph of the sub-class Cirripedia, Vol. 2. the Balanidae., London.

Darwin, C. (1859). On the origin of species by means of natural selection or the preservation of favoured races in the struggle for life. John Murray, London.

Darwin, C. (1871). The descent of man and selection in relation to sex. John Murray, London.

Darwin, C. (1872). The Expression of the emotions in man and animals. John Murray, London.

Darwin, C. (1876). Sexual selection in relation to monkeys. *Nature Lond.*, **15**, 18–19.

Darwin, C. (1881). The formation of vegetable mould through the action of worms. John Murray, London.

Davenport, R. K. and Menzel, E. W. (1963). Stereotyped behavior of the infant chimpanzee. *Arch. Gen. Psychiat.*, **8**, 99–104.

David, G. F. X. and Ramaswami, L. S. (1969). Studies on menstrual cycles and other related phenomena in the langur (*Presbytis entellus entellus*). *Folia Primatol.*, **11**, 300–316.

David, G. F. X. and Ramaswami, L. S. (1971). Reproductive systems of the North Indian langur (*Presbytis entellus entellus Dufresne*). *J. Morphol.*, **135**, 99–129.

David, G. F. X. and Rao, C. A. P. (1969). Ovulation and the serum protein changes during ovulation in *Presbytis entellus entellus* Dufresne. *Gen. Comp. Endocr. Suppl.*, **2**, 197–202.

Davidson, J. M. (1966). Activation of the male rat's sexual behaviour by intracerebral implantation of androgens. *Endocrinology*, **79**, 783–794.

Davidson, J. M., Rodgers, C. H., Smith, E. R., and Bloch, G. J. (1968). Stimulation of female sex behavior in adrenalectomized rats with estrogen alone. *Endocrinology*, **82**, 193–195.

Davidson, J. M., Camargo, C. A., and Smith, E. R. (1979). Effects of androgens on sexual behavior of hypogonadal men. *J. Clin. Endoc. Metab.*, **48**, 955–958.

Davies, N. B. (1991). Mating systems. In *Behavioural ecology: an evolutionary approach.* (eds. J. R. Krebs and N. B. Davies), pp. 263–294. Blackwell Scientific Publications, Oxford.

Davies, T. S. (1974). Cyproterone acetate for male hypersexuality. *J. Int. Med. Res.*, **2**, 159–163.

Dawkins, R. (1976). *The selfish gene*. Oxford University Press.

Dawson, G. A. (1976). Behavioral ecology of the Panamanian tamarin, *Saguinus oedipus*. Unpublished Ph.D. thesis., Michigan State University, East Lansing.

Dawson, G. A. (1979). The use of time and space by the Panamanian tamarin, *Saguinus oedipus*. *Folia Primatol.*, **31**, 253–284.

De Bold, J. F. and Frye, C. A. (1994). Progesterone and the neural mechanisms of hamster sexual behavior. *Psychoneuroendocrinology*, **19**, 563–579.

De Groat, W. C. and Steers, W. D. (1988). Neuroanatomy and neurophysiology of penile erection. In *Contemporary management of impotence and infertility.* (eds. E. A. Tanagho, T. F. Lue, and R. D. McClure), pp. 3–27. Williams and Wilkins, Baltimore.

De Jong, F. H., Louwerse, A. L., Ooms, M. P., Evers, P., Endert, E., and Van de Poll, N. (1989). Lesions of the SDN–POA inhibit sexual behavior of male Wistar rats. *Brain Res. Bull.*, **23**, 483–492.

De Lacoste, M. C. and Woodward, D. J. (1988). The corpus callosum in non-human primates: determinants of size. *Brain Behav. Evol.*, **31**, 318–323.

De Lacoste, M. C., Holloway, R. L., and Woodward, D. J. (1986). Sex differences in the fetal human corpus callosum. *Human Neurobiol.*, **5**, 93–96.

De Lacoste-Utamsing, M. C. and Holloway, R. L. (1982). Sexual dimorphism in the human corpus callosum. *Science*, **216**, 1431–1432.

De Meyer, J. M. and De Sy, W. A. (1986). Intracavernous injection of noradrenaline to interrupt erections during surgical interventions. *Eur. Urol.*, **12**, 169–170.

De Ridder, C. M., Bruning, P. F., Zonderland, M. L., Thijssen, J. H. H., Bonfrer, J. M. G., Blankenstein, M. A., *et al.* (1990). Body fat mass, body fat distribution and plasma hormones in early puberty in females. *J. Clin. Endocr. Metab.*, **70**, 888–893.

De Ruiter, J. R. and Van Hooff, J. A. R. A. M. (1993). Male dominance rank and reproductive success in primate groups. *Primates*, **34**, 513–523.

De Ruiter, J. R. and Inoue, M. (1993). Paternity, male social rank and sexual behavior. *Primates*, **34**, 553–555.

De Ruiter, J. R., Scheffrahn, W., Trommelen, G. J. J. M., Uitterlinden, A. G., Martin, R. D. and Van Hooff, J. A. R. A. M. (1992). Male social rank and reproductive success in wild long-tailed macaques. In *Paternity in Primates: genetic tests and theories* (eds. R. D. Martin, A. Dixson and J. Wickings), pp. 175–190. Karger, Basel.

De Valois, R. L. and Jacobs, G. H. (1971). Vision. In *Behavior of non-human primates*; Vol. 3. (eds. A. Schrier and F. Stollnitz), pp. 107–157. Academic Press, New York.

De Waal, F. (1982). *Chimpanzee politics*. Jonathan Cape, London.

De Waal, F. B. M. (1986). The Brutal elimination of a rival among captive male chimpanzees. *Ethol. Sociobiol.*, 7, 237–251.

De Waal, F. B. M. (1989). *Peacemaking among primates*. Harvard University Press, Cambridge, Mass.

De Waal, F. B. M. (1992). Coalitions as part of reciprocal relations in the Arnhem chimpanzee colony. In *Coalitions and alliances in humans and other animals*. (eds. A. H. Harcourt and F. B. M. de Waal), pp. 233–257. Oxford University Press.

Deag, J. M. (1974). A study of the social behaviour and ecology of the wild barbary macaque, *Macaca sylvanus*. Unpublished Ph.D. Dissertation, University of Bristol.

Deag, J. M. and Crook, J. H. (1971). Social behaviour and 'agonistic buffering' in the wild Barbary macaque *Macaca sylvanus*. *Folia Primatol.*, 15, 183–200.

Defler, T. R. (1983). Some population characteristics of *Callicebus torquatus lugens* (Humbolt, 1812) (Primates: Cebidae) in eastern Colombia. *Loziana* (*Acta Zool.*) *Colombiana*, 38, 1–19.

Del Abril, A., Segovia, S., and Guillamon, A. (1987). The bed nucleus of the stria terminalis in the rat: regional sex differences controlled by gonadal steroids early after birth. *Devel. Brain Res.*, 32, 295–300.

Del Abril, A., Segovia, S., and Guillamon, A. (1990). Sexual dimorphism in the parastrial nucleus of the rat preoptic area. *Devel. Brain Res.*, 52, 11–15.

Delcomyn, F. (1980). Neural basis of rhythmic behavior in animals. *Science*, 210, 492–498.

Demeter, S., Ringo, J. L., and Doty, R. W. (1988). Morphometric analysis of the human corpus callosum and anterior commisure. *Human Neurobiol.*, 6, 219–226.

Demott, R. P. and Suarez, S. S. (1992). Hyperactivated sperm progress in the mouse oviduct. *Biol. Reprod.*, 46, 779–785.

Demski, L. S., Bauer, D. H., and Gerald, J. W. (1975). Sperm release evoked by electrical stimulation of the fish brain: a functional anatomical study. *J. Exp. Zoo.*, 191, 215–232.

Dennerstein, L. and Burrows, G. D. (1982). Hormone replacement therapy and sexuality in women. *Clinics in Endocr. Metab.*, 11, 661–679.

Denniston, R. H. (1964). Notes on the vaginal cornification cycle of captive squirrel monkeys. *J. Mammal.*, 45, 471.

Denton, D. (1993). The pinnacle of life: consciousness and self-awareness in humans and animals. Allen and Unwin, St. Leonards, Australia.

Deputte, B. L., Johnson, J., Hempel, M., and Scheffler, G. (1994). Behavioral effects of an antiandrogen in adult male rhesus macaques (*Macaca mulatta*). *Horm. Behav.*, 28, 155–164.

Desa, R. M. L., Pope, T. R., Struhsaker, T. T., and Glander, K. E. (1993). Sexual dimorphism in canine length of woolly spider monkeys (*Brachyteles arachnoides*, Geoffroy, E. 1806). *Int. J. Primatol.*, 14, 755–763.

Deslypere, J. P. and Vermeulen, A. (1984). Leydig cell function in normal men: effect of age, life style, residence, diet and activity. *J. Clin. Endocr. Metab.*, 59, 955–962.

Desmond, A. and Moore, J. (1991). *Darwin*. Penguin, London.

Després, J. P., Prudhomme, D., Pouliot, M., Tremblay, A., and Bouchard, C. (1991). Estimation of deep abdominal adipose-tissue accumulation from simple anthropometric measurements in men. *Am. J. Clin. Nutrition*, 54, 471–477.

Deutsch, J. and Larsson, K. (1974). Model-oriented sexual behavior in surrogate-reared rhesus monkeys. *Brain Behav. EVol.*, 9, 157–164.

Devine, M. C. (1984). Potential for sperm competition in reptiles: behavioral and physiological consequences. In *Sperm competition and animal mating systems*. (ed. R. L. Smith), pp. 509–521. Academic Press, New York.

DeVore, I. (1965). Dominance and mating behavior in baboons. In *Sex and Behavior* (ed. F. A. Beach), pp. 266–289. Wiley and Sons, New York.

DeVore, I. and Hall, K. R. L. (1965). Baboon ecology. In *Primate behavior: field studies of monkeys and apes*. (ed. I. DeVore), pp. 20–52. Holt, Rinehart and Winston, New York.

DeVries, G. J. (1995). Studying neurotransmitter systems to understand the development and function of sex differences in the brain: the case of vasopressin. In *Neurobiological effects of sex steroid hormones*. (eds. P. E. Micevych and R. P. Hammer), pp. 254–278. Cambridge University Press.

DeVries, G. J., Buijs, R. M., and Van Leeuwen, F W. (1984). Sex differences in vasopressin and other neurotransmitter systems in the brain. *Prog. Brain Res.*, 61, 185–203.

Dewsbury, D. A. (1972). Patterns of copulatory behavior in male mammals. *Q. Rev. Biol.*, 47, 1–33.

Dewsbury, D. A. (1982). Ejaculate cost and mate choice. *Am. Nat.*, 119, 601–610.

Dewsbury, D. A. (1988). A test of the role of copulatory plugs in sperm competition in deer mice. (*Peromyscus maniculatus*).

Dewsbury, D. A. and Baumgardner, J. D. (1981). Studies of sperm competition in two species of muroid rodents. *Behav. Ecol. Sociobiol.*, 9, 121–133.

Dewsbury, D. A. and Hodges, A. W. (1987). Copulatory behavior and related phenomena in spiny mice (*Acomys cahirinus*) and hopping mice (*Notomys alexis*). *J. Mammal.*, 69, 49–57.

Dewsbury, D. A. and Pierce, J. D. (1989). Copulatory patterns of primates as viewed in broad mammalian perspective. *Am. J. Primatol.*, 17, 51–72.

Diamond, E. J., Aksel, S., Hazelton, J. M., Jennings, R. A., and Abee, C. R. (1984). Seasonal changes of serum concentrations of estradiol and progesterone in Bolivian squirrel monkeys. (*Saimiri sciureus*). *Am. J. Primatol.*, 6, 103–113.

Diamond, M. (1970). Intromission and species vaginal code in relation to induction of pseudopregnancy. *Science*, 169, 995–997.

Diamond, M. (1993). Homosexuality and bisexuality in different populations. *Arch. Sex. Behav.*, 22, 291–310.

Diamond, M. and Sigmundson, H. K. (1997). Sex reassignment at birth: long term review and clinical implications. *Arch. Pediatr. Adolesc. Med.*, 151, 298–304.

Dittmann, R. W. (1998). Sexual behavior and sexual orientation in females with congenital adrenal hyperplasia. In *Sexual orientation: toward biological understanding* (eds. L.

Ellis and L. Ebertz), Ch. 4. Praeger, Westport, Connecticut.

Dittmann, R. W., Kappes, M. E., and Kappes, M. H. (1992). Sexual behavior in adolescent and adult females with congenital adrenal hyperplasia. *Psychoneuroendocr.*, **17**, 153–170.

Dittus, W. (1975). Population dynamics of the toque monkey, *Macaca sinica*. In *Socioecology and psychology of primates*. (ed. R. H. Tuttle). Mouton, The Hague.

Dittus, W. (1977). The social regulation of population density and age-sex distribution in the toque monkey. *Behaviour.*, **63**, 281–322.

Dittus, W. P. J. (1979). The evolution of behaviors regulating density and age-specific sex ratios in a primate population. *Behaviour*, **69**, 265–302.

Dittus, W. P. J. (1986). Sex differences in fitness following a group-takeover among toque macaques: testing models of social evolution. *Behav. Ecol. Sociobiol.*, **19**, 257–266.

Dixson, A. F. (1976). Effects of testosterone on the sternal cutaneous glands and genitalia of the male greater galago (*Galago crassicaudatus crassicaudatus*). *Folia Primatol.*, **26**, 207–213.

Dixson, A. F. (1977). Observations on the displays, menstrual cycles and sexual behaviour of the 'black ape' of Celebes (*Macaca nigra*). *J. Zool. Lond.*, **182**, 63–84.

Dixson, A. F. (1978). Effects of ovariectomy and estradiol replacement therapy upon the sexual and aggressive behavior of the greater galago (*Galago crassicaudatus crassicaudatus*). *Horm. Behav.*, **10**, 61–70.

Dixson, A. F. (1980). Androgens and aggressive behavior in primates: a review *Aggressive Behavior*, **6**, 37–67.

Dixson, A. F. (1981). *The natural history of the gorilla*. Weidenfeld and Nicolson, London.

Dixson, A. F. (1983a). The owl monkey. In *Reproduction in New World primates*. (ed. J. P. Hearn), pp. 69–113. MTP Press, Lancaster.

Dixson, A. F. (1983b). Observations on the evolution and behavioral significance of 'sexual skin' in female primates. *Adv. Study Behav.*, **13**, 63–106.

Dixson, A. F. (1983c). The hormonal control of sexual behaviour in primates *Oxford Revs. Repro. Biol.*, **5**, 131–218.

Dixson, A. F. (1986a). Genital sensory feedback and sexual behaviour in male and female marmosets (*Callithrix jacchus*). *Physiol. Behav.*, **37**, 447–450.

Dixson, A. F. (1986b). Plasma testosterone concentrations during postnatal development in the male common marmoset. *Folia Primatol.*, **47**, 166–170.

Dixson, A. F. (1986c). Proceptive displays of the female common marmoset (*Callithrix jacchus*): effects of ovariectomy and oestradiol 17β. *Physiol. Behav.*, **36**, 971–973.

Dixson, A. F. (1987a). Observations on the evolution of the genitalia and copulatory behaviour in male primates. *J. Zool. Lond.*, **213**, 423–443.

Dixson, A. F. (1987b). Baculum length and copulatory behavior in primates. *Am. J. Primatol.*, **13**, 51–60.

Dixson, A. F. (1987c). Effects of adrenalectomy upon proceptivity, receptivity and sexual attractiveness in ovariectomized marmosets (*Callithrix jacchus*). *Physiol. Behav.*, **39**, 495–499.

Dixson, A. F. (1988). Effects of dorsal penile nerve transection upon the sexual behaviour of the male marmoset (*Callithrix jacchus*). *Physiol. Behav.*, **43**, 235–238.

Dixson, A. F. (1989). Sexual selection, genital morphology and copulatory behaviour in male galagos. *Int. J. Primatol.*, **10**, 47–55.

Dixson, A. F. (1990a). The neuroendocrine regulation of sexual behaviour in female primates. *Ann. Revs. Sex Res.*, **1**, 197–226.

Dixson, A. F. (1990b). Medial hypothalamic lesions and sexual receptivity in the female common marmoset (*Callithrix jacchus*). *Folia Primatol.*, **54**, 46–56.

Dixson, A. F. (1990c). Observations on neuroendocrine control of sexual behavior in the male common marmoset (*Callithrix jacchus*). In *Hormones, brain and behaviour in vertebrates*, Vol. 2. (ed. J. Balthazart), pp. 63–74. Karger, Basel.

Dixson, A. F. (1991a). Penile spines affect copulatory behaviour in a primate (*Callithrix jacchus*). *Physiol Behav.*, **49**, 557–562.

Dixson, A. F. (1991b). Sexual selection, natural selection and copulatory patterns in male primates. *Folia Primatol.*, **57**, 96–101.

Dixson, A. F. (1992). Observations on postpartum changes in hormones and sexual behavior in callitrichid primates: do females exhibit post partum 'estrus'? In *Topics in primatology*, Vol. 2. (eds. N. Itoigawa, Y. Sugiyama, G. P. Sackett and R. K. R. Thompson), pp. 141–149. University of Tokyo Press, Tokyo.

Dixson, A. F. (1993a). Callitrichid mating systems: laboratory and field approaches to studies of monogamy and polyandry. In *Marmosets and Tamarins: systematics, behaviour and ecology* (ed. A B. Rylands), pp.164–175. Oxford University Press.

Dixson, A. F. (1993b). Sexual selection, sperm competition and the evolution of sperm length. *Folia Primatol.*, **61**, 221–227.

Dixson, A. F. (1993c). Sexual and aggressive behaviour of adult male marmosets (*Callithrix jacchus*) castrated neonatally, prepubertally or in adulthood. *Physiol. Behav.*, **54**, 301–307.

Dixson, A. F. (1993d). Effects of testosterone propionate upon the sexual and aggressive behavior of adult marmosets (*Callthrix jacchus*) castrated as neonates. *Horm. Behav.*, **27**, 216–230.

Dixson, A. F. (1994). Reproductive biology of the owl monkey. In *Aotus: the owl monkey* (eds. J. F. Baer, R. E. Weller and I. Kakoma), pp. 113–132. Academic Press, San Diego.

Dixson, A. F. (1995a). Sexual selection and the evolution of copulatory behaviour in nocturnal prosimians. In *Creatures of the dark: the nocturnal prosimians* (eds. L. Alterman, G. A. Doyle and M. K. Izard), pp. 93–118. Plenum, New York.

Dixson, A. F. (1995b). Sexual selection and ejaculatory frequencies in primates. *Folia Primatol.*, **64**, 146–152.

Dixson, A. F. (1995c). Baculum length and copulatory behaviour in carnivores and pinnipeds (Grand Order Ferae). *J. Zool. Lond.*, **235**, 67–76.

Dixson, A. F. (1997a). Evolutionary perspectives on primate mating systems and behavior. *Annals. N.Y. Acad. Sci.*, **807**, 42–61.

Dixson, A. F. (1997b). Sexual selection and evolution of the seminal vesicles in primates. *Folia Primatol.*

Dixson, A. F. and Fleming, D. (1981). Parental behaviour and infant development in owl monkeys (*Aotus trivirgatus griseimembra*). *J. Zool. Lond.*, **194**, 25–39.

Dixson, A. F. and Gardner, J. S. (1981). Diurnal variations in plasma testosterone in a male nocturnal primate, the owl monkey (*Aotus trivirgatus*). *J. Reprod. Fertil.*, **62**, 83–86.

Dixson, A. F. and Hastings, M. H. (1992). Effects of ibotenic acid-induced neuronal degeneration in the hypothalamus upon proceptivity and sexual receptivity in the female marmoset. *J. Neuroendocr.*, **4**, 719–726.

Dixson, A. F. and Herbert, J. (1974). The effects of testosterone on the sexual skin and genitalia of the male talapoin monkey. *J. Reprod. Fertil.*, **38**, 217–219.

Dixson, A. F. and Herbert, J. (1977). Gonadal hormones and sexual behavior in groups of adult talapoin monkeys (Miopithecus talapoin). *Horm. Behav.*, **8**, 141–154.

Dixson, A. F. and Lloyd, S. A. C. (1988). The hormonal and hypothalamic control of primate sexual behaviour. *Symp. Zool. Soc. Lond.*, **60**, 81–117.

Dixson, A. F. and Lloyd, S. A. C. (1989). Effects of male partners upon proceptivity in ovariectomized estradiol-treated marmosets (*Callithrix jacchus*). *Horm. Behav.*, **23**, 211–220.

Dixson, A. F. and Lunn, S. F. (1987). Post partum changes in hormones and sexual behaviour in captive marmoset groups. *Physiol. Behav.*, **41**, 577–583.

Dixson, A. F. and Mundy, N. I. (1994). Sexual behavior, sexual swelling and penile evolution in chimpanzees (*Pan troglodytes*). *Arch. Sex. Behav.*, **23**, 267–280.

Dixson, A. F. and Nevison, C. M. (1997). The socioendocrinology of adolescent development in male rhesus monkeys (*Macaca mulatta*). *Horm. Behav.*, **31**, 126–135.

Dixson, A. F. and Purvis, A. (in press). Sexual selection and genital morphology in male and female primates. *Folia Primatol.*

Dixson, A. F. and Van Horn, R. N. (1977). Comparative studies of morphology and reproduction in two subspecies of the greater bushbaby, *Galago crassicaudatus crassicaudatus* and *G. c. argentatus*. *J. Zool. Lond.*, **183**, 517–526.

Dixson, A. F., Everitt, B. J., Herbert, J., Rugman, S. M., and Scruton, D. M. (1973). Hormonal and other determinants of sexual attractiveness and receptivity in rhesus and talapoin monkeys. In *Symp. Int. Congr. Primatol.*, Vol. 2., *Primate reproductive behavior*. (ed. C. H. Phoenix), pp. 36–63. Karger, Basel.

Dixson, A. F., Scruton, D. M., and Herbert, J. (1975). Behaviour of the talapoin monkey (Miopithecus talapoin) studied in groups, in the laboratory. *J. Zool. Lond.*, **176**, 177–210.

Dixson, A. F., Moore, H. D. M., and Holt, W. V. (1980). Testicular atrophy in captive gorillas. *J. Zool. Lond.*, **191**, 315–322.

Dixson, A. F., Knight, J., Moore, H. D. M. and Carman, M. (1982). Observations on sexual development in male orangutans, *Pongo pygmaeus*. *Int. Zoo Yrbk.*, **22**, 222–227.

Dixson, A. F., Kendrick, K. M., Blank, M. A., and Bloom, S. R. (1984). Effects of tactile and electrical stimuli upon release of vasoactive intestinal polypeptide in the mammalian penis. *J. Endocr.*, **100**, 249–252.

Dixson, A. F., Anzenberger, G., Monteiro Da Cruz, M. A. O., Patel, I. and Jeffreys, A. J. (1992). DNA fingerprinting of free-ranging groups of common marmosets (*Callithrix jacchus jacchus*) in NE Brazil. In *Paternity in primates: genetic tests and theories* (eds. R. D. Martin, A. F. Dixson and E. J. Wickings), pp. 192–202. Karger, Basel.

Dixson, A. F., Bossi, T. and Wickings, E. J. (1993). Male dominance and genetically determined reproductive success in the mandrill (*Mandrillus sphinx*). *Primates*, **34**, 525–532.

Dockray, G. J. and Gregory, R. A. (1980). Relations between neuropeptides and gut hormones. In *Neuroactive peptides: a Royal Society discussion meeting*. (eds. A. Burgen, H. W. Kosterlitz and L. L. Iversen), pp. 151–164. The Royal Society, London.

Doherty, P. C., Bartke, A., and Smith, M. S. (1981). Differential effects of bromocryptine treatment on LH release and copulatory behavior in hyperprolactinaemic male rats. *Horm. Behav.*, **15**, 436–450.

Doherty, P. C., Baum, M. J., and Todd, R. B. (1986). Effects of chronic hyperprolactinaemia on sexual arousal and erectile function in male rats. *Neuroendocrinology*, **42**, 368–375.

Doherty, P. C., Lane, S. F., Pfeil, K. A., Morgan, W. W., Bartke, A., and Smith, M. S. (1989). Extra-hypothalamic dopamine is not involved in the effects of hyperprolactinaemia on male copulatory behavior. *Physiol. Behav.*, **45**, 1101–1105.

Dohler, K. D., Coquelin, A., and Davis, F. (1984). Pre- and postnatal influence of testosterone propionate and diethylstilbestrol on differentiation of the sexually dimorphic nucleus of the preoptic area in male and female rats. *Brain Res.*, **302**, 291–295.

Domanski, E., Prozekop, F., and Skubiszewski, B. (1972a). Interaction of progesterone and oestrogens on the hypothalamic centre controlling estrus behavior in sheep. *Acta Neurobiologica Experimenta*, **32**, 763–766.

Domanski, E., Przekop, F., and Skubiszewski, B. (1972b). The role of the anterior region of the medial basal hypothalamus in the control of ovulation and sexual behavior in sheep. *Acta Neurobiologica Experimenta*, **32**, 753–762.

Doogen, D. P. and Caillard, V. (1988). Setraline: a new antidepressant. *J. Clin. Psychiat.* (*Suppl.*), **49**, 46–51.

Dörner, G. (1976). *Hormones and brain differentiation*. Elsevier, Amsterdam.

Dörner, G. (1977). Hormone dependent differentiation, maturation and function of the brain and sexual behavior. *Endokrinologie*, **69**, 306–320.

Dörner, G. (1979). Hormones and sexual differentiation of the brain. *Ciba Found. Symp.*, **62**, 81–112. Excepta Medica, Amsterdam.

Dörner, G. (1988). Neuroendocrine response to estrogen and brain differentiation in heterosexuals, homosexuals and transsexuals. *Arch. Sex. Behav.*, **17**, 57–75.

Dörner, G., Geier, T., Ahrens, L., Krell, L., Münx, G., Sieler, et al. (1980). Prenatal stress as possible aetiogenetic factor of homosexuality in human males. *Endokrinologie*, **25**, 365–386.

Dorr, L. D. and Brody, M. J., (1967). Hemodynamic mechanisms of erection in the canine penis. *Am. J. Physiol.*, **213**, 1526–1531.

Dorsa, D. M., Smith, D. R., and Davidson, J. M. (1981). Endocrine and behavioral effects of continuous exposure of male rats to a potent luteinizing hormone-releasing hormone (LHRH) agonist: evidence for central nervous system actions of LHRH. *Endocrinology*, **109**, 729–735.

Doyle, G. A., Pelletier, A., and Bekker, T. (1967). Courtship, mating and parturition in the lesser bushbaby (*Galago senegalensis moholi*) under semi-natural conditions. *Folia Primatol.*, **7**, 169–196.

Drellich, M. G. and Waxenburg, S. E. (1966). Erotic and affectional components of female sexuality. In *Sexuality of women—Science and psychoanalysis*, Vol. 10. (ed. J. H. Masserman), pp. 45–55. Grune and Stratton, New York.

Drickamer, L. C. (1974a). Social rank, observability and sexual behaviour of rhesus monkeys. *J. Reprod. Fertil.*, **37**, 117–120.

Drickamer, L. C. (1974b). A ten year summary of reproductive data for free-ranging *Macaca mulatta*. *Folia Primatol.*, **21**, 61–80.

Drickamer, L. C. and Vessey, S. (1973). Group changing in free-ranging male rhesus monkeys. *Primates*, **14**, 359–368.

Drukker, B., Nieuwenhuigsen, K., Van Der Werff Ten Bosch, J. J., Van Hooff, J. A. R. A. M., and Slob, A. K. (1991). Harassment of sexual interactions among stumptail macaques, *Macaca arctoides*. *Anim. Behav.*, **42**, 171–182.

Dubey, A. K. and Plant, T. M. (1985). A suppression of gonadotropin secretion in castrated male rhesus monkeys (*Macaca mulatta*) by the interruption of hypothalamic gonadotropin-releasing hormone release. *Biol. Reprod.*, **33**, 423–431.

DuMond, F. V. (1968). The squirrel monkey in a seminatural environment. In *The*

Dumond, F. V. and Hutchinson, T. C. (1967). Squirrel monkey reproduction: the 'fatted' male phenomenon and seasonal spermatogenesis. *Science*, **158**, 1067–1070.

Dunbar, R. I. M. (1977). Age-dependent changes in sexual skin colour and associated phenomena of female gelada baboons. *J. Hum. Evol.*, **6**, 667–672.

Dunbar, R. I. M. (1978). Sexual behaviour and social relationships among gelada baboons *Anim. Behav.*, **26**, 167–178.

Dunbar, R. I. M. (1980). Determinants and evolutionary consequences of dominance among female gelada baboons. *Behav. Ecol. Sociobiol.*, **7**, 253–265.

Dunbar, R. I. M. (1984). Reproductive decisions: an economic analysis of gelada baboon social strategies. Princeton University Press, Princeton, New Jersey.

Dunbar, R. I. M. (1988). *Primate social systems*. Croom Helm, London.

Dunbar, R. I. M. (1992). Neocortex size as a constraint on group size in primates. *J. Hum. Evol.*, **20**, 469–493.

Dunbar, R. I. M. (1993). Social organization of the gelada. In *Theropithecus: the rise and fall of a primate genus*. (ed. N. G. Jablonski), pp. 425–439. Cambridge University Press.

Dunbar, R. I. M. and Dunbar, E. P. (1975). *Social dynamics of gelada baboons*. Karger, Basel.

Durham, N. M. (1969). Sex differences in visual threat displays of West African vervets. *Primates*, **10**, 91–95.

Dyer, R. G. and Bicknell, R. J. (eds) (1989). *Brain opioid systems and reproduction*. Oxford University Press.

East, M. L., Hofer, H., and Wickler, W. (1993). The erect 'penis' is a flag of submission in a female-dominated society: greetings in Serengetti spotted hyenas. *Behav. Ecol. Sociobiol.*, **33**, 355–370.

Eaton, G. G. (1973). Social and endocrine determinants of sexual behaviour in simian and prosimian females. In *Symp.*

IVth Int. Congr. Primatol., Vol. 2. (eds. C. H. Phoenix), pp. 20–35. Karger, Basel.

Eaton, G. G. (1978). Longitudinal studies of sexual behavior of the Oregon troop of Japanese macaques. In *Sex and behavior* (eds. T. E. McGill, D. A. Dewsbury, and B. D. Sachs), pp. 35–59. Plenum Press, New York.

Eaton, G. G. and Resko, J. A. (1974a). Ovarian hormones and behavior of *Macaca nemestrina*. *J. Comp. Physiol. Psychol.*, **86**, 919–925.

Eaton, G. G. and Resko, J. A. (1974b). Plasma testosterone and male dominance in a Japanese macaque (*Macaca fuscata*) troop compared to repeated measures of testosterone in laboratory males. *Horm. Behav.*, **5**, 251–259.

Eaton, G. G., Slob, A. K., and Resko, J. A. (1973a). Cycles of mating behaviour, oestrogen and progesterone in the thick-tailed bushbaby, *Galago crassicaudatus crassicaudatus*, under laboratory conditions. *Anim. Behav.*, **21**, 309–315.

Eaton, G. G., Goy, R. W., and Phoenix, C. H. (1973b). Effects of testosterone treatment in adulthood on sexual behaviour of female pseudohermaphrodite rhesus monkeys. *Nature Lond.*, **242**, 119–120.

Eaton, G. G., Worlein, J. M., and Glick, B. B. (1990). Sex differences in Japanese macaques (*Macaca fuscata*): effects of prenatal testosterone on juvenile social behavior. *Horm. Behav.*, **24**, 270–283.

Eberhard, W. G. (1985). *Sexual selection and animal genitalia*. Harvard University Press, Massachusetts and London.

Eberhard, W. G. (1996). *Female control: sexual selection by cryptic female choice*. Princeton University Press, New Jersey.

Eberhart, J. A. and Keverne, E. B. (1979). Influences of the dominance hierarchy on LH, testosterone and prolactin in male talapoin monkeys. *J. Endocr.*, **83**, 42–43.

Eckhard, C. (1869). *Beitrage zur anatomie und physiologie*, Vol. 4. Emil Roth, Giessen.

Eckstein, P. (1958). Internal reproductive organs. In *Primatologica* Vol. 3. (eds. H. Hofer, A. H. Schultz and D. Starck), pp. 543–629. Karger, Basel.

Edwards, R.G. (1955). Selective fertilization following the use of sperm mixtures in the mouse. *Nature Lond.*, **175**, 215–223.

Edwards, D. A. and Matthews, D. (1977). The ventromedial nucleus of the hypothalamus and the hormonal arousal of sexual behaviors in the female rat. *Physiol. Behav.*, **14**, 439–453.

Eerdenberg, F. V. (1991). Postnatal development of some nuclei in the pig hypothalamus. Unpublished Ph.D. thesis, University of Utrecht.

Ehardt, C. L. and Bernstein, I. S. (1986). Matrilineal overthrows in rhesus monkey groups. *Int. J. Primatol.*, **7**, 157–181.

Ehrhardt, A. A. and Meyer-Bahlburg, H. F. L. (1979). Psychosexual development: an examination of the role of prenatal hormones. In *Sex, hormones and behaviour. Ciba Found. Symp.*, **62**, pp. 41–50. Excerpta Medica, Amsterdam.

Ehrhardt, A. A. and Meyer-Bahlburg, H. F. L. (1981). Effects of prenatal sex hormones on gender-related behavior. *Science.*, **211**, 1311–1318.

Ehrhardt, A. A. and Money, J. (1967). Progestin-induced hermaphroditism: IQ and psychosexual identity in a study of ten girls. *J. Sex. Res.*, **3**, 83–100.

Ehrhardt, A. A., Epstein, R., and Money, J. (1968a). Fetal androgens and female gender identity in the early-treated adrenogential syndrome. *Johns Hopkins Med. J.*, **122**, 160–167.

Ehrhardt, A. A., Evers, K., and Money, J. (1968b). Influence of androgen and some aspects of sexually dimorphic behavior in women with the late-treated adrenogenital syndrome. *Johns Hopkins Med. J.*, **123**, 115–122.

Eibl-Eibesfeldt, I. (1970). *Ethology, the biology of behavior.* Holt, Rinehart and Winston, New York.

Eisenberg, J. F. (1976). Communication mechanisms and social integration in the black spider monkey, *Ateles fusciceps robustus*, and related species. *Smithson Contrib. Zool.*, **213**, 1–108.

Eisenberg, J. F. (1981). The mammalian radiations: an analysis of trends in evolution, adaptation and behaviour. Athlone Press, London.

Eisenberg, J. F. and Kuehn, R. (1966). The behavior of *Ateles geoffroyi* and related species. *Smithson. Misc. Coll.*, **151**, 1–63.

Eisenberg, J. F., Muckenhirn, N. A., and Rudran, R. (1972). The relation between ecology and social structure in primates. *Science*, **176**, 863–874.

Eisler, J. A., Tannenbaum, P. L., Mann, D. R., and Wallen, K. (1993). Neonatal testicular suppression with a GnRH agonist in rhesus monkeys: effects on adult endocrine function and behavior. *Horm. Behav.*, **27**, 551–567.

Ekman, P., Hager, J. C., and Friesen, W V. (1981). The symmetry of emotional and deliberate facial actions. *Psychophysiology*, **18**, 101–106.

Ellefson, J. O. (1968). Territorial behavior in the common white-handed gibbon, *Hylobates lar*. In *Primates: studies in adaptation and variability* (ed. P. Jay). pp. 180–199. Holt, Rinehart and Winston, New York.

Ellefson, J. O. (1974). A natural history of white-handed gibbons in the Malay peninsula. In *Gibbon and siamang*, Vol. 3 (ed. D. M. Rumbaugh), pp. Karger, Basel.

Ellis, L. (1995). Dominance and reproductive success among nonhuman animals: a cross-species comparison. *Ethol. Sociobiol.*, **16**, 257–333.

Ellis, L., Ames, M. A., Peckham, W., and Burke, D. (1988). Sexual orientation of human offspring may be altered by severe maternal stress during pregnancy. *J. Sex Res.*, **25**, 151–157.

Emery, D. E. and Sachs, B. D. (1976). Copulatory behavior in male rats with lesions in the bed nucleus of the stria terminalis. *Physiol. Behav.*, **17**, 803–806.

Emlen, S. T. and Oring, L. W. (1977). Ecology, sexual selection, and the evolution of mating systems. *Science*, **197**, 215–223.

Emory, L. E., Williams, D. H., Cole, C. M., Amparo, E. G., and Meyer, W. J. (1991). Anatomic variation of the corpus callosum in persons with gender dysphoria. *Arch. Sex Behav.*, **20**, 409–417.

Enomoto, T. (1974). The sexual behavior of Japanese macaques. *J. Hum. Evol.*, **3**, 351–372.

Enomoto, T., Seiki, K., and Haruki, Y. (1979). On the correlation between sexual behavior and ovarian hormone level during the menstrual cycle in captive Japanese monkeys. *Primates*, **20**, 563–570.

Epple, G. (1967). Soziale kommunikation bei *Callithrix jacchus* Erxleben, 1777. In *Progress in primatology* (eds. D. Starck, R. Schneider, and H. J. Kutin), pp. 247–254. Gustav Fischer, Stuttgart.

Epple, G. (1974). Olfactory communication in South American primates. *Ann. N.Y. Acad. Sci.*, **237**, 261–278.

Epple, G. (1975). The behavior of marmoset monkeys. In *Primate behavior*, Vol. 4. (ed. L. A. Rosenblum), pp. 195–239. Academic Press, New York.

Epple, G. (1976). Chemical communication and reproductive processes in nonhuman primates. In *Mammalian olfaction, reproductive processes and behavior*. (ed. R. L. Doty), pp. 257–282. Academic Press, New York.

Epple, G. (1980). Relationships between aggression, scent marking and gonadal state in a primate, the tamarin *Saguinus fuscicollis*. In *Chemical signals in vertebrates, 2. vertebrates and aquatic invertebrates*. (eds. D. Müller-Schwarze and R. M. Silverstein), pp. 87–105. Plenum, New York.

Epple, G. and Katz, Y. (1983). The saddle back tamarin and other tamarins. In *Reproduction in New World primates*. (ed. J. P. Hearn), pp. 115–148. MTP Press, Lancaster.

Epple, G. and Katz, Y. (1984). Social influences on oestrogen excretion and ovarian cyclicity in saddle-back tamarins (*Saguinus fuscicollis*). *Am. J. Primatol.*, **6**, 215–227.

Epple, G. and Lorenz, K. (1967). Vorkommen, morphologie und funktion der sternaldruse bei den platyrrhini. *Folia Primatol.*, **7**, 98–126.

Epple, G., Küderling, I., and Belcher, A. M. (1988). Some communicatory functions of scent marking in the cotton-top tamarin (*Saguinus oedipus oedipus*). *J. Cem. Ecol.*, **14**, 503–513.

Epple, G., Belcher, A. M., Kuderling, I., Zeller, U., Scolnik, L., Greenfield, K. L., and Smith, A. B. III. (1993). Making sense out of scents: species differences in scent glands, scent-marking behaviour, and scent-mark composition, the Callithrichidae. In *Marmosets and tamarins: systematics, behavior and ecology*. (ed. A. B. Rylands), pp. 123–151. Oxford University Press, London.

Epps, J. (1976). Social interactions of *Perodicticus potto* kept in captivity in Kampala, Uganda. In *Prosimian behaviour*. (eds. R. D. Martin, G. A. Doyle and A. C. Walker), pp. 233–244.

Erwin, J. and Maple, T. (1976). Ambisexual behavior with male-male anal penetration in male rhesus monkeys. *Arch. Sex. Behav.*, **5**, 9–14.

Estep, D. Q., Nieuwenhuijsen, K., Bruce, K. E. M., De Neef, K. J., Walters, P. A., Baker, S. C., and Slob, A. K. (1988). Inhibition of sexual behaviour among subordinate stumptail macaques, *Macaca arctoides*. *Anim. Behav.*, **36**, 854–864.

Estrada, A. and Sandoval, J. M. (1977). Social relations in a free-ranging troop of stumptail macaques *Macaca arctoides*: male care behaviour. *Primates.*, **18**, 793–813.

Etgen, A. M. and Karkanias, G. B. (1990). Estradiol regulates the number of α_1, but not β or α_2 noradrenergic receptors in the hypothalamus of female rats. *Neurochem. Intl.*, **16**, 1–9.

Etgen, A. M. and Karkanias, G. B. (1994). Estrogen regulation of noradrenergic signalling in the hypothalamus. *Psychoneuroendocrinology*, **19**, 603–610.

Etgen, A. M., Vathy, I., Petitti, N., Ungar, S., and Karkanias, G. B. (1990). Ovarian steroids, female reproductive behavior, and norepipephrine neurotransmission in the hypothalamus. In *Hormones, brain and behaviour in vertebrates*, Vol. 2. (ed. J. Balthazart), pp. 116–128. Karger, Basel.

Evans, C. S. and Goy, R. W. (1968). Social behaviour and reproductive cycles in captive ring-tailed lemurs (*Lemur catta*). *J. Zool. Lond.*, **156**, 181–197.

Evans, C. and Schilling, A. (1995). The accessory (vomeronasal) chemoreceptor system in some prosimians. In *Creatures of the dark: the nocturnal prosimians* (eds. L. Alterman, G. A. Doyle and M. K. Izard), pp. 393–411. Plenum, New York.

Evans, I. M. and Distiller, L. A. (1979). Effects of luteinizing hormone releasing hormone on sexual arousal in normal men. *Arch. Sex Behav.*, **8**, 385–395.

Everitt, B. J. (1977). Effects of clomipramine and other inhibitors of monoamine re-uptake on the sexual behaviour of female rats and rhesus monkeys. *Postgrad. Med. J. Suppl.*, **4**, 53, 202–210.

Everitt, B. J. (1990). Sexual motivation: a neural and behavioural analysis of the mechanisms underlying appetite and copulatory responses of male rats. *Neurosci. Biobehav. Revs.*, **14**, 217–232.

Everitt, B. J. and Herbert, J. (1969). The role of ovarian hormones in the sexual preference of rhesus monkeys. *Anim. Behav.*, **17**, 738–746.

Everitt, B. J. and Herbert, J. (1971). The effects of dexamethasone and androgens on sexual receptivity of female rhesus monkeys. *J. Endocr.*, **51**, 575–588.

Everitt, B. J. and Herbert, J. (1972). Hormonal correlates of sexual behaviour in sub-human primates. *Dan. Med. Bull.*, **19**, 246–255.

Everitt, B. J. and Herbert, J. (1975). The effects of implanting testosterone propionate into the central nervous system on the sexual behaviour of adrenalectomized female rhesus monkeys. *Brain Res.*, **86**, 109–120.

Everitt, B. J. and Stacey, P. (1987). Studies of instrumental behavior with sex reinforcement in male rats (*Rattus norvegicus*). II. Effects of preoptic area lesions, castration and testosterone. *J. Comp. Psychol.*, **101**, 407–419.

Everitt, B. J., Herbert, J., and Hamer, J. D. (1972). Sexual receptivity of bilaterally adrenalectomized female rhesus monkeys. *Physiol. Behav.*, **8**, 409–415.

Everitt, B. J., Fuxe, K., and Hökfelt, T. (1974). Inhibitory role of dopamine and 5-hydroxytryptamine in the sexual behaviour of female rats. *Eur. J. Pharmacol.*, **29**, 187–191.

Everitt, B. J., Keverne, E. B., Martensz, N. D., and Hansen, S. (1981). Hormones and social behaviour in rhesus and talapoin monkeys. In *Steroid hormone regulation of the brain*. (eds. K. Fuxe *et al.*), pp. 317–330. Pergamon Press, Oxford.

Ewer, R. F. (1973). *The carnivores*. Cornell University Press, Ithaca, New York.

Faiman, C. and Winter, J. S. D. (1971). Diurnal cycles in plasma F.S.H. testosterone and cortisol in men. *J. Clin. Endocr. Metab.*, **33**, 186–192.

Fairbanks, L. A. and McGuire, M. T. (1984). Determinants of fecundity and reproductive success in captive vervet monkeys. *Am. J. Primatol.*, **7**, 27–38.

Fairweather, H. (1982). Sex differences. In *Divided visual field studies of cerebral organization.* (ed. J. G. Beaumont), pp. 148–194. Academic Press, London.

Fargas, L. (1987). *Anthropometric facial proportions in medicine*. Charles C. Thomas, Springfield, Illinois.

Feder, H. H. (1971). The comparative action of testosterone propionate and 5α-androstan-17β-ol-3-one propionate on the reproductive behaviour, physiology and morphology of male rats. *J. Endocr.*, **51**, 242–252.

Fedigan, L. M. (1983). Dominance and reproductive success. *Yrbk. Phys. Anthropol.*, **26**, 91–129.

Fedigan, L. M., Fedigan, L., Gonzoules, S., Gonzoules, H., and Koyama, N. (1986). Lifetime reproductive success in female Japanese macaques. *Folia Primatol.*, **47**, 143–157.

Feldman, S. C and Bloch, E. (1977). Developmental pattern of testosterone synthesis by fetal rat testis in response to LH. *Endocrinology*, **102**, 999–1007.

Feneux, D., Serres, C., and Jouannet, P. (1985). Sliding spermatozoa: a dyskinesia responsible for human infertility. *Fertil. Steril.*, **44**, 508–511.

Ferrari, S. F. and Lopes Ferrari, M. A. (1989). A re-evaluation of the social organization of the *Callitrichidae* with special reference to the ecological differences between genera. *Folia Primatol.*, **52**, 131–147.

Ferstl, R., Eggert, F., Westphalt., E., Zarazava, N., and Müller-Ruchholtz, W. (1992). MHC-related odours in humans. In *Chemical signals in vertebrates* (ed. R. L. Doty), pp. 205–211. Plenum, New York.

Fine, M. L., Economos, D., Radtke, R., and McClung, J. R. (1984). Ontogeny and sexual dimorphism of the sonic motor nucleus in the oyster toadfish. *J. Comp. Neurol.*, **225**, 105–110.

Finn, C. A. and Booth, J. (1977). Physiological effects of oestrogen and progesterone. In *The Ovary*, Vol. 3. (2nd edition) (eds S. Zuckerman and B. J. Wier), pp. 151–225. Academic Press, New York.

Fisher, R. A. (1930). *The genetical theory of natural selection*. Clarendon Press, Oxford.

Fisher, R. A. (1965). Experiments in plant hybridisation. Mendel's original paper in English translation with commentary and assessment. Oliver and Boyd, Edinburgh and London.

Fisher, S. (1973). *The female orgasm*. Basic Books, New York.

Fletcher, T. J. (1978). The induction of male sexual behavior in red deer (*Cervus elaphus*) by the administration of testosterone to hinds and estradiol 17β to stags. *Horm. Behav.*, **11**, 74–88.

Flores, F., Naftolin, F., Ryan, K. J., and White, R. J. (1973). Estrogen formation by the isolated perfused rhesus monkey brain. *Science*, **180**, 1074–1075.

Foerg, R. (1982a). Reproduction in Cheirogaleus medius. *Folia Primatol.*, **39**, 49–62.

Foerg, R. (1982b). Reproductive behavior in *Varecia variegata*. *Folia Primatol.*, **38**, 108–121.

Fogden, M. P. L. (1976). A preliminary field study of the western tarsier, *Tarsius bancanus* Horsefield. In *Prosimian behavior*. (eds. R. D. Martin, G. A. Doyle and A. C. Walker), pp.151–165. Duckworth, London.

Foley, R. (1995). *Humans before humanity*. Blackwell, Oxford.

Folstad, I. and Karter, A. J. (1992). Parasites, bright males and the immunocompetence handicap. *Am. Nat.*, **139**, 603–622.

Fontaine, R. (1981). The uakaris, genus Cacajao. In *Ecology and behavior of neotropical primates*, Vol. 1. (eds. A. F. Coimbra-Filho and R. A. Mittermeier), pp. 443–493. Academia Brasileira de Ciências, Rio de Janeiro.

Fooden, J. (1966). Identification of the stumptail monkey, *Macaca speciosa*. I. Geoffroy, 1826. *Folia Primatol.*, **5**, 153–164.

Fooden, J. (1967). Complementary specialization of male and female reproductive structures in the bear macaque, *Macaca arctoides*. *Nature* 214 (5091), 939–941.

Fooden, J. (1969). Taxonomy and evolution of the monkeys of Celebes. *Biblio. Primatol.*, **10**, 1–148. Karger, Basel.

Fooden, J. (1972). Male external genitalia and systematic relationships of the Japanese macaque (*Macaca fuscata* Blyth 1875). *Primates*, **12**, 305–311.

Fooden, J. (1975). Taxonomy and evolution of liontail and pigtail macaques (Primates: *Cercopithecidae*). *Fieldiana Zool.*, **67**, 1–169.

Fooden, J. (1976). Provisional classification and key to living species of macaques (Primates: *Macaca*). *Folia Primatol.*, **25**, 225–236.

Fooden, J. (1979). Taxonomy and evolution of the *sinica* group of macaques. 1. Species and subspecies accounts of *Macaca sinica*. *Primates*, **20**, 109–140.

Fooden, J. (1980). Classification and distribution of living macaques (*Macaca*, lacépède 1799). In *The macaques, studies in ecology, behavior and evolution*. (ed. D. G. Lindburg), pp. 1–9. Van Nostrand Reinhold, New York.

Fooden, J. (1991). New perspectives on macaque evolution. In *Primatology today*. (eds. A. Ehara, T. Kimura, O. Takenaka, and M. Iwamoto), pp. 1–7. Elsevier, Amsterdam.

Forbes, T. R. (1949). A. A. Berthold and the first endocrine experiment: some speculation as to its origin. *Bull. Hist. Med.*, **23**, 263–267.

Ford, C. S. and Beach, F. A. (1952). *Patterns of Sexual behavior*. Eyre and Spottiswoode,

Foreman, M. M. and Doherty, P. C. (1993). Experimental approaches for the development of pharmacological therapies for erectile dysfunction. In *Sexual pharmacology* (eds. A. J. Riley, M. Peet, and C. Wilson), pp. 87–113. Clarendon Press, Oxford.

Foreman, M. M. and Hall, J. L. (1987). Effects of D_2-dopaminergic receptor stimulation on the lordotic response of female rats. *Psychopharmacology*, **91**, 96–100.

Foreman, M. M. and Moss, R. L. (1979). Role of hypothalamic dopaminergic receptors in the control of lordosis behavior in the female rat. *Physiol. Behav.*, **22**, 283–289.

Foreman, M. M., Hall, J. L., and Love, R. L. (1989). The role of the 5-HT_2 receptor in the regulation of sexual performance in male rats. *Life Sci.*, **45**, 1263–1270.

Forest, M. G. (1990). Pituitary gonadotrophin and sex steroid secretion during the first two years of life. In *Control of the onset of puberty*. (eds M. M. Grumbach, P. C. Sizonenko and M. L. Auberrt), pp. 451–477. Williams and Wilkins, Baltimore.

Forest, M. G., Deperetti, E., and Bertrand, J. (1976). Hypothalamic-pituitary-gonadal relationships from birth to puberty. *Clin. Endocr.*, **5**, 551–569.

Fossey, D. (1984). Infanticide in mountain gorillas (*Gorilla gorilla beringei*) with comparative notes on chimpanzees. In *Infanticide: comparative and evolutionary perspectives*. (eds. G. Hausfater and S. B. Hrdy), pp. 217–235. Aldine, New York.

Fox, C. A., Wolff, H. S., and Baker, J. A. (1970). Measurement of intra-vaginal and intra-uterine pressure during human coitus by radiotelemetry. *J. Reprod. Fertil.*, **22**, 243–251.

Fox, C. A., Meldrum, S. J., and Watson, B. W. (1973). Continuous measurement by radiotelemetry of vaginal pH during human coitus. *J. Reprod. Fertil.*, **33**, 69–75.

Fragaszy, D. M., Schwartz, S. and Schinosaka, D. (1982). Longitudinal observation of care and development of infant titi monkeys (*Callicebus moloch*). *Am. J. Primatol.*, **2**, 191–200.

Franks, S., Jacobs, H. S., Martin, N., and Nabarro, J. D. N. (1978). Hyperprolactinaemia and impotence. *Clin. Endocr.*, **8**, 277–287.

Freeman, M. E. (1988). The ovarian cycle of the rat. In *The physiology of reproduction*, Vol. 2. (eds. E. Knobil and J. D. Neill), pp. 1891–1928. Raven Press, New York.

Freese, C. (1976). Censusing *Alouatta palliata*, *Ateles geoffroyi* and *Cebus capucinus* in the Costa Rican dry forest. In *Neotropical primates: field studies and conservation*. (eds. R. W. Thorington and P. G. Heltne), pp. 4–9. National Academy of Sciences, Washington, D. C.

Freese, C. H. and Oppenheimer, J. R. (1981). The capuchin monkeys, genus *Cebus*. In *Ecology and behavior of neotropical primates*, Vol. 1. (eds. A. F. Combra-Filho and R. A. Mittermeier), pp. 331–390. Academia Brasileira de Ciências, Rio de Janeiro.

French, J. A., Abbott, D. H., and Snowdon, C. T. (1984). The effect of social environment on oestrogen secretion, scent making and socio-sexual behavior in tamarins (*Saguinus oedipus*). *Am. J. Primatol.*, **6**, 155–167.

French, J. A., Inglett, B. J., and Dethlefs, T. M. (1989). The reproductive status of nonbreeding group members in captive golden lion tamarin social groups. *Am. J. Primatol.*, **18**, 73–86.

Freud, S. (1977). On sexuality: three essays on the theory of sexuality, and other works (English translation of the original, 1905, essays). Pelican Books, London.

Frisch, R. E. and McArthur, J. W. (1974). Menstrual cycles: fatness as a determinant of minimum weight for height necessary for their maintenance or onset. *Science*, **185**, 949–951.

Fulkerson, W. J., Synott, A. L., and Lindsay, D. R. (1982). Numbers of spermatozoa required to effect a normal rate of conception in naturally mated Merino ewes. *J. Reprod. Fertil.*, **66**, 129–132.

Furuichi, T. (1987). Sexual swelling, receptivity and grouping of wild pigmy chimpanzee females at Wamba, Zaire. *Primates*, **28**, 309–318.

Furuichi, T. (1992). The prolonged oestrus of females and factors influencing mating in a wild group of bonobos (*Pan paniscus*) in Wamba, Zaire. In *Topics in Primatology*, Vol. 2. (eds. N. Itoigawa, Y. Sugiyama, G. P. Sackett and R. K. R. Thompson), pp. 179–190. University of Tokyo Press, Tokyo.

Gaffan, D., Murray, E. A., and Fabrethorpe, M. (1993). Interaction of the amygdala with the frontal lobe in reward memory. *Eur. J. Neurosci.*, **5**, 968–975.

Gage, M. J. G. (1994). Associations between body size, mating pattern, testis size and sperm lengths across butterflies. *Proc. Roy. Soc. Lond. B.*, **258**, 247–254.

Gage, M. J. G. (1998). Mammalian sperm morphometry. *Proc. Roy. Soc. Lond. B.*, **265**, 97–103.

Gagneux, P., Woodruff, D. S., and Boesch, C. (1997). Furtive mating in female chimpanzees. *Nature Lond.*, **387**, 358–359.

Gajdusek, D. (1964). Congenital absence of the penis in Muniri and Simbari Kukukuku people of New Guinea. *Program and Abstracts of the American Pediatric Society*, 74th annual meeting, Washington, DC.

Galdikas, B. M. F. (1978). Orang-utan adaptation at Tanjung Puting Reserve, Central Borneo. Ph.D. dissertation, University of California, Los Angeles.

Galdikas, B. M. F. (1979). Orang-utan adaptation at Tanjung Puting Reserve: mating and ecology. In *The great apes* (eds. D. A. Hamburg and E. R. McCown). Benjamin/Cummings, Menlo Park, California.

Galdikas, B. M. F. (1981). Orang-utan reproduction in the wild. In *Reproductive biology of the great apes*. (ed. C. E. Graham), pp. 281–300. Academic Press, New York.

Galdikas, B. M. F. (1983). The orangutan long call and snag crashing at Tanjung Puting Reserve. *Primates.*, **24**, 371–384.

Galdikas, B. M. F. (1985). Subadult male orangutan sociality and reproductive behavior at Tanjung Puting. *Am. J. Primatol.*, **8**, 87–99.

Galdikas, B. M. F. (1995). Social and reproductive behavior of wild adolescent female orangutans. In *The neglected ape*. (eds. R. D. Nadler, B. F. M. Galdikas, L. K. Sheeran and N. Rosen), pp. 163–182, Plenum Press, New York.

Galenson, E. and Roiphe, H. (1974). The emergence of genital awareness during the second year of life. In *Sex differences in behavior*. (eds. R. C. Friedman, R. M. Richart and R. L. Van de Wiele), pp. 233–258. Wiley, New York.

Gallup, G. G. Jr (1975). Towards an operational definition of self-awareness. In *Socio-ecology and psychology of primates*. (ed. R. H. Tuttle). Mouton, The Hague.

Garber, P. A. and Teaford, M. F. (1986). Body weights in mixes species troops of *Saguinus mystax mystax* and *Saguinus fuscicollis nigrifrons* in Amazonian Peru. *Am. J. Phys. Anthropol.*, **71**, 331–336.

Garber, P., Moya, L., and Malaga, C. (1984). A preliminary field study of the moustached tamarin monkey (*Saguinus mystax*) in northeastern Peru: questions concerned with the evolution of a communal breeding system. *Folia Primatol.*, **42**, 17–32.

Garber, P. A., Moya, L., Encarnacion, F., and Pruetz, J. (1991). Demography, patterns of dispersal, and polyandrous matings in an island population of moustached tamarin monkeys. *Amer. J. Primatol.*, **24**, 102. Abstract.

Garber, P. A., Moya, L., Pruetz, J. D., and Ique, C. (1996). Social and seasonal influences on reproductive biology in male moustached tamarins (*Saguinus mystax*). *Am. J. Primatol.*, **38**, 29–46.

Gardner, B. T. and Gardner, R A. (1969). Teaching sign language to a chimpanzee. *Science.*, **165**, 664–672.

Gartlan, J. S. (1964). Dominance in East African monkeys. *Proc. E. Afr. Acad.*, **2**, 75–79.

Gartlan, J. S. (1969). Sexual and maternal behaviour in the vervet monkey (*Cercopithecus aethiops*). *J. Reprod. Fertil. Suppl.*, **6**, 137–150.

Gartlan, J. S. (1970). Preliminary notes on the ecology and behaviour of the drill (*Mandrillus leucophaeus*), Ritgen 1824. In *Old World Monkeys*; Evolution, systematics and behavior. (eds. J. R. Napier and P. H. Napier), pp. 445–480. Academic Press, New York.

Gautier, J. P. (1971). Etude morphologique et fonctionelle des annexes extra-laryngées des Cercopithecinae; liaison avec les cris d'espacement. *Biol. Gabon.*, **7**, 229–267.

Gautier, J. P. (1988). Interspecific affinities among guenons as deduced from vocalizations. In *A primate radiation: evolutionary biology of the African guenons.* (eds. A. Gautier-Hion, F. Boulière, J. P. Gautier and J. Kingdon), pp. 194–226. Cambridge University Press.

Gautier, J. P. and Gautier, A. (1977). Communication in Old World monkeys. In *How animals communicate.* (ed. T. A. Sebeok), pp. 890–964. Indiana University Press, Bloomington, Indiana.

Gautier, J. P. and Gautier-Hion, A. (1969). Les associations polyspécifiques chez les *Cercopithecidae* du Gabon. *Terre et Vie*, **23**, 164–201.

Gautier, J. P. and Gautier-Hion, A. (1983). Comportement vocal des mâles adultes et organisation supraspécifique dans les troupes polyspécifiques de cercopithèques. *Folia Primatol.*, **40**, 161–174.

Gautier-Hion, A. (1968). Etude du cycle annuel de reproduction du talapoin (*Miopithecus talapoin*), vivant dans son Milieu naturel. *Biol. Gabon.*, **4**, 163–173.

Gautier-Hion, A. (1971). Repertoire comportemental du talapoin (*Miopithecus talapoin*). *Biol. Gabon.*, **7**, 295–391.

Gautier-Hion, A. (1988). Polyspecific associations among forest guenons: ecological, behavioural and evolutionary aspects. In *A primate radiation: evolutionary biology of the African guenons*. (eds. A. Gautier-Hion, F. Boulière, J. P. Gautier and J. Kingdon), pp. 452–476. Cambridge University Press.

Gautier-Hion, A., Boulière, F., Gautier, J. P., and Kingdon, J. (1988). (eds.). *A primate radiation: evolutionary biology fo the African guenons*. Cambridge University Press.

Gebhard, P. H. and Johnson, A. B. (1979). The Kinsey data: marginal tabulation of the 1938–1963 interviews conducted by the Institute for Sex Research. W. B. Saunders, Philadelphia.

Geissmann, T. (1987). A sternal gland in the siamang gibbon (*Hylobates syndactylus*). *Int. J. Primatol.*, **8**, 1–15.

Gessa, G. L., Paglietti, E., and Pellegrini-Quarantotti, B. (1979). Induction of copulatory behavior in sexually inactive rats by naloxone. *Science*, **204**, 203–205.

Gilbert, A. N., Yamazaki, K., Beauchamp, G. K., and Thomas, L. (1986). Olfactory discrimination of mouse strains (*Mus musculus*) and major histocompatablity types by humans (*Homo sapiens*). *J. Comp. Psychol.*, **100**, 262–265.

Gilbert, H. W. and Gingell, J. C. (1991). The results of an intracorporal papaverine clinic. *Sex. and Mar. Therapy*, **6**, 49–56.

Gillman, J. (1940). The effect of multiple injections of progesterone on the tumescent perineum of the baboon (*Papio porcarius*). *Endocrinology*, **26**, 1072–1077.

Gillman, J. and Gilbert, C. (1946). The reproductive cycle of the chacma baboon with special reference to the problem of menstrual irregularities as assessed by the behaviour of the sex skin. *S. Afr. J. Med. Sci.*, **11**, 1–54.

Gillman, J. and Stein, H. B. (1941). A quantitative study of the inhibition of oestradiol benzoate by progesterone in the baboon (*Papio porcarius*). *Endocrinology*, **28**, 274–282.

Ginsberg, J. R. and Huck, U. W. (1989). Sperm competition in mammals. *Trends Ecol. Evol.*, **4**, 74–79.

Girolami, L. (1989). The female's role in primate socio-sexual communication: a study of the vervet monkey (*Cercopithecus aethiops pygerythrus*) and the chacma baboon (*Papio ursinus*). Unpublished Ph.D. thesis, University of Witwatersrand, South Africa.

Girolami, L. and Bielert, C. (1987). Female perineal swelling and its effects on male sexual arousal: on apparent sexual releaser in the chacma baboon (*Papio ursinus*). *Int. J. Primatol.*, **8**, 651–661.

Gittins, S. P. and Raemakers, J. J. (1980). Siamang, lar and agile gibbons. In *Malayan forest primates, ten year's study in tropical rainforest*. (ed. D. J. Chivers), pp. Plenum Press, New York.

Giuliano, F., Bernabe, J., Jardin, A., and Rousseau, J. P. (1993). Antierectile role of the sympathetic nervous system in rats. *J. Urol.*, **150**, 519–524.

Gladue, B. A., Green, R., and Hellman, R. E. (1984). Neuroendocrine response to estrogen and sexual orientation. *Science*, **225**, 1496–1499.

Glander, K. E. (1980). Reproduction and population growth in free-ranging mantled howling monkeys. *Am. J. Phys. Anthropol.*, **53**, 25–36.

Glick, B. (1979). Testicular size, testosterone level, and body weight in male *Macaca radiata*. *Folia Primatol.*, **32**, 268–289.

Glick, B. B. (1980). Ontogenetic and psychobiological aspects of the mating activities of male *Macaca radiata*. In *The macaques: studies in ecology, behavior and evolution*. (ed. D. G. Lindburg), pp. 345–369. Van Nostrand Reinhold, New York.

Glick, B. B., Brenner, R. M., Jensen, J. J., and Phoenix, C. H. (1982). Moxestrol (R2858), estradiol benzoate, and sexual behavior of cynomolgus macaques (*Macaca fascicularis*). *Horm. Behav.*, **16**, 59–65.

Godec, C. J. and Bates, H. (1984). Cholinergic receptors in corpora cavernoa. *Urology*, **24**, 31–33.

Godfrey, L. R. (1988). Adaptive diversification of Malagasy strepsirhines. *J. Hum. Evol.*, **17**, 93–134.

Going, J. J. and Dixson, A. F. (1990). Morphometry of the adult human corpus callosum: lack of sexual dimorphism. *J. Anat. Lond.*, **171**, 163–167.

Goldfoot, D. A. (1971). Hormonal and social determinants of sexual behavior in the pigtail monkey (*Macaca nemestrina*). In *Normal and abnormal development of brain and behavior*. (eds. G. B. A. Stoelinga and J. J. Van der Werff ten Bosch), pp. 325–342. University of Leiden Press, Leiden.

Goldfoot, D. A. (1977). Sociosexual behaviors of nonhuman primates during development and maturity: social and hormonal relationships. In *Behavioural primatology: advances in research and theory*, Vol. 1. (ed. A. M. Schrier), pp. 139–184. Lawrence Erlbaum Assoc. Hillsdale, New Jersey.

Goldfoot, D. A. (1981). Olfaction, sexual behavior, and the pheromone hypothesis in rhesus monkeys: a critique. *Am. Zool.*, **21**, 153–164.

Goldfoot, D. A. and Goy, R. W. (1970). Abbreviation of behavioral estrus in guinea pigs by coital and vagino-cervical stimulation. *J. Comp. Physiol. Psychol.*, **72**, 426–434.

Goldfoot, D. A., Slob, A. K., Scheffler, G., Robinson, J. A., Wiegand, S. J., and Cords. J. (1975). Multiple ejaculations during prolonged sexual tests and lack of resultant serum testosterone increases in male stumptail macaques (*M. arctoides*). *Arch. Sex. Behav.*, **4**, 547–560.

Goldfoot, D. A., Kravetz, M. A., Goy, R. W., and Freeman, S. K. (1976). Lack of effect of vaginal lavages and aliphatic acids on ejaculatory responses in rhesus monkeys: behavioral and chemical responses. *Horm. Behav.*, **7**, 1–27.

Goldfoot, D. A., Essock-Vitale, S. M., Asa, C. S., Thornton, J. E., and Leshner, A. I. (1978a). Anosmia in male rhesus monkeys does not alter copulatory activity with cycling females. *Science*, **199**, 1095–1096.

Goldfoot, D. A., Wiegend, S. J., and Scheffler, G. (1978b). Continued copulation in ovariectomized adrenal-suppressed stumptail macaques (*Macaca arctoides*). *Horm. Behav.*, **11**, 89–99.

Goldfoot, D. A., Westerborg-Van Loon, H., Groeneveld, W., and Slob, A. K. (1980). Behavioral and physiological evidence of sexual climax in the female stump-tailed macaque (*Macaca arctoides*). *Science.*, **208**, 1477–1479.

Goldizen, A. W. (1987). Tamarins and marmosets: communal care of offspring. In *Primate societies* (eds. B. Smuts, D. Cheney, R. Seyfarth, R. Wrangham and T. Struhsaker), pp. 34–43. University of Chicago Press.

Goldman-Rakic, P. S. (1987). Circuitry of primate prefrontal-cortex and regulation of behavior by representational memory. In *Handbook of physiology, the nervous system*, Vol. 5. (ed. F. Plum), pp. 373–417. American Physiological Society, Bethesda.

Gombe, S., Hall, W. C., McEntee, K., Hansel, W., and Pickett, B. W. (1973). Regulation of blood levels of LH in bulls: influence of age, breed, sexual stimulation and temporal fluctuations. *J. Reprod. Fertil.*, **35**, 493–503.

Gomendio, M. (1990). The influence of maternal rank and infant sex on maternal investment trends in rhesus macaques: birth sex ratios, inter-birth intervals and suckling patterns. *Behav. Ecol. Sociobiol.*, **27**, 365–375.

Gomendio, M. and Roldan, E. R. S. (1991). Sperm competition influences sperm size in mammals. *Proc. Roy. Soc. Lond. B.*, **243**, 181–185.

Gomendio, M. and Roldan, E. R. S. (1993). Coevolution between male ejaculates and female reproductive biology in eutherian mammals. *Proc. Roy. Soc. Lond. B.*, **252**, 7–12.

Goodall, J. (1964). Tool-using and aimed throwing in a community of free-living chimpanzees. *Nature Lond.*, **201**, 1264–1266.

Goodall, J. (1965). Chimpanzees of the Gombe Stream Reserve. In *Primate behavior: field studies of monkeys and apes*. (ed. I. DeVore), pp. 425–473. Holt, Rinehart and Winston, New York.

Goodall, J. (1968). The behaviour of free-living chimpanzees in the Gombe stream area. *Anim. Behav. Monogr.*, **1**, 161–311.

Goodall, J. (1973). Cultural elements in chimpanzees community. In *Precultural Primate behaviour*, Vol. 1 (ed. E. W. Menzel), pp. 144–184. Karger, Basel.

Goodall, J. (1986). *The chimpanzees of Gombe: patterns of behavior*. Belknap Press, Harvard.

Goodman, E. D., Bunnell, B. L., Dewsbury, D. A., and Boland, B. (1969). Septal lesions and male rat copulatory behavior. *Psychon. Sci.*, **16**, 123–124.

Goodman, E. R. L., Hotchkiss, J., Karsch, F. J., and Knobil, E. (1974). Diurnal variations in serum testosterone concentrations in the adult male rhesus monkey. *Biol. Reprod.*, **11**, 624–630.

Goodman, L. and Wislocki, G. B. (1935). The cyclical uterine bleeding in a New World monkey (*Ateles geoffroyi*). *Anat. Rec.*, **61**, 379–387.

Goodman, R. L. (1988). Neuroendocrine control of the ovine oestrous cycle. In *The physiology of reproduction*, Vol. 2. (eds. E. Knobil and J. D. Neill), pp. 1929–1969. Raven Press, New York.

Goodwin, F. K. (1971). Psychiatric side effects of levodopa in man. *J. Am. Med. Assoc.*, **218**, 1915–1920.

Gooren, L. J. G. (1985). Human male sexual functions do not require aromatization of testosterone: a study using tamoxifen, testolactone and dihydrotestosterone. *Arch. Sex. Behav.*, **14**, 539–548.

Gooren, L. J. G. (1986). The neuroendocrine response of luteinizing hormone to estrogen administration in heterosexual, homosexual and transsexual subjects. *J. Clin. Endocr. Metab.*, **63**, 583–589.

Gooren, L. and Cohen-Kettenis, P. T. (1991). Development of male gender identity/role and a sexual orientation towards women in a 46,XY subject with an incomplete form of the andogens insensitivity syndrome. *Arch. Sex. Behav.*, **20**, 459–470.

Gordon, T. P. and Bernstein, I. S. (1973). Seasonal variation in sexual behavior of all-male rhesus troops. *Am. J. Phys. Anthropol.*, **38**, 221–226.

Gordon, T. P., Rose, R., and Bernstein, I. S. (1976). Seasonal rhythm in plasma testosterone levels in the rhesus monkey (*Macaca mulatta*): a three-year study. *Horm. Behav.*, **7**, 229–243.

Gordon, T. P., Bernstein, I. S., and Rose, R. M. (1978). Social and seasonal influences on testosterone secretion in the male rhesus monkey. *Physiol. Behav.*, **21**, 623–627.

Gorski, R. A., Gordon, J. H., Shryne, J. E., and Southam, A. M. (1978). Evidence for a morphological sex difference within the medial preoptic area of the rat brain. *Brain Res.*, **148**, 333–346.

Gorski, R. A., Hanlan, R. E., Jacobson, C. D., and Shryne, J. E. (1980). Evidence for the existence of a sexually dimorphic nucleus in the preoptic area of the rat. *J. Comp. Neurol.*, **193**, 529–539.

Gorynski, G. and Katz, J. L. (1997). The polycystic ovary syndrome: psychosexual correlates. *Arch. Sex. Behav.*, **6**, 215–222.

Gorzalka, B B., Luck, K. A., and Tanco, S. A. (1991). Effects of the oxytocin fragment prolyl-leucyl-glycinamide on sexual behavior in the rat. *Pharmacol. Biochem. Behav.*, **38**, 273–279.

Gosling, L. M. (1986). Selective abortion of entire litters in the coypu: adaptive control of offspring production in relation to quality and sex. *Am. Nat.*, **127**, 772–795.

Gould, J. E., Overstreet, J. W., and Hanson, F. W. (1984). Assessment of human sperm function after recovery from the female reproductive tract. *Biol. Reprod.*, **31**, 888–894.

Gould, S. J. (1983). *Hens teeth and horses toes*. Pelican Books, Bungay, Suffolk.

Gouzoules, H. (1974). Harassment of sexual behavior in the stumptailed macaque (*Macaca arctoides*). *Folia Primatol.*, **22**, 208–217.

Gouzoules, H. (1980). A description of genealogical rank changes in a troop of Japanese monkeys (*Macaca fuscata*). *Primates*, **21**, 262–267.

Gouzoules, H., Gouzoules, S., and Fedigan, L. (1982). Behavioural dominance and reproductive success in female Japanese monkeys (*Macaca fuscata*). *Anim. Behav.*, **30**, 1138–1150.

Goy, R. W. (1968). Organizing effects of androgen on the behaviour of rhesus monkeys. In *Endocrinology, and human behaviour*. (ed. R. P. Michael), pp. 12–31. Oxford University Press, London.

Goy, R. W. (1978). Development of play and mounting behaviour in female rhesus virilized prenatally with esters of testosterone and dihydrotestosterone. In *Recent advances in primatology*, Vol. 1. *Behaviour*. (eds. D. J. Chivers and J. Herbert), pp. 449–462. Academic Press, London.

Goy, R. W. (1981). Differentiation of male social traits in female rhesus macaques by prenatal treatment with androgens: variation in type of androgen, duration and timing of treatment. In *Fetal endocrinology*. (eds. M. J. Novy and J. A. Resko), pp. 319–339. Academic Press, New York.

Goy, R. W. (1992). The mating behavior of primates: estrus or sexuality? In *Topics in primatology, Vol. 2. behavior, ecology and conservation*. (eds. N. Itiogawa, Y. Sugiyama, G. P. Sackett and R. K. R. Thompson), pp. 151–161. University of Tokyo Press, Tokyo.

Goy, R. W. and Goldfoot, D. A. (1974). Experiential and hormonal factors influencing development of sexual behavior in the male rhesus monkey. In *The neurosciences: third study program*. (eds. F. O. Schmitt and F. G. Worden), pp. 571–581. MIT Press, Cambridge, Massachusetts.

Goy, R. W. and McEwen, B. S. (1980). *Sexual differentiation of the brain*. MIT Press, Cambridge, Massachusetts.

Goy, R. W. and Resko, J. A. (1972). Gonadal hormones and behavior of normal and pseudohermaphroditic nonhuman female primates. *Rec. Prog. Horm. Res.*, **28**, 707–33.

Goy, R. W. and Robinson, J. A. (1982). Prenatal exposure of rhesus monkeys to patent androgens: morphological, behavioral and physiological consequences. *Banbury Report*, **11**, 355–378.

Goy, R. W., Wallen K., and Goldfoot, D. A. (1974). Social factors affecting the development of mounting behavior in male rhesus monkeys. In *Reproductive behavior*. (eds. W. Montagna and R. Sadler), pp. 223–247. Plenum, New York.

Goy, R. W., Wolf, J. E., and Eisele, S. G. (1978). Experimental female hermaphroditism in rhesus monkeys: anatomical and psychological characteristics. In *Handbook of sexology*, Vol. 2. (eds. J. Money and H. Mustaph), pp. 141–156. Elsevier, New York.

Goy, R. W., Bercovitch, F. B., and McBrair, M. (1988). Behavioral masculinization is independent of genital masculinization in prenatally androgenized female rhesus macaques. *Horm. Behav.*, **22**, 552–571.

Gradwell, P. B., Everitt, B. J., and Herbert, J. (1975). 5-Hydroxytryptamine in the central nervous system and sexual receptivity of female rhesus monkeys. *Brain Res.*, **88**, 281–293.

Grady, K. L., Phoenix, C. H., and Young, W. C. (1965). Role of the developing rat testis in differentiation of the neural tissues mediating mating behavior. {*J. Comp. Physiol. Psychol.*, **59**, 176–182.

Grf, K. J., Brotherton, J., and Neumann, F. (1974). Clinical use of antiandrogens. In *Handbook of experimental pharmacology*, Vol. 35. (eds. O. Eichler, A. Farah, H. Herken, and A. D. Welch), pp. 485–542. Springer, Berlin.

Graham, C. E. and Nadler, R. D. (1990). Socioendocrine interactions in great ape reproduction. In *Socioendocrinology of primate reproduction*. (eds. T. E. Zeigler and F. B. Bercovitch), pp. 33–58. Wiley-Liss, New York.

Graham, C. E., Keeling, M., Chapman, C., Cummins, C. B., and Haynie, J. (1973). Method of endoscopy in the chimpanzee: relations of ovarian anatomy, endometrial histology and sexual swelling. *Am. J. Phys. Anthropol.*, **38**, 211–216.

Grant, E. C. G. and Pryse-Davies, J. (1968). Effect of oral contraception on depressive mood changes and on endometrial oxidase and phosphatases. *Br. Med. J.*, iii, 777–780.

Grant, L. D. and Stumpf, W. E. (1975). Hormone uptake sites in relation to CNS biogenic amine systems. In *Anatomical neuroendocrinology*. (eds. W. E. Stumpf and L. D. Grant), pp. 445–463. Karger, Basel.

Gray, G. D., Smith, E. R., and Davidson, J. M. (1980). Hormonal regulation of penile erection in castrated male rats. *Physiol. Behav.*, **24**, 463–468.

Green, J. D., Clemente, C. D., and DeGroot, J. (1957). Rhinencephalic lesions and behavior in cats. *J. Comp. Neurol.*, **108**, 505–545.

Green, R. (1985). Gender identity in childhood and later sexual orientation: follow-up of seventy-eight males. *Am. J. Psychiat.*, **142**, 339–341.

Green, R. (1987). The 'sissy-boy syndrome' and the development of homosexuality. Yale University Press, New Haven.

Green, S. and Minkowski, K. (1977). The lion-tailed monkey and south Indian rainforest habitat. In *Primate Conservation*. (ed. Prince Rainier III and G. H. Bourne). Academic Press, New York.

Greenough, W. T., Carter, C. S., Steerman, C., and DeVoogd, T. J. (1977). Sex differences in dendritic patterns in hamster preoptic area. *Brain Res.*, **126**, 63–72.

Greer, W. E., Roussel, J. D., and Austin, C. R. (1968). Prevention of coagulation in monkey semen by surgery. *J. Reprod. Fertil.*, **15**, 153–155.

Grierson, J. P., James, M. D., Pearson, J. R., and Wilson, C. A. (1988). The effect of selective D_1 and D_2 dopaminergic agents on sexual receptivity in the female rat. *Neuropharmacology*, **27**, 181–189.

Gross, C. G., Rocha-Miranda, C E., Bender, D. B. (1972). Visual properties of neurons in inferotemporal cortex of the macaque. *J. Neurophysiol.*, **35**, 96–111.

Groves, C. P. (1980). Speciation in Macaca: the view from Sulawesi. In *The macaques: studies in ecology, behavior and evolution*. (ed. D. G. Lindburg), pp. 84–124. Van Nostrand Reinhold, New York.

Grubb, P. (1973). Distribution, divergence and speciation of the drill and mandrill. *Folia Primatol.*, **20**, 161–177.

Grubb, P. and Jewell, P. A. (1973). The rate and the occurrence of oestrus in the Soay sheep on St. Kilda. *J. Reprod. Fertil. Suppl.*, **19**, 491–502.

Grumbach, M. M., Ducharme, J. R., and Moloshok, R. E. (1959). On the foetal masculinizing action of certain oral progestins. *J. Clin. Endocr.*, **19**, 1368.

Grundemar, L. and Hkkansson, R. (1993). Multiple neuropeptide Y receptors are involved in cardiovascular regulation. Peripheral and central mechanisms. *Gen. Pharmacol.*, **24**, 785–796.

Grunt, J. A. and Young, W. C. (1952). Differential reactivity of individuals and the response of the male guinea pig to testosterone propionate. *Endocrinology*, **51**, 237–248.

Grunt, J. A. and Young, W. C. (1953). Consistency of sexual behavior patterns in individual male guinea pigs following castration and androgen therapy. *J. Comp. Physiol. Psychol.*, **46**, 138–144.

Gu, J., Polak, J. M., Probert, L., Islam, K. N., Marangos, P.J., Mina, S.,et al. (1983). Peptidergic innervation of the human male genital tract. *J. Urol.*, **130**, 386–391.

Guillamón, A. and Segovia, S. (1993). Sexual dimorphism in the accessory olfactory system. In *The development of sex differences and similarities in behavior*. NATO. ASI Series, Vol. 3. (eds. M. Haug, R. E. Whalen, C. Aron and K. L. Olsen), pp. 363–376. Kluwer Academic, Dordrecht/Norwell, MA.

Guldner, F. H. (1984). Suprachiasmatic nucleus: numbers of synaptic appositions and various types of synapses, a morphogenetic study on male and female rats. *Cell. Tiss. Res.*, **235**, 449–452.

Gunnett, J. W., Lookinland, K. L., and Moore, K. E. (1986). Comparison of the effects of castration and steroid replacement on incerto-hypothalamic dopaminergic neurones in male and female rats. *Neuroendocrinology*, **44**, 269–275.

Gust, D. A., Gordon, T. P., Gergits, W. F., Casna, N. J., Gould, K. G., and McClure, H. M. (1997). Male dominance rank and offspring-initiated affiliative behaviors were not predictors of paternity in a captive group of pigtail macaques. *Primates*, **37**, 271–278.

Haimoff, E. H. (1984). Acoustic and organizational features of gibbon songs. In *The lesser apes: evolutionary and behavioural biology*. (eds. H. Preuschoft, D. Chivers, W. Brockelman and N. Creel), pp. 333–353. Edinburgh University Press.

Halata, Z. and Munger, B. L. (1986). The neuroanatomical basis for the protopathic sensibility of the human glans penis. *Brain Res.*, **317**, 205–230.

Hall, K. R. L. (1962). The sexual, agonistic, and derived social behaviour patterns of the wild chacma baboon, *Papio ursinus*. *Proc. Zool. Soc. Lond.*, **13**, 283–327.

Hall, K. R. L. (1965). Behaviour and ecology of the wild patas monkey, (*Erythrocebus patas*), in Uganda. *J. Zool. Lond.*, **148**, 15–87.

Hall, K. R. L. (1967). Social interactions of the adult male and adult females of a patas monkey group. In *Social communication among primates*. (ed. S. A. Altmann), pp. 261–280. University of Chicago Press.

Hall, K. R. L. and DeVore, I. (1965). Baboon social behavior. In *Primate behaviour: field studies of monkeys and apes*. (ed. I. DeVore), pp. 53–110. Holt, Rinehart and Winston, New York.

Hall, K. R. L. and Mayer, B. (1967). Social interactions in a group of captive patas monkeys (*Erythrocebus patas*) Folia Primatol., **5**, 213–236.

Hall, K. R. L., Boelkins, R. C., and Goswell, M. J. (1965). Behaviour of patas monkeys, *Erythrocebus patas*, in captivity, with notes on the natural habitat. *Folia Primatol.*, **3**, 22–49.

Hall-Craggs, E. C. B. (1962). The testis of *Gorilla gorilla beringei*. *Proc. Zool. Soc. Lond.*, **139**, 511–514.

Halpern, D. F. (1987). *Sex differences in cognitive abilities*. Erlbaum, Hillsdale, New Jersey.

Hamai, M., Nishida, T., Takasaki, H., and Turner, L. A. (1992). New records of within-group infanticide and cannibalism in wild chimpanzees. *Primates*, **33**, 151–162.

Hamer, D. H., Hu, S., Magnuson, V. L., Hu, N., and Pattattuci, A. M. L. (1993). A linkage between DNA markers on the X chromosome and the male sexual orientation. *Science*, **261**, 321–327.

Hamilton, D. W. (1990). Anatomy of mammalian male accessory reproductive organs. In *Marshall's Physiology of reproduction*, Vol. 2. (Fourth edition). (ed. G. E. Lamming), pp. 691–746. Churchill Livingstone, England.

Hamilton, D. W. and Cooper, T. G. (1978). Gross and histological variations along the length of the rat vas deferens. *Anat. Rec.*, **190**, 795.

Hamilton, G. V. (1914). A study of sexual tendencies in monkeys and baboons. *J. Anim. Behav.*, **4**, 295–318.

Hamilton, W. D. (1964). The genetical evolution of social behaviour. I, II. *J. Theoret. Biol.*, **7**, 1–52.

Hamilton, W. D. and Zuk, M. (1982). Heritable true fitness and bright birds: a role for parasites? *Science.*, **218**, 384–387.

Hamilton, W. J. III. and Arrowood, P. (1978). Copulatory vocalizations of chacma baboons (*Papio ursinus*), gibbons (*Hylobates hoolock*), and humans. *Science*, **200**, 1405–1409.

Hampton, J. K. Jr, Hampton, S. H., and Landwehr, B. T. (1966). Observations on a successful breeding colony of the marmoset *Oedipomidas oedipus*. *Folia Primatol.*, **4**, 265–287.

Hanby, J. P. (1977). Social factors affecting primate reproduction. In *Handbook of sexology*. (eds. J. Money and H. Mustaph), pp. 461–484. Excerpta Medica, Amsterdam.

Hanby, J. P. and Brown, C. E. (1974). The development of sociosexual behaviors in Japanese macaques (*Macaca fuscata*). *Behaviour*, **49**, 151–196.

Hanby, J. P., Robertson, L. T., and Phoenix, C. H. (1971). The sexual behavior of a confined troop of Japanese macaques. *Folia Primatol.*, **16**, 123–143.

Hancock, R. J. T. (1981). Immune responses to sperm. *Oxford Revs. Repro. Biol.*, **3**, 182–208.

Hansen, S. (1982a). Hypothalamic control of motivation: the medial preoptic area and masculine sexual behaviour. *Scand. J. Psychol. Suppl.*, **1**, 121–126.

Hansen, S. (1982b). Spinal control of sexual behavior: effects in intrathecal administration of lisuride. *Neurosci. Lett.*, **33**, 329–332.

Hansen, S. and Gummersson, B. M. (1982). Participation of the lateral midbrain tegmentum in the neuroendocrine control of sexual behavior and lactation in the rat. *Brain Res.*, **251**, 319–325.

Hansen, S., Stanfield, E. S., and Everitt, B. J. (1980). The role of ventral bundle noradrenergic neurones in sensory components of sexual behaviour and coitus-induced pseudopregnancy. *Nature Lond.*, **286**, 151–154.

Hansen, S., Kohler, Ch., Goldstein, M., and Steinbusch, H. V. M. (1982). Effects of ibotenic acid-induced neuronal degeneration in the medial preoptic area and the lateral hypothalamic area on sexual behavior in the male rat. *Brain Res.*, **239**, 213–232.

Hanson, F. W. and Overstreet, J. W. (1981). The interaction of human spermatozoa with cervical mucus *in vitro*. *Am. J. Obstet. Gynecol.*, **140**, 173–177.

Harcourt, A. H. (1987). Dominance and fertility among female primates. *J. Zool. Lond.*, **213**, 471–487.

Harcourt, A. H. (1989). Deformed sperm are probably not adaptive. *Anim. Behav.*, **37**, 863–865.

Harcourt, A. H. (1991). Sperm competition and the evolution of non-fertilizing sperm in mammals. *Evolution*, **45**, 314–328.

Harcourt, A. H. and De Waal, F. B. M. (eds) (1992). *Coalitions and alliances in humans and other animals*. Oxford University Press.

Harcourt, A. H. and Gardiner, J. (1994). Sexual selection and genital anatomy of male primates. *Proc. Roy. Soc. Lond. B.*, **255**, 47–53.

Harcourt, A. H. and Stewart, K. J. (1981). Gorilla male relationships: can differences during immaturity lead to contrasting reproductive tactics in adulthood. *Anim. Behav.*, **29**, 206–210.

Harcourt, A. H., Stewart, K. J. and Fossey, D. (1976). Male emigration and female transfer in wild mountain gorillas. *Nature, Lond.*, **263**, 226–227.

Harcourt, A. H., Fossey, D., Stewart, K., and Watts, D. P. (1980). Reproduction in wild gorillas and some comparisons with chimpanzees. *J. Reprod. Fertil. Suppl.*, **28**, 59–70.

Harcourt, A. H., Harvey, P. H., Larson, S. G., and Short, R. V. (1981a). Testis weight, body weight and breeding system in primates. *Nature, Lond.*, **293**, 55–57.

Harcourt, A. H., Stewart, K. J., and Fossey, D. (1981b). Gorilla reproduction in the wild. In *Reproductive biology of the great apes: comparative and biomedical perspectives* (ed. C. E. Graham), pp. 265–318. Academic Press, New York.

Harcourt, A. H., Purvis, A., and Liles, L. (1995). Sperm competition: mating system, not breeding season, affects testes size of primates. *Funct. Ecol.*, **9**, 468–476.

Harcourt, C. S. (1984). The behaviour and ecology of galagos in Kenyan coastal forest. Unpublished Ph.D. thesis, University of Cambridge, U. K.

Harcourt, C. S. and Nash, L. T. (1984). Social organization of galagos in Kenyan coastal forests. *Am. J. Primatol.*, **10**, 339–355.

Harding, C. F. and Feder, H. H. (1976). Relation of uptake and metabolism of (1, 2, 6, 7, ^{3}H) testosterone to individual differences in sexual behavior in male guinea pigs. *Brain Res.*, **105**, 137–149.

Harding, R. D., Hulme, M. J., Lunn, S. F., Henderson, C., and Aitken, R. J. (1982). Plasma progesterone levels throughout the ovarian cycle of the common marmoset (*Callithrix jacchus*). *J. Med. Primatol.*, **11**, 43–51.

Harding, R. S. O. and Olson, D. K. (1986). Patterns of mating among male patas monkeys (*Erythrocebus patas*) in Kenya. *Am. J. Primatol.*, **11**, 343–358.

Harlow, H. F. (1965). Sexual behavior in the rhesus monkey. In *Sex and behavior* (ed. F. A. Beach), pp. 234–265. Wiley, New York.

Harlow, H. F. (1971). *Learning to love*. Ballantine Books, New York.

Harlow, H. F. and Harlow, M. K. (1966). Learning to love. *Am. Sci.*, **54**, 244–272.

Harlow, H. F., Joslyn, W. D., Senko, M. G., and Dopp, A. (1966). Behavioral aspects of reproduction in primates. *J. Anim. Sci.*, **25**, 49–67.

Harman, S. M. (1983). Relation of the neuroendocrine system to reproductive decline in men. In *Neuroendocrinology of aging*. (ed. J. Meites), pp. 203–219. Plenum, New York.

Harman, S. M. and Tsitouras, P. D. (1980). Reproductive hormones in aging men. I. Measurement of sex steroids, basal luteinizing hormone and Leydig cell response to human chorionic gonadotrophin. *J. Clin. Endocr. Metab.*, **51**, 35–40.

Harms, J. W. (1956). Fortplflanzungsbiologie. In *Primatologia* Vol. 1. (eds. H. Hofer, A. Schultz and D. Starck), Karger, Basel.

Harper, M. J. K. (1982). Sperm and egg transport. In *Reproduction in mammals*, Vol. 1. *Germ cells and fertilization*. (eds. C. R. Austin and R. V. Short), pp. 101–127. Cambridge University Press.

Harper, M. J. K. (1994). Gamete and zygote transport. In *The physiology of reproduction*, Vol. 1. (2nd edition). (eds. E. Knobil and J. D. Neill), pp. 123–187. Raven Press, New York.

Harrington, J. E. (1977). Discrimination between males and females by scent in *Lemur fulvus*. *Anim. Behav.*, **25**, 147–151.

Harris, G. W. (1965). *The neural control of the pituitary gland*. Edward Arnold, London.

Harris, V. H. and Sachs, B. D. (1975). Copulatory behavior in male rats following amygdaloid lesions. *Brain Res.*, **86**, 514–518.

Harrison, M. (1988). New guenon from Gabon. *Oryx*, **22**, 190–191.

Harrison, R. M. and Lewis, R. W. (1986). The male reproductive tract and its fluids. In *Comparative primate biology*, Vol. 3. (eds. W. R. Dukelow and J. Erwin), pp. 101–148. Alan R. Liss, New York.

Harris Poll (1988). *Survey for project hope*. Louis Harris and Associates, New York.

Hart, B. L. (1968). Alteration of quantitative aspects of sexual reflexes in spinal male dogs by testosterone. *J. Comp. Physiol. Psychol.*, **66**, 726–730.

Hart, B. L. (1972a). The action of extrinsic penile muscles during copulation in the male dog. *Anat. Rec.*, **173**, 1–5.

Hart, B. L. (1972b). Manipulation of neonatal androgen: effects on sexual responses and penile development in male rats. *Physiol. Behav.*, **8**, 841–845.

Hart, B. L. (1974). Medial preoptic-anterior hypothalamic area and sociosexual behavior of male dogs: a comparative neuropsychological analysis. *J. Comp. Physiol. Psychol.*, **86**, 328–349.

Hart, B. L. (1978). Hormones, spinal reflexes and sexual behaviour. In *Biological determinants of sexual behaviour*. (ed. J. B. Hutchison), pp. 319–347. John Wiley, Chichester.

Hart, B. L. (1986). Medial preoptic-anterior hypothalamic lesions and sociosexual behavior of male goats. *Physiol. Behav.*, **36**, 301–305.

Hart, B. L. and Jones, T. O. A. C. (1978). Effects of castration on sexual behavior of tropical male goats. *Horm. Behav.*, **6**, 247–258.

Hart, B. L. and Kitchell, R. L. (1966). Penile erection and contraction of penile muscles in the spinal and intact dog. *Am. J. Physiol.*, **210**, 257–261.

Hart, B. L. and Leedy, M. G. (1983). Female sexual responses in male cats facilitated by olfactory bulbectomy and medial preoptic/anterior hypothalamic lesions. *Behav. Neurosci.*, **97**, 608–614.

Hart, B. L. and Leedy, M. G. (1985). Neurological bases of male sexual behavior. In *Handbook of behavioral neurobiology*, Vol. 7. (eds. N. Adler, D. Pfaff, and R. W. Goy), pp. 373–422. Plenum, New York.

Hart, B. L. and Melese-d'Hospital, P. (1983). Penile mechanisms and the role of the striated penile muscles in penile reflexes. *Physiol. Behav.*, **31**, 807–813.

Hart, B. L., Haugen, C. M., and Peterson, D. M. (1973). Effects of medial preoptic-anterior hypothalamic lesions on mating behavior of male cats. *Brain Res.*, **54**, 177–191.

Hart, D. V. (1968). Homosexuality and transvestism in the Philippines. *Behav. Sci. Notes*, **3**, 211–248.

Hartman, C. G. (1931). On the relative sterility of the adolescent organism. *Science*, **74**, 226–227.

Hartung, T. G. and Dewsbury, D. A. (1978). A comparative analysis of copulatory plugs in muroid rodents and their relationship to copulatory behavior. *J. Mammal.*, **59**, 717–723.

Hartz, A. J., Rupley, D. C., and Rimm, A. A. (1984). The association of girth measurements with disease in 32, 856 women. *Am. J. Epidem.*, **119**, 71–80.

Harvey, P. H. and May, R. M. (1989). Out for the sperm count. *Nature Lond.*, **337**, 508–509.

Harvey, P. H., Kavanagh, M., and Clutton-Brock, T. H. (1978). Sexual dimorphism in primate teeth. *J. Zool. Lond.*, **186**, 475–485.

Harvey, P.H., Martin, R.D., and Clutton-Brock, T.H. (1987). Life Histories in Comparative Perspective. In *Primate Societies*. (ed. Smuts *et al.*), pp. 181–196. University of Chicago Press.

Hasegawa, T. and Hiraiwa-Hasegawa, M. (1990). Sperm competition and mating behavior. In *The Chimpanzees of the Mahale mountains: sexual and life history strategies*. (ed. T. Nishida), pp. 115–132. University of Tokyo Press, Tokyo.

Haug, M., Brain, P. F., and Aron, C. (1991), (eds.). *Heterotypical behaviour in man and animals*. Chapman and Hall, London.

Hauser, M. D. (1990). Do female chimpanzee copulation calls incite male-male competition? *Anim. Behav.*, **39**, 596–596.

Hausfater, G. (1975). Dominance and reproduction in baboons (*Papio cynocephalus*): a quantitative analysis. Karger, Basel.

Hausfater, G. and Hrdy, S. B. (eds.) (1984). *Infanticide: comparative and evolutionary perspectives*. Aldine, New York.

Hausfater, G. and Takacs, D. (1987). Structure and function of hindquarter presentations in yellow baboons (*Papio cynocephalus*). *Ethol.*, **74**, 297–319.

Havelock Ellis, H. (1902). Studies in the psychology of sex: the evolution of modesty, the phenomena of sexual periodicity, auto-erotism. F. A. Davis Co., Philadelphia.

Heape, W. (1900). The 'sexual season' of mammals and the relation of the 'pro-oestrus' to menstruation. *Quar. J. Microscop. Sci.*, **44**, 1–70.

Hearn, J. P. (1983). The common marmoset (*Callithrix jacchus*). In *Reproduction in New World primates*. (ed. J. P. Hearn), pp. 181–215. MTP Press, Lancaster.

Hearn, J. P. and Lunn, S. F. (1975). The reproductive biology of the marmoset monkey (*Callithrix jacchus*). *Lab. Anim. Handb.*, **6**, 191–202.

Hediger, H. (1965). Environmental factors influencing the reproduction of zoo animals. In *Sex and behavior*. (ed. F. A. Beach), pp. 319–354. Wiley and Sons, New York.

Hedlund, H. and Andersson, K. E. (1985). Comparison of the responses to drugs acting on adrenoceptors and muscarinic receptors in human isolated corpus cavernosum and cavernous artery. *J. Auton. Pharmacol.*, **5**, 81–88.

Heimer, L. and Larsson, K. (1966). Impairment of mating behavior in male rats following lesions in the preoptic-anterior hypothalamic continuum. *Brain Res.*, **3**, 248–263.

Heimer, L. and Larsson, K. (1967). Mating behavior of male rats after olfactory bulb lesions. *Physiol. Behav.*, **2**, 207–209.

Heltne, P. G., Wojcik, J. F., and Pook, A. G. (1981). Goeldi's monkey, genus *Callimico*. In *Ecology and behavior of neotropical primates*, Vol. 1. (eds. A. F. Coimbra-Filho and R. A. Mittermeier), pp. 169–209. Academia Brasileira de Ciências, Rio de Janeiro.

Hendrickx, A. and Kraemer, D. C. (1969). Observation of the menstrual cycle, optimal mating time and preimplantation embryos of the baboon. *J. Reprod. Fertil. Suppl.*, **6**, 119–128.

Hendricks, S. E., Graber, B., and Rodriguez-Sierra, J. F. (1989). Neuroendocrine response to exogenous estradiol: no difference between heterosexual and homosexual men. *Psychoneuroendocrinology*, **14**, 176–185.

Henkin, R. I. (1974). Sensory changes during the menstrual cycle. *Biorhythms and human reproduction*. (eds. M. Ferin, F. Hallberg, R. M. Richart, and R. L. Vande Wiele), p. 227. Wiley, New York.

Hennessey, A. C., Wallen, K., and Edwards, D. A. (1986). Preoptic lesions increase the display of lordosis by male rats. *Brain Res.*, **370**, 21–28.

Hepp-Reymond, M. C. (1988). Functional organization of motor cortex and its participation in voluntary movements. In *Comparative primate biology, Vol. 4. Neurosciences* (eds. H. D. Steklis and J. Erwin), pp. 501–624. Alan R. Liss, New York.

Herbert, D. C., Schuppler, J., Poggel, A., Gunzel, P., and El Etrepy, M. F. (1977). Effect of cyproterone acetate on prolactin secretion in the female rhesus monkey. *Cell. Tiss. Res.*, **183**, 51–60.

Herbert, J. (1966). The effect of oestrogen applied directly to the genitalia upon the sexual attractiveness of the female rhesus monkey. *Int. Cong. Ser. Excerpta Med.*, **111**, 212.

Herbert, J. (1968). Sexual preference in the rhesus monkey (*Macaca mulatta*) in the laboratory. *Anim. Behav.*, **16**, 120–128.

Herbert, J. (1970). Hormones and reproductive behavior in rhesus and talapoin monkeys. *J. Reprod. Fertil.*, **11**, 119–140.

Herbert, J. (1973). The role of the dorsal nerves of the penis in the sexual behavior of the male rhesus monkey. *Physiol. Behav.*, **10**, 293–300.

Herbert, J. (1974). Some functions of hormones and the hypothalamus in the sexual activity of primates. *Prog. Brain Res.*, **41**, 331–348.

Herbert, J. (1977). Hormones and behaviour. *Proc. Roy. Soc. Lond. B.*, **199**, 425–443.

Herbert, J. (1981). Hormones and the sexual strategies of primates. *Symp. Zool. Soc. Lond.*, **46**, 337–359.

Herbert, J. (1993). Peptides in the limbic system: neurochemical codes for co-ordinated adaptive responses to behavioural and physiological demand. *Progr. in Neurobiol.*, **41**, 721–791.

Herbert, J. and Trimble, M. (1967). The effect of oestradiol and testosterone on the sexual receptivity and attractiveness of the female rhesus monkey. *Nature Lond.*, **216**, 165–166.

Herbert, J., Martensz, N. D., Umberkoman Witta, B., and Hansen, S. (1982). Distribution of prolactin and cortisol between serum and CSF in rhesus monkeys. In *Frontiers in hormone research*. (ed. Tj. B. Van Wimermsa Greidarus), pp. 159–172, Karger, Basel.

Herdt, G. H. (1981). *Guardians of the flutes*. McGraw-Hill, New York.

Herdt, G. H. (1982). Sambia nose-bleeding rites and male proximity to women. *Ethos.*, **10**, 189–231.

Herdt, G. H. (1984). *Ritualized homosexuality in Melanesia*. University of California Press, Berkeley.

Herdt, G. H. (1987). *The Sambia: ritual and gender in New Guinea*. Holt, Rinehart and Winston, New York.

Herdt, G. H. and Davidson, J. (1988). The Sambia 'turnimman': sociocultural and clinical aspects of gender formation in male pseudohermaphrodites with 5 alpha-reductase deficiency in Papua New Guinea. *Arch. Sex. Behav.*, **17**, 33–56.

Herdt, G. H. and Stoller, R. J. (1985). Sakulambei—a hermaphrodite's secret: an example of clinical ethnography. *Psychoanal. Stud. Soc.*, **11**, 117–158.

Herdt, G. H. and Stoller, R. J. (1989). Commentary to 'The socialization of homosexuality and heterosexuality in a non-western society'. *Arch. Sex. Behav.*, **18**, 31–34.

Herndon, J. G. (1983). Seasonal breeding in rhesus monkeys: influence of the behavioral environment. *Am. J. Primatol.*, **5**, 197–204.

Herndon, J. G., Turner, J. J., and Collins, D. C. (1981). Ejaculation is important for mating-induced testosterone increases in male rhesus monkeys. *Physiol. Behav.*, **27**, 873–877.

Hershkovitz, P. (1977). *Living New World monkeys*, Vol. 1. University of Chicago Press.

Hershkovitz, P. (1984). Taxonomy of squirrel monkeys genus *Saimiri* (*Cebidae Platyrrhini*): a preliminary report with description of a hitherto unnamed form. *Am. J. Primatol.*, **7**, 155–210.

Hertig, A. T., Barton, B. R., and Mackay, J. J. (1976). The female reproductive tract of the owl monkey (*Aotus trivirga-*

tus) with special reference to the ovary. *Lab. Anim. Sci.*, **26**, 1041–1067.

Hess, D. L. and Resko, J. A. (1973). The effects of progesterone on the patterns of testosterone and oestradiol concentrations in the systemic plasma of the female rhesus monkey during the intermenstrual period. *Endocrinology*, **92**, 446–453.

Hess, D. L., Hendrickx, A. G., and Stabenfeldt, G. H. (1979). Reproductive and hormonal patterns in the African green monkey (*Cercopithecus aethiops*). *J. Med. Primatol.*, **8**, 273–281.

Hess, J. P. (1973). Some observations on the sexual behaviour of captive lowland gorillas, *Gorilla g. gorilla* (Savage and Wyman). In *Comparative ecology and behaviour of primates*. (eds. R. P. Michael and J. H. Crook), pp. 507–581. Academic Press, London.

Heymann, E. W., Zeller, U., and Schwibbe, M. H. (1989). Muzzle-rubbing in the moustached tamarin, *Saguinus mystax* (Primates, Callitrichidae)—behavioural and histological aspects. *Z. Säugetierk.*, **54**, 265–275.

Hill, D. A. (1987). Social relationships between adult male and female rhesus macaques: 1. Sexual consortships. *Primates*, **28**, 439–456

Hill, K. and Kaplan, H. (1988). Trade offs in male and female productive strategies among the Ache, Part 2. In *Human reproductive behaviour: a Darwinian perspective*. (eds. L. Betzig, M. Borgerhoff Mulder and P. Turke), pp. 291–305. Cambridge University Press.

Hill, W. C. O. (1952). The external and visceral anatomy of the olive colobus monkey (*Procolobus verus*). *Proc. Zool. Soc. Lond.*, **122**, 127–186.

Hill, W. C. O. (1953). *Primates, comparative anatomy and taxonomy*, Vol. 1. *strepsirhini*. Edinburgh University Press.

Hill, W. C. O. (1955a). *Primates, comparative anatomy and taxonomy*, Vol. 2. *haplorhini: Tarsioidea*. Edinburgh University Press.

Hill, W. C. O. (1955b). A note on integumental colours with special references to the genus *Mandrillus*. *Säugetierkd Mitt.*, **3**, 145–151.

Hill, W. C. O. (1957). Primates, comparative anatomy and taxonomy, Vol. 3. Pithecoidea, platyrrhini, family *Hapalidae*. Edinburgh University Press.

Hill, W. C. O. (1958). External genitalia. In *Primatologia. Handbook of primatology*, Vol. 3. (eds. H. Hofer, A. H., Schultz and D. Starck), pp. 630–704. Karger, Basel.

Hill, W. C. O. (1960). *Primates, comparative anatomy and taxonomy*, Vol. 4, *Cebidae*, Part A. Edinburgh University Press.

Hill, W. C. O. (1962). *Primates, comparative anatomy and taxonomy*, Vol. 5. *Cebidae*, Part B. Edinburgh University Press.

Hill, W. C. O. (1966). *Primates, comparative anatomy and taxonomy*, Vol. 6. *catarrhini, Cercopithecoidea, Cercopithecinae*. Edinburgh, University Press, Edinburgh.

Hill, W. C. O. (1970). Primates, comparative anatomy and taxonomy, Vol. 8. *Cynopithecinae, Papio, Mandrillus, Theropithecus*. Edinburgh University Press.

Hill, W. C. O. (1972). *Evolutionary biology of the primates*. Academic Press, London.

Hill, W. C. O. (1974). Primates, comparative anatomy and

taxonomy, Vol. 7. *Cynopithecinae, Cercocebus, Macaca, Cynopithecus*. Edinburgh University Press.

Hinde, R. A. (1965). Interaction of internal and external factors in integration of canary reproduction. In *Sex and behavior*. (ed. F. A. Beach), pp. 381–415. Wiley, New York.

Hinde, R. A. and Rowell, T. E. (1962). Communication by postures and facial expressions in the rhesus monkey (*Macaca mulatta*). *Proc. Zool. Soc. Lond.*, **138**, 1–21.

Hinde, R. A. and Simpson, M. J. A. (1975). Qualities of mother-infant relationships in monkeys. *Ciba. Found. Symp.*, **33**, 39–67.

Hines, M. (1981). Prenatal diethystilbestrol (DES) exposure, human sexually dimorphic behavior and cerebral lateralization. Unpublished Ph.D. thesis, University of California, Los Angeles.

Hines, M. (1982). Prenatal gonadal hormones and sex differences in human behavior. *Psychol. Bull.*, **92**, 56–80.

Hines, M. and Kaufman, F. R. (1994). Androgen and the development of human sex-typical behavior: rough-and-tumble play and sex of preferred playmates in children with congenital adrenal hyperplasia (CAH). *Child Devel.*, **65**, 1041–1053.

Hines, M. and Sandberg, E. C. (1996). Sexual differentiation of cognitive abilities in women exposed to diethylstilbestrol (DES) prenatally. *Horm. Behav.*, **30**, 354–363.

Hiraiwa-Hasegawa, M. (1993). Skewed birth sex ratios in primates: should high-ranking mothers have daughters or sons? *Trends in Ecol. and Evol.*, **11**, 395–400.

Hodges, A. K., Eastman, S. A. K., and Jenkins, N. (1983). Sex steroids and their relationship to binding proteins in the serum of the marmoset monkey (*Callithrix jacchus*). *J. Endocr.*, **96**, 443–450.

Hodges, J. K., Czekala, N. M., and Lasley, B. L. (1979). Estrogen and luteinizing hormone secretion in diverse species from simplified urinary analysis. *J. Med. Primatol.*, **8**, 349–364.

Hoenig, J. (1977). Dramatis personae: selected biographical sketches of 19th century pioneers in sexology. In *Handbook of Sexology* (ed. J. Money and H. Mustaph), pp. 21–43 Elsevier/North Holland Biomedical Press, Amsterdam.

Hoffman, G. E., Knigge, K. M., Moynihan, J. A., Melynk, V., and Arimura, A. (1978). Neuronal fields containing luteinizing hormone releasing hormone (LHRH) in mouse brain. *Neuroscience*, **3**, 219–231.

Hofman, M. A. and Swaab, D. F. (1993). Diurnal and seasonal rhythms of neuronal activity in the suprachiasmatic nucleus of humans. *J. Biol. Rhythms.*, **8**, 283–295.

Hogg, J. T. (1984). Mating in bighorn sheep: multiple creative male strategies. *Science*, **225**, 526–529.

Hohmann, G. (1989). Comparative study of vocal communication in two Asian leaf monkeys, *Presbytis johnii* and *Presbytis entellus*. *Folia Primatol.*, **52**, 27–57.

Hökfelt, T. (1991). Neuropeptides in perspective: the last ten years. *Neuron*, **7**, 867–879.

Hökfelt, T., Johansson, O., Ljungdahl, A., Lundberg, J. M., and Schultzberg, M. (1980a). Peptidergic neurones. *Nature Lond.*, **284**, 515–521.

Hökfelt, T., Johansson, O., Ljungdahl, A., Lundberg, J. M., and Schultzberg, M. (1980b). Cellular localization of peptides in neural structures. *Proc. Roy. Soc. Lond.*, **210**, 63–77.

Holekamp, K. E. and Smale, L. (1995). Rapid change in offspring sex—ratios after class fission in the spotted hyena. *Am. Nat.*, **145**, 261–278.

Hollihn, U. (1973). Remarks on the breeding and maintenance of colobus monkeys *Colobus guereza*, proboscis monkeys *Nasalis larvatus* and douc langurs *Pygothrix nemaeus* in zoos. *Int. Z. Yrbk.*, **13**, 185–188.

Holloway, R. L. and De Lacoste, M. C. (1986). Sexual dimorphism in the human corpus callosum: a replication study. *Human Neurobiol.*, **5**, 87–91.

Holman, S. D. and Rice, A. (1996). Androgenic effects on hypothalamic asymmetry in a sexually differentiated nucleus related to vocal behavior in Mongolian gerbils. *Horm. Behav.*, **30**, 662–672.

Holmes, G. M., Chapple, W. D., Leipheimer, R. E., and Sachs, B. D. (1991). Electromyographic analysis of male rat perineal muscles during copulation and reflexive erections. *Physiol. Behav.*, **49**, 1235–1246.

Holmquist, F., Andersson, K. E., and Hedlund, H. (1990). Actions of endothelin on isolated corpus cavernosum from rabbit and man. *Acta Physiol. Scand.*, **139**, 113–122.

Holmquist, F., Kirkeby, H. J., Larsson, B., Forman, A., and Andersson, K. E. (1992). Functional effects, binding sites and immunolocalization of endothelin-1 in isolated penile tissues from man and rabbit. *J. Pharmacol. Exp. Ther.*, **261**, 795–802.

Homonnai, Z. T., Shilon, M., and Paz, G. F. (1984). Phenoxybenzamine—an effective male contraceptive pill. *Contraception*, **29**, 479 481.

Van Hooff, J. A. R. A. M. (1967). The facial displays of the catarrhine monkeys and apes. In *Primate ethology*. (ed. D. Morris), pp. 7–68. Weidenfeld and Nicolson, London.

Hoon, P. W., Bruce, K., and Kinchloe, B. (1982). Does the menstrual cycle play a role in sexual arousal? *Psychophysiology*, **19**, 21–27.

Hooper, E. and Musser, G. (1964). The glans penis in neotropical cricetines (*Muridae*) with comments on classification of muroid rodents. *Misc. Publ. Mus. Zool. Univ. Mich.*, **123**, 1–57.

Horel, J. A. (1988). Limbic neocortical interrelations. In *Comparative primate biology, Vol. 4. neurosciences*. (eds. J. Erwin and H. G. Steklis), pp. 81–97. Alan R. Liss, New York.

Horr, D. A. (1972). The Bornean orang-utan: population structure and dynamics in relationship to ecology and reproductive strategy. In *Primate behavior*, Vol. 4. (ed. L. A. Rosenblum), pp. 25–80. Academic Press, New York.

Horwich, R. H. (1983). Breeding behaviors of the black howler monkey, *Alouatta pigra*, of Belize. *Primates*, **24**, 222–230.

Hoskins, D. D. and Patterson, D.L. (1967). Prevention of coagulum formation with recovery of motile spermatozoa from rhesus monkey semen. *J. Reprod. Fertil.*, **13**, 337–340.

Hotchkiss, J. and Knobil, E. (1994). The menstrual cycle and its neuroendocrine control. In *The physiology of reproduction*, Vol. 2. (2nd edition) (eds. E. Knobil and J. D. Neill), pp. 711–749. Raven Press, New York.

Howell, N. (1979). *Demography of Dobe !Kung*. Academic Press, New York.

Hrdy, S. B. (1974). Male-male competition and infanticide

among the langurs (*Presbytis entellus*) of Abu, Rajasthan. *Folia Primatol.*, **22**, 19–58.

Hrdy, S. B. (1977). *The langurs of Abu*. Harvard University Press, Cambridge, Massachusetts.

Hrdy, S. B. (1979). Infanticide among animals: a review, classification and examination of the implications for the reproductive strategies of females. *Ethol. Sociobiol.*, **1**, 13–40.

Hrdy, S. B. (1981). *The woman that never evolved*. Harvard University Press, Cambridge, Massachusetts.

Hrdy, S. B. (1988). The primate origins of human sexuality. In *Nobel Conference XXIII: the evolution of sex.* (eds. R. Bellig and G. Stevens), pp. 101–136. Harper and Row, San Francisco.

Hrdy, S. B. and Whitten, P. L. (1987). Patterning of sexual activity. In *Primate societies.* (eds. B. Smuts, D. Cheney, R. Seyfarth, R. Wrangham and T. Struhsaker), pp. 370–384. University of Chicago Press.

Hu, S., Pattatucci, A. M. L., Patterson, C., Li, L., Fulker, D. W., Cherny, S. S., Kruglyak, L. and Hamer, D. H. (1995). Linkage between sexual orientation and chromosome Xq28 in males but not in females. *Nature Genetics*, **11**, 248–256.

Huber, E. (1931). Evolution of facial musculature and facial expression. Oxford University Press.

Hubrecht, R. C. (1984). Field observation on group size and composition of the common marmoset (*Callithrix jacchus jacchus*) at Tapacura, Brazil. *Primates*, **25**, 13–21.

Hughes, A. M., Everitt, B. J., and Herbert, J. (1987a). Selective effects of β-endorphin infused into the hypothalamus, preoptic area and bed nucleus of the stria terminalis on the sexual and ingestive behaviour of male rats. *Neuroscience*, **23**, 1063–1073.

Hughes, A. M., Everitt, B. J., Lightman, S. L., and Todd, K. (1987b). Oxytocin in the central nervous system and sexual behavior in male rats. *Brain Res.*, **41**, 133–137.

Hughes, A. M., Everitt, B. J., and Herbert, J. (1988). The effects of simultaneous or separate infusions of some pro-opiomelanocortin-derived peptides (β-endorphin, melanocyte stimulating hormone and corticotrophin-like intermediate polypeptide) and their acetylated derivatives upon sexual and ingestive behavior of male rats. *Neuroscience*, **2**, 289–298.

Hughes, A. M., Herbert, J., and Everitt, B. J. (1990). Comparative effects of preoptic area infusions of opioid peptides, lesions and castration on sexual behavior in male rats: effects of instrumental behavior, conditioned place preference and partner preference. *Psychopharmacology*, **102**, 243–256.

Huhtaniemi, I., Dunkel, L., and Perheentupa, J. (1986). Transient increase in postnatal testicular activity is not revealed by longitudinal measurements of salivary testosterone. *Pediatr. Res.*, **20**, 1324–1327.

Hull, E. M., Bazzett, T. J., Warner, R. K., Eaton, R. C. and Thomson, J. T. (1990). Dopamine receptors in the ventral tegmental area modulate male sexual behavior in rats. *Brain Res.*, **512**, 1–6.

Hull, E. M., Du, J., Lorrain, D. S., and Matuszewich, L. (1995). Extracellular dopamine in the medial preoptic area: implications for sexual motivation and hormonal control of copulation. *J. Neurosci.*, **15**, 7465–7471.

Hunt, M. (1974). *Sexual behavior in the 1970's*. Playboy Press, New York.

Hunt, R. D, Chalifoux, L. V., King, N. W., and Trum, B. F. (1975). Arrested spermatologenesis in owl monkeys (*Aotus trivirgatus*). *J. Med. Primatol.*, **4**, 399–400.

Hunter, A. J., Fleming, D., and Dixson, A. F. (1984). The structure of the vomeronasal organ and nasopalatine ducts in *Aotus trivirgatus* and some other primate species. *J. Anat. Lond.*, **138**, 217–225.

Hunter, R. H. F. (1988). Transport of gametes, selection of spermatozoa and gamete lifespan in the female tract. In *The fallopian tubes*. (ed. R H. F. Hunter), pp. 53–80. Springer, New York.

Hunter, R. H. F., Flechon, B., and Flechon, J. E. (1991). Distribution, morphology and epithelial interactions of bovine spermatozoa in the oviduct before and after ovulation: a scanning election microscope study. *Tissue Cell*, **23**, 641–656.

Hutchinson, T. C. (1970). Vaginal cytology and reproduction in the squirrel monkey (*Saimiri sciureus*). *Folia Primatol.*, **12**, 212–223.

Hutchison, J. B. (1978). Hypothalamic regulation of male sexual responses to androgen. In *Biological determinants of sexual behaviour*. (ed. J. B. Hutchison), pp. 227–317. John Wiley, Chichester.

Huws, R. (1993). Antihypertensive medication and sexual problems. In *Sexual pharmacology* (eds. A. J. Riley, M. Peet and C. Wilson), pp. 146–158. Oxford Medical Publications, Clarendon Press, Oxford.

Huws, R. and Sampson, G. (1993). Recreational drugs and sexuality. In *Sexual pharmacology*. (eds. A. J. Riley, M. Peet and C. Wilson), pp. 197–210. Clarendon Press, Oxford.

Huxley, T. H. (1863). *Evidence as to man's place in nature*. Williams and Norgate, London.

I'Anson, H., Foster, D.L., Foxcroft, G. R., and Booth, P. J. (1991). Nutrition and reproduction. *Oxford Revs. Repro. Biol.*, **13**, 239–311.

Idani, G. (1995). Function of peering behavior among bonobos (*Pan paniscus*) at Wamba, Zaire. *Primates*, **36**, 377–383.

Iggo, A. (1982). Cutaneous sensory mechanisms. In *The senses*. (eds. H. B. Barlow and J. D. Mollon), pp. 369–408. Cambridge University Press.

Ignarro, L. J., Buga, G. M., Wood, K. S., Byrns, R. E., and Chaudhury, G. (1987). Endothelium-derived relaxing factor produced and released from artery and vein is nitric oxide. *Proc. Nat. Acad. Sci. USA*, **84**, 9265–9269.

Imperato-McGinley, J., Guerrero, L., Cautier, T., and Peterson, R. E. (1974). Steroid 5α—reductase deficiency in man: an inherited form of male pseudohermaphroditism. *Science*, **186**, 1213–1215.

Imperato-McGinley, J., Peterson, R E., Gautier, T., and Sturla, E. (1979). Androgens and the evolution of male gender identity among male pseudohermaphrodites with 5 alpha-reductase deficiency. *N. Eng. J. Med.*, **300**, 1231–1237.

Imperato-McGinley, J., Peterson, R. E., Gautier, T., Cooper, G., Danner, R., Arthur, A., Morris, P. L., Sweeney, W. J., and Shackleton, C. (1982). Hormonal evaluation of a large kindred with complete androgen insensitivity: evidence for secondary 5 alpha-reductase deficiency. *J. Clin. Endocr. Metab.*, **54**, 931–941.

Ingram, J. C. (1977). Interactions between parents and infants, and the development of independence in the common marmoset (*Callithrix jacchus*). *Anim. Behav.*, **25**, 811–827.

Inoue, M., Mitsunaga, F., Nozaki, M., Ohsawa, H., Takenaka, A., Sugiyama, Y., et al. (1993). Male dominance rank and reproductive success in an enclosed group of Japanese macaques: with special reference to post-conception mating. *Primates*, **34**, 503–511.

Ioannou, J. M. (1966). The oestrous cycle of the potto. *J. Reprod. Fertil.*, **11**, 455–457.

Ismail, A. A. A., Davidson, D. W., Loraine, J. A., and Fox, C. A. (1972). Relationship between plasma cortisol and human sexual activity. *Nature Lond.*, **237**, 288–289.

Izard, M. K. (1990). Social influences on the reproductive success and reproductive endocrinology of prosimian primates. In *Socioendocrinology of primate reproduction*. (eds. T. E. Ziegler and F. B. Bercovitch), pp. 159–186. Wiley-Liss, New York.

Izard, M. K. and Rasmussen, D. T. (1985). Reproduction in the slender loris (*Loris tardigradus malabaricus*). *Am. J. Primatol.*, **8**, 153–165.

Izor, R. J., Walchuk, S. L., and Wilkins, L. (1981). Anatomy and systematic significance of the penis of the pigmy chimpanzee, *Pan paniscus*. *Folia Primatol.*, **35**, 218–224.

Jacubczak, L. F. (1964). Effects of testosterone propionate on age differences in mating behavior. *J. Gerontol.*, **19**, 458–461.

James, M. D., Hole, D. R., and Wilson, C. A. (1989). Differential involvement of 5-hydroxytroyptamine (5HT) in specific hypothalamic areas in the mediation of steroid-induced changes in gonadotrophin release and sexual behavior in female rats. *Neuroendocrinology*, **49**, 561–569.

James, W. H. (1971). The distribution of coitus within the human intermenstruum. *J. Biosoc. Sci.*, **3**, 159–171.

Janson, C. (1984). Female choice and the mating system of the brown capuchin monkey, Cebus apella (Primates: Cebidae). *Z. Tierpsychol.*, **65**, 177–200.

Janus, S. S. and Janus, C. L. (1993). *The Janus Report*. Wiley, New York.

Jasczak, S. and Hafez, E. S. E. (1972). The cervix uteri and sperm transport in female macaques. In *Medical Primatology, Proc. 3rd Conf. exp. Med. Surg.* (eds. E. I. Goldsmith and J. Moor-Jankowski), pp. 273–270. Karger, Basel.

Jay, P. (1965). The common langur of North India. In *Primate behavior: field studies of monkeys and apes*. (ed. I. De Vore), pp. 197–249. Holt, Rinehart and Winston, New York.

Jeffreys, A. J. Wilson, V., and Thein, S. L. (1985). Hypervariable 'Minisatellite' regions in human DNA. *Nature Lond.*, **314**, 67–73.

Jenni, D. A. and Collier, G. (1972). Polyandry in the American jacana. *Auk*, **89**, 743–765.

Jit, I. and Sanjeev, (1988). The volume of the testes in North-West Indian children. *Hum. Biol.*, **60**, 945–951.

Johns, M. A. (1979). The role of the vomeronasal system in mammalian reproductive physiology. In *Chemical signals: Vertebrates and aquatic invertebrates*. (D. Müller-Schwarze and R. M. Silverstein), pp. 341–364. Plenum, New York.

Johnson, D. F. and Phoenix, C. H. (1976). Hormonal control of female sexual attractiveness, proceptivity and receptivity in rhesus monkeys. *J. Comp. Physiol. Psychol.*, **90**, 473–483.

Johnson, D. F. and Phoenix, C. H. (1978). Sexual behavior and hormone levels during the menstrual cycles of rhesus monkeys. *Horm. Behav.*, **11**, 160–174.

Johnson, J. (1968). *Disorders of sexual potency in the male*. Pergamon Press, Oxford.

Johnson, M. and Everitt, B. J. (1988). *Essential reproduction*, third edition. Blackwell Scientific Publications, Oxford.

Johnson, R. D. and Halata, Z. (1991). Topography and ultra structure of sensory nerve endings in the glans penis of the rat. *J. Comp. Neurol.*, **312**, 299–310.

Johnson, R. D., Kitchell, R. I., and Gilanpour, H. (1986). Rapidly and slowly adopting mechanoreceptors in the glans penis of the cat. *Physiol. Behav.*, **37**, 69–78.

Johnston V. S. and Franklin, M. (1993). Is beauty in the eye of the beholder? *Ethol. Sociobiol.*, **14**, 183–199.

Jolly, A. (1966). *Lemur behavior*. University of Chicago Press.

Jolly, A. (1972). *The evolution of primate behavior*. Macmillan, New York.

Jolly, A. (1984). The puzzle of female feeding priority. In *Female primates: studies by women primatologists*. (ed. M. Small), pp. . Alan Liss, New York.

Jolly, C. J. (1970a). The seed eaters: a new model of hominid differentiation based on a baboon analogy. *Man*, **5**, 5–26.

Jolly, C. J. (1970b). The large African monkeys as an adaptive array. In *Old World monkeys: evolution, systematics and behaviour*. (eds. J. R. Napier and P. N. Napier), pp. 139–174. Academic Press, New York.

Jones, C. B. (1983). Social organization of captive black howler monkeys (*Alouatta caraya*): social competition and the use of non-damaging behavior. *Primates*, **24**, 25–39.

Jones, C. B. (1985). Reproductive patterns in mantled howler monkeys: estrus mate choice and competition. *Primates.*, **26**, 130–142.

Jones, C. B. and Sabater, PI, J. (1971). Comparative ecology of *Gorilla gorilla* (Savage and Wyman) and Pan troglodytes (Blumenbach) in Rio Muni, West Africa. Bibliotheca Primatologica **13**, 1–96. Karger, Basel.

Jones, D. (1995). Sexual selection, physical attractiveness and facial neoteny. *Curr. Anthropol.*, **36**, 723–748.

Jost, A. (1961). The role of fetal hormones in prenatal development. *Harvey Lec.*, **55**, 201–226.

Jost, A. (1970). Hormonal factors in sex differentiation of the mammalian foetus. *Phil. Trans. Roy. Soc. B.*, **259**, 119–130.

Jouventin, P. (1975). Observations sur la socio-ecologie du mandrill. *Terre et vie*, **29**, 493–532.

Judd, H. L. and Yen, S. S. C. (1973). Serum androstenedione and testosterone levels during the menstrual cycle. *J. Clin. Endocr. Metab.*, **36**, 475–481.

Jünemann, K. P., Ehmke, H., Kummer, W., Mayer, B., Bührle, C. P., and Alken, P. (1993). Coexistence of nitric oxide and vasoactive intestinal polypeptide in nerve terminals of human corpus cavernosum. *J. Urol.*, **149**: 245A (abstract).

Jungers, W. L. (1990). Problems and methods in reconstructing body size in fossil primates. In *Body size in mammalian paleobiology: estimation and biological implications*. (eds. J. Damuth and B. J. MacFadden), pp. 103–118. Cambridge University Press.

Jurke, M. H., Pryce, C. R., and Dobelli, M. (1995). An investigation into sexual motivation and behavior in female Goeldi's monkeys (*Callimico goeldii*): effect of ovarian state,

mate familiarity and mate choice. *Horm. Behav.*, **29**, 531–553.

Kalkstein, T. (1991). The occurrence and function of sexual swelling in stumptail macaques. *Am. J. Phys. Anthropol. Suppl.*, **12**, 101–102.

Kallmann, F. J. (1952). Comparative twin study on the genetic aspects of male homosexuality. *J. Nerv. Ment. Dis.*, **115**, 283–298.

Kalra, S. P. and Kalra, P. S. (1983). Neural regulation of luteinising hormone secretion in the rat. *Endocr. Revs.*, **4**, 311–351.

Kamel, F., Mock, E. J., Wright, W. W., and Frankel, A. I. (1975). Alterations in plasma concentrations of testosterone, LH, and prolactin associated with mating in the male rat. *Horm. Behav.*, **6**, 277–288.

Kandel, E. R. (1985). Central representation of touch. In *Principles of neural science* (2nd edition). (eds. E. R. Kandel and J. H. Schwartz), pp. 316–330. Elsevier, New York.

Kaneko, N., Debski, E. A., Wilson, M. C., and Whitten, W. K. (1980). Puberty acceleration in mice: 2. Evidence that the vomcronasal organ is receptor for the primer pheromone in male mouse urine. *Biol. Reprod.*, **22**, 873–878.

Kano, T. (1982). The social group of pygmy chimpanzees (*Pan paniscus*) of Wamba. *Primates*, **23**, 171–188.

Kano, T. (1992). *The last ape: pygmy chimpanzee behavior and ecology*. Stanford University Press, Stanford, California.

Kaplan, J. N., Chen, J., Smith, E., and Davidson, J. M. (1981). The development of seasonal variation in gonadal hormones and body weight in the maturing squirrel monkey. *Int. J. Primatol.*, **2**, 369–380.

Kappeler, P. M. (1987). Reproduction in the crowned lemur (*Lemur coronatus*) in captivity. *Am. J. Primatol.*, **12**, 497–503.

Kappeler, P. M. (1991). Patterns of sexual dimorphism in body weight among prosimian primates. *Folia Primatol.*, **57**, 131–146.

Kappeler, P. M. (1993). Sexual selection and lemur social systems. In *Lemur social systems and their ecological basis*. (eds. P. M. Kappeler and J. U. Ganzhorn), p. 223–240. Plenum, New York.

Kappeler, P. M. (1996). Intrasexual selection and phylogenetic constraints in the evolution of sexual canine dimorphism in strepsirhine primates. *J. Evol. Biol.*, **9**, 43–65.

Kappeler, P. M. (1997a). Intrasexual selection in *Mirza coquereli*: evidence for scramble competition polygyny in a solitary primate. *Behav. Ecol. Sociobiol.*, **45**, 115–127.

Kappeler, P. M. (1997b). Intrasexual selection and testis size in strepsirhine primates. *Behav. Ecol.*, **8**, 10–19.

Karsch, F. J., Dierschke, D., and Knobil, E. (1973). Sexual differentiation of pituitary function: apparent difference between primates and rodents. *Science*, **179**, 484–486.

Kashon, M. L., Ward, O. B., Grisham, W., and Ward, I. L. (1992). Prenatal β-endorphin can modulate some aspects of sexual differentiation in rats. *Behav. Neurosci.*, **106**, 555–562.

Katongole, C. B., Naftolin, F., and Short, R. V. (1971). Relationship between blood levels of luteinizing hormone and testosterone in bulls, and effects of sexual stimulation. *J. Endocr.*, **50**, 457–466.

Kaufman, I. C. and Rosenblum, L. A. (1966). A behavioral

taxonomy for *Macaca nemestrina* and *Macaca radiata*: based on longitudinal observation of family groups in the laboratory. *Primates*, **7**, 206–258.

Kaufman, J. H. (1965). A three year study of mating behavior in a free ranging band of rhesus monkeys. *Ecology*, **40**, 500–512.

Kavanagh, M. (1983). A complete guide to monkeys, apes and other primates. Jonathan Cape, London.

Kawabe, M. and Mano. T. (1972). Ecology and behavior of the wild proboscis monkey, *Nasalis larvatus* (Wurmb) in Sabah, Malaysia. *Primates*, **13**, 213–227.

Kawai, M. (1979). *Ecological and sociological studies of gelada baboons*. Contrib. Primatol. Vol. 16. Karger, Basel.

Kawai, M., Azuma, S., and Yoshiba, K. (1967). Ecological studies of reproduction in Japanese monkeys (*Macaca fuscata*). 1. *Problems of the birth season. Primates*, **8**, 35–74.

Kawai, M., Dunbar, R. I. M., Ohsawa, H. and Mori, U. (1983). Social organization of gelada baboons: social units and definitions. *Primates*, **24**, 13–24.

Kay, R. F., Plavcan, J. M., Glander, K. E., and Wright, P. C. (1988). Sexual selection and canine dimorphism in New World monkeys. *Am. J. Phys. Anthropol.*, **77**, 385–397.

Kaye, S. A., Folson, A. R., Prineas, R. J., Potter, J. D., and Gapstur, S. M. (1990). The association of body fat distribution with lifestyle and reproductive factors in a population study of postmenopausal women. *Int. J. Obesity*, **14**, 583–591.

Keddy, A. C. (1986). Female mate choice in vervet monkeys (*Cercopithecus aethiops*). *Am. J. Primatol.*, **10**, 125–143.

Kellett, J. (1993). The nature of human sexual desire and its modification by drugs. In *Sexual pharmacology* (eds. A. J. Riley, M. Peet and C. Wilson), pp. 130–145. Clarendon Press, Oxford.

Kelly, J. P. (1985). Principles of the functional and anatomical organization of the nervous system. In *Principles of neural science* (2nd edition). (eds. E. R. Kandel and J. H. Schwartz), pp. 211–221. Elsevier, New York.

Kenagy, G. J. and Trombulak, G. C. (1986). Size and function of mammalian testes in relation to body size. *J. Mammal.*, **67**, 1–22.

Kendrick, K. M. (1983). Electrophysiological effects of testosterone on the medial preoptic-anterior hypothalamus of the rat. *J. Endocr.*, **96**, 35–42.

Kendrick, K. M. (1994). Neurobiological correlates of visual and olfactory recognition in sheep. *Behav. Proc.*, **33**, 89–112.

Kendrick, K. M. and Baldwin, B. A. (1987). Cells in temporal cortex of conscious sheep can respond preferentially to the sight of faces. *Science*, **236**, 448–450.

Kendrick, K. M. and Dixson, A F. (1983). The effect of the ovarian cycle on the sexual behaviour of the common marmoset (*Callithrix jacchus*). *Physiol. Behav.*, **30**, 735–742.

Kendrick, K. M. and Dixson, A. F. (1984a). A quantitative description of copulatory and associated behaviors of captive marmosets (*Callithrix jacchus*). *Int. J. Primatol.*, **5**, 199–212.

Kendrick, K. M. and Dixson, A. F. (1984b). Ovariectomy does not abolish proceptive behaviour cyclicity in the common marmoset (*Callithrix jacchus*). *J. Endocr.*, **101**, 155–162.

Kendrick, K. M. and Dixson, A. F. (1984c). Hypoprolacti-

naemia abolishes sensitivity of stria terminalis neurones to testosterone. *J. Endocr.*, **102**, 109–113.

Kendrick, K. M. and Dixson, A. F. (1985a). Effects of oestradiol 17β, progesterone and testosterone upon proceptivity and receptivity on ovariectomized common marmosets (*Callithrix jacchus*). *Physiol. Behav.*, **34**, 123–128.

Kendrick, K. M. and Dixson, A. F. (1985b). Luteinizing hormone releasing hormone enhances proceptivity in a primate. *Neuroendocrinology*, **41**, 449–453.

Kendrick, K. M. and Dixson, A. F. (1986). Anteromedial hypothalamic lesions block proceptivity but not receptivity in the female common marmoset (*Callithrix jacchus*) *Brain Res.*, **375**, 221–229.

Kendrick, K. M. and Drewett, R. F. (1979). Testosterone reduces refractory period of stria terminalis neurons in the rat brain. *Science*, **204**, 877–879.

Kendrick, K. M., Drewett, R. F., and Wilson, C. A. (1981). Effect of testosterone on neuronal refractory periods, sexual behaviour and luteinizing hormone: a comparison of time courses. *J. Endocr.*, **89**, 147–155.

Kennard, M. A. and Willner, M. D. (1941). Findings at autopsies of seventy anthropoid apes. *Endocrinology*, **28**, 967–976.

Kenyon, A. J. (1969). The sea otter in the eastern Pacific Ocean. *N. Am. Fauna.*, **68**, 1–352.

Keverne, E. B. (1970). Investigation of the sexual behavior of rhesus monkeys by operant techniques combined with free cage studies. Unpublished Ph.D. thesis, University of London.

Keverne, E. B. (1976). Sexual receptivity and attractiveness in the female rhesus monkey. *Adv. Study Behav.*, **7**, 155–196.

Keverne, E. B. (1977). Hormonal regulation of reproductive behaviour in the female primate. In *Use of non-human primates in biomedical research*. (eds. M. R. N. Prasad and T. C. Anand Kumar), pp. 183–204. Indian National Science Academy, New Delhi.

Keverne, E. B. (1979a). The dual olfactory projections and their significance for behavior. In *Chemical ecology: odor communication in animals* (ed. F. Fitter), pp. 75–83. Elsevier, Amsterdam.

Keverne, E. B. (1979b). Sexual and aggressive behaviour in social groups of talapoin monkeys. *Ciba Found. Symp.*, **62**, 271–297.

Keverne, E. B. (1981). Do Old World primates have oestrus? *Malay. Appl. Biol.*, **10**, 119–126.

Keverne, E. B. (1985). Reproductive behaviour. In *Reproduction in mammals*, Vol. 4. (2nd edition) (eds. C. R. Austin and R. V. Short), pp. 133–175. Cambridge University Press.

Keverne, E. B. (1987). Processing of environmental stimuli and primate reproduction. *J. Zool. Lond.*, **213**, 395–408.

Keverne, E. B. (1992). Primate social relationships: their determinants and consequences. *Adv. Study Behav.*, **21**, 1–37.

Keverne, E. B. (1995). Neurochemical changes accompanying the reproductive process: their significance for maternal care in primates and other mammals. In *Motherhood in human and nonhuman primates*. (eds. C. R. Pryce, R. D. Martin and D. Skuse), pp. 69–77. Karger, Basel.

Keverne, E. B., Leonard R. A., Scruton, D. M., and Young, S. K. (1978). Visual monitoring in social groups of talapoin monkeys (Miopithecus talapoin). *Anim. Behav.*, **26**, 933–944.

Keverne, E. B., Meller, R. E., and Eberhart, A. (1982). Dominance and subordination: concepts or physiological states. In *Advanced views in primate biology*. (eds. A. B. Chiarelli and R. S. Corruccini), pp. 81–94. Springer, New York.

Keverne, E. B., Fundele, R., Narasimha, M., Barton, S. C., and Surani, M. A. (1996). Genomic imprinting and the differential roles of parental genomes in brain development. *Dev. Brain Res.*, **92**, 91–100.

Kiersch, T. A. (1990). Treatment of sex offenders with Depo-Provera. *Bull. Am. Acad. Psychiatr. Law*, **18**, 179–187.

Kimura, D. (1992). Sex differences in the brain. *Sci. Am.* September 1992, 81–87.

Kindon, H. A., Baum, M. J., and Paredes, R. J. (1996). Medial preoptic/anterior hypothalamic lesions induce a female-typical profile of sexual partner preference in male ferrets. *Horm. Behav.*, **30**, 514–527.

King, C. (1989). *The natural history of weasels and stoats*. Christopher Helm, London.

Kingdon, J. (1980). The role of visual signals and face patterns in African forest monkeys (guenons) of the genus *Cercopithecus. Trans. Zool. Soc. Lond.*, **35**, 425–475.

Kingdon, J. (1988). What are face patterns and do they contribute to reproductive isolation in guenons. In *A primate radiation: evolutionary biology of the African guenons*. (eds. A. Gautier-Hion, F. Boulire, J. P. Gautier and J. Kingdon), pp. 227–245. Cambridge University Press.

Kingdon, J. (1992). Facial patterns as signals and masks. In *The Cambridge Encyclopedia of human evolution*. (eds. S. Jones, R. Martin and D. Pilbeam), pp. 161–165. Cambridge University Press.

Kingsley, S. K. (1982). Causes of non-breeding and the development of the secondary sexual characteristics in the male orangutan: a hormonal study. In *The orangutan: its biology and conservation*. (ed. L. M. de Boer), pp. 215–229. Dr. W. Junk, The Hague.

Kingsley, S. R. (1988). Physiological development of male orang-utans and gorillas. In *Orang-utan biology*. (ed. J. H. Schwartz), pp. 123–131. Oxford University Press, New York.

Kinsey, A. C., Pomeroy, W. B., and Martin, C. E. (1948). *Sexual behavior in the human male*. W. B. Saunders Co., Philadelphia.

Kinsey, A. C., Pomeroy, W. B., Martin, C. E., and Gebhard, P. H. (1953). *Sexual behavior in the human female*. W. B. Saunders Co., Philadelphia.

Kinzey, W. G. The titi monkeys, genus *Callicebus*. In *Ecology and behavior of neotropical primates*, Vol. 1. (eds. A. F. Coimbra-Filho and R. A. Mittermeier), pp. 241–276. Academia Brasileria de Ciências, Rio de Janeiro.

Kleiman, D. G. (1977). Monogamy in mammals *Quart. Rev. Biol.*, **52**, 34–69.

Kleiman, D. G. (1978). Characteristics of reproduction and socio-sexual interactions in pairs of lion tamarins (*Leontopithecus rosalia*). In *The biology and conservation of the callitrichidae*. (ed. D. Kleiman), pp.181–190. Smithsonian University Press, Washington.

Kleiman, D. G. and Mack, D. S. (1977). A peak in sexual activity during mid-pregnancy in the golden lion tamarin, *Leontopithecus rosalia* (Primates: Callitrichidae). *J. Mammal.*, **58**, 657–660.

Kleiman, D. G., Hoage, R. J., and Green, K. M. (1988). The lion tamarins, genus *leontopithecus*. In *Ecology and behavior of neotropical primates*, Vol. 2. (eds. R. A. Mittermeier, A. B. Rylands, and G. A. B. Fonseca), pp. 299–347. World Wildlife Fund, Washington DC

Klein, L. L. (1971). Observations on copulation and seasonal reproduction of two species of spider monkeys, *Ateles belzebuth* and *A. geoffroyi. Folia Primatol.*, **15**, 233–248.

Klein, L. L. and Klein, D. J. (1971). Aspects of social behavior in a colony of spider monkeys, *Ateles geoffroyi*, at the San Francisco Zoo. *Int. Z. Yb.*, **11**, 175–181.

Klima, M. (1972). Ossa suprasternalia der primaten und die specialanpassungen des manubrium sterni bei den brüllaffen (*Alouatta*). *Folia Primatol.*, **17**, 421-433.

Kling, A., Lancaster, J., and Benitone, J. (1970). Amygdalectomy in the free-ranging vervet. *J. Psychiat. Res.*, **7**, 191–199.

Klüver, N. and Bucy, P. C. (1939). Preliminary analysis of functions of the temporal lobes in monkeys. *Archs. Neurol. Psychiat.*, **42**, 979–1000.

Knapp, L. A., Ha, J. C., and Sackett, G. P. (1996). Parental MHC antigen sharing and pregnancy wastage in captive pigtailed macaques. *J. Repro. Immunol.*, **32**, 73—88.

Knight, T. W., Gherardi, S., and Lindsay, D. R. (1987). Effects of sexual stimulation on testicular size in the ram. *Anim. Reprod. Sci.*, **13**, 105–115.

Knobil, E. (1974). On the control of gonadotropin secretion in the rhesus monkey. *Rec. Progr. Horm. Res.*, **30**, 1–46.

Koering, M. J. (1974). Comparative morphology of the primate ovary. In *Contributions to primatology*, Vol. 3 (ed. W. P. Luckett), pp. 38–81. Karger, Basel.

Koering, M. J. (1979). Folliculogenesis in primates: process of maturation and atresia. In *Animal models for research on contraception and fertility*. (ed. N. J. Alexander), pp. 187–199. Harper and Row, Hagerstown.

Koering, M. J. (1986). Ovarian architecture during follicle maturation. In *Comparative primate biology*, Vol. 3. reproduction and development. (eds. W. R. Dukelow and J. Erwin), pp. 215–262. Alan R. Liss, New York.

Koering, M. J. (1987). Follicle maturation and atresia: morphological correlates. In *The primate ovary*. (ed. R. L. Stouffer), pp. 3–23. Plenum, New York.

Koford, C. B. (1965). Population dynamics of rhesus monkeys on Cayo Santiago. In *Primate behavior: field studies of monkeys and apes*. (ed. I. DeVore), pp. . Holt, Rinehart and Winston, New York.

Kolodny, R. C. and Bauman, J. E. (1979). Letter to editor. Female sexual activity at ovulation. *New Eng. J. Med.*, **300**, 626.

Komisaruk, B. R. (1978). The nature of the neural substrate of female sexual behaviour in mammals and its hormonal sensitivity: review and speculations. In *Biological determinants of sexual behaviour*. (ed. J. B. Hutchison), pp. 349–393. John Wiley and Sons, Chichester.

Konishi, M. and Gurney, M. E. (1982). Sexual differentiation of brain and behaviour. *Trends in Neurosci.*, **5**, 20–23.

Koraitim, M., Schafer, W., Melchior, H., and Lutzeyer, W. (1977). Dynamic activity of bladder neck and external sphincter in ejaculation. *Urology*, **10**, 130–132.

Koyama, N. (1971). Observations on the mating behaviour of wild siamang gibbons at Fraser's Hill, Malaysia. *Primates*, **12**, 183–189.

Koyama, N. (1988). Mating behavior of ring-tailed lemurs (*Lemur catta*) at Berenty, Madagascar. *Primates.*, **29**, 163–175.

Koyama, N., Takahata, Y., Huffman, M A, Norikoshi, K., Suzuki, H. (1992). Reproductive parameters of female Japanese macaques: 30 years data from the Arashiyama troops, Japan. *Primates*, **33**, 33–47.

Koyama, Y., Fujita, I., Aou, S., and Oomura, Y. (1988). Proceptive presenting elicited by electrical stimulation of the female monkey hypothalamus. *Brain Res.*, **446**, 199–203.

Krackow, S. (1995). Potential mechanisms for sex ratio adjustment in mammals and birds. *Biol. Rev.*, **70**, 225–241.

Kraemer, H. C., Becker, H. B., Brodie, H. K. H., Doering, C. H., Moos, R. H., and Hamburg, D. A. (1976). Orgasmic frequency and plasma testosterone levels in normal human males. *Arch. Sex. Behav.*, **5**, 125–132.

Krohn, P. L. and Zuckerman, S. (1937). Water metabolism in relation to the menstrual cycle. *J. Physiol. Lond.*, **88**, 369–387.

Kruijver, F. P. M., de Jonge, F. H., Van den Brock, W. T., Van der Woude, T., Endert, E., and Swaab, D. F. (1993). Lesions of the suprachiasmatic nucleus do not disturb sexual orientation of the adult male rat. *Brain Res.*, **624**, 342–346.

Kruuk, H. (1972). The spotted hyena: a study of predation and social behavior. University of Chicago Press.

Kuester, J. and Paul, A. (1989). Reproductive strategies of subadult Barbary macaque males at Affenberg Salem. In *Sociobiology of reproductive strategies in animals and humans.* (eds. A. E. Rasa, C. Vogel and E. Voland), pp. 93–109. Chapman and Hall, London.

Kuester, J., Paul, A., and Arnemann, J. (1994). Kinship, familiarity and mating avoidance in barbary macaques, *Macaca sylvanus*. *Anim. Behav.*, **48**, 1183–1194.

Kuester, J., Paul, A., and Arnemann, J. (1995). Age-related and individual differences of replacement success in male and female Barbary macaques. *Primates*, **36**, 461–476.

Kuhn, H.-J. (1967). Zur systematik der Cercopithecidae. In *Neue Ergebnisse der primatologie.* (eds. D. Starck, R. Schneider and H.-J. Kuhn), pp. 25–46. Gustav Fischer, Stuttgart.

Kulin, H. E. and Reiter, E. O. (1976). Gonadotropin and testosterone measurement after estrogen administration to adult men, prepubertal and pubertal boys, and men with hypogonadism: evidence for maturation of positive feedback in the male. *Ped. Res.*, **10**, 46–51.

Kumar A. and Kurup, G. U. (1985). Sexual behavior of the lion-tailed macaque (*Macaca silenus*). In *The Lion-tailed macaque: status and conservation.* (ed. P. Heltne) pp. 109–130. Alan R. Liss, New York.

Kumar, D. (1967). Hormonal regulation of myometrial activity: clinical implications. In *Cellular biology of the uterus.* (ed. R. M. Winn), pp. 449–474. Appleton-Century-Crofts, New York.

Kumar, R., Harper, M. J. K., and Hanahan, D. J. (1988). Occurrence of platelet-activating factor in rabbit spermatozoa. *Arch. Biochem. Biophys.*, **260**, 497–502.

Kummer, H. (1968). Social organization of hamadryas baboons: a field study. Bibliotheca Primatologia., **6**, 1–189. Karger, Basel.

Kummer, H. (1971). *Primate Societies.* Aldine-Atherton, Chicago.

Kummer, H. (1984). From laboratory to desert and back: a social system of hamadryas baboons. *Anim. Behav.*, **32**, 965–971.

Kummer, H. (1990). The social system of hamadryas baboons and its presumable evolution. In *Baboons: behavior and ecology, use and care.* (eds. M. T. de Mello, A. Whtten and R. W. Byrne). pp. 43–60. Brasilia, Brasil.

Kurland, J. A. (1977). Kin selection in the Japanese monkey. *Contributions to Primatology*, Vol. 12. Karger, Basel.

Laidler, L. (1982). *Otters in Britain.* David and Charles, London.

Lal, S., Tesfaye, Y., Tharundayil, J., Thompson, T. R., Kiely, M. E., Nair, N. P. V., *et al.* (1989). Apomorphine: clinical studies on erectile impotence and yawning. *Prog. Neuropsychopharm. Biol. Psychiat.*, **13**, 329–339.

Lamano Carvalho, T. L., Hudson, N. P., Blank, M. A., Watson, P. F., *et al.* (1986). Occurrence, distribution and origin of peptide-containing nerves of guinea pig and rat male genitalia and the effects of denervation on sperm characteristics. *J. Anat.*, **149**, 121–141.

Lancaster, J. B. and Lee, R. B. (1965). The annual reproductive cycle in monkeys and apes. In *Primate behavior: field studies of monkeys and apes* (ed. I. DeVore), pp. 486–513. Holt, Rinehart and Winston, New York.

Langevin, R., Hucker, S. J., Handy, L., Purins, J. E., Russon, A. E., and Hook, H. J. (1985a). Erotic preference and aggression in pedophilia: a comparison of heterosexual, homosexual and bisexual types. In *Erotic preference, gender identity and aggression in men: new research studies.* (ed. R. Langerin), pp. 137–160. Lawrence Erlbaum, Hillsdale, New Jersey.

Langevin, R., Paitich, D., and Russon, A. E. (1985b). Are rapists sexually anomalous, aggressive or both? In *Erotic preference, gender identity and aggression in men: new research studies.* (ed. R. Langerin), pp. 17–38. Lawrence Erlbaum, Hillsdale, New Jersey.

Langfeldt, T. (1981). Sexual development in children. In *Adult sexual interest in children.* Academic Press, London.

Langlois, J. R. and Roggman, L. A. (1990). Attractive faces are only average. *Psychol. Sci.*, **1**, 115–121.

Langlois, J. R., Roggman, L. A., and Musselman, L. (1994). What is average and what is not average about attractive faces? *Psychol. Sci.*, **5**, 214–220.

Larsson, K. (1956). Conditioning and sexual behavior in the male albino rat. Almquist and Wiksell, Stockholm.

Larsson, K. (1979). Features of the neuroendocrine regulation of masculine sexual behavior. In *Endocrine control of sexual behavior.* (ed. C. Beyer), pp. 77–163. Raven Press, New York.

Larsson, K. and Södersten, P. (1973). Mating in male rats after section of the dorsal penile nerve. *Physiol. Behav.*, **10**, 567–571.

Larsson, K., Södersten, P., and Beyer, C. (1973). Sexual behavior in male rats treated with estrogen in combination with dihydrotestosterone. *Horm. Behav.*, **4**, 289–299.

Larsson, K., Fuxe, K., Everitt, B. J., Holmgrem, M., and Södersten, P. (1978). Sexual behavior in male rats after intracerebral infusion injection of 5,7-dihydroxytryptamine. *Brain Res.*, **131**, 293–303.

Laschet, U. and Laschet, L. (1975). Antiandrogens in the

treatment of sexual deviations in men. *J. Steroid Biochem.*, **6**, 821–826.

Latta, J., Hop, S., and Ploog, D. (1967). Observation on mating behavior and sexual play in the squirrel monkey (*Saimiri sciureus*). *Primates*, **8**, 229–246.

Lau, L. C., Adaikan, P. G., and Ratnam, S. S. (1991). Effect of endothelin-1 on the human corpus cavernosum and penile vasculature. *Asian Pac. J. Pharmacol.*, **6**, 287–292.

Law, T. and Meagher, W. (1958). Hypothalamic lesions and sexual behavior in the female rat. *Science*, **128**, 1626–1627.

Le Boeuf, B. J. (1974). Male-male competition and reproductive success in elephant seals. *Am. Zool.*, **14**, 163–176.

Le Magnen, J. (1952). Les phénoménes olfacto-sexuels chez le rat blanc. *Arch. Sci. Physiol.*, **6**, 295–331

Le Vay, S. (1991). A difference in hypothalamic structure between heterosexual and homosexual men. *Science*, **253**, 1034–1037.

Le Vay, S. (1993). *The sexual brain*. MIT Press, Cambridge, Massachusetts.

Lehrman, D. S. (1965). Interaction between internal and external environments in the regulation of the reproductive cycle of the ring dove. In *Sex and behavior*. (ed. F. A. Beach), pp. 355–380. Wiley, New York.

Leipheimer, R. E. and Sachs, B. D. (1993). Relative androgen sensitivity of the vascular and striated-muscle systems regulating penile erection in rats. *Physiol. Behav.*, **54**, 1085–1090.

Leland, L., Strusaker, T. T., and Butynski, T. (1984). Infanticide by adult males in three primate species of Kibale Forest, Uganda: a test of hypotheses. In *Infanticide: comparative and evolutionary perspectives*. (eds. G. Hausfatcr and S. B. Hrdy), pp. 151–171. Adline, New York.

Lemmon, W. B. and Oakes, E. (1967). Tieing between stumptailed macaques during mating. *Lab. Primate Newsletter*, **6**, 14–15.

Lepor, H. and Kuhar, M. J. (1984). Characterization of muscarinic cholinergic receptor binding in the vas deferens, bladder, prostate and penis of the rabbit. *J. Urol.*, **132**, 392–396.

Lernould, J. M. (1988). Classification and geographical distribution of guenons: a review. In *A primate radiation: evolutionary biology of the African guenons*. (eds. A. Gautier-Hion, F. Boulière, J. P. Gautier and J. Kingdon), pp. 54–78. Cambridge University Press.

Leslie, J. F. and Dingle, H. (1983). Interspecific hybridization and genetic divergence in milkweed bugs (*Oncopeltus: Hemiptera: LygaeidaeA*). *Evol.*, **37**, 583–591.

Leutenneger, W. and Kelly, J. T. (1977). Relationship of sexual dimorphism in canine size and body size to social, behavioral and ecological correlates in anthropoid primates. *Primates*, **18**, 117–136.

Levin, R. J. and Wagner, G. (1985). Orgasm in women in the laboratory-quantitative studies on duration, intensity, latency and vaginal blood flow. *Arch. Sex. Behav.*, **14**, 439–450.

Levin, R. M. and Wein, A. J. (1980). Adrenergic alpha-receptors outnumber beta-receptors in human penile corpus cavernosum. *Invest. Urol.*, **18**, 225–226.

Lewontin, R. C. (1983). Gene, organism and environment. In *Evolution from molecules to men*. (ed. D. S. Bendall), pp. 273–285. Cambridge University Press.

Liers, E. E. (1951). Notes on the river otter (*Lutra canadensis*). *J. Mammal.*, **32**, 1–9.

Lightfoot-Klein, H. (1989). The sexual experience and marital adjustment of genitally circumcised and infibulated females in the Sudan. *J. Sex. Res.*, **26**, 375–392.

Lincoln, G. A. (1974). Luteinizing hormone and testosterone in man. *Nature Lond.*, **252**, 232–233.

Lincoln, G. A. and Ssewannyana, E. (1989). Opioid peptides and seasonal reproduction. In *Brain opioid systems and reproduction*. (eds. R. G. Dyer and R. J. Bicknell), pp. 52–69. Oxford University Press.

Lindburg, D. G. (1967). A field study of the reproductive behavior of the rhesus monkey (*Macaca mulatta*). Unpublished Ph.D. thesis, University of California.

Lindburg, D. G. (1969). Rhesus monkeys: mating season mobility of adult males. *Science*, **166**, 1176–1178.

Lindburg, D. G. (1971). The rhesus monkey in North India: an ecological and behavioral study. In *Primate behavior*, Vol. 2. (ed. L. A. Rosenblum), pp. 1–106. Academic Press, New York.

Lindburg, D. G. (1983). Mating behavior and estrus in the Indian rhesus monkey. In *Perspectives in primate biology*. (ed. P. K. Seth), pp. 45–61. Today and Tomorrow, New Delhi.

Lindburg, D. G. (1986). Seasonality of reproduction in primates. In *Comparative primate biology*, Vol. 2. Part B (eds. G. Mitchell and J. Erwin), pp. 167–218. Alan R. Liss, New York.

Lindburg, D. G., Shideler, S., and Fitch, H. (1985). Sexual behavior in relation to the time of ovulation in the lion-tailed macaque. In *The Lion-tailed macaque: status and conservation*. (ed. P. G. Heltne), pp. 131–148. Alan R. Liss, New York.

Lindsay, D. R. (1984). Quantitative requirements of females for spermatozoa and output of spermatozoa by rams. In *The male in farm animal reproduction*. (ed. M. Courot), pp. 324–338. Martinus Nighoff, Dordrecht, Netherlands.

Lindsay, D. R. and Fletcher, I. C. (1972). Ram seeking activity associated with oestrous behaviour in ewes. *Anim. Behav.*, **20**, 452–456.

Linn, G. S., Mase, D., Lafrancois, D., O'Keeffe, R. T., and Lifshitz, K. (1995). Social and menstrual group influences on the behavior of group-housed *Cebus apella*. *Am. J. Primatol.*, **35**, 41–57.

Lippert, W. (1974). Beobachtungen zum Schwangerschafts und Geburtsverhalten beim Orang-utan (*Pongo pygmaeus*) im Tierpark Berlin. *Folia Primatol.*, **21**, 108–134.

Lippert, W. (1977). Erfaurungen bei der Aufzucht Von Orang-utans (*Pongo pygmaeus*) im Tierpark Berlin. *Zool. Garten* (NF), **47**, 209–225.

Lipschitz, D. W. (1988). Endocrine factors determining female *Galago senegalensis moholi* mating behaviour. Unpublished Master of Science thesis, University of Witwatersrand, South Africa.

Lipschitz, D. W. (1994). Endocrine factors influencing female *Galago moholi* mating behaviour. Unpublished Ph.D. Thesis, University of Witwatersrand, South Africa.

Lish, J. D., Meyer-Bahlburg, H. F. L., Ehrhardt, A. A., Travis, B. G, and Vendiano, N. P. (1992). Prenatal exposure to diethylstilbestrol (DES): childhood play behavior and adult gender-role behavior in women. *Arch. Sex. Behav.*, **21**, 423–441.

Lisk, R. D. (1978). The regulation of sexual 'heat'. In *Biological determinants of sexual behaviour*. (ed. J. B. Hutchison), pp. 425–466. John Wiley and Sons, Chichester.

Lloyd, H. G. (1980). *The red fox*. Batsford, London.

Lloyd, J. E. (1979). Mating behavior and natural selection. *Florida Entomologist.*, **62**, 17–23.

Lloyd, S. A. C. and Dixson, A. F. (1988). Effects of hypothalamic lesions upon the sexual and social behaviour of the male common marmoset (*Callithrix jacchus*). *Brain Res.*, **463**, 317–329.

Lloyd, S. A. C. and Dixson, A. F. (1989). Effects of hypothalamic lesions upon sexual and aggressive behaviour in the common marmoset (*Callithrix jacchus*). *Primate Report*, **23**, 41–50.

Loireau, J. N. and Gautier-Hion, A. (1988). Olfactory marking behaviour in guenons and its implications. In *A primate radiation: evolutionary biology of the African guenons*. (eds. A. Gautier-Hion, F. Boulière, J. P Gautier, and J. Kingdon), pp. 246–253. Cambridge University Press.

Long, C. A. and Frank, T. (1968). Morphometric function and variation in the baculum with comments on the correlation of parts. *J. Mammal.*, **49**, 32–43.

Lorenz, K. (1950). The comparative method in studying innate behaviour patterns. *Symp. Soc. Exp. Biol.*, **4**, 221–268.

Lovejoy, J. C. and Wallen, K. (1990). Adrenal suppression and sexual initiation in group-living female rhesus monkeys. *Horm. Behav*.

Lovejoy, O. (1981). The origins of man. *Science*, **211**, 241–250.

Lowsley, O. S., Hinman, F., Smith, D. R., and Gutierrez, R. (1942). *The sexual glands of the male*. New York.

Loy, J. (1971). Estrous behavior of free-ranging rhesus monkeys. (*Macaca mulatta*). *Primates*, **12**, 1–31.

Loy, J. (1987). The sexual behavior of African monkeys and the question of estrus. In *Comparative behavior of African monkeys*. (ed. E. Zucker), pp. 175-195. Alan R. Liss, New York.

Loy, J. and Loy, K. (1977). Sexual harassment among captive patas monkeys (*Erythrocebus patas*). *Primates*, **18**, 691–699.

Lue, T. and Tanagho, E. (1988). Functional anatomy and mechanism of penile erection. In *Contemporary management of impotence and infertility*. (eds. E. A. Tangho, T. Lue and R. D. McClure), pp. 39–50. Williams and Wilkins, Baltimore.

Lundberg, J. M., Höfelt, T., Anggard, A., Uvnas-Wallensten, K., Brimijoin, S., Brodin, E., and Fahrenkrug, J. (1980). Peripheral peptide neurons: distribution, axonal transport, and some aspects of possible function. In *Neural peptides and neuronal communication* (eds E. Costa and M. Trabucchi), pp. 63–82. Raven Press, New York.

Lunn, S. F., Recio, R., Morris, K., and Fraser, H. M. (1994). Blockade of the neonatal rise in testosterone by a gonadotrophin-releasing hormone antagonist: effects on timing of puberty and sexual behaviour in the male marmoset monkey. *J. Endocr.*, **141**, 439–447.

Luttge, W. G. (1979). Endocrine control of mammalian male sexual behavior: an analysis of the potential role of testosterone metabolites. In *Endocrine control of sexual behavior* (ed. C. Beyer), pp. 341–363. Raven Press, New York.

Luttge, W. G. and Hall, N. R. (1973). Differential effectiveness of testosterone and its metabolites in the induction of male sexual behavior in two strains of albino mice. *Horm. Behav.*, **4**, 31–43.

Maccoby, E. E. and Jacklin, C. N. (1974). *The Psychology of sex differences*. Stanford University Press, Stanford, California.

Machida, H. and Giacometti, L. (1967). The anatomical and histochemical properties of the skin of the external genitalia of the primates. *Folia Primatol.*, **6**, 48–69.

Mack, D. and Kafka, H. (1978). Breeding and rearing of woolly monkeys at the National Zoological Park, Washington. *Int. Z. Yrbk.*, **18**, 117–122.

Mackinnon, J. (1974). The ecology and behaviour of wild orang-utans (*Pongo pygmaeus*). *Anim. Behav.*, **22**, 3–74.

Mackinnon, J. and Mackinnon, K. (1980). The behaviour of wild spectral tarsiers. *Int. J. Primatol.*, **1**, 361–379.

Maclean, P. D. (1964). Mirror display in the squirrel monkey. *Science.*, **146**, 950–952.

Maclean, P. D. and Ploog, D. W. (1962). Cerebral representation of penile erection. *J. Neurophysiol.*, **25**, 29–55.

Macrides, F., Fernandez, F., and D'Angelo, W. (1974). Effects of exposure to vaginal odor and receptive females on plasma testosterone in the male hamster. *Neuroendocrinology*, **15**, 355–364.

Macrides, F., Bartke, A., and Dalterio, S. (1975). Strange females increase testosterone levels in male mice. *Science*, **189**, 1104–1105.

Malinowski, B. (1932). The sexual life of savages in North Western Melanesia. Routeledge and Sons, London.

Mlmnas, C. O. (1977). Short-latency effects of testosterone on copulatory behaviour and ejaculation in sexually experienced intact male rats. *J. Reprod. Fertil.*, **51**, 351–354.

Mangat, H. K., Chhina, G. S., Singh, B., and Anand, B. K. (1978a). Effect of testosterone propionate on electrical activity of brain in intact and gonadectomized rhesus monkeys. *Indian. J. Exp. Biol.*, **16**, 893–896.

Mangat, H. K., Chhina, G. S., Singh, B., and Anand, B. K. (1978b). Influence of gonadal hormones and genital afferents on E.E.G. activity of the hypothalamus in adult male rhesus monkeys. *Physiol. Behav.*, **20**, 377–384.

Manley, G. H. (1966). Reproduction in lorisoid primates. *Symp. Zool. Soc. Lond.*, **15**, 493–509.

Manley, G. H. (1967). Gestation periods in the Lorisidae. *Int. Z. Yrbk.*, **1**, 80–81.

Mann, D. R. and Fraser, H. M. (1996). The neonatal period: a critical interval in male primate development. *J. Endocr.*, **149**, 191–197.

Mann, D. R., Gould, K. G., Collins, D. C. , and Wallen, K. (1989). Blockade of neonatal activation of the pituitary testicular axis: effect on peripubertal luteinizing hormone and testosterone secretion and on testicular development in male monkeys. *J. Clin. Endocr. Metab.*, **68**, 600–607.

Mann, T. (1964). The biochemistry of semen and of the male reproductive tract. Methuen, London.

Mann, T. and Lutwak-Mann, C. (1981). *Male reproductive function and semen*. Springer, Berlin.

Manning, A. and Thompson, M. L. (1976). Postcastration retention of sexual behaviour in the male BDF mouse: the role of experience. *Anim. Behav.*, **24**, 523–533.

Manning, J. T. and Chamberlain, A. T. (1993). Fluctuating

asymmetry, sexual selection and canine teeth in primates. *Proc. Roy. Soc. Lond. B.*, **251**, 83–87.

Manning, J. T. and Chamberlain, A. T. (1994). Fluctuating asymmetry in gorilla canines: a sensitive indicator of environmental stress. *Proc. Roy. Soc. Lond. B.*, **255**, 189–193.

Manning, J. T. and Hartley, M. A. (1991). Symmetry and ornamentation are correlated in the peacock's train. *Anim. Behav.*, **42**, 1020–1021.

Manogue, K. R., Leshner, A. I., and Candland, D. K. (1975). Dominance status adrenocortical reactivity to stress in squirrel monkeys (*Saimiri sciureus*). *Primates*, **16**, 457–463.

Manson, J. H. (1992). Measuring female mate choice in Cayo Santiago rhesus macaques. *Anim. Behav.*, **44**, 405–416.

Manson, J. H. (1993). Sons of low-ranking female rhesus macaques can attain high dominance rank in their natal groups. *Primates*, **34**, 285–288.

Manson, J. H. (1994). Mating patterns, mate choice and birth season heterosexual relationships in free-ranging rhesus macaques. *Primates*, **35**, 417–433.

Maple, T. (1977). Unusual sexual behavior of nonhuman primates. In *Handbook of sexology*. (eds. J. Money and H. Mustaph), pp. 1167–1186. Excerpta Medica, Amsterdam.

Markham, R. (1995). Doing it naturally: reproduction in captive orang-utans (*Pongo pygmaeus*). In *The neglected ape*. (eds. R. D. Nadler, B. F. M. Galdikas, L. K. Sheeran and N. Rosen), pp. 273–278. Plenum Press, New York.

Marler, P. (1969). *Colobus guereza*: territoriality and group composition. *Science*, **163**, 93–95.

Marler, P. (1972). Vocalizations of East African monkeys, 2. black and white colobus. *Behaviour*, **42**, 175–197.

Marler, P. and Hobbett, L. (1975). Individuality in a long-range vocalization of chimpanzees. *Z. Tierpsychol.*, **38**, 97–109.

Marlow, B. J. (1961). Reproductive behaviour of the marsupial mouse, *Antechinus flavipes*. (Waterhouse) (Marsupialia) and the development of the pouch young. *Aust. J. Zool.*, **9**, 203–217.

Marmor, J. and Green, R. (1977). Homosexual behavior. In *Handbook of sexology*. (eds. J. Money and H. Mustaph), pp. 1051–1068. Excerpta Medica, Amsterdam.

Marshal, E. A. and Tanner, J. M. (1969). Variations in patterns of pubertal changes in girls. *Arch. Dis. Childhood*, **44**, 291–303.

Marshall, F. H. A. (1911). The male generative cycle in the hedgehog. *J. Physiol.*, **43**, 247.

Marshall, P. E. and Goldsmith, P. C. (1980). Neuroregulatory and neuroendocrine GnRH pathways in the hypothalamus and forebrain of the baboon. *Brain Res.*, **193**, 353–372.

Marson, J., Gervais, D., Cooper, R. W., and Jouannet, P. (1989). Influence of ejaculation frequency on semen characteristics in chimpanzees (*Pan troglodytes*). *J. Reprod. Fertil.*, **85**, 43–50.

Martan, J. and Shepherd, B. A. (1976). The role of copulatory plugs in reproduction in the guinea pig. *J. Exp. Zool.*, **196**, 79–83.

Martel, F. L., Nevison, C. M., Rayment, F. D., Simpson, M. J. A., and Keverne, E. B. (1993). Opioid receptor blockade reduces maternal affect and social grooming in rhesus monkeys. *Psychoneuroendocrinology*, **18**, 307–321.

Martensz, N. D. and Everitt, B. J. (1982). Effects of passive immunisation against testosterone on the sexual activity of female rhesus monkeys. *J. Endocr.*, **94**, 271–282.

Martensz, N. D., Vellucci, S. V., Keverne, E. B., and Herbert, J. (1986). β-endorphin levels in the cerebrospinal fluid of male talapoin monkeys in social groups related to dominance status and luteinizing hormone response to naloxone. *Neuroscience*, **18**, 651–658.

Martin C. E. (1981). Factors affecting sexual functioning in 60–79 year old married males. *Arch. Sex. Behav.*, **10**, 399–420.

Martin, D. E. and Gould, K. G. (1981). The male ape genital tract and its secretions. In *Reproductive biology of the great apes* (ed. C. E. Graham), pp. 127–162. Academic Press, New York.

Martin, J. H. (1985). Receptor physiology and submodality coding in the somatic sensory system. In *Principles of neural science*. (2nd edition). (eds. E. R. Kandel and J. H. Schwartz), pp. 287–300. Elsevier, New York.

Martin, P. A., Reimers, T. J., Lodge, J. R., and Dzuik, P. (1974). The effect of ratios and numbers of spermatozoa mixed from two males on proportions of offspring. *J. Reprod. Fertil.*, **39**, 251–258.

Martin, R. D. (1972). Adaptive radiation and behaviour of the Malagasy lemurs. *Phil. Trans. Roy. Soc. Lond. Ser. B.*, **264**, 295–352.

Martin, R. D. (1973). A review of the behaviour and ecology of the lesser mouse lemur (*Microcebus murinus* J. F. Miller, 1777). In *Comparative ecology and behaviour of primates*. (eds. R. P. Michael and J. H. Crook), pp. 1–68. Academic Press, London.

Martin, R. D. (1981). Field studies of primate behaviour. *Symp. Zool. Soc. Lond.*, **46**, 287–336.

Martin, R. D. (1990). Primate origins and evolution: a phylogenetic reconstruction. Chapman and Hall, London.

Martin, R. D. (1993). Primate origins—plugging the gaps. *Nature Lond.*, **363**, 223–234.

Martin, R. D. (1995). Phylogenetic aspects of primate reproduction: the context of advanced maternal care. In *Motherhood in human and non-human primates*. (eds. C. R. Pryce, R. D. Martin and D. Skuse), pp. 16–26. Karger, Basel.

Martin, R. D., Dixson, A. F., and Wickings, E. J. (1992). Paternity in primates, genetic testis and theories: implications of human DNA fingerprinting. Karger, Basel.

Martin, R. D., Willner, L. A., and Dettling, A. (1994). The evolution of sexual size dimorphism in primates. In *The differences between the sexes*. (eds. R. V. Short and E. Balaban), pp. 159–200. Cambridge University Press.

Maslow, A. H. (1936). The role of dominance in the social and sexual behavior of infra-human primates: III. a theory of sexual behavior of infra-human primates. *J. Genet. Psychol.*, **48**, 310–338.

Maslow, A. H. (1936–37). The role of dominance in the social and sexual behavior of infra-human primates. *J. Gen. Psychol.*, **48**, 261–277; **48**, 310–338; **49**, 161–198.

Mason, W. A. (1960). The effects of social restriction on the behavior of rhesus monkeys: 1. Free social behavior. *J. Comp. Physiol. Psychol.*, **53**, 582–589.

Mason, W. A. (1966). Social organization of the South American monkey, *Callicebus moloch*: a preliminary report. *Tulane Stud. Zool.*, **13**, 23–28.

Mason, W. A. (1968). Use of space by *Callicebus* groups. In *Primates: studies in adaptation and variability* (ed. P. C. Jay), pp. 200–216. Holt, Rinehart and Winston, New York.

Mason, W. A. and Kenney, M. D. (1974). Redirection of filial attachments in rhesus monkeys: dogs as mother surrogates. *Science*, **183**, 1209–1211.

Masters, W. H. and Johnson, V. E. (1966). *Human sexual response*. J. and A. Churchill, London.

Masters, W. H. and Johnson, V. E. (1979). *Homosexuality in perspective*. Little, Brown, Boston.

Mathews, D., Greene, S. B., and Hollingsworth, E. M. (1983). VMN lesion deficits in lordosis: partial reversal with pergolide mesylate. *Physiol. Behav.*, **31**, 745–748.

Matsumoto, A. and Arai, Y. (1980). Sexual dimorphism in wiring pattern in the hypothalamic arcuate nucleus and its modification by neonatal hormonal environment. *Brain Res.*, **190**, 238–242.

Matsumoto, A. and Arai, Y. (1983). Sex difference in volume of the ventromedial nucleus of the hypothalamus of the rat. *Endocr. Japan*, **30**, 277–280.

Matsumoto, A. and Arai, Y. (1986a). Development of sexual dimorphism in synaptic organization in the ventromedial nucleus of the hypothalamus in rats. *Neurosci. Lett.*, **68**, 165–168.

Matsumoto, A. and Arai, Y. (1986b). Male-female difference in synaptic organization of the ventromedial nucleus of the hypothalamus in the rat. *Neuroendocrinology*, **42**, 232–236.

Matsumura, S. (1993). Female reproductive cycles and the sexual behavior of moor macaques (*Macaca maurus*) in their natural habitat, South Sulawesi, Indonesia. *Primates*, **34**, 99–103.

Matthews, L. H. (1939). Reproduction in the spotted hyena, *Crocuta crocuta* (*Erxleben*). *Phil. Trans. Roy. Soc. B.*, **230**, 1–78.

Matthews, L. H. (1946). Notes on the genital anatomy and physiology of the gibbon (*Hylobates*). *Proc. Zool. Soc. Lond.*, **116**, 339–364.

Matthews, L. H. (1956). The sexual skin of the gelada baboon (*Theropithecus gelada*). *Trans. Zool. Soc. Lond.*, **28**, 543–552.

Matthews, M. K. and Adler, N. T. (1977). Facilitative and inhibitory influences of reproductive behavior on sperm transport in rats. *J. Comp. Physiol. Psychol.*, **91**, 727–741.

Matthews, M. K. and Adler, N. T. (1978). Systematic interrelationships of mating, vaginal plug position, and sperm transport in the rat. *Physiol. Behav.*, **20**, 303–309.

Matuszczyk, J. V., Silverin, B., and Larsson, K. (1990). Influence of environmental events immediately after birth on postnatal testosterone secretion and adult sexual behavior in the male rat. *Horm. Behav.*, **24**, 450–458.

Maynard Smith, J. (1982). *Evolution and the theory of games*. Cambridge University Press.

Mayr, E. (1963). *Animal species and evolution*. Harvard University Press, Cambridge, Massachusetts.

McArthur, J. W., Beitins, I. Z., Gorman, A., Collins, D. C. , Preedy, J. R. K., and Graham, C. E. (1981). The interrelationships between sex skin swelling and the urinary excretion of gonadotropins, estrone and pregnanediol by the cycling female chimpanzee. *Am. J. Primatol.*, **1**, 265–276.

McClintock, M. K. (1971). Menstrual synchrony and suppression. *Nature Lond.*, **229**, 244–245.

McClintock, M. K. (1978). Estrous synchrony and its mediation by airborne chemical communication in *Rattus norvegicus*. *Horm. Behav.*, **10**, 264–276.

McClintock, M. K. (1984). Estrous synchrony: modulation of ovarian cycle length by female pheromones. *Physiol. Behav.*, **32**, 701–705.

McComb, K. (1987). Roaring by red deer stags advances the date of reproduction in hinds. *Nature Lond.*, **330**, 648–649.

McEwen, B. (1991). Steroids affect neural activity by acting on the membrane and the genome. *Trends Pharmacol. Sci.*, **12**, 141–147.

McEwen, B. S. (1994). Steroid hormone action on the brain: when is the genome involved? *Horm. Behav.*, **28**, 396–405.

McFarland Symington, M. (1987). Sex ratio and maternal rank in wild spider monkeys: when daughters disperse. *Behav. Ecol. Sociobiol.*, **20**, 421–425.

McGill, T. E. (1977). Reproductive isolation, behavioral genetics and functions of sexual behavior in rodents. In *Reproductive behavior and evolution*. (eds. J. S. Rosenblatt and B. R. Komisaruk). Plenum, New York.

McGill, T. E. (1978). Genotype-hormone interactions. In *Sex and behavior: status and prospectus*. (eds. T. E. McGill, D. A. Dewsbury and B. D. Sachs), pp. 161–187. Plenum, New York.

McGill, T. E. and Manning, A. (1976). Genotype and relation of the ejaculation reflex in castrated male mice. *Anim. Behav.*, **24**, 507–518.

McGinnis, P. R. (1973). Patterns of sexual behaviour in a community of free-living chimpanzees. Unpublished Ph.D. thesis, University of Cambridge.

McGinnis, P. (1979). Sexual behavior in free-living chimpanzees: consort relationships. In *The great apes*. (eds. D. A. Hamburg and E. R. McCown), pp. 429–440. Benjamin Cummings, Menlo Park, California.

McGlone, J. (1980). Sex differences in human brain asymmetry: a critical survey. *Behav. Brain Sci.*, **3**, 215–264.

McGregor, A. and Herbert, J. (1992). Specific effects of β-endorphin infused into the amygdala on sexual behaviour in the male rat. *Neuroscience*, **46**, 165–172.

McGrew, W. C. (1992). *Chimpanzee material culture, implications for human evolution*. Cambridge University Press.

McMillan, C. A. (1989). Male age, dominance and mating success among rhesus macaques. *Am. J. Phys. Anthropol.*, **80**, 83–89.

McNeilly, A. S., Abbot, D. H., Lunn, S. F., Chambers, P., C., and Hearn, J. P. (1981). Plasma prolactin concentrations during the ovarian cycle and lactation and their relationship to return of fertility post partum in the common marmoset. *J. Reprod. Fertil.*, **62**, 253–260.

McWhirter, D. P. and Reinisch, J. (1988). (eds.) *Homosexuality / heterosexuality: the Kinsey scale and current research*. Oxford University Press, New York.

Mead, M. (1967). Male and female: a study of the sexes in a changing world. William Morrow and Co., New York.

Meier, B., Albinac, R., Peyrieras, A., Rumpler, Y. and Wright, P. (1987). A new species of *Hapalemur* (Primates) from south east Madagascar. *Folia Primatol.*, **48**, 211–215.

Meikle, D. B., Tilford, B. L., and Vessey, S. H. (1984). Dominance rank, secondary sex ratio and reproduction of offspring in polygynous primates. *Am. Nat.*, **124**, 173–188.

Meisel, R. L. and Sachs, B. D. (1994). The physiology of male sexual behavior. In *The physiology of reproduction*, Vol. 2. (2nd edition) (eds. E. Knobil and J. D. Neill), pp. 3–105. Raven Press, New York.

Meites, J., Bruni, J. F., Van Vugt, D. A., and Smith, A. F. (1979). Relation of endogenous opioid peptides and morphine to neuroendocrine functions. *Life. Sci.*, **24**, 1325.

Melis, M. R., Argiolas, A., and Gessa, G. L. (1987). Apomorphine-induced penile erection and yawning: site of action in the brain. *Brain Res.*, **415**, 98–107.

Meller, R. E., Keverne, E. B., and Herbert, J. (1980). Behavioural and endocrine effects of naltrexone in male talapoin monkeys. *Pharmacol. Biochem. Behav.*, **13**, 663–672.

Ménard, N., Scheffrahn, W., Vallet, D., Zidane, C., and Reber, C. (1992). Application of blood protein electrophoresis and DNA fingerprinting to the analysis of paternity and social characteristics of wild barbary macaques. In *Paternity in Primates: genetic tests and theories*. (eds. R. D. Martin, A. F. Dixson and E. J. Wickings), pp. 155–174. Karger, Basel.

Mendes, S. L. (1985). Uso do espago, padroes de atividades diarias e organizacao social de *Alouatta fusca* (Primates, Cebedae) em Caratinga-MG. Unpublished Masters Thesis, University of Brazilia, Brasilia.

Mendoza, S. P. and Mason, W. A. (1989). Behavioural and endocrine consequences of heterosexual pair formation in squirrel monkeys. *Physiol. Behav.*, **46**, 597–603.

Mendoza, S. P. and Mason, W. A. (1991). Breeding readiness in squirrel monkeys: female-primed females are triggered by males. *Physiol. Behav.*, **49**, 471–479.

Mendoza, S. P., Lowe, E. L., Resko, J. A., and Levine, S. (1978). Seasonal variations in gonadal hormones and social behavior in squirrel monkeys. *Physiol. Behav.*, **20**, 515–522.

Mendoza, S. P., Coe, C. L., Lowe, E. L., and Levine, S. (1979). The physiological response to group formation in adult male squirrel monkeys. *Psychoneuroendocrinology*, **3**, 221–229.

Merari, A. and Ginton, A. (1975). Characteristics of exaggerated sexual behavior induced by electrical stimulation of the medial preoptic area in male rats. *Brain Res.*, **86**, 97–108.

Mertl-Millhollen, A. S. (1979). Olfactory demarcation of territorial boundaries by a primate, *Propithecus verreauxi*. *Folia Primatol.*, **32**, 35–42.

Messenger, J. C. (1971). Sex and repression in an Irish folk community. In *Human sexual behavior*. (eds. D. S. Marshall, and R. C Suggs), pp. 3–37. Basic Books, New York.

Meyer-Bahlburg, H. F. L. (1993). Psychobiological research on homosexuality. *Child and Adolescent Psychiatric Clinics of North America.*, **2**, 489–500.

Meyer-Bahlburg, H. F. L., Ehrhardt, A. A., Rosen, L. R., Feldman, J. F., Veridiano, N. P., Zimmerman, I., *et al.* (1984). Psychosexual milestones in women prenatally exposed to diethylstilbestrol. *Horm. Behav.*, **18**, 359–366.

Meyer-Bahlburg, H. F. L., Ehrhardt, A. A., Rosen, L. R., Gruen, R. S., Veridiano, N. P, Vann, F. H., *et al.* (1995). Prenatal estrogens and the development of homosexual orientation. *Devel. Psychol.*, **31**, 12–21.

Meyerson, B. J. (1981). Comparison of the effects of beta-endorphin and morphine on exploratory and socio-sexual behavior in the male rat. *Eur. J. Pharmacol.*, **69**, 453–463.

Meyerson, B., Malmnas, C. O., and Everitt, B. J. (1985).

Neuropharmacology, neurotransmitters and sexual behavior in mammals. In *Handbook of behavioral neurobiology*, Vol. 7. (eds. N. Adler, D. Pfaff and R. W. Goy), pp. 495–536. Plenum, New York.

Michael, R. P. (1968). Gonadal hormones and the control of primate behaviour. In *Endocrinology and human behaviour*. (ed. R. P. Michael), pp. 69–93. Oxford University Press, London.

Michael, R. P. (1969). Neural and non-neural mechanisms in the reproductive behavior of primates. In *Progress in endocrinology*. (ed. C. Gjal), pp. 302–309. Excerpta Medica, Amsterdam.

Michael, R. P. (1971). Neuroendocrine factors regulating primate behaviour. In *Frontiers in neuroendocrinology* (eds. L. Martini and W. F. Ganong), pp. 359–398 Oxford University Press.

Michael, R. P. and Bonsall, R. W. (1977). Chemical signals and primate behavior. In *Chemical signals and vertebrates*. (eds. D. Muller-Schwarze and M. M. Mozell), pp. 251–271. Plenum, New York.

Michael, R. P. and Bonsall, R. W. (1990). Androgens, the brain and behavior in male primates. In *Hormones, Brain and behaviour in male vertebrates, Vol. 2. Behavioural activation in males and females—social interaction and reproductive endocrinology.* (ed. J. Balthazart), pp. 15–26. Karger, Basel.

Michael, R. P. and Keverne, E. B. (1968). Pheromones and the communication of sexual status in primates. *Nature Lond.*, **218**, 746–749.

Michael, R. P. and Keverne, E. B. (1970). Primate sex pheromones of vaginal origin. *Nature Lond.*, **225**, 84–85.

Michael, R. P. and Saayman, G. S. (1968). Differential effects on behaviour of sub-cutaneous and intravaginal administration of oestrogen in the rhesus monkey (*Macaca mulatta*). *J. Endocr.*, **41**, 231–246.

Michael, R. P. and Wilson, M. (1973). Effects of castration and hormone replacement in fully adult male rhesus monkeys (*Macaca mulatta*). *Endocrinology*, **95**, 150–159.

Michael, R. P. and Zumpe, D. (1977). Effects of androgen administration on sexual initiations by female rhesus monkeys (*Macaca mulatta*). *Anim. Behav.*, **25**, 936–944.

Michael, R. P. and Zumpe, D. (1983). Sexual violence in the United States and the role of season. *Am. J. Psychiat.*, **140**, 883–886.

Michael, R. P. and Zumpe, D. (1993). Medroxyprogesterone acetate decreases the sexual activity of male cynomolgus monkeys (*Macaca fascicularis*): an action on the brain? *Physiol. Behav.*, **53**, 783–788.

Michael, R. P., Herbert, J., and Wellegalla, J. (1966). Ovarian hormones and grooming behaviour in the rhesus monkey (*Macaca mulatta*) under laboratory conditions. *J. Endocr.*, **36**, 263–279.

Michael, R. P., Herbert, J., and Wellegalla, J. (1967). Ovarian hormones and the sexual behaviour of the male rhesus monkey (*Macaca mulatta*) under laboratory conditions. *J. Endocr.*, **39**, 81–98.

Michael, R. P., Saayman, G. S., and Zumpe, D. (1968). The suppression of mounting behaviour and ejaculation in male rhesus monkeys by administration of progesterone to their female partners. *J. Endocr.*, **41**, 421–431.

Michael, R. P., Plant, T. M., and Wilson, M. I. (1973). Preliminary studies on the effects of cyproterone acetate on sexual activity and testicular function in adult male rhesus monkeys (*Macaca mulatta*). In *Advances in Biosciences*, Vol. 10. (eds. G. Raspe and S. Bernhard), pp. 197–208. Pergamon, Press, Oxford.

Michael, R. P., Wilson, M. I., and Zumpe, D. (1974). The bisexual behavior of female rhesus monkeys. In *Sex differences in behavior*. (eds. R. C. Friedman, R. M. Richart., and R. L. Vande Wide), pp. 399–412. Wiley, New York.

Michael, R. P., Bonsall, R. W., and Kutner, M. (1975). Volatile fatty acids, 'copulins', in human vaginal secretions. *Psychoneuroendocrinology.*, **1**, 153–163.

Michael, R. P., Bonsall, R. W., and Zumpe, D. (1976). Chemical communication among primates. *Vitam. Horm. (N.Y.)*, **34**, 137–186.

Michael, R. P., Keverne, E. B., and Bonsall, R. W. (1977a). Pheromones: isolations of male sex attractants from a female primate. *Science*, **172**, 964–966.

Michael, R. P., Zumpe, D., Richter, M., and Bonsall, R. W. (1977b). Behavioral effects of a synthetic mixture of aliphatic acids in rhesus monkeys (*Macaca mulatta*). *Horm. Behav.*, **9**, 296–308.

Michael, R. P., Richter, M. C., Cairn, J. R., Zumpe, D., and Bonsall, R. W. (1978). Artificial menstrual cycles, behaviour and the role of androgens in female rhesus monkeys. *Nature Lond.*, **275**, 439–440.

Michael, R. P., Zumpe, D., and Bonsall, R. W. (1984a). Sexual behavior correlates with the diurnal plasma testosterone range in intact male rhesus monkeys. *Biol. Reprod.*, **30**, 652–657.

Michael, R. P., Bonsall, R. W., and Zumpe, D. (1984b). The behavioral thresholds of testosterone in castrated male rhesus monkeys (*Macaca mulatta*). *Horm. Behav.*, **18**, 161–176.

Michael, R. P., Zumpe, D., and Bonsall, R. W. (1986). Comparison of the effects of testosterone and dihydrotestosterone on the behavior of male cynomolgus monkeys (*Macaca fascicularis*). *Physiol. Behav.*, **36**, 349–355.

Michael, R. P., Rees, H. D., and Bonsall, R. W. (1989). Sites in the male primate brain at which testosterone acts as an androgen. *Brain Res.*, **502**, 11–20.

Michael, R. P., Zumpe, D., and Bonsall, R. W. (1990). Estradiol administration and the sexual activity of castrated male rhesus monkeys (*Macaca mulatta*). *Horm. Behav.*, **24**, 71–88.

Miller, G. S. (1933). The classification of gibbons. *J. Mammal.*, **14**, 158–159.

Milton, K. (1984). Habitat, diet, and activity patterns of free-ranging woolly spider monkeys. (*Brachyteles arachnoides*). *Int. J. Primatol.*, **5**, 491–514.

Milton, K. M. (1985a). Mating patterns of woolly spider monkeys, *Brachyteles arachnoides*: implications for female choice. *Behav. Ecol. Sociobiol.*, **17**, 53–59.

Milton, K. M. (1985b). Multimale mating and the absence of canine tooth dimorphism in woolly spider monkeys (*Brachyteles arachnoides*). *Am. J. Phys. Anthropol.*, **68**, 519–523.

Milton, K. and Johnston, P. (1984). Diet and reproduction in New World primates: a review. Paper presented at the 53rd annual meeting of the American Association of Physical Anthropologists, April 11–14, 1984.

Missakian, E. A. (1969). Reproductive behavior of socially deprived adult male rhesus monkeys (*Macaca mulatta*). *J. Comp. Physiol. Psychol.*, **69**, 403–407.

Missakian, E. (1973). Genealogical mating activity in free-ranging groups of rhesus monkeys (*Macaca mulatta*) on Cayo Santiago. *Behaviour*, **45**, 224–240.

Mitani, J. C. (1985). Mating behaviour of male orangutans in the Kutai Reserve, East Kalimantan, Indonesia. *Anim. Behav.*, **33**, 392–402.

Mitani, J. C. and Nishida, T. (1993). Contexts and social correlates of long-distance calling by male chimpanzees. *Anim. Behav.*, **45**, 735–746.

Mitani, J., Hasegawa, T., Gros-Louis, J., Marler, P., and Byrne, R. (1992). Dialects in wild chimpanzees? *Am. J. Primatol.*, **27**, 233–243.

Mitani, J. C., Gros-Louis, J., and Manson, J. H. (1996). Number of males in primate groups: comparative tests of competing hypotheses. *Am. J. Primatol.*, **38**, 315–332.

Mitchell, G. (1977). *Behavioral sex differences in nonhuman primates*. Van Nostrand Reinhold, New York.

Mitchell, J. B. and Stewart, J. (1989). Effects of castration, steroid replacement and sexual experience on mesolimbic dopamine and sexual behaviors in the male rat. *Brain Res.*, **491**, 116–127.

Mittermeier, R. A. (1977). The distribution, synecology and conservation of Surinam monkeys. Unpublished Ph.D. thesis, Harvard University, Cambridge, Massachusetts.

Mittermeier, R. A., Coimbra-Filho, A. F., Rylands, A. B., and Fonseca, G. A. B. (1988). (eds). *Ecology and behavior of neotropical primates*, Vol. 2. World Wildlife Fund, Washington, DC.

Mittermeier, R. A., Schwarz, M., and Ayres, J. M. (1992). A new species of marmoset, genus *Callithrix* Erxleben, 1777 (Callitrichidae, Primates) from the Rio Maues region, state of Amazonas, Central Brazilian Amazonia. *Goeldiana Zool.*, **14**, 1–17.

Mizukami, S., Nishizuka, M., and Arai, Y. (1983). Sexual difference in nuclear volume and its ontogeny in the rat amygdala. *Exp. Neurol.*, **79**, 569–575.

Mogenson, G. J. (1984). Limbic-motor integration, with emphasis on initiation of exploratory and goal-directed locomotion. In *Modulation of sensormotor activity during alterations in behavioral states*. (ed. R. Bandler), pp. 121–137. Alan R. Liss, New York.

Mogenson, G. J., Jones, D. L., and Yim, C. Y. (1980). From motivation to action: functional interface between the motor system and the limbic system. *Progr. in Neurobiol.*, **14**, 69–97.

Moguilewsky, M. and Raynaud, J. P. (1979). Estrogen-sensitive progestin-binding sites in the female rat brain and pituitary. *Brain Res.*, **164**, 165–175.

Mohnot, S. M. (1971). Some aspects of social changes and infant-killing in the Hanuman langur (*Presbytis entellus*) (Primates: Cercopithecidae) in western India. *Mammalia*, **35**, 175–198.

Møller, A. P. (1988a). Female choice selects for male sexual tail ornaments in the monogamous swallow. *Nature Lond.*, **322**, 640–642.

Møller, A. P. (1988b). Ejaculate quality, testes size and sperm competition in primates. *J. Hum. Evol.*, **17**, 479–488.

Møller, A. P. (1989). Ejaculate quality, testes size and sperm production in mammals. *Funct. Ecol.*, **3**, 91–96.

Møller, A. P. (1992). Parasites differentially increase the degree of fluctuating asymmetry in secondary sexual characters. *J. Evol. Biol.*, **5**, 691–699.

Møller, A. P. (1993). Morphology and sexual selection in the barn swallow *Hirundo rustica* in Chernobyl, Ukraine. *Proc. Roy. Soc. Lond. B.*, **252**, 51–57.

Møller, A. P. and Hoglund, J. (1991). Patterns of fluctuating asymmetry in avian feather adornments: implications for models of sexual selection. *Proc. Roy. Soc. Lond.*, **245**, 1–5.

Money, J. (1961). Sex hormones and other variables in human eroticism. In *Sex and internal secretions*, Vol. 2. (2nd edn) (ed. W. C. Young), pp. 1383–1400. Williams and Wilkins, Baltimore.

Money, J. (1970). Use of androgen-depleting hormone in the treatment of male sex offenders. *J. Sex. Res.*, **6**, 165–172.

Money, J. (1977a). Determinants of human gender identity/role. *Handbook of sexology*. (eds. J. Money and H. Mustaph), pp. 57–79. Excerpta Medica, Amsterdam.

Money, J. (1977b). Paraphilias. In *Handbook of sexology*. (eds. J. Money and H. Mustaph), pp. 917–928. Excerpta Medica, Amsterdam.

Money, J. (1988). Gay, straight and in-between: the sexology of erotic orientation. Oxford University Press, New York.

Money, J. (1991). The development of sexuality and eroticism in human kind. In *Heterotypical behaviour in man and animals*. (eds. M. Haug, P. F. Brain and C. Aron), pp. 127–166. Chapman and Hall, London.

Money, J. and Ehrhardt, A. A. (1972). *Man and woman, boy and girl*. Johns Hopkins University Press, Baltimore.

Money, J., Cawte, J. E., Bianchi, G. N., and Nurcombe, B. (1977). Sex training and traditions in Arnhem Land. In *Handbook of sexology*. (eds. J. Money and H. Mustaph), pp. 519–550. Excerpta Medica, Amsterdam.

Money, J., Schwartz, M., and Lewis, V. G. (1984). Adult erotosexual status and fetal hormonal masculinization and demasculinization: 46, XX congenital virilizing hyperplasia and 46XY androgen-insensitivity syndrome compared. *Psychoneuroendocr.*, **9**, 405–414.

Moore, C. L. and Morelli, G. A. (1979). Mother rats interact differently with male and female offspring. *J. Comp. Physiol. Psychol.*, **93**, 677–684.

Moore, C. L. and White, R. H. (1996). Sex differences in sensory and motor branches of the pudendal nerve of the rat. *Horm. Behav.*, **30**, 590–599.

Moore, C. L., Dou, H., and Juraska, M. (1992). Maternal stimulation affects the number of motor neurons in a sexually dimorphic nucleus of the lumbar spinal cord. *Brain Res.*, **572**, 52–56.

Moore, M. C. (1990). Application of organization—activation theory to alternative male reproductive male reproductive strategies: a review. *Horm. Behav.*, **25**, 154–179.

Morales, A., Condra, M., Owen, J., Surridge, D. H., Fenemore, J., and Harris, C. (1987). Is yohimbine effective in the treatment of organic impotence? Results of a controlled trial. *J. Urol.*, **137**, 1168–1172.

Morales, P., Overstreet, J. W., and Katz, D. F. (1988). Changes in human sperm motion during capacitation *in vitro*. *J. Reprod. Fertil.*, **83**, 119–128.

Morali, G. and Beyer, C. (1979). Neuroendocrine control of mammalian estrus behavior. In *Endocrine control of sexual behavior*. (ed. C. Beyer), pp. 33–75. Raven Press, New York.

Morimoto, Y., Oishi, T., Honasaki, N., Miyatake, A., Sato, B., Noma, K., *et al.* (1980). Interrelations among amenorrhea, serum gonadotrophins and body weight in anorexia nervosa. *Endocrinol. Jpn*, **27**, 191–200.

Morris, D. (1967). The naked ape: a zoologist's study of the human animal. Jonathan Cape, London.

Morris, N. M., Udry, R. J., Khan-Dawood, F., and Dawood, M. Y. (1987). Marital sex frequency and midcycle female testosterone. *Arch. Sex. Behav.*, **16**, 27–37.

Mortimer, C. H., Fisher, R. A., Murray, M. A. F., and Besser, G. M. (1974). Gonadotrophin-releasing hormone therapy in hypogonadal males with hypothalamic or pituitary dysfunction. *Br. Med. J.* IV, 617–621.

Mortimer, D. (1983). Sperm transport in the human female reproductive tract. *Oxford Revs. Repro. Biol.*, **5**, 30–61.

Moss, R. L. (1978). Effects of hypothalamic peptides on sex behavior in animals and man. In *Psychophamacology: a generation of progress*. (eds. M. A. Lipton, A. DiMascio and K. F. Killam), pp. 431–440. Raven Press, New York.

Moss, R. L. and Foreman, M. M. (1976). Potentiation of lordosis behavior by intrahypothalamic infusion of synthetic luteinizing hormone releasing hormone. *Neuroendocrinology*, **20**, 176–181.

Moss, R. L. and McCann, S. M. (1973). Induction of mating behavior in rats by luteinizing hormone-releasing factor. *Science*, **181**, 177–179.

Mossman, H. W. and Duke, K. L. (1973). *Comparative morphology of the Mammalian ovary*. University of Wisconsin Press, Madison.

Moynihan, M. (1964). Some behavior patterns of platyrrhine monkeys. 1. the night monkey (*Aotus trivirgatus*). *Smithson Misc. Coll.*, **146**, 1–84.

Muller, L. R. (1902). Klinische und experimentalle Studien über die Innervation der Blase, des Mastdarms, und des genital Apparates. *Dtsch. Z. Nervenheilkd.*, **21**, 86–155.

Murray, R. D., Bour, E. S., and Smith, E. O. (1985). Female menstrual cyclicity and sexual behavior in stumptail macaques (*Macaca arctoides*). *Int. J. Primatol.*, **6**, 101–113.

Myers, B. M. and Baum, M. J. (1979). Facilitation by opiate antagonists of sexual performance in the male rat. *Pharmacol. Biochem. Behav.*, **10**, 615–618.

Nadler, R. D. (1975). Sexual cyclicity in captive lowland gorillas. *Science*, **189**, 813–814.

Nadler, R. D. (1976). Sexual behavior of captive gorillas. *Arch. Sex. Behav.*, **5**, 487–502.

Nadler, R. D. (1977). Sexual behavior of captive orang-utans. *Arch. Sex. Behav.*, **6**, 457–475.

Nadler, R. D. (1980). Reproductive physiology and behaviour of gorillas. *J. Reprod. Fertil. Suppl.*, **28**, 79–89.

Nadler, R. D. (1982). Laboratory research on sexual behavior and reproduction of gorillas and orang-utans. *Am. J. Primatol. Suppl.*, **1**, 57–66.

Nadler, R. D. (1986). Sex-related behavior of immature wild mountain gorillas. *Devel. Phychobiol.*, **19**, 125–137.

Nadler, R. D. (1988). Sexual and reproductive behavior. In *Orang-utan biology*. (ed. J. H. Schwartz), pp. 105–116. Oxford University Press, New York.

Nadler, R. D. (1990). Homosexual behavior in nonhuman primates. In *Homosexuality/heterosexuality: concepts of sexual orientation* (eds. D. McWhirter, P. Sanders and J. M. Reinisch), pp. 138–170. Oxford University Press, New York.

Nadler, R. D. (1992). Sexual behavior and the concept of estrus in the great apes. In *Topics in primatology, Vol. 2. behavior, ecology and conservation.* (eds. N. Itiogawa, Y. Sugiyama, G. P. Sackett, and R. K. R. Thompson), pp. 191–206. University of Tokyo Press, Tokyo.

Nadler, R. D. (1994). Walter Heape and the issue of estrus in primates. *Am. J. Primatol.*, **33**, 83–87.

Nadler, R. D. and Rosenblum, L. A. (1972). Hormonal regulation of the 'fatted' phenomenon in male squirrel monkeys. *Anat. Rec.*, **173**, 181–187.

Nadler, R. D. and Rosenblum, L. A. (1973). Sexual behavior during successive ejaculations in bonnet and pigtail macaques *Am. J. Phys. Anthrop.*, **28**, 217–220.

Nadler, R. D., Graham, C. E., Collins, D. C., and Gould, K. G. (1979). Plasma gonadotropins, prolactin, gonadal steroids and genital swelling during the menstrual cycle of lowland gorillas. *Endocrinology*, **105**, 290–296.

Nadler, R. D., Collins, D. C., Miller, L. C., and Graham, C. E. (1983). *Horm. Behav.*, **17**, 1–17.

Nadler, R. D., Graham, C. E., Gosslin, R. E., and Collins, D. C. (1985). Serum levels of gonadotropins and gonadal steroids, including testosterone, during the menstrual cycle of the chimpanzee (*Pan troglodytes*). *Am. J. Primatol.*, **9**, 273–284.

Naftolin, F., Ryan, K. J., Davies, I. J., Reddy, V.V., Flores, F., Petro, Z., *et al.* (1975). The formation of estrogens by central neuroendocrine tissue. *Rec. Prog. Horm. Res.*, **31**, 295–316.

Nagel, U. (1973). A comparison of anubis baboons, hamadryas baboons and their hybrids at a species border in Ethiopia. *Folia Primatol.*, **19**, 104–165.

Nagle, C. A. and Denari, J. H. (1983). The *Cebus* monkey (*Cebus apella*). In *Reproduction in New World primates.* (ed. J. P. Hearn), pp. 39–67. MTP Press, Lancaster.

Nagle, C. A., Denari, J. H., Quiroga, S., Riarte, A., Merlo, A., Germino, I., *et al.* (1979). The plasma pattern of ovarian steroids during the menstrual cycle of the capuchin monkey (*Cebus apella*). *Biol. Reprod.*, **21**, 979–983.

Nakai, Y., Plant, T. M., Hess, D. L., Keogh, E. J., and Knobil, E. (1978). On the sites of the negative and positive feedback action of estradiol in the control of gonadotropin secretion in the rhesus monkey. *Endocrinology*, **102**, 1008–1014.

Nakatsuru, K. and Kramer, D. L (1982). Is sperm cheap? Limited male fertility and female choice in the lemon tetra (*Pisces, Charicidae*). *Science*, **216**, 753–755.

Nance, D. W., Shryne, J., and Gorski, R. A. (1975a). Facilitation of female sexual behavior in male rats by septal lesions: an interaction with estrogen. *Horm. Behav.*, **6**, 289–299.

Nance, D. W., Shryne, J., and Gorski, R. A. (1975b). Effects of septal lesions on behavioral sensitivity of female rats to gonadal hormones. *Horm. Behav.*, **6**, 59–64.

Nance, W. E. (1993). Back to the future. *Am. J. Hum. Genet.*, **53**, 6–15.

Napier, J. R. (1970). Paleoecology and catarrhine evolution. In *Old World monkeys: evolution, systematics and behavior.* (eds.

J. R. Napier and P. H. Napier), pp. 55–95. Academic Press, New York.

Napier, J. R. and Davis, P. R. (1959). The forelimb skeleton and associated remains of *Proconsul africanus*. *Fossil Mammals of Africa* No.16. British Museum of Natural History, London.

Napier, J. R. and Napier, P. H. (1967). A handbook of living primates: morphology, ecology and behaviour of nonhuman primates. Academic Press, London.

Napier, J. R. and Napier, P. H. (1985). *The natural history of the primates*. British Museum of Natural History, and Cambridge University Press.

Nash, L. T. (1974). Parturition in a feral baboon (*Papio anubis*). *Primates*, **15**, 279–285.

Nauta, W. J. H. (1971). The problem of the frontal lobe: a reinterpretation. *J. Psychiatr. Res.*, **8**, 167–187.

Nauta, W. J. H. and Haymaker, W. (1969). Hypothalamic nuclei and fibre connections. In *The hypothalamus.* (eds. W. Haymaker, E. Andersson and W. J. H. Nauta), pp. 136–209. Thomas, Springfield, Illinois.

Neal, E. (1986). *The natural history of badgers*. Croom Helm, London.

Neimeyer, C. L. and Anderson, J. R. (1983). Primate harassment of matings. *Ethol. Sociobiol.*, **4**, 205–220.

Neimeyer, C. L. and Chamove, A. S. (1983). Motivation of harassment of matings in stumptailed macaques. *Behaviour*, **87**, 298–323.

Neimitz, C., Nietsch, A., Warter, S., and Rumpler, Y. (1991). *Tarsius dianae*: a new primate species from central Sulawesi (Indonesia). *Folia Primatol.*, **56**, 105–116.

Nelson, R. J. (1995). *An introduction to behavioral endocrinology*. Sinauer, Sunderland, Massachusetts.

Neville, M. K. (1972). Social relations within troops of red howler monkeys (*Alouatta seniculus*). *Folia Primatol.*, **18**, 47–77.

Neville, M. K., Glander, K. E., Braza. F., and Rylands, A. B. (1988). The howling monkeys: genus *Alouatta*. In *Ecology and behavior of neotropical primates*, Vol. 2. (eds. R. A. Mittermeier, A B. Rylands, A. Coimbra-Filho and G. A. B. Fonseca), pp. 349–453. World Wildlife Fund, Washington DC.

Nevison, C. M., Rayment, F. D. G., and Simpson, M. J. A. (1996). Birth sex ratios and maternal rank in a captive colony of rhesus monkeys (*Macaca mulatta*). *Am. J. Primatol.*, **39**, 123–138.

Nevison, C. M., Brown, G. R., and Dixson, A. F. (1997). Effects of altering testosterone in early infancy on social behaviour in captive juvenile rhesus monkeys. *Physiol. Behav.*, **62**, 1397–1403.

Newman, H. F. and Northup, J. D. (1981). Mechanism of human penile erection: an overview. *Urology*, **17**, 399–408.

Newton, P.N. (1986). Infanticide in an undisturbed forest population of Hanuman langurs, *Presbytis entellus*. *Anim. Behav.*, **34**, 785–789.

Niemitz, C., Nietsch, A., Warter, S., and Rumpler, Y. (1991). *Tarsius dianae*: a new primate species from central Sulawesi (Indonesia). *Folia Primatol.*, **56**, 105–116.

Nieschlag, E. (1979). The endocrine function of the human testis in regard to sexuality. In *Sex, hormones and behaviour.*

Ciba Found. Symp., **62**, 183–197. Excerpta Medica, Amsterdam.

Nieuwenhuijsen, K., De Neef, K. J., Van der Werff Ten Bosch, J. J., and Slob, A. K. (1987). Testosterone, testis size, seasonality and behavior in group-living stumptail macaques. *Horm. Behav.*, **21**, 153–169.

Nieuwenhuijsen, K., Bonke-Jansen, M., Broekhuijzen, E., De Neef, K. J., Van Hoof, J. A. R. A. M ., Van der Werff Ten Bosch, J. J., *et al.* (1988). Behavioral aspects of puberty in group-living stumptail monkeys (*Macaca arctoides*). *Physiol. Behav.*, **42**, 255–264.

Nigi, H., Tiba, T., Yamamoto, S., Floescheim, Y., and Ohsawa, N. (1980). Sexual maturation and seasonal changes in reproductive phenomena of male Japanese monkeys (*Macaca fuscata*) at Takasakiyama. *Primates*, **21**, 230–240.

Nishida, T. (1966). A sociological study of solitary male monkeys. *Primates.*, **7**, 141–204.

Nishida, T. (1968). The social group of wild chimpanzees in the Mahale Mountains. *Primates*, **19**, 167–224.

Nishida, T. (1979). Social structure among wild chimpanzees of the Mahali Mountains. In *The great apes.* (eds. D. A. Hamburg and E. R. McCown), pp. 72–121. Benjamin/Cummings, Menlo Park, California.

Nishida, T. (ed.) (1990). The chimpanzees of the Mahale Mountains: sexual and life history strategies. University of Tokyo Press, Tokyo.

Nishida, T. and Hiraiwa-Hasegawa, M. (1987). Chimpanzees and bonobos: cooperative relationships among males. In *Primate Societies.* (eds. B. Smuts, D. Cheney, R. Seyfarth, R. Wrangham, and T. Struhsaker), pp. 165–177. University of Chicago Press.

Nishida, T., Takasaki, H., and Takahata, Y. (1990). Demography and reproductive profiles. In *The chimpanzees of the Mahale Mountains: sexual and life history strategies.* (ed. T. Nishida), pp. 63–97. University of Tokyo Press.

Nishimura, A. (1988). Mating behavior of woolly monkeys, *Lagothrix lagotricha*, at La Macarena, Colombia. *Field Stud. N.W. Monkeys*, **1**, 19–27.

Nishimura, A., da Fonseca, G. A. B., Mittermeier, R. A., Young, A. L., Strier, K. B., and Valle, C. M. C. (1988). The muriqui, genus *Brachyteles*. In *Ecology and behavior of neotropical primates*, Vol. 2. (eds. R. A. Mittermeier, A. B. Rylands, A. F. Coimbra-Filho and G. A. B. Fonseca), pp. 577–610. World Wildlife Fund, Washington, D. C.

Nishizuka, M. and Arai, Y. (1981). Organizational action of estrogen on synaptic pattern in the amygdala: implications for sexual differentiation of the brain. *Brain Res.*, **213**, 422–426.

Nissen, H. W. (1931). A field study of the chimpanzee. *Comp. Psychol. Monogr.*, **8**, 1–122.

Nissen, H. W. (1954). Development of sexual behavior in chimpanzees. Paper presented at the Amherst Symposium: *Genetics, psychological and hormonal factors in the establishment and maintenance of patterns of sexual behavior in mammals.* Amherst, 1954.

Noë, R. and Sluijter, A. A. (1990). Reproductive tactics of male savanna baboons. *Behaviour*, **113**, 117–170.

Notides, A. C. and Williams Ashman, H. C. (1967). The basic protein responsible for the clotting of guinea-pig semen. *Proc. Nat. Acad. Sci. U.S.A.*, **58**, 199.

Nottebohm, F. (1980). Testosterone triggers growth of vocal control nuclei in adult female canaries. *Brain Res.*, **189**, 229–236.

Nottebohm, F. (1981). A brain for all seasons: cyclical anatomical changes in song control nuclei of the canary brain. *Science*, **214**, 1368–1370.

Nottebohm, F. and Arnold, A. P. (1976). Sexual dimorphism in vocal control areas of the songbird brain. *Science*, **194**, 211–213.

Nottebohm, F., Stokes, T. M., and Leonard, C. M. (1976). Central control of song in the canary (*Serinus canarius*). *J. Comp. Neurol.*, **165**, 457–486.

Numan, M. (1988). Maternal behavior. In *The physiology of reproduction*, Vol. 2. (eds. E. Knobil and J. D. Neill), pp. 1569–1645. Raven Press, New York.

Numan, M. (1994). Maternal behavior. In *The Physiology of reproduction*, Vol. 2. (2nd edition) (eds. E. Knobil and J. D. Neill), pp. 221–302. Raven Press, New York.

O'Byrne, K. T., Lunn, S. F., and Dixson, A. F. (1988). Effects of acute stress on the patterns of LH secretion in the common marmoset (*Callithrix jacchus*). *J. Endocr.*, **118**, 259–264.

O'Byrne, K. T., Lunn, S. F., and Dixson, A. F. (1989). Naloxone reversal of stress-induced suppression of LH release in the common marmoset. *Physiol. Behav.*, **45**, 1077–1080.

O'Carroll, R., Shapiro, C., and Bancroft, J. (1985). Androgens, behaviour and nocturnal erections in hypogonadal men: the effect of varying the replacement dose. *Clin. Endocr.*, **23**, 527–538.

Oakley, K. P. (1949). *Man the tool-maker*. British Museum of Natural History, London.

Ober, C. and Van der Ven, K. (1996). HLA and fertility. In *HLA and the maternal-fetal relationship.* (ed. J. Hunt), pp. 133–146. R. G. Landes, Austin, Texas.

Oda, Y. and Masataka, N. (1992). Functional significance of female Japanese macaque copulatory calls. *Folia Primatol.*, **58**, 146–149.

Odling-Smee, F. J., Laland, K. N., and Feldman, M. W. (1996). Niche construction. *American Naturalist*, **147**, 641–648.

Oemedes, A. and Carroll, J. B. (1980). A comparative study of pair behaviours of four callitrichid species and the Goeldi's monkey. *Dodo: Jersey Wild. Preserv. Trust Ann. Rep.*, **17**, 51–62.

Ohsawa, H. (1991). Takeover of a one-male group and the subsequent promiscuity of patas monkeys. In *Primatology today* (eds. A. Ehara, T. Kimua, M. Iwamoto, and O. Takenata), pp. 221–224. Elsevier Science Publications, Amsterdam.

Ohsawa, H., Inoue, M., and Takenata, O. (1993). Mating strategy and reproductive success of male patas monkeys (*Erythrocebus patas*). *Primates*, **34**, 533–544.

Olsen, K. L. and Whalen, R. E. (1984). Dihydrotesterone activates male mating behavior in castrated King-Holtzman rats. *Horm. Behav.*, **18**, 380–392.

Olsson, M., Shine, R., Madsen, T., Gullberg, A., and Tegelström, H. (1996). Sperm selection by females. *Nature Lond.*, **383**, 585.

Onufrowicz, B. (1901). On the arrangement and function of the cell groups of the sacral region of the spinal cord in man. *Arch. Neurol. Psychopathol.*, **3**, 387–412.

Oomura, Y., Yoshimatsu, H., and Aou, S. (1984). Hypothalamic neuronal activity during sexual behaviour in male monkey. In *Modulation of sensorimotor activity during alterations in behavioral states.* (ed. R. Bandler), pp. 231–250. Alan R. Liss, New York.

Orbach, J. (1961). Spontaneous ejaculation in the rat. *Science*, **134**, 1072–1073.

Oring, L. W. and Knudson, M. L. (1973). Monogamy and polyandry in the spotted sandpiper. *The Living Bird*, **11**, 59–73.

Overstreet, J. W., Katz, D. F., and Yudin, A. I. (1991). Cervical mucus and sperm transport in reproduction. *Seminars in Perinatology*, **15**, 149–155.

Owens, N. (1973). The development of behaviour in free-living baboons, *Papio anubis*. Unpublished Ph.D. thesis, University of Cambridge.

Packer, C. (1979a). Inter-troop transfer and inbreeding avoidance in *Papio anubis*. *Anim. Behav.*, **27**, 1–36.

Packer, C. (1979b). Male dominance and reproductive activity in *Papio anubis*. *Anim. Behav.*, **27**, 37–45.

Packer, C., Scheel, D., and Pusey, a. E. (1990). Why lions form groups-food is not enough. *Am. Nat.*, **136**, 1–19.

Packer, C., Gilbert, D. A., Pusey, A. E., and O'Brien, S. J. (1991). A molecular genetic analysis of kinship and cooperation in African lions. *Nature Lond.*, **351**, 562–565.

Packer, C., Collins, D. A., Sindimwo, A., and Goodall, J. (1995). Reproductive constraints on aggressive competition in female baboons. *Nature Lond.*, **373**, 60–63.

Pagés-Feuillade, E. (1988). Modalités de l'occupation de l'espace at relations inter-individuelles chez un prosimien Malgache, *Microcebus murinus*. *Folia Primatol.*, **50**, 204–220.

Palmer, R. M. J., Ferrige, A. G., and Moncada, S. (1987). Nitric oxide release accounts for the biological activity of endothelium-derived relaxing factor. *Nature Lond.*, **327**, 524–526.

Palombit, R. A. (1994). Extra-pair copulations in a monogamous ape. *Anim. Behav.*, **47**, 721–723.

Pandya, I. J. and Cohen, J. (1985). The leucocytic reaction of the human cervix to spermatozoa. *Fertil. Steril.*, **43**, 417–421.

Panzica, G. C., Viglietti-Panzica, C., Calacagni, M., Anselmetti, G. C., Schumacher, M., and Balthazart, J. (1987). Sexual differentiation and hormonal control of the sexually dimorphic medial preoptic nucleus in the quail. *Brain Res.*, **416**, 59–68.

Pariente, G. (1979). The role of vision in prosimian behavior. In *The study of prosimian behavior.* (eds. G. A. Doyle and R. D. Martin), pp. 411–459. Academic Press, New York.

Parish, A. R. (1996). Female relationships in bonobos (*Pan paniscus*). *Human Nature*, **7**, 61–96.

Parker, G. A. (1970). Sperm competition and its evolutionary consequences in the insects. *Biol. Rev.*, **45**, 525–567.

Parker, S. L. and Crowley, W. R. (1993). Central stimulation of oxytocin release in the lactating rat: interaction of neuropeptide Y with α_1-adrenergic mechanisms. *Endocrinology*, **132**, 658–666.

Parr, D. (1976). Sexual aspects of drug abuse in narcotic addicts. *Brit. J. Addiction*, **71**, 261–268.

Parrott, R. F. (1978). Courtship and copulation in prepubertally castrated male sheep (wethers) treated with 17β estradiol, aromatizable androgens or dihydrotestosterone. *Horm. Behav.*, **11**, 20–27.

Parsons, P. A. (1992). Fluctuating asymmetry: a biological monitor of environmental and genomic stress. *Heredity*, **68**, 361–364.

Paton, J. A. and Nottebohm, F. N. (1984). Neurons generated in the adult brain are recruited into functional circuits. *Science*, **225**, 1046–1048.

Patterson, B. D. and Thaeler, C. S. (1982). The mammalian baculum: hypotheses on the nature of bacular variability. *J. Mammal.*, **63**, 1–15.

Paul, A. (1997). Breeding seasonality affects the association between dominance and reproductive success in non-human primates. *Folia primatologica.*, **68**, 344–349.

Paul, A. and Kuester, J. (1988). Life-history patterns of Barbary macaques (*Macaca sylvanus*) at Affenberg Salem. In *Ecology and behavior of food-enhanced primate groups.* (eds. J. E. Fa and C. H. Southwick), pp. 199–228. Alan R. Liss, New York.

Paul, A. and Kuester, J. (1990). Adaptive significance of sex ratio adjustment in semi-free-ranging Barbary macaques. (*Macaca sylvanus*) at Salem. *Behav. Ecol. Sociobiol.*, **27**, 287–293.

Paul, A. and Thommen, D. (1984). Timing of birth, female reproductive success and infant sex ratio in semi-free-ranging Barbary macaques. (*Macaca sylvanus*). *Folia Primatol.*, **42**, 2–16.

Paul, A., Kuester, J., and Arnemann, J. (1992). DNA fingerprinting reveals that infant care by male barbary macaques(*Macaca sylvanus*) is not paternal investment. *Folia Primatol.*, **58**, 93–98.

Paul, A., Kuester, J., and Arnemann, J. (1992). Maternal rank affects, reproductive success of male Barbary macaques: evidence by DNA fingerprinting. *Behav. Ecol. Sociobiol.*, **30**, 337–341.

Paul, A., Kuester, J., Timme, A., and Arnemann, J. (1993). The association between rank, mating effort and reproductive success in male Barbary macaques (*Macaca sylvanus*). *Primates*, **34**, 491–502.

Paul, A., Kuester, J., and Arnemann, J. (1996). The sociobiology of male-infant interactions in barbary macaques, *Macaca sylvanus*. *Anim. Behav.*, **51**, 155–170.

Paxinos, G. and Bindra, D. (1971). Hypothalamic knife cuts: effects on eating, drinking, irritability, aggression and copulation in the male rat. *J. Comp. Physiol. Psychol.*, **79**, 219–229.

Peaker, M. (1992). Chemical signalling systems: the rules of the game. *J. Endocr.*, **135**, 1–4.

Pedersen, C. A. (1997). Oxytocin control of maternal behavior. *Ann. N. Y. Acad. Sci.*, **807**, 126–145.

Pentland, B., Anderson, D. A., and Critchley, J. A. J. H. (1981). Failure of ejaculation with indoramine. *Br. Med. J.*, **282**, 1433–1444

Perachio, A. A. (1978). Hypothalamic regulation of behavioural and hormonal aspects of aggression and sexual performance. In *Recent advances in primatology, Vol 1, Behaviour.* (eds. D. J. Chivers and J. Herbert), pp. 449–465. Academic Press, London.

Perachio, A. A., Marr, L. D., and Alexander, M. (1979). Sexual behavior in male rhesus monkeys elicited by electrical stimulation of preoptic and hypothalamic areas. *Brain Res.*, **177**, 127–144.

Pereira, M. E. (1991). Asynchrony within estrous synchrony

among ringtailed lemurs (Primates: *Lemuridae*). *Physiol. Behav.*, **49**, 47–52.

Pereira, M. E. and Weiss, M. L. (1991). Female mate choice, male migration and the threat of infanticide in ringtailed lemurs. *Behav. Ecol. Sociobiol.*, **18**, 141–152.

Pereira, M. E., Sieligson, M. A., and Macedonia, J. M. (1988). The behavioral repertoire of the black and white ruffed lemur, *Varecia variegata variegata*. *Folia Primatol.*, **51**, 1–32.

Perloe, S. I. (1992). Male mating competition, female choice and dominance in a free ranging group of Japanese macaques. *Primates.*, **33**, 289–304.

Peroutka, S. J. (1995). 5-HT receptors: past, present and future. *Trends Neurosci.*, **18**, 68–69.

Perret, M. (1982). Influence du groupement social sur la reproduction de la femelle de *Microcebus murinus* (Miller, 1777). *Z. Tierpsychol.*, **60**, 47–65.

Perret, M. (1985). Influence of social factors on seasonal variations in plasma testosterone levels of *Microcebus murinus*. *Z. Tierpsychol.*, **69**, 265–280.

Perret, M. (1992). Environmental and social determinants of sexual function in the male mouse lemur (*Microcebus murinus*). *Folia Primatol.*, **59**, 1–25.

Perret, M. (1995). Chemocommunication in the reproductive function of mouse lemurs. In *Creatures of the dark: the nocturnal prosimians*. (eds. L. Alterman, G. A. Doyle, and M. K. Izard), pp. 377–392. Plenum Press, New York.

Perrett, D. I., Hietanen, J. K., Oram, M. W., and Benson, P. J. (1992). Organization and functions of cells responsive to faces in the temporal cortex. *Phil. Trans. Roy. Soc. Lond. B.*, **335**, 23–30.

Perrett, D. I., May, K. A., and Yoshikowa, S. (1994). Facial shape and judgements of female attractiveness. *Nature Lond.*, **368**, 239–242.

Perryman, R. L. and Thorner, M. O. (1981). The effects of hyperprolactinaemia on sexual and reproductive functions in man. *J. Androl.*, **2**, 233–242.

Persky, H., Charney, N., Lief, H. I., O'Brien, C. P., Miller, W. R., and Strauss, D. (1978a). The relationship of plasma oestradiol to sexual behavior in young women. *Psychosom. Med.*, **40**, 523–535.

Persky, H., Lief, H. I., Strauss, D., Miller, W. R., and O'Brien, C. P. (1978b). Plasma testosterone level and sexual behavior of couples. *Arch. Sex. Behav.*, **7**, 157–173.

Persson, E. and Rodriguez-Martinez, H. (1990). The ligamentum influndibulo-cornuale in the pig: morphological and physiological studies of the smooth muscle component. *Acta. Anat.*, **138**, 111–120.

Petrie, M. and Williams, A. (1993). Peahens lay more eggs for peacocks with larger trains. *Proc. Roy. Soc. Lond. B.*, **251**, 127–131.

Petrie, M., Halliday, T., and Sanders, C. (1991). Peahens prefer peacocks with elaborate trains. *Anim. Behav.*, **41**, 323–331.

Petter Rousseaux, A. (1964). Reproductive physiology and behavior of the Lemuroidea. In *Evolutionary and genetic biology of primates*. Vol. 2 (ed. J. Buettner-Janusch), pp. 91–132. Academic Press, New York.

Petter-Rousseaux, A. (1980). Seasonal activity rhythms, reproduction, and body weight variations in five sympatric nocturnal, in simulated light and climatic conditions. In *Nocturnal malagasy primates*. (eds. H. M. Cooper, A. Hladik, C. M.

Hladik, E. Pages, *et al.*), pp.137–152. Academic Press, New York.

Pfaff, D. W. (1980). *Estrogens and brain function*. Springer, New York.

Pfaff, D. W. and Keiner, M. (1973). Atlas of estradiol-concentrating cells in the central nervous system of the female rat. *J. Comp. Neurol.*, **151**, 121–158.

Pfaff, D. W. and Pfaffman, C. (1969). Olfactory and hormonal influences on the basal forebrain of the male rat. *Brain Res.*, **75**, 137–156.

Pfaff, D. W. and Schwartz-Giblin, S. (1988). Cellular mechanisms of female reproductive behaviors. In *The physiology of reproduction*, Vol. 2. (eds. E. Knobil and J. D. Neill), pp. 1487–1568. Raven Press, New York.

Pfaff, D. W., Gerlach, J., McEwen, B. S., Ferin, M., Carmel, P., and Zimmerman, E. (1976). Autoradiographic localization of hormone concentrating cells in the brain of the female rhesus monkey. *J. Comp. Neurol.*, **170**, 279–294.

Pfaff, D. W., Schwartz-Giblin, S., McCarthy, M., and Kow, L. M. (1994). Cellular and molecular mechanisms of female reproductive behaviors. In *The physiology of reproduction*, Vol. 2 (2nd edition) (eds. E. Knobil and J. D. Neill), pp. 107–220. Raven Press, New York.

Pfaus, J. G. and Everitt, B. J. (1995). The phychopharmacology of sexual behavior. In *Psychopharmacology: the fourth generation of progress*. (eds. F. E. Bloom and D. J. Kupfer), pp. 743–758. Raven Press, New York.

Pfaus, J. G., Damsma, G., Nomikos, G. G., Wenkstern, D. G., Blaha, C. D., Phillips, A. G., *et al.* (1990). Sexual behavior enhances central dopamine transmission in the male rat. *Brain Res.*, **530**, 345–348.

Phillips, D. M. (1970). Ultrastructure of spermatozoa of the woolly opossum, *Caluromys philander*. *J. Ultrastr. Res.*, **33**, 381–397.

Phoebus, E. C. (1982). Primate female orgasm. *Am. J. Primatol.*, **2**, 223–224.

Phoenix, C. H. (1973). The role of testosterone in the sexual behavior of laboratory male rhesus. In *Symp. IVth Int. Congr. Primatol.*, Vol. 2. (ed. C. H. Phoenix), pp. 99–122. Karger, Basel.

Phoenix, C. H. (1974a). Prenatal testosterone in the nonhuman primate and its consequences for behavior. In *Sex differences in behavior*. (eds. R. C Friedman, R. M. Richart, and R. L. Van der Wiele), pp. 19–32. Wiley, New York.

Phoenix, C. H. (1974b). Effects of dihydrotestosterone on sexual behavior of castrated male rhesus monkeys. *Physiol. Behav.*, **12**, 1045–1055.

Phoenix, C. H. (1976). Sexual behavior of castrated male rhesus monkeys treated with 19-hydroxytestosterone. *Physiol. Behav.*, **16**, 305–310.

Phoenix, C. H. and Chambers, K. C. (1982). Sexual behavior in aging male rhesus monkeys. In *Advanced views in primate biology*. (eds. A. B. Chiarelli and R. S. Corruccini), pp. 95–104. Springer, Berlin.

Phoenix, C. H. and Chambers, K. C. (1988). Testosterone therapy in young and old rhesus males that display low levels of sexual activity. *Physiol. Behav.*, **43**, 479–484.

Phoenix, C. H., Goy, R. W., Gerall, A. A., and Young, W. C. (1959). Organizational action of prenatally administered testosterone propionate on the tissues mediating behavior in the female guinea pig. *Endocrinology*, **65**, 369–382.

Phoenix, C. H., Slob, A. K., and Goy, R. W. (1973). Effects of castration and replacement therapy on sexual behavior of adult male rhesuses. *J. Comp. Physiol. Psychol.*, **84**, 472–481.

Phoenix, C. H., Copenhaver, K. H., and Brenner, R. M. (1976). Scanning election microscopy of penile papillae in intact and castrated rats. *Horm. Behav.*, **7**, 217–227.

Phoenix, C. H., Dixson, A. F., and Resko, J. A. (1977). Effects of ejaculation on levels of testosterone, cortisol and luteinizing hormone in peripheral plasma of rhesus monkeys. *J. Comp. Physiol. Psychol.*, **91**, 120–127.

Phoenix, C. H., Jensen, J. N., and Chambers, K. C. (1983). Female sexual behavior displayed by androgenized female rhesus macaques. *Horm. Behav.*, **17**, 146–151.

Phoenix, C. H., Chambers, K. C., Jensen, J. N., and Baughman, W. (1984). Sexual behavior of an androgenized female rhesus macaque with a surgically reconstructed vagina. *Horm. Behav.*, **18**, 393–399.

Pickford, M. and Genut, B. (1988). Habitat and locomotion in miocene cercopithecoids. In *A primate radiation: evolutionary biology of the African guenons*. (eds. A. Gautier-Hion, F. Boulire, J. P. Gautier, and J. Kingdon), pp. 35–53. Cambridge University Press.

Pinder, R. M., Brogan, R. N., Sawyer, P. R., Speight, T. M., and Avery, G. S. (1975). Fenfluramine: a review of its pharmacological properties and efficacy in obesity. *Drugs*, **10**, 241–323.

Pirke, K. M., Kockott, G., and Dittmar, F. (1974). Psychosexual stimulation and plasma testosterone in man. *Arch. Sex. Behav.*, **3**, 577–584.

Pirke, K. M., Wuttke, W., and Schweiger, U. (1989) (eds.). *The menstrual cycle and its disorders*. Springer, Berlin.

Pitnick, S., Spicer, G. S., and Markow, T. A. (1995). How long is a giant sperm? *Nature Lond.*, **375**, 109.

Plant, T. M. (1982). A striking diurnal variation in plasma testosterone concentrations in infantile male rhesus monkeys (*Macaca mulatta*). *Neuroendocr.*, **35**, 370–373.

Plant, T. M. (1994). Puberty in primates. In *The physiology of reproduction*, Vol. 2. (2nd edition) (eds. E. Knobil and J. D. Neill), pp. 453–485. Raven Press, New York.

Plant, T. M. and Dubey, A. K. (1984). Evidence in the rhesus monkey (*Macaca mulatta*) for the view that negative feedback control of luteinizing hormone secretion by the testis is mediated by a deceleration of hypothalamic gonadotrophin releasing pulse frequency. *Endocrinology*, **115**, 2145–2153.

Plant, T. M., Moor, Y., Hess, D. L., Nakai, Y., McCormack, P., and Knobil, E. (1979). Further studies on the effects of lesions in the rostral hypothalamus on gonadotrophin secretion in the female rhesus monkey *Macaca mulatta*. *Endocrinology*, **105**, 465–473.

Plaplinger, L. and McEwen, B. S. (1978). Gonadal steroid-brain interactions in sexual differentiation. In *Biological determinants of sexual behaviour*. (ed. J. B. Hutchison), pp. 193–218. John Wiley, New York.

Plavcan, J. M. and Van Schaik, C. P. (1992). Intrasexual competition and canine dimorphism in anthropoid primates. *Am. J. Phys. Anthropol.*, **87**, 461–477.

Plavcan, J. M., Van Schaik, C. P., and Kappeler, P. M. (1995). Competition, coalitions and canine size in primates. *J. Hum. Evol.*, **28**, 245–276.

Pleim, E. T., Matochik, J. A., Barfield, R. J., and Auerbach, S. J. (1990). Correlation of dopamine release in the nucleus accumbens with masculine sexual behavior in rats. *Brain Res.*, **524**, 160–163.

Ploog, D. W. (1967). The behavior of squirrel monkeys (*Saimiri sciureus*) as revealed by sociometry, bioacoustics and brain stimulation. In *Social communication among primates*. (ed. S. A. Altmann), p. 149–184. University of Chicago Press.

Plooij, F. X. (1984). The behavioral development of free-living chimpanzee babies and infants. Ablex, Norwood, New Jersey.

Poirier, F. E. (1970). Dominance structure of the Nilgiri langur (*Presbytis johnii*) of South India. *Folia Primatol.*, **12**, 161–186.

Poirier, F. E. (1979). The Nilgiri langur (*Presbytis johnii*) troop: its composition, structure, function and change. *Folia Primatol.*, 20–47.

Polak, J. M., Gu, J., Mina, S., and Bloom, S. R. (1981). VIPergic nerves in the penis. *Lancet* ii, 217–219.

Pollard, J. W., Plante, C., King, W. A., Hansen, P. J., Betteridge, K. J., and Suarez, S. S. (1991). Fertilizing capacity of bovine sperm may be maintained by binding to oviductal epithelial cells. *Biol. Reprod.*, **44**, 102–107.

Pollock, J. I. (1979). Spatial distribution and ranging behavior in lemurs. In *The study of prosimian behavior*. (eds. G. A. Doyle and R. D. Martin), pp. 359–409. Academic Press, New York.

Pomerantz, S. (1990). Apomorphine facilitates male sexual behavior of rhesus monkeys. *Pharmacol. Biochem. Behav.*, **35**, 659–664.

Pomerantz, S. M. (1991). Quinelorane (LY163502), a D2 dopamine receptor agonist, acts centrally to facilitate penile erections of male rhesus monkeys. *Pharmacol. Biochem. Behav.*, **39**, 123–128.

Pomerantz, S. M. and Goy, R. W. (1983). Proceptive behavior of female rhesus monkeys during tests with tethered males. *Horm. Behav.*, **17**, 237–248.

Pomerantz, S. M., Goy, R. W., and Roy, M. M. (1986). Expression of male-typical behavior in adult female pseudohermaphrodite rhesus: comparison with normal males and with neonatally gonadectomized males and females. *Horm. Behav.*, **20**, 483–500.

Pomerantz, S. M., Hepner, B. C., and Wertz, J. M. (1993). 5-HT$_{1A}$ and 5-HT$_{1C/1D}$ receptor agonists produce reciprocal effects on male sexual behavior of rhesus monkeys. *Eur. J. Pharmacol.*, **243**, 227–234.

Ponig, B. F. and Roberts, J. A. (1978). Seminal vesicles as organs of sperm storage. *Urology.*, **11**, 384–385.

Pope, T. R. (1990). The reproductive consequences of male cooperation in the red howler monkey: paternity exclusion in multi-male and single-male troops using genetic markers. *Behav. Ecol. Sociobiol.*, **27**, 439–446.

Potts, W. K., Manning, C. J., and Wakeland, E. K. (1991). Mating patterns in seminatural populations of mice influenced by MHC genotype. *Nature Lond.*, **352**, 619–621.

Powers, J. B. and Valenstein, E. S. (1972). Sexual receptivity: facilitation by medial preoptic lesions in the female rats. *Science*, **175**, 1003–1005.

Preslock. J. P., Hampton, S. H., and Hampton, J. K. Jr (1973).

Cyclic variations of serum progestins and immuno-reactive estrogens in marmosets. *Endocrinology*, **92**, 1096–1101.

Preti, G., Cutler, W. B., Garcia, C. R., Huggins, G. R., and Lawley, H. (1986). Human axillary secretions influence womens' menstrual cycles: the role of donor extract of females. *Horm. Behav.*, **20**, 474–482.

Preuschoft, H., Chivers, D. J., Brockelman, W. Y., and Creel, N. (1984). (eds.). *The lesser apes: evolutionary and behavioural biology*. Edinburgh University Press.

Price, D. (1963). Comparative aspects of development and structure in the prostate. *Nat. Cancer Inst. Monogr.*, **12**, 1–X.

Price, D. and Williams Ashman, H. G. (1961). The accessory reproductive glands of mammals. In *Sex and internal secretions*, Vol. 1. (third edition). (Ed. W. C. Young), pp. 366–448. Williams and Wilkins, Baltimore.

Price, J. L., Russchen, F. T., and Amaral, D. G. (1987). The limbic region II: the amygdaloid complex. In *Handbook of chemical neuroanatomy*. Vol. 5. (Eds. A. Bjorklund, T. Hokfelt, and L. W. Swanson), pp. 279–388. Elsevier, Amsterdam.

Price, J. S., Burton, J. L., Shuster, S., and Wolff, K. (1976). Control of scrotal colour in the vervet monkey. *J. Med. Primatol.*, **5**, 296–304.

Prins, G. S. and Zaneveld L. J. D. (1980). Radiographic study of fluid transport in the rabbit vas deferens during sexual rest and after sexual activity. *J. Reprod. Fertil.*, **58**, 311–319.

Profet, M. (1993). Menstruation as a defence against pathogens transported by sperm. *Q. Rev. Biol.*, **68**, 335–386.

Pugeat, M. M., Chrousos, G. P., Nisula, B. C., Loriaux, D. L., Brandon, D., and Lipsett, M. B. (1984). Plasma cortisol transport and primate evolution. *Endocrinology*, **115**, 357–361.

Purvis, K. and Haynes, N. B. (1974). Short term effect of copulation, human chorionic gonadotrophin injection and non-tactile association with a female on testosterone levels in the male rat. *Endocrinology*, **60**, 429–439.

Purvis, K., Landgren, B. M., Cekan, Z., and Diczfalusy, E. (1976). Endocrine effects of masturbation in man. *J. Endocr.*, **70**, 439–444.

Pusey, A. E. (1990). Mechanics of inbreeding avoidance in nonhuman primates. In *Pedophilia: biosocial dimensions*. (ed. J. R. Feierman), pp. 201–220. Springer, New York.

Pusey, A. E. and Packer, C. (1987). Dispersal and philopatry. In *Primate Societies* (eds. B. Smuts, D. Cheney, R. Seyfarth, R. Wrangham and T. Struhsaker), pp. 250–266, University of Chicago Press.

Pusey, A., Williams, J., and Goodall, J. (1997). The influence of dominance rank on the reproductive success of female chimpanzees. *Nature Lond.*, **277**, 828–831.

Quintin, L., Buda, M., Hilavic, G., Bardelay, C., Ghignone, M., and Piejol, J. F. (1986). Catecholamine metabolism in the rat locus coeruleus as studies by *in vivo* differential pulse voltammetry. III. Evidence for an alpha-2-adrenergic tonic inhibition in behaving rats. *Brain Res.*, **375**, 235–245.

Quris, R. (1976). Donnea3es comparatives sur la socioécologie de huit espèces de Cercopithecidae vivant dans une même zon de forêt primitive periodiquement inondée (Nord-est du Gabon) *Terre et Vie*, **30**, 193–209.

Racey, P. A. (1979). The prolonged storage and survival of spermatozoa in *Chiroptera. J. Reprod. Fertil.*, **56**, 403–416.

Racey, P. A. and Skinner, J. D. (1979). Endocrine aspects of sexual mimicry in spotted hyenas, *Crocuta crocuta. J. Zool. Lond.*, **187**, 315–326.

Rachman, S. and Hodgson, R. (1968). Experimentally induced 'sexual fetishism': replication and development. *Psychol. Rec.*, **18**, 25–27.

Raemakers, J. J. and Chivers, D. J. (1980). Socio-ecology of Malayan forest primates. In *Malayan forest primates: ten year's study in tropical rain forest*. (ed D. J. Chivers), pp. 279–316. Plenum Press, New York.

Rahaman, H. and Parthasarathy, M. D. (1969a). The role of olfactory signals in the mating behavior of bonnet monkeys (*Macaca radiata*). *Commun. Behav. Biol.*, **6**, 97–104.

Rahaman, H. and Parthasarathy, M. D. (1969b) Studies on the social behaviour of bonnet monkeys. *Primates*, **10**, 149–162.

Raisman, G. and Field, P. M. (1971). Sexual dimorphism in the preoptic area of the rat. *Science*, **173**, 731–733.

Raleigh, M. J. and McGuire, M. T. (1990). Social influences on endocrine function in male vervet monkeys. In *Socioendocrinology of primate reproduction*. (eds. T. E. Ziegler and F. B. Bercovich), pp. 95–111. Wiley-Liss, New York.

Ralls, K. (1976). Mammals in which females are larger than males. *Q. Rev. Biol.*, **51**, 245–276.

Ralls, K. and Ballou, J. (1982). Effects of inbreeding on infant mortality in captive primates. *Int. J. Primatol.*, **3**, 491–505.

Ramirez, M. (1988). The woolly spiders, genus *Lagothrix*. In *Ecology and behavior of neotropical primates*, Vol. 2. (eds. R. A. Mittermeier, A. B. Rylands, A. F. Coimbra-Filho and G. A. B. Fonseca), pp. 539–575. World Wildlife Fund, Washington, DC.

Ramos, A. S. Jr (1979). Morphologic variations along the length of the monkey vas deferens. *Arch. Androl.*, **3**, 187–196.

Ransom, T. W. (1981). *Beach troop of the Gombe*. Bucknell University Press, Lewisburg.

Ransom, T. W. and Rowell, T. E. (1972). Early social development of feral baboons. In *Primate Socialization* (ed. F. E. Poirier), pp. 105–144. Random House, New York.

Rasmussen, K. (1983). Age-related variation in the interactions of adult females with adult males in yellow baboons. In *Primate social relationships*. (ed. R. A. Hinde), pp. 47–53. Blackwell, Oxford.

Read, G. F. (1993). Status report on measurement of salivary estrogens and androgens. In *Salvia as a diagnostic fluid. Ann. N. Y. Acad. Sci.*, **694**, 146–160.

Rees, H. D., Bonsall, R. W., and Michael, R. P. (1986). Pre-optic and hypothalamic neurons accumulate [3H] medroxyprogesterone acetate in male cynomolgus monkeys. *Life Sci.*, **39**, 1353–1359.

Rees, H. D., Bonsall, R. W., and Michael, R. P. (1988). Estrogen binding and the actions of testosterone in the brain of the male rhesus monkey. *Brain Res.*, **452**, 28–38.

Rees, J. A. and Harvey, P. H. (1991). The evolution of animal mating systems. In *Mating and Marriage* (eds. V. Reynolds and J. Kellett), pp. 1–45. Oxford University Press.

Reichard, U. (1995). Extra-pair copulations in a monogamous gibbon (*Hylobates lar*). *Ethology*, **100**, 99–112.

Reichard, U., and Sommer, V. (1997). Group encounters in wild gibbons (*Hylobates lar*): agonism, affiliation and the concept of infanticide. *Behaviour.*, **134**, 1135–1174.

Reinberg, A., Lagoguey, M., Chauffournier, J. M., and Cesselin, F. (1975). Circannual and circadian rhythms in plasma testosterone in five healthy young Parisian males. *Acta. Endocr. Copenh.*, **80**, 732.

Reinisch, J. M. (1981). Prenatal exposure to synthetic progestins increases potential for aggression in humans. *Science*, **211**, 1171–1173.

Resko, J. A. (1970a). Androgen secretion by the fetal and neonatal rhesus monkey. *Endocrinology*, **87**, 680–687.

Resko, J. A. (1970b). Androgens in systemic plasma of male guinea pigs during development and after castration in adulthood. *Endocrinology*, **86**, 1444–1447.

Resko, J. A. (1971). Sex steroids in adrenal effluent plasma of the ovariectomized rhesus monkeys. *J. Clin. Endocr.*, **33**, 940–948.

Resko, J. A. (1974). The relationship between fetal hormones and the differentiation of the central nervous system in primates. In *Reproductive behavior*. (eds. W. Montagna and W. Sadler), pp. 211–222. Plenum, New York.

Resko, J. A., Feder, H. H., and Goy, R W. (1968). Androgen concentrations in plasma and testes of developing rats. *J. Endocrinol.*, **40**, 485–491.

Reyes, F. I., Winter, J. S. D., Faiman, C., and Hobson, W. C. (1975). Serial serum levels of gonadotropins, prolactin and sex steroids in the non-pregnant and pregnant chimpanzee. *Endocrinology*, **96**, 1447–1455.

Reynolds, J. and Keverne, E. B. (1979). Accessory olfactory system and its role in pheromonally mediated suppression of oestrus in grouped mice. *J. Reprod. Fertil.*, **57**, 31–35.

Reynolds, R. L. and Van Horn, R. N. (1977). Induction of estrus in intact *Lemur catta* under photoinhibition of ovarian cycles. *Physiol. Behav.*, **18**, 693–700.

Reynolds, V. (1991). The biological basis of human patterns of mating and marriage. In *Mating and marriage*. (eds. V. Reynolds and J. Kellett), pp. 46–90. Oxford University Press, New York.

Rhine, R. J. (1994). A twenty-one-year study of maternal dominance and secondary sex ratio in a colony group of stumptailed macaques (*Macaca arctoides*). *Am. J. Primatol.*, **32**, 145–148.

Rhine, R. J., Norton, G. W., Rogers, J., and Wasser, S. K. (1992). Secondary sex ratio and maternal dominance rank among wild yellow baboons (*Papio cynocephalus*) of Mikimi National Park, Tanzania. *Am. J. Primatol.*, **27**, 261–273.

Richard, A. (1974). Intra-specific variation in the social organization and ecology of *Propithecus verreauxi*. *Folia Primatol.*, **22**, 178–207.

Richard, A. (1976). Patterns of mating in *Propithecus verreauxi verreauxi*. In *Prosimian behaviour*. (eds. R. D. Martin, G. A. Doyle and A. C. Walker), pp. 49–74. Duckworth, London.

Richard, A. (1978). *Behavioral variation: case study of a Malagasy lemur*. Bucknell University Press, Lewisburg, Pennsylvania.

Richard, A. (1985). *Primates in nature*. W. H. Freeman, New York.

Richards, C.S., Watkins, S.C., Hoffman, E.P., Schneider, N.R., Milsark, I.W., Katz, K.S., Coor, J.D., Kunzel, L.M., and Contada, J.M. (1990). Skewed X inactivation in a female MZ twin results in Duchenne muscular dystrophy. *Am. J. hum. Genet.*, **46**, 672–681.

Ridley, M. (1986). The number of males in a primate group. *Anim. Behav.*, **34**, 1848–1858.

Rijksen, H. (1978). A field study on Sumatran orang-utans (*Pongo pygmaeus abelii*, Lesson 1827): ecology behaviour and conservation. Veenman and Zonen, Wageningen, Netherlands.

Riley, A. J. (1994). Yohimbine in the treatment of erectile disorder. *Br. J. Clin. Pharmacol.*, **38**, 133–136.

Riley, A. J. and Riley, E. J. (1993). Pharmacotherapy for sexual dysfunction: current status. In *Sexual pharmacology*. (eds. A. J. Riley, M. Peet, and C. Wilson), pp. 211–226. Clarendon Press, Oxford.

Riley, A. J., Peet, M., and Wilson, C. A. (1993). (eds.). *Sexual pharmacology*. Clarendon Press, Oxford.

Ripley, S. (1967). Intertroop encounters among Ceylon grey langurs (*Presbytis entellus*). In *Social communication among primates*. (ed. S. A. Altmann), pp. University of Chicago Press.

Robertson, G. S., Pfaus, J. G., Atkinson, L. J., Matsumura, H., Phillips, A. G., and Fibeger, H. C. (1991). Sexual behavior increases C-fos expression in the forebrain of the male rat. *Brain Res.*, **564**, 352–357.

Robinson, B. M. and Mishkin, M. (1966). Ejaculation evoked by stimulation of the preoptic area in monkeys. *Physiol. Behav.*, **1**, 269–272.

Robinson, J. A. and Bridson, W. E. (1978). Neonatal hormone patterns in the macaque. 1. steroids. *Biol. Reprod.*, **19**, 773–778.

Robinson, J. A. and Goy, R. W. (1986). Steroid hormones and the ovarian cycle. In *Comparative primate biology*, Vol. 3. reproduction and development. (eds. W. R. Dukelow and J. Erwin), pp. 63–91. Alan R. Liss, New York.

Robinson, J. A., Scheffler, G., Eisele, S. G., and Goy, R. W. (1975). Effects of age and season on sexual behavior and plasma testosterone and dihydrotestosterone concentrations of laboratory-housed male rhesus monkeys (*Macaca mulatta*). *Biol. Reprod.*, **13**, 203–210.

Robinson, J. E. and Short, R. V. (1977). Changes in human breast sensitivity at puberty, during the menstrual cycle and at parturition. *Br. Med. J.* i, 1188–1191.

Robinson, J. G. and Janson, C. H. (1987). Capuchins, squirrel monkeys and atelines: socioecological convergence with Old World primates. In *Primate societies* (eds. B. B. Smuts, D. L. Cheney, R. M. Seyfarth, R. W. Wrangham and T. T. Struhsaker), pp. 69–82. University of Chicago Press.

Robinson, S. M., Fox, T. O., Dikkes, P., and Pearlstein, R. A. (1986). Sex differences in the shape of the sexually dimorphic nucleus of the preoptic area and suprachiasmatic nucleus of the rat: 3-D computer reconstructions and morphometrics. *Brain Res.*, **371**, 380–384.

Rodger, J. C. and Bedford, J. M. (1982). Separation of sperm pairs and sperm-egg interaction in the opossum, *Didelphis virginiana*. *J. Reprod. Fertil.*, **64**, 171–179.

Rodman, P. S. and Mitani, J. C. (1987). Orang-utans: sexual dimorphism in a solitary species. In *Primate societies*. (eds. B. Smuts, D. Cheney, R. Seyfarth, R. Wrangham and T. Struhsaker), pp. 146–154. University of Chicago Press.

Rodriquez-Sierra, J. F., Crowley, W. R., and Komisaruk, B. R. (1975). Vaginal stimulation in rats induces prolonged lorosis responsiveness and sexual receptivity. *J. Comp. Physiol. Psychol.*, **89**, 79–85.

Rogers, C. M. and Davenport, R. K. (1969). Effects of restricted rearing on sexual behavior of chimpanzees. *Devel. Psychol.*, **1**, 200–204.

Rolls, E. T. (1986). Neural systems involved in emotion in primates. In *Emotion: theory, research and experience*, Vol. 3. (eds. R. Plutchik and H. Kellerman), pp. 125–143. Academic Press, New York.

Rolls, E. T. (1990). A theory of emotion, and its application to understanding the neural basis of emotion. *Cognition and Emotion*, **4**, 161–190.

Rolls, E. T. (1992). Neurophysiology and functions of the primate amygdala. In *The amygdala: neurobiological aspects of emotion, memory and mental dysfunction*. (ed. J. Aggleton), pp. 143–165. Wiley-Liss, New York.

Rolls, E. T. and Baylis, L. L. (1994). Gustatory, olfactory and visual convergence within the primate orbitofrontal cortex. *J. Neurosci.*, **14**, 5437–5452.

Rolls, E. T. and Williams, G. V. (1987). Sensory and movement-related neuronal activity in different regions of the primate striatum. In *Basal ganglia and behavior: sensory aspects and motor functioning*. (eds. J. S. Schneider and T. I. Lidsky), pp. 37–59. Hans Huber, Bern.

Romer, A. S. (1962). *The vertebrate body*, (3rd edn). Saunders, Philadelphia.

Roos, J., Roos, M., Schaeffer, C., and Aron, C. (1988). Sexual differences in the development of accessory olfactory bulbs in the rat. *J. Comp. Neurol.*, **270**, 121–131.

Root, W. S. and Bard, P. (1947). The mediation of feline erection through sympathetic pathways with some reference on sexual behavior after deafferentation of the genitalia. *Am. J. Physiol.*, **151**, 80–90.

Rose, J. D. and Michael, R. P. (1978). Facilitation by estradiol of midbrain and pontine unit responses to vaginal and somatosensory stimulation in the squirrel monkey (*Saimiri sciureus*). *Exp. Neurol.*, **58**, 46–58.

Rose, R. M., Gordon, T. P., and Bernstein, I. S. (1972). Plasma testosterone levels in the male rhesus: effects of sexual and social stimuli. *Science*, **178**, 643–645.

Rose, R. M., Bernstein, I. S., and Gordon, T. P. (1975). Consequences of social conflict on plasma testosterone levels in rhesus monkeys. *Psychosom. Med.*, **37**, 50–61.

Rose, R. M., Bernstein, I. S., Gordon, T. P., Lindsley, J. G. (1978). Changes in testosterone and behavior during adolescence in the male rhesus monkey. *Psychosom. Med.*, **40**, 60–70.

Rose, R. W., Nevison, C. M., and Dixson, A. F. (1997). Testes weight, body weight and mating systems in marsupials and monotremes. *J. Zool. Lond.*, **243**, 523–531.

Rosenberger, A. L. and Norconk, M. A. (1996). New perspectives on the pithecines. In *Adaptive radiation of neotropical primates*. (eds. M A. Norconk, A. L. Rosenberger, and P. A. Garber). Plenum, New York.

Rosenblatt, J. S. and Aronson, L. R. (1958). The decline of sexual behavior in male cats after castration with special reference to the role of prior sexual experience. *Behaviour*, **12**, 285–338.

Rosenblum, L. A. and Cooper, R. W. (1968). (eds.). *The squirrel monkey*. Academic Press, New York.

Rosenblum, L. A. and Nadler, R. D. (1971). The ontogeny of sexual behavior in male bonnet macaques. In *Influence of hormones on the nervous system*. (ed. D. H. Ford). Karger, Basel.

Rosenblum, L. A., Nathan, T., Nelson, J., and Kaufman, I. C. (1967). Vaginal cornification cycles in the squirrel monkey (*Saimiri sciureus*). *Folia Primatol.*, **6**, 83–91.

Rostal, D. C., Glick, B. B., Eaton, G. G., and Resko, J. A. (1986). Seasonality of adult male Japanese macaques (*Macaca fuscata*): androgens and behavior in a confined troop. *Horm. Behav.*, **20**, 452–462.

Rothe, H. (1975). Some aspects of sexuality and reproduction in groups of captive marmosets (*Callithrix jacchus*). *Zeitschrift fur Tierpsychol.*, **37**, 255–273.

Roussel, J. D. and Austin, C. R. (1967). Enzymic liquefaction of primate semen. *Int. J. Fertil.*, **12**, 288–290.

Rowe, N. (1996). *The pictorial guide to the living primates*. Pogonian Press, New York.

Rowell, T. E. (1963). Behaviour and reproductive cycles of female macaques. *J. Reprod. Fertil.*, **6**, 193–203.

Rowell, T. E. (1967a). A quantitative comparison of the behaviour of a wild and caged baboon group. *Anim. Behav.*, **15**, 499–509.

Rowell, T. E. (1967b). Female reproductive cycles and the behavior of baboons and rhesus macaques. In *Social Communication among primates*. (ed. S. A. Altmann), pp. 15–32. University of Chicago Press.

Rowell, T. E. (1970a). Baboon menstrual cycles affected by social environment. *J. Reprod. Fertil.*, **21**, 133–141.

Rowell, T. E. (1970b). Reproductive cycles of two *Cercopithecus* monkeys. *J. Reprod. Fertil.*, **22**, 321–338.

Rowell, T. E. (1972). Organization of caged groups of *Cercopithecus* monkeys. *Anim. Behav.*, **19**, 625–645.

Rowell, T. E. (1973). Social organization of wild talapoin monkeys. *Amer. J. Phys: Anthropol.*, **38**, 593–598.

Rowell, T. E. (1977a). Reproductive cycles of the talapoin monkey. *Folia Primatol.*, **28**, 188–202.

Rowell, T. E. (1977b). Variation in age at puberty in monkeys. *Folia Primatol.*, **27**, 284–290.

Rowell, T. E. (1988). Beyond the one male group. *Behaviour*, **104**, 188–201.

Rowell, T. E. and Chalmers, N. R. (1970). Reproductive cycles of the mangabey *Cercocebus albigena*. *Folia Primatol.*, **12**, 264–272.

Rowell, T. E. and Dixson, A. F. (1975). Changes in social organization during the breeding season of wild talapoin monkeys. *J. Reprod. Fert.*, **43**, 419–434.

Rowell, T. E. and Hartwell, K. M. (1978). The interaction of behaviour and reproductive cycles in patas monkeys. *Behav. Biol.*, **24**, 141–167.

Rowland, D. L., Kallan, K., and Slob, A. K. (1997). Yohimbine, erectile capacity, and sexual response in men. *Arch. Sex. Behav.*, **26**, 49–62.

Royal College of General Practitioners (1974). *Oral contraceptives and health*. Pitman, London.

Rubin, H. B. (1967). Temporal patterns of sexual behavior in rabbits as determined by an automatic recording technique. *J. Exp. Anal. Behav.*, **10**, 219–231.

Rubinstein, B. B., Strauss, H., Lazarus, M. L., and Hankin, H. (1951). Sperm survival in women. Motile sperm in the fundus and tubes of surgical cases. *Fertil. Steril.*, **2**, 15–19.

Rudran, R. (1973). The reproductive cycles of two subspecies of purple-faced langurs (*Presbytis senex*) with relation to environmental factors. *Folia Primatol.*, **19**, 41–60.

Rudran, R. (1979). The demography and social mobility of a red howler (*Alouatta seniculus*) population in Venezuela. In *Vertebrate ecology in the northern neotropics*. (ed. J. F. Eisenberg), Smithsonian Institution, Washington, D.C.

Ruiz de Elvira, M., Herndon, J., and Collins, D. (1983). Effect of estradiol-treated females on all female groups of rhesus monkeys during the transition between the nonbreeding and breeding seasons. *Folia Primatol.*, **41**, 191–203.

Rylands, A. B. (1981). Preliminary field observations on the marmoset *Callithrix humeralifer intermedius* (Hershkovitz, 1977) at Dardanalos, Rio Aripuanã, Mato Grosso. *Primates*, **22**, 46–59.

Rylands, A. B. (1982). The behaviour and ecology of three species of marmosets and tamarins (*Callithrichidae*, Primates) in Brazil. Unpublished Ph.D thesis, University of Cambridge.

Rylands, A. B. (1986). Infant-carrying in a wild marmoset group, *Callithrix humeralifer*: evidence for a polyandrous mating system. In *A primatologia no Brasil-2* (ed. M. T. de Mello), pp. 131–144. Sociedade Brasileira de Primatologia, Brasilia.

Ryser, J. (1992). The mating system and male mating success of the Virginia opossum (*Didelphis virginiana*) in Florida. *J. Zool. Lond.*, **228**, 127–139.

Saayman, G. S. (1970). The menstrual cycle and sexual behaviour in a troop of free-ranging chacma baboons (*Papio ursinus*). *Folia Primatol.*, **12**, 81–110.

Saayman, G. S. (1971). Behaviour of the adult males in a troop of free-ranging chacma baboons (*Papio ursinus*). *Folia Primatol.*, **15**, 36–57.

Saayman, G. S. (1973). Effects of ovarian hormones on the sexual skin and behavior of ovariectomized baboons (*Papio ursinus*) under free-ranging conditions. In *Proc. IVth Int. Congr. Primatol.*, Vol. 2 (ed. C. H. Phoenix), pp. 64–98. Karger, Basel.

Sachs, B. D. (1982). Role of the rat's striated penile muscles in penile reflexes, copulation and induction of pregnancy. *J. Reprod. Fertil.*, **66**, 433–443.

Sachs, B. D. (1983). Potency and fertility: hormonal and mechanical causes and effects of penile actions in rats. In *Hormones and behaviour in higher vertebrates*. (eds. J. Balthazart, E. Prove, and R. Gilles), pp. 86–110. Springer, Berlin.

Sachs, B. D. and Barfield, R. J. (1976). Functional analysis of masculine copulatory behavior in the rat. *Adv. Study Behav.*, **7**, 91–154.

Sachs, B. D. and Meisel, R. (1988). The physiology of male sexual behavior. In *The physiology of reproduction*, Vol. 2. (eds. E. Knobil and J. D. Neill), pp. 1393–1485. Raven Press, New York.

Sachs, B. D., Pollak, E. I., Krieger, M. A., and Barfield, R. J. (1973). Sexual behavior: normal male patterning in androgenized female rats. *Science*, **181**, 770–772.

Sade, D. (1964). Seasonal cycle in size of testes of free-ranging *Macaca mulatta*. *Folia Primatol.*, **2**, 171–180.

Saenz De Tejada, I., Carson, M. P., De Las Morenas, A., Goldstein, L., and Traish, A. M. (1991). Endothelin: localization, synthesis, activity and receptor types in human penile corpus cavernosum. *Am. J. Physiol (Heart Circ. Physiol. 30).*, **261**, H1078-H1085.

Saginor, M. and Horton, R. (1968). Reflex release of gonadotropin and increased plasma testosterone concentration in male rabbits during copulation. *Endocrinology*, **82**, 627–630.

Said, S. I. and Mutt, V. (1970). Polypeptide with broad biological activity: isolation from small intestine. *Science*, **169**, 1217–1218.

Saino, N. and Møller, A. P. (1994). Secondary sexual characters, parasites and testosterone in the barn swallow, *Hirundo rustica*. *Anim. Behav.*, **48**, 1325–1333.

Sakuma, Y. (1994). Estrogen-induced changes in the neural impulse flow from the female rat preoptic region. *Horm. Behav.*, **34**, 438–444.

Sakuma, Y. and Pfaff, D. W. (1980). LH-RH in the mesencephalic central gray can potentiate lordosis reflex of female rats. *Nature Lond.*, **283**, 566–567.

Salmon, V. J. and Geist, S. H. (1943). The effect of androgens upon libido in women. *J. Clin. Endocr.*, **3**, 235–238.

Sanders, D. and Bancroft, J. (1982). Hormones and the sexuality of women—the menstrual cycle. *Clinics in Endocr. and Metab.*, **11**, 639–659.

Sanders, D., Warner, P., Bäckström, T., and Bancroft, J. (1983). Mood, sexuality, hormones and the menstrual cycle. 1. Changes in mood and physical state: description of subjects and method. *Psychosom. Med.*, **45**, 487–501.

Sapolsky, R. M. (1982). The endocrine stress response and social status in the wild baboon. *Horm. Behav.*, **16**, 279–292.

Sapolsky, R. M. (1985). Stress-induced suppression of testicular function in the wild baboon: role of glucocorticoids. *Endocrinology*, **116**, 2273–2278.

Sapolsky, R. M. (1987). Stress, social status and reproductive physiology in free-living baboons. In *Psychobiology of reproductive behavior: an evolutionary perspective*. (ed. D. Crews), pp. 291–322. Prentice-Hall, Englewood Cliffs, New Jersey.

Sapolsky, R. M. (1989). Hypercortisolism among socially-subordinate wild baboons originates at the CNS level. *Arch. Gen. Psychiatr.*, **46**, 1047–1051.

Sapolsky, R. M. (1993). The physiology of dominance in stable versus unstable social hierarchies. In *Primate social conflict*. (eds. W. A. Mason and S. P. Mendoza), pp. 171–204. State University of New York Press, New York.

Sar, M. and Stumpf, W. E. (1981). Central noradrenergic neurones concentrate ^{3}H-oestradiol. *Nature Lond.*, **289**, 500–501.

Satou, M., Oka, Y., Kusunoki, M., Matsushima, T., Kato, M., Fujita, I., *et al.* (1984). Telecephalic and preoptic areas integrate sexual behavior in hime salmon (landlocked red salmon, *Oncorhynchus nerka*): results of electrical brain stimulation experiments. *Physiol. Behav.*, **33**, 441–447.

Sauther, M. L. (1991). Reproductive behavior of free-ranging *Lemur catta* at Beza Mahafaly special reserve, Madagascar. *Am. J. Phys. Anthrop.*, **84**, 463–477.

Savage, S. and Bakeman, R. (1978). Sexual morphology and

behavior in *Pan paniscus*. In *Recent advances in primatology*, Vol. 1. *Behaviour*. (eds. D. J. Chivers and J. Herbert), pp. 613–616. Academic Press, London.

Sayag, N., Hoffman, N. W., and Yahr, P. (1994). Telencephalic connections of the sexually dimorphic area of the gerbil hypothalamus that influence male sexual behavior. *Behav. Neurosci.*, **108**, 743–757.

Schaller, G. B. (1963). The mountain gorilla: ecology and behavior. University of Chicago Press.

Schantz, T. V., Wittzell, H., Göransson, G., Grahn, M., and Persson, K. (1996). MHC genotype and male ornamentation: genetic evidence for the Hamilton-Zuk model. *Proc. Roy. Soc. Lond. B.*, **263**, 265–271.

Schapiro, S. J. (1985). Reproductive seasonality: birth synchrony, female-female reproductive competition and cooperation in captive *Cercopithecus aethiops* and *C. mitis*. Unpublished Ph.D. dissertation, University of California, Davis.

Schenck, P. E. and Slob, A. K. (1986). Castration, sex steroids and heterosexual behavior in adult, laboratory housed stumptail macaques (*Macaca arctoides*). *Horm. Behav.*, **20**, 336–353.

Schilling, A. (1979). Olfactory communication in prosimians. In *The study of prosimian behavior*. (eds. G. A. Doyle and R. D. Martin), pp. 461–542. Academic Press, New York.

Schmidt, R. S. (1982). Masculinzation of toad pretrigeminal nucleus by androgens. *Brain Res.*, **244**, 190–192.

Schofield, S. P. M. and Dixson, A. F. (1982). Comparative anatomy of brain monoamingergic neurons in New World and Old World monkeys. *Am. J. Primatol.*, **2**, 3–19.

Schön, M. A. (1971). The anatomy of the resonating mechanism in howling monkeys. *Folia Primatol.*, **15**, 117–132.

Schreiner-Engel, P. (1980). Female sexual arousability: its relation to gonadal hormones and the menstrual cycle. Unpublished Ph.D. thesis, New York University.

Schreiner-Engel, P., Schiavi, R. C., Smith, H. and White, D. (1981). Sexual arousability and the menstrual cycle. *Psychosom. Med.*, **43**, 199–214.

Schubert, G. (1982). Infanticide by usurper Hanuman langur males: a sociobiological myth. *Soc. Sci. Inf.*, **21**, 199–244.

Schultz, A. H. (1938). The relative weights of the testes in primates. *Anat. Rec.*, **72**, 387–394.

Schultz, A. H. (1969). *The life of primates*. Weidenfeld and Nicolson, London.

Schultz, L. (1928). Zur Kenntnis des Körpers der Hottentotten und Buschmnner. *Denkaschr. Med.- Naturwiss. Ges. Jena* **17**, 145–227.

Schultze, H. G. and Gorzalka, B. B. (1991). Oxytocin effects on lordosis frequency and lordosis duration following infusion into the medial preoptic area and ventromedial hypothalamus of female rats. *Neuropeptides*, **18**, 99–106.

Schulze, H. and Meier, B. (1995). Behavior of captive *Loris tardigradus nordicus*: a qualitative description, including some information about morphological bases of behavior. In *Creatures of the dark, the nocturnal prosimians*. (eds. L. Alterman, G. A. Doyle and M. K. Izard), pp. 221–249. Plenum, New York.

Schumacher, M., Coirini, H., Pfaff, D. W., and McEwen, B. S. (1990). Behavioral effects of progesterone associated with rapid modulation of oxytocin receptors. *Science*, **250**, 691–694.

Schurmann, C. (1982). Mating behaviour of wild orangutans. In *The orangutan: its ecology and conservation*. (ed. L. E. M. de Boer), pp. 269–284. Dr. W. Junk, The Hague.

Schwartz, M. and Beach. F. A. (1954). Effects of adrenalectomy upon mating behavior in castrated male dogs. *Am. Psychol.*, **9**, 467–468.

Schwartz, M. F., Banman, J. E., and Masters, W. H. (1982). Hyperprolactinaemia and sexual dysfunction in men. *Biol. Psychiatr.*, **17**, 861–876.

Schwartz, S. M., Wilson, M. E., Walker, M. E., and Collins, D. C. (1985). Social and growth correlates of puberty onset in female rhesus monkeys. *Nutr. Behav.*, **2**, 225–232.

Schwartz, S. M., Wilson, M. E., Walker, M. E., and Collins, D. C. (1988). Dietary influences in growth and sexual maturation in premenarchial rhesus monkeys. *Horm. Behav.*, **22**, 231–251.

Scott, L. (1984). Reproductive behavior of adolescent female baboons (*Papio anubis*). In *Female primates: studies by women primatologists*. (ed. M. Small), pp. 77–102. Alan R. Liss, New York.

Scruton, D. M. and Herbert, J. (1970). The menstrual cycle and its effect upon behaviour in the talapoin monkey (*Miopithecus talapoin*). *J. Zool. Lond.*, **162**, 419–436.

Segraves, R. T., Bari, M., Segraves, K., and Spirnak, P. (1991). Effect of apomorphine on penile tumescence in men with psychogenic impotence. *J. Urol.*, **145**, 1174–1175.

Sekulic, R. (1982). Daily and seasonal patterns of roaring and spacing in four red howler (*Alouatta seniculus*) troops. *Folia Primatol.*, **39**, 22–48.

Sekulic, R. (1983). Male relationships and infant deaths in red howler monkeys (*Alouatta seniculus*). *Z. für Tierpsychol.*, **61**, 185–202.

Semans, J. H. and Langworthy, D. R. (1938). Observation on the neuropathy of sexual function in the male cat. *J. Urol.*, **40**, 836–846.

Setalo, G., Vigh, S., Schally, V., Arimura, A., and Flerki, B. (1976). Immunohistological study of the origin of LH-RH-containing nerve fibres in the rat hypothalamus. *Brain Res.*, **103**, 597–602.

Setchell, B P. (1982). Spermatogenesis and spermatozoa. In *Reproduction in Mammals* Vol. 1. *Germ cells and fertilization* (eds. C. R. Austin and R. V Short), pp. 63–101. Cambridge University Press.

Settlage, D. S. F. and Hendrickx, A. G. (1974). Observations on coagulation characteristics of the rhesus monkey electroejaculate. *Biol. Reprod.*, **11**, 619–623.

Settlage, D. S. F., Motoshrima, M., and Tredway, D. R. (1973). Sperm transport from the external cervical os to the fallopian tubes in women: a time and quantitation study. *Fertil. Steril.*, **24**, 655–661.

Seuánez, H. N. (1980). Chromosomes and spermatozoa of the African great apes. *J. Reprod. Fertil. Suppl.*, **28**, 91–104.

Seyfarth, R. M. (1978). Social relationships among adult male and female baboons, I: behaviour during sexual consortship. *Behaviour*, **64**, 204–226.

Shadle, A. R., Mirand, E. A., and Grace, J. T. Jr (165). Breeding responses in tamarins. *Lab. Anim. Care*, **15**, 1–10.

Shepher, J. (1971). Mate selection among second generation Kibbutz adolescents and adults: incest avoidance and negative imprinting. *Arch. Sex Behav.*, **1**, 293–308.

Sheridan, P. J. and Weaker, F. J. (1981). The primate spinal cord is a target for gonadal steroids. *J. Neuropath. Exp. Neurol.*, **40**, 447–453.

Sherwin, B. B. (1988). A comparative analysis of the role of androgen in human male and female sexual behavior: behavioral specificity, critical thresholds and sensitivity. *Psychobiology*, **16**, 416–425.

Sherwin, B. B. and Gelfand, M. M. (1987). The role of androgen in the maintenance of sexual functioning in oophorectomised women. *Phsychosom. Med.*, **49**, 397–409.

Sherwin, B. B., Gelfand, M. M., and Brender, W. (1985). Androgen enhances sexual motivation in females: a prospective, crossover study of sex steroid administration in the surgical menopause. *Psychosom. Med.*, **47**, 339–351.

Shideler, S. E., Lindburg, D. G., and Lasley, B. L. (1983). Estrogen-behavior correlates in the reproductive physiology and behavior of the ruffed lemur (*Lemur variegatus*). *Horm. Behav.*, **17**, 249–263.

Shimura, T. and Shimokochi, M. (1991). Modification of male rat copulatory behavior by lateral midbrain stimulation. *Physiol. Behav.*, **50**, 989–994.

Shively, C. and Kaplan, J. (1984). Effects of social factors on adrenal weight and related physiology of *Macaca fascicularis*. *Physiol. Behav.*, **33**, 777–782

Shively, C., Clarke, S., King, N., Schapiro, S., and Mitchell. G. (1982). Patterns of sexual behavior in male macaques. *Am. J. Primatol.*, **2**, 373–384.

Shopland, J. M. (1982). An intergroup encounter with fatal consequences in yellow baboons (*Papio cynocephalus*). *Am. J. Primatol.*, **3**, 263–266.

Short, R. V. (1976). The evolution of human reproduction. *Proc. Roy. Soc. Lond. B.*, **195**, 3–24.

Short, R. V. (1979). Sexual selection and its component parts, somatic and genital selection, as illustrated by man and the great apes. *Adv. Study Behav.*, **9**, 131–158.

Short, R. V. (1980). The origins of human sexuality. In *Reproduction in mammals*, Vol. 8. *Human sexuality*. (eds. C. R. Austin and R. V. Short), pp. 1–33. Cambridge University Press.

Short, R. V. (1984). Oestrous and menstrual cycles. In *Reproduction in mammals*, Vol. 3. hormonal control of reproduction. (eds. C. R. Austin and R. V. Short), pp. 115–152. Cambridge University Press.

Short, R. V. (1985). Species differences in reproductive mechanisms. In *Reproduction in mammals*, Vol. 4. *Reproductive fitness*. (eds C. R. Austin and R. V. Short), pp. 24–61. Cambridge University Press.

Siegelman, M. (1974). Parental background of male homosexuals and heterosexuals. *Arch. Sex. Behav.*, **3**, 3–18.

Siegelman, M. (1978). Psychological adjustment of homosexual and heterosexual men: a cross national replication. *Arch. Sex. Behav.*, **7**, 1–12.

Sigg, H., Stolba A., Abegglen, J.-J., and Dasser, V. (1982). Life history of hamadryas baboons: physical development, infant mortality, reproductive parameters and family relationships. *Primates*, **23**, 473–487.

Signoret, J. P., Cognie, Y., and Martin, G. B. (1984). The effect of males on female reproductive physiology. In *The male in farm animal reproduction*. (ed. M. Courot), pp. 290–304. Martinus Nijhoff, Boston.

Silk, J. B. (1983). Local resource competition and facultative adjustment of sex ratios in relation to competitive activities. *Am. Nat.*, **12**, 56–66.

Silk, J. B. (1988). Maternal investment in captive bonnet macaques, *Macaca radiata*. *Am. Nat.*, **132**, 1–19./Silk, J. B., Clark-Wheatley, C. B., Rodman, P. S., and Samuels, A. (1981). Differential reproductive success and facultative readjustment of sex ratios among captive female bonnet macaques (*Macaca radiata*). *Anim. Behav.*, **29**, 1106–1120.

Sillen-Tullberg, B. and Møller, A. P. (1993). The relationship between concealed ovulation and mating systems in primates: a phylogenetic analysis. *Am. Nat.*, **141**, 1–25.

Simonds, P. E. (1965). The bonnet macaque in South India. In *Primate behavior: field studies of monkeys and apes*. (ed. I. DeVore), pp. 175–196. Holt, Rinehart, and Winston, New York.

Simons, E. L. (1988). A new species of *Propithecus* (Primates) from Northeast Madagascar. *Folia Primatol.*, **50**, 143–151.

Simpson, A. E. and Simpson, M. J. A. (1985). Short-term consequences of different breeding histories for captive rhesus macaque mothers and young. *Behav. Ecol. Sociobiol.*, **18**, 83–89.

Simpson, G. G. (1945). The principles of classification and a classification of mammals. *Bull. Am. Mus. Nat. Hist.*, **85**, 1–350.

Simpson, M. J. A. and Simpson, A. E. (1982). Birth sex ratios and social rank in rhesus monkey mothers. *Nature Lond.*, **300**, 440–441.

Singh, D. (1993). Body shape and womens' attractiveness. *Human Nature.*, **4**, 297–321.

Singh, D. and Luis, S. (1995). Ethnic and gender consensus for the effect of waist-to-hip ratio on judgement of womens' attractiveness. *Human Nature*, **6**, 51–65.

Singh, D. and Young, R. K. (1995). Body weight, waist-to-hip ratio, breasts, and hips: role in judgements of female attractiveness and desirability for relationships. *Ethol. Sociobiol.*, **16**, 483–507.

Sirinathsinghji, D. J. S. (1983). GnRH in the spinal subarachnoid space potentiates lordosis behavior in the female rat. *Physiol. Behav.*, **33**, 717–723.

Sirinathsinghji, D. J. S., Whittington, P. E., Audsley, A., and Frazer, H. M. (1983). Beta-endorphin regulates lordosis in female rats by modulating LH-RH release. *Nature Lond.*, **301**, 62–64.

Sittitrai, W., Brown, T., and Virulrak, S. (1992). Patterns of bisexuality in Thailand. In *Biosexuality and HIV/AIDS*. (eds. R. Tielman, M. Carballo, and A. Hendrickx), pp. 97–117. Prometheus Books, Buffalo.

Sjöstrand, N. D. (1965). The adrenergic innervation of the vas deferens and the accessory male genital glands. *Acta Physiol Scand.* Suppl. 257.

Sjöstrand, N. O. and Klinge, E. (1979). Principal mechanisms controlling penile retraction and protrusion in rabbits. *Acta. Physiol. Scand.*, **106**, 199–214.

Skakerberg, G., Björklund, A., Lindvall, O., and Schmidt, RH. (1982). Origin and termination of the diencephalo-spinal dopamine system in the rat. *Brain Res. Bull.*, **9**, 237–244.

Skakkebaek, N. E., Bancroft, J., Davidson, D. W., and Warner, P. W. (1981). Androgen replacement with oral testosterone undecanoate in hypogonadal men: a double blind controlled study. *Clin. Endocr.*, **14**, 49–61.

Slikker, W., Jr, Hill, D. E., and Young, J. F. (1982). Comparison of the transplacental pharmacokinetics of 17β-estradiol and diethylstilbestrol in the subhuman primate. *J. Pharmacol. Exp. Ther.*, **221**, 173–182.

Slimp, J. C., Hart, B. L., and Goy, R. W. (1978). Heterosexual, autosexual and social behavior of adult male rhesus monkeys with medial preoptic-anterior hypothalamic lesions. *Brain Res.*, **142**, 105–122.

Slipjer, F. M. E. (1984). Androgens and gender role behaviour in girls with congenital adrenal hyperplasia (CAH). *Progr. Brain Res.*, **61**, 417–422.

Slob, A. K. and Nieuwenhuijsen, K. (1980). Heterosexual interactions of pairs of laboratory-housed stumptail macaques (*Macaca arctoides*) under continuous observation with closed-circuit video recording. *Int. J. Primatol.*, **1**, 63–80.

Slob, A. K. and Schenck, P. E. (1981). Chemical castration with cyproterone acetate (Androcur) and sexual behavior in laboratory housed stumptail macaques (*Macaca arctoides*). *Physiol. Behav.*, **27**, 629–636.

Slob, A. K., Wiegand, S. J., Goy, R. W., and Robinson, J. A. (1978). Heterosexual interactions in laboratory-housed stumptail macaques (*Macaca arctoides*): observations during the menstrual cycle and after ovariectomy. *Horm. Behav.*, **10**, 193–211.

Slob, A. K., Ooms, M. P., and Vreeberg, J. T. M. (1979). Annual changes in serum testosterone in laboratory-housed male stumptail macaques (*M. arctoides*). *Biol. Reprod.*, **20**, 981–984.

Slob, A. K., Groeneveld, W. H. and Van der Werff Ten Bosch, J. J. (1986). Physiological changes during copulation in male and female stumptail macaques (*Macaca arctoides*). *Physiol. Behav.*, **38**, 891–895.

Slob, A. K., Bax, C. M., Hop. W. C. J., Rowland, D. L., and Van der Werffe ten Bosch, J. J. (1996). Sexual arousability and the menstrual cycle. *Psychoneuroendocrinology*, **21**, 545–558.

Small, M. F. (1988). Female primate sexual behavior and conception. Are there really sperm to spare? *Curr. Anthropol.*, **29**, 81–100.

Small, M. F. (1990). Promiscuity in Barbary macaques (*Macaca sylvanus*). *Am. J. Primatol.*, **20**, 267–282.

Small, M. (1993). *Female choices: sexual behavior of female primates*. Cornell University Press, Ithaca, New York.

Small, M. F. and Hrdy, S. B. (1986). Secondary sex ratios by maternal rank, parity and age in captive rhesus macaques (*Macaca mulatta*). *Am. J. Primatol.*, **11**, 359–365.

Smith, A. H., Butler, T. M., and Pace, N. (1975). Weight growth of colony-reared chimpanzees. *Folia Primatol.*, **24**, 29–59.

Smith, D. G. (1993). A 15-year study of the association between dominance rank and reproductive success of male rhesus macaques. *Primates*, **34**, 471–480.

Smith, E., Damassa, D. A., and Davidson, J. M. (1977). Plasma testosterone and sexual behavior following intracerebral implantation of testosterone propionate in the castrated male rat. *Horm. Behav.*, **8**, 77–87.

Smith, J. D. and Madkour, G. (1980). Penial morphology and the question of chiropteran phylogeny. In *Proceedings of the fifth international bat research conference* (eds. D. E. Wilson and A. L. Gardner), pp. 347–365. Texas Tech. Press, Lubbock.

Smith, M., Harris, P. J., and Strayer, F. F. (1977). Laboratory methods for the assessment of social dominance among captive squirrel monkeys. *Primates*, **18**, 966–984.

Smith, R. L. (1979). Repeated copulation and sperm precedence: paternity assurance for a male brooding waterbug. *Science*, **205**, 1029–1031.

Smith, R. L. (1984). Human sperm competition. In *Sperm competition and the evolution of animal mating systems*. (ed. R. L. Smith), pp. 601–659. Academic Press, New York.

Smith, R. R. and Credland, P. F. (1977). Menstrual and copulatory cycles in the gelada baboon, *Theropithecus gelada*. *Int. Z. Yb.*, **17**, 183–185.

Smith, T. T. and Yanagimachi, R. (1991). Attach and release of spermatozoa from the caudal isthmus of the hamster oviduct. *J. Reprod. Fertil.*, **91**, 567–573.

Smuts, B. B. (1985). *Sex and friendship in baboons*. Aldine, New York.

Smuts, B. B. and Smuts, R. W. (1993). Male aggression and sexual coercion of females in nonhuman primates and other mammals: evidence and theoretical implications. *Adv. Study Behav.*, **22**, 1–63.

Snowdon, C. T. and Soini, P. (1988). The tamarins, genus Saguinus. In *Ecology and behavior of neotropical primates*, Vol. 2. (eds. R. A. Mittermeier, A. B. Rylands and G. A. B. Fonseca), pp. 223–298. World Wildlife Fund, Washington, D.C.

Södersten, P. and Hansen, S. (1978). Effects of castration and testosterone, dihydrotestosterone or oestradiol replacement treatment in neonatal male rats on mounting behaviour in the adult. *J. Endocr.*, **76**, 251–260.

Södersten, P., Eneroth, P., and Hansen, S. (1981). Induction of sexual receptivity in ovariectomized rats by pulse administration of oestradiol-17β. *J. Endocr.*, **89**, 55–62.

Södersten, P., Forsberg, G., Bednar, I., Eneroth, P, and Weisenfeld-Hallin, Z. (1989). Opioid peptide inhibition of sexual behaviour in female rats. In *Brain opioid systems in reproduction*. (eds R. G. Dyer and R. J. Bicknell), pp. 203–215. Oxford University Press.

Soini, P. (1987). Sociosexual behavior of a free-ranging Cebuella pygmaea (Callitrichidae, Platyrrhini) troop during post-partum oestrus of its reproductive female. *Am. J. Primatol.*, **13**, 223–230.

Soini, P. (1988). The pygmy marmoset, genus Cebuella. In *Ecology and behavior of neotropical primates*, Vol. 2 (eds. R. A. Mittermeier, A. B. Rylands and G. A. B Fonseca), pp. 79–129. World Wildlife Fund, Washington, D.C.

Sommer, V. (1987). Infanticide among free-ranging langurs (*Presbytis entellus*) at Jodhpur (Rajasthan/India): recent observations and a reconstruction of hypotheses. *Primates*, **28**, 163–197.

Sommer, V. (1989). Sexual harassment in langur monkeys (*Presbytis entellus*): competition for ova, sperm and nurture? *Ethology*, **80**, 205–217.

Sommer, V. (1993). Infanticide among the langurs of Jodhpur: testing the sexual selection hypothesis with a long-term record. In *Infanticide and parental care*. (eds. S. Parmigiani and F. vom Saal), pp. 155–198. Harwood Academic Publishers, London.

Sommer, V. and Ragpurohit, L. S. (1989). Male reproductive success in harem groups of Hanuman langurs (*Presbytis entellus*). *Int. J. Primatol.*, **10**, 293–317.

Sopchak, A. L. and Sutherland, A. M. (1960). Physiological impact of cancer and its treatment. VII. Exogenous sex hormones and their relation to life-long adaptations in women with metastatic cancer of the breast. *Cancer*, **13**, 528–531.

Southwick, C. H., Beg, M. H., and Siddiqi, M. R. (1965). Rhesus monkeys in north India. In *Primate behavior: field studies of monkeys and apes*. (ed. I. DeVore), pp. 111–159. Holt, Rinehart and Winston, New York.

Spies, H. G. and Niswender, G. D. (1971). Levels of prolactin, LH and FSH in the serum of intact and pelvic-neurectomized rats. *Endocrinology*, **88**, 937–943.

Spies, H. G., Norman, R. L., Clifton, D. K., Ochsner, A. J., Jensen, J. N., and Phoenix, C. H. (1976). Effects of bilateral amygdaloid lesions on gonadal hormones in serum and on sexual behavior in female rhesus monkeys. *Physiol. Behav.*, **17**, 985–992.

Srivastava, A., Borries, C., and Sommer, V. (1991). Homosexual mounting in free-ranging female Hanuman langurs (*Presbytis entellus*). *Arch. Sex. Behav.*, **20**, 487–512.

Stallings, J. R. and Mittermeier, R. A. (1983). The black-tailed marmoset (*Callithrix argentata melanura*) recorded from Paraguay. *Am. J. Primatol.*, **4**, 159–163.

Stammbach, E. (1987). Desert, forest and montane baboons: multilevel societies. In *Primate societies*. (eds. B. Smuts, D. Cheney, R. Seyfarth, R. Wrangham and T. Struhsaker), pp. 112–112. University of Chicago Press.

Starck, D. and Schneider, R. (1960). Respirationsorgane. In *Primatologia* Vol. 3. (eds. H. Hofer, A. H. Schultz and D. Starck), pp. 423–587. Karger, Basel.

Stavy, M. and Herbert, J. (1989). Differential effects of β-endorphin infused into the hypothalamic preoptic area at various phases of the male rat's sexual behavior. *Neuroscience*, **198**, 433–442.

Stearns, E. L., Winter, J. S. D., and Faiman, C. (1973). Effects of coitus upon gonadotropin, prolactin and sex steroid levels in man. *J. Clin. Endocr. Metab.*, **37**, 687–691.

Steimer, T. and Hutchison, J. B. (1981). Androgen increases formation of behaviourally effective oestrogen in the dove brain. *Nature Lond.*, **292**, 345–347.

Steinach, E. (1894). Untersuchungen zur vergleichenden Physiolgie der mannlichen Geschlechtsorgane. 1. Uber den Geschlechtstrieb und den Geschlechlstreib und den Geschlechtakt bei Froschen. *Pfluegers Arch.*, **56**, 304–330.

Steinach, E. (1910). Geschlechstieb und echt sekundare Geschlechsmerkmale als Folge der Innersekretoriechen funktion der Keimdrusen. 11. Uber der Enstehung des Umklammerunsreflexes bei Froschen. *Zentralbl Physiol.*, **24**, 551–570.

Stephenson, G. R. (1975). Social structure of mating activity in Japanese macaques. In *Proceedings of the fifth international congress of primatology*. (eds S. Kondo, M. Kawai, A. Ehara, and S. Kawamura), pp. 63–115. Japan Science Press, Tokyo.

Sterling, E. J. (1993). Patterns of range use and social organization in aye-ayes (*Daubentonia madagascariensis*) on Nosy

Mangabé. In *Lemur social systems and their ecological basis*. (eds P. Kappeler and J. U. Ganzhorn), pp. 1–10. Plenum, New York.

Sterling, E. J. (1994). Evidence for nonseasonal reproduction in wild aye-ayes (*Daubentonia madagascariensis*) in Madagascar. *Folia Primatol.*, **62**, 46–53.

Sterling, E. J. and Richard, A. F. (1995). Social organization in the aye-aye (*Daubentonia madagascariensis*) and the perceived distinctiveness of nocturnal primates. In *Creatures of the dark: the nocturnal prosimians*. (eds L. Alterman, G. A. Doyle and M. K. Izard), pp. 439–451. Plenum, New York.

Stevenson, M. F. and Rylands, A. B. (1988). The marmosets, genus *Callithrix*. In *Ecology and behavior of neotropical primates*, Vol. 2 (eds. R. A. Mittermeier, A. B. Rylands, A. Coimbra-Filho and G. A. B. Fonseca), pp. 131–222. World Wildlife Fund, Washington D. C.

Stief, C. G., Bernard, F., Bosch, R. J. L. H., Aboseif, S. R., Nunes, L., Lue, T., *et al.* (1989). Acetylcholine as a possible neurotransmitter in penile erection. *J. Urol.*, **14**, 1444–1448.

Stief, C. G., Wetterauer, U., Schaebsdau, F. H., and Jonas, U. (1991). Calcitonin-gene-related peptide: a possible role in human penile erection and its therapeutic application in impotent patients. *J. Urol.*, **146**, 1010–1014.

Stief, C. G., Benard, F., Bosch, R., Aboseif, S., Wetterauer, U., Lue, T. F., *et al.* (1993). Calcitonin-gene-related peptide: possibly neurotransmitter contributes to penile erection in monkeys. *Urology*, **41**, 397–401.

Stoleru, S. G., Annaji, A., Cournot, A., and Spira, A. (1993). LH pulsatile secretion and testosterone blood levels are influenced by sexual arousal in human males. *Psychoneuroendocrinology.*, **18**, 205–218.

Stribley, J. A., French, J. A., and Inglett, B. J. (1987). Mating patterns in the golden lion tamarin (*Leontopithecus rosalia*) : continuous receptivity and concealed oestrus. *Folia Primatol.*, **49**, 137–150.

Strier, K. B. (1986). The behavior and ecology of the woolly spider monkey or muriqui (*Brachyteles arachnoides* E. Geoffrey, 1806). Unpublished Ph.D. thesis, Harvard University, U.S.A.

Strier, K. B. (1994). Myth of the typical primate. *Yearbook Phys. Anthropol.*, **37**, 233–271.

Strier, K. B. and Ziegler, T E. (1994). Insights into ovarian function in wild muriqui monkeys. *Am. J. Primatol.*, **32**, 31–40.

Struhsaker, T. T. (1967a). Behavior of vervet monkeys (*Cercopithecus aethiops*). *Univ. Calif. Publ. Zool.*, **82**, 1–74.

Struhsaker, T. T. (1967b). Auditory communication among vervet monkeys (*Cercopithecus aethiops*). In *Social communication among primates*. (ed. S. A. Altmann), pp. 281–324. University of Chicago Press.

Struhsaker, T. T. (1969). Correlates of ecology and social organization among African cercopithecines. *Folia Primatol.*, **11**, 80–118.

Struhsaker, T. T. (1975). *The red colobus monkey*. University of Chicago Press.

Struhsaker, T. T. (1977). Infanticide and social organization in the redtail monkey (*Cercopithecus ascanius schmidti*) in the Kibale forest, Uganda. *Z. Tierpsychol.*, **45**, 75–84.

Struhsaker, T. T. (1988). Male tenure, multi-male influxes and reproductive success in redtail monkeys (*Cercopithecus ascanius*). In *A primate radiation: evolutionary biology of the African guenons*. (eds. A. Gautier-Hion, F. Boulière, J. P. Gautier and J. Kingdon), pp. 340–363. Cambridge University Press.

Struhsaker, T. T. and Gartlan, J. S. (1970). Observations on the behaviour and ecology of the patas monkey (*Erythrocebus patas*) in the Waza Reserve, Cameroon. *J. Zool. Lond.*, **161**, 49–63.

Struhsaker, T. T. and Leland, L. (1979). Socioecology of five sympatric monkey species in the Kibale forest, Uganda. *Adv. Study Behav.*, **9**, 159–228.

Struhsaker, T. T. and Leland, L. (1987). Colobines: Infanticide by Adult Males. In *Primate Societies* (eds. Smuts *et al.*), pp. 83–97. The University of Chicago Press.

Struhsaker, T. T. and Leland, L. (1988). Group fission in redtail monkeys (*Cercopithecus ascanius*) in the Kibale Forest, Uganda. In *A primate radiation: evolutionary biology of the African guenons* (eds. A. Gautier-Hion, F. Boulire, J. P. Gautier and J. Kingdon), pp. 364–388. Cambridge University Press.

Struhsaker, T. T., Butynski, T. M., and Lwanga, J. S. (1988). Hybridization between redtail (*Cercopithecus ascanius schmidti*) and blue (*Cercopithecus mitis stuhlmanni*) monkeys in the Kibale Forest, Uganda. In *A primate radiation: evolutionary biology of the African guenons*. (eds. A. Gautier-Hion, F. Boulière, J. P. Gautier and J. Kingdon), pp. 477–507. Cambridge University Press.

Sugawara, T. (1979). Stretch reception in the bursa copulatrix of the butterfly, *Pieris rapae crucivora* and its role in behaviour. *J. Comp. Physiol.*, **130**, 191–199.

Sugiyama, Y. (1965). On the social change of Hanuman langurs (*Presbytis entellus*) in their natural conditions. *Primates*, **6**, 381–418.

Sugiyama, Y. (1967). Social organization of Hanuman langurs. In *Social communication among primates*. (ed. S. A. Altmann), pp. 221–236. University of Chicago Press.

Sugiyama, Y. (1971). Characteristics of the social life of bonnet macaques (*Macaca radiata*). *Primates*, **12**, 247–266.

Sugiyama, Y. and Ohsawa, H. (1982). Population dynamics of Japanese monkeys with special reference to the effect of artificial feeding. *Folia Primatol.*, **39**, 238–263.

Suh, B. Y., Liu, J. H., Berga, S. L., Quigley, M. E., Lauglin, G. A., and Yen, S. S. (1988). Hypercortisolism in patients with functional hypothalamic amenorrhea. *J. Clin. Endocr. Metabol.*, **66**, 733–739.

Surbey, M. K. (1990). Family composition, stress, and the timing of human menarche. In *Socioendocrinology of primate reproduction*. (eds. T. E. Zeigler and F. B. Bercovitch), pp. 11–32. Wiley-Liss, New York.

Susset, J. G., Tessier, C. D., Wincze, J., Bansal, S., Malhotra, C., and Schwacha, M. G. (1989). Effects of yohimbine hydrochloride on erectile impotence: a double blind study. *J. Urol.*, **141**, 1360–1363.

Sussman, R. W. (1976). Ecological distinctions in sympatric species of *Lemur*. In *Prosimian behavior* (eds. R. D. Martin, G. A. Doyle and A. C. Walker), pp. 75–108. Duckworth, London.

Sussman, R. W. and Garber, P. A. (1987). A new interpretation of the social organization and mating system of the Callitrichidae. *Int. J. Primatol.*, **8**, 73–92.

Suzuki, A. (1971). Carnivority and cannibalism among forest-living chimpanzees. *J. Anthropol. Soc. Nippon*, **79**, 30–48.

Suzuki, Y., Ishii, H., Furuya, H., and Arai, Y. (1982). Developmental changes of the hypogastric ganglion associated with the differentiation of the reproductive tracts in the mouse. *Neurosci. Lett.*, **32**, 271–276.

Suzuki, Y., Ishii, H., and Arai, Y. (1983). Prenatal exposure of male mice to androgen increases neuron number in the hypogastric ganglion. *Devel. Brain Res.*, **10**, 151–154.

Swaab, D F. and Fliers, E. (1985). A sexually dimorphic nucleus in the human brain. *Science*, **228**, 1112–1115.

Swaab, D. F. and Hofman, M. A. (1988). Sexual differentiation of the human hypothalamus: ontogeny of the sexually dimorphic nucleus of the preoptic area. *Devel. Brain Res.*, **44**, 314–318.

Swaab, D. F. and Hofman, M. A. (1990). An enlarged suprachiasmatic nucleus in homosexual men. *Brain Res.*, **537**, 141–148.

Swaab, D. F., Gooren, L. J. G., and Hofman, M. A. (1992). The human hypothalamus in relation to gender and sexual orientation. *Progr. Brain Res.*, **93**, 205–219.

Swaab, D. F., Hofman, M. A., Lucassen, P. J., Purba, J. S., Raadsheer, F C., and Van de Nes, J. A. P. (1993). Functional neuroanatomy and neuropathology of the human hypothalamus. *Anat. Embryol.*, **187**, 317–330.

Swaddle, J. P. (1996). Reproductive success and symmetry in zebra finches. *Anim. Behav.*, **51**, 203–210.

Swaddle, J. P. and Cuthill, I. C. (1994). Female zebra finches prefer symmetric males. *Nature Lond.*, **367**, 165–166.

Swaddle, J. P. and Cuthill, I. C. (1995). Asymmetry and human facial attractiveness: symmetry may not always be beautiful. *Proc. Roy. Soc. Lond. B.*, **261**, 111–116.

Symons, D. (1979). *The evolution of human sexuality*. Oxford University Press, New York.

Synott, A. L., Fulkerson, W. J., and Lindsay, D. R. (1981). Sperm output by rams and distribution amongst ewes under conditions of continual mating. *J. Reprod. Fertil.*, **65**, 355–361.

Szechtman, H., Caggiula, A. R., and Wulkan, D. (1978). Preoptic knife cuts and sexual behavior in male rats. *Brain Res.*, **150**, 569–591.

Taggart, D. A., Johnson, J. L., O'Brien, H. P., and Moore, H. D. M. (1993). Why do spermatozoa of American marsupials form pairs? A clue from the analysis of sperm-pairing in the epididymis of the grey short-tailed possum, *Monodelphis domestica*. *Anat. Rec.*, **236**, 465–478.

Takahata, Y. (1980). The reproductive biology of a free-ranging troop of Japanese monkeys. *Primates*, **21**, 303–329.

Takahata, Y. (1982a). Social relations between adult males and females of Japanese monkeys in the Arashiyama B troop. *Primates*, **23**, 1–23.

Takahata, Y. (1982b). The socio-sexual behaviour of Japanese monkeys. *Z. Tierpsychol.*, **59**, 89–108.

Takahata, Y. (1985). Adult male chimpanzees kill and eat a newborn infant: newly observed intragroup infanticide and cannibalism in Mahale National Park, Tanzania. *Folia Primatol.*, **44**, 161–170.

Takahata, Y. (1990). Adult males' social relations with adult females. In *The chimpanzees of the Manhale Mountains: sexual and life history strategies*. (ed. T. Nishida), pp. 133–148. University of Tokyo Press, Tokyo.

Tanner, J. M. (1962). *Growth at adolescence*. Blackwell, Oxford.

Tanner, J. (1992). Human growth and development. In *The Cambridge encyclopedia of human evolution* (eds. S. Jones, R. Martin and D. Pilbeam), pp. 98–105. Cambridge University Press.

Taranger, J., Engström, I., Lichtenstein, H., and Svennberg-Redegren, I. (1976). The somatic development of children in a Swedish urban community. VI. Somatic pubertal development. *Acta. Paediatr. Scand. (Suppl.)*, **258**, 121–135.

Tarara, E. B. (1987). Infanticide in a chacma baboon troop. *Primates*, **28**, 267–270.

Tattersall, I. (1982). *The primates of Madagascar*. Columbia University Press, New York.

Taub, D. M. (1980). Female choice and mating strategies among wild Barbary macaques (*Macaca sylvanus*). In *The macaques: studies in ecology, behavior and evolution*. (ed. D. G. Lindburg), pp. 287–344. Van Nostrand Reinhold, New York.

Tauber, P. F., Propping, D., Schumacher, G. F. B., and Zaneveld, L. J. D. (1980). Biochemical aspects of the coagulation and liquefaction of human semen. *J. Androl.*, **1**, 280–288.

Taylor, G. T., Bardgett, M., and Weiss, J. (1990). Behavior and physiology of castrated rats with different episodic schedules of testosterone restoration. *Horm. Metab. Res.*, **22**, 57–59.

Temerlin, M. K. (1975). *Lucy: growing up human*. Science and Behavior Books, Palo Alto, California.

Temple-Smith, P. D. and Bedford, J. M. (1980). Sperm maturation and the formation of sperm pairs in the epididymis of the opossum, *Didelphis virginiana*. *J. Exp. Zool.*, **214**, 161–171.

Templeton, A. A., Cooper, I., and Kelly, R. W. (1978). Prostaglandin concentrations in the semen of fertile men. *J. Reprod. Fertil.*, **52**, 147–150.

Tenaza, R. R. (1989). Female sexual swellings in the Asian colobine *Simias concolor*. *Am. J. Primatol.*, **17**, 81–86.

Terborgh, J. and Goldizen, A. W. (1985). On the mating system of the cooperatively breeding saddle-backed tamarin (*Saguinus fuscicollis*). *Behav. Ecol. Sociobiol.*, **16**, 293–299.

Terzian, H. and Dalle Ore, G. (1955). Syndrome of Kluver and Bucy reproduced in man by bilateral removal of the temporal lobes. *Neurol. Minneap.*, **5**, 373–380.

Thompson, M. L., McGill, T. E., McIntosh, S. M., and Manning, A. (1976). The effects of adrenalectomy on the sexual behaviour of castrated and intact BDF1 mice. *Anim. Behav.*, **24**, 519–522.

Thompson-Handler, N., Malenky, R. K., and Badrian, N. (1984). Sexual behavior of *Pan paniscus* under natural conditions in the Lomako Forest, Equateur, Zaire. In *The pygmy chimpanzee: evolutionary biology and behavior*. (ed. R. L. Susman), pp. 347–368. Plenum, New York.

Thorner, M. O. (1977). Prolactin: clinical physiology and the significance and management of hyperprolactinaemia. In *Clinical neuroendocrinology*. (eds. L. Martin and G. M. Besser), pp. 319–361. Academic Press, New York.

Thorner, M. O. and Besser, G. M. (1997). Hyperprolactinaemia and gonadal function: results of bromocryptine. In *Prolactin and human reproduction*. (eds. P. G. Crosignani and C. Robyn), pp. 285–301. Academic Press, New York.

Thornhill, R. and Alcock, J. (1983). *The evolution of insect mating systems*. Harvard University Press, Cambridge, Massachusetts.

Thornill, R., Gangestad, S. W., and Comer, R. (1996). Human female orgasm and mate fluctuating asymmetry. *Anim. Behav.*, **50**, 1601–1615.

Tilson, R. L. and Tenaza, R. R. (1976). Monogamy and duetting in an Old World monkey. *Nature Lond.*, **263**, 320–321.

Tinbergen, N. (1951). *The study of instinct*. Oxford University Press, London.

Tinklepaugh, O. L. (1930). Occurrence of a vaginal plug in a chimpanzee. *Anat. Rec.*, **46**, 329–332.

Tissell, L. E. and Larsson, K. (1979). Unimpaired sexual behavior of male rats after complete removal of the prostate and seminal vesicles. *Invest. Urol.*, **16**, 274–275.

Tobet, S. A., Zahniser, D. J., and Baum, M. J. (1986). Sexual dimorphism in the preoptic/anterior hypothalamic area of ferrets: effects of adult exposure to sex steroids. *Brain Res.*, **364**, 249–257.

Tokuda, K. (1961). A study of sexual behavior in the Japanese monkey troop. *Primates*, **3**, 1–40.

Tokuda, K., Simms, R. C., and Jensen, J. D. (1968). Sexual behavior in a captive group of pigtail macaques (*Macaca nemestrina*). *Primates*, **9**, 283–294.

Tomasch, J. (1954). Size, distribution, and number of fibres in the human corpus callosum. *Anat. Rec.*, **119**, 119–135.

Tomkins, J. L. and Simmons, L. W. (1995). Patterns of fluctuating asymmetry in earwig forceps: no evidence for reliable signalling. *Proc. Roy. Soc. Lond. B.*, **259**, 89–96.

Tredway, D. R., Settlage, D. S. F., Nakamura, R. M., *et al.* (1975). Significance of timing for the post-coital evaluation of cervical mucus. *Am. J. Obstet. Gynecol.*, **121**, 287–293.

Treolar, O. L., Wolf, R. C., and Meyer, R. K. (1972). Failure of a single neonatal dose of testosterone to alter ovarian function in the rhesus monkey. *Endocrinology*, **90**, 281–284.

Trivers, R. L. (1972). Parental investment and sexual selection. In *Sexual selection and the descent of Man 1871–1971*. (ed. B. Campbell), pp. 136–179. Aldine, Chicago.

Trivers, R. L. (1985). *Social evolution*. Benjamin/Cummings Publishing Co., Menlo Park, California.

Trivers, R. L. and Willard, D. E. (1973). Natural selection of parental ability to vary the sex ratio of offspring. *Science*, **179**, 90–92.

Trollope, J. and Blurton-Jones, N. G. (1972). Age of sexual maturity in the stumptail macaque (*Macaca arctoides*): a birth from laboratory born parents. *Primates*, **13**, 229–230.

Tsibris, J. C. M. (1987). Cervical mucus. In *Gynecologic endocrinology*. (eds. J. J. Gold and J. B. Josimovich), pp. 175–183. Plenum, New York.

Tsingalia, H. M. and Rowell, T. E. (1984). The behaviour of adult male blue monkeys. *Z. Tierpsychol.*, **64**, 253–268.

Tsitouras, P. D. and Hagen, T. C. (1984). Testosterone, LH, FSHG, prolactin and sperm in aging healthy men. *Abstracts of 7th Int. Congr. Endocr.* No. 1951. Excerpta Medica. International Congress Series 652, Amsterdam.

Turner, C. D. (1960). *General endocrinology.* W. B. Saunders Co., Philadelphia.

Turner, C. D. and Bagnara, J. T. (1976). *General endocrinology* (6th edn). W. B. Saunders Co., Philadelphia.

Turner, W. J. (1994). Comments on discordant M2 twinning in homosexuality. *Arch. Sex. Behav.*, **23**, 115–119.

Turner, W. J. (1995). Homosexuality, type 1: an Xq28 phenomenon. *Arch. Sex. Behav.*, **24**, 109–134.

Tutin, C. E. G. (1979). Mating patterns and reproductive strategies in a community of wild chimpanzees (*Pan troglodytes schweinfurthii*). *Behav. Ecol. Sociobiol.*, **6**, 29–38.

Tutin, C. E. G. (1980). Reproductive behaviour of wild chimpanzees in the Gombe National Park, Tanzania. *J. Reprod. Fertil. Suppl.*, **28**, 43–57.

Tutin, C. E. G. and McGinnis, P. R. (1981). Chimpanzee reproduction in the wild. In *Reproductive biology of the great apes: comparative and biomedical perspectives.* (ed. C. E. Graham), pp, 239–264. Academic Press, New York.

Udry, J. R. and Morris, N. M. (1968). Distribution of coitus in the menstrual cycle. *Nature Lond.*, **227**, 593–596.

Uehara, S., Hiraiwa-Hasegawa, M., Hosaka, K., and Hamai, M. (1994). The fate of defeated alpha male chimpanzees in relation to their social networks. *Primates.*, **35** (1), 49–55.

Uetz, G. W., McClintock, W. J., Miller, D., Smith., E. I., and Cook, K. K. (1996). Limb regeneration and subsequent asymmetry in a male secondary sexual character influences sexual selection in wolf spiders. *Behav. Ecol. Sociobiol.*, **38**, 253–257.

Utian, W. H. (1975). Effect of hysterectomy, oophorectomy and estrogen therapy on libido. *Int. J. Obstet. Gynecol.*, **84**, 4314–4315.

Valderrama, X., Srikosamatrara, S., and Robinson, J. G. (1990). Infanticide in wedge-capped capuchin monkeys, *Cebus olivaceous. Folia Primatol.*, **54**, 171–176

Valenstein, E. S. and Young, W. C. (1955). An experiential factor influencing the effectiveness of testosterone propionate in eliciting sexual behavior in male guinea pigs. *Endocrinology*, **56**, 173–177.

Valenstein, E. S., Riss, W., and Young, W. C. (1955). Experiential and genetic factors in the organization of sexual behavior in male guinea pigs. *J. Comp. Physiol. Psychol.*, **48**, 397–403.

Van der Werff ten Bosch, J. J. (1982). The physiology of reproduction of the orang-utan. In *The orang-utan: its biology and conservation.* (ed. L. E. M. de Boer), pp. 201–214. Dr. W. Junk, The Hague.

Van Goozen, S. H. M., Wiegant, V. M., Endert, E., Helmond, F. A., and Van de Poll, N. E. (1997). Psychoendocrinological assessment of the menstrual cycle: the relationship between hormones, sexuality and mood. *Arch. Sex. Behav.*, **26**, 359–382.

Van Horn, R. N. (1980). Seasonal reproductive patterns in primates. In *Progress in reproductive biology*, Vol. 5. (ed. P. O. Hubinont), pp. 181–221. Karger, Basel.

Van Horn, R. N. and Eaton, G. G. (1979). Reproductive physiology and behavior in prosimians. In *The study of prosimian behavior.* (eds. G. A. Doyle and R. D. Martin), pp. 79–122. Academic Press, New York.

Van Horn, R. N. and Resko, J. A. (1977). The reproductive cycle of the ring-tailed lemur (*Lemur catta*): sex steroid levels and sexual receptivity under controlled photoperiods. *Endocrinology*, **101**, 1579–1586.

Van Horn, R. N., Beamer, N., and Dixson, A. F. (1976). Diurnal variations of plasma testosterone in two prosimian primates (*Galago crassicaudatus crassicaudatus* and *Lemur catta*). *Biol. Reprod.*, **15**, 523–528.

Van Krey, H. P., Balander, R. J., and Compton, M. M. (1981). Storage and evacuation of spermatozoa from the uterovaginal sperm host glands in the domestic fowl. *Poultry Sci.*, **60**, 871–878.

Van Lawick-Goodall, J. (1969). Some aspects of reproductive behaviour in a group of wild chimpanzees (*Pan troglodytes schweinfurthii*) at the Gombe Stream Reserve, Tanzania, East Africa. *J. Reprod. Fertil. Suppl.*, **6**, 353–355.

Van Noordwijk, M. A. (1985). Sexual behaviour of Sumatran long-tailed macaques (*Macaca fascicularis*). *Z. Tierpsychol.*, **70**, 277–296.

Van Noordwijk, M. A. and Van Schaik, C. P. (1985). Male emigration and rank acquisition in wild long-tailed macaques (*Macaca fascicularis*). *Anim. Behav.*, **33**, 849–861.

Van Noordwijk, M. A. and Van Schaik, C. P. (1988). Male careers in Sumatran long-tailed macaques (*Macaca fascicularis*). *Behaviour*, **107**, 24–43.

Van Pelt, L. F. and Keyser, P. E. (1970). Observations on semen collection and quality in macaques. *Lab. Anim. Care*, **20**, 726–733.

Van Roosmalen, M. G. M. and Klein, L. L. (1988). The spider monkeys, genus *Ateles.* In *Ecology and Behavior of neotropical primatesA*, Vol. 2 (eds. R. A. Mittermeier, A. B. Rylands, A. Coimbra-Filho, and G. A. B. Fonseca), pp. 455–537. World Wildlife Fund, Washington, D.C.

Van Roosmalen, M., Mittermeier, R. A., and Milton, K. (1981). The bearded sakis, genus *Chiropotes.* In *Ecology and behavior of neotropical primates*, Vol. 1 (eds. A F. Combra-Filho and R. A. Mittermeier), pp. 419–441. Academia Brasileira de Ciências, Rio de Janeiro.

Van Schaik, C. P. (1983). Why are diurnal primates living in groups? *Behaviour*, **87**, 120–144.

Van Schaik, C. P. and Dunbar, R. I. M. (1990). The evolution of monogamy in large primates : a new hypothesis and some crucial tests. *Behaviour*, **115**, 30–62.

Van Schaik, C. P. and Hostermann, N. (1994). Predation risk and number of adult males in a primate group: a comparative test. *Behav. Ecol. Sociobiol.*, **35**, 261–272.

Van Schaik, C. P., Netto, W. J., Amerongen, A. J. J Van, and Westland, H. (1989). Social rank and sex ratio of captive long-tailed macaque females (*Macaca fascicularis*). *Am. J. Primatol.*, **19**, 147–161.

Van Valen, L. (1962). A study of fluctuating asymmetry. *Evolution*, **16**, 125–142.

Van Wagenen, G. (1936). The coagulating function of the cranial lobe of the prostate gland in the monkey. *Anat. Rec.*, **118**, 231–251.

Van Zessen, G. and Sandfort, T. (1991). *Seksualiteit in Nederland* (Sex in the Netherlands). Swets and Zeitlinger, Amsterdam, The Netherlands.

Vance, E. B. and Wagner, N. N. (1976). Written descriptions of orgasms—a study of sex differences. *Arch. Sex. Behav.*, **5**, 87–89.

Vandenbergh, J. G. (1965). Hormonal basis of the sex skin in male rhesus monkeys. *Gen. Comp. Endocr.*, **5**, 31–34.

Vandenbergh, J. G. (1969). Endocrine coordination in monkeys: male sexual responses to the female. *Physiol. Behav.*, **4**, 261–264.

Vandenbergh, J. G. (1983). Pheromonal regulation of puberty. In *Pheromones and reproduction in mammals.* (ed. J. G. Vandenbergh), pp. 95–112. Academic Press, New York.

Vandenbergh, J. G. and Drickamer, L. (1974). Reproductive coordination among free-ranging rhesus monkeys. *Physiol. Behav.*, **13**, 373–376.

Vandenbergh, J. G. and Post, W. (1976). Endocrine coordination in rhesus monkeys: female responses to the male. *Physiol. Behav.*, **17**, 979–984.

Vandevoort, C. A., Tollner, T. L., Tarantal, A. F., and Overstreet, J. W. (1989). Ultrasound-guided transfundal uterine sperm recovery from *Macaca fascicularis. Gamete Res.*, **24**, 327–331.

Vasey, P. L. (1995). Homosexual behavior in primates: a review of evidence and theory. *Int. J. Primatol.*, **61**, 173–204.

Vendrely, E. (1985). Structure and histophysiology of the human vas deferens. In *Seminal vesicles and fertility.* (eds. C. Bollack and A. Clavert). Karger, Basel.

Ventura, W. P., Freund, M., Davis, J., and Pannuti, C. (1973). Influences of norepinephrine on the motility of the human vas deferens: a new hypothesis of sperm transport by the vas deferens. *Fertil. Steril.*, **24**, 68–77.

Verrell, P. A. (1992). Primate penile morphologies and social systems: further evidence for an association. *Folia Primatol.*, **59**, 114–120.

Vincent, P. A. and Etgen, A. M. (1993). Steroid priming promotes oxytocin-induced norepinephrine release in the ventromedial hypothalamus of female rats. *Brain Res.*, **620**, 189–194.

Virag, R. (1982). Intracavernous injection of papaverine for erectile failure. *Lancet* ii, 938.

Virag, R., Ottesen, B., Fahrenkrug, J., Levy, C., and Wagner, G. (1982). Vasoactive intestinal polypeptide release during penile erection in man. *Lancet* ii, 1166.

Von Koenigswald, W. F. (1979). Ein lemurenrest aus dem oezanen Olschiefer der grube messel bei Darmstadt. *Palaeontologische Zeitschift.*, **53**, 63–76.

Von Krafft-Ebing, R. (1965). *Psychopathia sexualis: with especial reference to the antipathic sexual instinct.* (12th edn. English translation). Stein and Day, New York.

Voss, R. (1979). Male accessory glands and the evolution of copulatory plugs in rodents. *Occ. Pap. Mus. Zools. Univ. Mich.*, **689**, 1–17.

Waage, J. K. (1979). Dual function fo the damselfly penis: sperm removal and transfer. *Science*, **203**, 916–918.

Wada, M. and Gorbman, A. (1977a). Relation of mode of administration of testosterone to evocation of male sex behavior in frogs. *Horm. Behav.*, **8**, 310–319.

Wada, M. and Gorbman, A. (1977b). Mate calling induced by electrical stimulation in freely moving leopard frogs (*Rana pipiens*). *Horm. Behav.*, **9**, 141–149.

Wagner, G. and Brindley, G. S. (1980). The effect of atropine, α- and β-blockers on human penile erection: a controlled pilot study. In *Proceedings of 1st. International Conference on corpus cavernosum revascularization* (eds. W. Zorgniotti and G. Rossi), pp. 77–81. Charles C. Thomas, Springfield.

Wagner, G. and Green, R. (1981). Impotence: physiological, psychological, surgical diagnosis and treatment. Plenum, New York.

Wagner, G. and Kaplan, H. S. (1993). The new injection treatment for impotence: medical and psychological aspects. Brunner/Mazel, New York.

Wainman, P. and Shipounoff, G. C. (1941). The effects of castration and testosterone propionate on the striated perineal musculature in thc rat. *Endocrinology*, **29**, 975–978.

Wallach, S. J. R. and Hart, B. L. (1983). The role of the striated penile muscles of the male rat in seminal plug dislodgement and deposition. *Physiol. Behav.*, **31**, 815–821.

Wallen, K. (1982). Influence of female hormonal state upon rhesus monkey sexual behavior varies with space for social interactions. *Science*, **217**, 375–377.

Wallen, K. (1990). Desire and ability: hormones and the regulation of female sexual behavior. *Neurosci. Biobehav. Revs.*, **14**, 233–241.

Wallen, K. (1997). Sexual behavior in same-sexed nonhuman primates: is it relevant to understanding human homosexuality. *Ann. Revs. Sex. Res.*, **8**, (in press).

Wallen, K. and Winston, L. A. (1984). Social complexity and hormonal influences on sexual behavior in rhesus monkeys (*Macaca mulatta*). *Physiol. Behav.*, **32**, 629–637.

Wallen, K., Winston, S., Gaventa, M., Davis-Dasilva, M., and Collins, D. C. (1984). Periovulatory changes in female sexual behavior and patterns of ovarian steroid secretion in group living rhesus monkeys. *Horm. Behav.*, **18**, 431–450.

Wallen, K., Eisler, J. A., Tannenbaum, P. L., Nagell, K. M., and Mann, D. R. (1991). Antide (NAL-LYS GnRH antagonist) suppression of pituitary-testicular function and sexual behavior in group-living rhesus monkeys. *Physiol. Behav.*, **50**, 429–435.

Wallen, K., Maestripieri, D., and Mann, D. R. (1995). Effects of neonatal testicular suppression with a GnRH antagonist on social behavior in group-living juvenile rhesus monkeys. *Horm. Behav.*, **29**, 322–337.

Wallis, J. (1985). Synchrony of estrous swelling in captive group-living chimpanzees (*Pan troglodytes*). *Int. J. Primatol.*, **6**, 335–356.

Wallis, J., King, B. J., and Roth-Meyer, C. (1986). The effect of female proximity and social interaction on the menstrual cycle of crab-eating monkeys (*Macaca fascicularis*). *Primates*, **27**, 83–94

Wallis, S. J. (1981). The behavioural repertoire of the grey-cheeked mangabey (*Cercocebus albigena johnstonii*). *Primates*, **22**, 523–532.

Wallis, S. J. (1983). Sexual behaviour and reproduction of *Cercocebus albigena johnstonii* in Kibale forest, Western Uganda. *Int. J. Primatol.*, **4**, 153–166.

Walters, J. R. (1987). Kin recognition in non-human primates. In *Kin recognition in animals.* (eds. D. J. C. Fletcher and C. D. Mitchener), pp. 359–393. John Wiley, New York.

Walton, A. (1962). Copulation and natural insemination. In *Marshall's physiology of reproduction.* (ed. A. S. Parkes), pp. 130–160. Little Brown and Co., Boston.

Ward, I. L. (1972). Prenatal stress feminizes and demasculinizes the behavior of males. *Science*, **175**, 82–84.

Ward, I. L., Ward, O. B., Winn, R. J., and Bielawski, D. (1994). Male and female sexual behavior potential of male rats prenatally exposed to the influence of alcohol, stress or both factors. *Behav. Neurosci.*, **108**, 1188–1195.

Warner, M. D., Peabody, C. A., and Whiteford, H. A. (1987). Trazadone and priapism. *J. Clin. Psychiat.*, **48**, 244.

Warren, D. C. and Kilpatrick, L. (1929). Fertilization in the domestic fowl. *Poultry Sci.*, **8**, 237–256.

Warren, R. P. and Aronson, L. R. (1957). Sexual behavior in adult male hamsters castrated-adrenalectomized prior to puberty. *J. Comp. Physiol. Psychol.*, **50**, 475–480.

Wasser, S. K. (1995). Costs of conception in baboons. *Nature, Lond.*, **376**, 219–220.

Wasser, S. K. (1996). Reproductive control in wild baboons measured by fecal steroids. *Biol. Reprod.*, **55**, 393–399.

Wasser, S. K. and Starling, A. K. (1988). Proximate and ultimate causation of reproductive suppression among female yellow baboons at Mikumi National Park, Tanzania. *Am. J. Primatol.*, **16**, 97–121.

Watson, J. W. (1964). Mechanisms of erection and ejaculation in the bull and ram. *Nature Lond.*, **204**, 95–96.

Waxenburg, S. E., Drellich, M. G., and Sutherland, A. M. (1959). The role of hormones in human behavior. I. Changes in female sexuality after adrenalectomy. *J. Clin. Endocr.*, **19**, 193–202.

Waxenburg, S. E., Frinkbeiner, J. A., Drellich, M. L., and Sutherland, A. M. (1960). The role of hormones in human behavior. II. Changes in sexual behavior in relation to vaginal smears of breast cancer patients after oophorectomy and adrenalectomy. *Psychom. Med.*, **22**, 435–442.

Weber, G. and Weis, S. (1986). Morphometric analysis of the human corpus callosum fails to reveal sex-related differences. *J. für Wirnforschung*, **27**, 237–240.

Wedekind, C. and Folstad, I. (1994). Adaptive or nonadaptive immunosuppression by sex hormones? *Am. Nat.*, **143**, 936–938.

Wedekind, C., Seebeck, T., Bettens, F., and Paepke, A. J. (1995). MHC-dependent mate preferences in humans. *Proc. Roy. Soc. Lond. B.*, **260**, 245–249.

Weick, P. F., Dierschke, D. J., Karsch, F. J., Butler, W. R., Hotchkiss, J., and Knobil, E. (1973). Periovulatory time courses of circulating gonadotropic and ovarian hormones in the rhesus monkey. *Endocrinology*, **93**, 1140–1147.

Weiskrantz, L. (1956). Behavioural changes associated with ablation of the amygdaloid complex in monkeys. *J. Comp. Physiol. Psychol.*, **49**, 381–391.

Weisner, J. B. and Moss, R. L. (1986). Suppression of receptive and proceptive behavior in ovariectomized, estrogen-progesterone-primed rats by intraventricular beta-endorphin. Studies of behavior specificity. *Neuroendocrinology*, **43**, 47–62.

Weisner, J. B. and Moss, R. L. (1989). A psychopharmacological characterization of the opioid suppression of sexual behaviour in the female rat. In *Brain opioid systems and reproduction* (eds. R. G. Dyer and J. R. Bicknell), pp. 187–202. Oxford University Press.

Welker, C., Höhmann, H., and Schäfer-Witt, C. (1990). Significance of kin relations and individual preferences in the social behaviour of *Cebus apella*. *Folia Primatol.*, **54**, 166–170.

Weller, L. and Weller, A. (1993). Multiple influences of menstrual synchrony: kibbutz room-mates, their best friends, and their mothers. *Am. J. Hum. Biol.*, **5**, 173–179.

Weller, L., Weller, A., and Avinir, O. (1995). Menstrual synchrony: only in roommates who are close friends? *Physiol. Behav.*, **58**, 883–889.

Wellings, K., Field, J., Johnson, A. M., and Wadsworth, J. (1990). *Sexual behaviour in Britain: the national survey of sexual attitudes and lifestyles*. Penguin Books, London.

Wells, K. D. (1977). The social behaviour of anuran amphibians *Anim. Behav.*, **25**, 666–693.

Westermarck, A. E. (1891). *The history of human marriage*. Macmillan, London.

Whalen, R. E. (1975). Cyclic changes in hormones and behavior. *Arch. Sex. Behav.*, **4**, 313–314

Wheatley, B. P. (1980). Feeding and ranging of East Bornean (*Macaca fascicularis*). In *The macaques: studies in ecology, behavior and evolution*. (ed. D. G. Lindburg), pp. 215–246. Van Nostrand Reinhold, New York.

Wheatley, B. P. (1982). Adult male replacement in *Macaca fascicularis* of East Kalimantan, Indonesia. *Int. J. Primatol.*, **3**, 203–219.

Whitam, F. L. and Mathy, R. M. (1986). Male homosexuality in four societies: Brazil, Guatemala, the Philippines and the United States. Praeger, New York.

Whitam, F. L., Diamond, M., and Martin, J. (1993). Homosexual orientation in twins: a report on 61 pairs and three triplet sets. *Arch. Sex. Behav.*, **22**, 187–206.

White, F. J. (1991). Social organization, feeding ecology and reproductive strategy of ruffed lemurs, *Varecia variegata*. In *Primatology today*. (eds. Ehara, A., Kimura, T., Takenaka, O. and Iwamoto, M.), pp. 81–84. Elsevier Science Publishers, Amsterdam.

Whitten, P. L. (1983). Diet and dominance among female vervet monkeys (*Cercopithecus aethiops*). *Am. J. Primatol.*, **5**, 139–159.

Wickings, E. J. and Dixson, A. F. (1992a). Testicular function, secondary sexual development and social status in male mandrills (*Mandrillus sphinx*). *Physiol. Behav.*, **52**, 909–916.

Wickings, E. J. and Dixson, A. F. (1992b). Development from birth to sexual maturity in a semi-free-ranging colony of mandrills (*Mandrillus sphinx*) in Gabon. *J. Reprod. Fertil.*, **95**, 129–138.

Wickings, E. J., Bossi T., and Dixson, A. F. (1993). Reproductive success in the mandrill (*Mandrillus sphinx*) : correlations of male dominance and mating success with paternity, as determined by DNA fingerprinting. *J. Zool. Lond.*, **231**, 563–574.

Wickler, W. (1967). Socio-sexual signals and their intra-specific imitation among primates. In *Primate ethology*. (ed. D. Morris), pp. 69–147. Weidenfeld and Nicolson, London.

Wiebe, R. H., Diamond, E., Akesel, S., Liu, P., Williams, L. E., and Abee, C. R. (1984). Diurnal variations of androgens in sexually mature male Bolivian squirrel monkeys (*Saimiri sciureus*) during the breeding season. *Am. J. Primatol.*, **7**, 291–297.

Wilcox, J. N., Barclay, S. R., and Feder, H. H. (1984). Administration of estradiol-17β in pulses to female guinea pigs: self-priming effects of estrogen on brain tissues mediating lordosis. *Physiol. Behav.*, **32**, 483–488.

Wildt, D. E. (1986). Spermatozoa: collection, evaluation, metabolism, freezing and artificial insemination. In *Comparative primate biology*, Vol. 3. (eds. W. R. Dukelow and J. Erwin), pp. 171–193. Alan R. Liss, New York.

Wildt, D. E., Doyle, U., Stone, S. C., and Harrison, R. M. (1977). Correlation of perineal swelling with serum ovarian hormone levels, vaginal cytology and ovarian follicular development during the baboon reproductive cycle. *Primates*, **18**, 261–270.

Wilen, R. and Naftolin, F. (1976). Age, weight and weight gain in the individual pubertal female rhesus monkey (*Macaca mulatta*). *Biol. Reprod.*, **15**, 356–360.

Wilkins, L., Jones, H. W., Holman, G. H., and Stempel, R. S. (1958). Masculinization of the human fetus associated with administration of oral and intramuscular progestins during pregnancy: non-adrenal female pseudohermaphroditism. *J. Clin. Endocr.*, **18**, 559.

Wilks, J. W. (1977). Endocrine characterization of the menstrual cycle of the stumptail monkey (*Macaca arctoides*). *Biol. Reprod.*, **16**, 474–478.

Williams, C. L. and Meck, W. H. (1991). The organizational effects of gonadal steroids on sexually dimorphic spatial ability. *Psychoneuroendocr.*, **16**, 155–176.

Williams, C. L., Barnett, A. M., and Meck, W. H. (1990). Organizational effects of early gonadal secretions on sexual differentiation in spatial memory. *Behav. Neurosci.*, **104**, 84–97.

Williams, G. C. (1966). Adaptation and natural selection: a critique of some current evolutionary thought. Princeton University Press, Princeton, New Jersey.

Williams-Ashman, H. G., Wilson, J. Beil, R. E., and Lorand, L. (1977). Transglutaminase reactions associated with rat semen clotting system: modulation by macromolecular polyanions. *Biochem. Biophys. Res. Commun.*, **79**, 1192.

Willis, E. A., Otteson, B., Wagner, G., Sundler, F., and Fahrenkrug, J. (1983). Vasoactive intestinal polypeptide (VIP) as a putative neurotransmitter in penile erection. *Life Sci.*, **33**, 383–391.

Wilson, A. P. and Boelkins, R. C. (1970). Evidence for seasonal variation in aggressive behaviour by *Macaca mulatta*. *Anim. Behav.*, **18**, 719–724.

Wilson, A. P. and Vessey, S. H. (1968). Behavior of free-ranging castrated rhesus monkeys. *Folia Primatol.*, **9**, 1–14.

Wilson, C A. (1993). Pharmacological targets for the control of male and female sexual behaviour. In *Sexual pharmacology*. (eds. A. J. Riley, M. Peet and C. A. Wilson), pp. 1–58. Oxford Scientific Publications, Clarendon Press, Oxford.

Wilson, E. O. (1978). *On human nature*. Harvard University Press, Cambridge, Massachusetts.

Wilson, M. E. (1981). Social dominance and female reproductive behaviour in rhesus monkeys (*Macaca mulatta*). *Anim. Behav.*, **29**, 472–482.

Wilson, M. E. and Gordon, T. P. (1989). Season determines timing of first ovulation in rhesus monkeys (*Macaca mulatta*) housed outdoors. *J. Reprod. Fertil.*, **85**, 583–591.

Wilson, M. E., Gordon, T. P., and Bernstein, I. S. (1978). Timing of births and reproductive success in rhesus monkey social groups. *J. Med. Primatol.*, **7**, 202–212.

Wilson, M. E., Walker, M. L., and Gordon, T. P. (1983).

Consequences of first pregnancy in rhesus monkeys. *Am. J. Phys. Anthropol.*, **63**, 103–110.

Wilson, M. I. (1977a). Characterisation of the oestrous cycle and mating season of squirrel monkeys from copulatory behaviour. *J. Reprod. Fertil.*, **51**, 57–63.

Wilson, M. I. (1977b). A note on the external genitalia of female squirrel monkeys (*Saimiri sciureus*). *J. Med. Primatol.*, **6**, 181–185.

Wilson, M. I., Daly, M., and Weghorst, S. J. (1980). Household composition and the risk of child abuse and neglect. *J. Biosoc. Sci.*, **12**, 333–340.

Wimmer, R. E. and Wimmer, C. (1985). Three sex dimorphisms in the granule cell layer of the hippocampus in house mice. *Brain Res.*, **328**, 105–109.

Winn, R. L. and Newton, N. (1982). Sexuality in aging: a study of 106 cultures. *Arch. Sex. Behav.*, **11**, 283–298.

Winn, R. M. (1994). Preliminary study of the sexual behaviour of three aye-ayes (*Daubentonia madagascariensis*) in captivity. *Folia Primatol.*, **62**, 63–73.

Winter, J. S. D., Faiman, C., Hobson, W. C., and Reyes, F. I. (1980). The endocrine basis of sexual development in the chimpanzee. *J. Reprod. Fertil. Suppl.*, **28**, 131–138.

Wislocki, G. B. (1932). On the female reproductive tract of the gorilla with a comparison with that of other primates. *Contrib. Embryol.*, **23**, 163–204.

Wislocki, G. B. (1933). The reproductive systems. In *The anatomy of the rhesus monkey*. (eds. C. G. Hartman and W. L. Straus), pp. 231–247. Baillière, Tindall, and Cox, London.

Witelson, S. F. (1985). The brain connection: the corpus callosum is larger in left-handers. *Science*, **229**, 665–668.

Witt, D. M. and Insel, T. R. (1991). A selective oxytocin antagonist attenuates progesterone facilitation of female sexual behavior. *Endocrinology*, **128**, 3269–3276.

Wolf, A. P. (1995). Sexual attraction and childhood association: a Chinese brief for Edward Westermarck. Stanford University Press, Stanford, California.

Wolf, D. P., Byrd, W., Dandekar, P., and Quigley, M. M. (1984). Sperm concentration and the fertilization of human eggs in vitro. *Biol. Reprod.*, **31**, 837–848.

Wolf, K. and Fleagle, J. G. (1977). Adult male replacement in a group of silvered leaf monkeys (*Presbytis cristata*) at Kuala Selangor, Malaysia. *Primates*, **18**, 949–955.

Wolf, R. C., O'Connor, R. F., and Robinson, J. A. (1977). Cyclic changes in plasma progestins and estrogens in squirrel monkeys. *Biol. Reprod.*, **17**, 228–231.

Wolfe, L. (1978). Age and sexual behavior of Japanese macaques (*Macaca fuscata*). *Arch. Sex. Behav.*, **7**, 55–68.

Wolfe, L. D. (1984a). Female rank and reproductive success among Arashiyama, B. Japanese macaques (*Macaca fuscata*). *Int. J . Primatol.*, **5**, 133–143.

Wolfe, L. D. (1984b). Japanese macaque female sexual behavior: a comparison of Arashiyama East and West. In *Female primates: studies by women primatologists*. (ed. M. F. Small), pp. 141–157. Alan R. Liss, New York.

Wolfe, L. D. (1991). Human evolution and the sexual behavior of female primates. In *Understanding behavior: what primate studies tell us about human behavior*. (eds. J. Loy and C. B. Peters), pp. 121–151. Oxford University Press, New York.

Woodruff, J. D. and Pauerstein, C. J. (1969). *The fallopian tube: structure, function, pathology, and management*. Williams and Wilkins, Baltimore.

Wrangham, R. (1977). Feeding behaviour of chimpanzees in Gombe National Park, Tanzania. In *Primate Ecology*. (ed. T. H. Clutton-Brock), pp. 503–538. Academic Press, London.

Wrangham, R. W. (1979). On the evolution of ape social systems. *Soc. Sci. Inform.*, **18**, 335–368.

Wrangham, R. W. (1980). An ecological model of female-bonded primate groups. *Behaviour*, **75**, 262–300.

Wrangham, R. W. and Smuts, B. B. (1980). Sex differences in the behavioural ecology of chimpanzees in the Gombe National Park, Tanzania. *J. Reprod. Fertil. Suppl.*, **28**, 13–31.

Wright, P. C. (1981). The night monkeys, genus *Aotus*. In *Ecology and behavior of neotropical primates*, Vol. 1. (eds. A. F. Coimbra-Filho and R. A. Mittermeier), pp. 211–240. Academia Brasileira de Ciencias, Rio de Janeiro.

Wright, P. C., Toyama, T. M., and Simons, E. I. (1986). Courtship and copulation in *Tarsius bancanus*. *Folia Primatol.*, **46**, 142–148.

Yahr, P. and Gregory, J. E. (1993). The medial and lateral cell groups of the sexually dimorphic area of the gerbil hypothalmus are essential for male sex behavior and act via separate pathways. *Brain Res.*, **631**, 287–296.

Yamanouchi, K. and Arai, Y. (1985). Presence of a neural mechanism for the expression of female sexual behaviors in the male rat brain. *Neuroendocrinology*, **40**, 393–397.

Yamanouchi, K. and Arai, Y. (1990). The septum as origin of a lordosis-inhibiting influence in female rats: effect of neural transection. *Physiol. Behav.*, **48**, 351–355.

Yanagimachi, R. (1994). Mammalian fertilization. In *The Physiology of reproduction*, Vol. 1. (2nd edition). (eds. E. Knobil and J. D. Neill), pp. 189–317. Raven Press, New York.

Yeager, C. P. (1989). Feeding ecology of the proboscis monkey (*Nasalis larvatus*) *Int. J. Primatol.*, **10**, 497–530.

Yeager, C. P. (1990a). Proboscis monkey (*Nasalis larvatus*) social organisation: group structure. *Am. J. Primatol.*, **20**, 95–106.

Yeager, C. P. (1990b). Notes on the sexual behavior of the proboscis monkey (*Nasalis larvatus*). *Am. J. Primatol.*, **21**, 223–227.

Yeager, C. P. (1995). Does intraspecific variation in social systems explain reported differences in the social structure of the proboscis monkey (*Nasalis larvatus*)? *Primates.*, **36**, 575–582.

Yeoman, R. R., Williams, L. E., Aksel, S., and Abee, C. R. (1985). Mating effects and diurnal changes in serum estradiol of the female squirrel monkey (*Saimiri boliviensis*). p. 378 in Program of the 10th Annual Meeting of the American Society of Primatologists.

Yeoman, R. R., Williams, L. E., Aksel, S., and Abee, C. R. (1987). Mating effects and diurnal changes in serum estradiol of the female squirrel monkey *Saimiri boliviensis*. *Am. J. Primatol.*

Yerkes, R. M. (1939). Sexual behavior in the chimpanzee. *Hum. Biol.*, **11**, 78–110.

Yerkes, R. M. (1943). *Chimpanzees: a laboratory colony*. Yale University Press, New Haven.

Yerkes, R. M. and Elder, J. H. (1936). Oestrus, receptivity and mating in the chimpanzee. *Com. Psychol. Monogr.*, **13**, 1–39.

Yoneda, M. (1981). Ecological studies of *Saguinus fuscicollis*

and *Saguinus labiatus* with reference to habitat segregation and height preference. In *Kyoto Univ. Overseas Res. Rep., New World Monkeys*, pp. 45–50.

Yoshi, I., Barker, W., Apicella, A., Chang, J., Sheldon, J., and Duara, R. (1986). Measurements of the corpus callosum (CC) on magnetic resonance (MR) scans: effects of age, sex, handedness and disease. *Neurol. Suppl.*, **1**, 133.

Yoshiba, K. (1968). Local and intertroop variability in ecology and social behavior of common Indian langurs. In *Primates: studies in adaptation and variability*. (ed. P. C. Jay), pp. 217–252. Holt, Rinehart and Winston, New York.

Yoshimatsu, H. (1983). Hypothalamic regulatory mechanism of sexual behavior in the male monkey. *Fukuoka Acta. Med.*, **74**, 89–100.

Young, W. C. (1961). The hormones and mating behavior. In *Sex and internal secretions*, Vol. 2. (ed. W. C. Young), pp. 1173–1239.

Young, W. C. and Orbison, W. D. (1944). Changes in selected features of behavior in pairs of oppositely sexed chimpanzees during the sexual cycle and after ovariectomy. *J. Comp. Psychol.*, **27**, 107–143.

Young, W. C. and Yerkes, R. M. (1943). Factors influencing the reproductive cycle in the chimpanzee: the period of adolescent sterility and related problems. *Endocrinology*, **31**, 121–154.

Young, W. C., Goy, R. W., and Phoenix, C. H. (1964). Hormones and sexual behavior. Broad relationships exist between the gonadal hormones and behavior. *Science*, **143**, 212–218.

Zachmann, M. A., Prader, A., Kind, H. P., Häflinger, H., and Budliger, H. (1974). Testicular volume during adolescence: cross sectional and longitudinal studies. *Helv. Pediatr. Acta.*, **29**, 61–72.

Zahavi, A. (1975). Mate selection-a selection for handicap. *J. Theor. Biol.*, **53**, 205–214.

Zamboni, L., Conaway, C. H., and Van Pelt, L. (1974). Seasonal changes in production of semen in free-ranging rhesus monkeys. *Biol. Reprod.*, **11**, 251–267.

Zastra, B. M., Seidell, J. C., Van Noord, P. A. H., te Velde, E. R., Habbema, J. D. F., Vrieswijk, B., and Karbaat, J. (1993). Fat and female fecundity: prospective study of effect of body fat distribution on conception rates. *Brit. Med. J.*, **306**, 484–487.

Zeh, J. A. and Zeh, D. W. (1994). Last male sperm precedence breaks down when females mate with three males. *Proc. Roy. Soc. Lond. B.*, **257**, 287–292.

Zhao, Q. K. (1993). Sexual behavior of Tibetan macaques at Mt. Emei, China. *Primates*, **34**, 431–444.

Zhao, Q. K. (1994). Mating competition and intergroup transfer of males in Tibetan macaques (*Macaca thibetana*) at Mt. Emei, China. *Primates*, **35**, 57–68.

Zhou, J. N., Hofman, M. A., Gooren, L. J. G., and Swaab, D. F. (1995a). A sex difference in the human brain and its relation to transsexuality. *Nature Lond.*, **378**, 68–70.

Zhou, J. N., Hofman, M. A., and Swaab, D. F. (1995b). VIP neurons in the human SCN in relation to sex, age and Alzheimer's disease. *Neurobiol. Aging.*, **16**, 571–576.

Ziegler, T. E., Bridson, W. E., Snowdon, C. T., and Eman, S. (1987). Urinary gonadotropin and estrogen excretion during the post-partum estrus, conception and pregnancy in the cotton-top tamarin (*Saguinus oedipus oedipus*). *Am. J. Primatol.*, **12**, 127–140.

Ziegler, T. E., Snowdon, C. T., and Warneke, M. (1989). Post-partum ovulation and conception in Goeldi's monkeys, *Callimico goeldii. Folia Primatol.*, **53**, 206–210.

Ziegler, T. E., Snowdon, C. T., and Bridson, W. E. (1990). Reproductive performance and excretion of urinary estrogens and gonadotropins in the female pygmy marmoset (*Cebuella pygmaea*). *Am. J. Primatol.*, **22**, 191–203.

Ziegler, T. E., Epple, G., Snowdon, C. T., Porter, T. A., Belcher, A. M., and Küderling, I. (1993). Detection of chemical signals of ovulation in the cotton-top tamarin, *Saguinus oedipus. Anim. Behav.*, **45**, 313–322.

Zimmermann, E. and Lerch, C. (1993). The complex acoustic design of an advertisement call in male mouse lemurs (*Microcebus murinus*) and sources of its variation. *Ethology*, **93**, 211–224.

Zinner, D., Schwibbe, M. H., and Kaumanns, W. (1994). Cycle synchrony and probability of conception in female hamadryas baboons. *Papio hamadryas. Behav. Ecol. Sociobiol.*, **35**, 175–183.

Zucker, K., Bradley, S. J., Oliver, J., Blake, J., Fleming, S., and Hood, J. (1996). Psychosexual development of women with congenital adrenal hyperplasia. *Horm. Behav.*, **30**, 300–318.

Zuckerman, S. (1932). *The social life of monkeys and apes.* Harcourt, Brace and Co., New York.

Zuckerman, S. and Parkes, A. S. (1939). Observations on the secondary sexual characters in monkeys. *J. Endocr.*, **1**, 430–439.

Zuckerman, S., Van Wagenen, G., and Gardner, R. H. (1938). The sexual skin of the rhesus monkey. *Proc. Zool. Soc. Lond.*, **108**, 385–401.

Zuk, M. (1992). The role of parasites in sexual selection: current evidence and future directions. *Adv. Study Behav.*, **21**, 39–68.

Zumpe, D. and Michael, R. P. (1968). The clutching reaction and orgasm in the female rhesus monkey (*Macaca mulatta*). *J. Endocr.*, **40**, 117–123.

Zumpe, D. and Michael, R. P. (1983). A comparison of the behavior of Macaca fascicularis and Macaca mulatta in relation to the menstrual cycle. *Am. J. Primatol.*, **4**, 55–72.

Zumpe, D. and Michael, R. P. (1984). Low potency of intact male rhesus monkeys after long-term visual contact with their female partners. *Am. J. Primatol.*, **6**, 241–252.

Zumpe, D. and Michael, R. P. (1985). Effects of testosterone on the behavior of male cynomolgus monkeys (*Macaca fascicularis*). *Horm. Behav.*, **19**, 265–277.

Zumpe, D. and Michael, R. P. (1988). Effects of medroxyprogesterone acetate on plasma testosterone and sexual behavior in male cynomolgus monkeys (*Macaca fascicularis*). *Physiol. Behav.*, **42**, 343–349.

Zumpe, D., Bonsall, R. W., and Michael, R. P. (1992). Some contrasting effects of surgical and 'chemical' castration on the behavior of male cynomolgus monkeys (*Macaca fascicularis*). *Am. J. Primatol.*, **26**, 11–22.

Zumpe, D., Bonsall, R. W., and Michael, R. P. (1993). Effects of the nonsteroidal aromatase inhibitor, Fadrazole, on the sexual behavior of male cynomolgus monkeys (*Macaca fascicularis*). *Horm. Behav.*, **27**, 200–215.

Index

Page entries referring to figures or tables are indicated in bold

abnormal sexual preferences; *see* paraphilias
Ache, hunter-gathers 71
adolescent 'growth spurt' 172
adolescent sterility 210, 214–15, 461–**2**
adrenocortical hormones
  and congenital adrenal hyperplasia 285 8
  metabolic pathways 365
  and sexual behaviour 363–**5**, **366–7**, 404, **465**
  and social stress 445 6, 448, 457–**8**, 450–1, 452
agonistic buffering hypothesis 88
Allenopithecus nigroviridis
  genitalia, male **255**
  sexual skin, in female **202**
  sexual skin, in male **190**
Allen's bushbaby; *see Galago alleni*
all male groups **15**, 32, 36, **66**
Alouatta belzebul
  hyoid volume **182**
Alouatta caraya
  hyoid volume **182**
  male dominance and mating success **153**
  retarded secondary sexual development in **197**
  sexual dichromatism **190**
Alouatta fusca
  hyoid volume **182**
  sexual dichromatism **190**
Alouatta palliata
  body weight, adult **40**
  consortships **58**
  genitalia, female 56, 324
  group composition **40**
  group size **40**
  intromission pattern **188, 120**
  male dominance and mating success **53**
  mating system **40**
  multiple-partner matings by females **38**
  ovarian cycle and sexual behaviour 324
  vocal anatomy and display 179–80, **182**
Alouatta pigra 12
Alouatta seniculus
  body weight, adult **30**
  DNA typing in **75**
  genitalia, female **268**
  group composition 30
  group size 30
  male dominance and reproductive success **75**
  mating system **30**
  'tongue-pumping' precopulatory display **95**, **96**, 97
  vocal anatomy and display 179–**81**, **182**
5-alpha-reductase
  deficiency of during development; *see* Imperato–McGinley
    syndrome
  see also dihydrotestosterone
alternative mating tactics 62–7
  coercive matings, by males 66–7
  in male anthropoids **65**, **66**
  in male mandrills 54
  in male nocturnal prosimians **64**

  nocturnal mating/mate-guarding **79**
  in solitary males 63, **65**, **66**
amygdala
  and behaviour 377–**9**, 423–4
androgen insensitivity syndrome **288**
androgenization
  and behavioural development 277–300
  and brain development 301–14
angwantibo; *see Arctocebus calabarensis*
anthropoids
  classification **5–7**
  description of extant forms **10–18**
  lack of oestrus, in 93–4
  sexual dimorphism in body size in **171–3**
antiandrogens
  and sexual behaviour in males 394–7, **398–9**
Aotus **28**
Aotus azarae
  body weight, adult **27**
  group size **27**
  mating system **27**
Aotus griseimembra
  intromission pattern **120**
  puberty **173**
Aotus lemurinus
  body weight, adult **27**
  ejaculatory frequency **240**
  group size **27**
  mating system **27**
  scent-marking behaviour **184**
  testes size **220**, **223**
  vocal anatomy 177–**8**
Aotus trivirgatus
  genitalia, male **254**
Arctocebus calabarensis **10**
  body weight, adult **46**
  genitalia, male **245**
  home range **46**
  intromission pattern **118, 120**
  mating system **46**
aromatization 284, 296–**8**, 310, 312, 397–400, **401**
Ateles belzebuth
  body weight, adult **40**
  copulatory posture, **111**
  genitalia, male **250**
  group composition **40**
  group size **40**
  intromission pattern **118, 120**
  mating system **40**
Ateles fusciceps
  body weight, adult **121**
  intromission pattern **120**
  mating system **121**
Ateles geoffroyi
  body weight, adult **121**
  intromission pattern **120**
  mating system **121**
Ateles paniscus
  consortships **58**
  genitalia, female **266**
  genitalia, male **266**

intromission pattern **118**
multiple-partner matings by females **38**
ovarian cycle and sexual behaviour 324–5
social organization 324–5
australopithecines 19, **21**
auto-erotism; *see* masturbation
Avahi laniger
genitalia, male **254**
testes size **220**
aye-aye; *see Daubentonia madagascariensis*

baboon; *see* Papio
baculum
in adapids 257–8
elongation of, and copulatory behaviour 253, **256**–9
evolution of 252–60
functions of, in reproduction 253, 256–60
length, comparative data on 254–5
length, and mating systems 248–**9**
radiographs, of the **253**
Barbary macaque; *see Macaca sylvanus*
Bateman's principle 80
Bercovitch–McMillan hypothesis 54, 79
bimaturism in body growth **172–3**
black and white colobus monkey; *see Colobus guereza*
black ape of Celebes; *see Macaca nigra*
black lemur; *see Lemur macaco*
black spider monkey; *see Ateles paniscus*
blue monkey; *see Cercopithecus mitis*
body weights
of adult primates **27**, **30**, **40**, **46**, **121**
bonnet macaques; *see Macaca radiata*
bonobo; *see Pan paniscus*
Brachyteles arachnoides **12**
body weight, adult **40**
group composition **40**
group size **40**
intromission pattern **118**, **120**
mating system **40**, 326
multiple-partner matings by females **38**, 39
ovarian cycle and sexual behaviour 325
proceptive displays **101**
brain
binding of steroid hormones **414–16**
defeminization 277–8, 310, 312
masculinization 277–8, 280–3, 310, 312
organization and behaviour 277–95, 301, 304–5, 312–314
and processing of visual stimuli 376–**9**
sex differences in structure 301–14, **302–3**, **308**, **311**, 312–14
and sexual behaviour in females 358–65, 365–88
and sexual behaviour in males 413–18, 418–43
brown capuchin; *see Cebus apella*
brown lemur; *see Lemur fulvus*
bulbocavernosus muscles; *see* penile muscles
bulbocavernosus reflex 430
bushbabies; *see* Galago

Cacajao calvus **10**
body weight, adult **121**
intromission pattern **120**
mating system **121**
retarded secondary sexual development in **197**
secondary sexual adornments of males **190**, **196**
Cacajao melanocephalus
genitalia, female **268**

Callicebus moloch
body weight, adult **27**
group size **27**
intromission pattern **120**
mating system **27**
vocal anatomy 177–**8**
Callicebus torquatus
body weight, adult **27**
group size **27**
mating system **27**
Callimico goeldii **14**
body weight, adult **27**
genitalia, male **254**
group size **27**
intromission pattern **120**
mating system **27**
orgasmic responses **130**
Callithrix argentata
genitalia, male **254**
group size **27**
mating system **27**
Callithrix flaviceps **13**
Callithrix humeralifer
body weight, adult **27**
genitalia, male **254**
group size **27**
mating system **27**
Callithrix jacchus **13**
adrenalectomy and sexual behaviour 365, **367**
brain mechanisms and female's sexual behaviour 370–4, 376, 384–**5**
body weight, adult **27**
chemical cues and sexual attractiveness 330–**1**
cutaneous glands **186**
ejaculatory frequency **240**
genitalia, female **268**
genitalia, male **251**, **254**, **261**, **264**
group size **27**
harassment/interruption of copulation **72**, **73**
intromission pattern **118**, **120**
mating system **27**
neuroendocrine control of sexual behaviour in the female 359–**64**, **367**, 370–4, **376**, 384–**5**
neuroendocrine control of sexual behaviour in the male 393–4, **410–12**, 418–21
neuroendocrine control of paternal behaviour 419–**21**
orgasmic responses **134**, **135**–6
ovarian cycle and sexual behaviour **326–31**
ovariectomy and hormone replacement 359–**64**
post-ejaculatory interval 105
postnatal testosterone surge 292–**4**
proceptive displays **95**, **96**, 326–30
puberty **172–3**
social organization 327
sociosexual behaviour **153**
sperm length **228**
suppression of ovarian cycle 189, **452–4**
testes size 220–1, 223
callitrichids
mating system 26–9, **27**
ovarian suppression in 189, **452–3**
paternal behaviour in 419–**21**
proceptivity in **95–6**, 326–30
sperm competition in 220–**1**
twinning in 29
capuchins; *see* Cebus

castration
  behavioural effects of 392–4, 400–1, 404–**5**
  peripheral effects of 409, 412–**13**
catarrhines
  nostril position **7**, 14
  body weight dimorphism in **171**–2
  sexual skin in **190**–5, **201–3**, 208–14
catecholamines; *see* dopamine and noradrenaline
cebids; *see* New World monkeys
Ceboidea
  classification **5, 6**, 7
  see also New World monkeys
Cebuella pygmaea
  body weight, adult **27**
  genitalia, male **251, 254**
  group size **27**
  intromission pattern **118, 120**
  male dominance and mating success **53**
  mating system **27**, 328
  orgasmic responses **130**
  ovarian cycle and sexual behaviour **328**
  testes size 220–**1**
Cebus **7**
Cebus albifrons
  body weight, adult **40**
  group composition **40**
  group size **40**
  mating system **40**
Cebus apella **12**
  body weight, adult **40**
  consortships **58**
  female choice 85
  genitalia, male **254**
  group composition **40**
  group size **40**
  male dominance and mating success **53**
  mating system **40**
  multiple-partner matings by females **38**
  ovarian cycle and sexual behaviour 325
  testosterone levels, in males **390**
Cebus capucinus
  body weight, adult **40**
  genitalia, male **254**
  group composition **40**
  group size **40**
  mating system **40**
Cebus nigrivittatus
  body weight, adult **40**
  group composition **40**
  group size **40**
  intromission pattern **120**
  mating system **40**
Cercocebus albigena
  body weight, adult **40**
  genitalia, female **268**
  genitalia, male **255**
  group composition **40**
  group size **40**
  intromission pattern **188, 120**
  mating system **40**
  menstrual cycle and sexual behaviour 338
  multiple-partner matings by females **38**
  orgasmic responses **130**
  sexual skin, in female **201–2**
Cercocebus aterrimus
  genitalia, female **268**
  genitalia, male **255**

Cercocebus atys
  sexual skin, in female **201–2**
Cercocebus torquatus
  genitalia, female **268**
  genitalia, male **255**
cercopithecines; *see* Old World monkeys
Cercopithecoidea
  classification **5, 6**, 7
  see also Old World monkeys
Cercopithecus aethiops **17**
  body weight, adult **40**
  female dominance and fertility **84**
  genitalia, male **253–4**
  group composition **40**
  group size **40**
  intromission pattern **118, 120**
  harassment of copulation **73**
  male dominance and mating success **53**
  mating season **390**
  mating system **40**
  menstrual cycle and sexual behaviour 338–**9**
  multiple-partner matings by females **38**
  red, white and blue display **193**
  sexual skin, in female **202**
  sexual skin, in male **190**
  sperm length **228**
  testes size **219**
Cercopithecus albogularis
  genitalia, male **252**
Cercopithecus ascanius
  genitalia, female **268**
  genitalia, male **264**
  group composition **30**
  group size **30**
  hybridization, with *C. mitis* 246
  mating system **30**, 34
  multimale influxes **34**
  mating system **34**
  social organization **34**
Cercopithecus cephus
  genitalia, female **268**
Cercopithecus diana
  sexual skin in adult male **190**
Cercopithecus hamlyni
  sexual skin in adult male **190**
Cercopithecus mitis
  body weight, adult **30**
  genitalia, female **268**
  genitalia, male **255**
  group composition **30**
  group size **30**
  hybridization with *C. ascanius* 246
  mating season **34**
  mating system **30**, 34
  multimale influxes **34**
  social organization **34**
Cercopithecus mona
  genitalia, male **252, 255**
Cercopithecus neglectus
  booming vocalization by adult male 179–**80**
  genitalia, female **266, 268**
  genitalia, male **252, 255**
  retarded secondary sexual development in **197**
  sexual skin, in male **190**
Cercopithecus petaurista
  genitalia, male **252**
cervix 203, 269–71, **270**

cervical mucus 123, 269–71
C-Fos mapping studies
  in sexually active male rats **417**
chacma baboon; *see Papio ursinus*
Cheirogaleus major
  genitalia, female **268**
  genitalia, male **254**
  intromission pattern **120**
  testes size **220**
Cheirogaleus medius **7**
  genitalia, male **254**
  ovarian cycle and behaviour 317
chimpanzee; *see Pan troglodytes*
Chiropotes satanas **11**
classification of primates 5–18, **5, 6**
clitoris
  evolution of 265, **266–7**
  hypertrophy of, in Atelinae **266–**7
coercive matings 66–7
colobines; *see* Old World monkeys
Colobus badius
  body weight, adult **40**
  copulatory vocalization 128
  genitalia, male **255**
  group composition **40**
  group size **40**
  harassment/interruption of copulation 72, **73**
  intromission pattern **120**
  mating system **40**
  menstrual cycle and sexual behaviour 338
  pseudo-sexual swelling in male **194–**5
  sexual activity at night 79
  sexual skin in female **194, 202**
Colobus guereza
  body weight, adult **30**
  genitalia, male **251, 253–4**
  group composition **30**
  group size **30**
  mating system **30**
  vocal anatomy and display **178–**9
Colobus polykomos
  genitalia, male **255**
  vocal anatomy **178**
Colobus verus
  genitalia, female **268**
  genitalia, male 255
  pseudo-sexual swelling in male **194–**5
  sexual skin in female **194**
common marmoset; *see Callithrix jacchus*
concealed ovulation **339**, 352–3
congenital adrenal hyperplasia **285–**8, 298–**9**
consciousness
  and human sexuality 136–8
  in great apes 137–9
consortships 37, 57–62
  and male reproductive success 61–2
  see also individual species
Coolidge effect 242–**3**
copulation
  hormonal consequences of 464–6
  and responses by females **135, 137–**8
copulatory
  facial expressions 126–27, **130, 133–34**
    see also orgasm
  frequencies 237–9, **240, 241**, 242–**3, 235**
  mechanism, in males; definition of 104
  plugs 235–7, 319, 466

postures
  dorso-ventral 111–12
  double-foot clasp **110**
  development of **154–**9
  evolution of **110–16**
  female superior 113, **116**
  inverted 111, **112**
  leg-lock **111**
  missionary position 113
  ventro-ventral 112–14, **115, 116**
  vocalizations 127–9, **130, 134**, 136
  see also orgasm
corpus callosum
  sexual dimorphism of **303, 305–7**
  splenium of **305–**7
cortisol; *see* adrenocortical hormones
cottontop tamarin; *see Saguinus oedipus*
critical periods 277–8, 296–**8**, 312–14
cutaneous glands 183–9, **185, 187–8**
cynomolgus monkey; *see Macaca fascicularis*

Daubentonia madagascariensis
  body weight, adult **46**
  copulatory posture **112**
  genitalia, male **254**
  intromission pattern **118, 120**
  mating system **46**
DeBrazza's monkey; *see Cercopithecus neglectus*
Demidoff's bushbaby; *see Galagoides demidoff*
Depo-provera; *see* medoxyprogesterone acetate
diethylstilbestrol
  prenatal exposure to 291–2
dihydrotestosterone
  and behavioural development **282–**3
  circannual changes in **392**
  and male genitalia 279, **415**
  and sexual behaviour, in males 392, 397–400, **402**
  and spinal motor nuclei 304–5
  uptake, in target organs **414–**17
dispersal patterns
  in anthropoids 23, **24**, 41–2, **65**, 66
  in prosimians 24, **64**
dispersed mating systems 45–50
  and alternative male tactics **64**
  and penile morphologies **245, 249**, 250–1
  and sperm competition 218–**20**, 227–**8**, 233–5, **236–**7
DNA typing 28, 36, 43, 74–80, 87
dominance
  and consortships 58–62
  definition of 52
  and female mating success 80, **81**
  and female reproductive success 81, **82, 83, 84**
  and fertility in females 81–3, **84, 449–53**
  and interrupted copulations 71
  and male mating success 52, **53–6, 125, 341–2**
  in male nocturnal prosimians 47, 48
  and male reproductive success 74–80
  between matrilines in macaques 83
  and sociosexual behaviour 146–9, **150–1**
  and testicular function 444–9, **447–8**
dopamine
  and female proceptivity/receptivity 381–3
  neuroanatomical distribution of 380–**1, 382**
  oestrogen binding and 380–**1**
  and sexual behaviour in males 441–2

dorsal penile nerves
  and sexual behaviour 409–11
dorsolateral spinal nucleus 304–5
dwarf lemurs; *see* Cheirogaleus

ejaculation
  frequency of 237–**40**
  inter-ejaculatory interval **105**
  multiple ejaculatory patterns 117–18
  neuroendocrine organization of 277–8, 392–4, 436–7
  in pseudohermaphroditic female monkeys 283
  post-ejaculatory interval 104, **105**, 107
erectile dysfunction
  causes of 434–5
  effects of age upon 406–**9**
  hyperprolactinaemia and 440–**1**
  testicular hormones and 393–**5**, **396–7**
  treatment of 435–6
erotosexual preferences
  of androgen-insensitive males 288
  of homosexuals 165
  Kinsey scale of **159**
  of men with Imperato–McGinley syndrome 289–90
  of paraphilic subjects 141–5
  of transsexuals 311
  of women exposed prenatally to DES 292
  of women with congenital adrenal hyperplasia 298–**9**
Erythrocebus patas
  body weight, adult **30**
  DNA typing 36, **75**
  ejaculatory frequency **240**
  genitalia, female **268**
  genitalia, male **251–2**, **255**
  group composition **30**
  group size **30**
  harassment/interruption of copulation **73**
  intromission pattern **120**
  male dominance and mating success **35**, **53**
  male dominance and reproductive success **75**
  mating season **390**
  mating system **30**, **35**, 36
  menstrual cycle and sexual behaviour 338
  multimale influxes **35**, 36
  proceptivity in **36**
  puberty **173**
  sexual skin, in female **202**
  sexual skin, in male **190**
  social organization **35**, 36
Eulemur 6
  see also Lemur
Euoticus elegantulus
  body weight, adult **46**
  genitalia, male **245**
  mating system **46**
evolution
  of the baculum 252–60
  of copulatory frequencies 237–**41**
  of copulatory postures **110–16**
  of copulatory vocalizations 128–9
  Darwinian theory of **1**, 2
  of fat distribution in women **214**–16
  of female genitalia 265–**76**
  of female orgasm 129–30, 133–35
  of female proceptivity 95–8
  of female sexual skin 206–14
  of intromission patterns 119–26, **211**, **213**

of male accessory sexual organs 228–35
of male secondary sexual adornments 192–8
of penile morphologies 245–65
of penile spines 260–3
of sperm morphologies 224–8
of testicular size and structure 217–23
evolutionary
  'costs' of high testosterone levels 447
  stable strategy 51, 64
  trees and divergence dates 18, 19, **20**, **21**
eye-contact, as a sexual invitation **96**, 97, **98**

fadrazole
  effects of, on male behaviour 398, **401**
fallopian tubes; *see* oviduct
fattness
  in male mandrills 53, 54
  in male squirrel monkeys 56
fetishism; *see* paraphilias
fluctuating asymmetry 198–201, **200**
forced copulations; *see* coercive matings

galagines
  description of extant forms 8, **9**, 10
Galago alleni
  body weight, adult **46**
  genitalia, female **266**
  genitalia, male **245**
  home range **46**
  mating system **46**
Galago crassicaudatus
  body weight, adult **46**
  genitalia, male **245**, **254**
  home range **46**
  intromission pattern **120**
  mating system **46**
  puberty in 460–1
  scent-marking behaviour **184**
Galago garnettii
  body weight, adult **46**
  copulatory posture **111**
  cutaneous glands **186–7**
  genitalia, female **208**
  genitalia, male **245**, **254**, **261**, **413**
  intromission pattern **118**, **120**
  home range **46**
  mating system **46**
  ovarian cycle and sexual behaviour 316–**17**
  ovariectomy/hormone replacement **358**
  puberty in 460–1
  testes size **220**
Galago moholi **9**
  body weight, adult **46**
  cutaneous glands **186**
  dispersal, in males **47**
  dominance, in males **47**
  home range, overlap **47**
  home range, size **46**
  intromission pattern **118**, **120**
  mating system **46**, 47
  ovarian cycle 317
  ovariectomy and hormone replacement 358
  puberty in 460–1
  retarded secondary sexual development in **197**
  social organization 47
  twinning in 47

Galago senegalensis
  genitalia, male **254**
Galagoides demidoff
  body weight, adult **46**
  genitalia, male **245**, **254**, **261**
  home range **46**
  intromission pattern **118**, **120**
  mating system **46**
  retarded secondary sexual development in **197**–8
  testes sizes **220**
gelada; *see Theropithecus gelada*
gender identity/role
  in great apes 137–9
  in human beings 136–8, 285, 289–90
genetic
  contribution to homosexuality 165–7
  variability and effects of castration 405–**6**
genital development
  effects of hormones upon 278–**80**, **283**, **285**, 288–**9**, **295**,
    296–7
genital displays
  in New World monkeys **152–3**
  see also penile erection
genital lock during copulation; *see* intromission patterns
gibbons; *see* Hylobates
glans penis
  measurements of **254–5**
  morphology of **245**, **249–50**, **251–3**, 263–**5**
  tactile sensitivity of 260
Goeldi's monkey; *see Callimico goeldii*
golden lion tamarin; *see Leontopithecus rosalia*
gonadotrophins 284, **300**, 328–**9**, 332–**3**, 343, 365–6, **368**,
    445–6, **451**, 452–**4**, 464–**5**
Gorilla gorilla
  adolescent sterility **462**
  body weight, adult **30**
  chest-beating display 180, **182**
  copulatory postures 113, **182**
  copulatory vocalizations 128
  cutaneous glands **186**, **188**
  ejaculatory frequency **240**
  female transfers 33, 64
  fluctuating asymmetry in 200
  genitalia, female **268**, **346**
  genitalia, male **230**, **251**, **255**
  group size and composition **30**
  harassment/interruption of copulation **72**
  interbirth interval 33
  inter-ejaculatory interval **105**
  intromission pattern **118**, **120**
  mating system **30**, 33
  menarche, age at **462**
  menstrual cycle and sexual behaviour 345–7
  orgasmic reponses **130**
  post-ejaculatory interval **105**
  proceptivity, in female **346–7**
  secondary sexual characteritics, adult male **19**, **191**
  social organization 33
  solitary males 64
  sperm competition 33, 217–**19**
  sperm morphology **226**, **228**
  testes size 219
  vocal anatomy **178**, **182**
great apes
  adolescent sterility **462–3**
  consortships in **59**
  copulatory behaviour in male 112–16, **118**, **120**

descriptions of extant forms 16–18, **19**, **20**
genitalia of **245**, **250–6**, **264**, **268–9**, 272–3
male dominance and mating success **53**
masturbation in 142
menarche, age at **462**
menstrual cycle and sexual behaviour 343–8
multi-partner matings by females **38**
orgasm, occurrence of **130**, **134**
precopulatory displays of **107**, 113–**14**
proceptive displays **96**
retarded secondary sexual development in **197**
scent-glands of **184**, **186**, **188**
secondary sexual adornments of male **191**
sexual skin, in female **201–2**
sociosexual behaviour of **146**, 148–**9**
sperm competition in **217–19**
greater (thick-tailed) bushbaby; *see Galago crassicaudatus*
grey-cheeked mangabcy; *see Cercocebus albigena*
guenons; *see Cercopithecus*

hamadryas (sacred) baboon; *see Papio hamadryas*
Hamilton–Zuk hypothesis 195, 207–8
Hanuman langur; *see Presbytis entellus*
Hapalemur griseus
  cutaneous glands **186**
  genitalia, male **254**
haplorhines **5**, 7
harassment of copulation 71
  comparative data on **72**, **73**
  male quality hypothesis 74
  parent infant conflict hypothesis 73
  sentinel hypothesis 74
  sexual competition hypothesis 71
Hominoidea
  classification **5**, **6**, 7
  see also great apes; lesser apes
Homo erectus 19
Homo habilis 19
Homo sapiens
  adolescent sterility **462–3**
  brain, anatomical sex differences **302–3**, **305–9**, **311**–13
  concealed ovulation 352–3
  copulatory postures 113, 114, **116**
  cutaneous glands **186**
  ejaculatory frequency **240**, 406–7
  evolution of 19, **21**
  fluctuating asymmetry in **200–1**
  gender identity/role 136–8
  genitalia, female **268**, **270**
  genitalia, male **230**, **255**
  homosexuality 163–9, 293, 295–300, 306–9, 312–14
  incest avoidance in 88
  infanticide in 70–1
  interbirth intervals 214
  intromission pattern **118**, **120**
  masturbation 139–40, **142**
  mating systems 33, 214
  menarche **462**, 3
  menstrual cycle and sexual behaviour 348–52
  menstrual synchrony 458, 466
  neuroendocrine control of sexual behaviour, in females
    360–**5**, 376–80
  neuroendocrine control of sexual behaviour, in males
    394–7, 399–401, 406–**8**, 425–**9**, 439–43, **465**
  orgasmic responses 129, 131, 133–**5**, 200, **272**
  penile erection, neural mechanisms 430–6

post-natal testosterone surge 280–**1**, 292
pre-menstrual tension 349–51
proceptive displays 97, **98**
pseudohermaphroditism **285**, **288–9**
psychosexual differentiation 284–95, 312–14
secondary sexual characteristics of female **214–16**
secondary sexual characteristics of male **191**, **196**
sexual attractiveness in **200**, 214–16
sexual differentiation of the genitalia 278–**80**
sociosexual behaviour 161–4
sperm competition 218–**19**, 243
sperm morphology **225–6**, **228**, **275**
sperm survival, in female tract 269–**70**, **271–2**
testes size 219, **223**
vocal anatomy/sex differences **178**
homosexual
    activities and preferences 165
homosexuality (and lesbianism)
    absence of, in monkeys and apes 159–61
    birth order and 167–**8**
    cross-cultural studies of **163**–5
    evolution of 161
    experiential contributions to 168–9, 313–14
    frequency of **163**–4
    genetic background to 165–7
    neuroendocrine background to 292, 295–300, 306–**9**, 312–14
    ritualization of, in the Sambia 164–5
hormones
    activation/maintenance effects, on sexual behaviour
        in adult females 315–53, 354–80
        in adult males 389–418
howler monkeys; *see Alouatta*
5-HT
    and female proceptivity/receptivity 383–4
    neuroanatomical distribution of **381–2**
    and sexual behaviour in males 443
human beings; *see Homo sapiens*
hybridization
    and evolution of complex penile morphologies 245–7
    and evolution of sexual skin 211–12
Hylobates agilis
    body weight, adult **27**
    genitalia, male **251**
    group size **27**
    mating system **27**
    vocal displays 176–7
Hylobates concolor
    sexual dichromatism **191**
    vocal displays 176–7
Hylobates hoolock
    body weight, adult **27**
    genitalia, female **268**
    group size **27**
    mating system **27**
    sexual dichromatism **191**
    vocal displays 176–7
Hylobates klossii
    body weight, adult **27**
    group size **27**
    mating system **27**
    vocal displays 176–7
Hylobates lar
    body weight, adult **27**
    genitalia, female **268**
    genitalia, male **255**
    group size **27**
    mating system 27, 28

sexual skin, in female **202–3**
social organization 27, 28
sperm length **228**
testis weight **219**
vocal displays 177–**8**
Hylobates muelleri
    vocal displays 176–7
Hylobates pileatus
    sexual dichromatism **191**
    vocal displays 176–7
hyperprolactinaemia
    and female infertility 450–**1**
    and sexual dysfunction in males 440–**1**
hypogonadism, in men 394, **396–7**
hypothalamic lesions
    and female proceptivity/receptivity
        in non-primate mammals 366–**9**
        in female primates 370–**4**
    and male sexual arousal/copulatory behaviour 418–23
    and paternal behaviour in marmosets 419–**21**
hypothalamus
    and electrophysiological studies of sexual behaviour 371–2, **375**, 420–**2**
    and sexual behaviour in females 365–76
    and sexual behaviour in males 418–23, **419–20**, 429–30
    sexual dimorphism of **302–3**, 306–12, **308**, **311**

immediate early genes; *see* C-Fos mapping studies
Imperato–McGinley syndrome 288–90
impotence; *see* erectile dysfunction
incest avoidance 85, 88–90
    in *Homo sapiens* 88
    in *Macaca sylvanus* 89, **90**
    in *Pan troglodytes* 88, **89**
Indri indri
    body weight, adult **27**
    group size **27**
    mating system **27**
    vocal apparatus **179**
infanticidal males
    in *Homo sapiens* 70–1
    in *Presbytis entellus* **67**, **68**, **69**, 70
    in other primate species **67**, **68**
inferotemporal cortex
    and processing of sexually relevant visual stimuli 376–**8**
innate releasing mechanism 141, 206
interruption of copulation
    comparative data on **72**, **73**
    in *Macaca arctoides* 71
interstitial nuclei of the anterior hypothalamus 306–**9**
intromission
    castration, effects upon **392–4**
    dorsal penile nerves and **411**
    duration of **118**, **120**
    penile mechanoreceptors and 411–12
    penile spines and **412**
intromission and ejaculatory mechanism; *see* copulatory mechanism
intromission patterns
    classification schemes for 116–20
    evolutionary significance of 119–26
    genital lock 114–15, **117**, **119**–20
    multiple brief 115, **117**, **119**–20
    pelvic thrusting and 115, **117–120**

Shively *et al.*'s hypothesis 124–6
single brief 116, **118**–20, 122, 124–6
single prolonged 116–17, **118–22**, 123
ischiocavernosus muscles; *see* penile muscles
isolation rearing, effects of
   in chimpanzees 158–9
   in rhesus monkeys 155–7, **158–9**

James–Lange theory of the emotions 413
Japanese macaque; *see Macaca fuscata*

Klüver–Bucy syndrome 377–8, 423
!Kung, hunter gatherers **462**–3

Lagothrix lagotricha 11
   body weights, adult **40**
   genitalia, female **266**, **268**
   group composition **40**
   group size **40**
   intromission pattern **118**, **120**
   male dominance and mating success **53**
   mating system **40**
   multiple-partner matings by females 39
   testes size **219**, **223**
langurs; *see Presbytis*
leks 24, 25
Lemur catta 8
   aggressive behaviour 39
   body weight **40**
   copulatory plug **237**
   cutaneous glands **185–6**
   DNA typing **75**
   ejaculatory frequency **240**
   female choice 41
   genitalia, female **268**
   genitalia, male **254**
   group composition **40**
   group size 39, **40**
   harassment/interruption of copulation **72–3**
   immigrant males, reproductive success **75**
   inter-ejaculatory interval **105**
   intromission pattern **120**
   mating season 39, 389–90, 459–60
   mating system 39–**40**
   multiple-partner matings by females **38**, 39
   oestrogen and female sexual behaviour 358–9
   oestrus 41, **103**, **319**
   ovarian cycle and sexual behaviour 318–**19**, 459–**60**
   scent-marking behaviour 39, **41**, **185**
   sexual activity at night 79
   testes size **220**
Lemur fulvus
   body weight, adult **40**
   copulatory plug **237**
   genitalia, male **250**, **254**
   group composition **40**
   group size **40**
   mating system **40**
   testes size **220**
Lemur macaco
   body weight, adult **40**
   group composition **40**
   group size **40**
   mating system **40**
   sexual dichromatism **190**
   testes size **220**

Lemur mongoz
   body weight, adult **27**
   group size **27**
   mating system **27**
   sexual dichromatism **190**
Lemuroidea
   classification **5**, **6**, 7
   see also Malagasy lemurs
Leontopithecus rosalia 14
   body weight, adult **27**
   genitalia, male **254**
   group size **27**
   intromission pattern **120**
   mating system **27**
   testes size **223**
Lepilemur leucopus 8
lesbianism; *see* homosexual/homosexuality
lesser apes
   descriptions of extant forms 15, 16, **18**
   genitalia of **251**, **255–6**, **268**
   mating systems 27, 28
   secondary sexual adornments, in male **191**
   sexual skin, in female **202–3**
   sperm competition **219**
   vocal displays 176–**7**
lesser galago; *see Galago moholi*
lion-tailed macaque; *see Macaca silenus*
long-tailed macaque; *see Macaca fascicularis*
Lophocebus 6
   see also Cercocebus
lordosis
   in female prosimians 101, **103**, 320
   in the female rat 101, **103**, 366–**9**
lorisines
   descriptions of extant forms 8, 9, **10**
Lorisoidea
   classification **5**, **6**, 7
   see also galagines; lorisines
Loris tardigradus
   body weight, adult **46**
   copulatory plug **237**
   copulatory posture 111–**12**
   genitalia, male **254**
   intromission pattern **118**, **120**
   mating system **46**
   ovarian cycle and behaviour 317
   precopulatory behaviour **102**
   testes size **220**
luteinizing hormone releasing hormone [LHRH]
   analogues of 292–**3**, **295**, 396–7, **399**
   and female proceptivity 384–**5**
   and sexual behaviour in males 439–40

Macaca arctoides
   body weight, adult **121**
   copulatory vocalization **132**
   DNA typing **75**
   ejaculatory frequency **240**
   genitalia, female **245**, **268**
   genitalia, male **245**, **250**, **255**
   harassment/interruption of copulation **72–3**
   inter-ejaculatory interval **105**
   intromission pattern **118**, **120**
   male dominance and mating success **53–5**

male dominance and reproductive success **75**
masturbation 54, 140, **144**
mating system **121**
menstrual cycle and sexual behaviour 338
orgasmic responses **130**, 131, **132–4**
post-copulatory pair-sit 71, **74**
post-ejaculatory interval **105**
puberty, in males 62
sex ratio at birth **91**
sexual skin, in female **202**
sociosexual postures **147–8**, **154**
sperm length **228**
testes size **219**
Macaca assamensis
genitalia, female **203**, **245**
genitalia, male **203**, **245**, **255**
sexual skin, in female **203**
sexual skin, in male **191**
Macaca fascicularis
body weight, adult **40**
consortship **58**, 60–**1**
DNA typing **75**, **80**
ejaculatory frequency **240**
female dominance and fertility **84**
genitalia, female **203**, **268**, **270**, **274**
genitalia, male **203**, **250**, **255**
group composition **40**
group size **40**
harassment of copulation **72**
intromission pattern **120**
male dominance and mating success **53**
male dominance and reproductive success **80**
mating system **40**
menstrual cycle and sexual behaviour 338
multiple-partner matings by females **38**
neuroendocrine control of sexual behaviour, in male 395–6, **398**, 399, **401–2**
retarded secondary sexual development in **197**
sex ratio at birth **91**
sexual skin, in female **202**
solitary males 63, 80, 198
sperm length **228**
sperm survival, in female tract **270–1**
Macaca fuscata
body weight, adult **40**
consortship **58**
female dominance and fertility **84**
friendships 87
genitalia, female **203**
genitalia, male **203**, **255**
group composition **40**
group size **40**
harassment/interruption of copulation **72–3**
intromission pattern **120**
male dominance and mating success **53**
masturbation 131, **141**
mating season 389–**92**
mating system **40**
menstrual cycle and sexual behaviour 338
multiple-partner matings by females **38**
orgasmic responses **134**
pseudohermaphroditism 297–8
sex ratio at birth **91**
sexual skin, in female **203**
sexual skin, in male **191**
sociosexual development **156**

Macaca maurus
body weight, adult **121**
genital morphology **203**
intromission pattern **120**
male dominance and mating success **53**
mating system **121**
menstrual cycle and sexual behaviour 338
sexual skin, in female **203**
Macaca mulatta
adolescent sterility 461–**2**
adrenal glands and sexual behaviour 364–**7**
body weight, adult **40**
consortship **58**, 61, **62**, 336
copulatory posture **110**, **393**
DNA typing **75**, 77, **78**
ejaculatory frequency **240**
female choice 77–8
female dominance and fertility **84**
female dominance and mating success **81**
female dominance and reproductive success **82**
friendships 87
genitalia, female **268**
genitalia, male **231**, **245**, **255**
group composition **40**
group size **40**
harassment/interruption of copulation **72–3**
inter-ejaculatory interval **105**
intromission pattern **120**, **123–4**
male dominance and mating success 42, **53**
male dominance and reproductive success **75**, 77, **78**
male precopulatory behaviour **108**
mating season 41, 78, 333, **390**, **454–6**
mating system **40–2**
menstrual cycle and sexual behaviour 332–8, 365–6, **368**
multiple-partner matings by females **38**
neuroendocrine control of sexual behaviour, in female 359–**60**, 363–**4**, 365–**7**, **368**, 383–4
neuroendocrine control of sexual behaviour, in male 392–**3**, 396–**9**, 402–**5**, 407–**9**, **410**, **414–16**, 420, **422**–3, 429–30, 464–**5**
observability of dominant males 55
orgasmic responses **130**, 131, **134**
penile erection, neurotransmitters and 441–3
post-ejaculatory interval **105**
postnatal testosterone surge 292–3, **295–7**
pregnancy, copulation during 333–4
proceptive displays **95**, **96**–7, 336–**7**
pseudohermaphroditic females 280–4, **283**
puberty 173, 461–4, **463**
sex ratio at birth **91**
sexually dimorphic behaviour, development of 153–9, 280–4
sexual preferences, in males 85–**6**
sexual skin, in female **203**, **209**
sexual skin, in male **191**
social organization 41–2
sociosexual development 153–**9**
sperm length **228**
testes size **219**
vaginal cues and female attractiveness 104, **335–7**, **355–7**
Macaca nemestrina
body weight, adult **40**
DNA typing **75**
genitalia, female **203**, **268**
genitalia, male **203**, **255**
group composition **40**
group size **40**
inter-ejaculatory interval **105**

intromission pattern **120**
mating system **40**
menstrual cycle and sexual behaviour 338
multiple-partner matings by females **38**
post-ejaculatory interval **105**, 242–3
sexual skin, in female **201–2**
sperm length **228**
testes size **219**
Macaca nigra 16
　body weight, adult **121**
　consortship **58**
　copulatory posture **110**
　ejaculatory frequency **240**
　genitalia, female **203**, **268**
　genitalia, male **203**, **255**
　harassment of copulation **73**
　intromission pattern **120**
　male dominance and mating success **53**
　mating system **121**
　menstrual cycle and sexual behaviour 338
　precopulatory behaviour in males **109**
　proceptive displays **99**
　sexual skin, in female **201–2**
　sociosexual postures **148**
Macaca radiata
　body weight, adult **40**
　consortship **59**
　ejaculatory frequency **240**
　genital morphology **203**
　group composition **40**
　group size **40**
　intromission pattern **118**, **120**
　mating season **390**
　mating system **40**
　multiple-partner matings by females **38**
　post-ejaculatory interval **105**
　sex ratio at birth **91**
　sexual skin in female **203**
　sociosexual postures **147**
　testes size **219**
Macaca silenus
　body weight, adult **40**
　consortship **58**
　copulatory vocalization 127–8
　genital morphology **203**
　group composition **40**
　group size **40**
　intromission pattern **120**
　mating system **40**
　menstrual cycle and sexual behaviour 338
　orgasmic responses **130**
　sexual skin, in female **202**
Macaca sinica
　body weight, adult **40**
　female dominance and fertility **84**
　genitalia, female **203**
　genitalia, male **203**, **255**
　group composition **40**
　group size **40**
　mating system **40**
　sexual skin, in female **202**
Macaca sylvanus 16
　body weight **40**
　consortship **59**
　DNA typing **75**, **78**, 79
　dispersal, in males 62–3
　ejaculatory frequency **240**

female dominance and fertility **84**
genital morphology, female **203**
genital morphology, male **203**
group composition **40**
group size **40**
incest avoidance 89–**90**
inter-ejaculatory interval **105**
intromission pattern **120**
male dominance and mating success **53**
male dominance and reproductive success **75**, **78**–80
male–infant associations 87–8
mating system **40**
menstrual cycle and sexual behaviour 338
multiple-partner matings by females **38**, **59**
proceptive displays **99**
puberty, in males 62–3
sex ratio at birth **91**
sexual activity at night **79**
sexual skin, in female **202**
Macaca thibetana
　body weight, adult **121**
　consortship **58**
　ejaculatory frequency **230**
　genital morphology, female **203**
　genital morphology, male **203**, **255**
　group composition **40**
　group size **40**
　harassment of copulation **72**
　inter-ejaculatory interval **105**
　intromission pattern **120**
　male dominance and mating success **53**
　mating system **40**
　post-ejaculatory interval **105**
　sexual skin, female **203**
major histocompatibility complex 90
Malagasy lemurs
　descriptions of extant forms 7, **8**
man; *see Homo sapiens*
Mandrillus leucophaeus
　cutaneous glands **186**
　genitalia, female **268**
　genitalia, male **255**
　secondary sexual adornments, in male **190**
　sexual skin, in female **202**
Mandrillus sphinx 15
　cutaneous glands **186**
　DNA typing **75–7**
　dominance hierarchy, males 53–4
　ejaculatory frequency **240**
　facial displays 108–9
　fatted and non-fatted males 53–4, **221**, 446–7
　female choice **77**
　female dominance and fertility 449–**50**
　genitalia, female **268**
　genitalia, male **253**, **255**
　inter-ejaculatory interval **105**
　interruption of copulation **72**
　intromission pattern **118**, **120**
　male dominance and mating success **53–4**
　male dominance and reproductive success 74, **76–7**
　mate-guarding **59**, **61**, **76–7**
　mating season **390**, **461**
　mating system 74, 76
　multiple-partner matings by females **38**
　menstrual cycle and sexual behaviuor 76–7, 338
　orgasmic responses **130**
　precopulatory behaviour 108–**9**

puberty **172–3**
  retarded secondary sexual development, in male **197**–8
  scent-marking 189
  secondary sexual adornments, in male **190**, 194, **196**
  sexual activity at night **79**
  sexual skin, in female **201–2**, 206–7
  sociosexual presentation and mounting 148
  testes size **221**–2, **390**
mangabeys; *see Cercocebus*
mantled howler; *see Alouatta palliata*
marmosets; *see Callithrix*
marriage
  and Kibbutzim in Israel 88
  and sim-pua in China 88
  and female waist-to-hip ratio **215**–16
masturbation 54, 131, **134**, 139–**42**, **144–5**
  effects of apomorphine **442**
  functions of 140–1, **144**
  object use for 140, **142**, **145**
  and the menstrual cycle 347–8, **351**–2
  in pseudohermaphroditic monkeys 283
mate choice
  female 84–5, 100–1, 323, 325, 342, 348
  and female preference for dominant males 85
  and fluctuating asymmetry 199–201
  and friendships in baboons 86–**7**
  major histocompatibility complex and 90
  and male sexual preferences 85–**6**, **215**–6
mate-guarding
  in chimpanzees 61
  in mandrills 61, 76, **77**
  prolonged copulations and **121**, 123
  see also consortship
mating seasonality 22, 34–6, 39, 41, 320, 323–4, 333, 389–**92**
  and numbers of males in groups 44–5
  reproductive synchrony and 453–61
  testicular function, copulation and **390** 2
mating success
  in females 80–90
  in males 52–67
  see also dominance and female mating success; dominance
    and male mating success; *under individual species*
mating systems
  classification of **25–6**
  comparative review of 24–50
  and copulatory frequencies **240**
  and copulatory patterns **118**, **121**–2
  and copulatory plugs **236**
  and female internal genitalia 267–9
  and penile morphologies **245**, 248–**52**
  and relative testes size 217–**23**
  and secondary sexual characteristics, in males **190–1**
  and seminal vesicles sizes **234**
  and sperm competition 217–43
  and sperm length 227–**8**
  and vocal anatomy **178**
  see also dispersed mating systems; monogamy; multimale–
    multifemale; polyandry; polygyny
mating tactics 51–90
  and alliances in male chimpanzees 57
  and coalitions in male baboons **57**
  and consortships 57–62
  and dominance, in males 52–6
  and harassment of copulation 71–4
  and infanticides by males **67**–71
  and interruption of copulation 71–4
  and mate-guarding 61, 76, **77**

medroxyprogesterone acetate
  and sexual behaviour, in males 395–6, **398**
menarche; *see* puberty
menstrual cycles
  in androgenized female primates 284, 287
  artificial, in rhesus macaques 363–4
  duration of 332
  hormonal events during 332–**3**, **336**
  neuroendocrine control of 365–6, **368**
  oviductal epithelium and 274–5
  suppression of **449**–53
  synchrony of 458–60, **461**
menstruation
  evolution of 272–3
  rarity of, in New World monkeys 322
  sexual abstinence during 351
Microcebus murinus
  body weight, adult **46**
  dominant males, mating success 48
  genitalia, female **266**
  genitalia, male **254**, **261**
  home range **46**
  intromission pattern **118**, **120**
  mating season **390**, 453–**5**
  ovarian cycle and behaviour **102**
  pheromones and reproduction 189, 191–2, 453–**5**
  scent marking **102**
  testes size **220**, **390**
  trill calls by oestrous females 99, **102**
  vocal anatomy **178**
Miopithecus talapoin **17**
  body weight, adult **40**
  copulatory posture **110**
  copulatory vocalization 133, **136**
  female dominance and fertility **450**–1
  genitalia, female **268**
  genitalia, male **255**
  group composition **40**
  group size **40**
  intromission pattern **118**, **120**
  male dominance and mating success 52, **53**, **55**
  mating season **390**, 453, **455**
  mating system **40**
  menstrual cycle and sexual behaviour 338
  multiple-partner matings by females **38**–9
  neuroendocrine control of sexual behaviour, in male 393–**5**,
    437–**8**
  orgasmic responses **130**, **134**
  precopulatory behaviour, in males **108–9**
  presentation postures **100**
  puberty 173, **209**
  sexual skin, in female **108**, **201–2**, **204**, **209**
  sexual skin, in male **190**
  sociosexual behaviour 146–7
monamines; *see* dopamine; 5-HT; noradrenaline
monogamy 26–29
  and penile morphology 249, 251–**2**
  and secondary sexual characteristics, in males **190–1**
  and sperm competition 217–**21**, 222–**3**, 227–**8**, 233–**5**,
    238–**40**, 241
  and vocal displays 176–8
mounting; *see* copulatory postures; sociosexual behaviour
mouse lemur; *see Microcebus murinus*
multimale–multifemale mating systems 37–45
  and alternative male tactics **65**
  and multi-partner matings by females 37, **38**, 39
  numbers of males in 44–5

and penile morphology **249, 250**–1
and secondary sexual characteristics, in males **190**–1
and sperm competition 217–19, 221–4, 227–8, 233–5, **236–40**, 241, 242–**3**
muriqui; *see Brachyteles arachnoides*

Nasalis larvatus
body weight, adult **30**
group composition **30**
group size **30**
intromission pattern **118, 120**
mating system **30**, 32
retarded secondary sexual development in **197**
secondary sexual adornments in males **190, 196**
social organization 32
testes size **219**
naloxone; *see* opioid peptides
needle-clawed bushbaby; *see Euoticus elegantulus*
neuroactive peptides and sexual behaviour
in female mammals 384–8
in male mammals 437–41
neurotransmitters and sexual behaviour
in female mammals 380–8
in male mammals 431–4, 436–43
New World monkeys
consortships in **58**
copulatory patterns **118, 120**
description of extant forms **10**, 14
male dominance and mating success **53**
mating systems 27, 28, 29, **40**
multiple-partner matings by females 38
orgasm, occurrence of **130, 134**
ovarian cycle and sexual behaviour in 321–32
precopulatory behaviour in males **106**–8
proceptive displays **96**
retarded secondary sexual development in 196–**7**
scent glands/marking in 183–**4, 186**–9
secondary sexual adornments in male **190**
sexual dimorphism in **171, 175**
sociosexual behaviour of **152**–3
sperm competition in **219–21**
nitric oxide
and penile erection 433
nocturnal sexual activity
in diurnal primates **79**–80
noradrenaline
and female proceptivity/receptivity 380–1
neuroanatomical distribution of 380–**1, 382**
oestrogen binding and 380–**1**
release of, during lordosis 380–1, **383**
and sexual behaviour in males 442–3
Nycticebus coucang **10**
body weight, adult 46
genitalia, male **245**
intromission pattern **118, 120**
mating system 46

oestrogens
and attractiveness of female primates 85–**6**, 354–7
circulating levels of, in female
great apes **354**
New World monkeys 324–5, **458**, 466
Old World monkeys 332–**3**
prosimians **317, 319, 320**–1
positive feedback effect of 284, **300**, 332–**3**, 450–**1**

effects on proceptivity/receptivity 358–**61, 362, 363, 364**
and sexual behaviour in males 397–400, **401**
and sexual behaviour in pseudohermaphroditic rhesus 283–4
uptake of, in target organs 414–16
oestrus
absence of, in anthropoids 93–4, 101–2, 331–2, 344–5
occurrence of, in prosimians 93, 316–21
Old World monkeys
consortships in **58, 59**
copulatory patterns in **118, 120**
descriptions of extant forms 11–**15, 16, 17**
genitalia of **245, 250**–7, 259, **264**, 265–**6**, 268–70, 272–**4**
lipsmacking displays of **95, 96**
male dominance and mating success **53**
mating system 27, **40**
menstrual cycle and sexual behaviour in 332–43
multiple-partner matings by females 38
precopulatory behaviour in males **106**–10
orgasm, occurrence of **130, 134**
proceptive displays 96
retarded secondary sexual development in **197**
scent glands/marking in 183–**4, 186**
secondary sexual adornments in male **190**–1
sexual skin in female **201–3**
sociosexual behaviour of 146–51
sperm competition in **219, 221**–2
olfactory communication 102–4, **106**–8, 266–7, 316–20, 325, 330–**1, 335**–7, 340–**1, 355**–7, 453–**5**
olive baboon; *see Papio anubis*
olive colobus; *see Colobus verus*
one-male units; *see* polygyny
Onuf's nucleus 430
opioid peptides
β-endorphin, effects on proceptivity/receptivity 386–**8**
β-endorphin and sexual behaviour in males and female fertility 437–9, 452–**4**
other pro-opiomelanocortin-derived peptides 440–1
oral sex 147, 160, 162, 164–5, 289–90
orang-utan; *see Pongo pygmaeus*
orbito-prefrontal cortex
and behaviour **378**, 379–80
organization hypothesis: brain development and behaviour 277–8, 279–95, 301–5, 312–14
orgasm 129–36, **132–5**, 200, **272**, 283
os penis; *see* baculum
Otolemur **6**
see also Galago
ovarian cycles
and locomotor activity 99, **102**
and sexual behaviour 315–32, 332–52
social suppression of **449–53, 54**
synchrony of 457–60, **61**
see also menstrual cycle
ovariectomy
behavioural effects of 354–64
oviduct **274–6**
owl monkeys; *see Aotus*
oxytocin
behavioural effects of 385–6, 440
neuroanatomical distribution of **386**

paedophilia; see paraphilias
Pan paniscus
body weight, adult **40**
copulatory postures 112, 113, **115**

copulatory vocalizations 128
ejaculatory frequency **240**
genitalia, male **255**
group composition **40**
group size **40**
harassment/interruption of copulation **72, 73**
intromission pattern **118, 120**
male dominance and mating success **53**, 54, **55**
mating system **40**, 43–4
menstrual cycle and sexual behaviour 345–**6**
multiple-partner matings by females **38**
proceptive displays **99**
sexual behaviour 44, 345
sexual skin, in female **202**, 345–**6**
social organization/fusion–fission 43–4
sociosexual behaviour 146, **148**
sperm morphology **226**
Pan troglodytes **20**
  adolescent sterility **462**
  alliances between males 57
  body weight, adult **40**
  consortship **59, 60**, 61
  copulatory plug **237**
  copulatory posture 112, **114**
  copulatory vocalization 128
  cutaneous glands **186**
  DNA typing 43, **75**
  ejaculatory frequency 9, **40**, 238
  female dominance and reproductive success 82, **83**
  genitalia, female **268–9**
  genitalia, male **218, 230, 250, 253, 255, 269**
  group composition **40**
  group size **40**
  harassment/interruption of copulation **72, 73**
  incest avoidance in 88, **89**
  interbirth intervals 43
  inter-ejaculatory interval **105**
  intromission pattern **118, 120**
  male dominance and mating success **53**, 344
  male dominance and reproductive success **75**
  masturbation 131, 140, **142**
  mating system **40**
  mating tactics, males 43
  menarche, age at **462**
  menstrual cycle and sexual behaviour **343–5**
  multiple-partner matings by females **38**, 39
  orgasmic responses 130, **134**
  pant-hooting displays 181–**3**
  puberty **173**
  sexual activity at night **79**
  sexual behaviour 43, 343–4
  sexual skin, in female **201–2, 269**, 343
  social organization/fusion-fission 42–3
  sociosexual genital contacts **149**
  sperm competition 217–**19**, 238–**9**
  sperm morphology **226, 228**
  testes size 217–**19, 221, 223**
  transfers by females 43
Papio anubis
  body weight, adult **40**
  consortship **79**
  friendships 86, **87**
  genitalia, male **255**
  group composition **40**
  group size **40**
  harassment/interruption of copulation **72, 73**
  intromission pattern **120**

male dominance and mating success **53**
male dominance and testicular function **445–6**
mating system **40**
menstrual cycle, effects of social stress **449**
multiple-partner matings by females **38**
sexual activity at night **79**
sexual skin, in female **202**
sperm length **228**
testes size **219**
Papio cynocephalus
  body weight, adult **40**
  coalitions between females 83, 449–50
  coalitions between males 57
  consortships **59**, 61, **62**
  DNA typing in **75**
  ejaculatory frequency **240**
  genitalia, male **250, 255**
  group composition **40**
  group size **40**
  female dominance and fertility 83, **84**, 449–50
  male dominance and mating success **53, 56**
  male dominance and reproductive success **75**
  mating system **40**
  menstrual cycle and sexual behaviour **56, 57**, 61, **62**, 338
  multiple-partner matings by females **38**
  puberty **173**
  sex ratio at birth **91**
  sexual skin, in female **202**
  sociosexual postures 147, 148, **150**
  sperm length **228**
  testes size **219, 221**
Papio hamadryas
  body weight, adult **30**
  follower male tactic **31**
  genitalia, female **268**
  genitalia, male **255**
  group composition **30**
  group size **30**
  intrasexual competition, between females **459**
  intromission pattern **120**
  mating system **30, 31**–3
  menstrual cycle and sexual behaviour 338
  multi-level troop structure **31**
  notifying behaviour in males 194–5
  secondary sexual adornments in male 190, **193**
  sexual skin, in female **202**, 206, **459**
  social organization 31–33
  testes size **219**
Papio papio
  genitalia, male **255**
  sexual skin, in female **202**
  sociosexual postures 147
Papio ursinus **15**
  body weight, adult **40**
  consortships **59**
  copulatory vocalization 127
  ejaculatory frequency **240**
  genitalia, female **268**
  genitalia, male **255**
  group composition **40**
  group size **40**
  intromission pattern **120**
  male dominance and mating success **53, 341–2**
  mating system **40**
  menstrual cycle and sexual behaviour 339, **340–2**
  multiple-partner matings in females **38**
  orgasmic responses 130, **134**

ovariectomy and hormone replacement 359
post-ejaculatory interval **105**
proceptive displays of **94, 96**
secondary sexual adornments, in male **191**
sexual skin, in female **201–2**, 204–6
testes size **219**
paraphilias 141–5
absence of, in monkeys and apes 142–4
in anthropological perspective 145
parental investment 23
patas monkey; *see Erythrocebus patas*
paternity confusion hypothesis 69, 80
paternity determination; *see* DNA typing
pelvic thrusting
by female primates 126
neurological control of, during copulation 425–9
see also intromission patterns
penile erection
conditioning of 141, 150, 206, **420**, 441–2
during sleep 151, 394
failure of; *see* erectile dysfunction
neural control of **430–4**, 439–43
as a precopulatory display **113–15**, 150, 344–5
during sociosexual communication 150–3
effects of testosterone upon **394–5, 397**
penile morphology
Eberhard's hypothesis 247–50
effects of castration upon 409, 412–**13**
evolution of 245–65
genitalic recognition hypothesis 247
lock and key hypothesis 244–7
mechanical conflict of interest hypothesis 247
pleiotropism hypothesis 247
penile muscles
functions of **263, 264–**5, 436
innervation of **304–**5
penile sensory feedback
processing within the spinal cord and brain **427–9**
penile spines
effects of castration upon 409, 412–**13**
functions of 261–3, 411–**12**
size of, and mating systems 248–50
types of 260–**1**
Perodicticus potto
body weight, adult **46**
dispersal, in males **48**
genitalia, male **254**
home range overlap **47**
home range, size **46**
mating system **46, 47**, 48
ovarian cycle and behaviour 317–**18**
social organization **47**, 48
testes size **220**
pheromones 189, 191–2, 452, 453–**5**, 457–8, 460
phylogenetic relationships
of extant primates **5**–7
pigtail macaque; *see Macaca nemestrina*
Pithecia monarchus
group size **27**
mating system **27**
Pithecia pithecia **28**
cutaneous glands **186**
genitalia, male **251**
group size **27**
mating system **27**
secondary sexual adornments in males **28, 190**
testes size **223**

vocal anatomy **178**
platyrrhines
body size dimorphism in **171**–2
nostril position **7**, 14
see also New World monkeys
polyandry 29
polygyny 29–37
and alternative male tactics **66**
in *Cercopithecus* species **34**
in *Erythrocebus patas* **35**, 36
and female competition for sperm 81
in hamadryas baboons **31**–3
in geladas 32, 33
in gorillas 33
in human beings 33
and multimale influxes 34
and penile morphology **249, 251**–2
in *Presbytis entellus* 36, **37**
in proboscis monkeys 32
and secondary sexual characteristics in males **190–1**
and sperm competition 30, 217–**19**, 222–4, 227–8, 234–5,
    238–**40, 241**
and vocal specializations
    in adult males **178, 180–1**, 182–3
polyspecific associations
in *Cercopithecus* species 212, 246
Pongo pygmaeus 19
adolescent sterility **462**
body weight, adult **46**
coercive matings in 49, 66
consortships **59**
copulatory postures **112, 113**
cutaneous glands **186**
female choice 85, 348
genitalia, male **218, 229**–30, **245, 255, 264**
great call, adult male 49, **180–1**
harassment of copulation **73**
home range **46**
interbirth interval 50
intromission pattern **118, 120**
masturbation **142, 145, 347**–8
mating system **46**, 49–50
menarche, age at **462**
menstrual cycle and sexual behaviour **347**–8
orgasmic responses **130**
proceptive behaviour 98–9, **347**–8
puberty **173**
retarded secondary sexual development in **197**–8, 447–**8**
secondary sexual characteristics, male 49, **191, 196**, 447–**8**
sexual activity at night **79**
social organization 49–50
sperm competition 217–**19**
sperm morphology **226, 228**
testes size 217–**19**
visual display, territorial males 50
post-ejaculatory refractory period **104, 105**, 107
see also ejaculation
potto; *see Perodicticus potto*
precopulatory behviour in male primates
facial displays **106, 108, 109**, 110
inspection of female genitalia **106**, 107
postural communication **107**
penile displays **107, 113–15**
pregnancy
sexual activity during 69, **71, 329**–30, 333–4
preoptic area
and sexual behaviour in females 370

and sexual behaviour in males **310**, 418–23, **422**, 429–30
sexual dimorphism of **302**, 309–11
Presbytis cristatus
 body weight, adult **30**
 group composition **30**
 group size **30**
 mating system **30**
Presbytis entellus **15**
 body weight, adult **30**
 genitalia, male **252**
 group composition **30**, **40**
 group size **30**, **40**
 group takeovers **36**
 harassment/interruption of copulation **72**, **73**, 81
 infanticide 37, **67–70**
 interbirth intervals **69**
 mating system **30**, 36, 37
 menstrual cycle and sexual behaviour 338
 pregnancy, sexual activity during 69, **71**
 proceptive displays 97
 social organization 36, **37**
 sociosexual mounting 149, **151**
 testes size **219**
Presbytis johnii
 body weight, adult **30**
 group composition **30**
 group size **30**
 mating system **30**
Presbytis potenziani
 body weight, adult **27**
 group size **27**
 mating system **27**
Presbytis senex
 body weight, adult **30**
 genitalia, male **252**
 group composition **30**
 group size **30**
 mating system **30**
Presbytis thomasi
 genitalia, male **252**
presentation postures; *see* proceptivity in female primates, sociosexual behaviour
proboscis monkey; *see Nasalis larvatus*
proceptivity in female primates **94–101**
 eye-contact and **95**, **96**, 97, **98**, 339–40, **376–7**
 facial displays **94–8**, 326–30
 neuroendocrine basis of 358–67, 370–**6**, 376–80, **378**
 during ovarian cycles 315–16, 324–32, 332–52
 presentation postures **94**, **96**, 97–8, **99**, 339–40
 pseudohermaphroditism and 283–4
 rarity of, in prosimians 321
 tactile elements **96**, 100
 vocalizations **96**, 99–100, **102**, 325
progestagens
 prenatal exposure to **286**, 290–1
progesterone
 and attractiveness in female primates 85, **86**, 354–7
 circulating levels of, in female
  great apes 343
  New World monkeys 324–5, **326**, **329**, 458
  Old World monkeys 332–3
  prosimians **317**, **319**, 320
 effects on proceptivity/receptivity 361–2, **364**
Propithecus verreauxi **8**
 body weight, adult **40**
 genitalia, female **268**
 genitalia, male **254**

group composition **40**
group size **40**
mating season **390**
mating system **40**
multiple-partner matings by females **38**
ovarian cycle and sexual behaviour 320
scent-marking behaviour **185**
prosimians
 classification of **5–7**
 copulatory patterns in **118**, **120**
 description of extant forms **7–10**
 genitalia of **208**–9, **245**, **250**, **254**, **256**, 260–3, **266**, **268**
 home ranges **46**
 mating aggregations, in nocturnal forms 48
 mating systems 26, **27**, **40**, **46**, **47**, 48
 multiple-partner mating by females **38**
 oestrus, in female 93, **102**, 103, 466
 ovarian cycle and sexual behaviour in 316–21, **460**
 pheromonal communication in 189, 191–2, 453–**5**, 460
 precopulatory behaviour in males **106**
 proceptive displays **96**
 retarded secondary sexual development in **197**
 scent glands/marking in 183–7, 189, 191–2
 secondary sexual adornments in males **190**
 sexual dimorphism in **171**, 173–**4**, 175–6
 sperm competition in **218–20**
prostate gland **229**, 231, 233
pseudohermaphroditism
 and androgen-insensitivity syndrome **288**
 in female rhesus monkeys 280–4, **283**
 and Imperato–McGinley syndrome 288–90
 progestin-induced **286–7**
pseudoswellings
 in juvenile male colobines **194–5**
psychosexual differentiation, in human beings 284–95, 312–14
puberty
 age at **173**, **461**–4
 growth during **172**
 and male dispersal 62, 63
 postnatal testosterone and 292–3
 and sexual swellings in females **209**–11
 social suppression/acceleration of 460–4, **462–3**
Pygathrix nemaeus
 genitalia, male **254**
pygmy marmoset; *see Cebuella pygmaea*

rape; *see* coercive matings; paraphilias
red colobus; *see Colobus badius*
red howler; *see Alouatta seniculus*
redtail monkey; *see Cercopithecus ascanius*
Rensch's rule 171
reproductive success
 in females **82**, **83**, **84**
 in males 74–80
rhesus monkey; *see Macaca mulatta*
Rhinopithecus roxellana
 secondary sexual adornments in males **190**, **196**
Ridley's hypothesis: numbers of males in primate groups 44–5
ringtailed lemur; *see Lemur catta*
ruffed lemur; *see Varecia variegata*

saddle-back tamarin; *see Saguinus fuscicollis*
Saguinus bicolor **13**
 genitalia, male **254**

Saguinus fuscicollis
  body weight, adult 27
  cutaneous glands 186–8
  genitalia, male 254
  group size 27
  mating system 27
  scent-marking behaviour 187
  testes size 223
Saguinus imperator 14
Saguinus labiatus
  genitalia, male 254
  group size 27
  mating system 27
Saguinus leucopus
  group size 27
  mating system 27
Saguinus midas
  body weight, adult 27
  genitalia, male 254
  group size 27
  mating system 27
Saguinus mystax
  female dominance and fertility 84
  genitalia, male 234
  group size 27
  mating system 27
Saguinus nigricollis
  genitalia, male 254
Saguinus oedipus
  body weight, adult 27
  cutaneous glands 186–7
  genitalia, male 251, 254
  group size 27
  intromission pattern 120
  mating system 27
  scent-marking behaviour 187
  testes size 220–1
Saimiri boliviensis
  genitalia, male 250, 254
  copulatory stimuli and oestradiol secretion, in females 466
Saimiri oerstedii
  fatted males and female mate choice 56
Saimiri sciureus
  body weight, adult 40
  genitalia, female 268
  genitalia, male 254
  group composition 40
  group size 40
  harassment of copulation 72, 73
  inter-ejaculatory interval 105
  intromission pattern 118, 120
  mating season 390, 457–8
  mating system 40, 323
  ovarian cycle and sexual behaviour 323–4
  puberty 173
  reproductive synchrony 457–8
  social organization 323
  sociosexual behaviour 152, 154
  sperm length 228
Sambia, of Papua New Guinea
  and Imperato–McGinley syndrome 289–90
  and ritualized homosexuality 164–5
scent-marking 102–4, 183–9, 191–2, 266–7, 318
secondary sexual adornments, in males
  and effects of castration 413
  fluctuating asymmetry in 199–201
  involving the face 190–8, 196

involving the pelage 190–7
retarded development of 65, 197–8, 446–9
sexual skin 192–5, 193–4
seminal vesicles 231–5
septum
  and sexual behaviour 424–5
serotonin; *see* 5-HT
sexology
  history of 2–4
sex play
  in human children 161–4
  see also sociosexual behaviour
sex ratio
  advantaged daughter hypothesis 91–2
  at birth 90–2
  local resource competition hypothesis 91–2
  Trivers Willard hypothesis 90–1
  operational 44, 459
  socioeconomic 172, 182
sexual arousal mechanism, in males
  definition of 104
sexual attractiveness
  and facial cues 200–1, 216
  in female primates 102–4, 204–6, 266–7, 315–16, 324–32, 332–48, 354–7
  in female rhesus monkeys 85, 86, 104, 355–7
  and human waist-to-hip ratios 215–16
  lack of, between close relatives 88–90
  and sexual skin swellings 204–5, 210–11, 339–42, 343–6, 354–5
  and vaginal odour 106, 107–8, 355–6, 357
sexual behaviour
  abnormal 140, 141–5, 155–9
  effects of age upon 406–10
  development of 153–9, 161–4, 168–9, 277–95
  in homosexuals 165
  individual variability of, in males 401–9, 410
  neuroendocrine basis of, in females 354–88
  neuroendocrine basis of, in males 389–433
  observation and measurement of 357–8
  and ovarian cycles 315–53
sexual differentiation
  of the brain 277–8, 279–95, 301–14
  of the genitalia 278–80, 283
sexually dimorphic behaviour 277, 280–2, 285–7, 290–1, 293–5, 296–8
sexual dimorphism
  in body weight 170–4
  in brain structure 301–12
  in canine tooth size 174–6
  in cutaneous glands 183–9, 185, 187–8
  in facial adornments 190–1, 195–7
  in pelage 190–1, 195–7
  in vocal anatomy 176–83, 178, 181–2, 301–4
  see also sexually dimorphic behaviour
sexual interference
  see harassment of copulation; interruption of copulation
sexual receptivity 101–2, 315–16
  effects of ovariectomy and hormone replacement 358–64
  in great apes 343–8
  in New World monkeys 321–32
  in Old World monkeys 332–52
  in prosimians 316–21
  in women 348–52
  neuroendocrine basis of 365–76
sexual selection 1, 2
  and body size 170–4

and canine tooth size **174**–6
and copulatory frequencies 237–**41**
by cryptic female choice 247–8, 267–**76**
and female genitalia 265–**76**
and female sexual skin 206–8
and fluctuating asymmetry 198–201
and infanticide 67–71
and intromission patterns 119–22, 124–6
and penile morphology 245–65
and secondary sexual features in women **214**–16
and seminal vesicles sizes 233–5
and sperm morphology 224–8
and testes sizes 217–**23**
and vocal displays **178**, 182–3
sexual skin, in females 201–14
comparative morphology of **201**–3
evolution of 206–14
relationship to menstrual cycle 201, **343**
secondary loss of **211**–14
and sexual behaviour 204–6, **205**, 339–**42**, **343**–6, 354–5
as a supernormal stimulus 210–11
see also secondary sexual adornments in males
siamang; *see Symphalangus syndactylus*
sifaka; *see Propithecus verreauxi*
Simias concolor
sexual skin in female **202**–3
slender loris; *see Loris tardigradus*
slow loris; *see Nycticebus coucang*
social deprivation
and abnormal sexual development 139, 143, 144–5, 155–7, **158**–9, 163 4
social organization
age-graded male units 36, 37
communities of chimpanzees/bonobos 42–4
evolution of 22–4
female-bonded societies 23, 24, 41–2, 51
fusion–fission societies 27, 42–4, 325
in non-gregarious species 45–50
socioendocrinology 444–64
see also dominance socioendocrinology
and female fertility **449**–**54**
and male dominance rank 444–**6**
and masculine secondary sexual traits 446–9
and pubertal development 460–4, **462**–**3**
and reproductive synchrony 453–**61**, **455**–6, **458**
and sexual activity 464–**5**, 466
sociosexual behaviour 146–64
as compared to homosexual behaviour 159–61
development of 153, **154**–**9**, 161–4
dominance and 146–9, **150**–**1**
in human beings 161, **162**–4
penile erection and 150–3
sociosexual mimicry **193**–5
solitary males 54, 63, **65**, 66
species-isolating mechanisms 244–6
sperm
acrosome reaction in **275**–6
capacitation of **275**–**6**
co-operative forms of **224**
count, and ejaculatory frequency 237–**9**
egg-getter 224–6
hyperactivated swimming of 274, **275**–**6**
kamikaze 225–7
as a limiting resource for females 81, 459
morphology of **224**–**8**, **225**, 227–**8**
pleiomorphism, in hominoid **226**
production and storage 223–4

survival times, in female tract 240–1, 269–**70**, **271**, **273**–6
transport in female tract 269–**76**
sperm competition 81, 217–43
and mating order effects 239–41, 242–**3**
raffle-principle of 242
spider monkeys; *see Ateles*
spinal nucleus of the bulbocavernosus **304**–5, 451–2
steatopygia **214**
stepfathers
and infanticide 70–1
squirrel monkeys; *see Saimiri*
strepsirhines **5**, 7
stumptail macaque; *see Macaca arctoides*
surrogate mothers
and behavioural development in rhesus monkeys 143–4, 156–**8**, **159**
Symphalangus syndactylus **18**
body weight, adult **27**
ejaculatory frequency **240**
group size **27**
mating system **27**
preputial tuft, in male **18**
vocal displays 176–**7**

talapoin monkey; *see Miopithecus talapoin*
tamarins; *see Saguinus; Leontopithecus*
tarsiers
description of extant forms **10**
Tarsioidea
classification **5**, **6**, 7
Tarsius bancanus **10**
intromission pattern **118**, **120**, 318
ovarian cycle and sexual behaviour 318
Tarsius spectrum
body weight, adult **27**
group size **27**
mating system **27**
testes
compartments of **222**–3
sizes, individual variations in 220–1
sizes, and sperm competition 217–22
weights of, in hominoids **219**
testosterone
and behavioural development 277–8, 280–4, 285–90, 292–**5**, 301, **304**–5, 312–14
and brain development 301–5
central effects of, upon sexual behaviour 413–18
circannual changes in **390**–**2**
and cutaneous glands 186–7
and dominance in males 444–**6**
diurnal rhythm of, and sexual behaviour 402–3
free fraction of 292, 407–8, **410**
and genital development 278–**80**, 281–**3**, **296**–**7**
and individual variability in sexual behaviour 401–10
metabolites of 397–**400**
peripheral effects of, upon sexual behaviour 409–**13**, **415**
postnatal surge of 280–1, 292–7
and proceptivity/receptivity in females **326**, **336**, **343**, 346, 352, 362–5, **366**–**7**
replacement and behaviour in males **294**, 392–403
sexual stimulation and secretion of 453–**5**, 464–**5**
uptake, in target organs **414**–16
Theropithecus gelada
body weight, adult **30**
ejaculatory frequency **240**
female dominance and fertility **84**
follower male tactic 32, 64

genitalia, female **268**
genitalia, male **255**
group composition **30**
group size **30**
group takeovers 32, 64
intromission pattern **120**
mating system **30**, 32, 33
menstrual cycle and sexual behaviour 338
secondary sexual adornments in male **190**, **193**
social organization 32, 33
sperm length **228**
testes size **219**
titi monkey; *see Callicebus*
toque monkey; *see Macaca sinica*
transsexuals
  brain structure in 306, **311**
transvestites 3
Trobriand islanders
  sexual attitudes of 161
Turner's syndrome 286–7
turnim-men; *see* Imperato–McGinley syndrome

uakari [bald, or red]; *see Cacajao calvus*
Urbach-Wiethe disease 378, 424
uterotubal junction 273–4
uterus **268**, 271–3, **276**

vaginal
  closure membrane **208**, 317, 319

cornification 317–19
  length **268–9**
  pH and sperm survival 269–**70**
vagino-plasty 283–**4**
Varecia variegata
  genitalia, male **254**
  intromission pattern **120**
  mating system 26–7
  ovarian cycle and sexual behaviour 319–**20**
  social organization 26–7
  testes size **220**
  vocal specializations 177–**8**
vas deferens 228–31
vasoactive intestinal polypeptide
  and penile erection **433–4**
vervet monkey; *see Cercopithecus aethiops*
vomeronasal organ 184, 330–1, 452, 460

waist-to-hip ratio, in *Homo sapiens* **215**–16
white-faced saki; *see Pithecia pithecia*
white-handed gibbon; *see Hylobates lar*
woolly monkey; *see Lagothrix lagotricha*
woolly spider monkey; *see Brachyteles arachnoides*

yellow [savannah] baboon; *see Papio cynocephalus*
Yolngu tribespeople of N. Australia
  sexual attitudes of 145, 162

Zahavi handicap hypothesis 195, 208